D1673228

PARTICLES AND NUCLEI

N. N. Bogolyubov

Editor-in-Chief
Director, Laboratory for Theoretical Physics
Joint Institute for Nuclear Research
Dubna, USSR

A Translation of Problemy Fiziki Élementarnykh Chastits i Atomnogo Yadra
(Problems in the Physics of Elementary Particles and the Atomic Nucleus)

Volume 1, Part 1

A SPECIAL RESEARCH REPORT / TRANSLATED FROM RUSSIAN

PARTICLES AND NUCLEI

Volume 1, Part 1

PARTICLES AND NUCLEI

Volume 1, Part 1

Elastic Scattering of Protons by Nucleons in the Energy Range 1-70 GeV
V. A. Nikitin
Probability Description of High-Energy Scattering and the Smooth Quasi-potential
A. A. Logunov and O. A. Khrustalev
Hadron Scattering at High Energies and the Quasi-potential Approach in Quantum Field Theory
V. R. Garsevanishvili, V. A. Matveev, and L. A. Slepchenko
Interaction of Photons with Matter
Samuel C. C. Ting
Short-Range Repulsion and Broken Chiral Symmetry in Low-Energy Scattering
V. V. Serebryakov and D. V. Shirkov
CP Violation in Decays of Neutral K-Mesons
S. M. Bilen'kii
Nonlocal Quantum Scalar-Field Theory
G. V. Efimov

Volume 1, Part 2

The Model Hamiltonian in Superconductivity Theory
N. N. Bogolyubov
The Self-Consistent-Field Method in Nuclear Theory
D. V. Dzholos and V. G. Solov'ev
Collective Acceleration of Ions
I. N. Ivanov, A. B. Kuznetsov, É. A. Perel'shtein, V. A. Preizendorf, K. A. Reshetnikov, N. B. Rubin, S. B. Rubin, and V. P. Sarantsev
Leptonic Hadron Decays
E. I. Mal'tsev and I. V. Chuvilo
Three-Quasiparticle States in Deformed Nuclei with Numbers between 150 and 190 (E/T)
K. Ya. Gromov, Z. A. Usmanova, S. I. Fedotov, and Kh. Shtrustnyi
Fundamental Electromagnetic Properties of the Neutron
Yu. D. Aleksandrov

Volume 2, Part 1

Self-Similarity, Current Commutators, and Vector Dominance in Deep Inelastic Lepton–Hadron Interactions
V. A. Matveev, R. M. Muradyan, and A. N. Tavkhelidze
Theory of Fields with Nonpolynomial Lagrangians
M. K. Volkov
Dispersion Relationships and Form Factors of Elementary Particles
P. S. Isaev
Two-Dimensional Expansions of Relativistic Amplitudes
M. A. Liberman, G. I. Kuznetsov, and Ya. A. Smorodinskii
Meson Spectroscopy
K. Lanius
Elastic and Inelastic Collisions of Nucleons at High Energy
K. D. Tolstov

PARTICLES AND NUCLEI

N. N. Bogolyubov

Editor-in-Chief

Director, Laboratory for Theoretical Physics
Joint Institute for Nuclear Research
Dubna, USSR

A Translation of Problemy Fiziki Élementarnykh Chastits i Atomnogo Yadra
(Problems in the Physics of Elementary Particles and the Atomic Nucleus)

Volume 1, Part 1

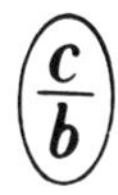

CONSULTANTS BUREAU • NEW YORK-LONDON • 1972

The original Russian text, published by Atomizdat in Moscow in 1970 for the Joint Institute for Nuclear Research in Dubna, has been revised and corrected for the present edition. This translation is published under an agreement with Mezhdunarodnaya Kniga, the Soviet book export agency.

PROBLEMS IN THE PHYSICS OF ELEMENTARY PARTICLES AND THE ATOMIC NUCLEUS
PROBLEMY FIZIKI ÉLEMENTARNYKH CHASTITS I ATOMNOGO YADRA
Проблемы физики элементарных частиц и атомного ядра

Library of Congress Catalog Card Number 72-83510
ISBN 0-306-17191-0

United Kingdom edition published by Consultants Bureau, London
A Division of Plenum Publishing Company, Ltd.
Davis House (4th Floor), 8 Scrubs Lane, Harlesden,
London NW10 6SE, England

Printed in the United States of America

CONTENTS

Volume 1, Part 1

	Eng.	Russ.
From the Editorial Board	1	5
Elastic Scattering of Protons by Nucleons in the Energy Range 1-70 GeV—V. A. Nikitin	2	7
Probability Description of High-Energy Scattering and the Smooth Quasi-potential —A. A. Logunov and O. A. Khrustalev	39	71
Hadron Scattering at High Energies and the Quasi-potential Approach in Quantum Field Theory—V. R. Garsevanishvili, V. A. Matveev, and L. A. Slepchenko	52	91
Interaction of Photons with Matter—Samuel C. C. Ting	80	131
Short-Range Repulsion and Broken Chiral Symmetry in Low-Energy Scattering — V. V. Serebryakov and D. V. Shirkov	106	171
CP Violation in Decays of Neutral K-Mesons — S. M. Bilen'kii	146	227
Nonlocal Quantum Scalar-Field Theory — G. V. Efimov	165	255

FROM THE EDITORIAL BOARD

With the present issue the Joint Institute for Nuclear Research initiates the series "Problems of the Physics of Elementary Particles and the Atomic Nucleus." The aim of the series is to make more widely known the most important theoretical and experimental results of investigations in the field of the physics of elementary particles and the atomic nucleus through the systematic publication of surveys on problems of topical interest written at a high scientific level. Four issues of the series are to be published each year. The individual issues will be devoted predominantly either to questions of elementary particle physics or questions of nuclear physics. The series will be referred to in the shortened form "Elementary Particles and the Atomic Nucleus (ÉChAYa) in Russian and "Particles and Nucleus" in English. The volumes and issues will be numbered.

The publication of the series "Particles and Nucleus" is needed to meet the increasing differentiation of the scientific branches of nuclear physics, a field in which thousands of highly qualified specialists are working. The publication of surveys devoted to specialized but important present-day branches of nuclear physics is made necessary to a large extent by the intensive development of these branches and the rapid accumulation of scientific information. The existing publications are not able to meet these requirements. The aim of the series is to publish such surveys as and when this becomes necessary. The series will contain review articles on subjects related to the investigations carried out at the Joint Institute for Nuclear Research. This includes theoretical and experimental problems of the physics of elementary particles, the nucleus, transuranic elements, neutron physics, problems of development of new accelerating methods and instruments, problems of automation and the mathematical processing of experimental data, and problems related to the construction of new experimental installations. The survey is also intended to acquaint the reader with new aspects of the theoretical and experimental physics of elementary particles and the atomic nucleus as they are developed in all parts of the world.

The series "Particles and Nucleus" is intended for qualified scientific researchers. Undoubtedly, it will also be very suitable for young scientific researchers, post-graduates, and students of higher courses of physics, mechanics and mathematics, and a number of engineering faculties since the surveys, written by specialists, will acquaint the reader with the problems of modern nuclear physics. It may also prove useful to other specialists not directly occupied with problems of nuclear physics in view of the considerable influence which the new experimental methods of investigation of the microscopic world exert on the technical progress of present-day society.

The publication of the series "Particles and Nucleus" will further the development of all the branches of nuclear physics that are so well represented at the Joint Institute for Nuclear Research, and have been extended to the countries that participate in the Joint Institute for Nuclear Research.

Translated from Problemy Fiziki Élementarnykh Chastits i Atomnogo Yadra, Vol. 1, No. 1, pp. 5-6, 1970.

ELASTIC SCATTERING OF PROTONS BY NUCLEONS IN THE ENERGY RANGE 1-70 GeV

V. A. Nikitin

Experimental methods for studying high-energy elastic scattering of protons by protons and nuclei are reviewed. The basic experimental results for pp and pd scattering for $E > 1$ GeV are given. The problem of checking dispersion relations for pN scattering is discussed. Models for high-energy nucleon–nucleon interactions are reviewed.

INTRODUCTION

Ever since Rutherford's famous experiments, elastic-scattering experiments have played an important role in the physics of elementary particles because they yield the information related most directly to the structure of the scattering center and the dynamics of the interaction.

A two-particle reaction is described completely by the scattering amplitude $A(s, t, \sigma)$ (where σ are the spin variables of the particles). At present it is not possible to completely reconstruct the function A at high energies [$s > 1$ (GeV/c)^2]. Experimentally, the following quantities can be observed: 1) $d\sigma/d\omega = |A|^2$ (measurement of the differential cross section), 2) $\mathrm{Im}\, A|_{t=0} = (k/4\pi)\sigma_{tot}$ (measurement of the total interaction cross section), and 3) $\mathrm{Re} A|_{t=0}$ (measurement of the real part of the nuclear amplitude through analysis of the differential scattering cross section for small t in the Coulomb-scattering region).

There has been little study of the dependence of A on the spins at high energies. The polarization of the reaction products has been measured in certain reactions. The experiments with polarized beams and polarized targets which are necessary for a detailed study of the function $A(\sigma)$ have only been begun.

Although the experiments which have been carried out in this field are not exhaustive [1-3], the data obtained in various laboratories on elastic scattering and on two-particle inelastic reactions constitute a rich store of material for checking and developing the basic directions of the theory. We will outline the most important of these directions which are based on information about the elastic scattering of particles.

1. Many limitations have been obtained on the behavior of $A(s, t)$ and $d^nA(s, t)/dt^n$ at fixed t and as $s \to \infty$ on the basis of the postulates of quantum field theory. Many relations follow directly from the basic axioms [12], so it is quite important to check them. An example is the familiar Pomeranchuk theorem and its generalizations. The cross sections for reactions of the type

$$a + b \to a + b \quad (\alpha);$$

$$\bar{a} + b \to \bar{a} + b \quad (\beta)$$

should be asymptotically equal, and the amplitudes should be related by

$$\mathrm{Re}\, A^{(\alpha)} = -\mathrm{Re}\, A^{(\beta)};$$

$$\mathrm{Im}\, A^{(\alpha)} = \mathrm{Im}\, A^{(\beta)}.$$

Joint Institute for Nuclear Research, Dubna. Translated from Problemy Fiziki Élementarnykh Chastits i Atomnogo Yadra, Vol. 1, No. 1, pp. 7-70, 1970.

An interesting restriction has been found on the behavior of the diffraction cone [13]:

$$b_J = \frac{d}{dt}(\ln|A|^2) \leqslant C\ln^2 s.$$

2. Much attention in both theory and experiment is devoted to checking the dispersion relations, which relate the real and imaginary parts of the elastic-scattering amplitude. In certain cases this relationship follows directly from the principles of causality, unitarity, and Lorentz invariance. The search for the applicability limits of these principles is of decisive importance for all science.

3. Since 1961 analysis of high-energy processes based on the analytic continuation of the scattering amplitude into the complex orbital-angular-momentum plane (the theory of Regge poles) has met with uneven success. In order to obtain by this approach results which can be checked experimentally, the basic axioms of field theory must be supplemented by certain specific assumptions about the analytic properties of the amplitude. The dynamics of the interaction is expressed in terms of a set of singularities of the analytic function. The problem here is to find the function having the fewest free parameters which is capable of describing the greatest number of phenomena. At present there is much optimism about this situation. The theory of complex angular momentum describes much experimental information on two-particle reactions. The hope has arisen that multiparticle reactions may be included in the studies, but this is far from a reality. A drawback of this approach is that many parameters must still be found experimentally. There is still uncertainty about the complete set of singularities which the amplitude has in the complex angular-momentum plane. There is no rigorous procedure available for calculating the contribution of branch-point singularities to observable quantities. The solution of these problems awaits progress both in theory and experiment.

One of the most interesting results of this approach is that an explicit relationship has been established between the properties of resonances and the scattering dynamics. Evidence of this is that the scattering is governed by the presence of the bound (or quasi-bound) states in the crossing channel (for pp scattering, these would be bound states in the $p\bar{p}$ system). If there are no bound states in the crossing channel, there will be no scattering in the forward channel. For example, the reaction

$$K^- p \to \pi^+ Y^{*-}$$

is strongly suppressed, at low momentum transfers, since K-π^- (the crossing channel of the reaction) has no bound states. This property of two-particle reactions is well supported experimentally. For this and many other reasons, we can conclude that the Regge-pole theory reflects the basic characteristics of hadron behavior. It may be that this theory will be the basis for a systematic theory of strong interactions [14-17].

4. Regge-pole theory describes scattering in the range $-t \ll s$; to completely describe reactions for any t is a complicated problem. The problem is formulated in this manner in the optical model, but here the classical potential turns out to depend strongly on the energy.

Several purely empirical attempts have been made to select a universal function $F = |A(s, t, \sigma)|$, which describes experiments for all t and s. Work has been begun in this direction by Krisch and Orear [18-21].

5. Many studies have been devoted to a relativistic generalization of the optical model. Logunov and Tavkhelidze [22] were the first to show that, according to the principles of quantum field theory, a system of two interacting particles may be described by a Schrödinger-type equation with a generalized complex potential which depends on the energy and momentum of the particles. Using this approach, one can both find the scattering amplitude and study the structure of the bound states. The potential, of course, cannot be determined theoretically. Only some of the properties which follow from the general axioms and from the form of the interaction are known. A quasi-potential approximation was used in [23] to describe experimental data on pp scattering in the momentum-transfer range 0.2-1 $(\mathrm{GeV}/c)^2$. The quasi-potential is written as a function with several parameters which are found empirically. This method is interesting because it is based on relativistic quantum field theory, and there is the possibility that it will yield a common description of scattering over the entire momentum-transfer range for a fixed energy. The quasi-potential found may be then used to solve problems not directly related to experiments (to study the analytic properties of amplitudes, form factors, etc.).

In general features, these are the questions which we will discuss below. Section 2 contains a more detailed examination of each of these directions along with experimental data. We turn now to the experimental procedures and a review of the results which have been obtained.

We will discuss here only experiments on proton–nucleon scattering, because of the unique methodological features and the great amount of experimental information available. The theory also deals with problems involving only the proton–nucleon interaction.

1. METHODS AND RESULTS OF STUDIES OF ELASTIC SCATTERING OF PROTONS BY PROTONS AND DEUTERONS

1.1. Spectrometry of Recoil Particles

There is much interest in the small-angle scattering of charged hadrons, in which case the Coulomb amplitude is large. Interference between electromagnetic and nuclear interactions yields information about the magnitude and sign of the real part of the nuclear scattering amplitude. However, it is difficult to carry out experiments at small angles.

Before 1962, the only way to measure the small-angle elastic scattering of protons with energies above 1 GeV was based on the photoemulsion method (perpendicular irradiation, as described in [11]).

The possibilities of this method are severely limited by the slow rate at which a statistically significant amount of information can be accumulated. For methodological reasons, experimental studies of nucleon–nucleon scattering based on electronic apparatus have excluded the small-angle range, in which Coulomb scattering occurs and in which interference phenomena are possible.

A method which permits a determination, within 1-3%, of the differential cross section for elastic scattering of protons by protons and nuclei, including the Coulomb-scattering range, was developed and successfully used in studies [1-10] carried out in the High-Energy Laboratory of the Joint Institute for Nuclear Research. The method consists essentially of: a) detecting the emission angle and momentum of the secondary particles created at the inner target of the accelerator in the angular range 70-90° and in the momentum range 40-500 MeV/c, and b) using a target which is of size and mass small enough to avoid distortion of the secondary-particle angles and momenta. The particle yield required can be obtained with a small target through the use of an irradiation setup in which the accelerated beam intersects the target several times.

The experimental setup is shown in Fig. 1. The inner accelerator beam passes through target T, a polymer film 0.5-2 μ in thickness. The film target is suspended against the target drive mechanism on glass filaments 7 μ in diameter. The recoil protons are detected by photoemulsion cameras 1 at the end of evacuated channel 2, three meters from the target. The photoemulsions are protected by concrete shielding 3 on the side toward the linear part of the accelerator, which is a background source. Telescopes of scintillation detectors S_i are used to select the irradiation conditions and to determine the proton flux incident on the target (beam monitoring).

In order to study elastic scattering through angles $\theta_{c.m.} \lesssim 1.5°$ (in which Coulomb scattering occurs), one must measure the angles and the momenta at which recoil protons having energies $\lesssim 1.5$ MeV are emitted. This places severe restrictions on the target thickness and its dimension along the beam direction. A simple calculation based on the kinematics of the elastic pp interaction and taking into account multiple scattering in the target shows that the target must be ≈ 1 μ thick and have a dimension of ≈ 1 cm along the beam. Its mass ($\approx 10^{-4}$ g) turns out to be 10^{-5} that of the targets ordinarily used, so repeated beam-target intersections are required to provide a sufficiently large number of interactions between primary protons and the target.

The system is designed so that the beam will pass through the target 6×10^3 times in 200 msec. Experimentally, knowing the elastic-scattering cross section and by measuring the flux of elastically scattered protons, one can find the same quantity; it turns out to be $\approx 7 \times 10^3$. In other words, each proton travels a distance of ≈ 15 mm through the 2-μ thick film. Figure 2 shows the range distribution of secondary particles in the emulsion. The maximum corresponds to recoil deuterons during elastic scattering of 10-GeV protons through a c.m. angle of 2.3°. Evaporated and cascade particles from carbon nuclei make up the background. Three factors make the primary contribution to the width of the maximum: 1) the

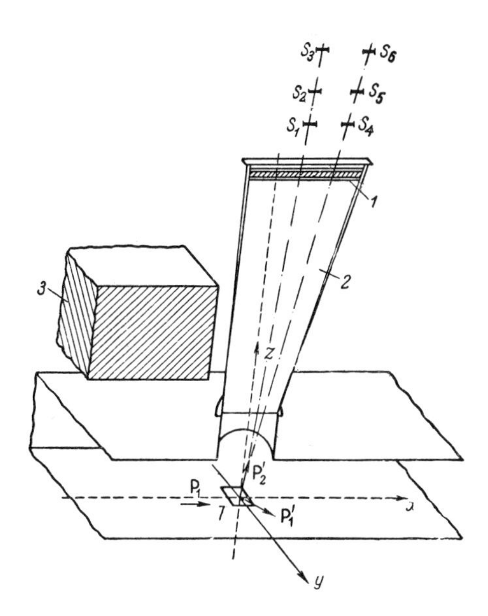

Fig. 1. Experimental setup for studying elastic scattering of protons by protons and nuclei by means of an inner film target. 1) Photoemulsion cameras; 2) ion guide; 3) radiation shielding; S_i) scintillation detectors; T) film targets; P_1) inner accelerator beam.

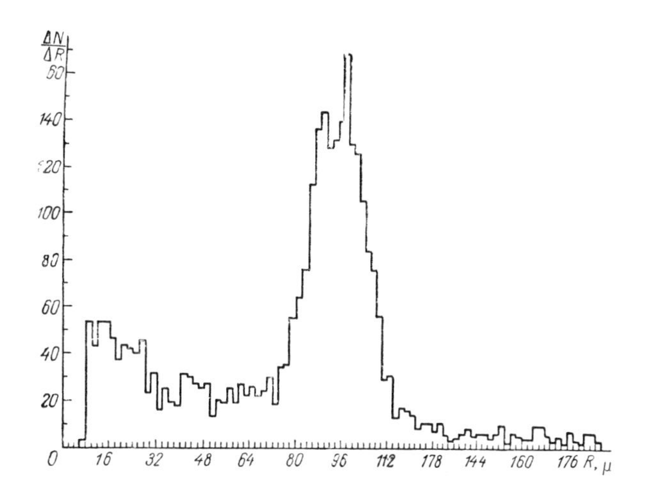

Fig. 2. Range distribution of recoil particles in the emulsion.

angular scatter in the primary proton beam and the finite extent of the target; 2) multiple Coulomb scattering of recoil protons in the target film; and 3) inaccuracy in the measurements of the range in the emulsion and straggling. For protons having a range of 1000 μ, the first factor is the basic one; for shorter-range protons, the second factor is.

The number of particles in the elastic-scattering peak is related in a definite manner to the differential scattering cross section. The absolute differential cross section, in millibarns per steradian, can be found by measuring the C^{11} activity induced in the target. The cross section for the reaction $C^{12}(p, pn)C^{11}$ in the energy range $E \geq 1$ GeV is known at present within $\approx 5\%$ [24].

In a discussion of the procedure described here, the question arises of the contribution from scattering by protons bound in the nucleus (so-called quasi-elastic scattering). Since there is carbon in the target, it is necessary to distinguish between elastic and quasi-elastic scattering; this can be done reliably because of the high resolution of this procedure. The accuracy with which the emission angle of the recoil particle can be measured is governed by the ratio of the target size to the target-detector separation: $\Delta\theta = 1\ \text{cm}/300\ \text{cm} \approx 3$ mrad.

The accuracy with which the particle momentum can be determined from its range in the emulsion is $\Delta p \approx 3$ MeV/c. These quantities must be compared with the angular and momentum scatter of the recoil protons which arise as the result of quasi-electric scattering:

$$\Delta p_{\text{quasi}} \approx p_{\text{fermi}}$$

$$\Delta\theta_{\text{quasi}} \approx \frac{p_{\text{fermi}}}{p_{\text{elastic}}} \approx \frac{150\ \text{MeV/c}}{300\ \text{MeV/c}} \approx 0.5\ \text{rad};$$

here $\Delta\theta \ll \Delta\theta_{\text{quasi}}$ and $\Delta p \ll \Delta p_{\text{quasi}}$, so there is essentially complete resolution of the elastic and quasi-elastic events.

In an experimental proof of this assertion, a thin aluminum target was irradiated by the inner beam of an accelerator, and an attempt was made to find in the recoil-particle spectrum a peak corresponding to elastic pp scattering. It was found that the admixture of quasi-elastic events within the diffraction cone was no greater than 2%.

Spectrometry of the recoil particles thus permits one to single out elastic scattering by protons or deuterons in the inner target. Figures 3a and 3b show how to determine the momentum range in which the spectrometry must be carried out in order to measure the cross sections in the Coulomb-interference range. The cross section for pp and pd scattering are shown as functions of t, the square of the momentum transfer. The scales at the bottom of the figure show other quantities characterizing the given experiment:

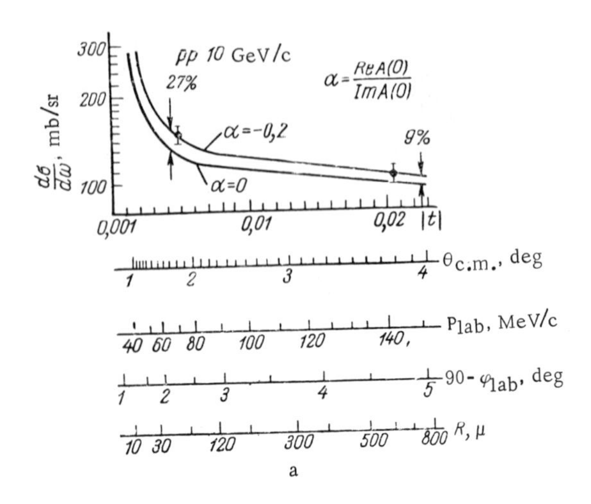

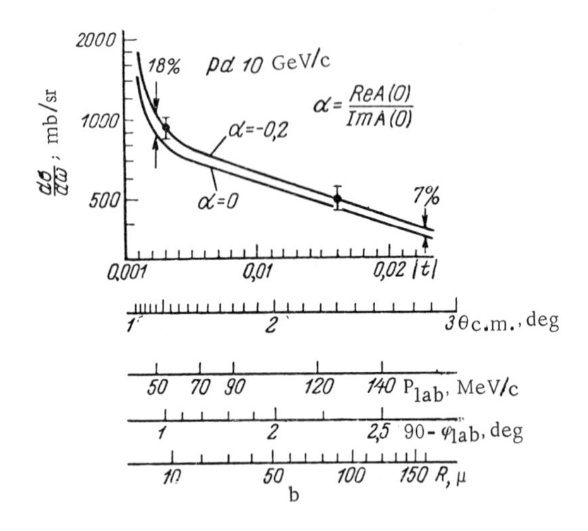

Fig. 3. Interference curves for: a) pp scattering; b) pd scattering.

the lab and c.m. scattering angles, the c.m. escape angle of the recoil particle, and its momentum. Two cases are treated: diffraction scattering and scattering in the case in which the real part of the nuclear scattering amplitude is 20% of the imaginary part. In current experimental practice, the relative error in the measurement of the differential cross section is 3-5%, and the absolute error is 7%. To distinguish between the two cases here, one must work in the momentum range 50-60 MeV/c (corresponding to a proton energy of ≈ 1 MeV and a deuteron energy of ≈ 0.5 MeV). The coordinate of each particle must be measured within ≈ 2 mm, and the momentum must be measured within $\approx 10\%$.

The simplest detector which satisfies these conditions is a nuclear emulsion. The use of a nuclear emulsion in this experimental geometry improves the productivity by a factor of about 100 over that of the method involving perpendicular irradiation of stacks in the inner accelerator beam. This is sufficient to resolve the "small-angle" problem; i.e., to determine whether the scattering corresponds to pure diffraction or whether there is a small-angle singularity which may be interpreted as interference of nuclear and Coulomb scattering.

Below (in Part 1.5) we examine a study in which a logical step has been taken toward improving the procedure for detecting elastic scattering in terms of recoil particles: the emulsions are replaced by semiconductor detectors connected to a computer. We will discuss the most important advantages and disadvantages of this procedure.

1. For an arbitrarily high primary-beam energy, the diffraction-cone region $|t| \leq 0.3$ $(\mathrm{GeV/c})^2$ corresponds to recoil particles in the ranges $90° \geq \theta \geq 70°$, $p \lesssim 500$ MeV/c, which are convenient for measurements. For comparison, we note that the method of detecting elastic coefficients on the basis of the fast scattered particle, widely used at present, runs into ever-increasing difficulties as the energy is increased, because of the reduction of the scattering angles.

2. Inelastic processes greatly disrupt the kinematic relationship between angle and momentum for the recoil particle. This provides an easy way to distinguish elastic events from the background – an especially important consideration in studying scattering by nuclei for which the level-excitation or nuclear-decay energies are small.

3. The momentum and emission angle of a slow particle may be measured highly accurately, so the systematic error in the calculation of the invariant t and the differential cross section may be kept small.

This method is comparatively easy to use with the inner accelerator beam, but it is difficult to use with the outer beams, since the low-mass target requires a highly intense beam. In [25], however, this difficulty was successfully overcome.

Because of the advantages of this procedure for recoil-particle spectrometry, it seems likely that it will be widely used in future high-energy accelerators ($\lesssim 100$ GeV).

1.2. Method of Magnetic Spectrometers

Magnetic spectrometers, in which the particle coordinates and angles are measured by spark chambers or scintillation hodoscopes, are currently widely used to study high-energy elastic scattering involving extracted beams. (The use of such a spectrometer in experiments with moderate-energy particles was

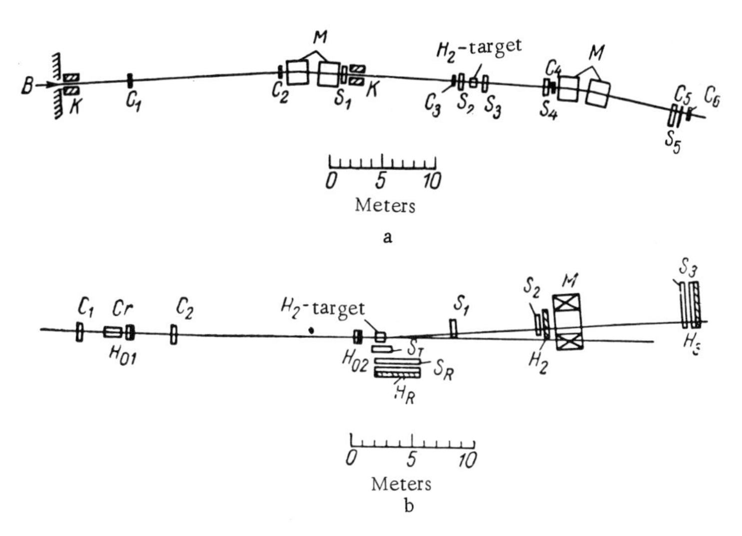

Fig. 4. Magnetic spectrometers of Bellettini et al. [26] (CERN) (a) and Foley et al. [27] (USA) (b). B) Monochromatic beam; K) collimator; M) analyzing magnets; C) control scintillation detectors; S) spark chambers; Cr) Cerenkov detector; H) scintillation hodoscopes.

described in [99].) An important part of the spectrometer is the computer, which receives digital information from the detecting electronics. A computer is used because the properties of large numbers of particles must be measured and analyzed. In the Coulomb-scattering range (corresponding to a lab angle of 0.5°), the particles which interact elastically in the target cannot be singled out in a simple manner from the direct beam, and there are relatively few of them. For example, the cross section for elastic πp scattering in the momentum-transfer range 40-70 MeV/c is ≈ 0.2 Mb. When a liquid-hydrogen target 20 cm long is used, only one beam particle in 10^4 will undergo the elastic scattering of interest here.

Typical experimental setups are shown in Fig. 4. The apparatus shown in Fig. 4a operates as follows. The proton beam obtained from the inner beryllium target of the accelerator has a diameter of 5 mm after the last collimator and an intensity of 300 particles per cycle. Elastic pp scattering in the hydrogen target is identified by measurement of the particle's angle and momentum before and after the target. The protons incident on the apparatus are detected by scintillation counters C_1 and C_2. Particles passing through the target (either scattered or not scattered) are detected by counters C_{3-5}. Scintillation counter C_4 has an aperture which allows most of the unscattered beam to pass. Scintillation counter C_6 detects most of the unscattered beam, and is connected in anticoincidence with all the other counters. This manner of connecting counters C_4 and C_6 significantly reduces the number of times the apparatus is triggered by particles passing unscattered through the target. The coordinates of particles are measured within ± 3 mm along the entire trajectory by acoustic spark chambers. Chambers S_1 and S_2 measure the angle before the target, while chambers S_3 and S_4 measure that after the target. The results are sufficient for calculating the scattering angle. Chambers S_{3-5} detect the angle through which a particle is deflected in the analyzing magnets M; this serves to determine the particle's momentum. The set of coordinates and indications of the scintillation counters are recorded on magnetic tape, and then the information is treated on an IBM 7090 computer. The angular and momentum resolution of the spectrometer are limited primarily by multiple scattering in the material of the counters, chambers, and the hydrogen target. Evacuation reduces the background in scattering in the system for producing and transporting the beam.

Figure 5a shows the momentum spectrum of the particles detected; the momentum resolution is seen to be ± 70 MeV/c. A certain fraction of the events involving pion creation fall in the elastic-scattering peak. In the case of the pp interaction, this fraction is small (4%) and may be taken into account by a suitable correction. For the case of pd scattering, however, for which the inelastic-channel threshold is 2.2 MeV (c.m.), this procedure is essentially incapable of resolving elastic from inelastic events.

A similar procedure has been used by Taylor et al. [28] on the Nimrod accelerator to study elastic pp scattering at a momentum of 7.85 GeV/c.

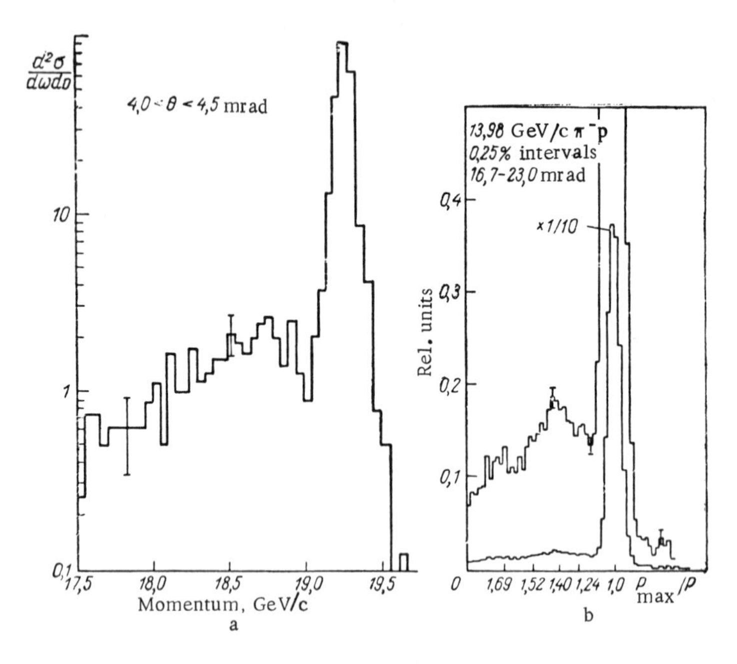

Fig. 5. Proton spectrum recorded by the magnetic spark spectrometer (a) and the spectrum of π^- mesons recorded by a magnetic scintillation spectrometer (b).

The group of Foley, Lindenbaum, et al. [27] at Brookhaven has long been using a spectrometer in which the coordinates of the beam particles and the scattered particles are measured by scintillation hodoscopes. An advantage of the latter over spark chambers is their speed: while a spark chamber detects 10-100 particles per accelerator cycle (at an ordinary beam intensity of ≈ 200 msec), a scintillation hodoscope may detect $\approx 10^4$ particles. However, the high detection rate is paid for by a tenfold accuracy loss in the measurement of the spatial coordinates. In 1967 Foley, Lindenbaum, et al. [27] obtained a spatial resolution of ± 1.5 mm (see Fig. 4b). The principle by which the elastic scattering was identified was exactly the same as that described above. Spark chambers were used only during the procedural stage of the study; the physical experiments were carried out without them. The hodoscope H_R was used to study scattering involving a momentum transfer above 200 MeV/c, in which case, the diffraction cone was studied. Studies in the interference angular range were carried out without the hodoscope H_R since the recoil particle did not leave the hydrogen target. In this case, the hodoscopes H_2 and H_3 fell within the direct beam. The hodoscope H_3 contains 120 scintillators for measuring the coordinates of the particle in the horizontal plane and 24 scintillators for measuring coordinates in the vertical plane. The other hodoscopes have fewer elements.

Figure 5b shows the momentum spectrum of the particles detected; the resolution is the same as in the experiments of the CERN group: ± 60 MeV/c at an energy of 14 GeV. The contribution of inelastic events amounts to a few percent. At small scattering angles the background due to the target walls is 30-50% of the effect (scattering by hydrogen). This background was measured in an experiment with an empty target.

Using this procedure, one can now measure the differential cross section with a relative accuracy of 5%. An obvious advantage of magnetic spectrometers is that they may be used with beams of various particles. The Brookhaven group has carried out experiments with p, π^+, π^-, and $\bar{p}$ particles. In principle it is possible to study the scattering of $K^\pm$ mesons. In evaluating the capabilities of this procedure, however, one should take note of its disadvantages: the spectrometer is a complicated set of measuring instruments and computational apparatus, considerable expenditures of time and money are required to produce such a system, and it is extremely difficult to find and analyze the systematic errors in the measurements. It is sufficient to state that the Brookhaven data on the real part of the pion scattering amplitude eventually go beyond the stated error limits, and the measured real part of the pp-scattering amplitude at 20 GeV contradicts the CERN experiments (see the discussion of the data in Part 2.3). The two CERN experiments carried out at a proton momentum of 10 GeV/c yielded results differing by two standard

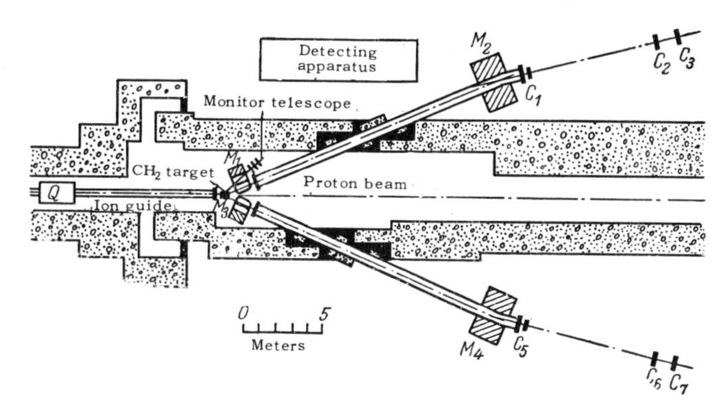

Fig. 6. Apparatus for studying large-angle pp scattering. M_i) Analyzing magnets; C_i) scintillation counters.

deviations [26]. All this implies that the investigators have not yet traced some important source of systematic errors in their procedures.

The momentum resolution of all the spectrometers which have been used up to the present is about $\pm$ 60 MeV/c at an energy of $\approx$10 GeV; this is not sufficient for identifying elastic scattering by nuclei. Sutter et al. [29] have developed a spectrometer based on wire-type spark chambers whose energy resolution is $\pm$ 1.7 MeV at a energy of 1 GeV for the primary proton beam. This resolution is achieved by reducing the pressure in the spark chamber to 70 torr. The chambers have needles 0.05 mm in diameter and Mylar walls 0.01 mm thick. This spectrometer has been used to study elastic scattering by nuclei at 1 GeV [30]. At higher energies (10 GeV and above), the absolute energy resolution of the instrument is again inadequate for discriminating between nuclear decay and excitation. The only adequate method remains that of detecting the recoil particles (see Part 1.1).

The differential cross section for elastic scattering decreases with increasing momentum transfer approximately according to $d\sigma/dt \sim e^{bt}$, where $b \approx 10\ (\mathrm{GeV/c})^{-2}$, while for $|t| \geq 2\ (\mathrm{GeV/c})^2$ it is proportional to $Ae^{c\sqrt{t}}$, where $1/c \approx 150$ MeV/c. For $|t| = 1\ (\mathrm{GeV/c})^2$ the cross section is $\approx 10^{-4}$ of $d\sigma/dt \cdot (t = 0)$. This estimate shows that in experiments with large-angle scattering one must strive to attain the maximum instrumental efficiency while simultaneously increasing the strictness of the criteria used to select elastic events, in order to reduce the background contribution. To illustrate the procedure for studying the range of large momentum transfer, we consider the study by Akerlof et al. [31] at Argonne National Laboratory (USA).

The apparatus (Fig. 6) consists of a combination of two magnetic channels. The extracted proton beam, with an intensity of $\approx 10^{11}$ particles per cycle, is directed onto a polyethylene target. The secondary particles pass along magnetic ducts M_1, M_2 and M_3, M_4 and are detected by telescopes C_{1-3} and C_{5-7}, connected in coincidence. The solid angle subtended by the apparatus is governed by only the last counter in the telescope, C_6: $\Delta\omega = 3 \cdot 10^{-5}$ sr. A calculation of $\Delta\omega$ must take into account the focusing properties of magnets M_1 and M_2. The angles between the channel axes and the primary beam are selected taking account of the kinematics of the pp $\to$ pp process being studied. The momentum resolution of each channel is 10%. The relative error in the measurement of the differential cross section by means of this apparatus is 3-7%, and the absolute error is 5%.

Many experiments at high $|t|$ have been carried out at CERN by means of a single-arm scintillation magnetic spectrometer [32]. The basic characteristics of this apparatus are as follows. The liquid-hydrogen target is 10 cm long. The magnet channel passes secondary particles having a momentum scatter of $\Delta p/p \approx 2.5\%$. The solid angle of the channel is $7 \cdot 10^{-6}$ sr. A hodoscope of ten counters singles out elastic events. The momentum resolution of the spectrometer is 0.5–0.9%. A peak corresponding to pp $\to$ pp elastic scattering in the target is observed in the spectrum of particles detected by the hodoscope. The background-to-signal ratio is $\approx$ 20%. About 8% of the particles are absorbed in the counters singling out the beam. The relative accuracy of the measurements is 5%, and the absolute calibration accuracy is 7%.

1.3. Description of the Differential Cross Section at Low Momentum Transfer

We can express the c.m. differential cross section in terms of the nuclear and Coulomb scattering amplitude:

$$\frac{d\sigma}{d\omega} = |\,iA_J + A_r + A_c\,|^2. \tag{1.1}$$

Inelastic processes contribute greatly at high energies. The de Broglie wavelength of the colliding particles is much smaller than the interaction region, so the diffraction approximation

$$A_J(\theta) = \frac{RJ_1(kR\sin\theta)}{k\sin\theta}$$

may be used for the imaginary part of the scattering amplitude as a function of the scattering angle. Here R is the radius of the interaction region, and $k = p_{c.m.}/\hbar$ is the wave vector. Because of the property

$$J_1(2x) \approx xe^{-x^2}; \quad x \ll 1$$

of the Bessel function at small θ, we have

$$A_J(\theta) \sim \frac{R}{2}\exp\left[-\left(\frac{kR}{2}\right)^2\theta^2\right]. \tag{1.2}$$

We can adopt the following similar parametrization of the imaginary part of the scattering amplitude:

$$A_J(t) = \sqrt{O}\exp\frac{1}{2}(b_J t + c_J t^2);$$
$$O = \left(\frac{k}{4\pi}\sigma_{tot}\right)^2, \quad t = -2p^2_{c.m.}(1-\cos\theta), \tag{1.3}$$

where σ_{tot} is the total cross section for the interaction in these particles, and $b_J = R^2/2$ is customarily called the "slope parameter of the diffraction cone."

In this case, the optical theorem for spinless particles is automatically satisfied:

$$\text{Im}\,A(0) = \frac{k}{4\pi}\sigma_{tot}.$$

The quadratic term $c_J t^2$ extends the applicability range of (1.3) beyond that of (1.2). For the real part A_r of the nuclear amplitude it is at present quite difficult to write down a plausible analytic expression. Experiments give us only the $A_r(0)$ value; we know nothing about the dependence on t. We postulate the equation

$$A_r(t) = \alpha\sqrt{O}\exp\frac{1}{2}(b_r t + c_r t^2);$$
$$\alpha = A_r(0)/A_J(0). \tag{1.4}$$

Depending on the parameters b_r and c_r, the function $A_r(t)$ may be constant, decreasing, or increasing; this is completely adequate for describing the experimental data available.

The role of the magnetic moments in the Coulomb interaction is small at small angles, so we can use a simple expression for the amplitude of the electromagnetic interaction:

$$A_c(t) = \frac{2nkF(t)}{t}e^{i\eta}; \quad n = \frac{1}{137\beta_{lab}}. \tag{1.5}$$

The electric form factor of the proton and deuteron is known from experiments on electron scattering. In the range $|t| \lesssim 0.01$, where Coulomb scattering is important ($A_c \approx A_J + A_r$), it is nearly one. It is approximated well by

$$F(t) = \exp\left(\frac{1}{2} b_J t\right). \tag{1.6}$$

The basic equation for $d\sigma/d\omega$ (1.1), is approximate, for it is known from quantum mechanics that scattering amplitudes combine nonadditively when potentials combine additively. As Bethe [33], Solov'ev [34], and Locher [35] have shown, however, Eq. (1.1) may be retained at high energies if the phase of the Coulomb scattering with respect to the nuclear scattering is chosen correctly. In the quasi-classical approximation and with a phenomenological description of the nuclear interaction, the following equation is obtained for η:

$$\eta = 2n \ln \frac{\varphi}{\theta_{c.m.}}; \quad \varphi = \frac{1.06}{ka}, \tag{1.7}$$

where a is the radius of the strong-interaction region.

Relativistic calculations lead to a slight change in the constant in the expression for φ. We can write (1.1) as

$$\frac{d\sigma}{d\omega} = C\left[A_J^2(1+\beta) + A_r^2 + A_c^2 - 2A_c\left(A_r + 2nA_J \ln \frac{\varphi}{\theta_{c.m.}}\right)\right]. \tag{1.8}$$

The factor C is introduced for convenience in describing the differential cross sections, specified in arbitrary (relative) units, by means of Eq. (1.8). The parameter β corrects the imaginary part of the nuclear amplitude for possible spin effects. We can clarify this point as follows.

We assume the scattering amplitude is imaginary but that it depends on the spin state of the pp system. Then the optical theorem

$$\frac{d\sigma}{d\omega}(0) = \left(\frac{d\sigma}{d\omega}\right)_{opt} = \left(\frac{k}{4\pi}\sigma_{tot}\right)^2 \tag{1.9}$$

must of course be altered:

$$\frac{d\sigma}{d\omega}(0) = \left(\frac{k}{4\pi}\right)^2 \left(\frac{1}{4}\sigma_1^2 + \frac{3}{4}\sigma_3^2\right), \tag{1.10}$$

where σ_1 and σ_3 are the total cross sections of the singlet and triplet spin states of the interacting protons. The total cross section for the pp interaction, measured in an unpolarized beam and for an unpolarized target, is evidently

$$\sigma_{tot} = \frac{1}{4}\sigma_1 + \frac{3}{4}\sigma_3. \tag{1.11}$$

Equation (1.10) [with an account of Eq. (1.11)] leads to Eq. (1.9) in the case $\sigma_1 = \sigma_3$. If, on the other hand, $\sigma_1 \neq \sigma_3$, we find from Eq. (1.10) that

$$4\left(\frac{d\sigma}{d\omega}\right)_{opt} \gg \frac{d\sigma}{d\omega}(0) \gg \left(\frac{d\sigma}{d\omega}\right)_{opt}, \tag{1.12}$$

i.e., the cross section $d\sigma/d\omega$ (0) may be significantly higher than the optical point (1.9) because of a difference in the interaction of protons in the singlet and triplet state. This is reflected by the parameter β, which must be determined experimentally. It follows from (1.12) that only positive values of β are permissible. If, on the other hand, $\beta < 0$, the implication is that there is an error in the experimental data.

All the experiments on elastic scattering at energies $E \gtrsim 1$ GeV and in the range $0.002 < |t| < 0.3$ $(\text{GeV}/c)^2$ are described well by Eq. (1.8) with the amplitude parameterization* (1.3-1.7).

---

* Scattering by complex nuclei may be described by Eq. (1.8), but in a narrower t range. The region of diffraction minima is of course not included in (1.8).

TABLE 1. Characteristics of Elastic pd Scattering

$E_{kin. lab}$, GeV	p_{lab}, GeV/c	α_{pd}	α_{pn}	b_j, $(GeV/c)^{-2}$	c_j, $(GeV/c)^4$	σ_{el}, mb.
1	1,70	$-0,14\pm0,08$	$-0,30\pm0,20$	$33,7\pm1$	42 ± 10	$10,7\pm0,7$
2	2,78	—	—	$37,5\pm1,5$	83 ± 12	$10,2\pm0,7$
4	4,85	—	—	$37,8\pm1,4$	72 ± 14	$9,5\pm0,7$
6	6,87	$-0,30\pm0,09$	$-0,38\pm0,17$	$36,0\pm0,8$	45 ± 6	$9,6\pm0,7$
8	8,89	$-0,26\pm0,09$	$-0,25\pm0,17$	$36,5\pm1,0$	40 ± 7	$9,3\pm0,7$
10	10,90	$-0,39\pm0,08$	$-0,47\pm0,17$	$34,3\pm0,9$	34 ± 6	$9,0\pm0,6$

TABLE 2. Characteristics of Elastic pp Scattering

$E_{kin. lab}$, GeV	p_{lab}, GeV/c	α_{pp}	b_j, $(GeV/c)^{-2}$	β
2	2,78	$-0,12\pm0,07$	$7,3\pm0,5$	$-0,1\pm0,09$
4	4,85	$-0,38\pm0,07$	$7,8\pm0,5$	$0,0\pm0,1$
6	6,87	$-0,30\pm0,07$	$9,2\pm0,4$	$0,0\pm0,1$
8	8,89	$-0,33\pm0,08$	$9,8\pm0,5$	—
10	10,90	$-0,26\pm0,05$	$9,2\pm0,5$	$0,0\pm0,1$

1.4. Experimental Results on Elastic pp and pd Scattering

Here we will review the experimental data obtained in studies of elastic pp and pd scattering. We will make some general comments.

At small angles, $|t| < 0.3\ (GeV/c)^2$, pp scattering in all the studies is analyzed on the basis of the interference equation (1.8). The parameters α, b_J, and β are calculated by the method of least squares. Taking account of the quadratic term in the exponential function in (1.3) does not reduce χ^2, so it is usually immediately assumed that $c_J = 0$ [in the case of pd scattering, we have $c_J \neq 0$ even at $|t| < 0.2\ (GeV/c)^2$]. The same is true of the spin parameter β at energies $E > 2$ GeV. The typical value of β is 0 ± 0.1; i.e., the spin-dependent amplitudes yield a contribution no greater than $\sqrt{\beta} \leq 30\%$. At lower energies, β is usually nonvanishing and must be taken into account. For example, with $p_{lab} = 1.3$ GeV/c, we have $\beta \approx 1$.

The data which have been published up till now are not sufficient for calculating the slope parameter b_r of the real part of the scattering amplitude, since information about the function $A_r(t)$ is essentially contained only in the Coulomb-scattering region, i.e., $0.003 \lesssim |t| \lesssim 0.02\ (GeV/c)^2$. In this narrow range of the argument, it is possible to find only the function; the derivative $b_r = 2(d/dt) \ln A_r(t)$ remains undetermined. For calculations, therefore, it is assumed that $b_r = b_J$ and $c_r = c_J$. A variation of b_r within $\pm 30\%$ does not affect the values of the other calculated quantities.

A series of studies has been carried out in the Joint Institute for Nuclear Research involving proton-deuteron elastic scattering at 1, 2, 4, 6, 8, and 10 GeV and proton-proton elastic scattering at 2, 4, 6, 8, and 10 GeV [1-10]. The differential cross section has been measured in the $|t|$ range from 0.003 to 0.2 $(GeV/c)^2$.

The systematic error in the measurements is $\approx 3\%$, while the absolute error is 7%. The range of small scattering angles, in which the interference of Coulomb and nuclear interactions is significant, yields the greatest amount of information for determining the quantity $\alpha_{pp, pd}$ – the ratio of the real part of the nuclear scattering amplitude to its imaginary part.

The results of an analysis on the basis of Eq. (1.8) are shown in Tables 1 and 2.

A large part of the error in $\alpha_{pp, pd}$ is due to the inaccuracy of the absolute calibration of the differential cross section ($\pm$ 7%). Figure 7 shows the c.m. differential cross sections for pd scattering, while Fig. 8 shows them for pp scattering, along with the data from other laboratories.

TABLE 3. Values of $\alpha_{pn} = \frac{\operatorname{Im} A_{pn}}{\operatorname{Re} A_{pn}}\Big|_{t=0}$, Obtained from Experiments on pp and pd Scattering in the Coulomb-Interference Range

p_{lab}, GeV/c	α_{pn}	Statistical error	Source
0,845	+0,05	0,2	[40]
1,29	−0,68	0,25	[41]
1,39	−0,48	0,13	
1,54	−0,36	0,18	
1,69	−0,50	0,15	
1,70	−0,30	0,20	[7, 8]
6,87	−0,38	0,17	
8,89	−0,25	0,17	
10,9	−0,47	0,17	
19,3	−0,35	Not indicated	[42]

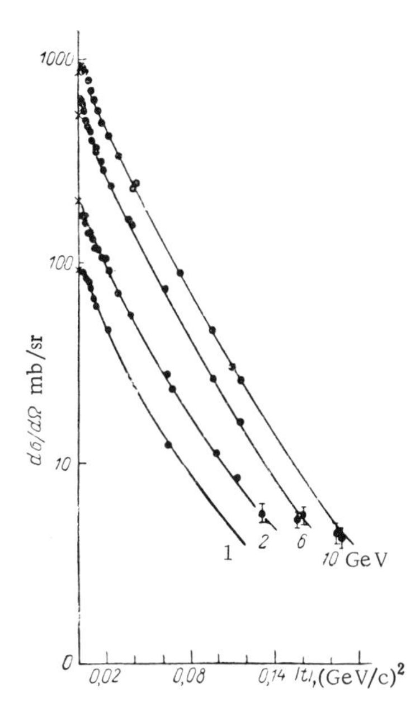

Fig. 7. Differential cross sections for pd scattering.

Most of the elastic pd scattering occurs in this momentum-transfer range [at $t = -0.2$ (GeV/c)^2 the differential cross section falls off by a factor of about 500]. Accordingly, the total cross section for elastic pd scattering can be found from the elastic distribution for the pd→pd reaction.

The error in the quantity σ_{el} indicated in Table 1 is governed completely by the calibration error. We note that the data on σ_{el} and the parameters b_J and c_J of the pd-scattering diffraction cone obtained in [7, 8] are unique, for the magnetic-spectrometer methods do not single out reactions involving deuteron decay. In Fig. 9 this is illustrated by a comparison of the data of [7, 8] with the data of [42].

Figure 10 shows the parameters b_J and c_J for pd scattering. Interestingly, the energy dependence $b_J(E)$ is similar to the energy dependence $\sigma_{tot.\,pd}(E)$ of the total cross section; this should be expected on the basis of the diffraction model for elastic scattering.

The quantity $\alpha_{pn} = (\operatorname{Re} A_{pn} | \operatorname{Im} A_{pn}) |_{t=0}$ is found from a joint analysis of data on pp and pd scattering on the basis of the Glauber model for the interaction of fast particles with a deuteron (see Part 1.6).

Data on the real part of the proton-neutron scattering amplitude are summarized in Table 3. The calculation procedure for pd-pd reactions was used in all studies. There have been no direct measurements of the real part of the pn scattering amplitude at high energies.

Table 4 summarizes data on the real part of the pp scattering amplitude. All the results were obtained by calculation of the parameter α in Eq. (1.8) on the basis of the differential cross section in the Coulomb-interference region.

There have been several studies of large-angle proton scattering [31, 32, 43-46]. We select for discussion here the results of [31, 32] as the most typical. (The characteristics of the apparatus used have been described above.)

Akerlof et al. [31] measured the differential cross section for pp scattering through a c.m. angle of 90° as a function of the energy of the primary particles. The function $d\sigma/dt\ (p_{c.m.}) | \theta_{c.m.} = 90°$ has the following features (Fig. 11).

1. There is a clearly defined break at the point $p^2_{c.m.} \approx 3.3$ (GeV/c)^2 ($p_{lab} \approx 8$ GeV/c) on the $\ln (d\sigma/dt)\ (p_{c.m.})$ curve. This can be explained on the basis of the optical model by the presence of two regions in the proton having different densities of hadron matter or displaying different types of interactions.

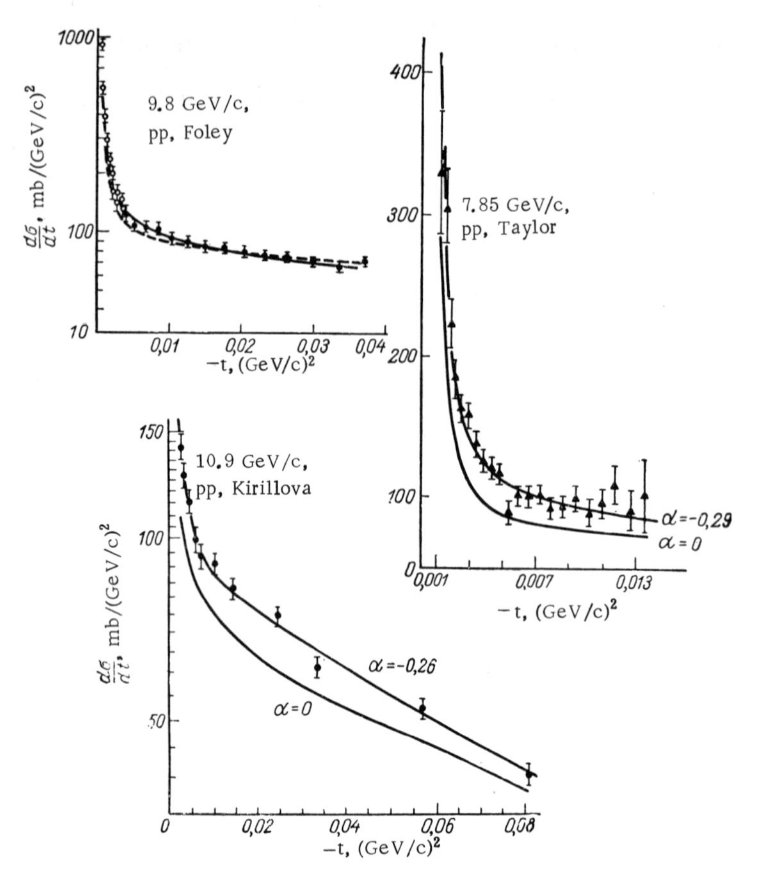

Fig. 8. Differential cross sections for pp scattering.

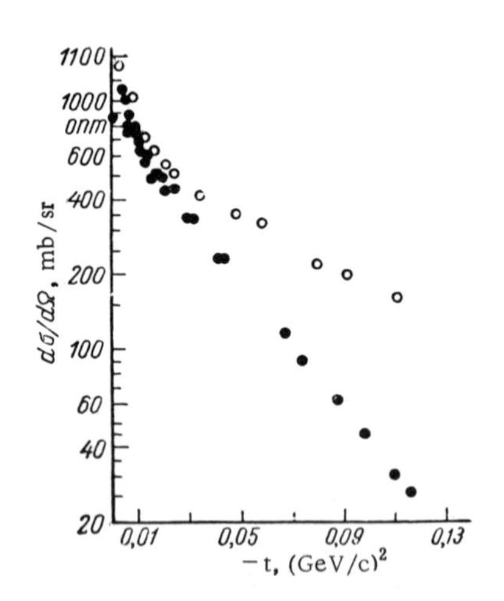

Fig. 9. Differential cross sections for pd scattering. ○) Data obtained by Bellettini with a magnetic spectrometer (the sum of the coherent and incoherent scattering); ●) data obtained by Kirillova by detection of recoil particles (coherent scattering only).

2. The function $d\sigma/dt$ ($p_{c.m.}$) does not display a resonance structure which could be introduced by means of a two-baryon quasi-bound state: pp → B → pp. If resonance B does exist, it is weak, and the contribution to the pp scattering cross section at $\theta_{c.m.} = 90°$ does not exceed 10%. It is interesting here to compare the situation with that of $\pi^- p$ scattering through an angle of 180°, for which the contribution of resonances is clearly defined.

Figure 12 shows the nature of the differential cross section for pp scattering at large momentum transfer and for various primary-beam energies [32].

At E > 10 GeV, the $d\sigma/dt$ (t) cross section displays an oscillatory structure. This interesting phenomenon is currently being widely discussed theoretically; the basic approach here is to take into account the higher orders of the interaction which a particle undergoes at the scattering center. In the optical model, for example, the scattering center is a complex potential (or quasi-potential). The n-th Born approximation in the solution of the Schrödinger equation may be treated as the n-th-order interaction at the potential.

1.5. Measurement of the Slope Parameter of the Diffraction Cone in the Energy Range 10-70 GeV

A group of investigators from the High-Energy Laboratory (Dubna) and the Institute of High-Energy Physics (Serpukhov) has measured the differential cross section for elastic pp scattering in the range $0.08 < |t| < 0.12$ $(GeV/c)^2$ and $10 < E_{lab} < 70$ GeV [49].

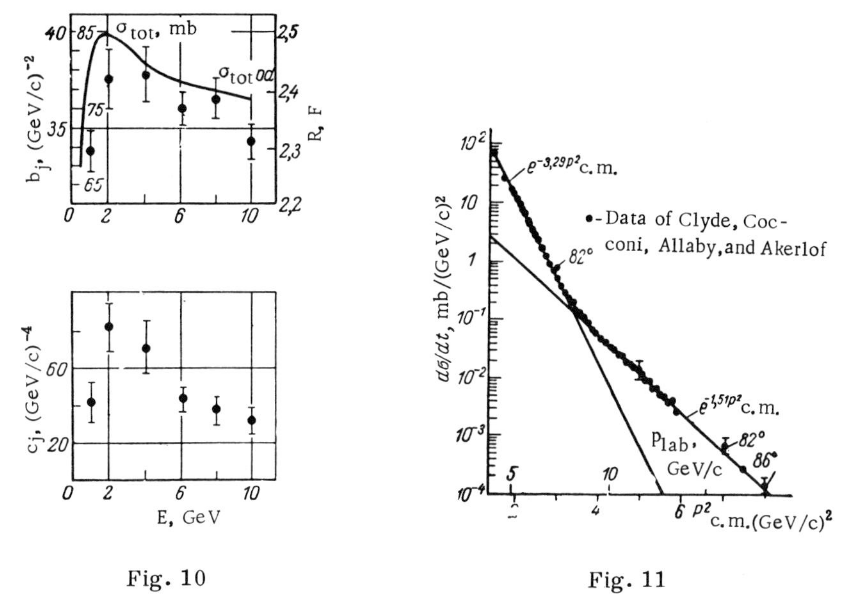

Fig. 10

Fig. 11

Fig. 10. Parameters of the pd scattering diffraction cone.

Fig. 11. Differential cross section for pp scattering through an angle $\theta_{c.m.} = 90°$.

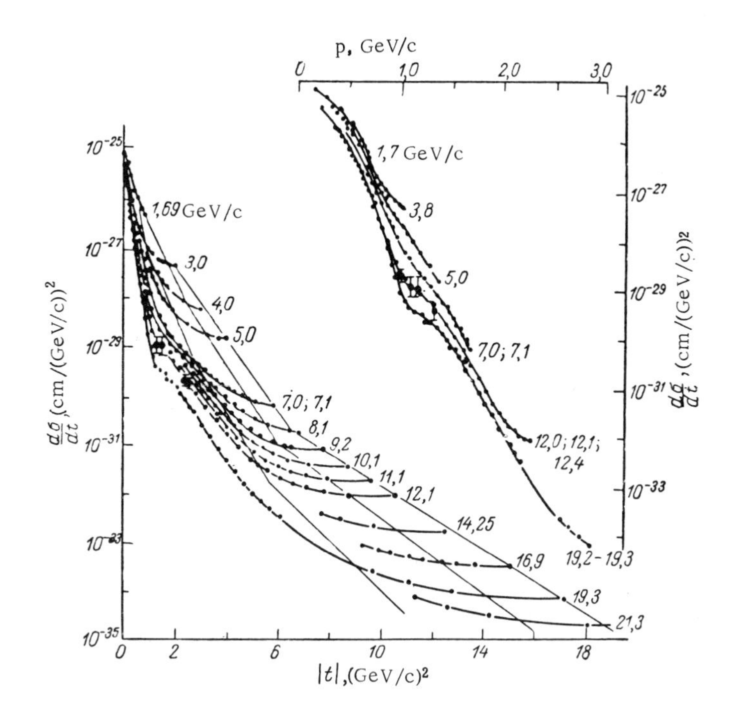

Fig. 12. Differential cross sections for large-angle pp scattering.

Figure 13 shows the apparatus, which is mounted at one of the linear gaps in the accelerator. During the acceleration a variation in the high frequency deflects the inner beam p_0 onto a thin polyethylene target M, 3 μ thick; the beam circulates through the target repeatedly for 2 sec. (The entire acceleration process requires 2.5 sec.) The target is suspended on quartz needles 7 μ in diameter and is brought into the working part of the chamber after the acceleration cycle is begun. Three independent scintillation telescopes T_1, T_2, and T_3, directed toward the target and having different solid angles, are used as monitors.

The secondary-particle detectors are eight silicon detectors $\approx 1\ cm^2$ in area, rectangular in shape, having a sensitive-layer thickness of 100–2000 μ and a resolution of 70–100 keV. They are placed 3.5 m from

TABLE 4. Values of $\alpha_{pp} = (\mathrm{Re}A/\mathrm{Im}\,A)|_{t=0}$ Obtained by the Interference Method

p_{lab}	α_{pp}	Statistical error	Systematic error	Source
2,78	0,12	0,07	—	[9, 10]
4,85	—0,38	0,10	—	
6,87	—0,30	0,07	—	
8,89	—0,33	0,08	—	
10,90	—0,26	0,05	—	
7,92	—0,247	0,023	0,059	[27]
9,94	—0,302	0,018	0,053	
12,14	—0,258	0,016	0,051	
17,82	—0,307	0,016	0,049	
7,81	—0,331	0,014	0,020	[27]
9,86	—0,345	0,018	0,020	
9,86	—0,343	0,009	0,020	
11,94	—0,290	0,013	0,020	
14,03	—0,272	0,013	0,020	
20,24	—0,205	0,013	0,020	
24,12	—0,157	0,018	0,020	
26,12	—0,154	0,025	0,020	
10,00	—0,330	0,035	—	[26]
10,11	—0,430	0,043	—	
19,33	—0,330	0,033	—	
26,42	—0,320	0,033	—	
7,85	—0,290	0,030	—	[28]
1,70	+0,010	0,090	—	[36]
1,54	—0,320	0,070	—	[37]
1,39	—0,580	0,060	—	
1,29	—0,760	0,130	—	
1,62	+0,100	0,160	—	
1,54	—0,300	0,090	—	
24,0	—0,190	0,090	—	[38]
27,5	—0,230	0,130	—	[39]

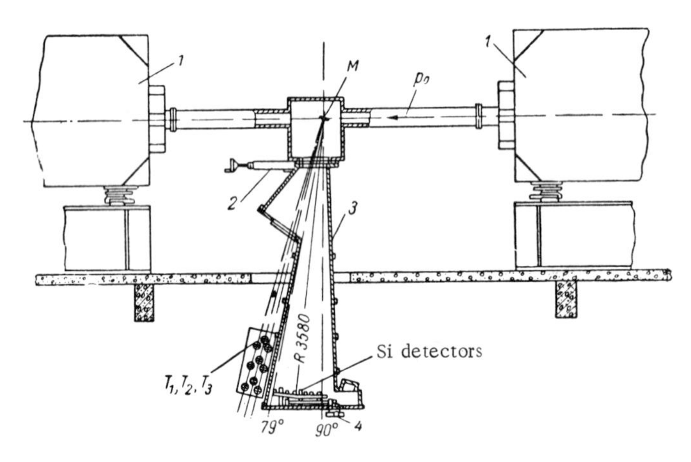

Fig. 13. Diagram of the apparatus used to study elastic pp scattering by means of semiconductor detectors. p_0) Direction of the beam of primary protons; M) film target; Si detectors) a movable carriage with semiconductor detectors; T_{1-3}) scintillation telescopes; 1) accelerator magnet; 2) slide valve covering the ion guide from the accelerator chamber; 3) ion guide; 4) mechanism for moving the detectors.

the target on the cover of the vacuum ion guide at an angle of 80-90° with respect to the beam. Six of the eight detectors can be displaced with respect to the target and used to detect recoil-particle spectra at various angles with respect to the vertical dropped from the target to the plane of the ion-guide cover. Two detectors remain fixed and can be moved only into the background position, where, according to the kinematics of the process, there are no recoil protons from elastic scattering events (at an angle of ≈ 90°). In this position, these two detectors can be used as additional monitors yielding information about the number of interactions in the target and about the overall accelerator background.

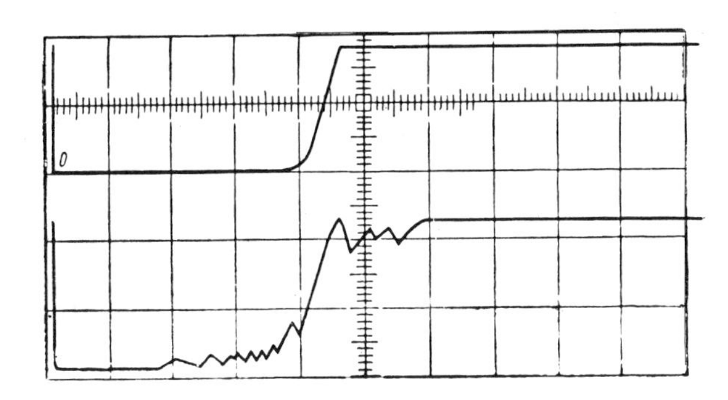

Fig. 14. Operation of the target and beam uniformity. Upper oscillogram) Path of the target as time elapses during the acceleration cycle; lower) uniformity of the beam at the target. The extreme left position on the scale coincides with the imposition of the trigger pulse. The sweep rate is 500 msec/cm.

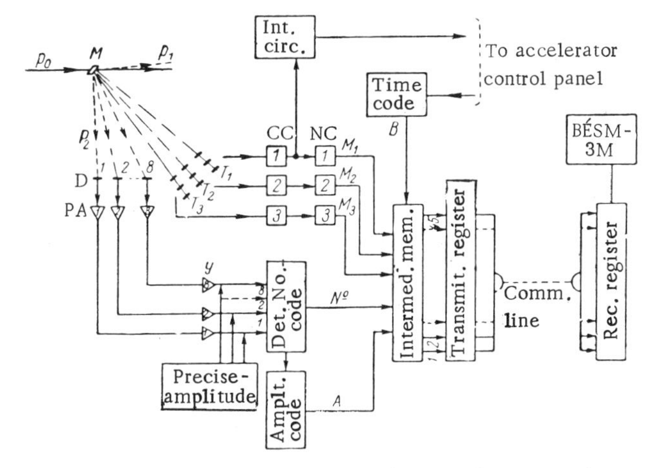

Fig. 15. Block diagram of the electronic apparatus. p_0) Beam of accelerated protons; p_1) scattered protons; D) semiconductor detectors; PA) preamplifiers; T_{1-3}) telescopes; CC) coincidence circuits; NC) counting circuits.

Figure 14 shows oscillograms characterizing the operation of the target and the uniformity of the beam over the target.

After 2 sec (one division corresponds to 500 msec) the target returns to its initial position. The deflection of the lower trace on the oscillogram from the horizontal is governed by a signal from one of the monitor telescopes T (Fig. 13), whose amplitude is proportional to the number of interactions in the target at a given time. The oscillogram shows that the beam is held uniform within ± 20% for 2 sec. The spikes on the lower oscillogram observed after the target leaves its working position are due to the count, with the beam on the target, of telescopes of the other channels, which operate later.

Constancy of the beam discharge onto the target is achieved by a feedback unit ("integral circuit" in Fig. 15) which generates a signal proportional to the count of one of the telescopes aligned toward the target. The feedback signal is supplied to the accelerator control room for connection to the system directing the beam onto the target. Information is received for the same length of time the beam is discharged onto the target: 2 sec. During each acceleration cycle, therefore, the apparatus receives information corresponding to a continuous energy spectrum of primary protons.

The reception of data over a wide energy range during a single acceleration cycle is a characteristic and important feature of this experiment: it reduces the count rate of the detectors and minimizes the systematic errors associated with the time drift of the apparatus characteristics.

The electronic apparatus operates on line with a computer (Fig. 15). The signals generated by the detectors are amplified by preamplifiers PA and amplifiers A to ≈ 5 V and are supplied to a detector-number coder which produces the code number of the detector from which the information was received; then the test pulse is sent to the input of an amplitude coder for measurement of the signal amplitude. The average dead time here is 30 μsec. The number code is a four-digit binary number, while the amplitude code is a seven-digit number. Together they constitute a "single event." Four events form a single word in the intermediate memory. The 45th digit in such a word is always 0, which indicates detector number-signal amplitude information.

All the amplitude information which comes from the detectors during each 16 msec is marked by time codes generated by a time coder. The time coder is triggered by a pulse "attached" to a specified accelerator magnetic field (i.e., to a certain energy of the inner proton beam). When the time code is delivered, information is removed from three counting circuits accumulating the count from the scintillation telescopes. The count from each telescope – the monitor number – takes up 11 digits. All three monitor numbers are combined in the intermediate memory with the time code and form a single word. Service information (the experiment number, etc.) is added to the same word. The 45th digit in such a word is assumed equal to 1 and indicates information of the monitor-time type. The energy corresponding to the position of the middle of a given time interval between two time codes is assigned to the information in the block of amplitude codes between these two time codes.

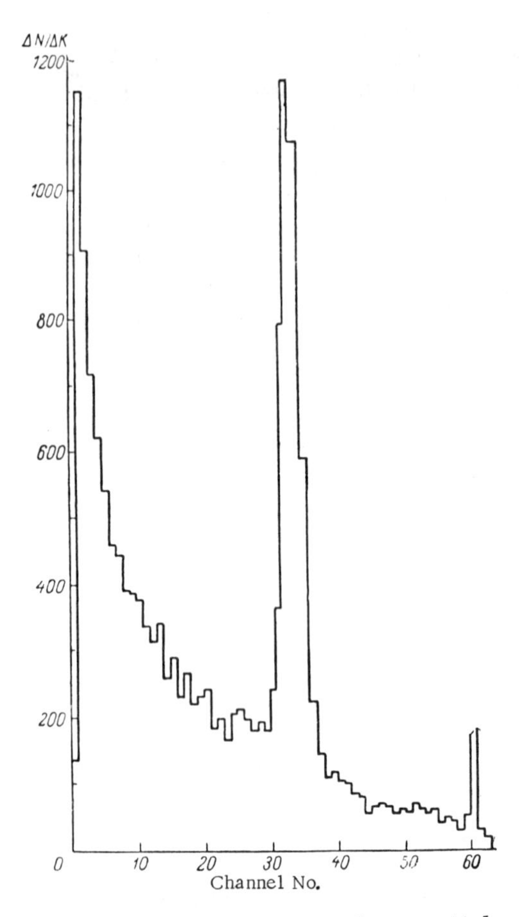

Fig. 16. Spectrum of secondary particles detected by a semiconductor detector. The energy of the primary proton beam is 17 GeV, and the energy of the recoil protons is 10 MeV.

The energy range corresponding to a time interval of 96 msec is 2.5-3.0 GeV. The BÉSM-3M computer questions the intermediate memory at intervals of 120 μsec. The intermediate memory is based on ferrite rings having a capacity of two machine words.

The following parameters of the system are the governing factors in the choice of the working beam intensity: 1) the capacity of the magnetic internal storage (1 cube or 4096 cells), 2) the dead time of the amplitude coder, and 3) the time required to transfer a single code from the intermediate memory to the magnetic internal storage of the computer.

The maximum possible frequency at the input to the apparatus is $\gamma = 4096 \times n/t = 8$ kHz, where $n = 4$ (four events per word), and $t = 2$ sec (the information-receiving time).

There is provision in the apparatus for blocking all types of coding during the "silent" time of the amplitude coder in order to eliminate the effect of counting errors on the relative detection efficiency in the various channels. Information was usually received at a frequency of 3 kHz. The counting error is 15%, the same for all channels.

The multiparameter spectrometric apparatus used in the experiment to directly extract information in the computer was described in [50].

A system of programs has been worked out for receiving, recording, and analyzing the information. Here, we will discuss only the functions of the program which directly tie the apparatus to the computer. This program controls the information flow and the operation of the individual units of the apparatus. It carries out the following analysis: 1) it checks the uniformity of the beam over the target; 2) it establishes the presence and correct order of the time markers; 3) it evaluates the temporal stability of each channel, by comparing the ratio of the count rate in any two channels at different times; 4) it constructs histograms of the spectra; 5) it periodically evaluates the spectral width in selected detectors and determines the signal-to-background ratio for all the channels in order to monitor the integrity of the target and the correctness of the beam aiming.

Figure 16 shows a typical amplitude spectrum recorded by one of the detectors. The maximum near the 34th channel corresponds to 10-MeV recoil protons. The width of the maximum is governed by the apparatus resolution, which depends primarily on the dimensions of the target and detectors in the beam direction. The target and detectors have lengths of 8 and 5 mm, respectively, along the beam. This leads, according to the kinematics of the process, to a distribution width of ≈ 6 MeV/c, independent of the recoil-particle energy.

The 60th channel contains a maximum which is used to check the operating stability of the amplitude coder. A signal from an oscillator at a frequency of ≈ 300 Hz is fed to the input of all eight spectrometer channels before the detector-number and amplitude coding. The operating stability of the amplitude coder, the presence of counting errors, and the correct operation of the programs are checked through a comparison of the number of counts in the oscillator peak in various spectrometric channels.

Figure 17 shows the operating stability of the various spectrometric channels in the worst possible case – that in which the beam distribution is extremely nonuniform (the lower curve). (As a rule, the beam nonuniformity is much less, no more than 20%.) The upper curves are the sum of counts from a reference-amplitude oscillator in the various spectrometric channels (from 0 to 7).

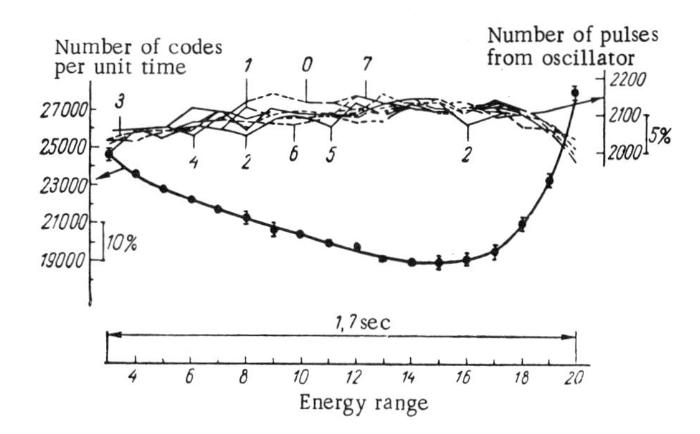

Fig. 17. Operating-stability diagram for the various spectrometric channels for the case of the nonuniform beam discharge onto the target.

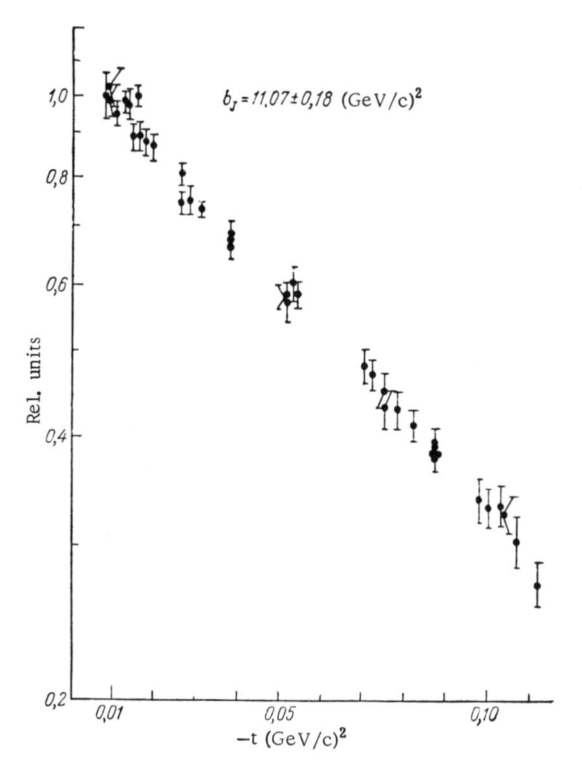

Fig. 18. Differential cross section for elastic pp scattering at a primary-beam of 58.1 GeV. One of 94 measurements carried out in the energy range 12–70 GeV is shown.

From Fig. 17 we see that the ratio of counts in the various spectrometric channels remain constant, regardless of the count rate, within ≈ 1%, although the count in the spectrometric channels falls off by about 5% as the beam discharge increases by a factor of about two.

The set of recoil-particles spectra obtained in seven detector positions (one of which is a background position) constitutes a single "experiment." Each experiment lasts 5–7 h and permits $3 \cdot 10^6$ particles to be detected, of which 10^6 turn out to be recoil protons from elastic pp → pp scattering in the energy range $\Delta E \approx 50$ GeV.

Measurements of the relative differential cross section for elastic pp scattering at 58.1 GeV are shown in Fig. 18. The groups of points near $|t| = 0.04$ and $|t| = 0.09$ $(\mathrm{GeV}/c)^2$ were obtained by means of fixed detectors. The measured relative differential cross sections are analyzed on the basis of Eq. (1.8).

The slope parameter b_J of the diffraction cone and the normalization factor C are found by the method of least squares. The α values are taken from the curve which follows from the dispersion relations. The optical point O is calculated from the total cross sections, obtained from the equation $\sigma_{tot} = c_1 + c_2/p^m$, which approximates the experimental data in the energy range up to 30 GeV. (In this equation, p is the proton momentum in the lab system, and c_1, c_2, and m are constants.) We note that O and α are not critical for determining b_J, while the factor C is, because the differential cross section is specified in relative units. The set of experimental data in the range 10–70 GeV permits the slope parameter to be determined at 94 energies.

Figure 19 shows 20 average values of the slope parameter obtained from the statistics of 10^7 elastic-scattering events, along with the results of earlier studies [6, 51, 52].

As this figure shows, the slope parameter b_J increases monotonically with increasing energy; in terms of the optical model, this is due to an increase of the interaction radius $r = \sqrt{2b_J}$ from 1.23 to 1.34 F in the energy range 10–70 GeV. Measured slope parameters are also shown in Table 5.

The main source of systematic errors is the uncertainty in the ratio of the areas of the various detectors and the uncertainty in the ratio of the real and imaginary parts of the amplitude for elastic pp scattering. It should be noted that, first, this error decreases significantly after measurement of α. Second, and very importantly, almost all the systematic errors may cause only a parallel rise or decent of the $b_J(E)$ curve, without affecting its slope.

The measured slope parameters obtained in this experiment are described by a function of the form

$$b_J = b_0 + 2b_1 \ln s,$$

TABLE 5. Measured Slope Parameter for Elastic pp Scattering for $0.008 \le |t| \le 0.12$ $(GeV/c)^2$

$E_{kin. lab}$, GeV	s, GeV^2	b_J, $(GeV/c)^{-2}$	R, F*
12,1	26,2	9,81±0,35	1,236±0,022
14,8	31,3	9,98±0,12	1,247±0,008
17,9	37,1	10,46±0,12	1,276±0,007
20,9	42,7	10,58±0,12	1,284±0,007
23,8	48,2	10,59±0,11	1,284±0,007
26,7	53,6	10,77±0,11	1,295±0,007
29,7	59,3	10,68±0,11	1,290±0,007
32,6	64,7	10,66±0,11	1,288±0,007
35,5	70,1	10,77±0,11	1,295±0,007
38,6	75,9	10,89±0,10	1,302±0,006
40,7	79,9	10,87±0,14	1,301±0,008
44,2	86,5	10,95±0,10	1,306±0,006
48,0	93,6	11,19±0,11	1,320±0,006
51,2	99,6	11,31±0,11	1,327±0,006
53,4	103,7	11,24±0,12	1,323±0,007
56,1	108,8	11,16±0,10	1,319±0,006
59,1	114,8	11,40±0,09	1,333±0,005
62,6	121,0	11,76±0,12	1,353±0,007
65,2	125,9	11,52±0,12	1,339±0,007
69,0	133,0	11,38±0,11	1,331±0,006

* The statistical errors are shown here. The systematic error is $\Delta b_J = \pm 0.3$ $(GeV/c)^{-2}$.

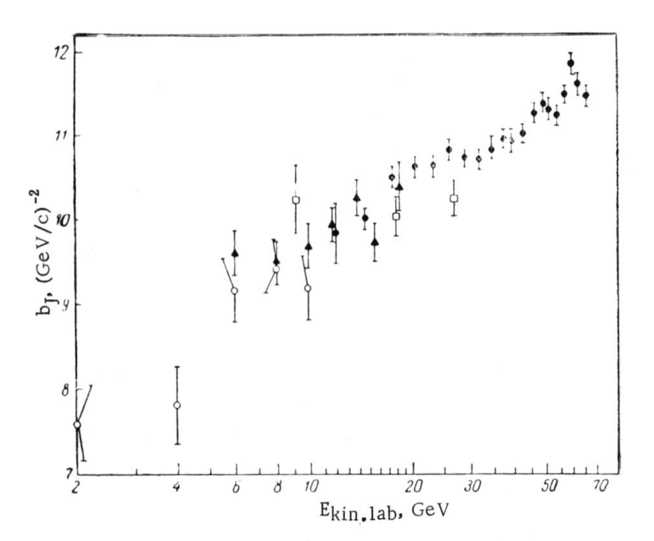

Fig. 19. Measured slope parameter for elastic pp scattering in the energy range 12-70 GeV: ●) Data of [49] (the error shown includes the statistical and random errors); ○) data of [6]; ▲) data of [51]; □) data of [47].

where s is the square of the total c.m. energy. It turned out that

$$\left.\begin{array}{l} b_1 = 0.47 \pm 0.09 \\ b_0 = 6.8 \pm 0.3 \end{array}\right\} \chi^2 = 24.8$$

for 20 experimental points.

1.6. Analysis of Data on the pd Interaction and Calculation of the Real Part of the Amplitude for pn Scattering

The problem of the interaction of particles with the nucleus is a many-body problem; it of course has no exact solution in either classical or quantum mechanics. Approximate methods exist in which two-body interactions are treated. In the optical model, e.g., the multiparticle system, the nucleus is replaced by a single object described by a complex potential. A more detailed picture of the process is obtained by means of the momentum approximation, in which the reaction is represented as a simple superposition of coherent scatterings of the incident particle by nucleons of the nucleus [53]. Various versions of this method have been used successfully at energies E < 1 GeV.

The model of multiple diffraction scattering posed by Glauber [54, 55] is a good approximation at high energies. In the Glauber approximation, the incident particle interacts with nucleons of the nucleus independent of other nucleons, as in the momentum approximation, but, in contrast to the latter, multiple scattering by several nucleons is taken into account. The Glauber procedure is currently used to treat high-energy hadron scattering. The particles are treated as extended objects whose individual elements interact independently.

In the Glauber theory, the amplitude for scattering by a deuteron is expressed in terms of the nucleon amplitudes:

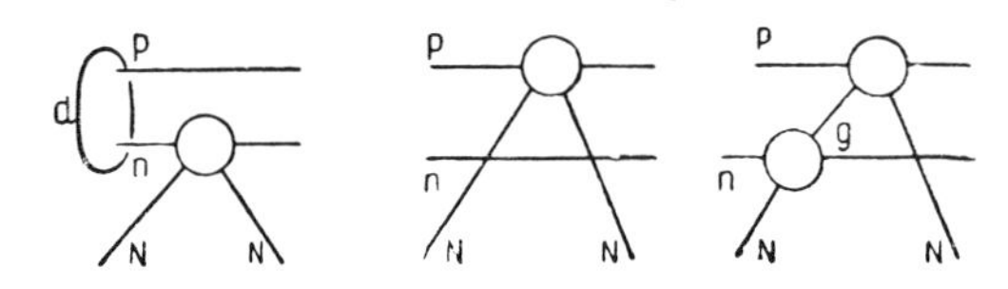

Fig. 20. Glauber-approximation diagrams for pd scattering.

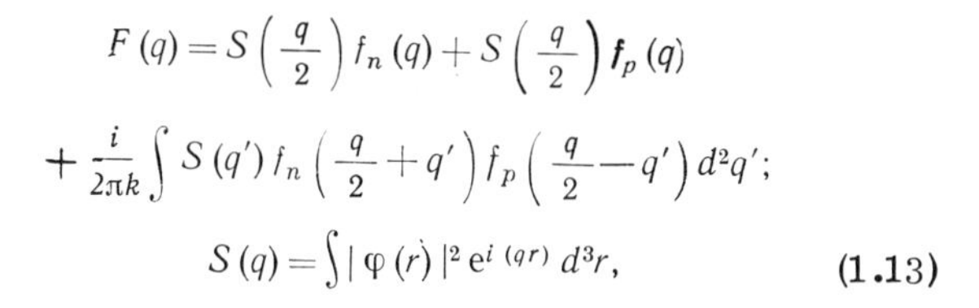

$$F(q)=S\left(\frac{q}{2}\right)f_n(q)+S\left(\frac{q}{2}\right)f_p(q)$$
$$+\frac{i}{2\pi k}\int S(q')f_n\left(\frac{q}{2}+q'\right)f_p\left(\frac{q}{2}-q'\right)d^2q';$$
$$S(q)=\int|\varphi(r)|^2 e^{i(qr)}d^3r, \qquad (1.13)$$

where S(q) is the deuteron form factor, and φ(r) is the ground-state wave function of the deuteron.

Equation (1.13) was derived in the quasi-classical approximation for small scattering angles, i.e., for $q/k \ll 1$. The nucleons in the nucleus are assumed to be on a mass surface [56]. The structure of function (1.13) clearly shows the essence of this approximation. The first and second terms are contributions from single scattering by a proton and by a neutron. The third term gives the contribution from double scattering of an incident particle by both deuteron nucleons. The integration is carried out over the intermediate momentum q'. The diagrams in Fig. 20 correspond to the amplitude in Eq. (1.13).

The vertex functions of the diagrams are expressed in terms of the corresponding nucleon amplitudes and the deuteron form factor. In this manner, the fact that the nucleons interacting with the primary particle are in a bound state is taken into account.

We can write down the amplitude for scattering through zero angle; by definition, we have S(0) = 1, and from (1.13) we find

$$F(0)=f_n(0)+f_p(0)+\frac{i}{2\pi k}\int S(q')f_n(q')f_p(q')d^2q'. \qquad (1.14)$$

Applying the optical theorem to this equation, we find a relation between total cross sections:

$$\sigma_d=\sigma_p+\sigma_n-\delta\sigma; \qquad (1.15)$$

$$\delta\sigma=\frac{2}{k^2}\int S(q)\left[\operatorname{Im}f_p\operatorname{Im}f_n-\operatorname{Re}f_p\operatorname{Re}f_n\right]d^2q. \qquad (1.16)$$

The screening correction (1.16) may be either positive or negative; at high energies, the real part of the scattering amplitude is small, and δ_σ in Eq. (1.15) is positive. At low energies, it changes sign. The screening effect is of a wave nature. The Glauber correction arises as a result of the coherent interaction of the incident wave with the nucleons making up the deuterons.

In the approximation $\operatorname{Re}f_{p,n}=\alpha_{p,n}\operatorname{Im}f_{p,n}$, Eq. (1.16) yields a convenient expression for $\delta\sigma$:

$$\delta\sigma=\frac{1}{4\pi}\langle r^{-2}\rangle\sigma_n\sigma_p(1-\alpha_{pp}\alpha_{pn}); \qquad (1.17)$$

$$\langle r^{-2}\rangle=\frac{1}{2\pi}\int\frac{S(q)\operatorname{Im}f_p(q)\operatorname{Im}f_n(q)}{\operatorname{Im}f_p(0)\operatorname{Im}f_n(0)}d^2q. \qquad (1.18)$$

Correction (1.17) is used in practice to calculate from Eq. (1.15) the total cross section for interaction with a neutron from data on the total cross sections $\sigma_{\text{tot}\,p}$ and $\sigma_{\text{tot}\,d}$. The constant $\langle r^{-2}\rangle$, which characterizes the ground state of the deuteron (the average reciprocal square radius) may be found from data on the total cross sections for scattering of pions by deuterons:

$$\langle r^{-2}\rangle=\left(\sigma_{p\pi^+}+\sigma_{p\pi^-}-\sigma_{d\pi}\right)\frac{4\pi}{\sigma_{p\pi^+}\sigma_{p\pi^-}}. \qquad (1.19)$$

Directly measurable quantities appear on the right side of this equation. Different authors [57, 58] using Eq. (1.19) have found values in the range 0.0239–0.0424 mb^{-1} for $\langle r^{-2}\rangle$. The typical measurement error is ≈ 0.009 (with an account of systematic errors). The average reciprocal square radius may be calculated from Eq. (1.18) through the use of a specific model for the deuteron. The nucleon amplitudes in

TABLE 6. Energy Dependence of the Glauber Correction

Primary-particle momentum, GeV/c	Glauber correction, mb	Source	Primary-particle momentum, GeV/c	Glauber correction, mb	Source
3	1,3±1,4	[85]	17,3	3,9±1,7	[62, 27]
6,5	3,0±1,7	[83]	21,6	5,0±1,5	[62, 27]
14,6	4,3±1,9	[62, 27]	27,0	8,1±0,9	[62, 27]

(1.18) are quite well known experimentally. Moreover, the result depends only slightly on the form of these amplitudes, since the function $f_p(q)f_n(q) \sim e^{-10q^2}$ varies slowly in comparison with the form factor $S(q) \sim e^{-40q^2}$. The value $\langle r^{-2}\rangle = 0.0311\ mb^{-1}$ was found in [59], in good agreement with experimental data.

The maximum uncertainty in the Glauber correction is therefore ≈ 40%. The error itself is ≈ 3.0 mb, i.e., ≈ 8%. This means that the accuracy of this model is no worse than 4% within the stated assumptions. We note that this uncertainty is large in comparison with that of contemporary measurements of total cross sections: ≈ 0.3%. The energy dependence of the correction for rescattering is an interesting topic. The total cross sections $\delta_{p,n}$ are essentially constant, and such a dependence may arise through the nucleon amplitude in (1.18). For a wide class of deuteron wave functions and for an exponential parametrization of the functions $f_{p,n}$

$$f_{p,n}(q) \sim e^{-b_{p,n}q^2},$$

we can calculate the integral in (1.18) [42]:

$$\langle r^{-2}\rangle = \frac{\pi^2}{b_d + b_p + b_n}, \tag{1.20}$$

where

$$b_d = \frac{d}{dq^2} \ln S(q)$$

is a parameter which depends on the form of the ground-state wave function of the deuteron.

Equation (1.20) is a good approximation for $\langle r^{-2}\rangle$, and it may be used to treat the energy dependence of the Glauber correction. As proton experiments have shown, the slope parameter b_p for the diffraction cone depends on the energy: $b_p = b_0 + b_1 \ln E_{lab}$, where $b_{p,n} \ll b_d$ for energies up to 70 GeV. The latter inequality is qualitatively obvious, since the deuteron radius is large in comparison with the nucleon radius. Accordingly, δ_σ should fall off with increasing energy, but very slowly:

$$\delta\sigma \sim \frac{1}{c_1 + c_2 \ln E_{lab}}.$$

This topic has been treated from the Regge-pole point of view [60, 61]. No rigorous conclusions have been obtained, but a power-law decrease of δ_σ is possible under certain circumstances. The experimental results are interesting in this connection. We have already seen that data on the total $\pi \pm d$ cross section can be used to determine $\langle r^{-2}\rangle$. It was asserted in [58] that $\langle r^{-2}\rangle = 0.042 \pm 0.003$ holds in the range 6–20 GeV. Data were obtained in [62] from the total cross sections σ_{pp} and σ_{np}, measured with a hydrogen target in proton and neutron beams. Figure 21 shows the cross section σ_{np} measured in the neutron beam. The results of [58], in which σ_{pn} was measured on the basis of deuteron measurements, are shown for comparison. The defect cross section is calculated from

$$\delta\sigma = \sigma_{pp} + \sigma_{np} - \sigma_{dp}.$$

The experimental results are shown in Table 6.

The increase in δ_σ with increasing energy is quite obvious, but the data are not yet adequate for final conclusions. Some caution is required, here, since δ_σ is sensitive to systematic errors in the total cross sections.

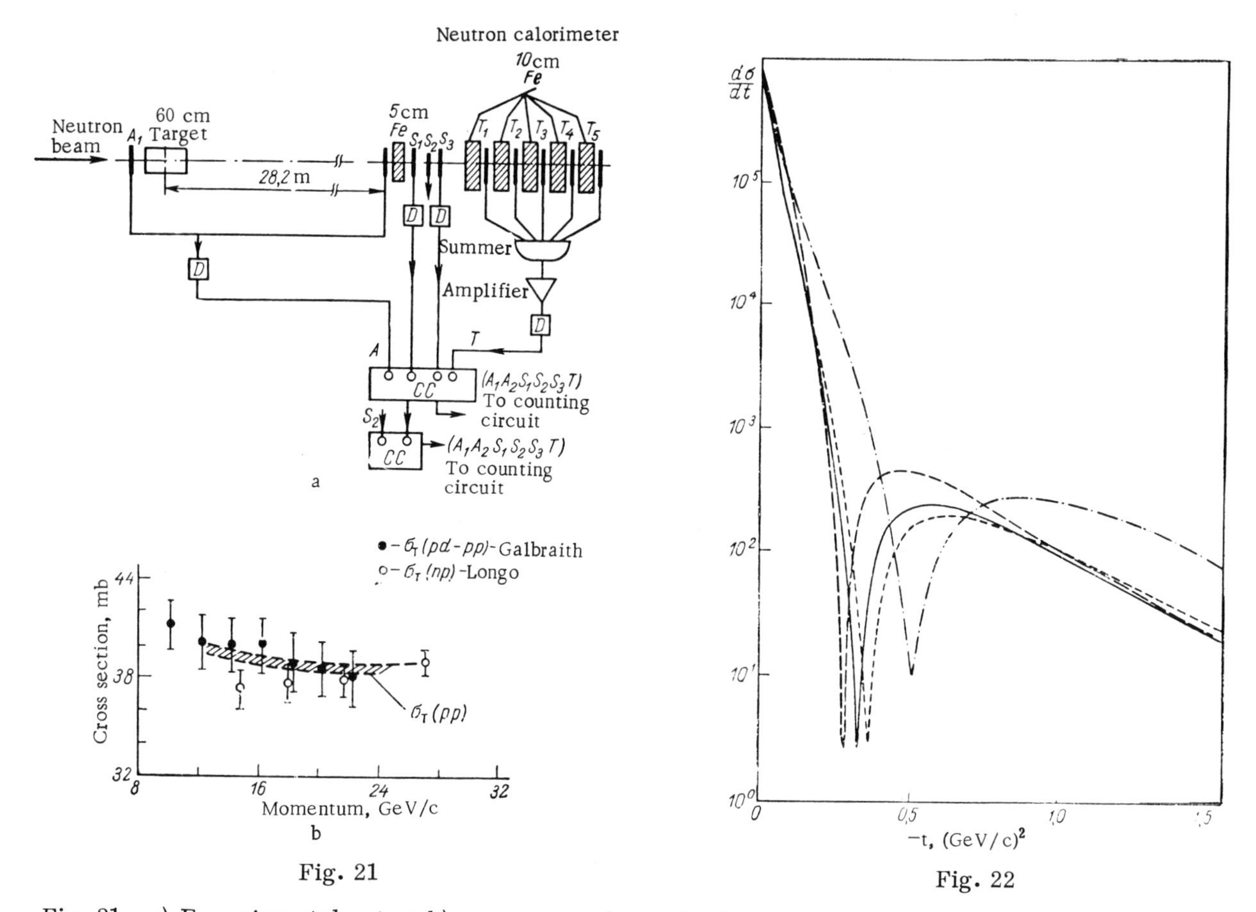

Fig. 21

Fig. 22

Fig. 21. a) Experimental setup; b) experimental results for measurement of the total cross section for the neutron–proton interaction.

Fig. 22. Differential cross section for pd scattering, obtained in the Glauber approximation with various deuteron wave functions: – – –) $\Phi(r) \propto e^{-\gamma r^2}$; ——) Moravcsik III; – – –) $\Phi(r) \propto (e^{\alpha r} - e^{-\beta r})/r$; — · — ·—) $\Phi(r) \propto e^{-\alpha r}/r$.

Finally, it may be asserted that the theory of multiple diffraction scattering is a reliable basis for analyzing deuteron reactions, with an accuracy of a few percent and under certain assumptions.

Equation (1.13) can be used to calculate the differential cross section for elastic scattering of protons by deuterons from the given nucleon amplitudes (Fig. 22) [64]. A characteristic feature of the cross section is a minimum at momentum transfers $t \approx -0.4$ $(GeV/c)^2$; it arises as the result of interference between the single-scattering and double-scattering waves. The position of the minimum is sensitive to the choice of deuteron wave functions. At small t the cross section depends weakly on the form of the wave function.

The interference phenomenon is governed by the phase difference between the constituent waves, so a study of scattering by deuterons yields information about the real part of the nucleon amplitudes at $t \neq 0$. At high energies this is as yet the only way to observe α_{pp} and α_{pn} with $t \neq 0$. For scattering by a more complicated nucleus, the interference pattern will be observed at a different value of t. Experiments with light nuclei may yield the function $\mathrm{Re} f_{pn}(E, t)$, but difficulties may arise here in connection with the proper choice of the nuclear wave function, when the dependence of these on the particle spins is taken into account. As Fig. 22 shows, the position of the minimum depends very strongly on the form of the wave function. The depth of the minimum in the theoretical curve may change when D, the wave in the deuteron ground state, is taken into account. This question has yet to be resolved.

Franco and Coleman [64] have numerically analyzed the dependence of the deuteron cross section on α_{pp} and α_{pn} (Fig. 23). The experimental data currently available near the interference minimum are not sufficient for judging the value of the real part of the nucleon amplitudes at $t \neq 0$, but, as the Franco-

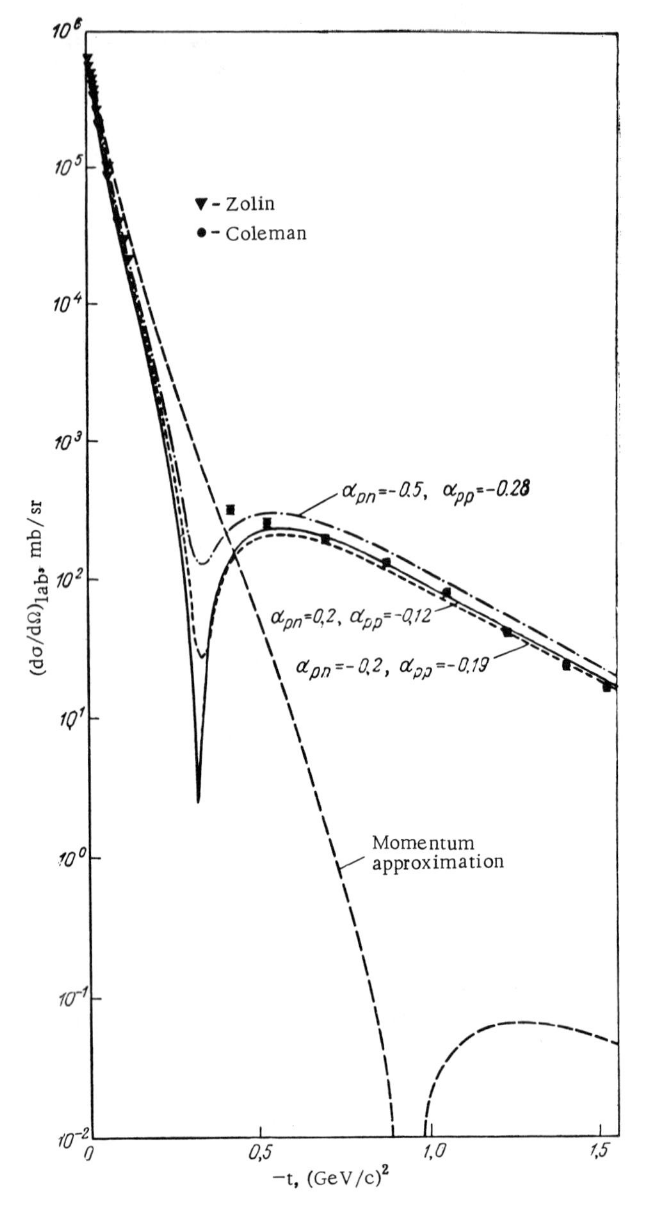

Fig. 23. Differential cross section for pd scattering, obtained in the Glauber approximation with various assumptions about the real part of the proton-nucleon amplitudes.

Coleman analysis shows, the problem of constructing the function $\mathrm{Re} f_{p,n}(F, t)$ can be solved with experimental techniques currently available.

Equation (1.13) may be used to treat the small-angle differential cross section for calculating α_{pn}. Using (1.14) and (1.18), we easily find a useful relationship among α_{pd}, α_{pp}, and α_{pn}. For zero scattering angle, we have

$$F(0) = f_p(0) + f_n(0) + \frac{i}{k} f_p(0) f_n(0) \langle r^{-2} \rangle.$$

Separating the real and imaginary parts, we find

$$\sigma_n = (\sigma_d - \sigma_{pp}) \left(1 + \langle r^{-2} \rangle \frac{\sigma_{pp}}{4\pi} \right) - (\sigma_d \alpha_{pd} - \sigma_{pp} \alpha_{pp}) \langle r^{-2} \rangle \frac{\sigma_{pp} \alpha_{pp}}{4\pi}; \tag{1.21}$$

$$\alpha_{pn}=\frac{1}{\sigma_{pn}}\left[(\sigma_d-\sigma_{pp})\langle r^{-2}\rangle\frac{\sigma_{pp}\,\alpha_{pp}}{4\pi}+(\sigma_d\,\alpha_{pd}-\sigma_{pp}\,\alpha_{pp})\left(1+\langle r^{-2}\rangle\frac{\sigma_{pp}}{4\pi}\right)\right]. \quad (1.22)$$

Equation (1.21) may be used to determine $<r^{-2}>$ or to check the consistency of the data used for a given $<r^{-2}>$. Equation (1.22) gives the desired value of α_{pn}. This simple method of calculating α_{pn} can be used in experiments in which coherent (elastic) scattering is easily distinguishable from scattering involving deuteron decay. In most experimental studies, the sum of the elastic and inelastic processes is measured, and it is considerably more difficult to analyze the data in this case. Harrington [65] has theoretically treated pd scattering with an account of the incoherent reaction.

2. CHECK OF CERTAIN CONSEQUENCES OF LOCAL QUANTUM FIELD THEORY AND MODELS FOR HADRON INTERACTIONS

2.1. Basic Postulates of Local Quantum Field Theory

Particle interactions are currently described on the basis of local quantum field theory, which has arisen as a natural generalization of quantum mechanics to the case of a system having a variable number of relativistic particles. The equations of this theory have not in general been solved yet, and it is still not clear which methods will lead to a systematic theory of elementary particles. The experimental data are explained on the basis of many models or particular hypotheses (e.g., the statistical theory for particle creation, the one-boson model for strong interactions, the Regge-pole hypothesis, etc.). The importance of each model depends on the results of an experimental check. If experiments confirm the conclusions which follow from a hypothesis, our knowledge is improved by another positive assertion. A negative result of an experimental check invalidates the particular hypothesis but it does not affect local quantum field theory as a whole. Experiments which check fundamental postulates of contemporary theory are very important.

The basic principles of relativistic local quantum field theory are as follows [48]: 1) invariance with respect to the inhomogeneous Lorentz group; 2) microscopic causality; 3) the spectrality condition, according to which there exists a complete system of physical states with positive energy; 4) the condition for unitarity of the scattering matrix.

The microcausality postulate sometimes permits a rigorous proof of the analyticity of the scattering amplitude A(E) as a function of the complex variable E (the energy). The Cauchy theorem gives a linear integral relationship between ReA and ImA which is called the "dispersion relation." By virtue of the unitarity condition, $\mathrm{ImA}|_{t=0}$ may be expressed in terms of the total interaction cross section. In this manner an equation is obtained in which experimentally observable quantities appear. Obviously, it is of primary importance to check the dispersion relations in order to establish the applicability limits of the axioms of modern theory.

Bogolyubov [66] was the first to rigorously derive dispersion relations for πp scattering. A detailed examination of this question may be found in [67, 68].

The analyticity of the scattering matrix in E and t permits one to establish other experimentally verifiable consequences. These are usually inequalities which only hold asymptotically as $E\to\infty$.

2.2. Some Specific Calculations by the Dispersion-Relation Method for Proton-Nucleon Scattering

The dispersion-relation method is one of the most highly developed fields of theoretical physics. Here we will compare theoretical and experimental data on pN scattering under the assumption that the basic positions of the dispersion-relation method are already familiar to the reader.

Calculation of the real part of the pN-scattering amplitude is based on the dispersion relations in the form given by Goldberger et al. [69]. In order to obtain a specific prediction for the quantity $\alpha_{p,n}(E)$, one must make several particular assumptions regarding the behavior of the scattering amplitude in the nonphysical energy range $2\mu - M$ (μ is the pion mass and M is the nucleon mass). The simplest approach to the problem [77] is to ignore the behavior of the integrand. We consider the series

$$\int_{E\,(2\mu)}^{M} \frac{A_{\bar{p}N}(E')}{E'+E}\,dE' = c_0 + \frac{c_1}{E} + \frac{c_2}{E^2} + \ldots$$

with arbitrary coefficients which must be determined from experimental data. A disadvantage of this solution is obvious: many free parameters arise, and the accuracy of the result turns out to be low.

In calculating $\mathrm{Re}A_{pp}(E)$, Levintov and Adelson-Velsky [70] postulated a definite parametrization of the nonphysical pole-type integral

$$\int \ldots dE' = \frac{E_1+\xi}{E+\xi}C.$$

The constant C is determined experimentally. The pole position is assumed completely indefinite. Since the result depends on ξ, its variation within the range $2\mu < \xi < M$ yields an estimate of the accuracy of this method.

Soding [71] started from a definite physical model for the $\bar{p}N$ system, according to which the nucleon–antinucleon system has a bound state in the energy range 0–M at the point μ (the pion) and a continuum of states in the range 2μ–M. Experiments on $\bar{p}p$ anhililation in the physical range indicate a predominant role of quasi-bound pion states of resonances. It follows from the theory of the electromagnetic form factor of the nucleon that quasi-bound states consisting of two pions, a ρ meson, predominate in the pion sheath of the nucleon. This implies that it may be possible to replace the continuum by a set of bound states with fixed energy (or mass). Mathematically, this means replacement of the cut in the analytic function $A_{\bar{p}N}(E)$ by a sum of poles:

$$\int_{E\,(2\mu)}^{M} \frac{A_{\bar{p}N}(E')}{E'+E}\,dE' = \sum_i \frac{R_i}{(M_i+E)\,2M}\,. \tag{2.1}$$

The constant R_i may be expressed in terms of the boson-nucleon coupling constant (Fig. 24).

The pole representation is a hypothesis which must be checked. Only qualitative arguments are given above. At small energies (E < 600 MeV), expansion of the nucleon scattering amplitude in terms of pole terms is successfully used to theoretically describe experimentally observed partial-wave phase shifts (the so-called one-boson exchange model [72, 73]). It has been suggested that π, σ, ρ, ω, η, and f mesons are the bosons responsible for the interaction. The boson-nucleon interaction constants are determined experimentally. All the experimental information available is consistent with this picture of the nucleon interaction; this is evidence that the nonphysical region may be represented by means of Eq. (2.1). Soding took the constants R_i corresponding to ρ and ω mesons from [72], where the scattering phase shifts were analyzed at energies up to 300 MeV on the basis of the one-boson exchange model. The coupling constant for the η meson was selected by comparison of the calculated $\mathrm{Re}A_{pp}(E)$ curve with experiment in the range 69–310 MeV. The narrow range of observable energies E < 50 MeV in which the cross section σ_{pp} is not yet known was assigned to the nonphysical integral.

The pp and pn amplitudes were calculated in [74, 75]; the nonphysical integral was replaced by only a single-pole term (in addition to the π meson and deuteron terms) corresponding to the ρ meson. The coupling constant and mass of the ρ meson were determined from low-energy experimental data.

Barashenkov et al. [76–78] have analyzed the dispersion relations for pp, pn, and πp scattering; the nonphysical region was treated on the basis of the one-boson approximation.

The higher the energy, the smaller the contribution of the unobservable region: according to estimates in [71], this contribution is 20% at 3 GeV, falling off thereafter as 1/p. The one-boson exchange model reproduces the experimental data at low energies with an accuracy generally no worse than 10-20%. The uncertainty in the dispersion calculations due to the nonphysical region at energies E > 3 GeV thus should not exceed a few percent.

Bialkowski and Pokorski [79] carried out calculations for the pp and $\bar{p}p$ amplitudes, explicitly making use of the smallness of the contribution from the nonphysical region. The integrand function $A_{\bar{p}p}(E')/(E' + E)$ was expanded in a series in powers of E. It turned out to be sufficient to use only two terms of the

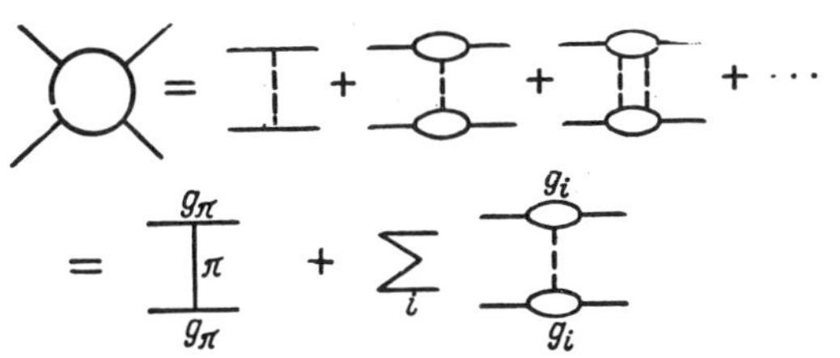

Fig. 24. Representation of the amplitude of nucleon-nucleon scattering as a sum of one-boson diagrams.

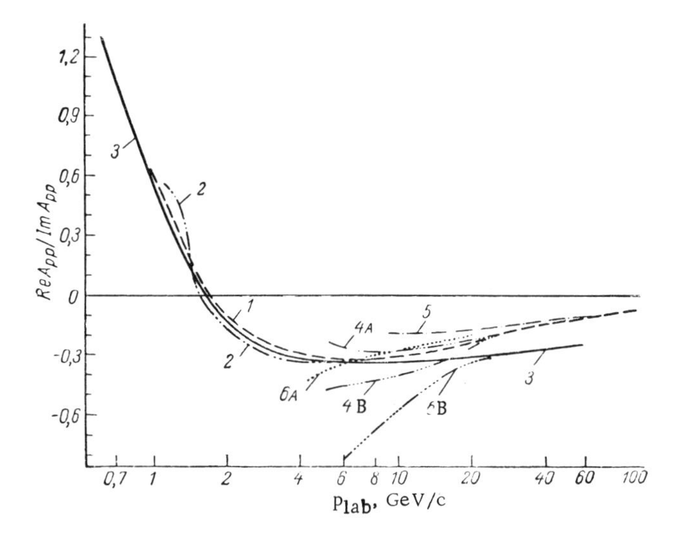

Fig. 25. Results of calculating the real part of the pp scattering amplitude from dispersion relations. 1) Data of [71]; 2) data of [75]; 3) data of [78]; 4A, 4B) data of [70] (the curves indicate the limits of the uncertainty in α_{pp}); 5) data of [79]; 6) A, B) data of [14]. Prediction of Regge-pole theory. (The curves show the boundaries of the uncertainty in α_{pp}.)

series for an accuracy of 1% in the range E > 10 GeV. One arbitrary constant was determined by comparison with experimental data, and the other was determined from the dispersion sum rule. The Bialkowski-Pokorski method differs significantly from others in that the parameter of the nonphysical region is found from high-energy experimental data. In a certain sense, this is equivalent to a subtraction at the point E_0 from the region E > 10 GeV.

The next important topic in a practical dispersion analysis is the assumption made regarding the behavior of the cross sections at energies E > 30 GeV, where there are no experimental data. It is currently common to use the extrapolation equation

$$\sigma_{pN,\,\bar{p}N} = \sigma^{(0)}_{p,\,\bar{p}} + c_{p,\,\bar{p}}\, p^{k_{p,\,\bar{p}}}, \tag{2.2}$$

which gives a good approximation of the available data, and follows from Regge-pole theory and from certain other models [80]. Bialkowski and Pokorski [79] have carried out a detailed numerical analysis of available data on the total pp and $p\bar{p}$ scattering cross sections, giving the parameters $\sigma^{(0)}$, c, and k.

The effect of the asymptotic behavior may be strongly suppressed by using experimental relations with a subtraction at a certain point E_0 from the high-energy region. This procedure is equivalent to normalizing the calculated curve in terms of the quantity $\mathrm{ReA}(E_0)_{\mathrm{exp}}$. In evaluating the agreement between theoretical and experimental data in this case, only the energy dependence of the compared curves should be taken into account.

Figures 25 and 26 show published results of a dispersion analysis of $\alpha_{p,n}$. At E < 10 GeV, the results of all the studies are in good agreement. Levintov's curves were obtained by a subtraction at the point 26 GeV; the experimental value of α_{pp} was the upper limit found for the real part of the pp scattering amplitude in [81]. The Soding curve was determined without subtractions in this region, so the coincidence of results here is fortuitous.

The data of Barashenkov differ greatly from the results of [71] because of a different choice of constants in extrapolation Eq. (2.2). Barashenkov used the most recent results (1968) on the total cross sections. In this sense, the Soding curve, calculated in 1964, should be considered obsolete. The same holds for the results on pn scattering; the data of Bugg and Carter [74, 75] are based on 1964 data, and they may be used only as rough estimates for the range E > 10 GeV.

The data of [79] on the energy dependence are analogous to the data of Barashenkov in that the corresponding studies were based on similar parameters in the extrapolation equation. In [79], however, as we have already noted, the constant for the nonphysical region was chosen on the basis of the experimental value of α_{pp} at 26 GeV. This explains the discrepancy with the data of Barashenkov.

Finally, we note that the problem of the nonphysical region in the dispersion relations for proton–nucleon scattering is largely solved. The solution is based on the many experiments and theoretical studies at low energies. The experimental data on the real part of the proton–nucleon scattering amplitude may be

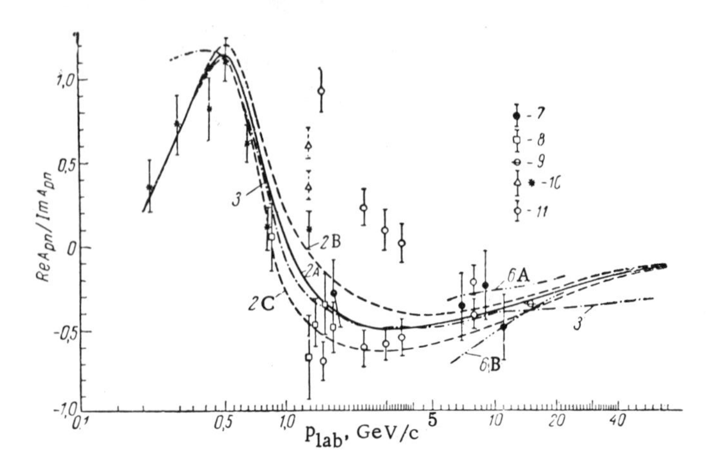

Fig. 26. Real part of the pn scattering amplitude. The curves show the results of theoretical calculations on the basis of dispersion relations and on the basis of the Regge-pole theory. 6A, 6B) Data of [15]; 7) data of [7, 8]; 8) data of [40, 41]; 9) data of [42] (the error is not indicated by the authors); 10) data from a phase analysis of pn scattering; 11) data following from pn → np charge exchange without an account of spins (two solutions); the other notation is the same as in Fig. 25.

adopted as a reliable basis for checking the basic axioms of relativistic quantum field theory. The inverse problem is also meaningful: the bases of the theory are assumed valid, and the dispersion relations give the high-energy behavior of the cross sections.

In this connection the dispersion relations are sometimes metaphorically called a crystal ball through which we can look into the asymptotic behavior.

Figure 26 and 27 show all published measurements of α_{pp} and α_{pn}. All the α_{pn} values are consistent with each other and with the theoretical curve.

The situation is more complicated in the case of pp scattering. For the range $E > 10$ GeV, we have the data of the Brookhaven group (Foley et al. [27]) and the data of the CERN group (Bellettini et al. [26]), which are in significant disagreement. The data of [26] agree with the Barashenkov curve. Interestingly, the curve of Bialkowski and Pokorski (curve 5 in Fig. 25) was obtained with a subtraction at the point $p = 26.1$ GeV, $\alpha_{pp} = -0.154 \pm 0.03$ (the last point of [27]). Nevertheless, agreement could not be reached with Foley's data [27]. Bialkowski and Pokorski [79] suggested that the procedure for obtaining the real part from the interference scattering was not correct because of the presence of two spin amplitudes in the pp scattering. If this is true, a rapid decrease in α_{pp} (according to the data of [27]) may primarily indicate the extinction of spin effects, rather than the behavior of the real part of the scattering amplitude. However, this point of view is denied by all experimentalists who have analyzed scattering in the Coulomb-interference range. For example, an attempt was made in [26, 27] to explicitly take into account spin effects: singlet and triplet amplitudes with different t dependences were introduced. It was shown that the interference pattern largely reflects the characteristic c/t dependence of the Coulomb amplitude on t. This could not be the case with nuclear amplitudes, since it would lead to an anomalously large effective range of the nuclear forces, as a detailed analysis by Levintov [82] has shown. Foley et al. [27] found the maximum possible correction for spin effects to α_{pp} to be $\Delta = -0.02$; this is not sufficient for achieving agreement with the theoretical curve of [79]. It is still too early to draw a serious conclusion from this situation, since there is little information in the energy range $E > 10$ GeV, and that which is available is contradictory.

In the energy range 2-10 GeV there is complete agreement between the experimental and theoretical data of various laboratories.

The results of Dutton and Van der Raay [37] are surprising: as Fig. 27 shows, their data fall off rapidly. So far, no explanation has been found for this. Repeated measurements should be expected in the

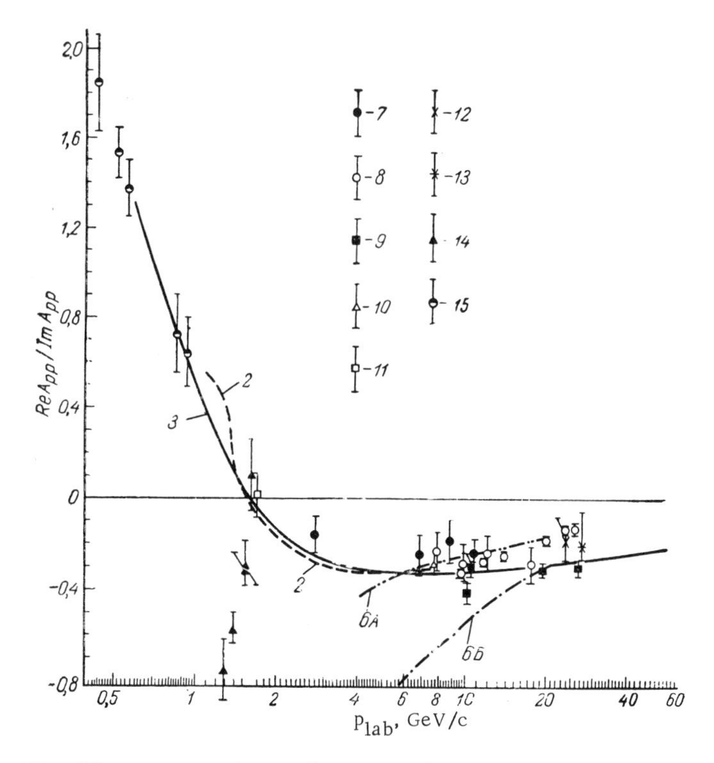

Fig. 27. Data on the real part of the pp scattering amplitude. 7) Data of [19, 10]; 8) data of [27]; 9) data of [26]; 10) data of [28]; 11) data of [36]; 12) data of [38]; 13) data of [39]; 14) data of [37]; 15) phase analysis of pp scattering [36] (the notation for the theoretical curves is given with Fig. 25).

range 1-2 GeV. If the results of [37] are valid, there is a significant contradiction with the dispersion-relation predictions. In this case, there are no variations of the nonphysical region or the total asymptotic cross sections which can explain the set of experimental data.

2.3. Method of Complex Momenta

The interactions of particles at low and moderate energies are extremely complicated. Hypotheses have been advanced according to which the interactions of hadrons would simplify with increasing energy. A characteristic example here is the Pomeranchuk theorem which asserts that the total cross sections for interactions of particles and antiparticles are equal at extremely high energies. The Pomeranchuk theorem has been generalized to differential cross sections and polarizations. It is natural to assume that the key to the understanding of the complete strong-interaction picture should be sought in the asymptotic behavior, where the basic assumptions on which modern field theory is constructed may be checked. A method of analyzing high-energy processes based on the analytic continuation of the scattering amplitude into the complex orbital angular momentum plane – the Regge-pole theory – has been developed with varying success over the last 10 years. According to this concept, the partial wave amplitude f_l (t) in the t channel has a pole in the complex l plane whose position depends on t (a moving pole). To each pole there corresponds a bound or quasi-bound state (resonance) in the t channel. The amplitude in the s channel at high energies is governed by the singularities of the function f_l (t).

The theory of complex angular momentum occupies an extremely important place in physics, for on it are pinned the hopes of constructing a strong-interaction theory for high energies. From the trajectories of poles of the function f_l (t), found from experiments on scattering for $t < 0$, extrapolated into the range $t > 0$, one can find the position of particles in the spin-mass plane. This gives rise to an interesting method for classifying particles according to their Regge-pole trajectories. One can treat the physics of particles and resonances and scattering dynamics from a common point of view. There is considerable experimental justification for constructing such a picture. Known particles often lie on a continuation of

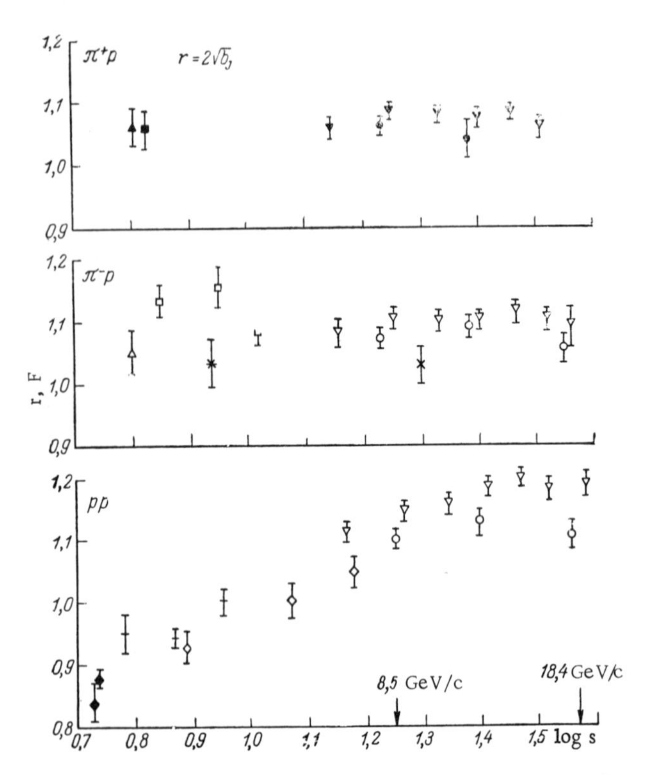

Fig. 28. Characteristics of the diffraction cone for $\pi^{\pm}p$ and pp scattering [91].

the trajectories of the corresponding poles. For example, the existence of trajectories associated with the ρ, ω, A_2, and π mesons has been well established, and the parameters of these trajectories have been found.

The data available at present (total cross sections, differential cross sections for two-particle reactions, polarizations, and real parts of scattering amplitudes) conform well to the multi-Regge-pole theory.

The data on the real parts of the amplitudes for pp and pn scattering shown in Figs. 26 and 27 and the results calculated from the Regge-pole theory [14, 15] agree with theory.

Some experiments which have been carried out, however, cannot be explained on the basis of the pole approximation. For example, it has not been possible to jointly describe the differential cross sections for the ($pp \to pp$), ($\bar{p}p \to \bar{p}p$), and ($\pi N \to \rho N$) reaction. The polarization of the neutron in the ($\pi^- p \to \pi^0 n$) reaction cannot be explained. It is difficult to understand the sharp forward peak in the angular distribution for the charge-exchange reactions ($np \to pn$) and ($pp \to \bar{n}n$).

These difficulties apparently show that the function f_l (t) has not only a pole, but also cuts in the l plane. The cut in the t channel of the two-particle reaction corresponds to an intermediate multiparticle state. This is completely plausible for strong interactions. At present only an approximate procedure for calculating the contribution of cuts to the s-channel amplitude has been developed [86, 87]. The results obtained show that account of cuts eliminates most of the contradictions between theory and experiment [88].

A very interesting aspect of the theory of complex angular momentum is the explanation of the nature of the pole with the vacuum quantum numbers whose trajectory assumes unit value at t = 0 (the Pomeranchuk pole). Because of this latter property [$\alpha(0) = 1$], the contribution of this pole does not appear as $s \to \infty$, because of the asymptotic constancy of the total cross sections and the fulfillment of the Pomeranchuk theorem. (The contributions of all other poles fall off as $s^{\alpha_i(0)-1}$ as $s \to \infty$.)

The problem is as follows: there is no reaction which could occur only as a result of a vacuum pole. At contemporary energies ($\ln s \sim 2$), it is always largely masked by the contribution of other poles. All this is complicated by the existence of reserves whose energy dependence is not completely clear. The pole trajectories at small t are assumed describable by a linear function: $\alpha(t) = \alpha(0) + \alpha' \cdot t$. In the one-pole model, the slope parameter α' of the trajectory is related to that of the diffraction cone in a simple manner:

$$b = \frac{d}{dt}\left(\ln \frac{d\sigma}{dt}\right), \quad b = b_0 + 2\alpha' \ln s. \tag{2.3}$$

This relation shows qualitatively the nature of the energy dependence of diffraction scattering; law (2.3) should be displayed more clearly with increasing energy, as the relative contribution of the vacuum pole increases. Figure 28 and 29 show data on the parameter b(s) for πp, $\bar{p}p$, and pp reactions. As we see, the picture is extremely foggy: $b_{\bar{p}p}$ falls off (the diffraction peak e^{bt} broadens), $b_{\pi p}$ remains constant, and b_{pp} increases, but there is an indication that the increase ends in the energy range 18-30 GeV. It has been suggested that $\alpha'_p = 0$; i.e., the Pomeranchuk pole differs greatly in nature from all other poles, for which $\alpha'_i \approx 0.6$. It is also important that apparently none of the known particles lie on the trajectory of the vacuum pole in region $t > 0$. There is also a qualitative argument: diffraction scattering is of a

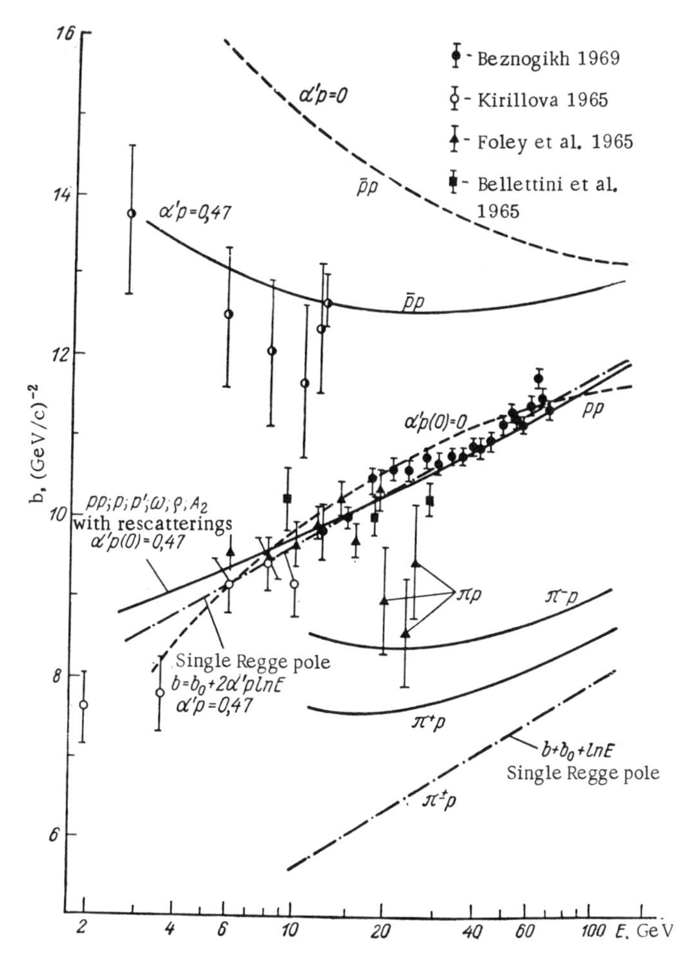

Fig. 29. Slope of the diffraction cone for $\pi^{\pm}$p, pp, and $\bar{p}$p scattering. The theoretical curves are calculated from Regge-pole theory [89]. The data on $\bar{p}$p scattering are from [52].

complicated nature, since it results from many inelastic processes, so it is difficult to expect it to be described by a simple pole model. If we have $\alpha'_p = 0$, the Pomeranchuk pole is a fixed (non-Regge) singularity in the l plane. A difficulty arises with the unitarity condition in the t channel. A revision of the entire Regge-pole concept may be required for this case.

It is therefore particularly important to obtain precise experimental data on b over the widest possible energy range. This has been precisely the purpose of the Dubna-Serpukhov group [49], which has measured the diffraction-cone slope parameter for pp scattering in the energy range 10-70 GeV. The measured experimental function $b_{pp}(s)$ has been analyzed on the basis of a five-pole model. All the parameters except α'_p were taken from [16] and fixed.

It turned out that $\alpha'_p = 0.40 \pm 0.09$. The statistical and systematic errors of the experiment were taken into account.

The experimental data are in agreement with the requirement $b_J < c \ln^2 s$ of axiomatic field theory [12, 13].

An interesting study was carried out by Ter-Martirosyan [89], who took into account, along with the poles, rescattering by a vacuum region (Fig. 29). The experimental data correspond well to the value $\alpha'_p = 0.47$. The value $\alpha'_p = 0$ may also be consistent with data on pp scattering, but there is a significant contradiction with the slope of the diffraction cone in the $p\bar{p}$ interaction.

Interestingly, construction of an asymptotic theory for strong interactions is a fundamentally new method for solving the question of particle structure. As all previous experience has shown, the monotonic behavior of the interaction cross section with increasing particle energy is unalterably disrupted. Moving upward along the energy scale, we pass several critical values when the incident wavelength is comparable to the characteristic dimension of the scatterer (molecule, atom, nucleus, or nucleon). A new scattering channel has been opened up, and we are dealing with a new class of interactions demonstrating the structure of the object. If it is possible to construct an asymptotic theory of the Regge-pole type, the structure problem in the sense above turns out to be exhausted. Everything should be reduced to the interaction laws of the elementary (structureless) fields.

2.4. Model for the Elastic Interaction of Hadrons

The ultimate goal of the theory is to describe the interaction of particles on the basis of certain dynamical principles which may be specified as equations for fields or as a postulate of the analytic properties of certain functions. Until this goal is reached, it is necessary to study particular models for particle interactions in which the missing units are replaced by phenomenological characteristics suggested by experiment.

In analyzing hadron behavior on the basis of models, we are trying to accumulate empirical material in a certain summary of laws (or rules) whose meaning may subsequently be revealed in terms of the basic postulates of field theory.

Let us consider the model for the elastic interaction of hadrons, primarily nucleon scattering, which is discussed most frequently in the literature. Empirical equations are given in many papers [20] to describe the differential cross section $d\sigma/dt\,(s,\,t)_{pp}$ over a wide range of energies and momentum transfer. The Orear equations [18] (two versions),

$$\frac{d\sigma}{d\omega} = Ae^{-ap_\perp},\quad A = 3\cdot 10^{-26}\ \text{cm/sr},\quad a^{-1} = 152\ \text{MeV/c}; \tag{2.4a}$$

$$\frac{d\sigma}{d\omega} = s^{-1} Ae^{-ap_\perp}, \tag{2.4b}$$

give an approximate description ($\approx 40\%$) of the experimental data at $t \gtrsim 2\ (\text{GeV/c})^2$. A detailed examination shows that the parameters A and a are not universal constants and must be functions of s and t.

A considerably more general law was found by Krisch [19]:

$$\left.\begin{aligned} &\frac{d\sigma^+}{dt} = I^{-1}\frac{d\sigma}{dt},\quad I = 1 + \exp\left(-2a\beta^2 p_l^2\right);\\ &\frac{d\sigma^+}{dt} = \sum_{i=1}^{3} A_i \exp\left(-a_i\,\beta^2 p_\perp^2\right),\end{aligned}\right\} \tag{2.5}$$

where A_i and a_i are arbitrary parameters, p_l and $p_\perp$ are the longitudinal and transverse momenta after the scattering, and β is the c.m. particle velocity. The function I takes into account the indistinguishability of protons.

Differential cross section (2.5) increases exponentially with increasing energy for any fixed angle $\theta \neq 0,\ \theta \neq \pi$, in contradiction of the Cerulus-Martin restriction [90], which follows from axiomatic field theory and the constancy of the total cross sections.

Fleming et al. [20, 21] corrected the shortcoming of Eq. (2.5) by replacing the sum of three terms by the infinite series

$$\left(\frac{d\sigma}{dt}\right)^+ \Big/ \left(\frac{d\sigma}{dt}\right)_{t=0} = C\sum_{n=0}^{\infty}\frac{a^{-n}}{(cn+1)^\nu}\exp\left(-\frac{bx^2}{cn+1}\right);\quad x = \beta p_\perp, \tag{2.6}$$

where a, b, c, and ν are free parameters. This Krisch-Fleming scheme gives a qualitative justification for the so-called incoherent-drop model for hadron scattering. According to this model, each term in series (2.6) corresponds to scattering by a certain structural region of the proton whose radius is given by the parameter $b' = (b/cn + 1)$, where n is the number of the region, and $r = \sqrt{2b'}$. The contributions of the various regions are combined incoherently.

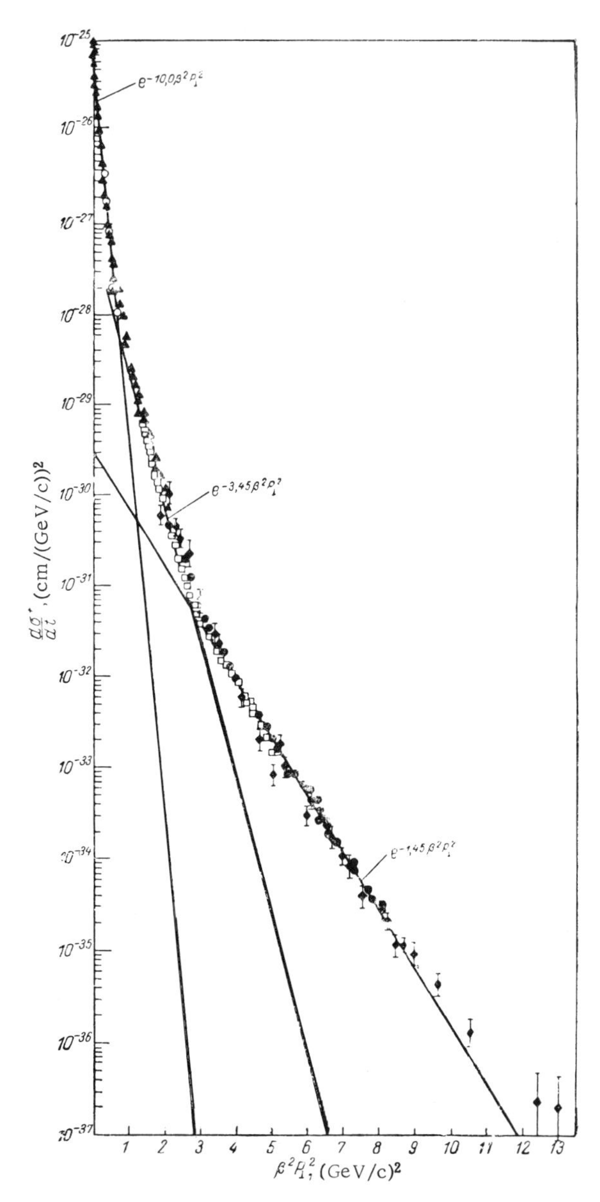

Fig. 30. Description of the differential cross section for pp scattering according to Krisch's equation, (2.5).

There is another possible interpretation of (2.6): the term of number n corresponds to n-fold scattering of the primary particle within the hadron target.

The agreement of Eqs. (2.5) and (2.6) with experimental data is evident from Figs. 30 and 31.

An interesting feature of the Krisch-Fleming equation is the use of the argument $(\beta p_\perp)^2$ instead of t or θ^2. The application of the classical optical model to the scattering of particles contracted along the scattering axis because of relativistic scale attraction ($l = l_0/\gamma$, $\gamma = 1/\sqrt{1-\beta^2}$) leads to a diffraction-cone shape $e^{-a(\beta p_\perp)^2}$. If this is true, we find a simple kinematic explanation for the contraction of the diffraction cone in proton-proton scattering. A contraction law is indicated. For small t, we may retain only the first term in Eq. (2.5) and (2.6):

$$\frac{d\sigma}{dt} \sim e^{-a(\beta p_\perp)^2} \sim e^{bt};\ b = a\beta^2. \tag{2.7}$$

For sufficiently high energies, we have $\beta \approx 1$, and the contraction of the cone ends. For π mesons, β is essentially constant at an energy E > 3 GeV, so the diffraction peak should be constant. The Krisch hypothesis is very alluring because of its visualizability and simplicity. However, in the energy range 30–70 GeV a contradiction is found with new data on pp scattering [49], since $b(\beta)$ curve (2.7) has a much smaller slope than the experimentally measured function $b_{J\,pp}$.

Huang [92] offered a different version of the incoherent-drop model and derived Orear's equation, (2.4a).

A characteristic feature of these approximations is that they predict a monotonic decrease of the cross section $d\sigma/dt$ with increasing t at all energies. The accumulation of experimental data and the increase in the accuracy of the data is gradually eating away at this simple picture. Interesting in this connection is the experiment carried out at CERN by the group of Allaby et al. [43]. This group measured the cross section for elastic pp scattering in the range $3 < E_{lab} < 21$ GeV, $0.2 < |t| < 20$ $(GeV/c)^2$. As Fig. 12 shows, the function is "modulated" at energies above 10 GeV: an oscillatory structure appears.

It was suggested in [93, 94] that the unitarity of the S matrix be used for a general description of the scattering at both small and large momentum transfer. Only a single term, corresponding to the two-particle intermediate state, was retained in the unitarity relation for the scattering amplitude. In this approximation, the problem may be solved explicitly. The model is found to be in good agreement with experimental data in the range $p\theta < 24$ (GeV/c)·rad. In particular, the structural features of the cross section $d\sigma/dt$ (t) are reproduced.

Logunov and Tavkhelidze [22] showed that according to relativistic field theory the scattering of two spinless particles is described by a Schrödinger-type equation with a complex potential which depends on the energy. In the loose-binding case, the potential may be constructed by the perturbation-theory method within the framework of relativistic field theory. For strong interactions there are no general methods for constructing the quasi-potential, so it may be assumed a phenomenological characteristic of the interaction.

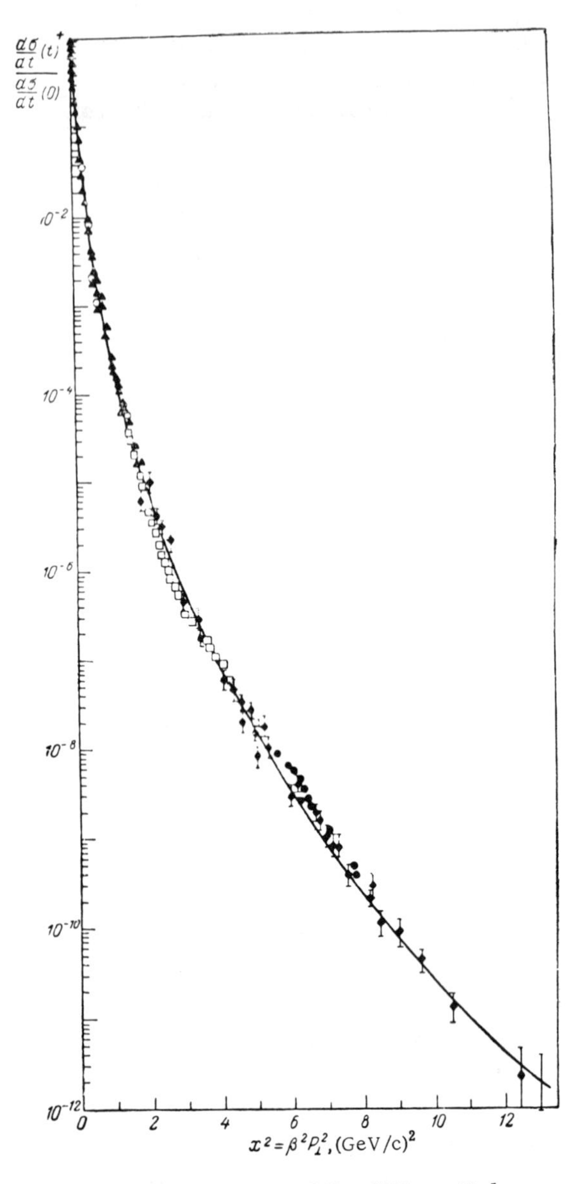

Fig. 31. Description of the differential cross section for pp scattering according to Fleming's equation, (2.6).

In [23] the (pp → pp) reaction was described by means of a purely imaginary smooth potential of the Gaussian type (Fig. 32).

The cross section falls off to zero at the diffraction minimum, since the solution was obtained only for the imaginary part of the scattering amplitude. Account of the real part partially fills in this minimum.

A characteristic feature of all elastic processes and many two-particle inelastic reactions in the energy range $E > 1$ GeV is a forward peak in the $d\sigma/dt \sim e^{bt}$ angular distribution, where the parameter b lies between 6 and 13 $(\mathrm{GeV}/c)^{-2}$ for all processes and all energies. It is therefore natural to represent a hadron as an extended object with a characteristic dimension $r = \sqrt{2b} \approx 1$ F. The length of the incidence wave is $\lambda = p/\hbar \approx 10^{-14}$ cm $= 10^{-1}$ F; i.e., the conditions of the quasi-classical approximation are satisfied, and one may use scattering theory in the impact-parameter approximation. The scattering amplitude is expressed in terms of the phase $\delta(b)$ (eikonal):

$$A(q) = \frac{1}{(2\pi)^2} \int e^{iqb} \left(1 - \exp(2i\delta(b)\right) d^2b;$$
$$2\delta(b) = -\frac{1}{hv} \int_{-\infty}^{+\infty} v(b^2 + z^2)\, dz. \tag{2.8}$$

It is natural to assume that the potential v is proportional to the density D(b) of hadron matter at point b:

$$2\delta(b) = KD(b). \tag{2.9}$$

The complex coefficient K relates the phase of the incidence wave and the matter density. Wu and Yang suggested that the function D(b) be described by means of the electric hadron form factor F(q):

$$D(b) = \frac{1}{(2\pi)^2} \int e^{iqb} F(q)\, d^2q. \tag{2.10}$$

If the colliding particles are of finite extent and are characterized by functions D_A and D_B, the natural generalization of (2.9) is

$$2\delta(b) = K_A \cdot K_B \int d^2b D_A(b - b') D_B(b'). \tag{2.11}$$

In the first Born approximation, we have $1 - e^{2i\delta} \approx 2i\delta$, and Eqs. (2.8)-(2.11) yield

$$A(q) = \frac{1}{(2\pi)^2} \int e^{iqb}\, 2i\delta(b)\, d^2b = CF_A(q) F_B(q). \tag{2.12}$$

In particular, for pp scattering we have

$$\frac{d\sigma}{dt} = CF_p^4(q). \tag{2.13}$$

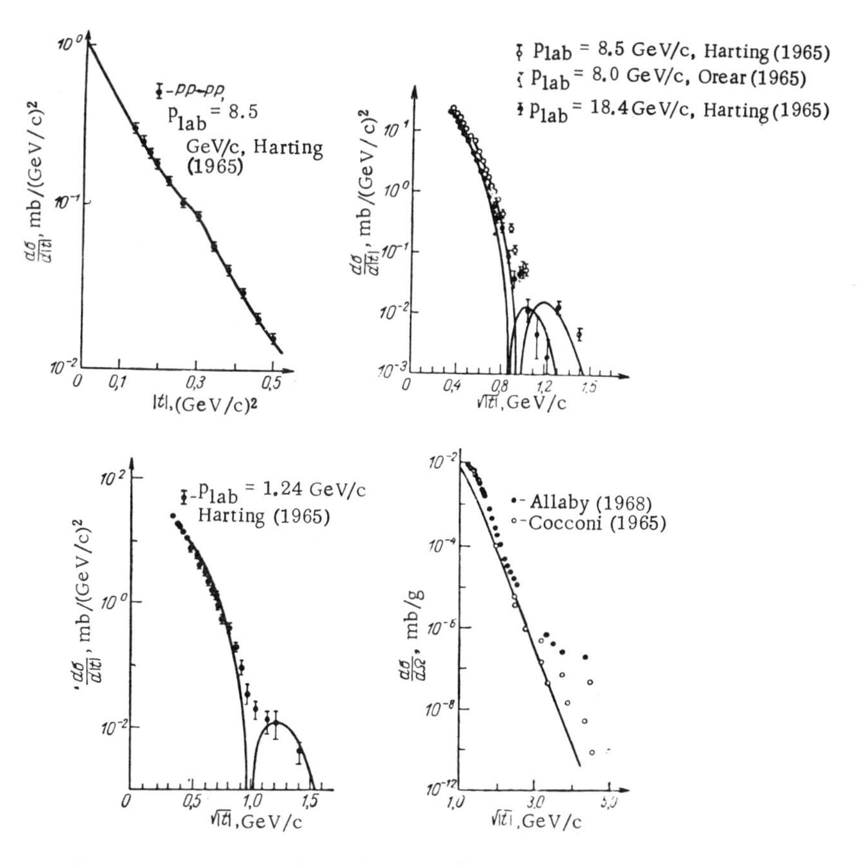

Fig. 32. Description of the differential cross section for pp scattering according to the quasi-potential model [23].

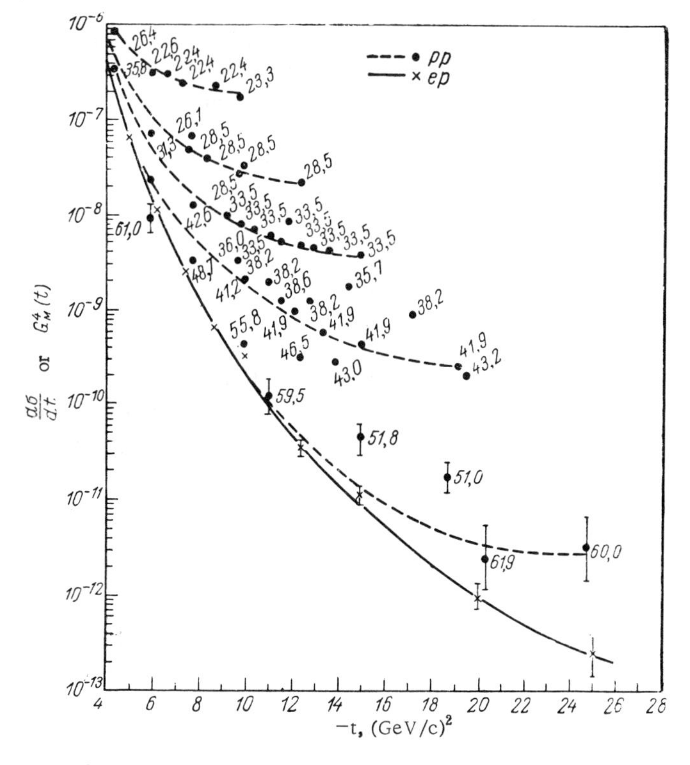

Fig. 33. Comparison of the differential cross sections for large-angle pp scattering with the magnetic form factor of the proton. The numbers show the square of the total c.m. energy of the pp system. The dashed curves connect points with the same s values. The solid curve is $G_m^4(t)$.

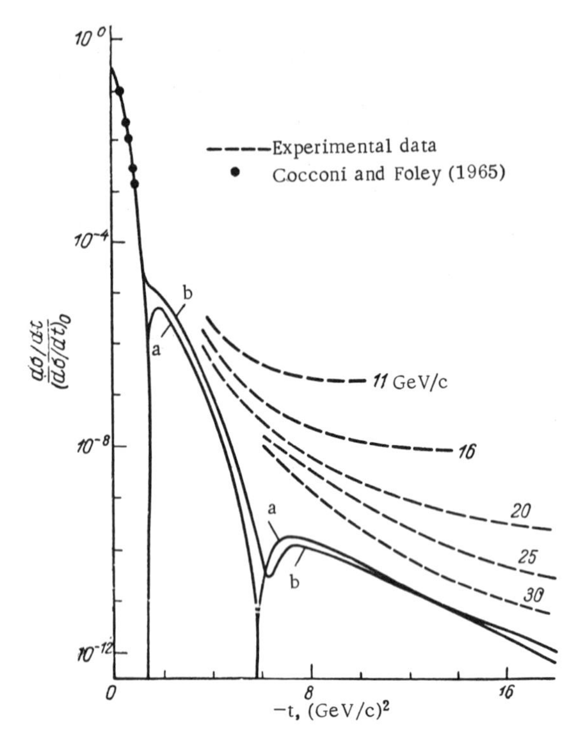

Fig. 34. Differential cross section for pp scattering calculated in the eikonal approximation. Versions a and b differ in the manner in which the spins are taken into account [96]. The dashed curves show experimental data.

Wu and Yang [95] obtained this equation in a slightly different manner. It also follows from the quark model. The experimental cross section lies systematically above curve (2.13), but the difference decreases with increasing energy (Fig. 33).

Durand and Lipes [96] carried out calculations on the basis of Eqs. (2.8)-(2.11) without resorting to simplification (2.12). The potential was assumed to be purely imaginary [the constant in Eq. (2.9) is purely imaginary]; the results are shown in Fig. 34. Interestingly, a diffraction-minimum structure appeared on the cross-section curve. This results from an account of rescattering, an explicit form for which may be found by a series expansion of the exponential function $e^{2i\delta}$ in Eq. (2.8). However, a significant discrepancy with experimental data remains.

The optical potential in (2.8) was interpreted in [63, 97, 98] as the result of an interaction which arises during the exchange of reggions. This approach is called the "hybrid model" or the "reggicized optical approximation." Experimental data on large-angle pp scattering were analyzed in [63], and a good agreement was found with experiment.

With slight variations, the reggicized optical model has been widely used to describe π and K scattering [84].

CONCLUSION

Further progress toward understanding the nature of hadron interactions depends largely on high-energy experiments. There has been considerable recent interest in data on the total cross sections for the interactions of π, K, and $\bar{p}$ with protons in the energy range 20-40 GeV, obtained on the 70-GeV proton synchrotron of the Institute of High-Energy Physics at Serpukhov.*

The new $\sigma^{\pi^-, K^-}(E)$ values for energies E > 30 GeV lie noticeably above the curves calculated from the Regge-pole model with an account of all the previous data from the region E < 20 GeV. This situation may be treated as an indirect indication of the violation of the Pomeranchuk theorem. If further experiments confirm this, the axiomatic field theory will receive a new interpretation.

Accurate study of elastic scattering is also quite important. For a further refinement of the Regge-pole parameters and for a check of the dispersion relations, the real part of the scattering amplitude and the slope of the diffraction cone must be measured with an accuracy no worse than 1-2%.

The accelerator experiments with colliding proton-proton and proton-antiproton beams, with a total c.m. energy of ≈50 GeV, currently being planned at CERN and at Novosibirsk, will be of decisive importance for asymptotic theories and for checking the axioms of local quantum field theory.

LITERATURE CITED

1. V. A. Nikitin et al., Pribory i Tekh. Éksperim., 6, 18 (1963).
2. V. A. Nikitin, Preprint of the Joint Institute for Nuclear Research R-1476, Dubna (1963).
3. L. F. Kirillov et al., Zh. Éksp. Teor. Fiz., 45, 1261 (1963).
4. V. A. Nikitin et al., Zh. Éksp. Teor. Fiz., 46, 1608 (1964).
5. Yu. K. Akimov et al., Zh. Éksp. Teor. Fiz., 48, 767 (1965).
6. L. F. Kirillov et al., Yadernaya Fizika, 1, 533 (1965).

*Preliminary data have been reported at the 1969 conference in Lund, Sweden.

7. Yu. K. Akimov et al., Yadernaya Fizika, 4, 88 (1966).
8. N. Dalkhazhav et al., Yadernaya Fizika, 8, 342 (1968).
9. L. S. Zolin et al., ZhÉTF Pis. Red., 3, 15 (1966).
10. L. F. Kirillova et al., Zh. Éksp. Teor. Fiz., 50, 77 (1966).
11. V. B. Dyubimov et al., Zh. Éksp. Teor. Fiz., 34, 310 (1959).
12. A. A. Logunov and Nguyen Van Hieu, International School on High-Energy Physics, Poprad, Czechoslovakia (1967), p. 3.
13. J. D. Bessis, Nuovo cimento, 45, 974 (1966).
14. R. J. N. Phillips and W. Rarira, Phys. Rev., B139, 1336 (1965); 165, 1615 (1968).
15. V. Barger and M. Olsson, Phys. Rev., 146, 1080 (1966).
16. K. A. Ter-Martirosyan, International School on High-Energy Physics, Poprad, Czechoslovakia (1967), p. 43.
17. L. B. Okun', Questions of the Physics of Elementary Particles [in Russian], Vol. 3, Izd. AN ArmSSR (1963), p. 133.
18. J. Orear, Phys. Letters, 13, 190 (1964).
19. A. D. Krisch, Phys. Rev. Letters, 19, 1149 (1967).
20. H. Fleming, A. Giovannini, and F. Predazzi, Ann. Phys., 54, 62 (1969).
21. H. Fleming, A. Giovannini, and F. Predazzi, Nuovo cimento, 56, 1131 (1968).
22. A. A. Logunov and A. N. Tavkhelidze, Nuovo cimento, 29, 380 (1963).
23. V. R. Garsevanishvili et al., Preprint of the Joint Institute of Nuclear Research E2-4361 (1969); V. R. Garsevanishvili et al., Preprint of the Joint Institute of Nuclear Research E2-4251 (1969).
24. B. I. Bekker et al., Preprint of the Joint Institute of Nuclear Research R-1358, Dubna (1963).
25. A. A. Nomofilov et al., Phys, Letters, 22, 350 (1966).
26. G. Bellettini et al., Phys. Letters, 19, 705 (1966); 14, 74 (1965).
27. K. J. Foley et al., Phys. Rev. Letters, 19, 857 (1967); 14, 74 (1965).
28. A. E. Taylor et al., Phys. Letters, 14, 54 (1965).
29. R. J. Sutter et al., Preprint BNL 11350 (1967).
30. G. J. Igo et al., Preprint BNL 11319 (1967).
31. C. W. Akerlof et al., Phys. Rev., 159, 1138 (1967).
32. J. V. Allaby, F. Binon, and A. N. Diddens et al., Phys. Letters, 28B, 67 (1968).
33. H. Bethe, Ann. Phys., 3, 190 (1958).
34. L. D. Solov'ev and A. V. Shchelkachev, Preprint of the Joint Institute of Nuclear Research R2-3670, Dubna (1968).
35. M. P. Locher, Nuclear Phys., B2, 525 (1967).
36. J. D. Dowell et al., Proceedings of the Sienna Conference, 1963, p. 683; Phys. Letters, 12, 252 (1964).
37. L. M. C. Dutton and H. B. Van der Raay, Phys. Letters, 26B, 11 (1968); 25B, 245 (1967).
38. E. Lohrmann, H. Meyer, and H. Winzeler, Phys. Letters, 13, 78 (1964).
39. G. Baroni, A. Manfredini, and V. Rossi, Nuovo cimento, 38, 95 (1965).
40. L. M. C. Dutton et al., Nuclear Phys., B9, 594 (1969).
41. L. M. C. Dutton and H. B. Van der Raay, Phys. Rev. Letters, 21, 1416 (1968).
42. G. Cocconi et al., Phys. Letters, 19, 341 (1965).
43. J. V. Allaby et al., Phys. Letters, 27B, 49 (1968).
44. J. V. Allaby et al., Phys. Letters, 25B, 156 (1967).
45. D. Harting, P. Blackall, et al., Nuovo cimento, 38, 60 (1965).
46. J. Orear et al., Phys. Rev., 152, 1162 (1966).
47. G. Bellettini, Phys. Letters, 14, 164 (1965).
48. A. A. Logunov, Nguyen Van Hieu, and I. T. Todorov, Usp. Fiz. Nauk, 88, No. 1, 51 (1966).
49. G. G. Beznogikh, A. Buyak, N. N. Zhidkov, et al., Preprint of the Joint Institute of Nuclear Research R1-4594, Dubna (1967).
50. G. I. Zabiyakin et al., Preprint of the Joint Institute of Nuclear Research 13-3397, Dubna (1967).
51. K. J. Foley et al., Phys. Rev. Letters, 11, 425 (1963).
52. L. Montanet, Proceedings of the Lund Conference, "Reporter talk on antiproton interactions," Lund (1969).
53. K. L. Kowalski and D. Feldman, Phys. Rev., 130, 276 (1963).
54. R. J. Glauber, Phys. Rev., 100, 242 (1955).
55. V. Franco and R. J. Glauber, J. Phys. Rev., 142, 1195 (1966).

56. V. S. Barashenkov and V. D. Toneev, Preprint of the Joint Institute of Nuclear Research R2-4292, Dubna (1969).
57. W. Galbraith, E. W. Lenkins, T. Kycia, et al., Phys. Rev., 138B, 913 (1965); W. F. Baker, E. W. Jenkins, et al., Proceedings of the Sienna Conference (1963), p. 634.
58. D. V. Bugg et al., Phys. Rev., 146, 980 (1966).
59. J. A. McIntyre and G. R. Burleson, Phys. Rev., 112, 2077 (1958).
60. E. Abers et al., Phys. Letters, 21, 339 (1966).
61. B. M. Udgaonkar and M. Gell-Mann, Phys. Rev. Letters, 8, 346 (1962).
62. M. J. Longo, Topical Conference on High Energy Collisions of Hadrons, Vol. 1, CERN (1968), p. 523.
63. A. A. Ansel'm and I. T. Dyatlov, Yadernaya Fizika, 9, 416 (1969).
64. V. Franco and E. Coleman, Phys. Rev. Letters, 17, 827 (1966).
65. D. R. Harrington, Phys. Rev., 135B, 358 (1964).
66. N. N. Bogolyubov, Report to the International Conference on Theoretical Physics, Seattle (1956).
67. N. N. Bogolyubov and D. V. Shirokov, Introduction to the Theory of Quantum Fields [in Russian], Gostekhizdat, Moscow (1957).
68. N. N. Bogolyubov, B. V. Medvedev, and M. K. Polivanov, Questions of the Theory of Dispersion Relations [in Russian], Fizmatgiz, Moscow (1958).
69. M. L. Goldberger, Y. Namba, and R. Oeme, Ann. Phys., 2, 226 (1957).
70. I. I. Levintov and G. M. Adelson-Velsky, Phys. Letters, 13, 185 (1964).
71. P. Soding, Phys. Letters, 8, 285 (1964).
72. M. Riazuddin and J. Moravcsik, Phys. Letters, 4, 243 (1963).
73. L. S. Azhgirei and V. I. Chizhikov, Preprint of the Joint Institute of Nuclear Research E1-3420, Dubna (1967).
74. A. A. Carter and D. V. Bugg, Phys. Letters, 20, 203 (1966).
75. D. V. Bugg et al., Phys. Rev., 146, 980 (1966).
76. V. S. Barashenkov, Forschr. Phys., 10, 205 (1962).
77. V. S. Barashenkov and V. I. Dedu, Nuclear Phys., 64, 636 (1965).
78. V. S. Barashenkov and V. D. Toneev, Preprint of the Joint Institute of Nuclear Research R2-3850, Dubna (1968).
79. G. Bialkowski and S. Pokorski, Nuovo cimento, 57A, 219 (1968).
80. V. Nambu and M. Sugawara, Phys. Rev., 132, 2724 (1962).
81. P. Breitenlohner, P. Egli, H. Hofer, et al., Phys. Letters, 7, 73 (1963).
82. I. I. Levintov, Phys. Letters, 19, 149 (1965).
83. L. Ozhdyain et al., Zh. Éksp. Teor. Fiz., 42, 392 (1962).
84. L. Blackmon and R. Goldstein, Phys. Rev., 179, 1480 (1969).
85. H. Palevsky et al., Proc. Congress International Phys. Nucl., Paris (1964), p. 162.
86. V. N. Gribov, Yadernaya Fizika, 9, 424 (1969).
87. V. N. Gribov and A. A. Migdal, Yadernaya Fizika, 8, 1002 (1968).
88. Jackson, Proceedings of the Conference on Elementary Particles, Lund (1969).
89. K. A. Ter-Martirosyan, Proceedings of the Conference on Elementary Particles, "Reporter talk by Lillethun," Lund (1969).
90. F. Cerulus and A. Martin, Phys. Letters, 8, 80 (1964).
91. D. O. Caldwell, B. Elsner, and D. Harting, et al., Phys. Letters, 8, 290 (1964).
92. K. Huang, Phys. Rev., 146, 1075 (1966).
93. I. V. Andreev, I. M. Dremin, and I. M. Gramenitsky, Nuclear Phys., 10B, 137 (1969).
94. I. V. Andreev and I. M. Dremin, Yadernaya Fizika, 8, 814 (1969).
95. T. T. Wu and C. N. Yang, Phys. Rev., 137B, 708 (1965).
96. L. Durand and R. Lipes, Phys. Rev. Letters, 20, 637 (1968).
97. R. C. Arnold, Phys. Rev., 153, 1523 (1967).
98. C. Chin and J. Finkelstein, Nuovo cimento, 59A, 92 (1969).
99. M. G. Meshcheryakov, V. P. Zrelov, et al., Zh. Éksp. Teor. Fiz., 31, 45 (1956).

PROBABILITY DESCRIPTION OF HIGH-ENERGY SCATTERING AND THE SMOOTH QUASI-POTENTIAL

A. A. Logunov and O. A. Khrustalev

The relation between the probability and quasi-potential descriptions of high-energy scattering is studied. It is shown that the probability description of scattering can be considered as a justification for the introduction of smooth quasi-potentials into quantum field theory.

1. INTRODUCTION

One of the most important experimental facts of the physics of strong interactions is the recently observed far-reaching similarity between elastic and two-particle exchange processes at high energies. For small momentum transfers ($-t \lesssim 0.5$ GeV^2) exchange processes have a sharply defined diffraction peak with a width and relative height close to those for elastic scattering, and weakly dependent on energy and the quantum numbers of the scattering particles.

The existence of a sharp peak in the cross section for elastic processes involving momentum transfers close to zero can be explained by assuming that at high energies the partial-wave scattering amplitudes have the maximum values consistent with analyticity in the large Lehmann ellipse, short of the unitary limit, and the scattering amplitude in this case is close to

$$f(s, t) = i \sum_{l=0}^{\tilde{l}} (2l + 1) P_l(\cos\theta), \tag{1.1}$$

with the limiting orbital angular momentum given by

$$\tilde{l} = q\varphi(q), \tag{1.2}$$

where q is the momentum in the center of mass system, and $\varphi(q)$ is a slowly varying function of energy.*

The fact that the amplitude of two-particle exchange processes differ in first approximation from the amplitude (1.1) only by an energy-dependent factor permits the assumption that at high energies the exchange quanta involving nonzero quantum numbers are spread out among a large number of quanta corresponding to internal vacuum quantum numbers and carrying momentum only. This picture of two-particle exchange processes is quite consistent with the intuitive notion that all high-energy processes are essentially many-particle processes. Within the limits of the phenomenological potential description of two-particle processes this indicates the inadequacy of Yukawa potentials for high-energy processes.

Alliluev et al. [2] noted the prospect of describing high-energy scattering by a quasi-potential equation [3] with smooth Gaussian potentials. Later Savrin et al. [4] showed that within the framework of the Schrödinger equation such potentials can be interpreted as having a variable range characterized by a constant having the dimensions of a length squared and leading to differential cross section curves whose shapes are determined by momentum transfer only. The shape of the diffraction peak thus becomes a

*The compatibility of the black sphere model with the analytic properties of the scattering amplitude is analyzed in [1].

Institute of High-Energy Physics, Serpukhov. Translated from Problemy Fiziki Élementarnykh Chastits i Atomnogo Yadra, Vol. 1, No. 1, pp. 71-90, 1970.

universal function of all two-particle processes. Thus, there is a basis for assuming that smooth potentials appear naturally in such a description of scattering when the contribution of many-particle intermediate states is taken into account explicitly, if only in a very crude approximation. The present paper is devoted to a justification of this assumption.

2. HIGH-ENERGY SCATTERING AS COHERENT EXCITATION

As a starting point for the formalism developed later we take the unitarity relation for the scattering amplitude in the form

$$\operatorname{Im} f(\omega, \alpha) = \sum_n \int dq_1 \ldots dq_n f_n^* (q; \omega) f_n (q; \alpha) \delta \left(\sum_{l=1}^{n} q_l - p_\alpha \right). \tag{2.1}$$

Here $f(\omega, \alpha)$ is the scattering amplitude in the center of mass system:

$$f(\omega, \alpha) = \delta(p' - p) \langle \omega | t | \alpha \rangle, \tag{2.2}$$

the matrix T is defined by

$$S = 1 + iT, \tag{2.3}$$

α and ω are unit vectors along **p** and **p'** and

$$f_n(q; p) = \langle q_1 \ldots q_n | t | p \rangle, \tag{2.4}$$

where $|q_1 \ldots q_n\rangle$ is the proper n-particle vector operator for the total momentum. The function $f_n(q; p)$ can be interpreted as a wave function in the momentum representation of a system of particles created in a two-particle collision, and each of the integrals of series (2.1), aside from the delta function, as an overlap integral of wave functions, containing as parameters two unit vectors α and ω, which determine a certain two-dimensional structure.

If it is assumed that the scattering amplitude is pure imaginary, and this assumption is completely justified by high-energy experiments, Eq. (2.1) determines the scattering amplitude as a sum of overlap integrals of n-particle wave functions. This representation of the amplitude permits one to consider high-energy scattering as the passage of the particle through an absorbing medium with possible coherent excitation of the medium.† Moreover, the idea of coherent excitation can be carried over to the process of momentum transfer. Actually, we consider the simplest overlap integral of two single-particle wave functions describing sufficiently accurately localized particles. In this case, one can take as a wave function a Gaussian packet in **x** space

$$\psi(\mathbf{x}, \mathbf{p}) = (\pi\delta)^{-3/4} \exp\left[-\frac{\mathbf{x}^2}{2\delta} + i\mathbf{x}\mathbf{p} \right]. \tag{2.5}$$

Then

$$\int d\mathbf{x} \psi^* (\mathbf{x}, \mathbf{p}_2) \psi(\mathbf{x}, \mathbf{p}_1) = \exp\left[-\frac{\delta}{4} (\mathbf{p}_2 - \mathbf{p}_1)^2 \right]. \tag{2.6}$$

The integral (2.6) shows that fraction of the state (2.5) with average momentum $\mathbf{p}_2$ is contained in the state with average momentum $\mathbf{p}_1$. It is natural that it decreases exponentially with the vector difference $\mathbf{p}_2 - \mathbf{p}_1$. Wave functions of the type (2.5) quite naturally appear in the integrals (2.1) since strong interactions have a sharply defined range. Because of this the analogy between momentum transfer and coherent excitation becomes clearer. One can suppose that the whole interaction, i.e., the formation of packets in the integral (2.1), occurs within a sphere of finite radius R. Then the momentum becomes a discrete variable as are all other characteristics of the particles taking part in the reaction. Both elastic and exchange scattering are now characterized by coherent exchange and a finite number of quantum numbers, and the separate role of momentum transfer is distinguished only by a larger separation of the initial and final quantum levels.

Let us now examine the role of the delta function in (2.1). We write the corresponding n-fold integral in the form

† This result is taken as the initial assumption in [5].

$$\frac{1}{2\pi}\int dv e^{-ivp_0}\int dq_1 \ldots$$

$$\ldots dq_n f_n^* (q; \omega) f_n (q; \alpha) e^{-iv\sum_{l=1}^{n}\sqrt{q_l^2+m^2}}\,\delta\left(\sum_{l=1}^{n} q_l\right). \qquad (2.7)$$

Since the wave functions $f_n(q_1 \ldots q_n; \rho)$ decrease rapidly with the difference between q_l and its average value, † the factor $\exp[iv\sqrt{q_l^2+m^2}]$ in the integral (2.7) can be considered a slowly varying function, and by replacing q_l^2 by its average value $\langle q_l^2\rangle$, the integration over v can be performed, setting

$$\sum_{l=1}^{n}\sqrt{\langle q_l^2\rangle + m^2} = p_0. \qquad (2.8)$$

Thus, taking account of conservation of energy does not spoil the concept of scattering as the passage of a particle through a two-dimensional structure. Equation (2.8) only supplements such a picture since the degree of inhomogeneity of this structure is determined by the initial energy and the number of particles in the intermediate state.

It should be noted that for statistical independence of individual particles in integral (2.1) the Gaussian wave packets (2.5) in x space, taking account of conservation of momentum, lead to isotropic scattering in the center of mass system. If we assume that $f_n(q_1 \ldots q_n; \rho)$ describes identically distributed particles with average momentum p, then after converting from the delta function to the Fourier integral, the inner integral of (2.7) can be written as

$$(2\pi)^{-3}\int du\int dq_1 \ldots dq_n \left(\frac{\delta}{\pi}\right)^{3n/2}\exp\left[-\frac{\delta}{2}\sum_l (q_l - p')^2 - \frac{\delta}{2}\sum_l (q_l - p)^2 + iu\sum q_l\right]. \qquad (2.9)$$

The exponent in this integral can be written in the form

$$-\delta\sum_{l=1}^{n}\left\{\left(q_l - \frac{p'-p}{2} - i\frac{u}{2\delta}\right)^2 - \left(\frac{p'+p}{2} - i\frac{u}{2\delta}\right)^2 + \frac{p^2+p'^2}{2}\right\}. \qquad (2.10)$$

After appropriate translations integral (2.9) is reduced to an expression which does not involve the scalar product $p \cdot p'$. It is easy to understand the meaning of this result. The dispersion of the radius vector to a particle, distributed according to the law (2.5), is proportional to the unit tensor $\delta_{\alpha\beta}$. It is natural that an isotropic distribution of scatterers does not lead to anisotropic scattering. Therefore, scattering can become anisotropic because of a correlation between individual particles in the intermediate state or because of an anisotropy of the distributions of individual particles (even statistically independent) in the intermediate state.

Thus, the laws of energy and momentum refine the picture of scattering as the passage of a particle through a two-dimensional structure; the structure must be sufficiently anisotropic and the degree of its nonuniformity varies with energy and with the number of particles in the intermediate state, determining each partial structure in (2.1). Equation (2.8) shows that if we again turn to plane waves in a sphere of radius R, then in the case of statistically independent particles there will be a decrease in the distance between the levels the system occupies before and after scattering as n increases and the momentum transfer remains constant. This means that for a given momentum transfer the relative weights of higher intermediate states in Eq. (2.1) increase with increasing n, and this leads to a flattening out of the differential cross section with increasing momentum transfer.

The explicit dependence of the overlap integral on n for large n has been found [6, 7] under the condition that individual particles in an n-particle state are either statistically independent or sufficiently

† The Fourier transform of the wave function (2.5) is

$$\left(\frac{\pi}{\delta}\right)^{3/4}\exp\left[-\frac{\delta}{2}(q-p)^2\right].$$

weakly correlated. It is convenient in this case, to apply the central limit theorem of probability theory and to describe the angular dependence of the differential cross section by a certain quadratic form $\varphi(s, \theta) = \varphi(p_l, p_t)$ of the momentum components along the perpendicular directions

$$\sigma = \frac{\varkappa + \omega}{2\cos\theta/2}, \quad \pi = \frac{\varkappa - \omega}{2\sin\theta/2} \tag{2.11}$$

(for small scattering angles this quadratic form is proportional to the momentum transfer, or, what amounts to the same thing, to the square of the transverse momentum). If it is assumed that each of the statistically independent particles carries a definite fraction of the transverse momentum, the n-particle overlap integral is proportional to

$$\exp\left(-\frac{\varphi(p_l, p_t)}{2n}\right). \tag{2.12}$$

If $\sqrt{\varphi(p_l, p_t)}$ is identified with the transverse momentum (the validity of this identification is discussed in [7]) the expression (2.12) can be identified with the probability that n of the independent particles, each of which carries a definite transverse momentum $\langle p_t \rangle$, together carry a transverse momentum $\sqrt{\varphi(p_l, p_t)}$.

3. DEPENDENCE OF $\varphi(s, \theta)$ ON THE SCATTERING ANGLE

We now give a more detailed discussion of the dependence of $\varphi(s, \theta)$ on the scattering angle. For statistically independent particles $\varphi(s, \theta)$ is given by [7]

$$\varphi(s, \theta) = A_{\alpha\beta}^{-1} \langle p_\alpha \rangle \langle p_\beta \rangle, \tag{3.1}$$

where $\langle p \rangle$ is the average momentum of an intermediate state, and $A_{\alpha\beta}$ is the dispersion of the momentum of an individual particle. Since there are only two independent vectors $\boldsymbol{\sigma}$ and $\boldsymbol{\pi}$ at our disposal, the averages $\langle p \rangle$ and $\langle p_s p_t \rangle$ must be resolved into components along $\boldsymbol{\sigma}$ and $\boldsymbol{\pi}$ in the following way:

$$\langle \mathbf{p} \rangle = \langle \mathbf{p}\sigma \rangle \sigma + \langle \mathbf{p}\pi \rangle \pi; \tag{3.2}$$

$$\langle p_l p_t \rangle = \gamma_1 \delta_{lt} + \gamma_2 \sigma_l \sigma_t + \gamma_3 \pi_l \pi_t + \gamma_4 (\sigma_l \pi_t + \pi_l \sigma_t). \tag{3.3}$$

It is easy to show that

$$\gamma_1 = \langle \mathbf{p}^2 \rangle - \langle (\mathbf{p}\sigma)^2 \rangle - \langle (\mathbf{p}\pi)^2 \rangle; \tag{3.4}$$

$$\gamma_2 = 2\langle (\mathbf{p}\sigma)^2 \rangle + \langle (\mathbf{p}\pi)^2 \rangle - \langle \mathbf{p}^2 \rangle; \tag{3.5}$$

$$\gamma_3 = 2\langle (\mathbf{p}\pi)^2 \rangle + \langle (\mathbf{p}\sigma)^2 \rangle - \langle \mathbf{p}^2 \rangle; \tag{3.6}$$

$$\gamma_4 = \langle (\mathbf{p}\sigma)(\mathbf{p}\pi) \rangle. \tag{3.7}$$

For an isotropic distribution, when $\langle p \rangle = 0$, $\langle (\mathbf{p}\pi) \times (\mathbf{p}\sigma) \rangle = 0$, $\langle (\mathbf{p}\sigma)^2 \rangle = \langle (\mathbf{p}\pi)^2 \rangle = 1/3 \langle p^2 \rangle$, all the coefficients in (3.3) are zero except $\gamma_1 = 1/3 \langle p^2 \rangle$. The dispersion of the momentum A_{lt} and the inverse matrix A_{lt}^{-1} are similar in form to (3.3):

$$A_{lt} = a_1 \delta_{lt} + a_2 \sigma_l \sigma_t + a_3 \pi_l \pi_t + a_4 (\sigma_l \pi_t + \pi_l \sigma_t); \tag{3.8}$$

$$A_{lt}^{-1} = b_1 \delta_{lt} + b_2 \sigma_l \sigma_t + b_3 \pi_l \pi_t + b_4 (\sigma_l \pi_t + \pi_l \sigma_t), \tag{3.9}$$

and the quadratic form $\varphi(s, \theta)$ in this notation is

$$\varphi(s, \theta) = (b_1 + b_2)\langle (\mathbf{p}\sigma)^2 \rangle + (b_1 + b_3)\langle (\mathbf{p}\pi)^2 \rangle + 2b_4 \langle (\mathbf{p}\sigma)(\mathbf{p}\pi) \rangle. \tag{3.10}$$

A straightforward calculation shows that $\varphi(s, \theta)$ can be expressed as a function of two random variables

$$\xi_1 = (\mathbf{p}\sigma), \quad \xi_2 = (\mathbf{p}\pi) \tag{3.11}$$

the components of momentum along the orthogonal vectors σ and π, namely:

$$\varphi(s, \theta) = \Delta^{-1}_{\lambda\varkappa} \xi_\lambda \xi_\varkappa, \tag{3.12}$$

where

$$(\Delta_{\lambda\varkappa}) = (\langle \xi_\lambda \xi_\varkappa \rangle - \langle \xi_\lambda \rangle \langle \xi_\varkappa \rangle). \tag{3.13}$$

We note that the average components of **p** (and also ξ_λ) are calculated by using a probability density of a very special form. Its dependence on the scattering angle is given by*

$$f(\omega \mathbf{n}) f(\mathbf{n}\alpha), \tag{3.14}$$

where $\vec{n}$ is a unit vector with the components

$$\mathbf{n} = (\sin\tilde{\theta}\cos\tilde{\varphi},\ \sin\tilde{\theta}\sin\tilde{\varphi},\ \cos\tilde{\theta}). \tag{3.15}$$

If we introduce a coordinate system in which the vectors α and ω have the components

$$\alpha = (0, 0, 1), \quad \omega = (\sin\theta, 0, \cos\theta), \tag{3.16}$$

the vectors $\vec{\sigma}$ and $\vec{\pi}$ can be written in the form

$$\sigma = (z_1, 0, z_2), \quad \pi = (-z_2, 0, z_1), \tag{3.17}$$

where

$$z_1 = \sin\theta/2, \quad z_2 = \cos\theta/2, \tag{3.18}$$

and the density of the angular distribution (3.14) is

$$f\left(2z_1 z_2 \sin\tilde{\theta}\cos\tilde{\varphi} + (z_2^2 - z_1^2)\cos\tilde{\theta}\right) f(\cos\tilde{\theta}). \tag{3.19}$$

The densities of the distributions of individual momenta (3.14) must be chosen in such a way that the total momentum of the intermediate state is zero. There is no loss in generality, however, if we give up this condition and consider $\langle\xi\rangle_\theta - \langle\xi\rangle_0$ instead of the averages $\langle\xi\rangle_\theta$ calculated by using the density (3.14). This then permits us to consider only individual particles and not the whole collection of particles in an n-particle intermediate state.

Since in the coordinate system (3.16)

$$\mathbf{n}\sigma = z_1 \sin\tilde{\theta}\cos\tilde{\varphi} + z_2\cos\tilde{\theta}; \tag{3.20}$$

$$\mathbf{n}\pi = -z_2 \sin\tilde{\theta}\cos\tilde{\varphi} + z_1\cos\tilde{\theta}, \tag{3.21}$$

and the density of the angular distribution for small scattering angles, when $z_1 \ll 1$ and $z_2 \sim 1$, is approximately

$$2f'(\cos\tilde{\theta}) f(\cos\tilde{\theta}) \sin\tilde{\theta}\cos\tilde{\varphi} z_1, \tag{3.22}$$

to first order in z_1

$$\langle \mathbf{n}\sigma \rangle = 0; \tag{3.23}$$

$$\langle \mathbf{n}\pi \rangle = A z_1, \tag{3.24}$$

where

$$A = 2\int d\mathbf{n} (\sin\tilde{\theta}\cos\tilde{\varphi})^2 f'(\cos\tilde{\theta}) f(\cos\tilde{\theta}). \tag{3.25}$$

As was noted in [6], the dispersions of the momentum distributions of intermediate particles vary slowly with energy and with the scattering angle, and the absolute magnitude of the average momentum is proportional to $\sqrt{s}$, and therefore at small scattering angles

* For simplicity we consider a real function $f(\mathbf{n}\,\alpha)$.

$$\varphi(s, \theta) = -c(s)\, t, \tag{3.26}$$

where t is the momentum transferred and c(s) is a slowly varying function of energy.

Such a dependence of $\varphi(s, \theta)$ on the scattering angle is most likely to be valid only for small scattering angles, however, since the value of $\langle \mathbf{n}\sigma \rangle$ increases with angle. This is easy to see by considering the limiting case of scattering at 90°, when $z_1 = z_2 = 1/\sqrt{2}$. In this case, the density of the angular distribution (3.16) is

$$f(\sin\tilde{\theta}\cos\tilde{\varphi})\, f(\cos\tilde{\theta}), \tag{3.27}$$

and

$$\mathbf{n}\sigma = \sin\tilde{\theta}\cos\tilde{\varphi} + \cos\tilde{\theta}; \tag{3.28}$$

$$\mathbf{n}\pi = -\sin\tilde{\theta}\cos\tilde{\varphi} + \cos\tilde{\theta}. \tag{3.29}$$

Going over to Cartesian coordinates for the averaging we find that

$$\langle \mathbf{n}\sigma \rangle \approx \int \frac{dx\, dy\, dz}{r}\, \delta(r^2 - 1)(x + z)\, f\left(\frac{x}{r}\right) f\left(\frac{z}{r}\right), \tag{3.30}$$

and

$$\langle \mathbf{n}\pi \rangle \approx \int \frac{dx\, dy\, dz}{r}\, \delta(r^2 - 1)(x - z)\, f\left(\frac{x}{r}\right) f\left(\frac{z}{r}\right). \tag{3.31}$$

Thus, for scattering at 90° it is apparent that the magnitude of the longitudinal component of momentum is of fundamental importance.

The following can be said about the properties of $\varphi(s, \theta)$ as a function of the scattering angle. At small scattering angles $\varphi(s, \theta)$ can be identified with momentum transfer; $\varphi(s, \theta)$ is small in this region, but increases rapidly. At large scattering angles $\varphi(s, \theta)$ is large, but almost certainly does not reduce to a function of $\sin\theta/2$ only, since in this case, the longitudinal component of momentum becomes increasingly important. The following should be noted: for $\theta < \pi/2$ the variable $z_2 = \cos\theta/2$, which we relate to the longitudinal component of momentum, varies more slowly than $z_1 = \sin\theta/2$, which is related to the momentum transfer t. Their rates of change are comparable only at $\theta \simeq (\pi/2)$. Therefore, there is a possible range of angles where $\varphi(s, \theta)$ is large enough, but still can be considered a function of z_1 only. For larger angles the rate of change of $\varphi(s, \theta)$ can be very much smaller than the rate of change of t. At the same time $\varphi(s, \theta)$ in this case, will increase with increasing energy s even for fixed t.

These considerations of the dependence of $\varphi(s, \theta)$ on energy and scattering angle are qualitatively confirmed by the experimental data on the differential cross sections for elastic scattering [8]. For small scattering angles ($-t \lesssim 0.5$ GeV^2) the scattering cross section varies approximately as $\exp(a, t)$, where a varies slowly with the scattering energy s. One can associate a $\varphi(s, \theta)$ of the form (3.26) with this region, where $\varphi(s, \theta)$ is still so small that the saddle point of the integrand approximating series (2.1) is outside the contour of integration.* In the region adjoining the diffraction cone the scattering cross section behaves as $\exp(-b\sqrt{-t})$, where b has a tendency to increase with energy. The cross section must behave this way if $\varphi(s, \theta)$, although large, may be considered a function of z_1 only, with the dependence of $\varphi(s, \theta)$ on z_2 manifesting itself only in the fact that $\varphi(s, \theta)$ begins to increase more rapidly with energy. Finally, for larger scattering angles the cross section varies relatively slowly with the scattering angle but decreases rapidly with s for a fixed momentum transfer. This dependence appears to be close to $\exp(-c\sqrt{s})$. This feature of the cross section is also in complete agreement with the assumption that at larger scattering angles z_2 is at least as important a variable as z_1.

4. THE EIKONAL REPRESENTATION OF THE IMAGINARY PART OF THE SCATTERING AMPLITUDE

The well-known eikonal representation can be obtained by starting from the representation of the imaginary part of the scattering amplitude in the form of series (2.1). We use the equation

*A detailed discussion of the estimate of series (2.1) is given in [6].

$$\frac{1}{n}\exp\left[-\frac{\varphi(s,\theta)}{2n}\right]=\int_0^\infty \xi\, d\xi J_0\left(\xi\sqrt{\varphi(s,\theta)}\right)\exp\left[-n\frac{\xi^2}{2}\right]. \tag{4.1}$$

By substituting this expression into series (2.1) and interchanging the order of summation and integration we find that

$$\operatorname{Im} T(s,\theta)=\int_0^\infty \xi\, d\xi J_0\left(\xi\sqrt{\varphi(s,\theta)}\right)\rho(\xi), \tag{4.2}$$

where

$$\rho(\xi)=\sum\frac{c(n)}{n}\sqrt{\frac{n}{2\pi}}\exp\left[-n\frac{\xi^2}{2}\right]. \tag{4.3}$$

For small scattering angles $\varphi(s,\theta) = -at$ and therefore, (4.2) goes over into the usual eikonal representation of the scattering amplitude

$$\operatorname{Im} T(s,t)=\int_0^\infty b\, db J_0\left(b\sqrt{-t}\right)\rho(b), \tag{4.4}$$

where ξ in the integral (4.2) differs from the usual impact parameter b only by a factor which may vary slowly with energy. We call ξ the impact parameter, although at large scattering angles Eq. (4.2) may differ considerably from (4.4).

The quantity $\rho(\xi)$ has a simple probability meaning. We can extend the integration in (4.2) from $-\infty$ to ∞ and consider negative values of ξ. We assume that each of the particles in the n-th intermediate state has a certain distribution over the impact parameter with an average value ξ and that this distribution is such that the central limit theorem holds for the sum of the impact parameters. Then

$$\sqrt{\frac{n}{2\pi}}\exp\left[-n\frac{\xi^2}{2}\right] \tag{4.5}$$

will be the probability density for zero total impact parameter in an n-particle intermediate state. The relative weight of each n-particle state is determined by the C(n)/n distribution.

Following Chou and Yang [8] we turn from the representation (4.2) to the two-dimensional Fourier representation, replacing the Bessel function by the integral*

$$J_0\left(\xi\sqrt{\varphi(s,\theta)}\right)=\int_0^{2\pi} dv e^{i\xi\sqrt{\varphi(s,\theta)}\cos v} \tag{4.6}$$

and transforming from polar to Cartesian coordinates in the integral (4.2). Then

$$\operatorname{Im} T(s,\theta)=\int d\xi e^{i\xi\varkappa(s,\theta)}\rho(\xi), \tag{4.7}$$

where

$$\xi=(\xi_x,\xi_y); \tag{4.8}$$

$$\varkappa(s,\theta)=\left(\varkappa_x(s,\theta),\quad \varkappa_y(s,\theta)\right),\quad \varkappa^2=\varphi(s,\theta). \tag{4.9}$$

In order to give an analogous relation obtained from (4.4) a clear physical meaning, Chou and Yang [8] proposed considering the scattering of strongly interacting particles as the penetration of two spheres through one another, each of which appears to the other as a disk. Therefore, the scattering amplitude is written in the form of the Fourier transform of some two-dimensional structure.

*Chou and Yang wrote representation (3.7) for the scattering amplitude, although they then assumed that it is purely imaginary.

It is easy to see that Eq. (4.7) can be given exactly such a meaning. In doing this we find just what must be understood by the internal structure of the colliding particles. To do this we recall that the terms of series (2.1) were obtained from the asymptotic estimate of the integral [6]

$$w_n(\omega, \alpha) = \frac{1}{2\pi} \int dv e^{-ivp^0_\alpha} \rho_n(\mathrm{p}_\alpha, v), \tag{4.10}$$

where

$$\rho_n(\mathrm{p}_\alpha, v) = \int d\mathrm{q} f_n^*(\mathrm{q}, \omega) f_n(\mathrm{q}, \alpha) e^{iv \sum_{l=1}^{n} \sqrt{\mathrm{q}_l^2 + m^2}} \delta(\sum \mathrm{q}_l - \mathrm{p}_\alpha). \tag{4.11}$$

The factor exp $\left[iv \sum_{l=1}^{n} \sqrt{\mathrm{q}_l^2 + m^2}\right]$ guarantees conservation of energy and is not important in what follows.

If the scattering angle $\theta = 0$, the integrand contains the square of the absolute magnitude of the wave function of a system of n particles, and $w_n(\omega, \alpha)$ can be interpreted as the contribution of the n-particle state to the meson exchange of the colliding particles. If the scattering angle is not zero the integral (4.11) should rather be identified with the overlap integral of the wave functions.

The unit vectors $\boldsymbol{\alpha}$ and $\boldsymbol{\omega}$ along the initial and final momenta determine the two-dimensional structure of the overlap integral. The transition probability in this case, as it should be according to quantum mechanics, is proportional to the square of the absolute magnitude of the overlap integral. Therefore, if we seek analogies closer to Eq. (4.7) than those of high-energy physics, the most likely are the σ- and π-bonds of quantum chemistry. Our σ-bond is determined by the magnitude of $\xi_1 = (\mathrm{p}\,\sigma)$ (cf. Eq. (3.13)) and gives the main effect – forward scattering. The π-bond, which is small in comparison, gives rise to a small but rapidly increasing effect – the deviation of the particle from its rectilinear path. In this case, the scattering amplitude is in good agreement with the eikonal representation (3.4). With an increase in the effect of the π-bond, however, the old basis of σ and π vectors turns out to be inadequate to describe the interaction, and the quadratic form $\varphi(s, \theta)$ (cf. Eq. (3.14)) deviates more and more from diagonal form. In particular this effect can lead to the previously described flattening out of the differential cross section at larger scattering angles.

The integral (4.7), evaluated at $\varkappa = 0$, gives the area of our two-dimensional structure, and by the optical theorem this integral is proportional to the total scattering cross section. Thus, we are led to the formula $\sigma_{tot} \sim r^2$, which is well-known from semiclassical considerations. Here r is the characteristic range of the interaction. Chou and Yang [8] identify integral (4.4) at small t with the convolution of the two form factors of the scattering particles and conclude that for small momentum transfers the differential proton-proton scattering cross section is

$$\frac{d\sigma}{dt} \sim [F(\mathrm{q})]^4, \tag{4.12}$$

where F(q) is the proton form factor. This conclusion is valid in our case also, but for large momentum transfers it appears that (4.12) must be violated since $\varkappa^2(s, \theta)$ from integral (4.7) can differ appreciably from $-t$, and by measuring the differential proton-proton scattering cross section we obtain information on the overlap integral which can differ appreciably from the Fourier transform of the meson exchange density distribution at large scattering angles.

The difference between our interpretation of $\rho(\xi)$ and that given $\rho(b)$ by Chou and Yang [8] should be noted. They assume that

$$\rho(b) = 1 - s(b), \tag{4.13}$$

with s given by

$$-\ln s(b) = \int d\mathbf{b}' D_A(\mathbf{b} - \mathbf{b}') D_B(\mathbf{b}'), \tag{4.14}$$

The integral (4.14) is the convolution of the two-dimensional Fourier transforms of the density of the disks corresponding to particles A and B. This identification was made on the basis of an optical analogy, since it is known that the absorption of light passing through a body that is not completely transparent is proportional to exp (−g), where g is the opaqueness of the material.

In this case, Eq. (4.4) leads to the expansion of the scattering amplitude in an alternating series in which the relative weights of the higher terms of the expansion increase with increasing momentum transfer. This leads to the vanishing of the differential cross section for certain sufficiently large momentum transfers. Actually this is true only for asymptotically large energies. Durand and Lipes [9] showed that the correction for finite scattering energy partly fills up the dip in the differential cross section. Nevertheless the presence of a dip in the differential cross section for sufficiently high momentum transfers is an intrinsic feature of the Chou and Yang theory. The phenomenon of the dip is due to the interference between single and double scattering, well-known in the scattering of fast particles by nuclei [10], and the expansion of the scattering amplitude given in [8] makes the scattering of high-energy particles similar to the scattering of particles by a nucleus.

The expansion of the amplitude (4.7) does not require asymptotically large scattering energies. In addition the internal structure of the scattering particles, described by the series (4.3), is more complicated than the simple multiple reproduction of a single structure used in [8]. Therefore, the phase relations between terms of the series (4.3) require special study.

5. EIKONAL REPRESENTATION OF THE QUASI-POTENTIAL SCATTERING AMPLITUDE

To study the detailed behavior of the differential cross section in the region directly adjacent to the diffraction cone it is convenient to go over to the quasi-potential equation for the scattering amplitude [3]. Garsevanishvili et al. [11] studied the scattering of relativistic particles by a complex Gaussian quasi-potential. In particular they showed that for moderate momentum transfers the scattering amplitudes have an eikonal structure. In this section, we obtain this result another way which, in the case of a pure imaginary scattering amplitude, permits the reduction of the quasi-potential by the representation (4.2) for the scattering amplitudes obtained from other considerations.

Calogero [12] showed that the quasi-potential equation for the wave function in **x** space has the form of a nonlocal Schrödinger equation

$$(\nabla^2+k^2)\,u(\mathbf{x})=\int F(\mathbf{x},\mathbf{y})\,V(\mathbf{y},k^2)\,u(\mathbf{y})\,d\mathbf{y}, \tag{5.1}$$

where

$$F(\mathbf{x},\mathbf{y})=\frac{2m^2}{(2\pi)^2}\cdot\frac{K_1(m\,|\mathbf{x}-\mathbf{y}|)}{m\,|\mathbf{x}-\mathbf{y}|}. \tag{5.2}$$

The equation for the radial wave functions, defined by the relation

$$u(\mathbf{x})=\frac{1}{x}\sum_{l,m}c(l,m)\,u_l(x)\,Y_{lm}(\hat{x}), \tag{5.3}$$

can be written in the form

$$u_l''(x)+\left[k^2-\frac{l(l+1)}{x^2}\right]u_l(x)=\int V_l(x,y)\,u_l(y)\,dy, \tag{5.4}$$

where

$$V_l(x,y)=\int_0^\infty d\xi\,\varphi(l,\xi)\,K_{i\xi}(mx)\,K_{i\xi}(my)\,V(y), \tag{5.5}$$

and the functions $\varphi(l,\xi)$ are defined as the expansion coefficients of the Gegenbauer functions in spherical harmonics

$$\frac{2}{(2\pi)^2}\,\xi C^1_{-1+i\xi}(-\hat{x}\,\hat{y})=\sum_{l,m}\varphi(l,\xi)\,Y_{lm}(\hat{x})\,Y^*_{lm}(\hat{y}). \tag{5.6}$$

In order to find the scattering phases we use the method of phase functions developed in [13]. We define the functions

$$S_\lambda(x)=\left(\frac{\pi kx}{2}\right)^{1/2} J_{l+1/2}(kx),\quad C_\lambda(x)=-\left(\frac{\pi kx}{2}\right)^{1/2} Y_{l+1/2}(kx), \tag{5.7}$$

which satisfy the equation

$$J''+\left[k^2-\frac{l(l+1)}{x^2}\right]y=0, \tag{5.8}$$

and for asymptotically large kx go over into

$$S_\lambda\sim\sin\left(kx-\frac{\pi l}{2}\right),\quad C_\lambda\sim\cos\left(kx-\frac{\pi l}{2}\right). \tag{5.9}$$

In view of the analogy between S_λ and C_λ and the sine and cosine we employ the notation [13]:

$$S_\lambda(x)=\hat{D}_\lambda(x)\sin\hat{\delta}_\lambda(x),\quad C_\lambda(x)=\hat{D}_\lambda(x)\cos\hat{\delta}_\lambda(x). \tag{5.10}$$

We seek a solution of Eq. (5.4) in the form

$$u_l(x)=\alpha_\lambda(x)\left[S_\lambda(x)\cos\delta_\lambda(x)+C_\lambda(x)\sin\delta_\lambda(x)\right], \tag{5.11}$$

under the condition that the functions $\alpha_\lambda(x)$ and $\delta_\lambda(x)$ are related by

$$\alpha_\lambda(x)=\alpha_\lambda\exp\left[\int_0^\infty dy\,\delta_\lambda'(y)\,\mathrm{ctg}\,(\delta_\lambda(y)+\hat{\delta}_\lambda(y))\right]. \tag{5.12}$$

Then Eq. (5.4) reduces to a nonlinear integro-differential equation for $\delta_\lambda(x)$:

$$\delta_\lambda'(x)=-\frac{1}{k}\hat{D}_\lambda(x)\sin(\hat{\delta}_\lambda(x)+\delta_\lambda(x))\int_0^\infty V_l(x,y)\frac{\alpha_\lambda(y)}{\alpha_\lambda(x)}\hat{D}_\lambda(y)\sin(\hat{\delta}_\lambda(y)+\delta_\lambda(y)). \tag{5.13}$$

The physical meaning of $\delta_\lambda(x)$ becomes clear when the wave function $u_l(x)$ is written in the form (5.11): the limit of $\delta_\lambda(x)$ as $x\to\infty$ is the scattering phase. Suppose the potential $V(\mathbf{x}, k^2)$ is given by

$$V(\mathbf{x}, k^2)=g(\mathbf{x}^2)\exp(-\varphi(\mathbf{x}^2)), \tag{5.14}$$

where the functions g and φ depend on $\mathbf{x}^2$ (and possibly contain the energy as a parameter) and have no singularities along a radial line $0\le x<\infty$. We take the following as first approximations to the functions $\alpha_\lambda(x)$ and $\delta_\lambda(x)$: $\alpha_\lambda(x)$ = const and $\delta_\lambda(x)$ is a solution of the equation

$$\delta_\lambda(x)=-\frac{1}{k}S_\lambda(x)\int_0^\infty V_l(x,y)S_\lambda(y)\,dy \tag{5.15}$$

with the initial condition $\delta_\lambda(0)=0$. It is easy to show that in this approximation the phase will be of the order $V(0)/p^2$ which can be considered a small quantity provided that the total scattering cross section does not increase at sufficiently high energies. Equation (5.13) is now transformed into a construction — a definite iteration procedure for constructing the scattering phase. In the second approximation it will contain the function

$$\alpha_\lambda(x)=\alpha_\lambda\exp\left\{-\frac{1}{k}\int_x^\infty C_\lambda(y)\int_0^\infty V_l(y,z)S_\lambda(t)\,dy\,dz\right\}. \tag{5.16}$$

So long as x is not too small the exponent in (5.16) is proportional to V/k^2 because of the oscillating character of the functions S_λ and C_λ, and when x and y are not small the ratio $\alpha_\lambda(x)/\alpha_\lambda(g)$ in the integral in Eq. (5.13) can be set equal to unity. The integral in (5.13) also contains values of $\alpha_\lambda(y)$ for small y, however, and in this case, the singularity of the function $C_\lambda(x)$ at zero raises a doubt about the previous estimate of $\alpha_\lambda(x)$. The specific properties of the potentials (5.14) must be taken into account in this case. The fact that φ is an even function of x shows that when the integral in (5.16) is evaluated by the saddle-point method the main contribution comes from a range of complex x values rather far from zero. Therefore, for potentials which are smooth in the sense of (5.14) a good approximation at high energies is the expression

$$\delta_\lambda = \frac{1}{k}\int_0^\infty S_\lambda(x)\,V_l(x, y)\,S_\lambda(y)\,dx\,dy. \tag{5.17}$$

Substituting the expression for the potential $V_l(x, y)$ from (5.5) into (5.17) and using the formulas for products of K functions [14] reduces (5.17) to the form

$$\delta_\lambda = -\frac{1}{k}\int_0^\infty dx\,dy\,S_\lambda(x)\,S_\lambda(y)\,V(y)\int_0^\infty \frac{dt}{t}\exp\left[-\frac{t}{2} - m^2\frac{(x-y)^2}{2t}\right]\int_C d\xi\tilde{v}_l\left(\xi, \frac{xy}{t}\right), \tag{5.18}$$

where C is a contour in the ξ plane parallel to the imaginary axis and $v_l(\xi, xy/t)$ is a slowly varying function of l and xy/t. The Bessel functions in (5.18) determine a region of characteristic values of x and y ~ 1/k, and therefore, the exponent in the integral over t is actually proportional to $(m^2/k^2)\,(x-y)^2$. In the nonrelativistic case when $m^2/k^2 \gg 1$, the integral (5.18) goes over into an integral determining the phase of the scattering by the local potential. In this case, the oscillations of the Bessel functions are not so important and the determining factor in the estimate of the integral is the Gaussian exponential, degenerating into a δ function. These considerations show that for scattering by potentials of the form (5.14) the integral for the scattering phase is also simplified. In this case, the potential V(y) localizes the significant values of y in a certain portion of the complex plane which does not approach zero even at large energies. The Gaussian exponential can be relatively large only if the x values fall in this same region, and this leads to a reduction of (5.18) to a one-dimensional integral which agrees with the expression for the nonrelativistic scattering phase.

The structure of the scattering amplitude with the phase given by

$$\delta_\lambda = -\frac{1}{k}\int_0^\infty V(x)\,S_\lambda^2(x), \tag{5.19}$$

was studied in detail by Savrin et al. [4]. They showed that for small momentum transfers the scattering amplitude can be written in the form

$$f(k, \theta) = -ik\int_0^\infty b\,db J_0(b\sqrt{-t})\left(e^{i\chi(b)} - 1\right), \tag{5.20}$$

where

$$i\chi(b) = -\frac{i}{\pi k}V\left(\sqrt{b^2+z^2}\right)dz. \tag{5.21}$$

A comparison of Eqs. (5.20) and (5.21) with the representation of the scattering amplitude (4.4) obtained in the preceding section enables one to find the effective quasi-potential by starting from the imaginary part of the scattering amplitude.

6. THE UNITARITY RELATION AND THE EFFECTIVE QUASI-POTENTIAL

We limit ourselves to momentum transfers which are small in comparison with the energy. In this case, the unitarity relation for the elastic-scattering amplitude is conveniently written in the impact-parameter representation

$$\operatorname{Im} f(b) = \frac{1}{2} |f(b)|^2 + \rho(b), \tag{6.1}$$

where $f(b)$ is the spectral density of the impact parameter, i.e.,

$$F(s, t) = \int 2q^2 b\, db J_0 (b\sqrt{-t}) f(b), \tag{6.2}$$

q is the momentum in the center of mass system, and $\rho(b)$ is the spectral density of the impact parameter contribution of inelastic channels to the imaginary part of the scattering amplitude.*

If the scattering amplitude is assumed to be pure imaginary Eq. (6.1) can be regarded as an equation determining the spectral density of the impact parameter in terms of the given contribution of inelastic channels [15]:

$$f(b) = i(1 - \sqrt{1 - 2\rho(b)}). \tag{6.3}$$

We now employ the eikonal representation of the quasi-potential scattering amplitude obtained in the preceding section and the eikonal representation of the contribution of inelastic channels from Sec. 4 within the framework of our hypothesis on the correlation of particles in higher intermediate relations and in the relative probabilities of the creation of n particles in two-particle collisions.

By combining Eqs. (5.21) and (6.3) we obtain an integral equation for the quasi-potential,

$$-i \ln \sqrt{1 - 2\rho(b)} = -\frac{2}{\pi q} \int_b^\infty \frac{V(r)\, r\, dr}{\sqrt{r^2 - b^2}}, \tag{6.4}$$

where $\rho(b)$, after appropriate normalization, is given by Eq. (4.3). Hence, $\rho(b)$ is a function of b^2. Equation (6.4) is the well-known Abel's equation which has the solution†

$$V(r) = iq \int_r^\infty \frac{\rho'(b)}{1 - 2\rho(b)} \cdot \frac{db}{\sqrt{b^2 - r^2}}. \tag{6.5}$$

Specifying the function $\rho(b)$ in the form

$$\rho(b) = \psi(b) \exp(-\varphi(b)), \tag{6.6}$$

where $\psi(b)$ and $\varphi(b)$ depend on b^2, it is easy to estimate the dependence of the quasi-potential on r. In particular it is convenient to expand the integral (6.5) in an asymptotic series. The first term in the expansion of (6.5) gives

$$V(r) = ig(r) \exp(-\varphi(r)), \tag{6.7}$$

where

$$g(r) = q \sqrt{\frac{\pi}{2r\varphi'(r)}} \cdot \frac{\psi'(r) - \varphi'(r)\psi(r)}{1 - 2\rho(r)}. \ddagger \tag{6.8}$$

Thus, the assumption of the eikonal nature of the quasi-potential scattering amplitude for small momentum transfers and the hypothesis of the statistical character of the scattering amplitude make it possible to describe high-energy scattering with momentum transfers which are small in comparison with the energy by a smooth quasi-potential of the form (5.14).

This result has a simple physical meaning. It has already been noted that since potentials of the form (5.14) have a variable range they are characterized by quantities quite different from the mass of an exchange quantum, as is usual with the Yukawa potential. In describing scattering in terms of many-particle

* For the normalization $\sigma_{tot} = (2\pi/q^2) \operatorname{Im} F(s, 0)$.
† More general potentials are considered in [16].
‡ It is easy to see that V(r) depends on r^2.

intermediate states we took as a first approximation a beam of uncorrelated particles. In this approximation it is easy to find the global dimensional characteristic of each n-particle state, independently of the number of particles. This is a weakly energy-dependent dispersion of the total momentum of the intermediate state. Since this quantity has the dimensions of momentum transfer, it naturally leads to a potential depending on the square of the distance. Thus, the quasi-potential and the probability methods of describing scattering naturally supplement one another. The potential description of scattering with a given quasipotential permits a description of the phase shifts of individual iterations which are essential in the transition from the diffraction region of momentum transfer to the Orear region. The probability description can be considered as a justification for the introduction of smooth quasi-potentials into field theory, and in addition it appears to be more promising for describing scattering with momentum transfers comparable with the energy.

LITERATURE CITED

1. A. A. Logunov, Nguyen Van Hieu, and O. A. Khrustalev, Problems in Theoretical Physics [in Russian], Nauka, Moscow (1969).
2. S. P. Alliluev, S. S. Gerstein, and A. A. Logunov, Phys. Lett., 18, 195 (1965).
3. A. A. Logunov and A. N. Tavkhelidze, Nuovo Cimento, 29, 380 (1963).
4. V. I. Savrin and O. A. Khrustalev, Preprint Institute of High-Energy Physics 68-19-K, Serpukhov (1968); O. A. Khrustalev, V. I. Savrin, and N. E. Tyurin, Communications JINR E2-4479 (1969).
5. N. Byers and C. N. Yang, Phys. Rev., 142, 976 (1966).
6. A. A. Logunov and O. A. Khrustalev, Preprint Institute of High-Energy Physics 69-20, Serpukhov (1969).
7. A. A. Logunov and O. A. Khrustalev, Preprint Institute of High-Energy Physics 69-21, Serpukhov (1969).
8. T. T. Chou and C. N. Yang, Phys. Rev., 170, 1591 (1968).
9. L. Durand and R. Lipes, Phys. Rev. Lett., 20B, 637 (1968).
10. R. Glauber, Lectures in Theoretical Physics, Vol. 1, Interscience, New York (1959).
11. V. R. Garsevanishvili et al., Phys. Lett., 29B, 191 (1969).
12. O. A. Khrustalev, Preprint Institute of High-Energy Physics 69-24, Serpukhov (1969).
13. F. Calogero, Nuovo Cimento, 28, 66 (1963).
14. G. N. Watson, Theory of Bessel Functions, Macmillan, New York (1944).
15. V. I. Savrin, N. E. Tyurin, and O. A. Khrustalev, Preprint Institute of High-Energy Physics 69-23, Serpukhov (1969).
16. V. I. Savrin, N. E. Tyurin, and O. A. Khrustalev, Preprint Institute of High-Energy Physics 69-65, Serpukhov (1969).

HADRON SCATTERING AT HIGH ENERGIES AND THE QUASI-POTENTIAL APPROACH IN QUANTUM FIELD THEORY

V. R. Garsevanishvili, V. A. Matveev,
and L. A. Slepchenko

The paper gives a review of methods and certain results for the quasi-potential approach to hadron collision processes at high energies.

INTRODUCTION

The theoretical foundation of an understanding of strong interaction processes at high energies is made up of such general principles of quantum field theory as causality, unitarity, and the proof on the basis of these notions of the dispersion relationships for the scattering amplitude [1].

Based on general propositions concerning analyticity and unitarity of the scattering amplitude, one can obtain a number of rigorous corollaries and constraints for physical observables [2]. The study of the analytic properties within the framework of the double spectral representation forms the base for various approximate schemes in the theory of strong interactions [3, 4].

The successful application of the dispersion relationships served as the impetus for fundamental papers on the asymptotic approach in quantum field theory. Specific assumptions concerning the character of the asymptotic behavior allow a number of important corollaries to be derived which can be checked experimentally. The first relationship of this kind is the equality between the total cross sections for interactions of particles and antiparticles at high energies [5]. On the basis of the Phragmen-Lindelöf theorem a number of important relationships between the cross sections of various processes have been derived [6].

Experiments on modern accelerators are establishing a number of regularities in the interaction of high-energy hadrons. A detailed analysis of the data shows a tendency of the total cross sections toward a fixed limit with increasing energy. At small angles the scattering is of a diffraction character, and the differential cross sections of the elastic processes are characterized by a fast (exponential) falloff with an increase in the square of the transferred momentum. The results of recent measurements of the differential cross sections of elastic pp-scattering on the accelerator of the Institute of High-Energy Physics [7] show the monotonic growth of the parameter of the diffraction-peak slope for increasing energy in the 12 to 70 GeV^2 range. Such behavior is in agreement with the general principles of quantum field theory and substantiates the idea of growth of the effective radius of strong interactions [8]. The diffraction character of high-energy scattering allows methods to be resorted to which are close to quasi-classical methods, the eikonal approximation [9], and quantum-mechanical scattering by smooth potentials [10]. Many different models were proposed which are based on modifications of the optical formalism, the impact-parameter representation, and multipole Reggeon models [11-17].

In terms of multiple scattering using certain simple propositions concerning the dynamics of the interaction of hadrons, model explanations were obtained for various phenomena in small- and large-angle scattering: the structure in the angular distributions, the characteristic change of operating mode with increasing transferred momentum, and others. In a number of papers [18] the Reggeon technique is

---

Joint Institute for Nuclear Research, Dubna. Translated from Problemy Fiziki Élementarnykh Chastits i Atomnogo Yadra, Vol. 1, No. 1, pp. 91-130, 1970.

developed, and consideration of the cuts which develop in the complex l-plane as a result of multiple exchange of Regge poles is discussed in connection with various problems of scattering at high energies [19].

Below we shall concentrate our attention on the use of the quasi-potential description of hadron scattering at high energies [20–23]. The approach developed in [20–23] is based on the quasi-potential equation derived by Logunov and Tavkhelidze for the scattering amplitude and quantum field theory [24, 25].

In concept the quasi-potential approach is close to the optical model of the nucleus (i.e., to the consideration of nucleon scattering by nuclei according to the type of light scattering by a semitransparent optical medium). Under these conditions the scattering problem is treated not as a many-body problem but as a problem in the motion of a nucleon in a field which can be described by a complex potential.

The quasi-potential equation as it applies to high-energy hadron scattering allows a unified description to be given of the principal regularities of small- and large-angle scattering, the treatment of inelastic processes, etc.

Note the heuristic role which was played by the conventional potential approach in the study of hadron interaction processes at high energies. Within the framework of the theory of potential scattering the theory of Regge poles [26] was formulated, the presence of cuts in the complex plane of the angular momentum was indicated for the first time [27], and the double dispersion representations were proved. The eikonal representation, which is currently being studied in relativistic field theory [12, 28], has long been known in potential scattering.

§ 1. THE QUASI-POTENTIAL EQUATION AND THE PROPERTIES OF THE LOCAL QUASI-POTENTIAL

Let us consider the simplest case of scattering of two spinless particles of equal mass. In this case, the quasi-potential equation for the scattering amplitude has the form

$$T(\mathbf{p}, \mathbf{k}; E) = V[(\mathbf{p}-\mathbf{k})^2; E] + \int \frac{d\mathbf{q}}{\sqrt{m^2+\mathbf{q}^2}} \frac{V[(\mathbf{p}-\mathbf{q})^2; E]\, T(\mathbf{q}, \mathbf{k}; E)}{\mathbf{q}^2+m^2-E^2-i0}, \tag{1.1}$$

where E is the energy; $\mathbf{p}$, $\mathbf{k}$, and $\mathbf{q}$ are the relative momenta of the center of inertia system in the initial, final, and intermediate states, respectively.

The physical relativistically invariant amplitude is determined from the condition

$$T(s, t) = 32\pi^3 T(E; \mathbf{p}, \mathbf{k})\big|_{\substack{s=4(\mathbf{p}^2+m^2)=4(\mathbf{k}^2+m^2)=4E^2. \\ t=-(\mathbf{p}-\mathbf{k})^2}} \tag{1.2}$$

Equation (1.1) represents the generalization of the Lipman-Schwinger equation for the case of quantum field theory. However, unlike quantum mechanics, the quasi-potential V in Eq. (1.1) is in general a complex function which depends parametrically on energy. The imaginary part of the quasi-potential is caused by inelastic processes in the two-particle system and is a positive-definite quantity.

It can be proved rigorously that the condition of positiveness of the imaginary part of the quasi-potential ensures fulfillment of the condition

$$SS^{+} \leqslant 1 \tag{1.3}$$

for the two-particle scattering matrix S [29]. For a purely real quasi-potential, for example, we have the relativistic condition for two-particle unitarity.

As was shown in [24, 25], for the case of weak coupling a regular method exists for constructing the local quasi-potential; this method uses the expansion of the scattering amplitude into a perturbation theory series.

The quasi-potential constructed by this method yields a solution of Eq. (1.1) which coincides with the physical scattering amplitude on the energy surface. The method indicated was used in [30] for investigating the energy levels for positronium and hydrogenic atoms within the framework of quantum electrodynamics.

In the case of strong interactions no regular method exists for constructing the quasi-potential. Therefore, the approach which we have developed to hadron scattering at high energies is based on a phenomenological selection of the quasi-potential in Eq. (1.1).

In reality we shall make use of the following general principles.

A. The existence of the local quasi-potential $V(s, \mathbf{r})$ which provides an adequate description of hadron scattering at high energies.

B. Positive-definiteness of the imaginary part of the quasi-potential

$$\operatorname{Im} V(s, \mathbf{r}) \geqslant 0. \tag{1.4}$$

C. Smooth and nonsingular behavior of the quasi-potential as a function of the relative two-particle coordinate.

The latter principle evidently concerns the dynamics of hadron interaction at high energies and signifies that hadrons behave as extended objects having finite dimensions in high-energy collisions [31-33].

An important corollary of item C is the quasi-classical character of the scattering when the wavelengths of the colliding particles become considerably shorter than the interaction range. The results of such a consideration are the eikonal representation for the small-angle scattering amplitude [20-23] and the exponential drop of the large-angle scattering amplitude with increasing energy [33]. Further, item C allows the scattering amplitude to be found in the form of a convergent series of Born approximations.

It is of interest to note that under specified conditions only the first several terms of the Born series provide a substantial contribution to the scattering amplitude at large transferred momenta. On the other hand, experimental data indicate the fact that elastic scattering of hadrons at high energies and low momentum transfers is of a diffraction character and can be described approximately by a purely imaginary amplitude in the following form:

$$T(s, t) \approx is\sigma_{\text{tot}}(\infty)\, e^{a(s)\, t}. \tag{1.5}$$

The amplitude (1.5) ensures a constant total cross section and an exponential falloff of the differential cross section with an increase of the square of the transferred momentum. The quantity $a(s)$ associated with the slope parameter of the diffraction peak in general varies slowly with energy. It is natural to consider Eq. (1.5) as the Born approximation for the scattering amplitude. The corresponding local quasi-potential can be found by Fourier transformation and has a Gaussian form

$$V(s, \mathbf{r}) = isg_0 \left(\frac{\pi}{a}\right)^{3/2} e^{-\frac{r^2}{4a}}. \tag{1.6}$$

Here

$$\sigma_{\text{tot}}(\infty) = 32\pi^3 g_0 \tag{1.7}$$

and g_0 is a positive parameter.

The local quasi-potential (1.6) has a positive-definite imaginary part and is a smooth nonsingular function of $\mathbf{r}$. Let us treat (1.6) as the simplest form of quasi-potential which satisfies all of the principles enumerated above and also the requirement of diffractive behavior at small transferred momenta. Further on we shall show that the quasi-potential (1.6) conveys the principal properties of small- and large-angle hadron scattering.

Let us emphasize the fact that certain results do not depend on the specific sum of the quasi-potential and have a general character. Let us briefly discuss the relationship between the quasi-potential approach and the approach based on the Regge-pole hypothesis.

Several years ago it was proposed [34] that diffraction behavior develops in high-energy hadron scattering having an amplitude

$$T_P(s, t) = -\beta(t)\, s^{\alpha_p(t)} \frac{1+e^{-i\pi\alpha_p(t)}}{\sin \pi\alpha_p(t)} \tag{1.8}$$

as a result of exchange of a Pomeranchuk Regge pole in the t-channel. Actually, if it is assumed that the trajectory has a linear growth and the behavior of the residue function is smooth, i.e.,

$$\left.\begin{aligned} \alpha_p(t) &= 1+\alpha' t; \\ \beta(t) &= \beta(0)\, e^{bt}, \end{aligned}\right\} \tag{1.9}$$

we obtain Eq. (1.5) with

$$\left.\begin{aligned} \sigma_{\text{tot.}}(\infty) &= \beta(0); \\ a(s) &= b + \alpha' \ln(s). \end{aligned}\right\} \tag{1.10}$$

However, as is well known, a solitary Pomeranchuk pole, if it exists, yields a very rough approximation of the scattering amplitude as high energies. In general, an infinite sequence of Regge cuts appears in the angular-momentum plane, and this series must be considered [18]. Regrettably, notwithstanding the considerable efforts which have been made in the investigation of Regge cuts, there is as yet no unique way of calculating their contributions.

For this reason the quasi-potential approach has an essential advantage, since it allows the corrections to the Born term of the type (1.5) to be found in a simple way.

It should likewise be noted that here we are considering a fairly simplified model of purely elastic scattering. In the intermediate energy range it is necessary to consider secondary effects such as, for example, exchange of nonprincipal Regge poles or exchange of resonances in the forward channel. In order to carry this out in a self-consistent manner it is necessary to use the requirements of analyticity and crossing-symmetry in the form of the finite-energy sum rules [35, 36].

§ 2. ELASTIC SMALL-ANGLE SCATTERING

Let us consider elastic small-angle scattering at high energies:

$$\left|\frac{t}{s}\right| \ll 1, \quad as \gg 1,$$

where the parameter a is associated with the effective dimension of the interaction range by the relationship $4a = R^2$.

We shall assume that the Gaussian quasi-potential in Eq. (1.1) is purely imaginary, while below we shall discuss the situation which develops for other parametrizations of the quasi-potential.

2.1. SMALL TRANSFERRED MOMENTA

Let us begin by considering the small case of small transferred momenta at high energies:

$$a|t| < 1, \quad as \gg 1. \tag{2.1}$$

In this case, Eq. (1.1) can be solved by the method of iteration.

$$T(\Delta^2; E) = V(\Delta^2; E) + \delta T(\Delta^2; E) + \ldots; \; \Delta^2 = -t. \tag{2.2}$$

The first correction to the Born approximation has the following form:

$$\delta T(\Delta^2; E) = \int \frac{d\mathbf{q}}{\sqrt{m^2+\mathbf{q}^2}} \cdot \frac{V[(\mathbf{p}-\mathbf{q})^2; E]\, V[(\mathbf{q}-\mathbf{k})^2; E]}{\mathbf{q}^2+m^2-E^2-i0} = (isg_0)^2 e^{at/2} A(\Delta^2; E), \tag{2.3}$$

where

$$A(\Delta^2; E) = \int \frac{d\mathbf{q}}{\sqrt{m^2+\mathbf{q}^2}} \cdot \frac{e^{-2a(\mathbf{q}-\boldsymbol{\lambda})^2}}{\mathbf{q}^2+m^2-E^2-i0}; \quad \boldsymbol{\lambda} = \frac{\mathbf{p}+\mathbf{k}}{2}. \tag{2.4}$$

Integrating over the angles, we have

$$A(\Delta^2; E) = \frac{\pi}{2a\lambda} \int_{-\infty}^{\infty} \frac{q\,dq}{\sqrt{m^2+q^2}} \cdot \frac{e^{-2a(q-\lambda)^2}}{q^2+m^2-E^2-i0}, \tag{2.5}$$

where $\lambda = \left|\frac{\mathbf{p}+\mathbf{k}}{2}\right|$, or on the energy surface

$$\lambda = \sqrt{p^2+t/4} = p\cos\frac{\theta}{2}. \tag{2.6}$$

Let us realize (2.5) in the following form:

$$A = R + iI, \tag{2.7}$$

where

$$I = \frac{\pi^2}{4a\lambda\sqrt{p^2+m^2}} \left\{e^{-2a(p-\lambda)^2} - e^{-2a(p+\lambda)^2}\right\}; \tag{2.8}$$

R is determined by the principal value of the integral (2.5). In the limit of high energies and small transfers

$$I = \frac{\pi^2}{as} + O\left(\frac{1}{s^2}\right); \tag{2.9a}$$

$$R = \frac{\pi^2}{as} \cdot \frac{1}{\sqrt{2\pi as}} + O\left(\frac{1}{s^2}\right). \tag{2.9b}$$

Thus, the first two terms of the series (2.2) yield

$$T(\Delta^2; E) = isg_0 e^{at} - isg_0 \frac{\pi^2 g_0}{a}(1+i\alpha)e^{at/2} + \dots, \tag{2.10}$$

where $\alpha = (1/\sqrt{2\pi as})$.

The conditions for applicability of the Born approximation are

$$\frac{\pi^2 g_0}{a} < 1; \quad a|t| < 2\ln\frac{a}{\pi^2 g_0}. \tag{2.11}$$

Under these conditions the amplitude (2.10) describes diffraction scattering at small momentum transfers with a slope $A \approx 2a$ at the diffraction peak. Near the point

$$t \approx -\frac{2}{a}\ln\frac{a}{\pi^2 g_0}, \tag{2.12}$$

where the correction becomes comparable with the first Born approximation in the differential scattering cross section, a minimum can be observed. From (2.10) we obtain

$$\sigma_{tot} = 32\pi^3 g_0\left(1 - \frac{\pi^2 g_0}{a}\right) \tag{2.13}$$

for the total cross section for $t = 0$.

If we assume the parameter a to be a logarithmically rising function of energy, then it can easily be seen that the total cross section will tend from below to its asymptotic limit [37, 18]

$$\sigma_{tot}(\infty) = 32\pi^3 g_0. \tag{2.14}$$

For fixed g_0 this result remains valid when the entire series of Born approximations is considered. Without contradicting the unitarity condition which derives from Eq. (1.1), one may assume that the parameter g_0 is a decreasing function of energy and tends to a certain fixed value β_0 for $s \to \infty$; for example,

$$g_0 = \beta_0 \left(1 + \frac{\gamma_0}{\ln s/s_0}\right).$$

As can easily be seen, in this case, one can obtain a falloff of the cross section in the preasymptotic region for specified constraints on the constant γ_0.

Further, in the approximation (2.10) the ratio between the real part of the amplitude and the imaginary part is determined by the equation

$$\rho(s, t) = \frac{\operatorname{Re} T(s, t)}{\operatorname{Im} T(s, t)} = \frac{\pi^2 g_0}{a\sqrt{2\pi a s}} e^{-at/2} \tag{2.15}$$

and is a small positive-definite quantity.

Note that the experimentally observed behavior of the total cross section and the real part of the amplitude does not in general contradict the consideration presented and can be explained as the effect of the influence of nonleading Regge poles or resonances in the forward channel. In any case, the smallness of the real part of the amplitude at high energies and small transferred momenta does not contradict experiment. Therefore, in the subsequent consideration we shall neglect the real part of the amplitude at high energies and small scattering angles.

2.2. LARGE TRANSFERRED MOMENTA

Let us now consider scattering at large-momentum transfers outside the diffraction domain when

$$\left|\frac{t}{s}\right| \ll 1; \quad a|t| > 1. \tag{2.16}$$

In this case, the scattering amplitude can be found in the form of a convergent series of Born approximations

$$T(\Delta^2; E) = \sum_{n=0}^{\infty} T^{(n+1)}, \tag{2.17}$$

where

$$T^{(n)} = isg_0 \frac{e^{at/n}}{n \cdot n!} \left(-\frac{4\pi^2 g_0}{a}\right)^{n-1}. \tag{2.17a}$$

(The calculation of the n-th Born approximation $T^{(n)}$ is given in Appendix A.)

It can easily be seen that at large transferred momenta the principal contribution to the sum (2.17) is made by terms having a large number n.

Let us study the asymptotic behavior of the series (2.17) in the limit $a|t| \gg 1$. Using the Watson–Sommerfeld transform, we rewrite the series (2.17) in the form

$$T(\Delta^2; E) = \frac{as}{8\pi^2} \int_C \frac{dz\, e^{-\sigma z - \omega/z}}{2z\Gamma(z+1)\sin \pi z}, \tag{2.18}$$

where

$$\sigma = \ln \frac{a}{4\pi^2 g_0}; \quad \omega = a|t|. \tag{2.19}$$

The integration contour C envelops the positive real half-axis in the counterclockwise direction and incorporates the integer points z = 1, 2, ...

It can be shown that in domain (2.16) the basic contribution to the integral (2.18) is made by the first term in the expansion of the function $1/\sin\pi z$ into the series

$$\frac{1}{\sin \pi z} = \mp 2i \sum_{n=0}^{\infty} e^{\pm 2i\pi z(n+1/2)} \tag{2.20}$$

on the upper and lower bounds of the contour C, respectively. Using the method of steepest descent, we obtain

$$T(\Delta^2; E) \to -\frac{isa}{4\pi^2} \operatorname{Re}\left\{\frac{e^{-2\sqrt{2a|t|}\,\mathrm{sh}\,\gamma+\frac{\gamma}{2}}}{\sqrt{a|t|}\,\mathrm{ch}\,\gamma}\right\} \quad a|t| \gg 1. \tag{2.21}$$

The parameter γ is associated with the saddle point z_0 by the relationship

$$z_0 = \sqrt{2a|t|}\,e^{-\gamma} \tag{2.22}$$

and is determined from the following condition:

$$e^{2\gamma} + 2\gamma = 2(\sigma - i\pi + \ln\sqrt{2a|t|}). \tag{2.23}$$

Equation (2.21) can be rewritten in the form

$$T(\Delta^2; E) \to \frac{i\,s\alpha}{\sqrt{a|t|}}\,e^{-\beta\sqrt{a|t|}}\cos\psi(s, t), \tag{2.24}$$

$a|t| \gg 1$, where the parameters α and β are slowly varying functions of s and t.

Thus, in the domain (2.16) the scattering amplitude is an exponentially falling function of the momentum transfer $\sqrt{|t|}$ with possible oscillations [19] near points where

$$\psi(s, t) = \pi(k + 1/2);\ \text{k is an integer.} \tag{2.25}$$

However, as is evident from (2.23), for fairly large transfers the oscillations, if they exist, must be smoothed and the scattering amplitude takes the form

$$T(\Delta^2; E) \to -\frac{isa}{2\pi^2}(\ln 2\omega)^{1/4}\,\frac{e^{-\sqrt{2\omega\ln 2\omega}}}{\sqrt{2\omega\ln 2\omega}}, \quad \omega = a|t| \to \infty. \tag{2.26}$$

2.3. EIKONAL DESCRIPTION OF SMALL-ANGLE SCATTERING

The eikonal approximation for the scattering amplitude, which has long been known in nonrelativistic quantum mechanics [9], has become very popular lately. It is based on the quasi-classical picture of high-energy scattering in which the wavelengths of the colliding particles are considerably shorter than the effective dimensions of the interaction range.

We see that the quasi-classical character of small-angle hadron scattering at high energies is intrinsically associated with the nonsingular, or smooth, behavior of the quasi-potential as a function of the relative two-particle coordinate.

Let us begin by discussing the properties of the solution of (2.17). Substituting the equation

$$\frac{e^{at/n}}{n} = \frac{1}{4\pi a}\int e^{i\rho\Delta_\perp}\,e^{\frac{-\rho^2 n}{4a}}\,d^2\rho \tag{2.27}$$

(where $t = -\Delta_\perp^2$; $\Delta_\perp = \mathbf{p} - \mathbf{k}$ varies in the plane perpendicular to the vector 1/2 $(\mathbf{p} + \mathbf{k})$) into (2.17), we obtain

$$T(\Delta^2; E) = \frac{s}{(2\pi)^3}\int d^2\rho\,e^{i\Delta_\perp\rho}\left(\frac{e^{2i\chi} - 1}{2i}\right), \tag{2.28}$$

where

$$2i\chi = -\frac{4\pi^2 g_0}{a} e^{-\rho^2/4a}. \quad (2.29)$$

Equation (2.28) is none other than the eikonal representation of the amplitude in the high-energy region having the phase function $\chi(\rho)$ which is associated with the quasi-potential by the relationship

$$\chi(\rho) = \frac{1}{s} \int_{\infty}^{+\infty} V(s, \sqrt{\rho^2 + z^2})\, dz. \quad (2.30)$$

It is evident that the representation will hold for a broad class of smooth quasi-potentials.

In order to show this it is convenient to consider the quasi-potential equation for the wave function of two particles in **r**-space.

$$(E^2 - m^2 + \nabla^2)\psi(\mathbf{r}) = -\frac{1}{\omega} V(s, \mathbf{r})\psi(\mathbf{r}), \quad (2.31)$$

where

$$\omega = \sqrt{m^2 - \nabla^2}. \quad (2.32)$$

The presence of the operator ω makes Eq. (2.31) nonlocal. However, for the condition of nonsingular, or smooth, behavior of the quasi-potential $V(s, \mathbf{r})$ Eq. (2.31) takes an effectively local form in the high-energy limit. Actually, we shall seek the solutions of Eq. (2.31) in the form

$$\psi(\mathbf{r}) = e^{ipz}\varphi(\mathbf{r}), \quad (2.33)$$

where $\varphi(\mathbf{r})$ is assumed to be a slowly varying function and

$$E = \sqrt{p^2 + m^2}.$$

It can easily be seen that on the space of slowly varying functions in the high-energy limit $p \to \infty$ we have

$$e^{-ipz}\omega e^{ipz} \to p - i\nabla_z + O(1/p) \quad (2.34a)$$

or

$$e^{-ipz}\frac{1}{\omega} e^{ipz} \to \frac{1}{p} + O(1/p^2). \quad (2.34b)$$

Thus, the function $\varphi(\mathbf{r})$ satisfies the equation

$$-2ip\frac{\partial\varphi(\mathbf{r})}{\partial z} = \frac{1}{p} V(s, \mathbf{r})\varphi(\mathbf{r}), \quad (2.35)$$

which coincides with the corresponding equation that derives from the Klein-Gordon equation with the effective potential $1/p\, V(s, \mathbf{r})$. As a result we obtain the eikonal representation (2.28) with the phase function $\chi(\rho)$ determined by Eq. (2.30). Thus, there is no need to consider the eikonal (or, as it is usually called, the Glauber) representation as the primary dynamic principle (see, for example, [11]). It is a consequence of the assumption that hadron interaction at high energies has a nonsingular character.

In recent years many attempts have been made to formulate the eikonal representation for the scattering amplitude, based solely on general principles such as relativistic invariance, unitarity, and analyticity, without restriction to small scattering angles [12]. One should think, however, that additional assumptions are introduced in implicit form into these derivations.

2.4. THE APPROACH BASED ON THE UNITARITY CONDITION

The approach to the description of hadron scattering at high energies on the basis of the unitarity equation in quantum field theory [38, 39] is very attractive. Here, we shall briefly discuss this approach and its connection with the quasi-potential method.

The unitarity equation for the scattering amplitude of two spinless equal-mass particles has the form

$$\mathrm{Im}\, T(s,t) = \int d\omega T(s,t')\, T^{+}(s,t'') + F(s,t), \tag{2.36}$$

where in the center of inertia system

$$\left.\begin{aligned} t &= -(\mathbf{p}-\mathbf{k})^2;\\ t' &= -(\mathbf{p}-\mathbf{q})^2;\\ t'' &= -(\mathbf{q}-\mathbf{k})^2;\\ s &= 4(m^2+\mathbf{p}^2) = 4(m^2+\mathbf{q}^2) = 4(m^2+\mathbf{k}^2) \end{aligned}\right\} \tag{2.37}$$

and

$$d\omega = \frac{1}{8\pi^2}\cdot\frac{d\mathbf{q}\, d\mathbf{q}'}{2q_0\, 2q_0'}\, \delta(p+p'-q-q') \tag{2.38}$$

is associated with the two-particle phase volume. The quantity F(s, t) is known under the name of the Van Hove overlap function and represents the contribution of the inelastic states of Eq. (2.36).

It is convenient to choose the following directions of the momenta:

$$\begin{aligned} \mathbf{p}+\mathbf{k} &= (2p_z,\ 0,\ 0);\\ \mathbf{k}-\mathbf{p} &= (0,\ \mathbf{\Delta}_\perp), \end{aligned} \tag{2.39a}$$

or

$$\mathbf{p} = \left(p_z,\ \frac{\mathbf{\Delta}_\perp}{2}\right);\quad \mathbf{k} = \left(p_z,\ -\frac{\mathbf{\Delta}_\perp}{2}\right). \tag{2.39b}$$

From Eqs. (2.39) it follows that

$$\left.\begin{aligned} s &= 4(m^2+p_z^2)+\mathbf{\Delta}_\perp^2;\\ t &= -\mathbf{\Delta}_\perp^2. \end{aligned}\right\} \tag{2.40}$$

At high energies and small scattering angles, when $|t/s| \ll 1$, we have

$$s \approx 4p_z^2,\quad t = -\mathbf{\Delta}_\perp^2 = -(\mathbf{k}_\perp - \mathbf{p}_\perp)^2. \tag{2.41}$$

Having assumed that the principal contribution to the integral in Eq. (2.36) is made from the small-angle region, one can substitute

$$t' = -(\mathbf{p}_\perp - \mathbf{q}_\perp)^2;\quad t'' = -(\mathbf{q}_\perp - \mathbf{k}_\perp)^2 \tag{2.42}$$

into Eq. (2.36). For the same reason we have

$$\int d\omega \to \frac{1}{8\pi^2 s}\int_{4\mathbf{q}_\perp^2 \leqslant s} d^2 q_\perp. \tag{2.43}$$

Thus, in this approximation the unitarity equation (2.36) takes the following form:

$$\mathrm{Im}\, T\left[s_1, -(\mathbf{p}_\perp-\mathbf{k}_\perp)^2\right] = \frac{1}{8\pi^2 s}\int_{4\mathbf{q}_\perp^2\leqslant s} d^2 q_\perp\, T\left[s_1, -(\mathbf{p}_\perp-\mathbf{q}_\perp)^2\right]\, T^{+}\left[s_1, -(\mathbf{q}_\perp-\mathbf{k}_\perp)^2\right] + F[s_1, -(\mathbf{p}_\perp-\mathbf{k}_\perp)^2]. \tag{2.44}$$

If it is assumed that the domain of integration in Eq. (2.44) can be extended beyond the circle $4q_\perp^2 \leq s$ over the entire transverse-momentum plane, then we obtain the well-known convolution formula

$$\operatorname{Im} T = \frac{1}{8\pi^2 s} T * T^+ + F. \tag{2.45}$$

Note, however, that such an expansion is nontrivial and requires continuation beyond the mass surface. In any case, it is associated with the assumption of the nonsingular character of hadron interaction at high energies. Let us now consider the approximation of the purely imaginary amplitude

$$T(s, t) = isA(s, t) \tag{2.46}$$

and let us assume [38] that the overlap function has a Gaussian form

$$F(s, t) = s\sigma_{\text{inel}} e^{at}. \tag{2.47}$$

Equation (2.45) can easily be solved using the Fourier transform

$$A(s, t) = \int d^2\rho e^{i\rho\Delta_\perp} a(s, \rho), \tag{2.48}$$

where $t = -\Delta_\perp^2$. As a result, we obtain the equation

$$a(s, \rho) = \frac{1}{2} a^2(s, \rho) + \frac{\sigma_{\text{inel}}}{4\pi a} e^{-\rho^2/4a} \tag{2.49}$$

with the following solution:*

$$a(s, \rho) = 1 - \sqrt{1 - \eta(s, \rho)}, \tag{2.50}$$

where

$$\eta(s, \rho) = \frac{\sigma_{\text{inel}}}{2\pi a} e^{\rho - 2/4a}. \tag{2.51}$$

Assuming that

$$\eta(s, \rho) < 1, \tag{2.52}$$

we expand the solution (2.50) into a series in powers of $\eta(s, \rho)$:

$$a(s, \rho) = \sum_{n=1}^{\infty} \frac{(2n)!}{(n!)^2} \cdot \frac{e^{-n\rho^2/4a}}{2n-1} \left(\frac{\sigma_{\text{inel}}}{8\pi a}\right)^n. \tag{2.53}$$

Substituting Eq. (2.53) into Eq. (2.48), we obtain

$$T(s, t) = is\,\sigma_{\text{inel}} \sum_{n=1}^{\infty} \frac{(2n)!}{(n!)^2} \cdot \frac{e^{at/n}}{2n(2n-1)} \left(\frac{\sigma_{\text{inel}}}{8\pi a}\right)^{n-1} \tag{2.54}$$

for the scattering amplitude.

Thus, we obtain a series of the type (2.17) for the scattering amplitude at high energies and small scattering angles, but the series has sign-constant coefficients which decrease considerably more slowly with increasing n.

The fact that the terms of this series are sign-constant was noted in certain papers as a shortcoming of this approach, since it complicates the explanation of the dips and oscillations in the differential scattering cross section.

Another difficulty consists in the following. It can be shown that the sum in the series (2.54) falls off as $e^{-\sqrt{|t|}}$ with increasing transfer, whereas the overlap function (2.47) falls off as e^t. It is well known,

*We choose the solution which vanishes for $F(s, t) \to 0$.

however, that at large transfers there is no substantial difference between the behavior of the elastic and inelastic processes. In contradiction to this the quasi-potential equation (1.1) corresponds to a unitarity equation having a fairly complex overlap function:

$$F(s, t) = s\beta e^{at} + s\beta \sum_{n=2}^{\infty} \frac{e^{at/n}}{n \cdot n!} (2^n - 3) \left(\frac{-4\pi^2 g_0}{a} \right)^{n-1}, \tag{2.55}$$

where

$$\beta = 32\pi^3 g_0.$$

Equation (2.55) can be obtained by direct calculation using the series (2.17) and the unitarity condition.

It can easily be seen that the series (2.55) has the same properties as does the elastic-scattering amplitude.

2.5. APPLICATION TO ELASTIC pp-SCATTERING

Let us compare the results obtained on the basis of the quasi-potential equations (1.1) with the experimental data on high-energy small-angle pp-scattering.

In considering specific physical processes it is necessary, in general, to consider the spin structure of the scattering amplitude.

For example, in the case of elastic pp-scattering there are five independent invariant amplitudes which can be chosen as follows on a spiral basis:

$$\left.\begin{aligned}
T_1 &= \left\langle \frac{1}{2}, \frac{1}{2} \middle| T \middle| \frac{1}{2}, \frac{1}{2} \right\rangle; \\
T_2 &= \left\langle \frac{1}{2}, \frac{1}{2} \middle| T \middle| -\frac{1}{2}, -\frac{1}{2} \right\rangle; \\
T_3 &= \left\langle \frac{1}{2}, -\frac{1}{2} \middle| T \middle| \frac{1}{2}, -\frac{1}{2} \right\rangle; \\
T_4 &= \left\langle \frac{1}{2}, -\frac{1}{2} \middle| T \middle| -\frac{1}{2}, \frac{1}{2} \right\rangle; \\
T_5 &= \left\langle \frac{1}{2}, \frac{1}{2} \middle| T \middle| \frac{1}{2}, -\frac{1}{2} \right\rangle.
\end{aligned}\right\} \tag{2.56}$$

However, only two of them, T_1 and T_2, which correspond to spin-nonflip processes, yield a nonvanishing contribution for forward scattering. For nonzero scattering angles the relative magnitude of the spin-flip amplitudes T_2, T_4, and T_5 may be determined from the value of the polarization parameter which does not exceed 10% at high energies and decreases with rising energy [40].

The remaining spin-nonflip amplitudes T_1 and T_3 are approximately equal to one another, which is a consequence of the truly elastic character of high-energy scattering which occurs mainly with exchange of zero quantum numbers in the cross channels.

Thus, in describing the scattering of unpolarized protons by high-energy protons one may limit consideration to one amplitude $T \approx T_1 \approx T_3$ within the framework of the quasi-potential equation (1.1) for spinless particles.* The total differential cross sections for scattering of unpolarized protons by protons are associated with the amplitude $T(\Delta^2; F)$ which satisfies the quasi-potential equation as follows:

$$\sigma_{\text{tot}} = \frac{8\pi^3}{pE} T(E; \quad \Delta^2 = 0); \tag{2.57a}$$

$$\frac{d\sigma}{dt} = -4\pi \left| \frac{\pi^2 T(E, \Delta^2)}{pE} \right|^2, \quad t = -\Delta^2. \tag{2.57b}$$

* This proposition may, however, turn out to be invalid in the region of large scattering angles $\theta \approx 90°$, where the requirement of cross symmetry in general makes it necessary to consider the spin-flip amplitudes also.

The solution (2.17) of Eq. (1.1) with the quasi-potential (1.6) in the region of small scattering angles depends on the two real parameters a and g_0, which are included in the definition of the quasi-potential.

The numerical values of these parameters can be found from experimental data at zero and small transferred momenta (i.e., from the total cross section and the width of the diffraction peak 1/A) by means of the following equations:

$$\sigma_{tot} = 8\pi a I(x), \quad x = \frac{4\pi^2 g_0}{a}; \tag{2.58a}$$

$$A = \frac{d}{dt} \ln \left[\frac{d\sigma}{dt}\right]_{t=0} = 2a \frac{1}{I(x)} \int_0^x \frac{d\xi}{\xi} I(\xi); \tag{2.58b}$$

$$I(x) = -\sum_{n=1}^{\infty} \frac{(-x)^n}{n \cdot n!} = \int_0^x \frac{d\xi}{\xi} (1 - e^{-\xi}). \tag{2.59}$$

We compared the results obtained above with experimental data on elastic pp-scattering at p_L = 8.5; 12.4, and 18.4 GeV/c [41-42]. The theoretical curves in Figs. 1 and 2 correspond to the following values of the parameters g_0 and a:

$$p_L = 8.5 \frac{\text{GeV}}{c}, \quad g_0 = 0.13 \left(\frac{\text{GeV}}{c}\right)^{-2}, \quad a = 2.6 \left(\frac{\text{GeV}}{c}\right)^{-2};$$
$$p_L = 12.4 \frac{\text{GeV}}{c}, \quad g_0 = 0.12 \left(\frac{\text{GeV}}{c}\right)^{-2}, \quad a = 2.8 \left(\frac{\text{GeV}}{c}\right)^{-2};$$
$$p_L = 18.4 \frac{\text{GeV}}{c}, \quad g_0 = 0.14 \left(\frac{\text{GeV}}{c}\right)^{-2}, \quad a = 3.8 \left(\frac{\text{GeV}}{c}\right)^{-2},$$

calculated using the experimental values of the total cross sections [43] and the widths of the diffraction peaks [41] at the corresponding energies.* As is evident from Figs. 1 and 2, the theoretical curves reproduce the behavior of the differential cross section of elastic pp-scattering fairly well, as well as the positions of the diffraction minima and their energy dependence.

It is of interest to note that the numerical value of the parameter a for the energy range considered corresponds to an effective interaction radius $R \approx (1/2\mu) = 0.71$ F or $a = 3.3$ $(\text{GeV/c})^{-2}$, which is determined by the position of the nearest singularity in the t-channel in the case of elastic scattering. Moreover, at these energies the total cross section is very close to its geometric value:

$$\sigma_{tot} \simeq 2\pi R^2 = \frac{\pi}{2\mu^2} = 32 \text{ mb}.$$

Note that a qualitative analysis of elastic pp-scattering at high energies from various points of view has likewise been carried out in [11-17, 38, 39]. Near points at which the sum (2.17) vanishes it is in general necessary to consider the next terms of the expansion of the amplitude in inverse powers of the momentum p. This leads to so-called "filling-in of the minima." Further on (Fig. 3) the behavior of the differential elastic pp-scattering cross section is shown at p_L = 8.5 GeV/c in the region of transfers $0 \leq |t| < 0.6$ $(\text{GeV/c})^2$. The existence of a large "arm" at $|t| \approx 0.3$ $(\text{GeV/c})^2$ is in agreement with the results of the theoretical calculations. An analogous behavior is also observed at other energies.

§ 3. LARGE-ANGLE SCATTERING

Let us now consider the scattering of two particles at high energies and a fixed scattering angle:

$$as \gg 1, \quad \left|\frac{t}{s}\right| \approx \sin^2 \frac{\theta}{2} = \text{fixed}. \tag{3.1}$$

* The numerical values of these and other parameters calculated in the paper contain uncertainties caused by errors in the corresponding experimental data.

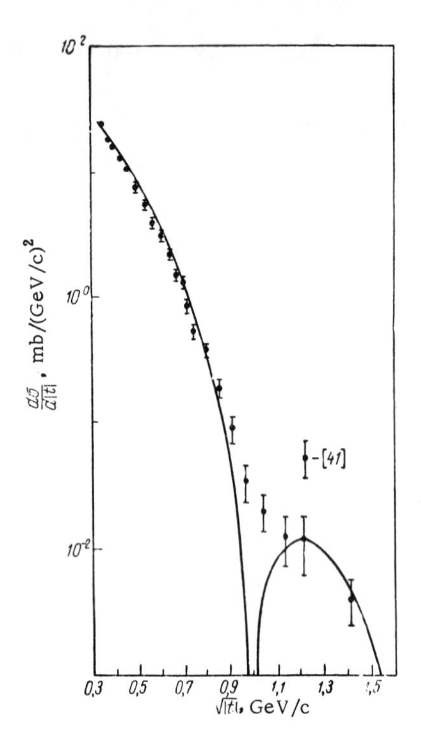

Fig. 1. pp-Scattering at $p_L = 12.4$ GeV/c.

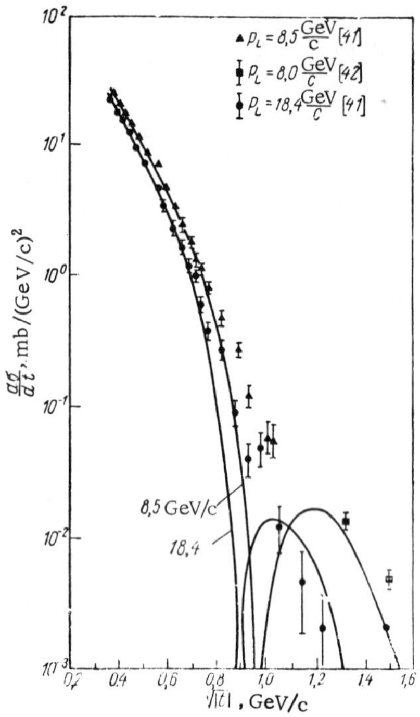

Fig. 2. pp-Scattering at $p_L = 8.5$; 8.0, and 18.4 GeV/c.

In this case, the series of Born approximations for the scattering amplitudes has the form (see Appendix B)

$$T(\Delta^2;\ E) = \mathrm{i}\, sg_0 \sum_{n=1}^{\infty} \frac{n^{2n}}{(n!)^2} \cdot \frac{e^{at/n}}{n^{3/2}} \left(\frac{\mathrm{i}\, sg_0 \pi \sqrt{\pi}\, e^{\varphi(\theta)}}{tpa\sqrt{a}} \right)^{n-1}, \tag{3.2}$$

where

$$\varphi(\theta) = 1 - \frac{\theta}{2\,\mathrm{tg}\,\theta/2}. \tag{3.3}$$

The function $\varphi(\theta)$ is fairly small, and we shall neglect it. It can easily be seen that for $a|t| \gg 1$ the principal contribution to the sum (3.2) is made by terms with $n \gg 1$, and the Stirling formula $n! \approx \sqrt{2\pi n}\,(n/e)^n$ may be used. As a result,

$$\underset{\substack{\theta = \text{fixed} \\ s \to \infty}}{T(\Delta^2;\ E)} = \mathrm{i}\, sg_0 \frac{e^2}{2\pi} \sum_{n=1}^{\infty} \frac{e^{at/n}}{n^{5/2}} (-\mathrm{i}\gamma)^{n-1}. \tag{3.4}$$

Here

$$\gamma = \frac{g_0 e^2}{p \sin^2\theta/2} \left(\frac{\pi}{a} \right)^{3/2} = \frac{sg_0 e^2}{|t|\, p} \left(\frac{\pi}{a} \right)^{3/2}. \tag{3.5}$$

Let us use the representation

$$\frac{e^{at/n}}{n^{5/2}} = \frac{1}{(4\pi a)^{5/2}} \int d^5 r\, e^{\mathrm{i}\mathbf{rq}} \left[e^{-r^2/4a} \right]^n, \tag{3.6}$$

where $t = -\Delta^2 = -q^2$, $\mathbf{r}$ is the vector of certain five-dimensional auxiliary space.

Having used (3.6), we rewrite the series (3.4) in integral form

$$\underset{\substack{\theta = \text{fixed} \\ s \to \infty}}{T(\Delta^2, E)} = \mathrm{i}\, sg_0 \frac{e^2}{2\pi (4\pi a)^{5/2}} \int d\mathbf{r}\, e^{\mathrm{i}\mathbf{rq}} \frac{e^{-r^2/4a}}{1 + \mathrm{i}\gamma e^{-r^2/4a}}. \tag{3.7}$$

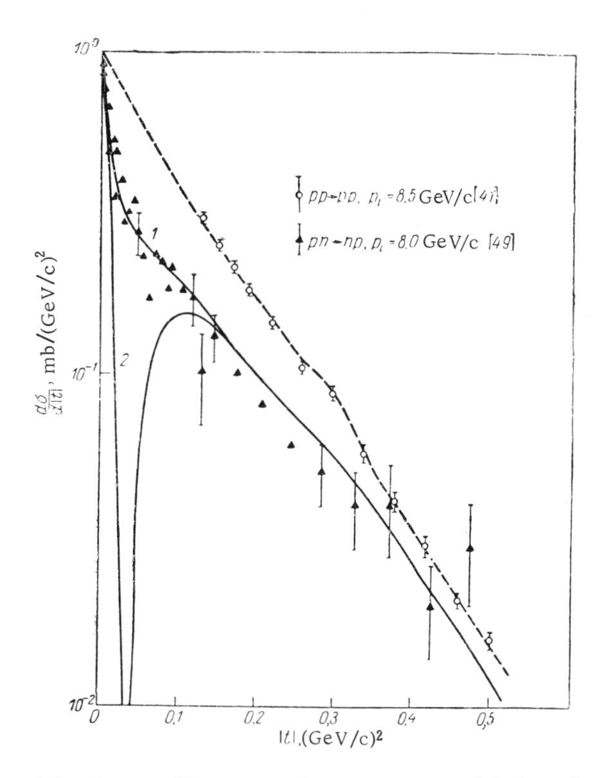

Fig. 3. pn-Charge exchange at p_L = 8.0 GeV/c, and pp-elastic scattering at p_L = 8.5 GeV/c. The cross section for pp-scattering has been normalized to 1 mb/(GeV/c)^2 for t=0.

After integration over the angles we obtain

$$T(E;\ q^2) = i\, s g_0 \frac{2\pi e^2}{3(4\pi a)^{5/2}} \frac{1}{iq} \int_{-\infty}^{+\infty} r^3 dr\, e^{iqr} \frac{e^{-r^2/4a}}{1 + i\gamma e^{-r^2/4a}}. \quad (3.8)$$

The integral in (3.8) can be taken using the theorem of residues; as a result, we have

$$T(E,\, q^2) \rightarrow \frac{qpr_0^3}{12\pi^2} e^{iqr_0}, \quad \text{where } q = \sqrt{|t|}. \quad (3.9)$$

The positions of the poles of the integrand are determined from the equation

$$1 + i\gamma e^{-r_0^2/4a} = 0 \quad (3.10)$$

or

$$r_0^2 = -\,2\pi i a\left(1 + \frac{2i}{\pi}\ln\gamma\right). \quad (3.11)$$

The pole in the upper half-plane, which makes the principal contribution, is situated at the point

$$r_0 = i\sqrt{2\pi i a\left(1 + \frac{2i}{\pi}\ln\gamma\right)}. \quad (3.12)$$

At intermediate energies, at which the second term in Eq. (3.11) can be discarded (i.e., $r_0^2 = -2\pi i a$), we obtain the following equation for the differential cross section at large angles:

$$\frac{d\sigma}{d\Omega} = \left(\frac{\pi a}{3}\right)^2 q^2 e^{-2q\sqrt{\pi a}}, \quad q = \sqrt{|t|}. \quad (3.13)$$

An interesting peculiarity of the result (3.13) is the fact that at large momentum transfers corresponding to large angles $d\sigma/d\Omega$ depends weakly on energy. The only energy dependence enters into the equation via the parameter a which is associated with the width of the forward-scattering peak.

For fairly high energies at which the principal contribution in Eq. (3.11) is made by the first term we obtain the following asymptotic behavior for the differential cross section:

$$\left.\frac{d\sigma}{d\Omega}\right|_{\substack{\theta=\text{fixed}\\ s\to\infty}} \rightarrow s\left(\frac{a\sin\frac{\theta}{2}\ln s}{3}\right)^2 e^{-c(\theta)\sqrt{2as\ln s}}, \quad (3.14)$$

where $c(\theta) = 2\sin\theta/2$.

It is interesting that for a logarithmic growth of the parameter a with energy we obtain the estimate on the lower bound of the decrease in fixed-angle scattering amplitude at high energies deriving from the rigorous results of field theory [44, 45].

Let us briefly discuss the application of these results to a description of elastic large-angle pp-scattering. As is evident from the preceding consideration in the p_L = 10–20 GeV/c range of laboratory momenta, the parameters a and g_0 are approximately equal to

$$a \approx 3.0\left(\frac{\text{GeV}}{c}\right)^{-2};\ g_0 \approx 0.13\left(\frac{\text{GeV}}{c}\right)^{-2}. \quad (3.15)$$

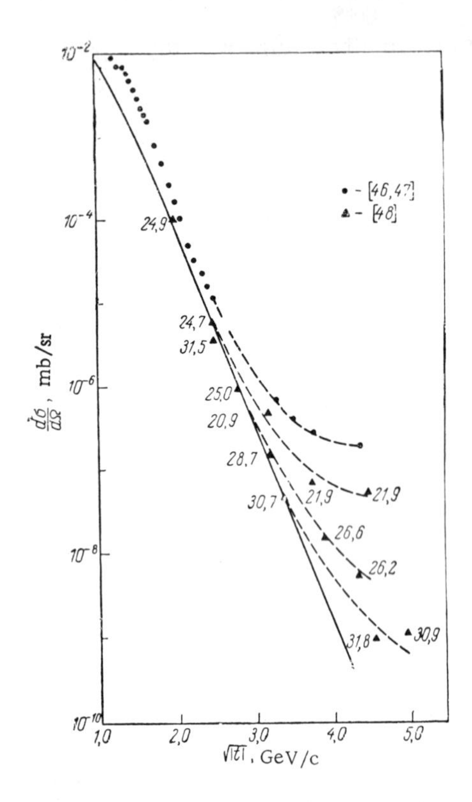

Fig. 4. Large-angle pp-scattering. The numbers on the figure denote the energies of the impinging particles in the laboratory frame of reference. The dashed lines connect points having approximately equal energies.

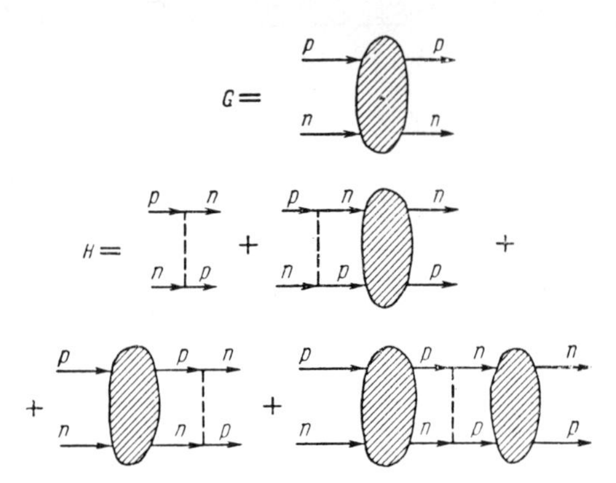

Fig. 5. Schematic representation of the amplitudes G and H.

The condition $|\gamma| < 1$ for convergence of the series (3.4) appears as follows for these numerical values of the parameters a and g_0:

$$|t| > 0.3\sqrt{s} \text{ GeV}. \tag{3.16}$$

The theoretical curve in Fig. 4, which was plotted using Eq. (3.13) and the parameters (3.15), conveys the character of the falloff and absolute value of the differential cross section in the region of large scattering angles [46–48] which are constrained by the condition (3.16). Let us emphasize the fact that, in accordance with the footnote on p. 62, these results are in general inapplicable in the region of angles $\theta \approx 90°$.

§ 4. BACKWARD SCATTERING

As is evident from § 3, the large-angle scattering amplitude (3.9) falls off exponentially with increasing energy. Thus, the solution of Eq. (1.1) with the quasi-potential (1.6) leads to an exponentially small cross section for backward scattering of high-energy particles, which contradicts experiment in a number of cases. As was indicated in [21], this fact can be explained by neglect of the exchange forces in the two-particle system.

Below we shall show how the exchange forces can be taken into account within the framework of the quasi-potential equation, and we shall use the results to analyze the experimental data on elastic backward np-scattering [42].

With allowance for the exchange forces, the scattering amplitude can be represented in the form of the sum of two quantities:

$$T(\mathbf{p}, \mathbf{k}; E) = G(\mathbf{p}, \mathbf{k}; E) + H(\mathbf{p}, \mathbf{k}; E) \tag{4.1}$$

which satisfy the following system of quasi-potential equations:*

$$G = g + g \times G + h \times H, \tag{4.2a}$$

$$H = h + h \times G + g \times H. \tag{4.2b}$$

The multiplication operation in Eqs. (4.2) denotes integration according to Eq. (1.1). The quantities g and h are the Fourier-transforms of the "direct" and "exchange" parts of the quasi-potential, respectively:

$$g(s, t) = \frac{1}{(2\pi)^3}\int d\mathbf{r} e^{i\mathbf{p}\mathbf{r}} V(s, r) e^{-i\mathbf{k}\mathbf{r}}, \quad t = -(\mathbf{p}-\mathbf{k})^2; \tag{4.3a}$$

* The system of equations (4.2) is equivalent to the pair of equations for amplitudes having a definite parity which were considered in [24].

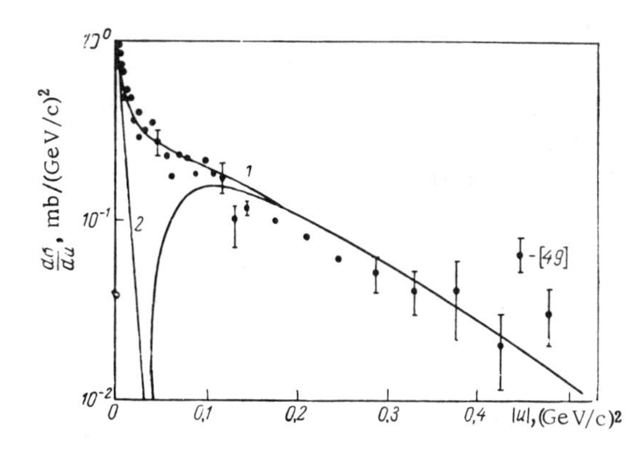

Fig. 6. Backward np-scattering at p_L = 8.0 GeV/c.

$$h(s,u)=\frac{1}{(2\pi)^3}\int d\mathbf{r}e^{i\mathbf{p}\mathbf{r}}V_e(s,r)\hat{P}e^{-i\mathbf{k}\mathbf{r}},\quad u=-(\mathbf{p}+\mathbf{k})^2, \tag{4.3b}$$

where $\hat{P}$ is the coordinate communication operator.

We shall make use of Eq. (1.6) as the direct-interaction quasi-potential, whence, it follows that $g(s, t) = isg_0e^{at}$. The exchange part of the quasi-potential is defined by the contributions of the cross u-channel. Taking account of the condition

$$\left|\frac{h(s,0)}{sg_0}\right|\ll 1 \quad \text{for} \quad s\to\infty, \tag{4.4}$$

one may neglect the last term in Eq. (4.2a). Iterating the system of equations obtained, we find

$$H=h+h\times G+G\times h+G\times h\times G, \tag{4.5}$$

where G is defined by the solution of Eq. (1.1) with the quasi-potential (1.6). The expression for the amplitude A is depicted symbolically in Fig. 5.

Let us now assume that the exchange quasi-potential may be represented in the form of the sum

$$h(s,u)=\sum_i h_i(s)e^{b_iu}, \tag{4.6}$$

where $|h_i(s)/sg_0| \ll 1$ at high energies.

In this case, the exchange-scattering amplitude H can be found in the following form in the $|u/s| \ll 1$ region:

$$H(\mathbf{q}'^2;\ E)=\sum_i H_i(\mathbf{q}'^2;\ E), \tag{4.7}$$

where

$$H_i(\mathbf{q}'^2;\ E)=h_i(s)\sum_{n=0}^{\infty}\frac{ae^{\frac{ab_i}{a+nb_i}u}}{(a+nb_i)n!}\left(-\frac{4\pi^2g_0}{a}\right)^n, \tag{4.8}$$
$$u=-q'^2.$$

The results obtained were used in analyzing elastic backward pn-scattering at an energy p_L = 8 GeV/c and $|u| < 0.6$ (GeV/c)^2. Under these conditions an expression of the type (4.6) in which only two terms were considered was used for the exchange potential. The parameters h_1 and h_2 were assumed real for simplicity, the cases of both identical and opposite signs of the quantities h_i being considered. The parameters a and g_0 included in the definition of the elastic part of the potential were determined on the basis of the experimental data on forward pp-scattering at p_L = 8.5 GeV/c:

$$g_0=0.13\left(\frac{\text{GeV}}{c}\right)^{-2};\quad a=2.6\left(\frac{\text{GeV}}{c}\right)^{-2}.$$

The theoretical curves 1 and 2 in Fig. 6, which correspond to identical and opposite signs of the quantities h_1 and h_2, were calculated for the following values of the parameters h_i and b_i.

Identical signs (curve 1):

$$|h_1|=0.07;\quad b_1=110\ (\text{GeV/c})^{-2};$$
$$|h_2|=0.30;\quad b_2=1.8\ (\text{GeV/c})^{-2}.$$

Opposite signs (curve 2):

$$|h_1|=0.29;\quad b_1=34\ (\text{GeV/c})^{-2};$$
$$|h_2|=0.30;\quad b_2=1.8(\text{GeV/c})^{-2}.$$

In Fig. 3 those same curves are depicted for comparison purposes along with the curve for the differential cross section of elastic pp-scattering at $p_L = 8.5$ GeV/c, normalized to 1 mb/(GeV/c)^2 at $t = 0$. A comparison with experiment shows that the case of identical signs is preferable. Note, however, that the given analysis, without taking account of spin dependence, has only a qualitative character.

§ 5. CHARGE EXCHANGE PROCESSES AND INELASTIC SCATTERING

Let us generalize the quasi-potential approach for the case of inelastic quasi-two-particle processes, and let us compare the results with data on the production of isobars in pp-collisions.

Let us consider the multichannel problem of quasi-two-particle scattering of spinless particles with mass m_i ($i = 1, 2, \ldots n$). The quasi-potential equations for the amplitudes of the transition $T_{ik \to jl}(E, \mathbf{p}, \mathbf{k})$ can be written symbolically in the matrix form

$$T = V + \int d\mathbf{q}\, [VF]\, T, \tag{5.1}$$

where

$$[VF]_{ik \to jl} = V_{ik \to jl}(E, (\mathbf{p} - \mathbf{q})^2)\, F_{jl}(E, \mathbf{q}). \tag{5.2}$$

The quantities $F_{ik}(E, \mathbf{q})$ are associated with the masses of the intermediate particles in the following manner:

$$F_{ik}(E, \mathbf{q}) = \frac{\omega_i + \omega_k}{\omega_i \omega_k} \cdot \frac{2}{(\omega_i + \omega_k)^2 - s - i0}, \tag{5.3}$$

where $\omega_i = \sqrt{m_i^2 + q^2}$ and $s = (E_1 + E_2)^2$.

Under the condition of smoothness of the generalized quasi-potential at high energies one can neglect the deviation of the masses m_i from a certain average mass m. As a result, Eq. (5.1) takes the form [50]

$$T = V + V \times T, \tag{5.4}$$

where

$$[V \times T]_{ik \to jl} = \sum_{i'k'} \int \frac{d\mathbf{q}}{\sqrt{\mathbf{q}^2 + m^2}} \cdot \frac{V_{ik \to i'k'}(E, (\mathbf{p} - \mathbf{q})^2)\, T_{i'k' \to jl}(E, \mathbf{q}, \mathbf{k})}{(\mathbf{q}^2 + m^2 - E^2 - i0)}. \tag{5.5}$$

Let us consider our problem as it applies to the processes of isobar production in pp-collisions.

We introduce the following notation for the elements of the transition-amplitude matrix:

$$T = T_{pp \to pp}; \quad T^* = T_{pp \to pN^*} \tag{5.6}$$

and use analogous notation for the generalized quasi-potential. Hereafter in our equations we shall neglect terms associated with "doubly inelastic" transitions in the scattering of a proton or isobar by a proton. The possibility of such neglect is indicated by the experimental fact that the isobar excitation cross sections are small compared with the cross section of elastic pp-scattering at high energies in the diffraction region [51]. It can be shown that if the so-called reduced system of equations with the notation of (5.6) is used, the system (5.4) may be written efficiently in the form of a two-channel equation:

$$T = V + V \times T + V^* \times T^*, \tag{5.7a}$$

$$T^* = V^* + V^* \times T + V \times T^*. \tag{5.7b}$$

Iterating the system obtained, we find the following solution for the isobar production amplitude:

$$T^* = V^* + T \times V^* + V \times T^* + T \times V^* \times T. \tag{5.8}$$

Here, the quantity T is the pp-scattering amplitude which satisfies the one-channel equation with a quasi-potential of the Gaussian type [20, 22]:

$$V(E, (\mathrm{p}-\mathrm{k})^2) = isg_0 e^{at};$$
$$s = 4E^2; \quad t = -(\mathrm{p}-\mathrm{k})^2. \tag{5.9}$$

Let us use the results obtained here to describe the processes of production of isobars having isotopic spin I = 1/2. In view of the quasi-elastic character of these processes, which take place without exchange of nonzero quantum numbers in the t-channel, we juxtapose these processes with the quasi-potential V* of the type (5.9); i.e.,

$$V^*(E, (\mathrm{p}-\mathrm{k})^2) = isg^* e^{bt}, \tag{5.10}$$

where g* is in general a complex number.

Let us give the relationship between the quasi-potential amplitude and the differential production cross section (with allowance for the normalization of the quasi-potential amplitude):

$$\left(\frac{d\sigma}{dt}\right)_{\text{prod}} = \frac{4\pi}{s}\left|\frac{32\pi^3 T^*}{8\pi\sqrt{s}}\right|^2.$$

We consider the properties of the nucleon isobars having isotopic spin I = 1/2. These include the following three resonances: N*(1.4), N*(1.52), and N*(1.69) having masses of 1400, 1581, and 1688 MeV, respectively. The resonance N*(1.4) is a strongly excited nucleon state at very small transferred momenta (in the region $|t| \sim 10^{-2}$ $(\mathrm{GeV/c})^2$), and N*(1.52) and N*(1.69) are weakly excited states.

The measurements carried out by the Brookhaven group [52] yield the following values of diffraction-peak slopes* in the $|t| \sim (0.01\text{–}0.8)$ $(\mathrm{GeV/c})^2$ region:

$$B = (14.4 \pm 2.5)\ (\mathrm{GeV/c})^{-2} \text{ for } N^*\ (1.4);$$
$$B = (3.83 \pm 0.37)(\mathrm{GeV/c})^{-2} \text{ for } N^*\ (1.52);$$
$$B = (5.25 \pm 0.48)(\mathrm{GeV/c})^{-2} \text{ for } N^*\ (1.69).$$

As is evident, the lightest isobar has a slope twice that of the elastic one $A \approx (8\text{–}9)$ $(\mathrm{GeV/c})^{-2}$, while the two others have slopes which are approximately equal to one another and amount to one-half of the slope of the elastic pp-scattering cross section.

Let us consider Eq. (5.8). In the region of small transferred momenta one can write out the solution in the following form:

$$T^* = \varphi(s) \sum_{n=0}^{\infty} \frac{a e^{\frac{ab}{a+nb}t}}{(a+nb)\, n!} \left(-\frac{4\pi^2 g_0}{a}\right)^n,$$
$$\varphi(s) = isg^*. \tag{5.11}$$

Let us rewrite this solution in the impact-parameter representation (see Appendix C):

$$T^* = \frac{\varphi(s)}{(2\pi)^3} \int d^2\rho e^{i\Delta\rho} \chi^*(\rho, b) e^{2i\chi(\rho, a)}, \tag{5.12}$$

where

$$\chi(\rho, a) = \frac{1}{s} \int_{-\infty}^{+\infty} V\left(s, \sqrt{\rho^2 + z^2}\right) dz;$$

$$\chi^*(\rho, b) = \frac{1}{\varphi(s)} \int_{-\infty}^{+\infty} V^*\left(s, \sqrt{\rho^2 + z^2}\right) dz \tag{5.13}$$

---

* Parametrization of the differential cross section

$$\frac{d\sigma}{dt} = A e^{Bt}$$

is used.

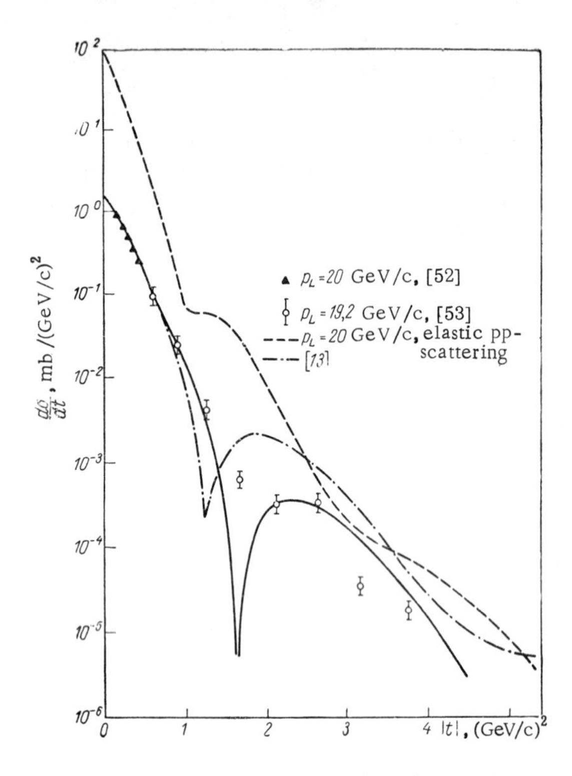

Fig. 7. Production of a N* (1.69) isobar in pp-collisions at p_L = 20 GeV/c.

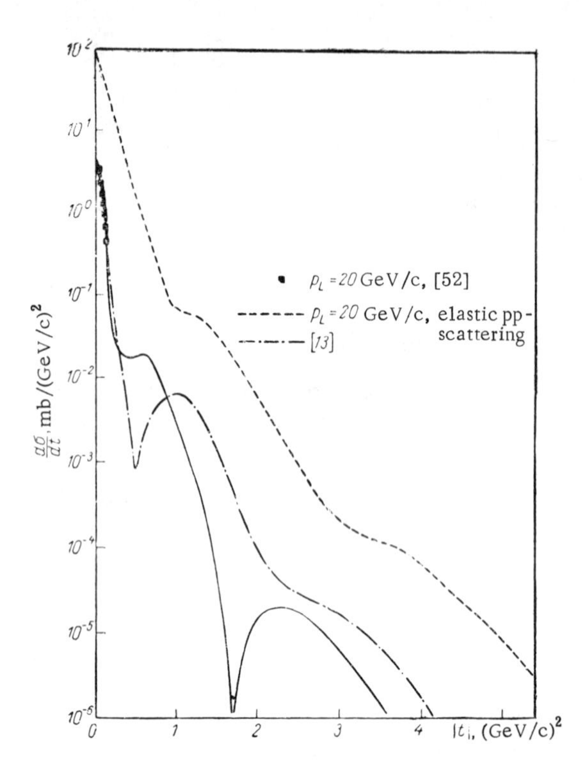

Fig. 8. Production of a N* (1.4) isobar in pp-collisions at p_L = 20 GeV/c.

are the so-called eikonal functions, while V(s, **r**), V*(s, **r**) are the Fourier transforms of the quasi-potentials (5.9) and (5.10). Equation (5.12) gives a clear picture of the process of isobar production with subsequent multiple elastic rescattering.

In Eq. (5.11) we make use of the experimental values of the slope parameters in isobar production [52]. The parameters of elastic pp-scattering, as well as the analog of coupling with the measured quantities for inelastic interaction, have been taken from [22].

Let us give the values of the parameters of the theoretical curves, which were calculated from the experimental slopes for p_L = 20 GeV/c:

$$\text{for } N^*(1.4)\, b = 7 \ (\text{GeV/c})^{-2};$$
$$\text{for } N^*(1.52)\, b = 1.4 \ (\text{GeV/c})^{-2};$$
$$\text{for } N^*(1.69)\, b = 1.5 \ (\text{GeV/c})^{-2};$$

The results of the calculations are displayed by the graphs in Figs. 7–9 in which a comparison is made with experimental data [55] in the region of transfers $|t|$ = 0–5 (GeV/c)^2 at an impinging-proton energy of 20 GeV/c in the laboratory system. The averaged curve for elastic pp-scattering has been plotted in Figs. 7 and 8 for comparison purposes.

As is evident from the graphs, the theoretical curves fit the experimental points well in the region of applicability of the solution (5.11), which corresponds to relatively small angles. The minima corresponding to the experimental dips lie in the region $|t| \approx$ 1.6–1.8 (GeV/c)^2.* The subsequent emergence into a new mode indicates transition from single to double, etc., rescattering.

Thus, the model with elastic rescattering in isobar production is in good agreement with the corresponding experimental situations. It is of interest to note the experimental indications which substantiate such a picture for these processes [51].

The total cross section of the pp $\to pN^*_{I=1/2}$ excitation is almost constant and does not depend on energy above p_L = 10 GeV/c. This is analogous to the behavior of σ_{el} in elastic pp-scattering and substantiates

* The problem of the presence of a zero in the amplitude was discussed in [22].

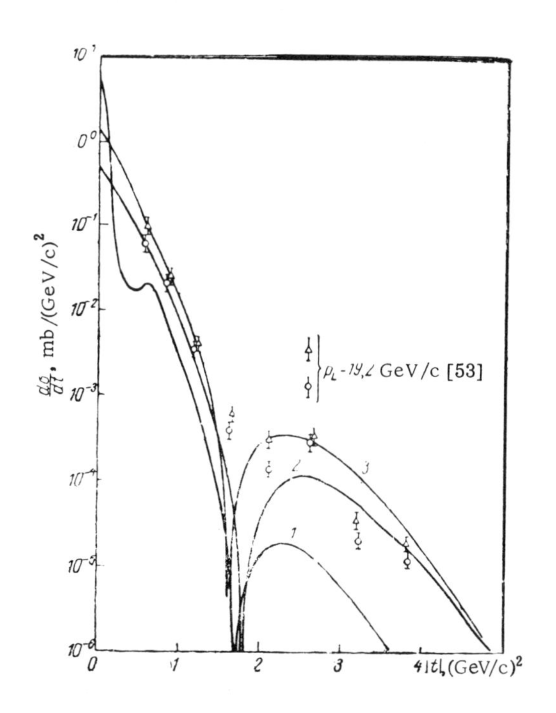

Fig. 9. Comparative graph of the cross sections of isobar productions: 1) N* (1.4); 2) N* (1.52); 3) N* (1.69).

the fact that the processes given go on without any change of the internal quantum numbers.

For comparison purposes the theoretical curves of [13] are plotted in Figs. 7 and 8; these curves were calculated on the basis of the Glauber formula in the model with rescattering by a Pomeranchuk pole.

Note that a number of papers [54] have likewise been devoted to model descriptions of isobar production at high energies.

Let us consider Eq. (5.11) in the region of large fixed angles, i.e., for

$$\left|\frac{t}{s}\right| = \sin^2\frac{\theta}{2} = \text{fixed}. \tag{3.1}$$

The solution of the equation in this region can be written in the form

$$T^* \to c(s)\sum_{n=0}^{\infty}\frac{(n+1)(n+\gamma)^{2n}}{n!\,\Gamma(n+\gamma)}\cdot\frac{e^{\frac{at}{n+\gamma}}}{(n+\gamma)^{3/2}}\left(\frac{isg_0\pi\sqrt{\pi}}{atp\sqrt{a}}\right)^n, \tag{5.14}$$

where

$$c(s) = sg^* e^{\gamma\varphi(0)}\gamma^{3/2}\Gamma(\gamma), \tag{5.15}$$

$n\gamma = a/b$, and $\varphi(\theta)$ is defined in [20]. Let us consider pp $\to$pN* processes in the region of high energies and large momentum transfers for the isobars N* (1.52) and N* (1.69). Then $\gamma = a/b \approx 2$ in a good approximation, and Eq. (5.14) takes the form

$$T^* \approx c\sum_{n=0}^{\infty}\frac{(n+1)(n+2)^{2n}}{n!\,\Gamma(n+2)(n+2)^{3/2}}e^{\frac{at}{n+2}}(-ix)^n,$$
$$x = \frac{sg_0}{|t|p}\left(\frac{\pi}{a}\right)^{3/2}. \tag{5.16}$$

since in the region of small angles the numbers n $\gg$ 1 play the dominant role in the sum, one can use the Stirling formula for factorials and transform the amplitude to the form

$$T^* \approx c\frac{e^2}{2\pi}\sum_{n=1}^{\infty}\frac{e^{\frac{at}{n+1}}}{(n+1)^{5/2}}(-ixe^2)^{n-1}. \tag{5.17}$$

Comparing (5.17) with the results for the amplitude of fixed-angle elastic scattering [22] and carrying out analogous calculations, we obtain the differential production cross section in the region (3.1):

$$\left(\frac{d\sigma}{d\Omega}\right) \to cq^2 e^{-2q\sqrt{\pi a}},$$
$$\theta = \text{fixed},\ s\to\infty. \tag{5.18}$$

The constant c in (5.18) has been determined for the value $\gamma \simeq 2$. We see that the formula for the differential cross section of large-fixed-angle production includes only the value of the elastic-scattering slope parameter a. Thus, the analogy between the angular distributions of production of the N* isobar and elastic pp-scattering indicates that a general mechanism must exist which is responsible for these processes at large angles independently of the details of final state. Experimental facts are available [53] which indicate this interesting property: namely, that the elastic and inelastic angular distributions (for the pp-system) have an identical slope in the region of large angles which does not depend on their relationship at small-momentum transfers.

§ 6. SCATTERING OF A SCALAR PARTICLE BY A SPINOR PARTICLE

In studying high-energy hadron scattering it is of interest to consider spin effects. Within the framework of the quasi-potential approach this can be reduced to the derivation and investigation of the quasi-potential equations for particles having spin.

Let us consider the case of interacting particles having spins 0 and 1/2, and let us investigate the corresponding equation in the high-energy limit. Without dwelling in detail on the derivation, we write out the quasi-potential equations for the two-particle wave function in the case considered:*

$$\left[E\gamma_0-(\gamma\mathbf{p}+M)\left(1+\frac{\omega}{W}\right)\right]\Psi(\mathbf{p})=\frac{1}{\omega}\int V_E(\mathbf{p},\ \mathbf{k})\Psi(\mathbf{k})\,d\mathbf{k}, \tag{6.1}$$

where $\omega = \sqrt{\mu^2 + \mathbf{p}^2}$; $W = \sqrt{M^2 + \mathbf{p}^2}$; μ and M are the masses of the scalar and spinor particles, respectively; E is the total energy; $\mathbf{p}$ is the momentum in the center of inertia system; $(\gamma_0,\ \boldsymbol{\gamma})$ are Dirac matrices.

Carrying out reasoning similar to that presented above, it can be shown that in the high-energy limit the quasi-potential equation (6.1) goes over into an equation of the Dirac type with the effective quasi-potential $1/\mathbf{p}\cdot V(s,\ \mathbf{r})$ in the coordinate representation:

$$[\varepsilon+i\boldsymbol{\alpha}\nabla-\beta M-\beta V_{\mathrm{eff}}\ (s,\ \mathbf{r})]\,\Psi(\mathbf{r})=0, \tag{6.2}$$

where $\varepsilon = E/2$. In the case of a scalar quasi-potential this equation leads to the following scattering amplitude [56]:

$$T(\mathbf{p},\ \mathbf{k})=\Phi_0^*(\mathbf{p})\,[a+\sigma_y b]\,\Phi_0(\mathbf{k}). \tag{6.3}$$

Here a and b are the so-called spin-nonflip and spin-flip amplitudes, respectively; σ_y is a Pauli matrix. The amplitudes a and b have the form

$$a=-ip\int_0^\infty \rho d\rho J_0(\rho\Delta)\,\{e^{\chi_0}\cos\chi_1-1\}, \tag{6.4a}$$

$$b=-ip\int_0^\infty \rho d\rho J_1(\rho\Delta)\,e^{\chi_0}\sin\chi_1. \tag{6.4b}$$

The functions $\chi_0(\rho)$ and $\chi_1(\rho)$, which are called eikonal functions, are associated with the quasi-potential:

$$\chi_0(\rho)=\frac{1}{2ip}\int_{-\infty}^{+\infty} dz\,[V_{\mathrm{eff}}^2+2mV_{\mathrm{eff}}], \tag{6.5a}$$

$$\chi_1(\rho)=\frac{1}{2ip}\int_{-\infty}^{+\infty} dz\,\frac{\partial V_{\mathrm{eff}}}{\partial\rho}. \tag{6.5b}$$

The results obtained can be used, for example, for the analysis of experimental data on πN- and KN-scatterings.

Note that in [57] this group of problems was considered on the basis of the Schrödinger equation on the assumption of the nonsingular behavior of the potential.

CONCLUSION

We have discussed certain results of the quasi-potential approach to hadron collision processes at high energies.

The close relationship between the equations of the quasi-potential method in quantum field theory and the quantum-mechanical formalism of the Schrödinger equation combined with the physical assumption of the

* A number of papers [55] has been devoted to the consideration of the quasi-potential equations for two spin particles.

nonsingular character of hadron interaction at high energies forms the basis of the given approach and makes it a convenient instrument in the study of the regularities governing hadron collision processes at high energies.

The theoretical results obtained on the basis of the quasi-potential equations with a quasi-potential of the Gaussian type are in good agreement with experimental data.

The authors deeply thank N. N. Bogolyubov for his stimulating discussions of the problems considered here and his valuable comments, A. N. Tavkhelidze for his constant interest in the work, his useful discussions, and his fruitful collaboration, D. I. Blokhintsev, S. S. Gershtein, S. V. Goloskokov, V. G. Kadyshevskii, A. A. Logunov, M. A. Markov, M. A. Mestvirishvili, R. M. Muradyan, Nguyen Van Hieu, M. K. Polivanov, V. I. Savrin, Ya. A. Smorodinskii, L. D. Solov'ev, N. E. Tyurin, O. A. Khrustalev, V. P. Shelest', and D. V. Shirkov for their numerous fruitful discussions.

APPENDIX A

Let us calculate the contribution of n + 1 iterations to the scattering amplitude (2.1) T(p, k; E) in the region of high energies and small scattering angles:

$$T^{(n+1)}(\mathbf{p}, \mathbf{k}; E) = (i s g_0)^{n+1} \iint \cdots \int \frac{d\mathbf{q}_1 \ldots d\mathbf{q}_n}{\varepsilon_1 \ldots \varepsilon_n} \; \frac{\exp\left\{-a\left[(\mathbf{p}-\mathbf{q}_1)^2 + \sum_{l=1}^{n-1}(\mathbf{q}_l - \mathbf{q}_{l+1})^2 + (\mathbf{q}_n - \mathbf{k})^2\right]\right\}}{\prod_{l=1}^{n}\left(\mathbf{q}_l^2 - p^2 - i0\right)}, \tag{A.1}$$

where $\varepsilon_l = \sqrt{\mathbf{q}_l^2 + m^2}$ $l = 1, 2, \ldots, n$. We perform the substitution of variables $\Delta_l = \mathbf{q}_l - \lambda_l$, $l = 1, 2, \ldots, n$, where

$$\lambda_l = \mathbf{q}_l^{\text{extrm}} = \frac{(n+1-l)\,\mathbf{p} + l\mathbf{k}}{n+1}, \quad l = 1, 2, \ldots, n \tag{A.2}$$

are the extremal momenta of the quadratic form in the exponent of the exponential in Eq. (A.1). We introduce the orthogonal vectors $\mathbf{l} = (\mathbf{p} + \mathbf{k})/2$, $\mathbf{r} = (\mathbf{p} - \mathbf{k})/2$, $(\mathbf{lr} = 0)$ and rewrite Eq. (A.2):

$$\lambda_l = \mathbf{l} + \frac{n-2l+1}{n+1}\mathbf{r} = \mathbf{l} + \mathbf{r}_l, \quad l = 1, 2, \ldots, n, \tag{A.3}$$

where $\mathbf{r}_l = [(n-2l+1)]/(n+1)\,\mathbf{r}$. Then the exponent of the exponential can be split into the contribution at the extremum point and a remainder:

$$(\mathbf{p}-\mathbf{q}_1)^2 + (\mathbf{q}_n-\mathbf{k})^2 + \sum_{l=1}^{n-1}(\mathbf{q}_l - \mathbf{q}_{l+1})^2 = \frac{(\mathbf{p}-\mathbf{k})^2}{n+1} + 2\left[\sum_{l=1}^{n}\Delta_l^2 - \sum_{l=1}^{n-1}\Delta_l\Delta_{l+1}\right] = \frac{(\mathbf{p}-\mathbf{k})^2}{n+1} + \Delta_1^2 + \Delta_n^2 + \sum_{l=1}^{n-1}(\Delta_l - \Delta_{l+1})^2.$$

In the new variables (A.1) takes the form

$$T^{(n+1)}(\mathbf{p}, \mathbf{k}; E) = (isg_0)^{n+1} e^{\frac{at}{n+1}} \iint \cdots \int \frac{d\Delta_1 \ldots d\Delta_n}{\varepsilon_1 \ldots \varepsilon_n} \times \frac{\exp\left\{-2a\left[\sum_{l=1}^{n}\Delta_l^2 - \sum_{l=1}^{n-1}\Delta_l\Delta_{l+1}\right]\right\}}{\prod_{l=1}^{n}\left[(\Delta_l + \lambda_l)^2 - p^2 - i0\right]}.$$

Let us decompose integration over Δ into longitudinal and transverse components in the vector $\mathbf{l}$:

$$\Delta = (\Delta_\perp; \Delta), \quad (\Delta_\perp \mathbf{l}) = 0;$$

$$T^{(n+1)}(\mathbf{p}, \mathbf{k}; E) = (i s g_0)^{n+1} e^{\frac{at}{n+1}} \iint \cdots \int d^2\Delta_\perp^{(1)} \ldots d^2\Delta_\perp^{(n)} \times \exp\left\{-2a\left[\sum_{l=1}^{n}\Delta_\perp^{(l)2} - \sum_{l=1}^{n-1}\Delta_\perp^{(l)}\Delta_\perp^{(l+1)}\right]\right\} I_n; \tag{A.4}$$

$$I_n = \iint \cdots \int \frac{d\Delta_1 \ldots d\Delta_n}{\varepsilon_1 \ldots \varepsilon_n} \cdot \frac{\exp\left\{-2a\left[\sum_{l=1}^{n} \Delta_l^2 - \sum_{l}^{n-1} \Delta_l \Delta_{l+1}\right]\right\}}{\prod_{l=1}^{n} \left[\Delta_l^2 + 2\Delta_l l + (\Delta_\perp^{(l)} + \mathbf{r})^2 - i0\right]}.$$

We represent each denominator in the form of a decomposition into two poles:

$$\frac{1}{[\ldots]_k} = \frac{1}{2l}\left(\frac{1}{\Delta_k + \frac{(\mathbf{r}+\Delta_\perp^k)^2}{2l} - i0} - \frac{1}{\Delta_k + 2l - \frac{(\mathbf{r}+\Delta_\perp^k)^2}{2l} + i0}\right).$$

It can be shown that in the region of large energies and small angles

$$|\mathbf{l}| \gg 1, \quad |\mathbf{r}| \ll 1, \quad \mathbf{l}^2 = p^2 + t/4$$

, the principal contribution will be made by the first pole. Then

$$I_n \simeq \frac{1}{(2l^2)^n}\left\{\iint \cdots \int d\Delta_1 \ldots d\Delta_n \frac{\exp\left\{-2a\left[\sum_1^n \Delta_l^2 - \sum_1^{n-1} \Delta_l \Delta_{l+1}\right]\right\}}{\prod_{l=1}^{n}\left(\Delta_l + \frac{\Delta_\perp^{(l)2}}{2l} - i0\right)} - J_n\right\},$$

where consideration has been given to the fact that for the same assumptions $\varepsilon_k \simeq l$ (k = 1, 2, ..., n) and J_n is the contribution of the nonprincipal poles. Let us represent each Gaussian factor in spectral form $e^{-a\Delta^2} = \int_{+\infty}^{-\infty} e^{i\Delta z} v(z)\, dz$. Taking account of the definition of the θ-function in the limit $\Delta_\perp^{(l)2}/2l \to 0$, it is not difficult to obtain

$$I_n \cong \frac{1}{(2l^2)^n}\left\{\frac{(2\pi i)^n}{(n+1)!} - J_n\right\}.$$

Thus, the principal contribution to the scattering amplitude in the given region is

$$T^{(n+1)}(\mathbf{p}, \mathbf{k}; E) \approx (i s g_0)^{n+1} e^{\frac{at}{n+1}} \left(\frac{2\pi i}{2l^2}\right)^n \frac{1}{(n+1)!} \iint \cdots \int d^2\Delta_1 \ldots d^2\Delta_n \exp\left\{-2a\left[\sum_1^n \Delta_\perp^{l^2} - \sum_1^{n-1} \Delta_\perp^l \Delta_\perp^{l+1}\right]\right\}.$$

Making use of the well-known equation

$$\iint \cdots \int d^2\Delta_1 \ldots d^2\Delta_n \exp\left\{-a \sum_{\alpha,\beta}^{n} C_{\alpha\beta} \Delta_\alpha \Delta_\beta\right\} = \left(\frac{\pi}{a}\right)^n \frac{1}{\operatorname{Det} C} \tag{A.5}$$

and taking account of the fact that in our case Det C = n + 1, we finally obtain

$$T^{(n+1)}(\mathbf{p}, \mathbf{k}; E) \approx (i s g_0)^{n+1} \frac{e^{\frac{at}{n+1}}}{(n+1)!} \left(\frac{2\pi i}{2l^2}\right)^n \left(\frac{\pi}{a}\right)^n \frac{1}{n+1} = i s g_0 \left(-\frac{4\pi^2 g_0}{a}\right)^n \frac{e^{\frac{at}{n+1}}}{(n+1)(n+1)!}. \tag{A.6}$$

APPENDIX B

Let us calculate $T^{(n+1)}(\mathbf{p}, \mathbf{k}; E)$ in the region (3.1). Hereafter, we shall use the notation of Appendix A. We consider

$$T^{(n+1)}(\mathbf{p}, \mathbf{k}; E) = (i s g_0)^{n+1} e^{\frac{at}{n+1}} \iint \cdots \int \frac{d^3\Delta_1 \ldots d^3\Delta_n}{\varepsilon_1 \ldots \varepsilon_n} \frac{\exp\left\{-2a\left[\sum_{l=1}^{n} \Delta_l^2 - \sum_{l=1}^{n-1} \Delta_l \Delta_{l+1}\right]\right\}}{\prod_{l=1}^{n} [(\Delta_l + \lambda_l)^2 - p^2 - i0]}. \tag{B.1}$$

In the high-energy limit, without assuming that the scattering angle is small, we have

$$\varepsilon_l = \sqrt{(\Delta_l+\lambda_l)^2+m^2} \approx |\lambda_l| = p\sqrt{1-4\sin^2\frac{\theta}{2}\,\frac{(n+1-l)\,l}{(n+1)^2}} \tag{B.2}$$

$$l = 1, 2, \ldots, n,$$

where λ_l is defined in (A.2), while θ is the scattering angle in the center of inertia system, since

$$\lambda_l^2 - p^2 = -4p^2\sin^2\frac{\theta}{2}\cdot\frac{l\,(n+1-l)}{(n+1)^2};$$

$$\prod_{l=1}^{n}(\Delta_l^2+2\Delta_l\lambda_l+\lambda_l^2-p^2-i0) \approx \prod_{l=1}^{n}(\lambda_l^2-p^2-i0) = \frac{(-4p^2\sin^2\theta/2)^n}{(n+1)^{2n}}\prod_{l=1}^{n} l\,(n+1-l) = \frac{t^n\,(n!)^2}{(n+1)^{2n}}. \tag{B.3}$$

Taking (B.2), (B.3) into account, we obtain

$$T^{(n+1)}(\mathbf{p},\mathbf{k};E) = \frac{(i\,sg_0)^{n+1}\,(n+1)^{2n}}{p^n\,t^n\,(n!)^2}\cdot\frac{e^{\frac{at}{n+1}}}{\prod_{l=1}^{n}\sqrt{1-4\sin^2\frac{\theta}{2}\cdot\frac{(n+1-l)l}{(n+1)^2}}}\iint\cdots\int d\Delta_1\ldots d\Delta_n\exp\left\{-2a\left[\sum_{l=1}^{n}\Delta_l^2-\sum_{l=1}^{n-1}\Delta_l\Delta_{l+1}\right]\right\} \tag{B.4}$$

from (B.1). The integration in (B.4) is carried out over the three-momenta Δ. Let us use the three-dimensional analog of Eq. (A.5). Then

$$T^{(n+1)}(\mathbf{p},\mathbf{k};E) = i\,sg_0\left(\frac{i\,sg_0\,\pi\sqrt{\pi}}{pta\sqrt{a}}\right)^n\frac{(n+1)^{2n}e^{\frac{at}{n+1}}}{(n!)^2\,(n+1)^{3/2}}\,\frac{1}{\prod_{l=1}^{n}\sqrt{1-4\sin^2\frac{\theta}{2}\,\frac{(n+1-l)l}{(n+1)^2}}}. \tag{B.5}$$

Let us consider the function

$$f_n(\gamma) = \prod_{l=1}^{n}\left(1-\gamma\frac{l}{n+1}\right), \text{where } \gamma = 2i\sin\frac{\theta}{2}\,e^{-i\theta/2},$$

$$|f_n(\gamma)| = \prod_{l=1}^{n}\sqrt{1-4\sin^2\frac{\theta}{2}\cdot\frac{(n+1-l)\,l}{(n+1)^2}},$$

and let us calculate $\ln f_n(\gamma)$ for $n \gg 1$:

$$\ln f_n(\gamma) = \sum_{l=1}^{n}\ln\left(1-\gamma\frac{l}{n+1}\right) \approx \int^{n}\ln\left(1-\gamma\frac{l}{n}\right)dl = -n\left[1+\frac{1-\gamma}{\gamma}\ln(1-\gamma)\right].$$

Thus,

$$\prod_{l=1}^{n}\sqrt{1-4\sin^2\frac{\theta}{2}\cdot\frac{(n+1-l)\,l}{(n+1)^2}} \simeq e^{-n\varphi(\theta)}, \quad n \gg 1,$$

where

$$\varphi(\theta) = 1+\mathrm{Re}\,\frac{1-\gamma}{\gamma}\ln(1-\gamma) = 1-\frac{\theta}{2\,\mathrm{tg}\,\theta/2}.$$

Substituting this equation into (B.5), we obtain the contribution to $T^{(n+1)}$ in the region (3.1):

$$T^{(n+1)}(\mathbf{p},\mathbf{k};E) \simeq i\,sg_0\left(\frac{i\,sg_0\,e^{\varphi(\theta)}\,\pi\sqrt{\pi}}{a\sqrt{a}\,pt}\right)^n\frac{(n+1)^{2(n+1)}}{[(n+1)!]^2}\cdot\frac{e^{\frac{at}{n+1}}}{(n+1)^{3/2}}. \tag{B.6}$$

APPENDIX C

Let us calculate $H=\sum_{n=0}^{\infty}\sum_{k=0}^{n} H^{(n;\,k)}$ in the region (4.4), where

$$H^{(n;\,k)}=\underbrace{g\otimes g\otimes\ldots}_{k-1}\otimes g\otimes h\otimes\underbrace{g\otimes\ldots g}_{n-k+1}, \tag{C.1}$$

while g(s, t), h(s, u) are defined by Eq. (4.3). We write out $H^{(n;k)}$ in explicit form:

$$H^{(n;\,k)}(\mathbf{p},\,\mathbf{k};\,E)=(i\,sg_0)^n\,h_0(s)\iint\ldots\int\frac{d\mathbf{q}_1\ldots d\mathbf{q}_n}{\varepsilon_1\ldots\varepsilon_n}$$

$$\times\frac{\exp\{-a[(\mathbf{p}-\mathbf{q}_1)^2+\ldots(\mathbf{q}_{k-2}-\mathbf{q}_{k-1})^2]-b\,(\mathbf{q}_{k-1}+\mathbf{q}_k)^2\;-a\,[(\mathbf{q}_h-\mathbf{q}_{k+1})^2+\ldots+(\mathbf{q}_n-\mathbf{k})^2]\}}{\prod\limits_{l=1}^{n}(\mathbf{q}_l^2-p^2-i0)}$$

By analogy with (A.2) we introduce the system of extremal momenta and perform the substitution

$$\Delta_i=\mathbf{q}_i-\lambda_i';\quad i=1,\,2,\,\ldots,\,k-1;$$
$$\Delta_i=\mathbf{q}_i-\lambda_i,\quad i=k,\,k+1,\,\ldots,\,n;$$

where

$$\lambda_l'=\frac{[a+(n-l)\,b]\,\mathbf{p}-lb\mathbf{k}}{a+nb}=\mathbf{l}+\mathbf{r}_l',\quad l=1,\,2,\,\ldots\,k-1; \tag{C.2}$$

$$\lambda_l=\frac{-(n+1-l)\,b\mathbf{p}+[a+b\,(l-1)]\,\mathbf{k}}{a+nb}=-\mathbf{l}+\mathbf{r}_l,\quad l=k,\,\ldots,\,n \tag{C.2'}$$

and

$$\mathbf{r}_l'=\frac{a+(n-2l)\,b}{a+nb}\,\mathbf{r},\quad l=1,\,2,\,\ldots,\,k-1;$$

$$\mathbf{r}_l=\frac{a-[n-2\,(l-1)]\,b}{a+nb}\,\mathbf{r},\quad l=k,\quad k+1,\,\ldots,\,n.$$

Note that unlike Appendix A, we use the following notation here:

$$\mathbf{l}=\frac{\mathbf{p}-\mathbf{k}}{2},\quad \mathbf{r}=\frac{\mathbf{p}+\mathbf{k}}{2}.$$

The expression for the exponent of the exponential is transformed as follows after the substitution of variables:

$$a\,[(\mathbf{p}-\mathbf{q}_1)^2+\ldots+(\mathbf{q}_{k-2}-\mathbf{q}_{k-1})^2]+b\,(\mathbf{q}_{k-1}+\mathbf{q}_k)^2\;+a\,[\mathbf{q}_h-\mathbf{q}_{k+1})^2+\ldots+(\mathbf{q}_n-\mathbf{k})^2]=\frac{ab}{a+bn}(\mathbf{p}+\mathbf{k})^2$$

$$+a\left[\sum_{\substack{l=1\\ l\neq k-1}}^{n-1}(\Delta_l-\Delta_{l+1})^2+\Delta_1^2+\Delta_n^2\right]+b\,(\Delta_{k-1}+\Delta_k)^2. \tag{C.3}$$

Taking (C.2) and (C.2') into account,

$$H^{(n;\,k)}=(i\,sg_0)^n\,h_0(s)\,e^{\frac{ab}{a+bn}u}\iint\ldots\int\frac{d\Delta_1\ldots d\Delta_n}{\varepsilon_1\ldots\varepsilon_n}\cdot\frac{1}{\prod\limits_{i=1}^{k-1}[(\Delta_i+\lambda_i')^2-p^2-i0]}$$

$$\times\frac{1}{\prod\limits_{j=k}^{n}[(\Delta_j+\lambda_j)^2-p^2-i0]}\exp\left\{-a\left[\sum_{\substack{l-1\\ l\neq k-1}}^{n-1}(\Delta_l-\Delta_{l+1})^2+\Delta_1^2+\Delta_n^2\right]-b\,(\Delta_{k-1}+\Delta_k)^2\right\}.$$

Let us partition Δ into components along l and $\perp$l:

$$\Delta=(\Delta_{\perp},\ \Delta),\quad (\Delta_{\perp}\cdot l)=0$$

and let us calculate the corresponding integral over the longitudinal components

$$D^{(n;\,k)}=\iint\ldots\int\frac{d\Delta_1\ldots d\Delta_n}{\varepsilon_1\ldots\varepsilon_n}\,\frac{\exp\left\{-a\left[\sum_{1}^{n-1}(\Delta_l-\Delta_{l+1})^2+\Delta_1^2+\Delta_n^2\right]-b\,(\Delta_{k-1}+\Delta_k)^2\right\}}{\prod_{i=1}^{k-1}[\ldots]\prod_{j=k}^{n}[\ldots]}.$$

Isolating the contributions of the principal poles in the region $\theta\approx\pi$ by analogy with Appendix A, taking account of $l^2 = p^2 + u/4$, and using the spectral representations

$$e^{-a\Delta^2}=\int_{-\infty}^{+\infty}e^{i\Delta z}\,v(z)\,dz;\quad e^{-b\Delta^2}=\int_{-\infty}^{+\infty}e^{i\Delta z}\,u(z)\,dz$$

and the definition of the θ-function, we obtain

$$D^{(n;\,k)}\simeq\left(\frac{2\pi i}{2l^2}\right)^n\frac{1}{(n-k+1)!\,(k-1)!}\int_{-\infty}^{+\infty}u(z)dz\left[\int_z^{\infty}v(x)\,dx\right]^{n-k+1}\left[\int_{-\infty}^{z}v(y)\,dy\right]^{k-1}.$$

Having taken the Gaussian quadrature over the transverse components of Δ with allowance for the fact that Det C = $a^{n-1}(a+nb)$, we obtain

$$H^{(n;\,k)}\simeq(i\,sg_0)^n\,h_0(s)\,e^{\frac{ab}{a+bn}u}\left(\frac{\pi}{a}\right)^n\frac{a}{a+nb}\,\frac{\int_{-\infty}^{+\infty}u(z)\,dz\left[\int_z^{\infty}v(x)\,dx\right]^{n-k+1}\left[\int_{-\infty}^{z}v(y)\,dy\right]^{k-1}}{(n-k+1)!\,(k-1)!}.\qquad\text{(C.4)}$$

Finally, the result of summation over k can be represented in the form

$$H\approx h_0(s)\sum_{n=0}^{\infty}\frac{a\,e^{\frac{ab}{a+bn}u}}{(a+nb)\,n!}\left(-\frac{4\pi^2 g_0}{a}\right)^n\qquad\text{(C.5)}$$

or in eikonal form

$$H\cong\frac{h_0(s)}{(2\pi)^3}\int e^{i\Delta\rho}\,d^2\rho\,\overline{\chi}(\rho,\ b)\,e^{2i\chi(\rho,\ a)},\qquad\text{(C.6)}$$

where

$$\chi(\rho,\ a)=\frac{1}{s}\int_{-\infty}^{+\infty}V(s,\sqrt{\rho^2+z^2})\,dz;$$
$$\overline{\chi}(\rho,\ b)=\frac{1}{h_0(s)}\int_{-\infty}^{+\infty}V_e(s,\sqrt{\rho^2+z^2})\,dz\qquad\text{(C.7)}$$

are the eikonal functions.

LITERATURE CITED

1. N. N. Bogolyubov, B. V. Medvedev, and M. K. Polivanov, Problems in the Theory of Dispersion Relationships [in Russian], Gostekhizdat, Moscow (1958).
2. H. Epstein, "Rigorous theoretical considerations on high-energy scattering," in: Topical Conference on High-Energy Collisions of Hadrons, CERN, Geneva (1968).

2a. A. A. Logunov and Nguyen Van Hieu, "On some consequences of analyticity and unitarity," in: Topical Conference on High-Energy Collisions of Hadrons, CERN, Geneva (1968).

3. New Method in the Theory of Strong Interactions, Collection of Papers [Russian translation], Izd. Inostr. Lit., Moscow (1960).
4. D. V. Shirkov, V. V. Serebryakov, and V. A. Meshcheryakov, Dispersion Theories of Strong Interactions at Low Energies [in Russian], Nauka, Moscow (1967).
5. I. Ya. Pomeranchuk, Zh. Éksperim. i Teor. Fiz., 34, 725 (1958).
6. A. A. Logunov, Nguyen Van Hieu, and I. T. Todorov, Usp. Fiz. Nauk, 88, 51 (1966).
7. G. G. Beznogykh et al., Preprint of the Joint Institute for Nuclear Research, F1-4628, Dubna (1969).
8. A. A. Logunov and Nguyen Van Hieu, Preprint of the Technical Physics Council 69-4, Serpukhov (1969).
9. G. Moliere, Z. für Naturforsch., 2A, No. 3 (1947). R. J. Glauber, Lectures in Theoretical Physics, Vol. 1, Interscience, New York (1959).
10. A. A. Logunov and M. A. Mestvirishvili, Phys. Lett., 24B, 620 (1967). V. I. Savrin and O. A. Khrustalev, Preprint of the Institute of High-Energy Physics 68-19-K, Serpukhov (1968).
11. R. C. Arnold, Phys. Rev., 153, 1523 (1967); R. Torgerson, Phys. Rev., 143, 1194 (1966).
12. M. M. Islam, Impact Parameter Description of High-Energy Scattering. Lectures in Theoretical Physics, Vol. 15 (1968).
13. S. Frautschi and B. Margolis, Nuovo Cimento, 56A, 1155 (1968); 57A, 427 (1968); F. Frautschi, O. Kofoed-Hansen, and B. Margolis, Nuovo Cimento, 61A, 41 (1969).
14. G. Cohen-Tannoudji, A. Morel, and H. Navelet, Nuovo Cimento, 48A, 1075 (1967); R. J. Rivers and L. M. Saunders, Nuovo Cimento, 58A, 385 (1968).
15. C. B. Chiu and J. Finkelstein, Nuovo Cimento, 57A, 649 (1968).
16. J. Finkelstein and M. Jacob, Nuovo Cimento, 56A, 681 (1968).
17. T. T. Chou and C. N. Yang, Phys. Rev., 170, 1591 (1968); Phys. Rev., 175, 1832 (1968); L. Durand and R. Lipes, Phys. Rev. Lett., 20, 637 (1968).
18. V. N. Gribov, Zh. Éksperim. i Teor. Fiz., 53, 654 (1968); V. N. Gribov, I. Ya. Pomeranchuk, and K. A. Ter-Martirosyan, Phys. Rev., 139B, 184 (1965).
19. A. A. Ansel'm and I. T. Dyatlov, Nuclear Fusion, 6, 591 (1967); V. N. Gribov and A. A. Migdal, Nuclear Fusion, 8, 1002 (1968); K. A. Ter-Martirosyan, in: Transactions of the International Theoretical School of High-Energy Physics, Poprad, Czechoslovakia (1967), p. 43.
20. V. R. Garsevanishvili, V. A. Matveev, L. A. Slepchenko, and A. N. Tavkhelidze, Preprint of the Joint Institute for Nuclear Research E2-4251, Dubna (1969). Talk Given at the Coral Gables Conference, Miami, Gordon and Breach Publishers (1969), p. 74.
21. V. R. Garsevanishvili, V. A. Matveev, L. A. Slepchenko, and A. N. Tavkhelidze, Phys. Lett., 29B, 191 (1969).
22. V. R. Garsevanishvili, S. V. Goloskokov, V. A. Matveev, and L. A. Slepchenko, Nuclear Fusion, 10, 627 (1969); Preprint of the Joint Institute for Nuclear Research E2-4361, Dubna (1969).
23. V. R. Garsevanishvili, V. A. Matveev, L. A. Slepchenko, and A. N. Tavkhelidze, Preprint of the ICTP, IC/69/87, Trieste (1969).
24. A. A. Logunov and A. N. Tavkhelidze, Nuovo Cimento, 29, 380 (1963).
25. A. N. Tavkhelidze, Lectures on the Quasi-Potential Method in Field Theory, Tata Institute of Fundamental Research, Bombay (1964); V. G. Kadyshevskii and A. N. Tavkhelidze, The Quasi-potential Method in the Relativistic Two-Body Problem, Collection of Papers Devoted to the Sixtieth Birthday of N. N. Bogolyubov [in Russian], Nauka, Moscow (1969), p. 261.
26. Collection: Theory of Strong Interactions at High Energies [Russian translation], Izd. Inostr. Lit., Moscow (1963).
27. V. A. Arbuzov, B. M. Barbashov, A. A. Logunov, Nguyen Van Hieu, A. N. Tavkhelidze, R. N. Faustov, and A. T. Filippov, Phys. Lett., 4, 272 (1963).
28. B. M. Barbashov, S. P. Kuleshov, et al., Preprint of the Joint Institute for Nuclear Research E2-4612, Dubna (1969).
29. P. N. Bogolyubov, Preprint of the Joint Institute of Nuclear Research E2-4417 (1969); Int. Rep. ICTP, IC/69/77, Trieste (1969).
30. R. N. Faustov, Lectures at the International Winter School of Theoretical Physics [in Russian], Joint Institute for Nuclear Research, Dubna (1964).
31. D. I. Blokhintsev, Nuovo Cimento, 30, 1094 (1963); D. I. Blokhintsev, V. S. Barashenkov, and B. M. Barbashov, Usp. Fiz. Nauk, 68, 417 (1969).
32. T. T. Wu and C. N. Yang, Phys. Rev., 137B, 708 (1965).

Teiler und Vielfache

Zu Seite 1

1 a) ist Teiler
kein Teiler
ist Teiler

b) kein Teiler
kein Teiler
ist Teiler

c) ist Teiler
ist Teiler
kein Teiler

d) kein Teiler
kein Teiler
ist Teiler

2 a) 6, 66, 24, 90, 84, 48

b) 72, 24, 144, 108, 240, 132

3 a) MEISE b) REISE c) KREIS d) REIS
e) RIESE f) EIS

4 a) Teiler von 18: 1, 2, 3, 6, 9, 18
Teiler von 24: 1, 2, 3, 4, 6, 8, 12, 24

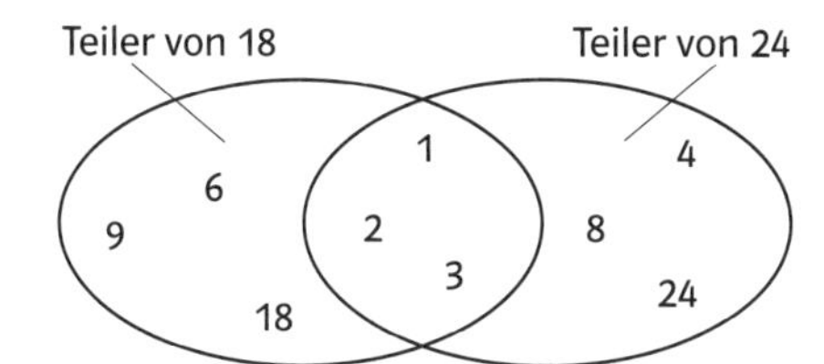

b) Teiler von 50: 1, 2, 5, 10, 25, 50
Teiler von 30: 1, 2, 3, 5, 6, 10, 15, 30

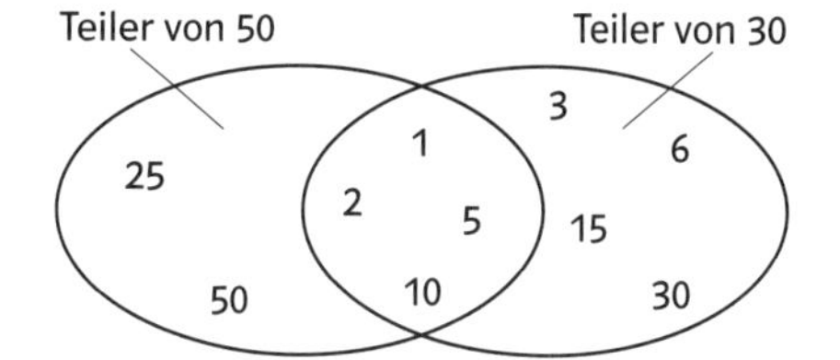

5 a)

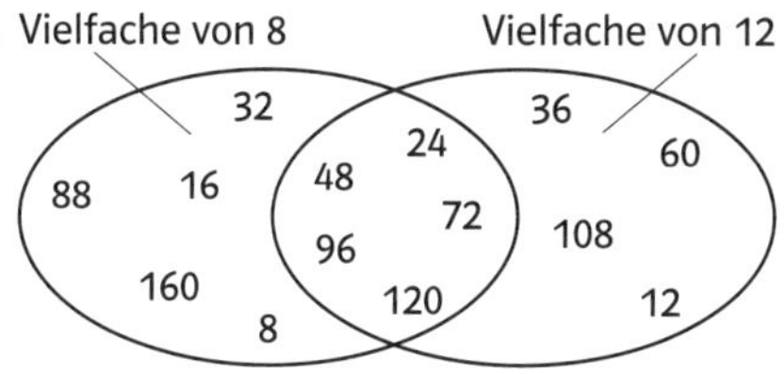

b)

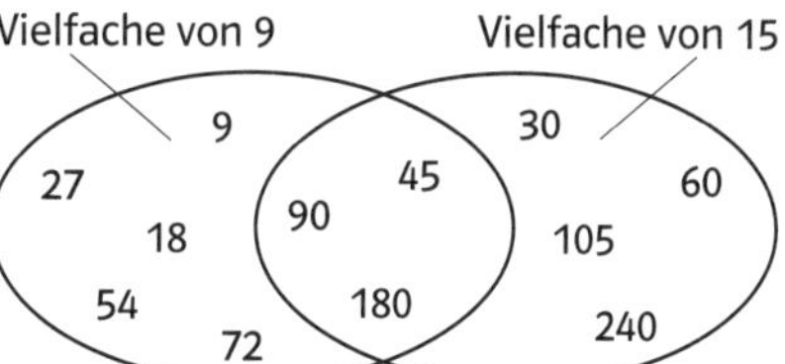

Zu Seite 2

6 a) $T_{12} = \{1, 2, 3, 4, 6, 12\}$
b) $T_{36} = \{1, 2, 3, 4, 6, 9, 12, 18, 36\}$
c) $T_{40} = \{1, 2, 4, 5, 8, 10, 20, 40\}$
d) $T_{50} = \{1, 2, 5, 10, 25, 50\}$

7 a) $V_6 = \{6, 12, 18, 24, 30, \ldots\}$
b) $V_9 = \{9, 18, 27, 36, 45, \ldots\}$
c) $V_{12} = \{12, 24, 36, 48, 60, \ldots\}$
d) $V_{13} = \{13, 26, 39, 52, 65, \ldots\}$

8 a) $T_{18} = \{1, 2, 3, 6, 9, 18\}$
b) $T_{50} = \{1, 2, 5, 10, 25, 50\}$
c) $T_{15} = \{1, 3, 5, 15\}$

9 a) $V_9 = \{9, 18, 27, 36, 45, \ldots\}$
b) $V_{23} = \{23, 46, 69, 92, 115, \ldots\}$
c) $V_{13} = \{13, 26, 39, 52, 65, \ldots\}$
d) $V_{15} = \{15, 30, 45, 60, 75, \ldots\}$

10 4 Teiler: T_{15}, T_8
5 Teiler: T_{16}, T_{81}
6 Teiler: T_{20}, T_{18}

11 T_{40}, T_{16}, T_{24}

12

32	4	20	10
16	8	40	5
1	48	2	30
12	24	6	3

Gemeinsame Teiler und gemeinsame Vielfache, Teilbarkeitsregeln

Zu Seite 3

1 a) $T_{18} = \{\underline{1}, \underline{2}, \underline{3}, \underline{6}, 9, 18\}$
$T_{12} = \{\underline{1}, \underline{2}, \underline{3}, 4, \underline{6}, 12\}$
ggT(12,18) = 6

b) $V_{10} = \{10, 20, \underline{30}, 40, 50, \underline{60} \ldots\}$
$V_{15} = \{15, \underline{30}, 45, \underline{60}, 75, 90 \ldots\}$
kgV(10,15) = 30

2 a) 8 b) 4 c) 48 d) 60
e) 12 f) 48 g) 24 h) 18

3

a)	b)	c)	d)
f	w	w	w
w	f	f	f
w	w	w	w
w	w	w	w
f	w	f	w

4

a)	Die Zahl soll durch 3 teilbar sein.	3□48	3348, 3648, 3948
		27□79	27279, 27579, 27879
		15□22	15222, 15522, 15822
		117□82	117282, 117582, 117882
b)	Die Zahl soll durch 4 teilbar sein.	12□	120, 124, 128
		17□4	1704, 1744, 1784
		3□8	308, 348, 97
		134□	1340, 1344, 1348

5 a) 3348 b) 8757 c) 30942 d) 27279
e) 985788 f) 2691

Primzahlen

Zu Seite 4

1 a) Primzahlen: 23, 37, 13, 41, 53, 7, 47, 61, 67, 17
b) Primzahlen: 59, 19, 11, 79, 97, 43, 83, 71, 73, 67, 29
c) Primzahlen: 71, 31, 17, 103, 29, 89, 13, 2, 101, 5

2 Primzahlen: 11, 13, 17, 19, 23, 29, 31, 37, 41, 43, 47, 53, 59, 61, 67, 71, 73, 79, 83, 89, 97
Alle Primzahlen liegen in der zweiten oder vierten Zeile.

3 a)

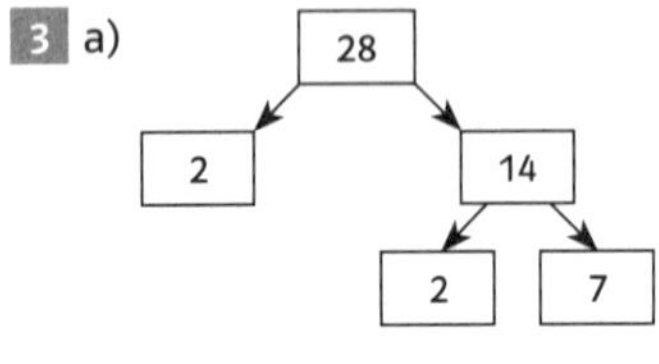

28 = 2 · 2 · 7

b)

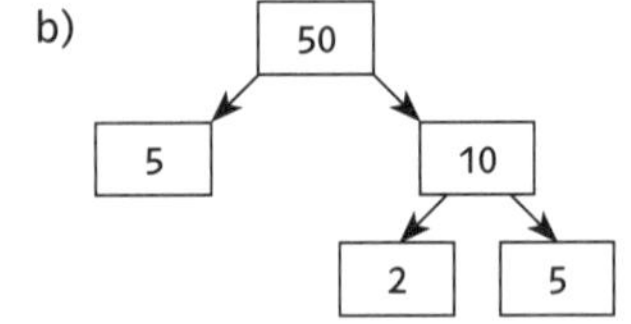

50 = 5 · 2 · 5

c)

140
10 14
2 5 2 7

140 = 2 · 5 · 2 · 7

d)

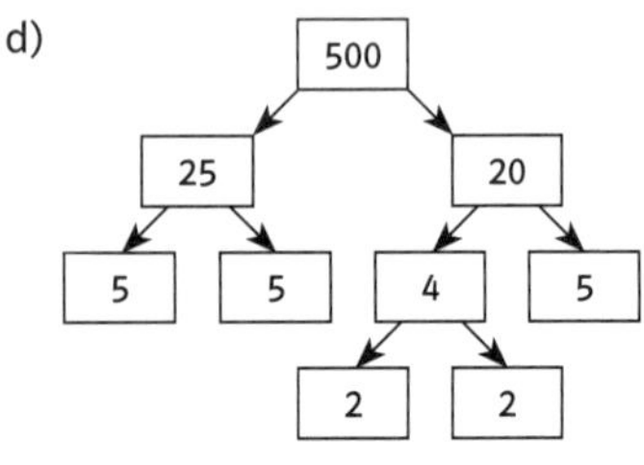

500 = 5 · 5 · 2 · 2 · 5

e)

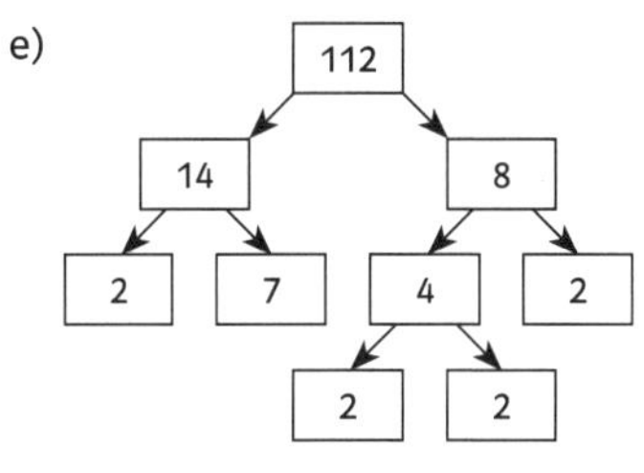

$112 = 2 \cdot 7 \cdot 2 \cdot 2 \cdot 2$

f)

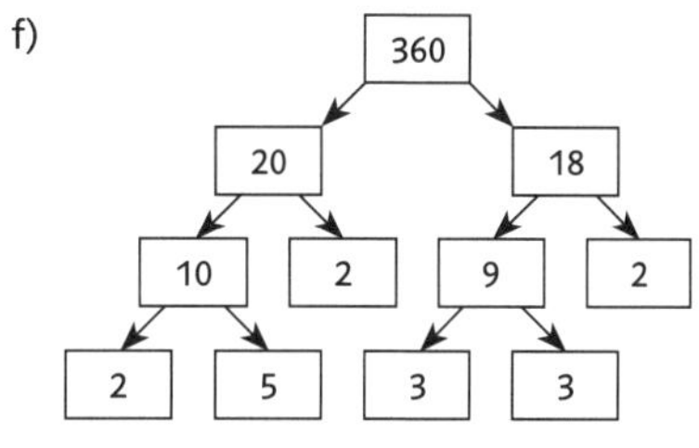

$360 = 2 \cdot 5 \cdot 2 \cdot 3 \cdot 3 \cdot 2$

4 (A) f, (B) f, (C) w, (D) w, (E) f, (F) f, (G) w, (H) w

Brüche darstellen

Zu Seite 5

1

Figur	A	B	C	D	E	F	G
Bruchteil gefärbt	$\frac{1}{6}$	$\frac{1}{2}$	$\frac{2}{5}$	$\frac{1}{4}$	$\frac{3}{4}$	$\frac{2}{8}$	$\frac{7}{12}$
Bruchteil weiß	$\frac{5}{6}$	$\frac{1}{2}$	$\frac{3}{5}$	$\frac{3}{4}$	$\frac{1}{4}$	$\frac{6}{8}$	$\frac{5}{12}$

2

Buchstabe	M	A	L	T	E
Bruchteil gefärbt	$\frac{4}{11}$	$\frac{5}{13}$	$\frac{3}{7}$	$\frac{5}{9}$	$\frac{4}{10}$
Bruchteil weiß	$\frac{7}{11}$	$\frac{8}{13}$	$\frac{4}{7}$	$\frac{4}{9}$	$\frac{6}{10}$

3 a) b) c) d)

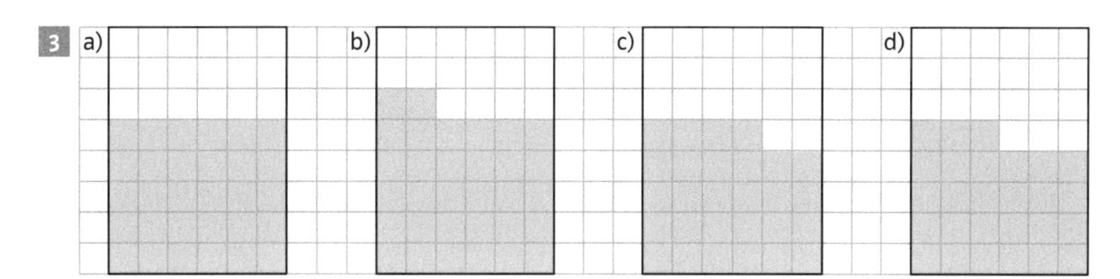

4 a) b) c) d)

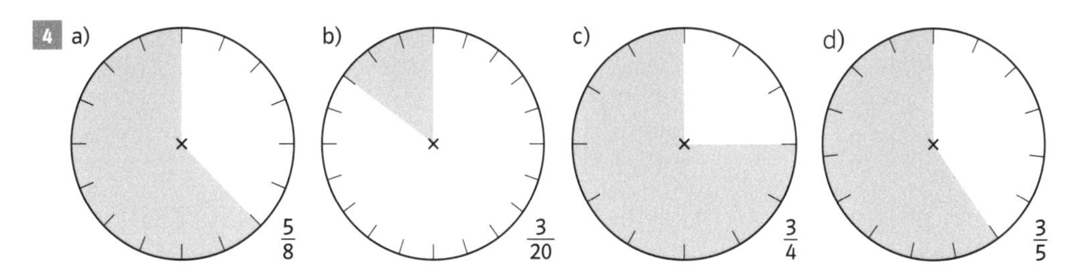

Bruchteile berechnen, das Ganze bestimmen

Zu Seite 6

1 a)

b)

c)

2 a) 30 kg
16 m
90 €
60 m²

b) 44 m
40 l
350 g
63 t

3

	Strecke in einer Sekunde
Schwalbe	54 m
Brieftaube	18 m
Pferd	10 m
Finnwal	5 m
Mensch	9 m

4 a) 325 m
80 kg
63 €
99 l
108 m²

b) 38,40 €
16,90 m
9 hl
10 m
2,070 km

Erweitern und Kürzen

Zu Seite 7

1 a) $\frac{2}{3} = \frac{16}{24}$ erweitert mit 8 b) $\frac{3}{4} = \frac{12}{16}$ erweitert mit 4

2 a) $\frac{16}{24} = \frac{4}{6}$ gekürzt durch 4 b) $\frac{4}{12} = \frac{2}{6}$ gekürzt durch 2

3 a) $\frac{4}{28}$ b) $\frac{6}{27}$ c) $\frac{9}{15}$ d) $\frac{63}{99}$
e) $\frac{60}{96}$ f) $\frac{44}{187}$ g) $\frac{18}{27}$ h) $\frac{21}{119}$

4 a) $\frac{4}{7}$ b) $\frac{12}{15}$ c) $\frac{8}{10}$ d) $\frac{7}{12}$
e) $\frac{8}{9}$ f) $\frac{2}{5}$ g) $\frac{2}{3}$ h) $\frac{2}{3}$

5 a) $\frac{2}{9}$ b) $\frac{5}{8}$ c) $\frac{2}{3}$ d) $\frac{5}{11}$
e) $\frac{1}{7}$ f) $\frac{3}{5}$ g) $\frac{2}{3}$ h) $\frac{11}{18}$
i) $\frac{1}{7}$ k) $\frac{11}{19}$

6 a) > b) < c) > d) <
e) = f) > g) > h) >
i) > k) > l) = m) >
n) < o) <

Brüche, Dezimalbrüche und Prozente

Zu Seite 8

1 A: $0{,}05 = \frac{1}{20}$ B: $0{,}2 = \frac{1}{5}$ C: $0{,}35 = \frac{7}{20}$ D: $0{,}5 = \frac{1}{2}$ E: $0{,}75 = \frac{3}{4}$ F: $0{,}9 = \frac{9}{10}$

2 A: $2{,}4 = \frac{12}{5}$ B: $2{,}7 = \frac{27}{10}$ C: $3{,}2 = \frac{16}{5}$ D: $3{,}5 = \frac{7}{2}$ E: $3{,}8 = \frac{19}{5}$

3 a) $\frac{4}{10} = 0{,}4$ b) $\frac{5}{10} = 0{,}5$ c) $\frac{75}{100} = 0{,}75$ d) $\frac{125}{1000} = 0{,}125$
e) $\frac{15}{100} = 0{,}15$ f) $\frac{34}{100} = 0{,}34$ g) $\frac{28}{100} = 0{,}28$ h) $\frac{8}{1000} = 0{,}008$

4

	a)	b)	c)	d)	e)	f)	g)	h)
Bruch	$\frac{1}{4}$	$\frac{3}{10}$	$\frac{4}{5}$	$\frac{1}{20}$	$\frac{1}{25}$	$\frac{2}{25}$	$\frac{3}{4}$	$\frac{1}{2}$
Bruch mit dem Nenner Hundert	$\frac{25}{100}$	$\frac{30}{100}$	$\frac{80}{100}$	$\frac{5}{100}$	$\frac{4}{100}$	$\frac{8}{100}$	$\frac{75}{100}$	$\frac{50}{100}$
Dezimalbruch	0,25	0,3	0,8	0,05	0,04	0,08	0,75	0,5
Prozente	25 %	30 %	80 %	5 %	4 %	8 %	75 %	50 %

5 a) 48 kg b) 32 € c) 228 € d) 90 kg
e) 0,050 kg f) 5 € g) 5 kg

Gleichnamige Brüche addieren und subtrahieren

Zu Seite 9

1 a) $\frac{4}{8} + \frac{3}{8} = \frac{7}{8}$ b) $\frac{6}{16} + \frac{3}{16} = \frac{9}{16}$ c) $\frac{13}{18} - \frac{4}{18} = \frac{9}{18}$ d) $\frac{2}{9} + \frac{3}{9} = \frac{5}{9}$

e) $\frac{5}{12} - \frac{3}{12} = \frac{2}{12}$

2 a) $\frac{1}{2}$ b) $\frac{1}{2}$ c) $\frac{3}{5}$ d) $\frac{1}{3}$

$\frac{1}{4}$ $\frac{4}{9}$ $\frac{1}{5}$ $\frac{1}{2}$

3 a) $1\frac{3}{7}$ b) $1\frac{1}{9}$ c) $1\frac{1}{2}$ d) 1

$1\frac{4}{5}$ $1\frac{7}{15}$ $1\frac{1}{8}$ $1\frac{2}{9}$

4 a) $2\frac{5}{8} + \frac{3}{8} = 3$ b) $2\frac{1}{8} - \frac{5}{8} = 1\frac{1}{2}$

5 a) $3\frac{18}{12} = 4\frac{6}{12} = 4\frac{1}{2}$ b) $6\frac{30}{24} = 7\frac{6}{24} = 7\frac{1}{4}$ c) $3\frac{20}{15} = 4\frac{5}{15} = 4\frac{1}{3}$

6 a) $9\frac{2}{7}$ b) $5\frac{3}{4}$ c) $\frac{5}{11}$

$22\frac{8}{19}$ $4\frac{2}{3}$ $8\frac{5}{6}$

7 a) $3\frac{1}{4}$ b) $4\frac{1}{3}$ c) $3\frac{3}{5}$

$1\frac{4}{5}$ $10\frac{1}{2}$ 18

Ungleichnamige Brüche addieren und subtrahieren

Zu Seite 10

1 a) $\frac{1}{2} + \frac{8}{16} = 1$ b) $\frac{1}{2} - \frac{1}{6} = \frac{1}{3}$

2 a) $\frac{19}{20}$ b) $\frac{19}{30}$ c) $\frac{7}{12}$ d) $\frac{5}{14}$

$\frac{7}{8}$ $\frac{13}{40}$ $\frac{11}{18}$ $\frac{7}{20}$

$\frac{11}{14}$ $\frac{33}{50}$ $\frac{19}{24}$ $\frac{11}{25}$

3 a) $\frac{1}{3}$ b) $\frac{1}{2}$ c) $\frac{4}{7}$ d) $\frac{2}{3}$

$\frac{3}{10}$ $\frac{1}{2}$ $\frac{7}{8}$ $\frac{3}{10}$

$\frac{2}{5}$ $\frac{1}{16}$ $\frac{7}{8}$ $\frac{1}{9}$

4 Leonie und Emily haben unterschiedlich erweitert. Leonie hat nicht auf den kgV der Nenner erweitert und muss daher zum Schluss noch einmal kürzen.

5 a) $\frac{9}{10}$ b) $\frac{9}{40}$ c) $\frac{19}{30}$ d) $\frac{11}{36}$

$\frac{14}{15}$ $\frac{13}{36}$ $\frac{5}{24}$ $\frac{1}{20}$

$\frac{29}{40}$ $\frac{13}{21}$ $\frac{11}{24}$ $\frac{11}{75}$

Zu Seite 11

6 a) $\frac{1}{6} + \frac{7}{12} = \frac{3}{4}$

b)

$\frac{8}{15}$	$\frac{1}{15}$	$\frac{2}{5}$
$\frac{1}{5}$	$\frac{1}{3}$	$\frac{7}{15}$
$\frac{4}{15}$	$\frac{3}{5}$	$\frac{2}{15}$

7 a) $1\frac{19}{30}$ b) $6\frac{5}{6}$ c) $1\frac{1}{60}$

$1\frac{7}{24}$ $1\frac{5}{44}$ $1\frac{7}{30}$

$1\frac{7}{12}$ $1\frac{5}{24}$ $1\frac{1}{45}$

8 a) $7\frac{7}{40}$ b) $8\frac{1}{24}$ c) $10\frac{31}{60}$

$6\frac{13}{30}$ $2\frac{7}{15}$ $2\frac{11}{26}$

$9\frac{17}{30}$ $7\frac{9}{40}$ $6\frac{11}{60}$

$5\frac{17}{28}$ $3\frac{5}{8}$ $7\frac{17}{30}$

9 a) $2\frac{13}{20}$ b) $7\frac{8}{15}$ c) $9\frac{19}{24}$

$3\frac{9}{14}$ $5\frac{5}{6}$ $3\frac{17}{24}$

$5\frac{23}{24}$ $4\frac{23}{40}$ $8\frac{29}{60}$

$2\frac{7}{12}$ $8\frac{19}{30}$ $10\frac{28}{45}$

10 a) $4\frac{1}{2}$ b) $6\frac{2}{3}$ c) $6\frac{2}{3}$ d) $6\frac{7}{10}$ e) $2\frac{3}{8}$

$5\frac{1}{12}$ $4\frac{3}{10}$ $6\frac{9}{10}$ $5\frac{33}{40}$ $6\frac{2}{3}$

$4\frac{3}{5}$ $5\frac{1}{3}$ $4\frac{1}{2}$ $4\frac{3}{10}$ $4\frac{1}{2}$

Namen der Städte: Sofia, Minsk, Paris

Brüche multiplizieren

Zu Seite 12

1 a) $\frac{1}{2} \cdot \frac{1}{5} = \frac{1}{10}$ b) $\frac{3}{4} \cdot \frac{3}{5} = \frac{9}{20}$

2 a) b) c)

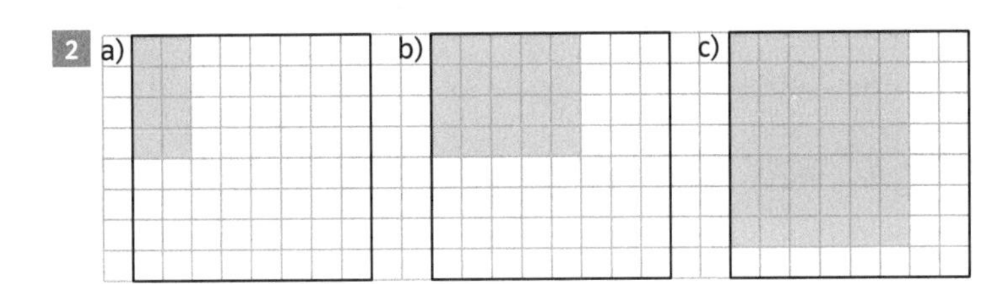

3 a) $\frac{3}{2}$ b) $\frac{3}{10}$ c) $\frac{1}{8}$

$\frac{1}{2}$ $\frac{1}{6}$ $\frac{4}{15}$

$\frac{1}{4}$ $\frac{1}{14}$ $\frac{1}{6}$

$\frac{72}{175}$ $\frac{5}{9}$ $\frac{1}{2}$

4 a) $\frac{1}{11}$ b) $\frac{1}{25}$

5 a) $6\frac{3}{4}$ b) $10\frac{1}{2}$ c) 51

8 $7\frac{1}{2}$ $7\frac{1}{5}$

$25\frac{1}{2}$ 8 $3\frac{1}{3}$

Zu Seite 13

6 a) $\frac{4}{7} \cdot \frac{7}{12} = \frac{1}{3}$ b) $\frac{6}{7} \cdot \frac{1}{8} = \frac{3}{28}$

$\frac{6}{11} \cdot 55 = 30$ $\frac{8}{13} \cdot 156 = 96$

$\frac{4}{5} \cdot \frac{3}{8} = \frac{3}{10}$ $\frac{54}{63} \cdot \frac{45}{72} = \frac{15}{28}$

7 a) $2\frac{2}{5}$ km = 2 400 m b) $\frac{3}{4}$ kg = 750 g c) 8 l = 80 dl d) $11\frac{1}{4}$ m = $112\frac{1}{2}$ dm

8 a) $\frac{13}{4} \cdot \frac{2}{3} = 2\frac{1}{6}$ b) $\frac{48}{7} \cdot \frac{35}{48} = 5$

$\frac{18}{7} \cdot \frac{14}{15} = 2\frac{2}{5}$ $\frac{6}{23} \cdot \frac{46}{3} = 4$

$\frac{59}{6} \cdot \frac{3}{4} = 7\frac{3}{8}$ $\frac{11}{4} \cdot \frac{8}{33} = \frac{2}{3}$

$\frac{18}{19} \cdot \frac{38}{5} = 7\frac{1}{5}$ $\frac{76}{9} \cdot \frac{18}{19} = 8$

9 a) $4\frac{8}{9}$ b) 15 c) $4\frac{19}{20}$ d) $51\frac{4}{5}$

10 Der Apotheker hat $\frac{3}{10}$ Liter verkauft.

11 Fahrschüler: 135
Fahrradfahrer: 80
Fußgänger: 140
Beifahrer: 5

Brüche durch natürliche Zahlen dividieren

Zu Seite 14

1 a) $\frac{1}{8}$ b) $\frac{1}{9}$ c) $\frac{3}{16}$

2 a) $\frac{1}{4} : 4 = \frac{1}{16}$ b) $\frac{5}{6} : 2 = \frac{5}{12}$ c) $\frac{2}{5} : 3 = \frac{2}{15}$

3 a)

b)

c)

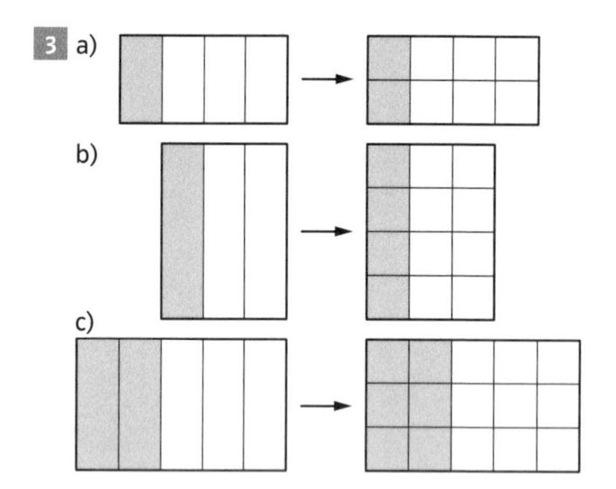

4 a) $\frac{1}{12}$ b) $\frac{3}{11}$ c) $\frac{1}{14}$ d) $\frac{5}{21}$

e) $\frac{2}{35}$ f) $\frac{1}{12}$ g) $\frac{2}{25}$ h) $\frac{5}{98}$ i) $\frac{8}{55}$ k) $\frac{3}{32}$

5 a) $\frac{3}{11}$ b) $\frac{3}{5}$ c) $\frac{4}{7}$ d) $1\frac{2}{3}$ e) $\frac{6}{7}$ f) $1\frac{2}{5}$

Durch Brüche dividieren

Zu Seite 15

1

a) 2	b) 1	c) $\frac{1}{2}$
$1\frac{1}{9}$	$\frac{2}{3}$	2
$7\frac{7}{8}$	$\frac{11}{20}$	$\frac{3}{7}$
$\frac{25}{27}$	$\frac{3}{4}$	$\frac{1}{2}$

2 a) $\frac{1}{2}$ b) 10 c) $1\frac{1}{3}$ d) $1\frac{1}{2}$ e) $\frac{1}{2}$ f) $1\frac{5}{6}$

3 a) $x = \frac{6}{7}$ b) $x = \frac{5}{6}$ c) $x = \frac{2}{3}$ d) $x = \frac{1}{2}$ e) $x = \frac{2}{5}$ f) $x = \frac{1}{15}$

4 $2\frac{2}{3} \cdot x = 3\frac{1}{5}$. Paulas Zahl ist $1\frac{1}{5}$.

5 a) $\frac{1}{2} \cdot \frac{1}{5} = \frac{1}{10}$. Rabia benötigt $\frac{1}{10}$ Liter des Konzentrats.

b) $\frac{7}{10} \cdot 5 = \frac{7}{2}$. Max kann $\frac{7}{2}$ Liter Getränk damit herstellen.

Grundrechenarten bei Brüchen

Zu Seite 16

1 [1] $\frac{1}{4}$ [2] 7 [3] $1\frac{3}{5}$ [4] 10 [5] 3 [6] 40 [7] $\frac{1}{2}$

2 a) $1\frac{3}{4}$ b) $\frac{48}{49}$ c) $1\frac{3}{4}$ d) $1\frac{17}{30}$

Man muss Punkt- vor Strichrechnung beachten.

Dezimalbrüche addieren und subtrahieren

Zu Seite 17

1 a) 0,4
0,66
0,95
0,85
0,425
0,63

b) 8,4
3,1
6,71
2,52
5,75
8,9

c) 61,3
24,7
56,1
0,9
48,8
16,25

2 a) 6,8
2,2
19,9
5,5

b) 18,2
21,4
21,2
31,0

c) 21,2
21,8
17,0
20,3

3 a) 6,011 b) 5,194 c) 213,649

4 18,59
15,19
21,201
12,3
10,29
11,3

5 a) 2,5
19,6
14,7

b) 12,3
135,08
0,004

Zu Seite 18

6 a) $1500 - x = 285,34$
$x = 1\,214,66$

b) $x - 39,15 = 905,7$
$x = 944,85$

c) $x + 6,459 = 39,6$
$x = 33,141$

d) $x + 15,38 + 17,095 = 38,145$
$x = 5,67$

7 a) beispielhaft:
10,5 + 46,7 + 35,25
34,56 + 12,34 + 45,55
1,08 + 2,57 + 88,8

b) 182,94

8 a) 16,8 b) 0,1 c) 3,07 d) 7,21 e) 1,15 f) 8

9 a) Bei Lösungsweg A wird die Aufgabe Schritt für Schritt berechnet. Bei Lösungsweg B werden zunächst alle Additionen zusammen berechnet und dann mit dem Ergebnis alle Subtraktionen durchgeführt.
b) 26,476
108,8977
6,207
762,983

Dezimalbrüche multiplizieren

Zu Seite 19

1

	· 10	· 100	· 1000
24,67	246,7	2467	24 670
376,4	3764	37 640	376 400
0,489	4,89	48,9	489
0,025	0,25	2,5	25
8,023	80,23	802,3	8023
0,000 6	0,006	0,06	0,6

2 a) 3,6
0,28
7,5

b) 0,18
0,024
0,065

3 a) 24,44 b) 2,592 c) 0,348

4 a) 5,796
63 · 0,092
630 · 0,0092

b) 27,76
3,47 · 8
0,347 · 80

c) 1,001
38,5 · 0,026
385 · 0,0026

d) 87,6125
70,09 · 1,25
700,9 · 0,125

5 a) 87,696 b) 2,196 c) 0,021

Dezimalbrüche dividieren

Zu Seite 20

1

	: 10	: 100	: 1000
11,8	1,18	0,118	0,011 8
3,4	0,34	0,034	0,003 4
2,202	0,220 2	0,022 02	0,002 202
546,7	54,67	5,467	0,546 7
0,256	0,025 6	0,002 56	0,000 256
0,004 7	0,000 47	0,000 047	0,000 004 7

2 a) 1,2 b) 0,04 c) 1,09 d) 2,2
e) 3,2 f) 0,36 g) 0,71 h) 0,26
i) 0,06

3 a) 2
0,2

b) 0,3
0,03

c) 250
0,025

d) 1
0,1

e) 0,0003
30000

f) 70
0,07

4 a) 3,05
5,9
6,8

b) 5,3
6,7
0,34

c) 8,3
9,6
1,29

5 a) 6,1 b) 110,6 c) 3900

Verbindung der Grundrechenarten

Zu Seite 21

1 a) 31,7 b) 0,7 c) 14,9 d) 12,212
e) 24 f) 14,062 5
Lösungswort: Schule

2 a) $28{,}2 \cdot (16{,}5 + 3{,}9) = 575{,}28$
b) $(26{,}2 - 4{,}7) \cdot 12{,}2 = 262{,}3$
c) $12 : 0{,}75 \cdot 12 : 0{,}3 = 640$
d) $14 : 2{,}5 - 6 : 1{,}5 = 1{,}6$
e) $(3{,}5 \cdot 3{,}8 - 1{,}2 \cdot 0{,}8) : 1{,}6 = 7{,}7125$

Periodische Dezimalbrüche

Zu Seite 22

1 a) 0,5 b) 0,75 c) 0,4 d) 0,625 e) 0,35

2 a) $\frac{2}{5}$ b) $\frac{11}{20}$ c) $\frac{7}{25}$ d) $\frac{7}{40}$

3 a) $0{,}\overline{1}$ b) $0{,}\overline{6}$ c) $0{,}\overline{45}$ d) $0{,}4\overline{6}$
e) $0{,}4\overline{09}$ f) $0{,}91\overline{6}$

4 a) $0{,}5 < 0{,}55 < 0{,}\overline{5} < 0{,}556 < 0{,}5\overline{56}$
b) $10{,}70\overline{56} < 10{,}756 < 10{,}\overline{765} < 10{,}7\overline{60}$
c) $2{,}37 < 3{,}373 < 2{,}\overline{37} < 2{,}377 < 2{,}378$
d) $45{,}015 < 45{,}1\overline{5} < 45{,}11 < 45{,}1\overline{51}$

5 a) $\frac{1}{3}$ b) $\frac{2}{3}$ c) $3\frac{5}{33}$ d) $6\frac{5}{11}$
e) $\frac{7}{111}$ f) $5\frac{30}{37}$

Sachaufgaben

Zu Seite 23

1 a) Juni: 24,501 m^3 Juli: 28,808 m^3
b) Zählerstand am 1. Mai: 34 192,004 m^3
c) Gasverbrauch im Mai: 26,499 m^3

2 a)

	Yesmin			Paul		
Anzahl der Umdrehungen	10	100	1000	10	100	1000
Länge der Strecke insgesamt	21 m	210 m	2100 m	22,3 m	223 m	2230 m

b) Yesmins Räder drehen sich ca. 5238-mal.
c) Paul legt 6,5 km zurück. Die Durchschnittsgeschwindigkeit beträgt 19,5 km/h.
d) Die Länge der Strecke beträgt 23,1 km.

Zu Seite 24

3 a) Aufgabe: Wie viel Wechselgeld bekommt sie?
Lösung: 13,52 €
b) Aufgabe: Reicht das Geld?
Lösung: Nein, er muss 104,58 € bezahlen, also 4,58 € mehr.

4 a) Familie Klever erhält 640 Pfund Sterling und 10 536 Schwedische Kronen.
b) Herr Büchner muss 930 € bezahlen.

5 Unter der Annahme, dass der Klassenraum komplett leer ist, entfallen auf jeden Schüler 4,63 m^3 Luft.

6 Länge des Kopfes: 0,18 m
Länge des Oberkörpers: 0,72 m
Länge des Unterarms: 0,36 m
Länge der Hand: 0,144 m
Länge des Fußes: 0,24 m

Muster

Zu Seite 25

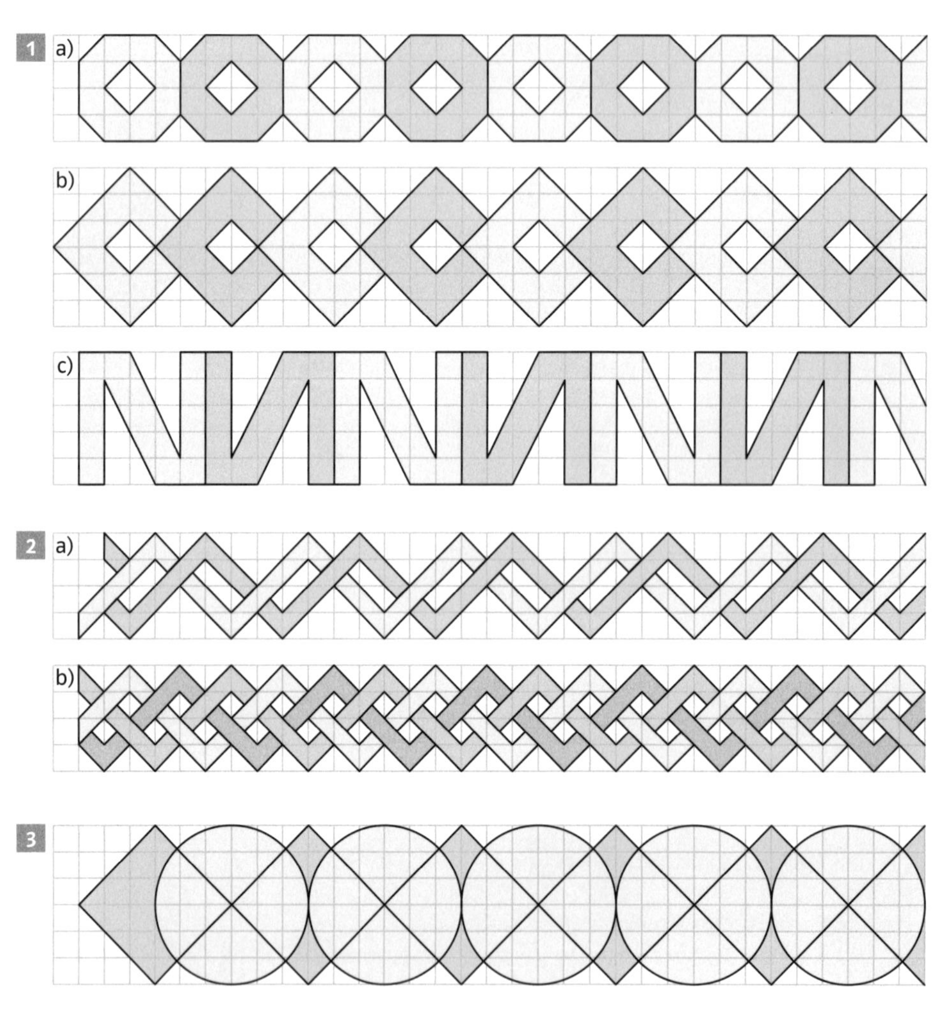

Verschiebung

Zu Seite 26

1 Der Fehler ist eingekreist und die richtige Position in die Grafik eingezeichnet.

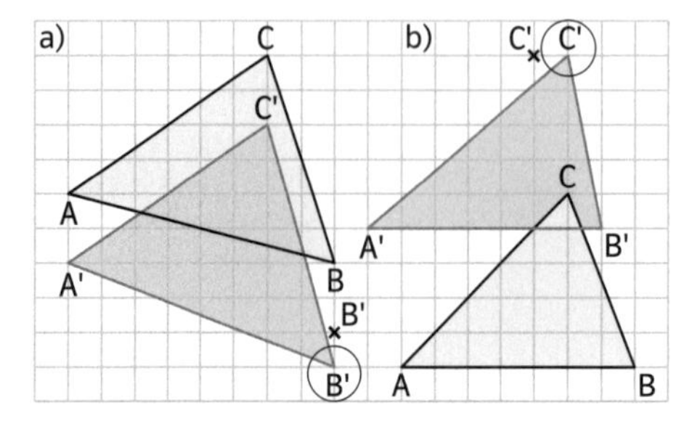

2

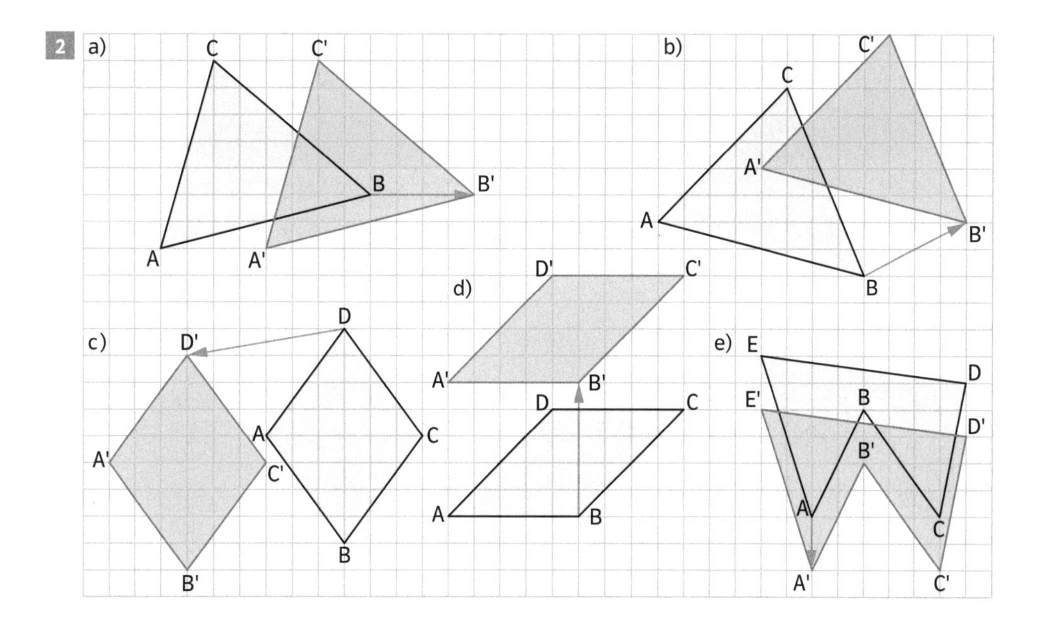

3

	a)	b)	c)
Eckpunkte der Originalfigur	A(3|4), B(7|1), C(9|7)	A(5|9), B(9|6), C(11|12)	A(5|5), B(9|3), C(10|11)
Eckpunkte der Bildfigur	A′(7|7), B′(11|4), C′(13|10)	A′(3|4), B′(7|1), C′(9|7)	A′(6|10), B′(10|8), C′(11|16)
Verschiebungsvorschrift	4 Kästchen nach rechts und 3 nach oben.	2 Kästchen nach links und 5 nach unten.	1 Kästchen nach rechts und 5 nach oben.

Achsenspiegelung

Zu Seite 27

1

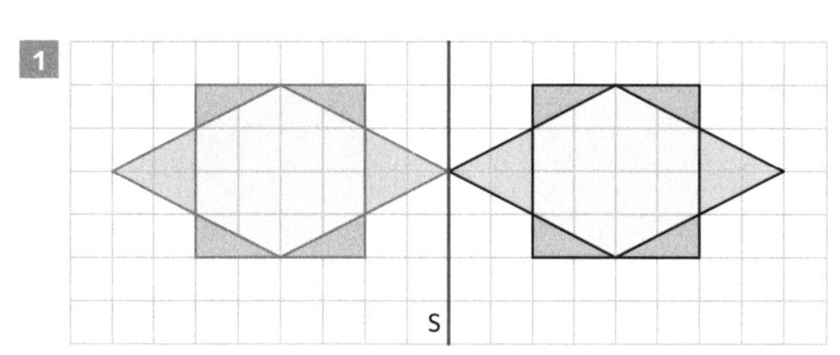

2

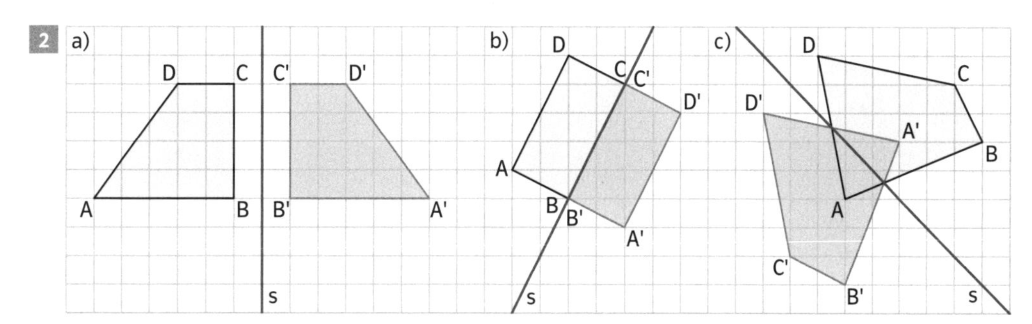

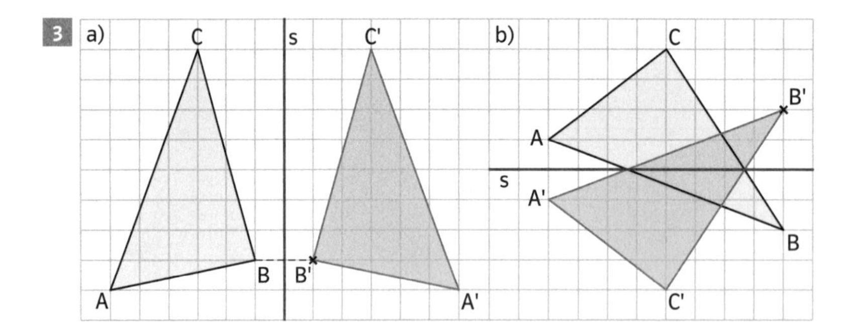

4

Quadrat	
Original	Bild
A(3|1)	A'(13|1)
B(8|3)	B'(8|3)
C(6|8)	C'(10|8)
D(1|6)	D'(15|6)
P(8|6), R(8|8)	

Rhombus	
Original	Bild
A(5|3)	A'(5|9)
B(9|1)	B'(9|11)
C(13|3)	C'(13|9)
D(9|5)	D'(9|7)
P(6|6), R(14|6)	

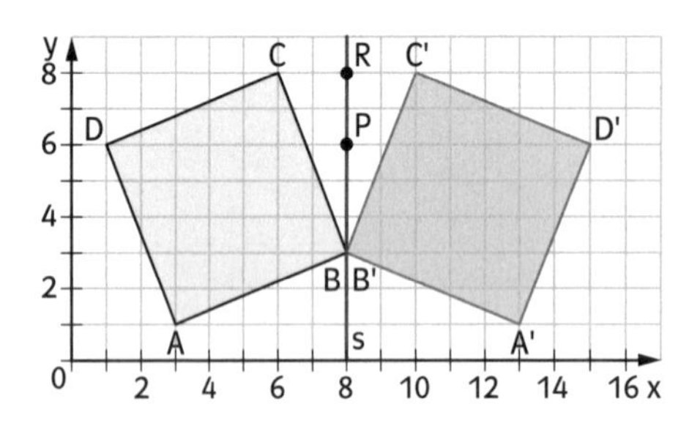

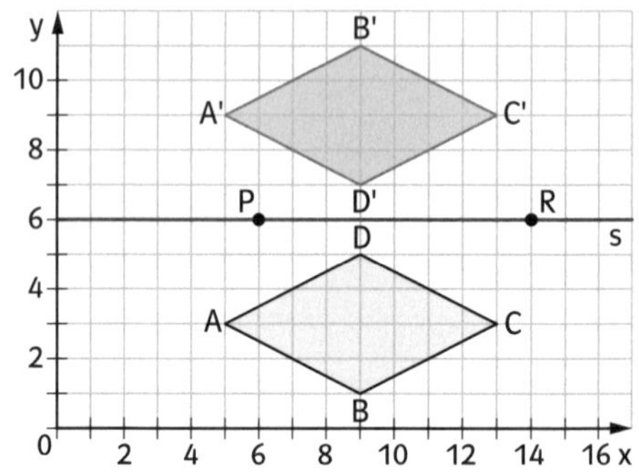

Mittelsenkrechte, Lot fällen

Zu Seite 28

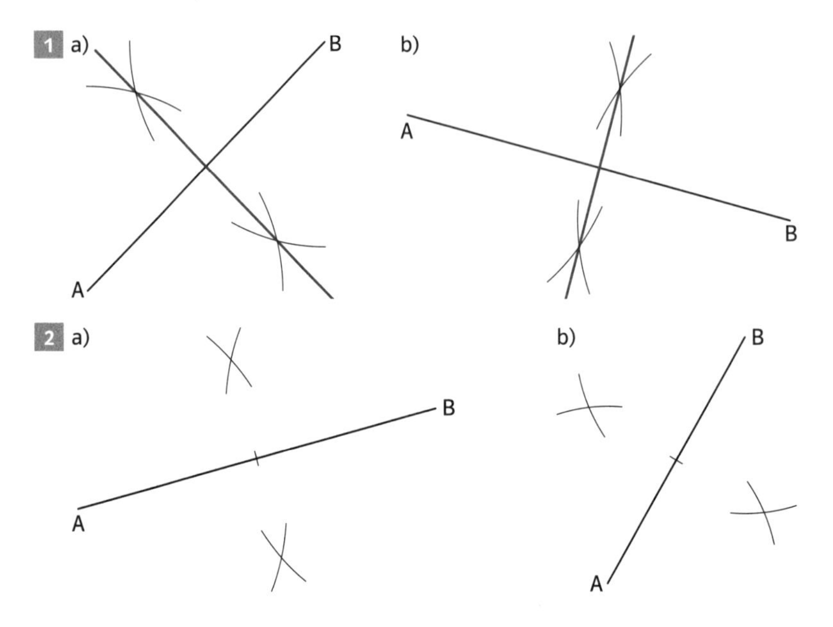

3

Lot von	Koordinaten des Fußpunktes
P	(6|6)
Q	(10|8)
R	(8|7)
S	(4|5)

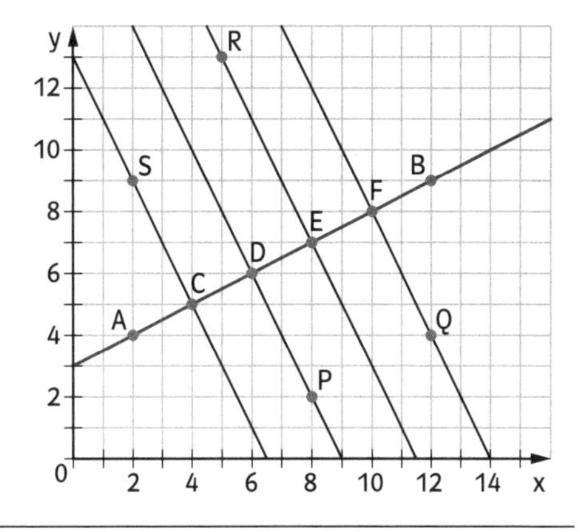

Senkrechte errichten, Winkelhalbierende

Zu Seite 29

1 a) b)

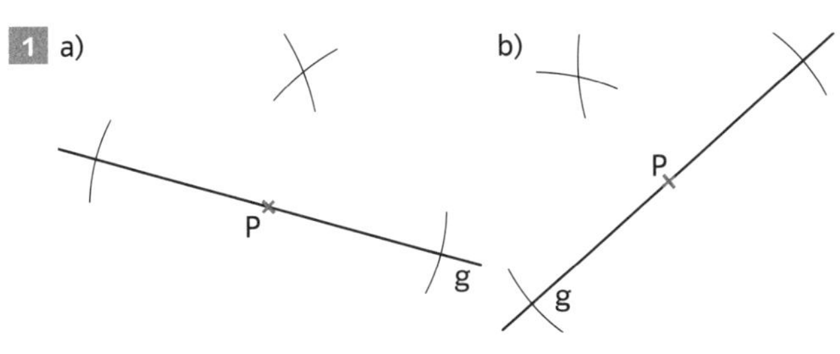

2 a) b)

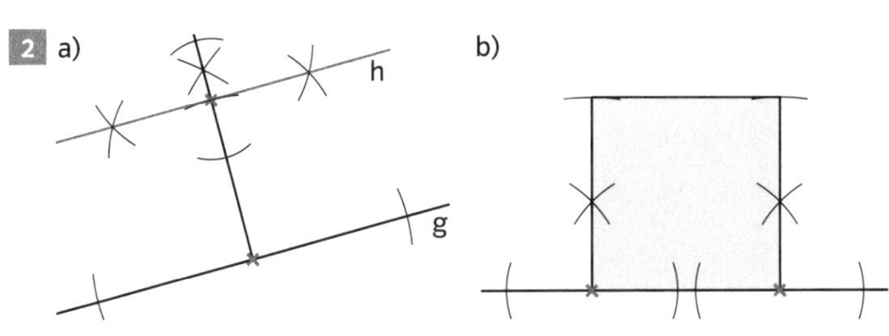

3 a) b) c)

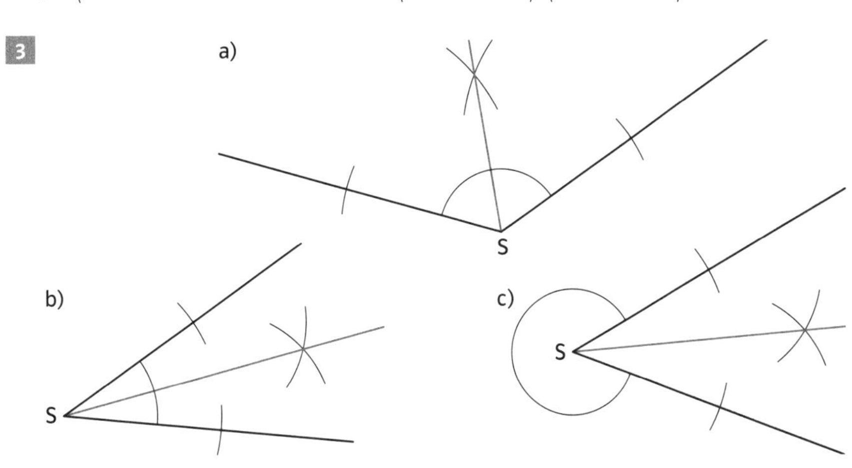

Drehung

Zu Seite 30

1

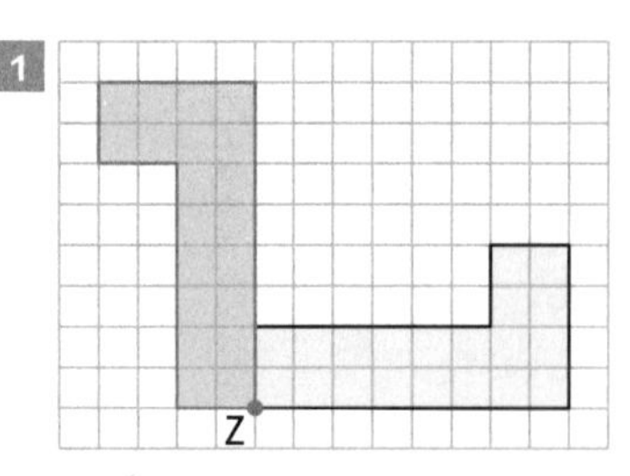

2

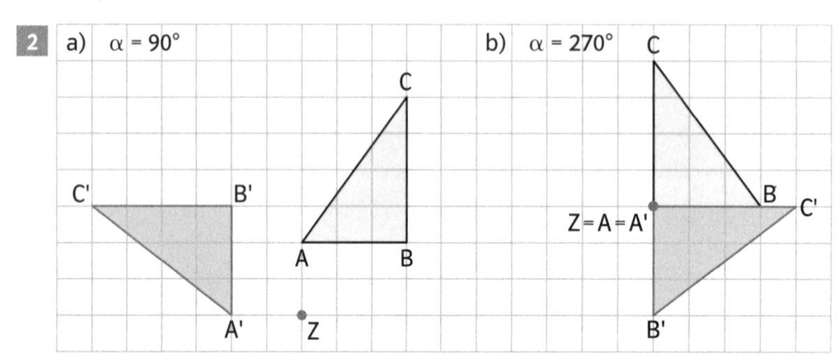

3

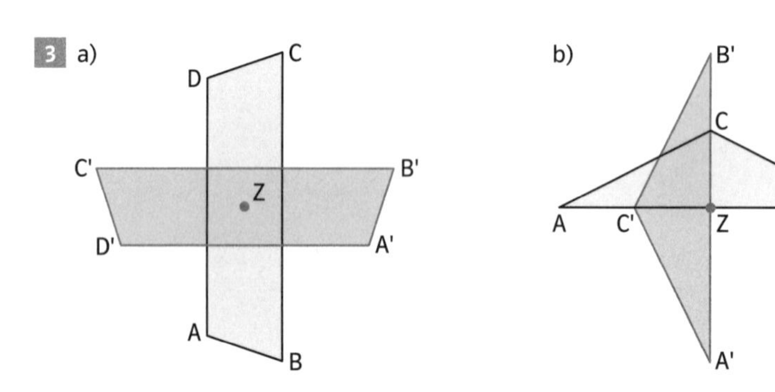

Punktspiegelung

Zu Seite 31

1

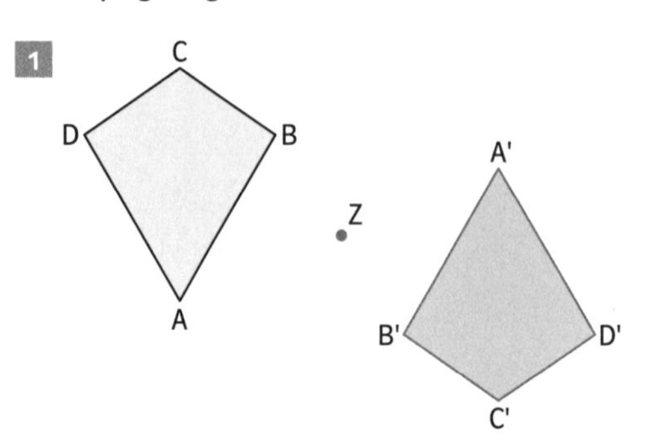

2

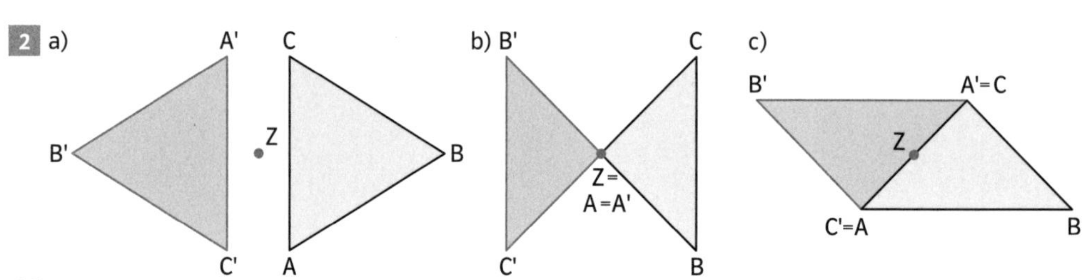

3 a)

Originalpunkt	Bildpunkt
A(2|4)	A'(8|12)
B(5|3)	B'(5|13)
C(6|6)	C'(4|10)
D(3|7)	D'(7|9)
Symmetriezentrum Z(5|8)	

b)

Originalpunkt	Bildpunkt
A(13|8)	A'(11|8)
B(15|12)	B'(9|4)
C(13|14)	C'(11|2)
D(11|12)	D'(13|4)
Symmetriezentrum Z(12|8)	

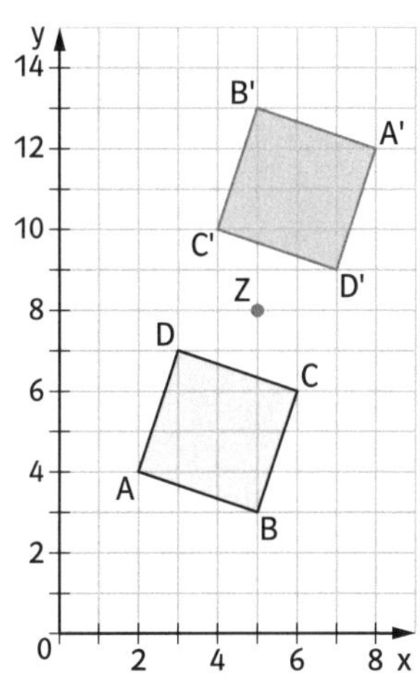

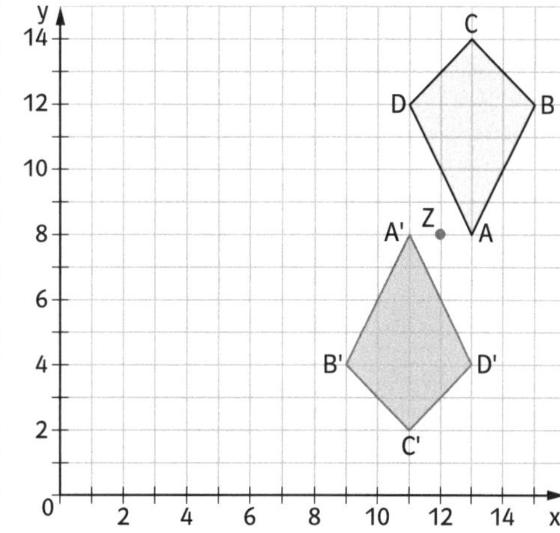

Symmetrische Figuren

Zu Seite 32

1

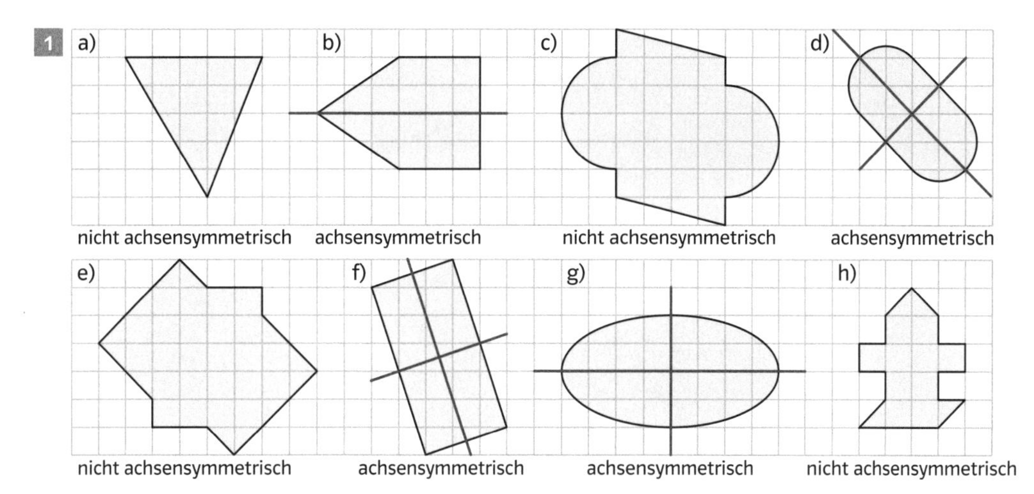

a) nicht achsensymmetrisch
b) achsensymmetrisch
c) nicht achsensymmetrisch
d) achsensymmetrisch
e) nicht achsensymmetrisch
f) achsensymmetrisch
g) achsensymmetrisch
h) nicht achsensymmetrisch

2

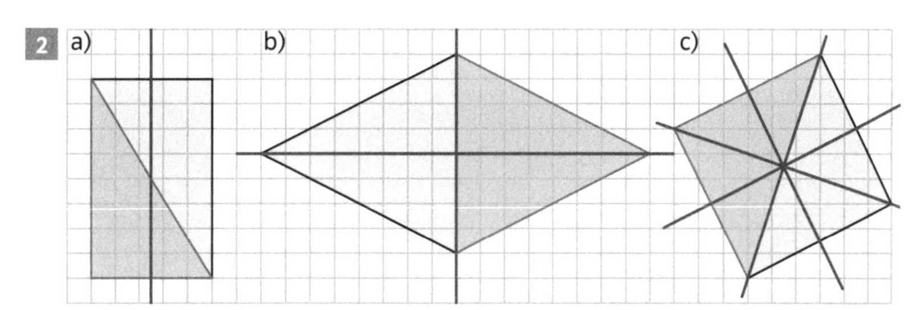

3

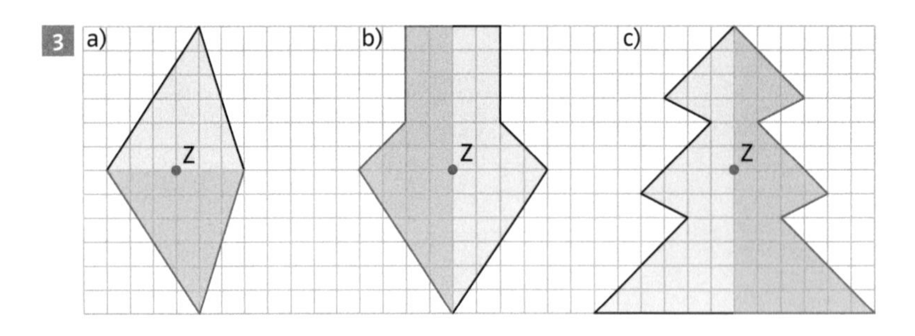

Zufallsexperimente durchführen und auswerten

Zu Seite 33

Die Aufgaben behandeln Zufallsexperimente, die von Schülerinnen und Schülern durchgeführt werden müssen – keine Lösungen.

Zu Seite 34

3 a) und b)

Ergebnis	absolute Häufigkeit	relative Häufigkeit als Bruch	relative Häufigkeit %	Winkel
Rollenspiele	12	$\frac{12}{150} = \frac{8}{100}$	8	28,8° ≈ 29°
Adventure-Spiele	9	$\frac{9}{150} = \frac{6}{100}$	6	21,6° ≈ 22°
Sport- und Simulationsspiele	33	$\frac{33}{150} = \frac{22}{100}$	22	79,2° ≈ 79°
Actionspiele	57	$\frac{57}{150} = \frac{38}{100}$	38	136,8° ≈ 137°
Strategie- und Denkspiele	39	$\frac{39}{150} = \frac{26}{100}$	26	93,6° ≈ 94°
Summe	150	$\frac{150}{150} = 1$	100	361° (aufgrund von Rundungen)

c)

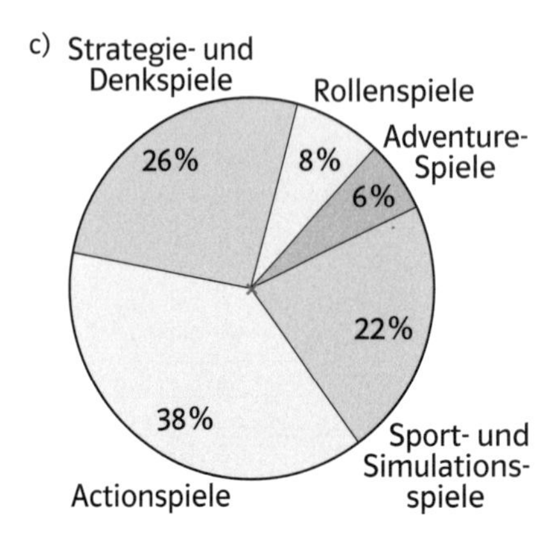

4

Ergebnis	absolute Häufigkeit	relative Häufigkeit		Winkel
		als Bruch	%	
rot	56	$\frac{56}{200} = \frac{28}{100}$	28	100,8°
schwarz	34	$\frac{34}{200} = \frac{17}{100}$	17	61,2°
blau	22	$\frac{22}{200} = \frac{11}{100}$	11	39,6°
weiß	88	$\frac{88}{200} = \frac{44}{100}$	44	158,4°
Summe	200	$\frac{200}{200} = 1$	100	360°

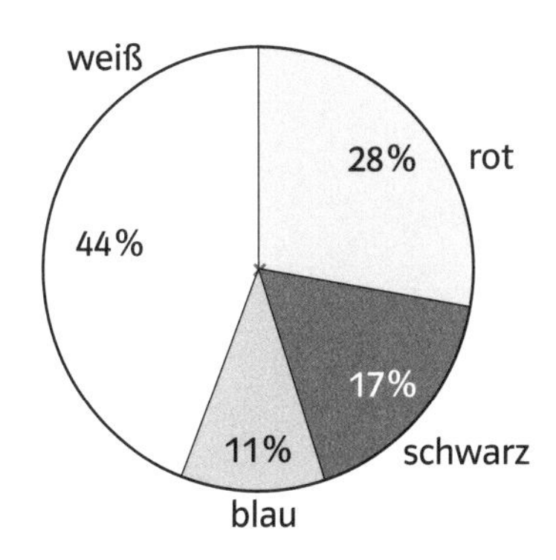

Arithmetisches Mittel und Median

Zu Seite 35

1 $\bar{x} = \frac{38 + 45 + 65 + 73 + 45 + 29 + 30 + 41 + 62 + 18}{10} = \frac{446}{10} = 44{,}6$ min

2 $\bar{x} = \frac{1 \cdot 28 + 2 \cdot 44 + 3 \cdot 19 + 4 \cdot 9}{100} = \frac{209}{100} = 2{,}09$ Fernseher

3 a) $\bar{x} = 30{,}4$ m b) $\tilde{x} = 33$ m

c) Der Median beschreibt ihre Wurfleistungen besser, da er den schwachen Wurf von 18 m nicht berücksichtigt. Beim Median fallen Ausreißer aus der Wertung heraus.

4 a) Jana: $\bar{x} = 33{,}417$ m b) $\tilde{x} = 35{,}75$ m

Lea: $\bar{x} = 34$ m $\tilde{x} = 34$ m

c) Die Frage kann nicht sicher beantwortet werden. Lea wirft konstant gut, während Jana in Einzelfällen eine sehr gute Leistung bringt, aber auch schwache Weiten wirft.

Zu Seite 36

5 Mädchen: $\bar{x} = 151{,}87$ cm $\tilde{x} = 152$ cm

Jungen: $\bar{x} = 147{,}64$ cm $\tilde{x} = 147$ cm

6 $\bar{x} = 3{,}66$ Personen pro Haushalt

7 a)

Augenzahl	absolute Häufigkeit
1	8
2	12
3	13
4	7
5	14
6	6
Summe	60

b) $\bar{x} = 3{,}417$ c) $\tilde{x} = 3$

Wahrscheinlichkeiten bestimmen

Zu Seite 37

1 $P(7) = \frac{1}{9}$ $P(\text{rot}) = \frac{4}{9}$ $P(\text{blau}) = \frac{5}{9}$

2 a) $\frac{1}{2}$ b) $\frac{1}{5}$ c) $\frac{3}{10}$ d) $\frac{1}{10}$

e) $\frac{2}{5}$ f) $\frac{1}{5}$ g) $\frac{3}{10}$ h) 0

3

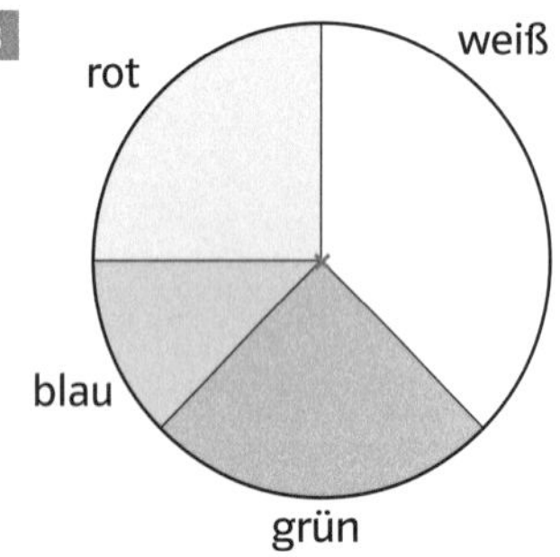

4 Summe: 500

a) $\frac{248}{500} = \frac{62}{125} = 49{,}6\ \%$ b) $\frac{188}{500} = \frac{47}{125} = 37{,}6\ \%$

c) $\frac{44}{500} = \frac{11}{125} = 8{,}8\ \%$ d) $\frac{252}{500} = \frac{63}{125} = 50{,}4\ \%$

e) $\frac{64}{500} = \frac{16}{125} = 12{,}8\ \%$ f) $\frac{480}{500} = \frac{24}{25} = 96\ \%$

Zu Seite 38

5 a)

1 — 1 (1; 1)
1 — 3 (1; 3)
1 — 5 (1; 5)
3 — 1 (3; 1)
3 — 3 (3; 3)
3 — 5 (3; 5)
5 — 1 (5; 1)
5 — 3 (5; 3)
5 — 5 (5; 5)

b) $P((5,5)) = \frac{1}{3} \cdot \frac{1}{3} = \frac{1}{9}$

6 a) (1,1), (1,2), (1,3), (1,4), (2,1), (2,2), (2,3), (2,4), (3,1), (3,2), (3,3), (3,4), (4,1), (4,2), (4,3), (4,4)

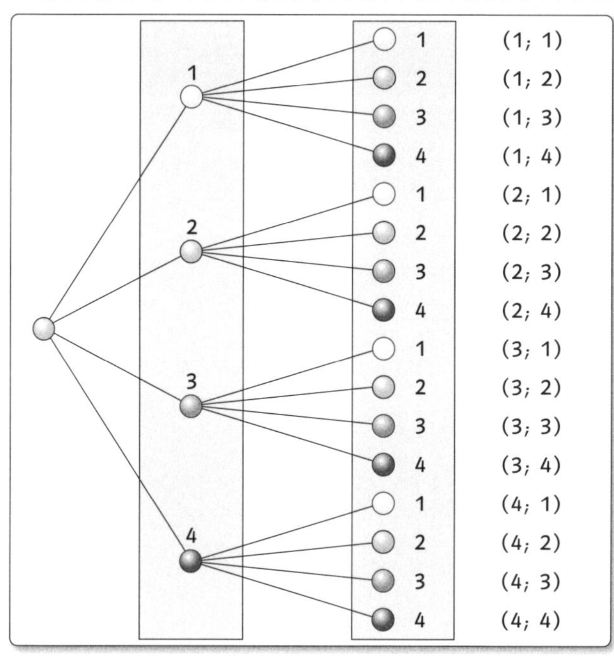

b) $P((3,3)) = \frac{1}{4} \cdot \frac{1}{4} = \frac{1}{16}$

7 a) 27 Möglichkeiten b) $P(r,r,r) = 1/3 \cdot \frac{1}{3} \cdot \frac{1}{3} = \frac{1}{27}$

Seiten und Winkel eines Dreiecks, Dreiecksformen

Zu Seite 39

1

	Seitenlängen	Winkelgrößen	Dreiecksform
a)	a = 6,5, b = 7,4, c = 3,5 cm	$\alpha = 62°, \beta = 90°, \gamma = 28°$	rechtwinklig
b)	a = 6,0, b = 6,0, c = 6,0 cm	$\alpha = 60°, \beta = 60°, \gamma = 60°$	gleichseitig
c)	a = 6,0, b = 6,0, c = 8,5 cm	$\alpha = 57°, \beta = 80°, \gamma = 43°$	kein spezielles Dreieck
d)	a = 5,5, b = 6,5, c = 4,5 cm	$\alpha = 45°, \beta = 45°, \gamma = 90°$	rechtwinklig, gleichschenklig

2 a) möglich

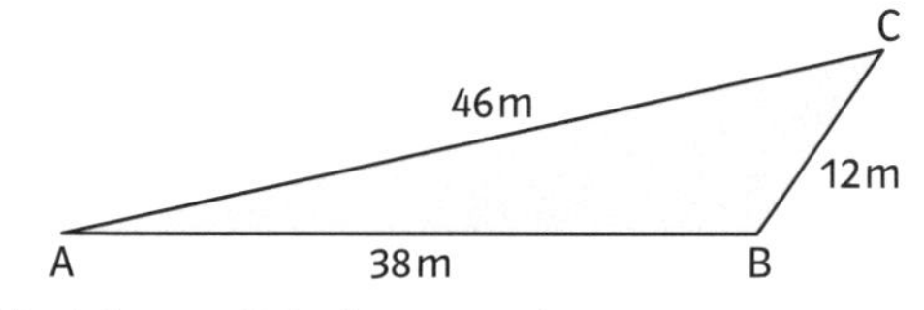

b) nicht möglich, da $a + c < b$
c) nicht möglich, da $b + c = a$

Zu Seite 40

3 a) länger b) größere c) 180°

4

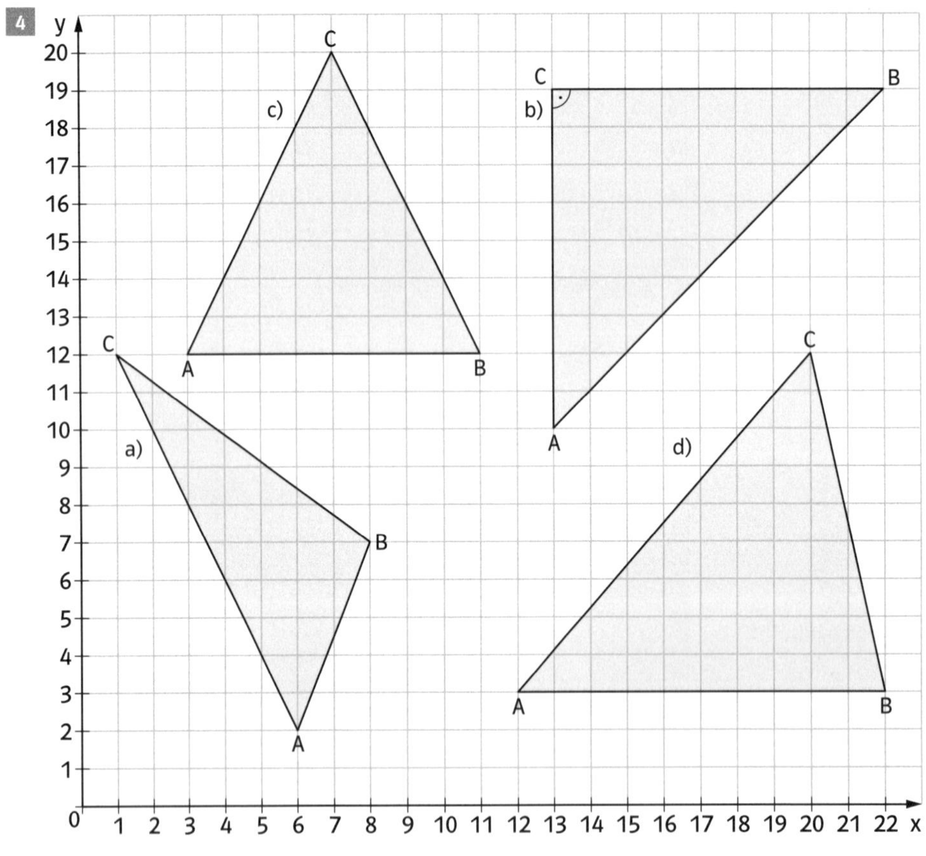

a) Dreieck mit stumpfem Winkel
b) gleichschenklig, rechtwinklig
c) gleichschenklig
d) Dreieck mit spitzen Winkeln

Dreiecke konstruieren

Zu Seite 41

1 a)

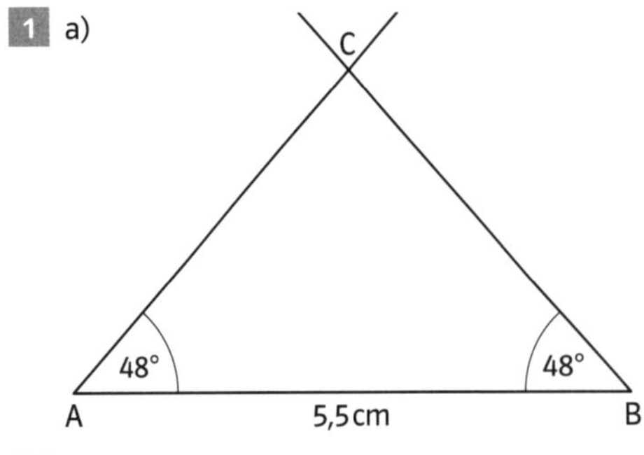

b)

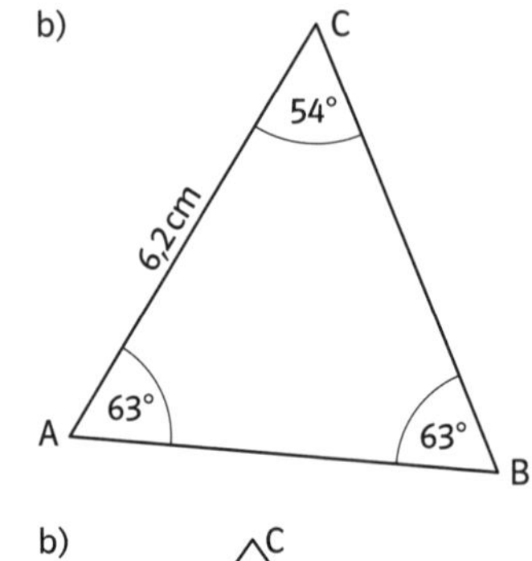

2 a)

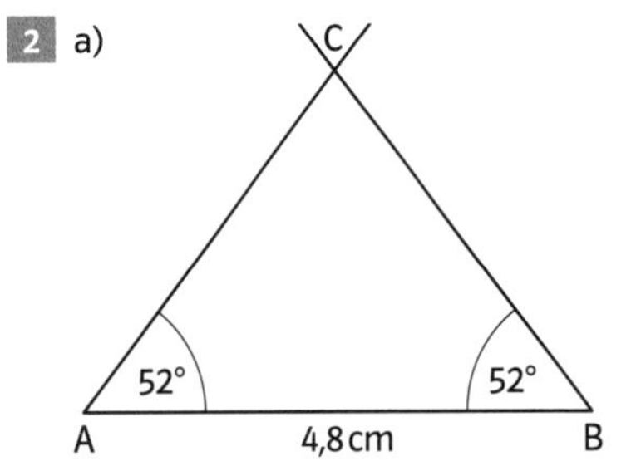

b)

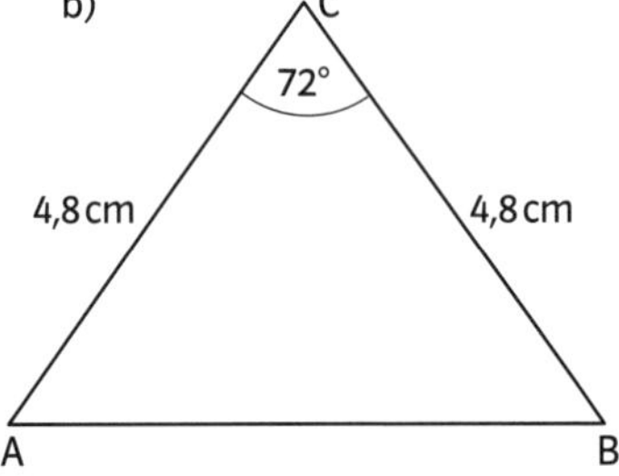

3 a)

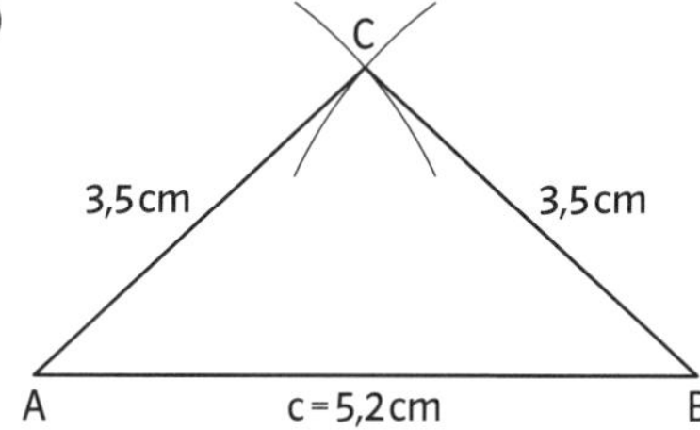

4 In einem gleichseitigen Dreieck sind alle Winkel 60°.

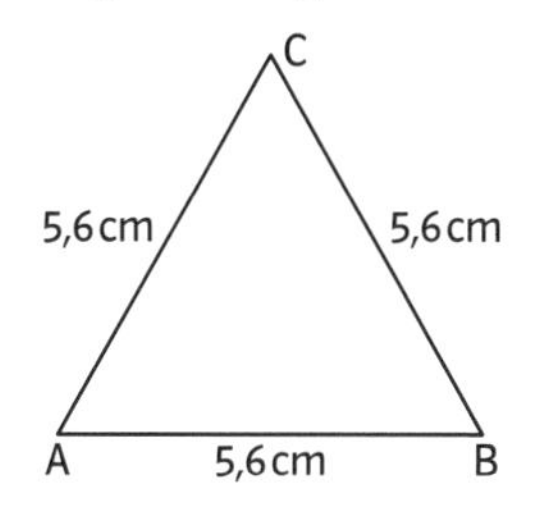

Winkel am Dreieck

Zu Seite 42

1

	α	β	γ
Dreieck I	45°	55°	80°
Dreieck II	60°	25°	95°
Dreieck III	100°	20°	60°

2

	a)	b)	c)	d)
α	50°	35°	50°	75°
β	50°	35°	50°	75°
γ	80°	110°	80°	30°

3 Die Dreiecke sind gleichseitig, da jede Seite dem Radius des Kreises entspricht.
Die Winkel sind also jeweils 60°.

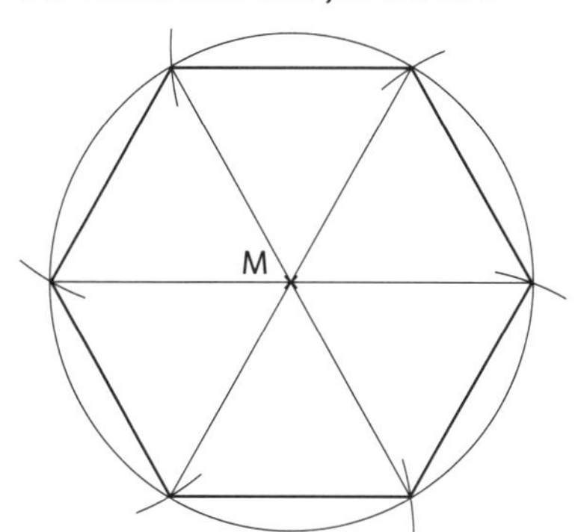

Winkel berechnen

Zu Seite 43

1 a) $\alpha = 32°, \beta = 48°$ b) $\delta = 108°$ c) $\delta = 35°$

2 $\alpha = 80°, \beta = 35°$

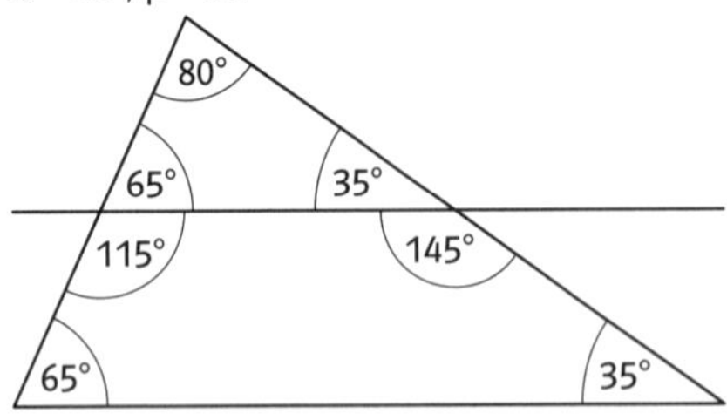

3 a) $\alpha = 150°, \beta = 70°$ b) $\alpha = 56°, \beta = 68°, \gamma = 12°$

Umkreis eines Dreiecks

Zu Seite 44

1 a)

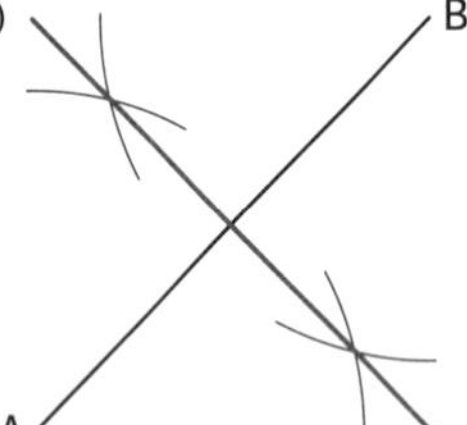

b)

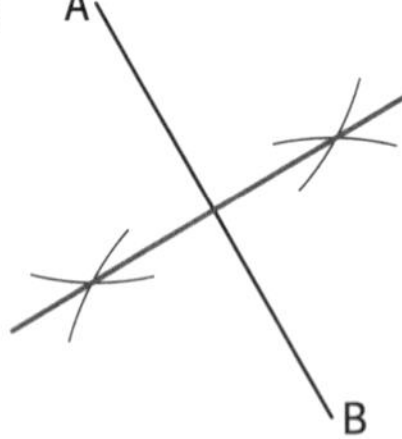

2 a)

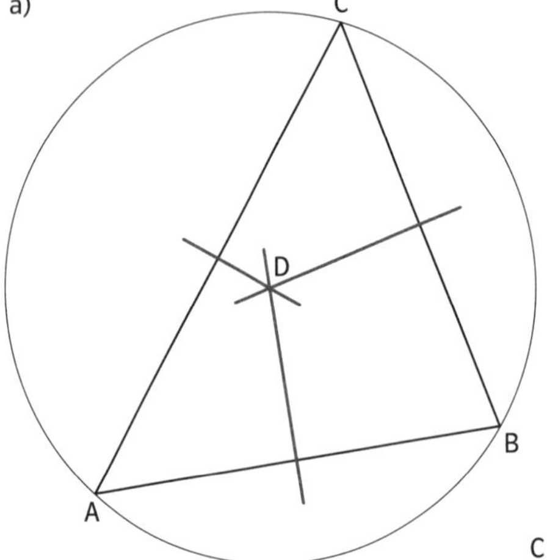

b)

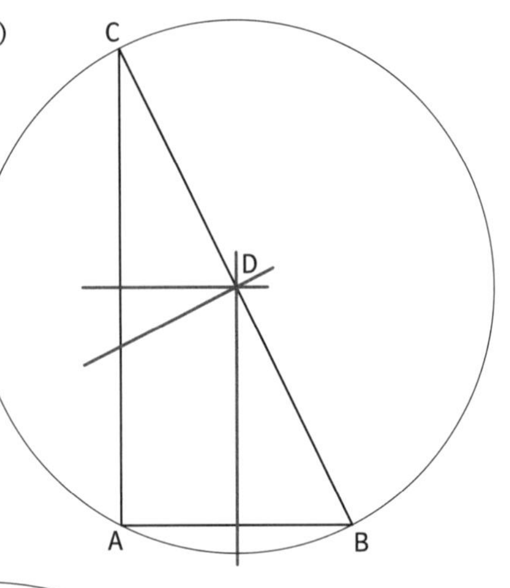

c) A C D B

Zu Seite 45

3 a) M(5|10) b) M(3|10) c) M(7|8) d) M(10|8)

4 a)

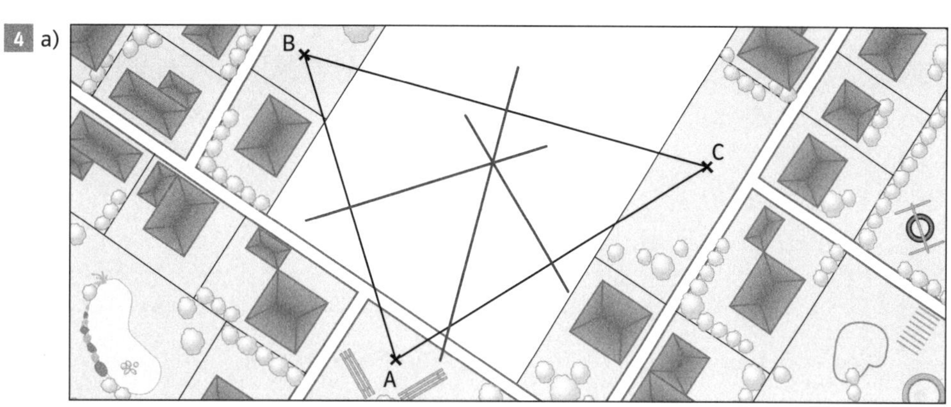

b) 4,2 cm entsprechen 8 400 cm = 84 m

5 Wähle drei beliebige Punkte auf dem Kreis, der Kreis ist somit der Umkreis des Dreiecks. Der Schnittpunkt der Mittelsenkrechten des Dreiecks ist der Mittelpunkt des Kreises.

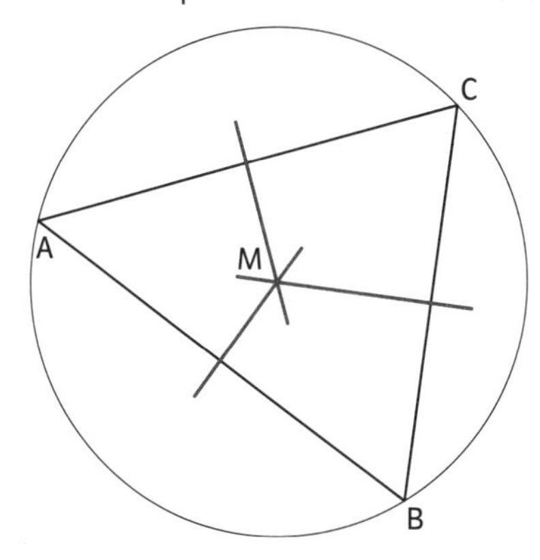

Inkreis des Dreiecks

Zu Seite 46

1 a) b) c)

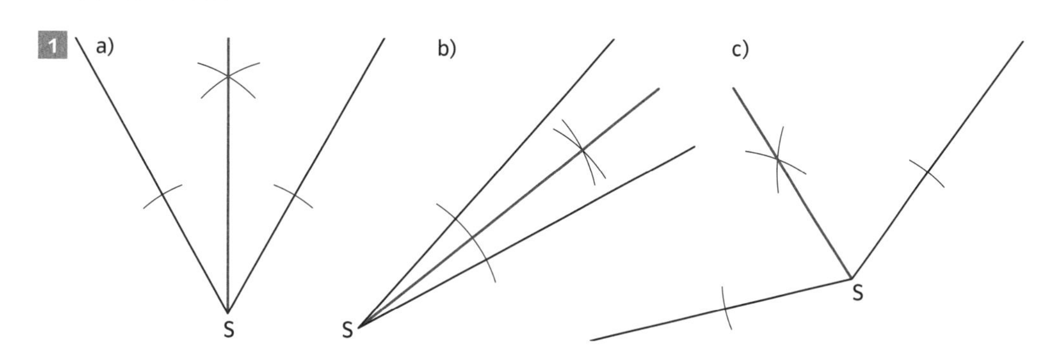

2

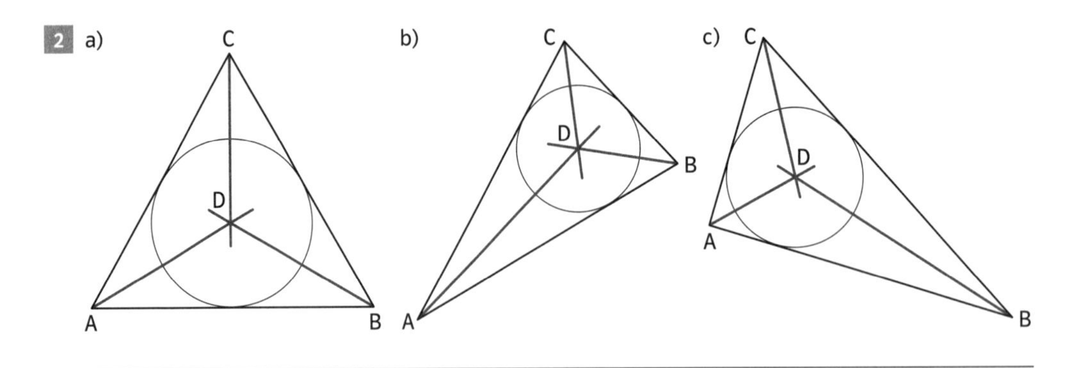

Zu Seite 47

3 a) r = 1,7 cm b) r = 1,7 cm

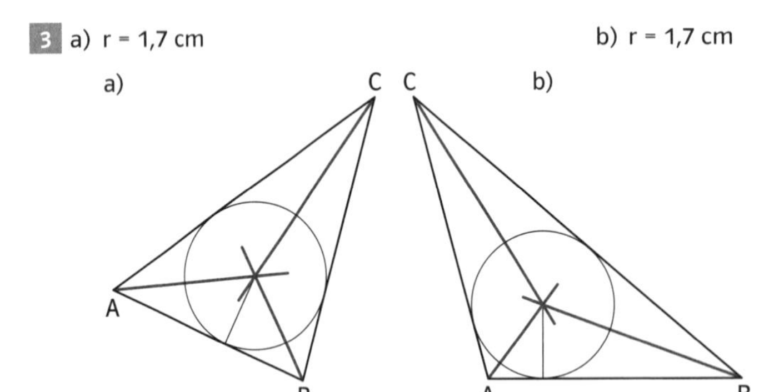

4 a) M(6|6) r = 5 b) M(11|6) r = 5

5

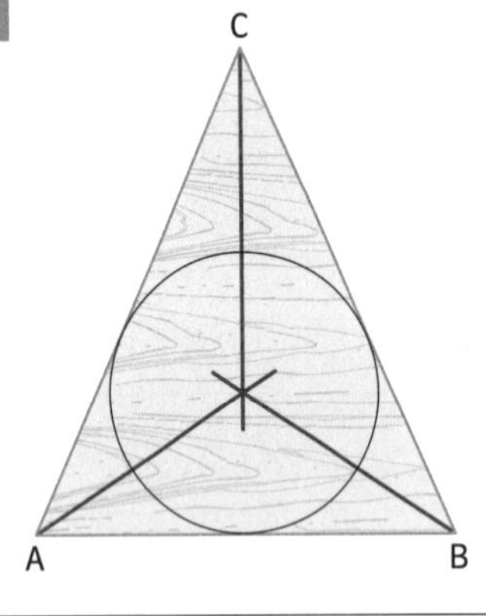

Radius: 2,5 cm entsprechen 125 cm
125 cm = 1,25 m

Höhen und Seitenhalbierende eines Dreiecks

Zu Seite 48

1

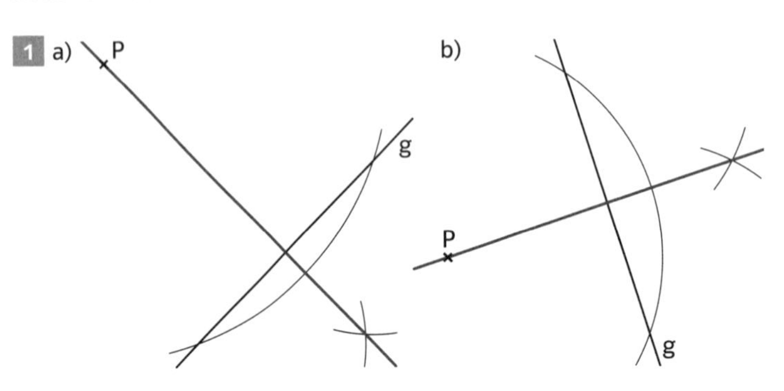

2 a) b) c)

Der Höhenschnittpunkt liegt
a) bei einem Dreieck mit spitzen Winkeln innerhalb des Dreiecks,
b) bei einem rechtwinkligen Dreieck auf einem Eckpunkt,
c) bei einem Dreieck mit einem stumpfen Winkel außerhalb des Dreiecks.

3 a) S(4|4)
b) Die Seitenhalbierenden verlaufen alle durch das Dreieck, da der Mittelpunkt der gegenüberliegenden Seite zwischen den Eckpunkten liegt. Daraus folgt, dass der Schnittpunkt der Seitenhalbierenden, also der Schwerpunkt, innerhalb des Dreiecks liegt.

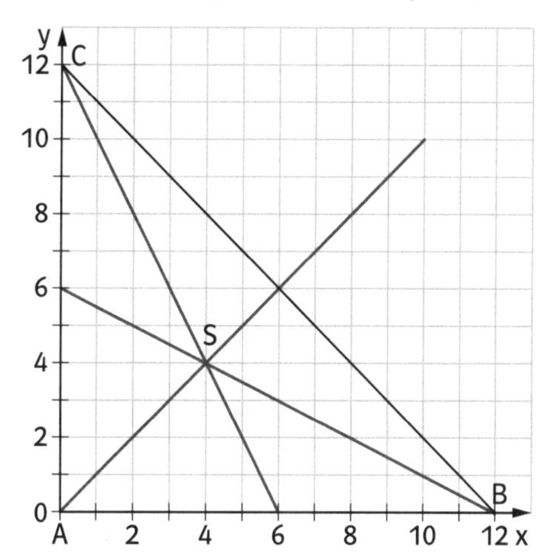

Terme aufstellen und berechnen

Zu Seite 49

1 a) y = 26 · x + 19 y = Gesamtlänge (in Meter), x = Anzahl der IC-Reisezugwagen
b) y = 45 · x + 87 y = Gesamtgewicht (in Tonnen)
c)

Anzahl der Wagen	1	4	7	9	10
Gesamtlänge (m)	45	123	201	253	279
Gesamtmasse (t)	132	267	402	492	537

d) 6 Wagen e) 5 Wagen

2 a) Johanna: ein Fruchtsaft und eine unbekannte Anzahl von Müsliriegeln
Sara: 2 Tafeln Schokolade, 2 Gummibärchenpackungen, unbekannte Anzahl von Kaugummi
Paul: Fruchtsaft, gleiche Anzahl von Schokolade und Müsliriegel, andere Anzahl von Schokoriegel
b) 0,30 · x + 0,80 · y + 1,90 · z
c) 10 – (3 · 1,20 + 2 · 0,90 + 0,80 · x) 5 Müsliriegeln

Zu Seite 50

3

	a)	b)	c)	d)	e)	f)	g)
x	x + 25	x − 27	65 + (x + 12)	400 − (x − 20)	5x + 25	x : 5 − 6	100 + x − (x + 10)
35	60	8	112	385	200	1	90
125	150	98	202	295	650	19	90
45	70	18	122	375	250	3	90

4 a) $56 - x$ b) $x \cdot 20 + 8$ c) $x : 12$ d) $56 + x$
e) $50 - 2 \cdot x$ f) $5 \cdot x \cdot 30$ g) $60 + 4 \cdot x$ h) $x : 9 + 10$
i) $5 \cdot x - 5$

5 a) Subtrahiere von einer Zahl 10.
b) Addiere zu 60 das Doppelte einer Zahl.
c) Dividiere eine Zahl durch 20.
d) Subtrahiere von dem Achtfachen einer Zahl 35.
e) Subtrahiere von dem Produkt von einer Zahl mit 25 die Zahl 50.
f) Subtrahiere von 100 eine Zahl.
g) Addiere zum Quotienten von einer Zahl und 7 die Zahl 11.

Gleichungen lösen

Zu Seite 51

1 a) 22 Wagen b) $11 \cdot x + 19 = 129$; 10 Wagen
c) Term: $11 \cdot x + 20$, entweder 97 m mit 7 Wagen oder 108 m mit 8 Wagen

2 a) L = {3} b) L = {15} c) L = {4} d) L = {4}
e) L = { } f) L = {48}

3 a) $5 \cdot x + 12 = 67$ L = {11}
b) $x \cdot 8 - 14 = 58$ L = {9}
c) $12 \cdot x - 5 = 91$ L = {8}
d) $x : 5 = 12 : 6$ L = {10}

Ungleichungen lösen

Zu Seite 52

1 a) 3, 5, 10 b) 2, 5, 7 c) 4, 2, 1 d) 5, 3, 1

2

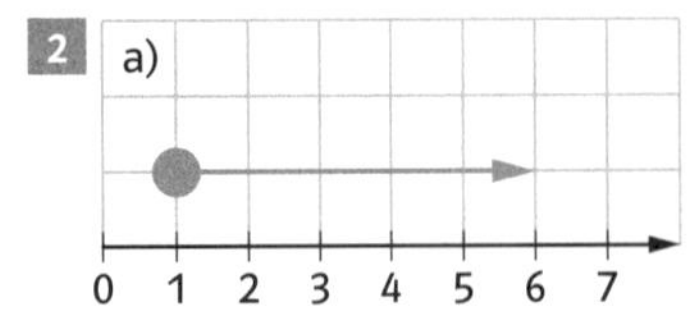

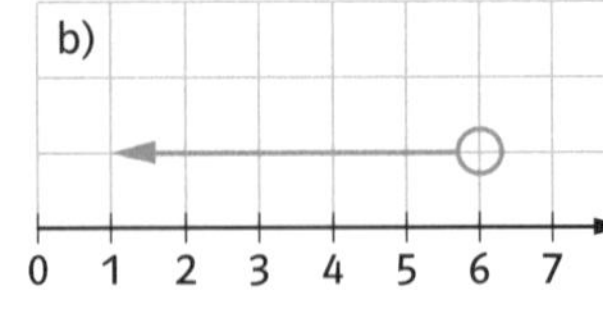

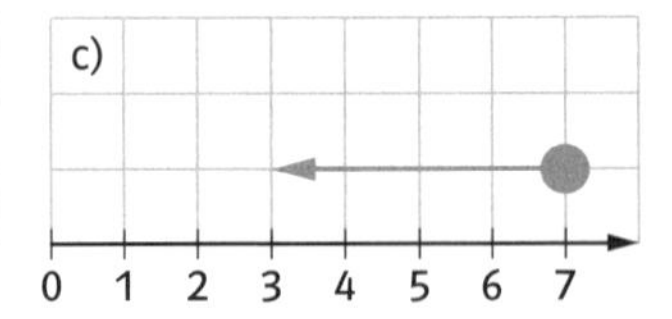

3 a)

	Umgeformte Ungleichung	Lösungsmenge L
a)	$x < 4$	{0, 1, 2, 3}
b)	$x \geq 4$	{4, 6, 8, 10, 12}
c)	$x > 20$	{25, 30}
d)	$x > 6$	{7, 8, 9, 10}
e)	$x \geq \frac{9}{4}$	{3, 4, 5, 6}
f)	$x > 31\frac{1}{4}$	{40, 50, 60}
g)	$x \leq 6\frac{2}{3}$	{0, 2, 4, 6}

4 a) $4 \cdot x + 4 \cdot x + x + x = 10 \cdot x \leq 40$
$x \leq 4$
Antwort: Die Seiten des Rechtecks dürfen höchstens a = 4 cm und b = 16 cm lang sein.
b) $x + 8 + x + 8 + x + x = 4 \cdot x + 16 \geq 60$
$x \geq 11$
Antwort: Die Seiten des Rechtecks sind mindestens a = 11 cm und b = 19 cm lang.

Sachprobleme erfassen und erkunden

Zu Seite 53

1 3 Aufgaben:
Wann muss Johanna das Haus verlassen, wann kommt sie wieder zu Hause an?
Wie teuer wird die Fahrt insgesamt?
Wie lang ist die Strecke, die längs der Weser zurückgelegt wird?

Informationen:
Bahn: 7,50 € + 4,50 € = 12 €
Übernachtungen + Frühstück: 3 · 15,80 € = 47,40 €
Lunchpakete: 3 · 4,60 € = 13,80 €
Gesamt: 73,20 €

Abfahrt- und Ankunftszeiten: Abfahrt: 10 Minuten vor 11.40 Uhr
Ankunft: 10 Minuten nach 17.39 Uhr

Johanna muss am Sonntag spätestens um 11.30 Uhr das Haus verlassen und kommt am Mittwoch frühstens um 17.49 Uhr dort wieder an.

Strecke:		
	Hann. Münden – Karlshafen:	42,0 km
	Karlshafen – Beverungen:	10,4 km
	Beverungen – Höxter:	13,6 km
	Höxter – Holzminden:	9,8 km
	Holzminden – Bodenwerder:	26,4 km
	Bodenwerder – Hameln:	23,6 km
	Hameln – Rinteln:	26,8 km
	Rinteln – Porta Westfalica:	15,4 km

Johanna legt am ersten Tag von Hann. Münden aus 42,0 km zurück, am zweiten Tag 33,8 km, am dritten Tag 50,0 km und am vierten Tag bis Porta Westfalica 42,4 km. Insgesamt legt sie 168,2 km zurück.

Sachprobleme durch Schätzen, Messen und Überschlagen lösen

Zu Seite 54

1 Fußballfeld
minimal: $90\ m \cdot 45\ m = 4050\ m^2 = 40\,500\,000\ cm^2$
maximal: $120\ m \cdot 90\ m = 10\,800\ m^2 = 108\,000\,000\ cm^2$

18 Monate = 540 Tage

Annahme: ein Mann rasiert sich einmal pro Tag
Minimale Fläche: $220\ cm^2 \cdot 7 \cdot 540 = 831\,600\ cm^2$
Maximale Fläche: $280\ cm^2 \cdot 7 \cdot 540 = 1\,058\,400\ cm^2$

Die Behauptung kann nicht stimmen. Selbst die minimale Fußballfeldfläche ist ca. 40-mal so groß wie die maximale rasierte Fläche.

2 -

Zu Seite 55

3 Flächeninhalt einer Spielkarte: 6 cm · 9,5 cm = 57 cm²
Beispielhaftes Klassenzimmer: 10 m · 7 m · 3 m
1 Wandlänge wird nicht berechnet: Fenster, Tafel, Heizkörper, Tür, Schrank
Fläche der Wände (ohne Decke und Fußboden): ca. 70 m² = 700 000 cm²
Anzahl der Karten: 700 000 : 57 ≈ 12 280

4 Die Rechnung beruht nur auf Schätzungen:
Ein Einheitsquadrat entspricht einem Flächeninhalt von 10 000 m². Die Fläche erstreckt sich über ungefähr 33 Einheitsquadrate. Bei einer mittleren Wassertiefe von 6,5 m ergibt sich ein Volumen von 2 145 000 m³ Wasser.

Sachprobleme durch Rückwärtsrechnen lösen

Zu Seite 56

1

Preis für einen Müsliriegel		Preis für sechs Müsliriegel		Preis für sechs Müsliriegel und die Flasche Saft		Gesamtbetrag mit Rückgeld
0,35	· 6 →	2,10	+ 85 →	2,95	+ 2,05 →	5,00
0,35	← : 6	2,10	← − 85	2,95	← − 2,05	5,00

2

Preis für eine Batterie		Preis für vier Batterien		Preis für vier Batterien und die Zeitschrift		Gesamtbetrag mit Rückgeld
0,95	← : 4	3,80	← − 3,50	7,30	← − 2,70	10,00

3 a) Frage: Wie lange muss Nina sparen, um sich den MP3-Player leisten zu können?
Antwort: 13 Monate
b) Frage: Wie viel Liter Benzin benötigt der Wagen pro 100 Kilometer?
Antwort: 6 Liter/100 km

Oberflächeninhalt von Quader und Würfel

Zu Seite 57

1 A_O = 164 cm²

2 a) A_O = 664 cm² b) A_O = 864 cm² c) A_O = 502 cm²

Zu Seite 58

3 A_O = 72 cm²

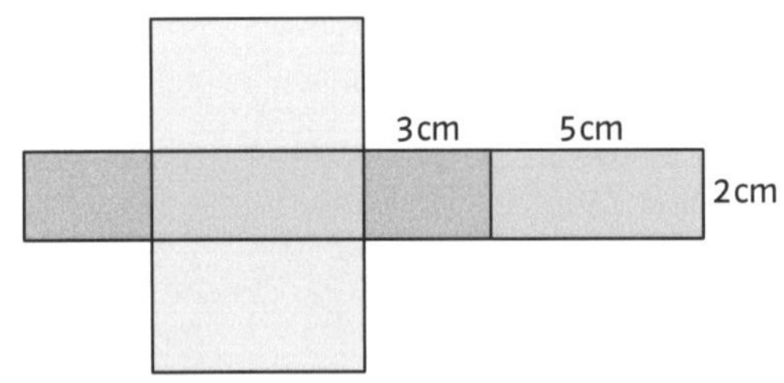

4 a) A_O = 184 cm² b) A_O = 1350 m² c) A_O = 1060 m²

Volumen vergleichen

Zu Seite 59

1 a) Schachtel A: 54 Schachtel B: 60 Schachtel C: 144
b) größtes Volumen: Schachtel C kleinstes Volumen: Schachtel A

2 In jede Kiste kann das gleiche Volumen geschüttet werden. Das Volumen beträgt 24 cm³.

3 a) Quader A: Kantenlänge 48 cm Quader B: Kantenlänge 44 cm
b)

Länge	12	12	48	16	6	2	2
Breite	2	4	1	3	8	8	1
Höhe	2	1	1	1	1	3	24

Volumeneinheiten

Zu Seite 60

1 Mülltonne: l Stecknadel: mm³ Wassertropfen: mm³ Kochtopf: l
See: m³

2 a) 10 Fruchtsafttüten passen in die Tasche.
b) 100 000 Tüten passen in die Lagerhalle.
c) 10 000 Würfel passen in die Tasche.
d) 1 000 000 000 000 Stecknadelköpfe passen in die Lagerhalle.

3

25 m³ = 25 000 dm³
0,025 m³ = 25 dm³
2,5 m³ = 2500 dm³
25 cm³ = 25 000 mm³
0,250 cm³ = 250 mm³
250 l = 250 dm³

2,5 l = 2500 cm³
25 dm³ = 25 000 cm³
250 ml = 0,250 l
250 cm³ = 0,250 dm³
2500 l = 2,5 m³
2,5 dm³ = 2500 ml

4 a) 22 cm³
45 000 dm³
4 m³
62 dm³
70 hl

b) 6,7 m³
2500 mm³
300 ml
120 l
500 ml

c) 0,095 m³
4000 cm³
2800 ml
4500 l
0,045 l

Volumen von Quader und Würfel

Zu Seite 61

1

	I	II	III
Volumen einer Stange	12 cm³	5 cm³	6 cm³
Anzahl der Stangen	8	9	6
Volumen einer Schicht	96 cm³	45 cm³	36 cm³
Anzahl der Schichten	10	3	6
Volumen des Quaders	960 cm³	135 cm³	216 cm³

2 a) V = 220 cm³ b) V = 125 cm³ c) V = 900 dm³ d) V = 36 cm³
e) V = 126 m³

Oberflächeninhalt und Volumen zusammengesetzter Körper

Zu Seite 62

1 a) $V = 864\ cm^3$ b) $V = 11\,664\ cm^3$ c) $V = 30\,500\ cm^3$

2 a) $A_O = 416\ cm^2$ $V = 384\ cm^3$
b) $A_O = 4300\ cm^2$ $V = 8400\ cm^3$

Sachaufgaben

Zu Seite 63

1 Welches Volumen hat ein Quader?
$V = 1{,}26\ m^3$
Wie viele Quadratmeter Folie werden für einen Stapel benötigt?
Es werden 7,12 m^2 Folie benötigt.

2 Es gibt die folgenden Möglichkeiten bezüglich der Maßzahlen der Schachtel:
12 cm × 2 cm × 1 cm, 8 cm × 3 cm × 1 cm, 6 cm × 4 cm × 1 cm, 6 cm × 2 cm × 2 cm,
4 cm × 3 cm × 2 cm

3 a) Die Halle fasst 8064 m^3 Luft. b) Es müssen 64 260 l Wasser eingefüllt werden.

4 80 000 000 Bundesbürger verbrauchen im Jahr rund 3 504 000 000 m^3 Wasser.

Zu Seite 64

5 Das Schwimmbad fasst 278 608 Liter Wasser.

6 Nein, das abgebildete Aquarium fasst nur 240 Liter.

7 a) Maßstab: 1 cm entspricht 10 cm also 1:10

b) 144 l c) 5 cm

Beilage zum Arbeitsheft Mathematik 6 Mecklenburg-Vorpommern

ISBN 978-3-14-121905-0

Westermann Schroedel Diesterweg Schöningh Winklers GmbH, Braunschweig
www.westermann.de

Zeichnungen: Technische Grafik Westermann (Frau Wohlt), Braunschweig
Satz: media service schmidt, Hildesheim
Druck und Bindung: westermann druck GmbH, Braunschweig

33. S. P. Alliluev, S. S. Gershtein, and A. A. Logunov, Phys. Lett., 18, 195 (1965).
34. V. N. Gribov and I. Ya. Pomeranchuk, Zh. Éksperim. i Teor. Fiz., 42, 1682 (1962).
35. A. A. Logunov, L. D. Soloviev, and A. N. Tavkhelidze, Phys. Lett., 24B, 181 (1967).
36. R. J. Rivers, Phys. Rev. Lett., 22, 85 (1969); L. L. Zhenkovskii, A. I. Lend'el, and R. S. Tutik, Preprint of the ITF 63-30, Kiev (1969).
37. V. I. Akimov, I. M. Drëmin, I. I. Roizen, and D. S. Chernavskii, Nuclear Fusion, 7, 629 (1968).
38. D. Amati, M. Cini, and A. Stanghellini, Nuovo Cimento, 30, 193 (1963); L. Van Hove, Lectures Given at the Cargese Summer School (1963).
39. V. I. Savrin and O. A. Khrustalev, Nuclear Fusion, 8, 1016 (1968); I. V. Andreev and I. M. Drëmin, Nuclear Fusion, 8, 814 (1968).
40. S. Nilsson, Nucleon–Nucleon High-Energy Interaction, International School on Elementary Particle Physics, Herceg Novi (1968).
41. D. Harting et al., Nuovo Cimento, 38, 60 (1965).
42. J. Orear, et al., Phys. Rev., 152, 1162 (1966).
43. W. Galbraith, et al., Phys. Rev., 138B, 913 (1965).
44. F. Cerulus and A. Martin, Phys. Lett., 8, 80 (1964).
45. A. A. Logunov and M. A. Mestvirishvili, Phys. Lett., 24B, 583 (1967).
46. J. V. Allaby, et al., Phys. Lett., 25B, 156 (1967).
47. J. V. Allaby, et al., Phys. Lett., 28B, 67 (1968).
48. G. Cocconi, et al., Phys. Rev., 138B, 165 (1965).
49. G. Manning, et al., Nuovo Cimento, 41, 167 (1966).
50. A. A. Logunov, Nguyen Van Hieu, and O. A. Khrustalev, Nucl. Phys., 50, 295 (1964); A. A. Khelashvili, Soobshch. Akad. Nauk Gruz. SSR, 42, 555 (1966).
51. G. Belletini, Two-Body Intermediate- and High-Energy Collisions, Rapporteur's Talk, Fourteenth International Conference on High-Energy Physics, Vienna (1968).
52. E. W. Anderson, et al., Phys. Rev. Lett., 16, 855 (1966).
53. J. V. Allaby, et al., Phys. Lett., 28B, 229 (1968); C. N. Ankenbrandt, et al., Phys. Rev., 170, 1233 (1968).
54. M. Jacob and S. Pokorski, Nuovo Cimento, 61A, 233 (1969); C. B. Chiu and J. Finkelstein, Nuovo Cimento, 59A, 92 (1969); R. C. Arnold, Phys. Rev., 157, 1292 (1967).
55. G. M. Desimirov and D. Ts. Stoyanov, Preprint of the Joint Institute for Nuclear Research R-1568, Dubna (1964); V. A. Matveev, R. M. Muradyan, and A. N. Tavkhelidze, Preprint of the Joint Institute for Nuclear Research E-2-3498, Dubna (1967); V. G. Kadyshevskii and M. D. Matveev, Nuovo Cimento, 55A, 275 (1968); A. A. Khelashvili, Preprint of the Joint Institute for Nuclear Research R2-4327, Dubna (1969).
56. S. P. Kuleshov, V. A. Matveev, and A. N. Sissakian, Preprint of the Joint Institute for Nuclear Research E2-4455, Dubna (1969).
57. O. A. Khrustalev, V. I. Savrin, and N. E. Tyurin, Preprint of the Joint Institute for Nuclear Research E2-4479, Dubna (1969).

INTERACTION OF PHOTONS WITH MATTER

Samuel C. C. Ting

Recent experimental work on the interaction of photons with matter is reviewed. The review consists of three parts devoted to: 1) the validity of quantum electrodynamics at distances of 10^{-14}–10^{-15} cm (interaction of photons with electrons); 2) leptonic decays of vector mesons (interaction of photons with photon-like particles); 3) photoproduction of vector mesons (interaction of photons with nuclei). The review deals only with the most important and most recent experiments. A list and details of other similar experiments can be found in the Proceedings of the Vienna Conference [1].*

1. VALIDITY OF QUANTUM ELECTRODYNAMICS (QED) AT SMALL DISTANCES: PHOTOPRODUCTION OF e^+e^- PAIRS AT LARGE MOMENTUM TRANSFERS†

Three first-order diagrams (Fig. 1) contribute to pair production. The first two, the Bethe-Heitler (BH) diagrams, can be calculated exactly. The last, the Compton diagram, cannot be calculated exactly, but experimental conditions can be chosen so that its contribution is small. This is possible because the e^+e^- pairs of the BH diagram behave under charge conjugation as two photons (C = +1) and the BH cross section varies rapidly with angle (proportional to $\theta^{-6}, \ldots, \theta^{-8}$), whereas the Compton contribution has the charge properties of one photon (C = −1) and the Compton cross section decreases more slowly with angle (proportional to θ^{-3}).

If a symmetrical detector with a small opening angle is used, the interference between the BH and Compton amplitudes is eliminated and the contribution from the pure Compton diagram is reduced to a few percent.

For symmetrical pairs with angles $\theta_+ = \theta_- \leq 10°$ the momentum transfer to the recoil nucleus is limited to $|q| \simeq E\theta^2 \leq 100$ MeV/c, while the mass of the virtual electron propagator $|t| \approx \sqrt{2}E\theta \simeq 1000$ MeV/c. Thus, under these conditions a heavy nuclear target can be used with relatively small form factor corrections. The yield increases as Z^2 and the measured cross section of e^+e^- pair production can thus be compared with the predictions of QED at a momentum transfer of 1 GeV/c.

The DESY-MIT‡ group has reported new results on pair production with a precision of ± 5% and up to a pair invariant mass of 1 GeV/c^2. The experiment was conducted on the 7.5 GeV synchrotron and the

* The author of the review is the director of a broad program of experiments on the Hamburg electron synchrotron (DESY). Most of the review is devoted to photoproduction experiments — Editor's note.

† This part of the review deals very briefly (on the basis of DESY results) with only one aspect of the test of QED at small distances — the photoproduction of e^+e^- pairs at large momentum transfers. A full review of the situation in QED can be found in: Electromagnetic Interactions and the Structure of Elementary Particles [Russian translation], Mir, Moscow (1969) — Editor's note.

‡ DESY is the Deutsches Elektronen-Synchrotron in Hamburg. MIT is Massachusetts Institute of Technology, U.S.A. Later on we encounter SLAC — Stanford Linear Accelerator Center, U.S.A. — Editor's note.

Department of Physics and Laboratory for Nuclear Science, Massachusetts Institute of Technology. Deutsches Elektronen-Synchrotron, Hamburg. Translated from Problemy Fiziki Élementarnykh Chastits i Atomnogo Yadra, Vol. 1, No. 1, pp. 131-170, 1970.

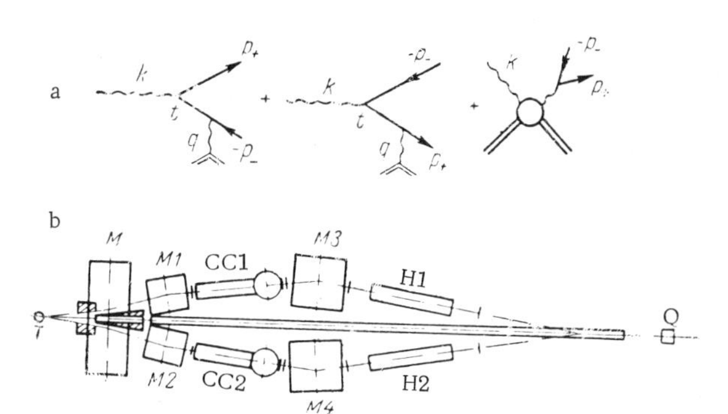

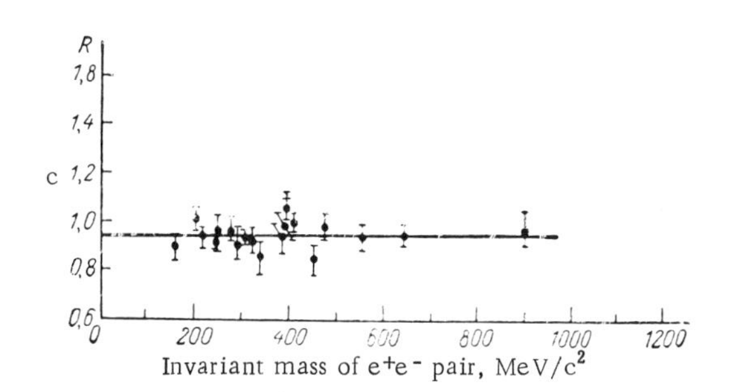

Fig. 1. Wide angle e^+e^--pair photoproduction experiment: a) lower-order diagrams (first two — Bethe-Heitler, third — Compton); b) setup of DESY-MIT experiment; T) target, M-M4) spectrometer magnets, CC1, CC2) Cerenkov counters, H1, H2) hodoscopes, Q) quantameter; c) main result of DESY-MIT experiment — ratio of measured pair yield to theoretical yield (R).

pair production angle was restricted to $\leq 7.7°$. A carbon target, four high-resolution Cerenkov counters for rejection of pions, and fast electronics to reduce the number of chance coincidences were used. The results of the experiment, based on 400–1000 events for each pair mass, are in good agreement with the predictions of QFD. These results, together with the results obtained on the same apparatus earlier, are shown in Fig. 1.

According to Kroll, the correct application of the Ward identities of higher orders requires a modification of the BH cross section in the form $\sigma_{exp}/\sigma_{BH} = 1 \pm (m/\Lambda)^n$, $n \geq 4$, where Λ is a cutoff parameter which can be used as a standard of comparison between various experiments to test QFD. Following Kroll's expression, the DESY-MIT group obtained from their experimental data $\Lambda > 2$ GeV at a 68% confidence level ($n = 4$). This experiment now provides the most accurate test of QED.

2. LEPTONIC DECAYS OF VECTOR MESONS*

Having obtained some experimental evidence on the validity of quantum electrodynamics at momentum transfers up to 1 GeV/c, we turn to experiments on the detailed understanding of the nature of light.

The rest of the discussion will be devoted to experiments designed to measure the coupling between photons and vector mesons ("massive" photons, which have the same quantum numbers as the photon J = 1, C = −1, P = −1, but with nonzero rest mass).

§ 2.1. PURPOSE OF INVESTIGATIONS

The investigation of leptonic decays of vector mesons is of interest for the following reasons [1].

1. Measurement of the branching ratio (B)

$$B = \frac{V^0 \to e^+ + e^-}{V^0 \to \text{All modes}}$$

is the only direct way to determine the coupling constant between γ rays and the vector mesons ρ, ω, and φ.

The coupling constant γ_V is related to the partial decay width $\Gamma(V^0 \to e^+e^-)$ in the following way:

$$\frac{\gamma_V^2}{4\pi} = \frac{\alpha^2}{12} \cdot \frac{m_V}{B \cdot \Gamma_{tot}} = \frac{\alpha^2}{12} \cdot \frac{m_V}{\Gamma(V^0 \to e^+e^-)} .$$

Exact determination of γ_V or $\Gamma(V^0 \to e^+e^-)$ enables us to:

a) calculate the $\varphi - \omega$ mixing angle from the relationships

$$\operatorname{tg}\theta = \frac{m_\omega}{m_\varphi}\operatorname{tg}\theta_Y = \frac{m_\varphi}{m_\omega}\operatorname{tg}\theta_B;$$

*In recent years there has been a wide range of investigations of leptonic decays of vector mesons: in π-meson beams, in colliding e^+e^- beams, and γ-ray beams. This review is devoted to γ-ray investigations. For the other questions reference will have to be made to the original papers — Editor's note.

$$\frac{\Gamma(\omega \to e^+ e^-)}{\Gamma(\varphi \to e^+ e^-)} = \frac{m_\omega}{m_\varphi} \operatorname{tg}^2 \theta_Y = \frac{m_\varphi}{m_\omega} \operatorname{tg}^2 \theta = \frac{\gamma_\varphi^2}{\gamma_\omega^2};$$

b) check Weinberg's first sum rule, which is based on the current mixing model and predicts

$$\frac{1}{3} \cdot \frac{m_\rho^2}{\gamma_\rho^2} = \frac{m_\omega^2}{\gamma_\omega^2} + \frac{m_\varphi^2}{\gamma_\varphi^2};$$

c) compare the experimental data with the quark model calculations of Dar and Weisskopf [2]:

$$\Gamma(\rho \to e^+ e^-) = 5.8 \text{ keV/c}^2, \ \Gamma(\varphi \to e^+ e^-) = 0.95 \text{ keV/c}^2,$$

etc. In particular, the quantity γ_V appears directly in the vector dominance model approximation, which relates the electromagnetic current $J_\mu(x)$ of the hadrons to the phenomenological fields $\rho_\mu(x)$, $\omega_\mu(x)$, and $\varphi_\mu(x)$ of the vector mesons in the following way:

$$J_\mu(x) = -\left[\frac{m_\rho^2}{2\gamma_\rho} \rho_\mu(x) + \frac{m_\omega^2}{2\gamma_\omega} \omega_\mu(x) + \frac{m_\varphi^2}{2\gamma_\varphi} \varphi_\mu(x)\right]. \tag{1}$$

Thus, in the vector dominance model a knowledge of γ_V is essential to an understanding of the electromagnetic form factors of nucleons and pseudoscalar mesons, and to an understanding of the electromagnetic decays of mesons. For example, in the simple vector dominance model the pion form factor is found from the diagram [3],

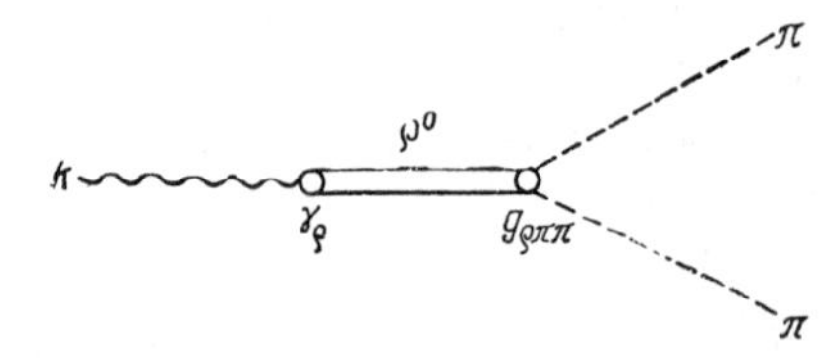

giving the result $F_\pi(k^2) = (g_{\rho\pi\pi}/2\gamma_\rho) \times (m_\rho^2/m_\rho^2 - k^2)$, where $g_{\rho\pi\pi}$ can be calculated from the decay width $\rho \to \pi\pi$, and the normalization condition $F_\pi(0) = 1$ gives $g_{\rho\pi\pi} = 2\gamma_\rho$.

2. A comparison of the probabilities of the decays $V^0 \to e^+e^-$ and $V^0 \to \mu^+ + \mu^-$* gives us a direct check of μ, e universality in the time-like region at high momentum transfers $q^2 = m_V^2 > 0$. These reactions can be used to detect any differences in the form factors of the electron $F_e(q^2)$ and the muon $F_\mu(q^2)$ in a domain which is inaccessible to elastic scattering experiments ($\mu + p \to \mu + p$ and $e + p \to e + p$, where $q^2 < 0$) or low momentum transfer experiments, like measurement of the gyromagnetic ratios $(g-2/2)_e$ and $(g-2/2)_\mu$.

3. An investigation of the e^+e^- mass spectrum from reactions

$$\gamma + C \to C + V^0 \quad (V^0 \to e^+ e^-)$$

and

$$e^+ e^- \to \pi^+ \pi^- \quad \text{or} \quad e^+ e^- \to K^+ K^-$$

is the best way to determine the mass m_V and the width Γ_V of the vector mesons,† since the background contribution to the peak corresponding to $V^0 \to e^+e^-$ can be calculated exactly.

4. It follows from (1) that the photoproduction cross section of vector mesons can be related directly to the vector meson nucleon cross sections [4] by the relationship

* Questions of μ, e universality are not discussed in the review.

† The opinion of the author, who did not include the reaction $\pi^- p \to n e^+ e^-$ in this list, is not generally accepted – Editor's note.

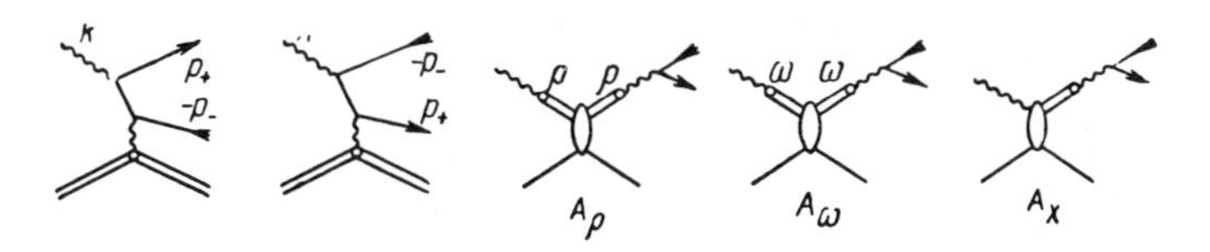

Fig. 2. Feynman diagrams for the process of photoproduction of pair (2). The first two are BH diagrams.

$$\sigma(\gamma + A \to B + C) = \sum_V \frac{\alpha\pi}{\gamma_V^2} \sigma_{tot} \ (V + A \to V + A).$$

We will discuss the physical sense of this situation later.

§ 2.2. EXPERIMENTAL RESULTS: BRANCHING RATIOS OF $\rho \to e^-e^+$, $\omega \to e^-e^+$ AND OBSERVATION OF COHERENT INTERFERENCE BETWEEN ρ AND ω DECAYS

The DESY-MIT group carried out an experiment to determine the branching ratios of $\rho \to e^-e^+$ and $\omega \to e^-e^+$, and also investigated the rare, previously unobserved, phenomenon of coherent interference of $\rho \to e^-e^+$ and $\omega \to e^+e^-$ in the photoproduction of vector mesons. Interference was revealed by the results of measurement of the yield of e^+e^- pairs from the reaction

$$\gamma + \mathrm{Be} \to \mathrm{Be} + V^0(\rho, \omega) \qquad \qquad (2)$$
$$\qquad\qquad\qquad\qquad \hookrightarrow e^+e^-.$$

In the region of values of mass $m_{e^+e^-} \simeq m_\rho \simeq m_\omega$ five diagrams (Fig. 2) make a contribution to reaction (2). The first two are BH diagrams, the contribution of which can be calculated [5]; the next two are diffractive production diagrams, where the e^+e^- pairs are the result of leptonic decays of diffractively produced ρ and ω mesons. The last diagram represents the e^+e^- pairs produced by the decay of vector mesons formed as a result of one-pion exchange (OPE henceforth), or incoherent production, or from both these processes.

The total amplitude of e^+e^- pair production is equal to the sum of the amplitudes:

$$A = A_{\mathrm{BH}} + A_\rho + A_\omega + A_X. \qquad (3)$$

The aim of the described experiment was to measure the contribution of $A = A_\rho + A_\omega$ to e^+e^--pair production and to investigate interference between these two amplitudes.

Physical Sense. The e^+e^- pair yield associated with the amplitudes of coherent ρ and ω production is determined in the following way [6]:

$$A_\rho = g_{\gamma\rho} A(\rho A \to \rho A)\, g_{\gamma\rho} \frac{1}{m_\rho^2 - m^2 - i m_\rho \Gamma_\rho} A(\gamma \to e^+e^-);$$
$$|A|^2 = g_{\gamma\rho}^4 |A(\rho A \to \rho A)|^2 \left(\frac{m_\rho}{m}\right)^4 \left| \frac{\sqrt{\delta(m)}}{m_\rho^2 - m^2 - i m_\rho \Gamma_\rho} + \frac{g_{\gamma\omega}^2}{g_{\gamma\rho}^2} |R| e^{i\varphi} \frac{1}{m_\omega^2 - m^2 - i m_\omega \Gamma_\omega} \right|^2, \qquad (4)$$

where $g_{\gamma\rho}$ and $g_{\gamma\omega}$ are the coupling constants between vector mesons and photons; Γ_ρ and Γ_ω are the resonance widths; $\delta(m)$ is a possible correction for the yield from the mass shape, equal to unity for the normal Breit-Wigner form, and to $(m_\rho/m)^4$ in the remaining cases; $(A_\omega A \to \omega A)/(A_\rho A \to \rho A) = |R| e^{i\varphi}$, where φ is the relative phase of ρ- and ω-meson production. Despite the fact that the absolute value of the ω amplitude is probably small, the contribution of interference of the ρ and ω amplitudes may be fairly large near the ω peak.

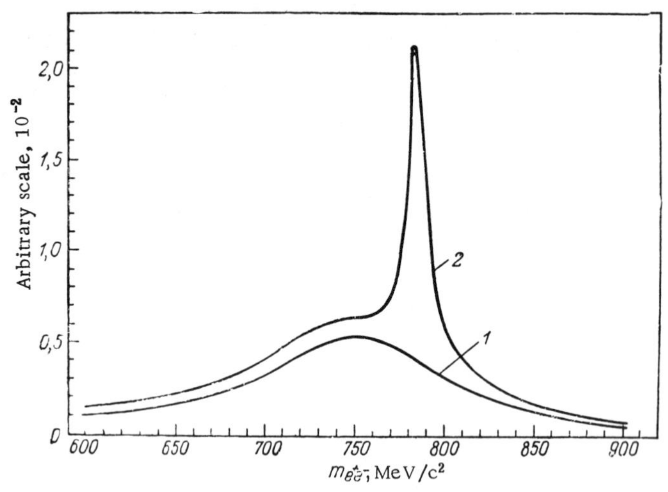

Fig. 3. Spectra of effective masses from $\rho \to e^+e^-$ decays (curve 1) and from their interference with $\omega \to e^+e^-$ decays (curve 2), calculated from Eq. (4) with the following values of parameters [7]: $\varphi = 0°$, $\delta(m) = 1$; $m_\rho = 765$ MeV/c²; $\Gamma_\rho = 128$ MeV/c²; $m_\omega = 783.3$ MeV/c²; $\Gamma_\omega = 12.2$ MeV/c²; $(\gamma_\rho^2/\gamma_\omega^2) = 1/9$.

The contribution of interference of the ρ and ω amplitudes to the $\rho \to e^+e^-$ spectrum for reasonable values of the parameters [7] is shown in Fig. 3. It can be seen that for $\varphi = 0°$ (equal phases of ρ and ω meson production amplitudes) in the mass range close to the ω meson (750–800 MeV), the ascent of the curve due to interference is approximately twice as great as the corresponding contribution from the decay of the ρ meson alone.

It has been a puzzling feature for a long time that none of the experiments on $\rho \to e^+e^-$ photoproduction showed the expected rise in the curve [8]. This can be attributed partially to the poor statistics. For instance, the DESY data for $\rho \to e^+e^-$ decays had 12 events close to the ω meson mass. Since the expected peak was not observed in any of the experiments, however, it would seem that our views on vector dominance require radical revision.

The aim of the described experiment was to investigate the expected peak with large statistics (3000 events) and good mass resolution ($\Delta m = \pm 5$ MeV), and to compare the results of the measurements with the predictions of the vector dominance model.

Design of Experiment. It follows from Eq. (3) that in the design of an experiment to detect the narrow peak the following circumstances have to be taken into account.

a. Counting rate. For the attainment of the most efficient event-counting rate the contribution of QED, A_{BH}, must be kept small. Since coherent diffractive production of vector mesons on nuclei is given by the formula

$$N_\rho \sim \frac{d\sigma}{d\Omega} \sim A^{1.7} p^2 e^{at},$$

where $t = (k - p_+ - p_-)^2$, $a = a_0 A^{2/3}$, whereas $N_{BH} \sim |A_{BH}|^2 \sim Z^2 |G_E(q^2)|^2 p^{-2} \theta^{-7}$, then for reduction of the BH background the relationship

$$\frac{N_\rho}{\sqrt{N_\rho + N_{BH}}} = \frac{A^{1.7} p^2 e^{at}}{(p^2 A^{1.7} e^{at} + cZ^2 |G_E(q^2)|^2 p^{-2} \theta^{-7})^{1/2}}, \quad c = \text{const}. \tag{5}$$

has to be a minimum. In addition, since the final state of the e^+e^- pair in the BH diagrams behaves under the charge conjugation operation (C) like a two-photon system, while the charge properties of the e^+e^- pair in the other diagrams are similar to those of one photon, then it follows from the invariance against charge conjugation that in the case of symmetrical detection of e^+e^- pairs the interference term between the BH diagram and the other diagrams will disappear.

b. Mass resolution. Figure 3 shows that if the interference term is to be found the mass resolution has to be comparable with the small ω resonance width ($\Delta m = \pm 5$ MeV). For a prescribed target thickness multiple scattering and bremsstrahlung loss of e^+e^- pairs in the target are the main effects affecting the mass resolution. Hence, to obtain good resolution the variables p and X are chosen so that the deviation due to multiple scattering

$$\langle \theta_s \rangle = \frac{21\,\text{MeV}}{p\beta\,(\text{MeV})} \sqrt{\frac{X}{X_0}} \tag{6}$$

and the loss due to bremsstrahlung

$$W = \frac{1}{p} \cdot \frac{X}{X_0 \ln 2} \left(\ln \frac{p}{p_1} \right)^{\left(\frac{X}{X_0 \ln 2} - 1 \right)}$$

are as small as possible. Here (X/X_0) is the target thickness in radiation lengths, and p, p_1 is the momentum of the electrons before and after the bremsstrahlung act.

c. Incoherent contributions and one-pion exchange contributions. To facilitate the direct comparison of experimental data and theory the contributions of incoherent diffractive processes ($\sigma^i_{\rho,\omega}$) and the contributions of OPE (σ^0_ω) must be small. In fact, these terms are sufficiently small when

$$\sigma^0_\omega \sim E^{-1,6} \text{and } \sigma^i_{\rho,\omega} \to 0\ (\theta \to 0^\circ). \tag{7}$$

A combined analysis of expressions (5)-(7) leads to the following optimum conditions for distinguishing the narrow ρ, ω peak from the electrodynamic background and for limiting the undesirable contributions from the incoherent terms: $A = 9$ (Be), $7° \leq \theta \leq 9°$.

Experimental Apparatus. Figure 4 illustrates the new spectrometer (top view) designed for the described experiment. It has the following properties which are essential to experiments of this kind.

a. Counting rate. To obtain the maximum number of events (20 times greater than previously [8]) the scintillation counters, which determine the acceptance, are placed after four large deflecting magnets. The maximum field in the magnets is 18 kG. None of the counters is seen directly from the target. Corrections for dead time and chance coincidences in this experiment are less than 1.2%.for a current of 12 mA circulating in the synchrotron.

b. Pion rejection. Since the branching ratios of leptonic decays of ρ and ω mesons have values of the order of 10^{-5}, then an experiment to accuracy of one percent requires pion rejection of 10^7:1 or even better. This is achieved by using four large-aperture threshold Cerenkov counters and two shower counters. The total pion rejection obtained in the system during calibration was 10^{10} :1. This pion rejection was determined by comparing the triggering T(3C) from three Cerenkov counters with the triggering T(4C) from four Cerenkov counters. Since each counter has an efficiency of more than 99% for electrons and a measured pion rejection of 10^3 :1, the equality T(3C) = T(4C) indicates that the pion contamination in this experiment is much less than one percent.

c. Mass resolution. The effect of multiple scattering and bremsstrahlung on the resolution is reduced by using a thin beryllium target and placing all the counters in the distal part of the spectrometer after all the magnets. The use of small hodoscopes and an analysis* of 202,500 hodoscope combinations to determine the kinematic parameters of the e^+e^- pair resulted in a mass resolution $\Delta m = \pm 5$ MeV. The use of vacuum pipes and helium bags in the spectrometer reduced the count of the apparatus without a target to $\leq 10\%$.

Experimental Procedure and Consistency Checks. Numerous checks were made during the course of the experiment to ensure that the spectrometer behaved as designed and that all the systematic errors were correctly assessed. We cite the following examples.

a. To keep radiative corrections and bremsstrahlung losses constant the ratio k/k_{max} was fixed throughout the experiment by the values $k = 2p = 2 \times 2.56$ GeV and $k_{max} = 7.00$ GeV. The mass spectrum was obtained by varying the opening angle of the pair.

* The analysis was performed on an electronic computer operating "in line" with the experiment—Editor's note.

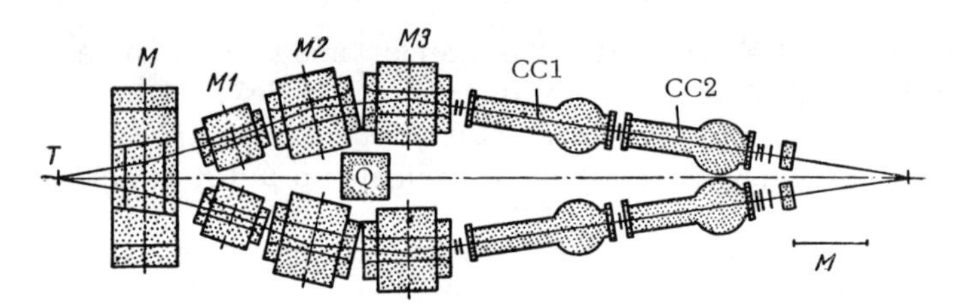

Fig. 4. Diagram of new spectrometer designed for the search for interference of $\rho \to e^+e^-$ and $\omega \to e^+e^-$: T) target; M, M1, M2, M3) magnets, CC1, CC2) Cerenkov counters; Q) quantameter.

b. Two sets of counters of different size were used to ensure that the acceptance of the spectrometer was not limited by the edges of the magnet or the shielding: the change in the counting rate agreed to within ± 3% with the expected change due to the different size of the counters.

c. The smallness of the contribution of second-order processes was checked by measuring the variation of the yield with target thickness. We found that to within ± 3% (1000 events) the yield of the reaction increased linearly with increase in target thickness from 0.5 to 3 cm Be.

d. The high degree of pion rejection required in this experiment, and also the dead time of the Cerenkov counters and the chance coincidences in the main triggering channel (not all completely independent), were determined by means of a two-dimensional trigger system $T_{i,j}$, where i = 2, 3, 4 is the number of Cerenkov counters in the system, and j = 4, 5, 6, 10 nsec is the resolving time between the two arms of the apparatus. The condition $T_{3j} = T_{4j}$ denotes the absence of pions and chance coincidences in the trigger.

e. To compensate for any possible asymmetry in the spectrometer and to reduce the interference contribution of the BH and Compton diagrams to zero, half of the data were obtained with one polarity of the spectrometer, and the other half with the opposite polarity.

f. The absolute normalization of the detecting system and the mass resolution were tested by measuring the yield from QED at production angles of 4°. The results, given in Fig. 5, contain 4000 events and are in good agreement with the QED predictions as regards shape (in the range $\Delta m = \pm 5$ MeV) and absolute normalization. This agreement means that the desired mass resolution is obtained and all the systematic corrections are introduced.

g. To eliminate the effect of any systematic errors in the acceptance calculations we used three different settings of the spectrometer ($\theta_\pm = 7.5, 8.4, 8.8°$, $p_\pm = 2.560$ MeV). The acceptance maximum was found either on the ascending side (8.8°) or the descending side (8.4°) of the expected peak. The data at 7.5° were used to construct the curve in the $\rho \to e^+e^-$ region.

Summary and Conclusions. The results of the measurements were corrected for yield without the target, loss due to bremsstrahlung, dead time, chance coincidence, etc.

In the analysis of the data the expected BH contribution must be subtracted from the total number of events recorded in each interval of the histogram. This contribution is calculated by using the elastic form factors measured for Be^9 and the corrections for inelasticity (of the order of 5%) calculated in accordance with the Drell–Schwartz sum rules [9, 10]. The total BH contribution close to the ω peak is $\simeq 40\%$ of the total yield in the experiment. The mass spectrum obtained after subtraction of the BH contribution and consideration of the production mechanism (4) is shown in Fig. 6. The graph shows that there is a distinct rise of the curve close to the ω-meson mass. This corresponds in general features to the predictions of the vector dominance model (4) (see Fig. 3).

To compare the spectrum (Fig. 6) with Eq. (4) we first subtract the contribution of the term $|A_X|^2$, which, according to the estimates of [11], is 5% of the term $|A_\rho|^2$. We use the following values: $m_\rho = 765$, $\Gamma_\rho = 130$, $m_\omega = 783.4$, $\Gamma_\omega = 12$ MeV/c^2. The unknown parameters are $(\gamma_\omega^2 \sigma_{\rho N}/\gamma_\rho^2 \sigma_{\omega N})$ and φ. The result of the approximation is shown in Fig. 7 and Table 1.

To avoid systematic errors the approximation is made in the mass range $m > m_c$, where m_c is a cutoff parameter, chosen so that a three percent change in absolute normalization does not significantly

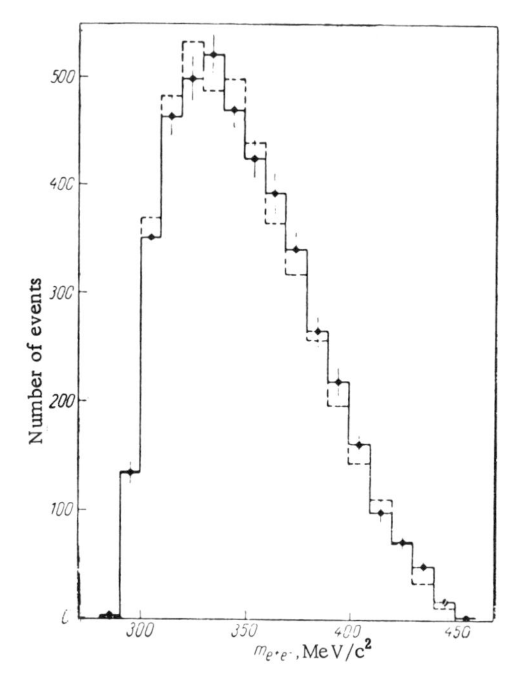

Fig. 5. Results of measurements (points and continuous curve) and calculation from QED (broken curve) for e^+e^- pair yield with $\theta_\pm = 4°$, $p_\pm = 2560$ MeV/c. The agreement confirms the correctness of the calculations of the absolute efficiency and mass resolution.

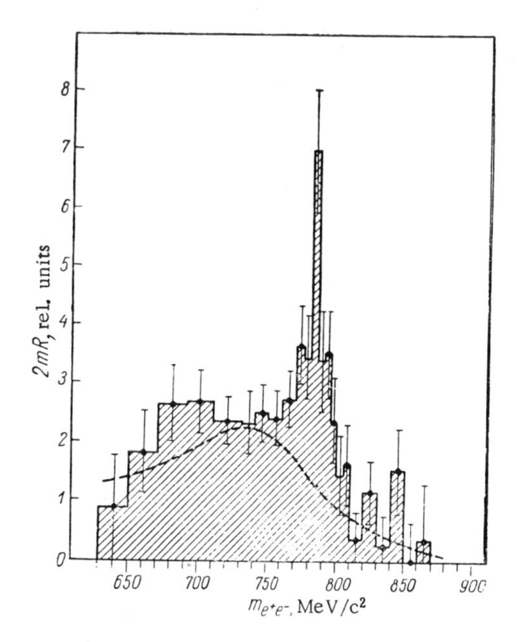

Fig. 6. Effective mass spectrum of e^+e^- pairs obtained from experimental yields after subtraction of BH contribution (about 40% of yield in region of ω peak). The dashed curve corresponds to ρ-meson decays alone.

affect the results of the approximation. The results are sensitive to a choice of the value of Γ_ω and are fairly insensitive to the choice of values of φ and Γ_ρ in the region $|\varphi| < 60°$ and $100 < \Gamma_\rho < 140$ MeV/c².

In approximation of the data by Eq. (4) we used two forms of $\delta(m)$, viz., $\delta(m) = 1$ and $\delta(m) = (m_\rho/m)^4$. As Table 1 and Fig. 7 show, the results are insensitive to the precise form of $\delta(m)$. The errors, indicated in Table 1, are due mainly to systematic errors and uncertainties in the parameters (Γ_ω, etc.).

Table 1 also gives the ratio (C) of the number of $\rho \to e^+e^-$ events expected from our measurements of the branching ratio of $\rho \to e^+e^-$ [8] and the ρ-meson production cross section [11] to the result of the fit. The value $C = 1.0 \pm 0.2$ indicates complete agreement between these experiments.

In conclusion we note that our data, as distinct from all previous measurements, reveal a rapid rise of the spectrum in the region of ω, superimposed on the peak due to leptonic decays of the ρ meson. If we assume further that $\sigma_{\omega N} = \sigma_{\rho N}$ then our value of $\gamma_\rho^2/\gamma_\omega^2$ is close to that measured with colliding beams [12]. Thus, the mass spectrum is reasonably consistent with the predictions of the vector dominance model.

Table 2 gives the results of two experiments with colliding beams, and also the earlier results obtained by the DESY-MIT group and the Harvard group.*

For the values of the coupling constant when the ρ meson is on the mass shell we obtain

$$\frac{\gamma_\rho^2}{4\pi} = \frac{\alpha^2}{12} \cdot \frac{m_\rho}{\Gamma_{\rho \to ee}} = 0.52^{+0.07}_{-0.06}\,.$$

*The first result in the world for $\rho \to e^+e^-$ decays ($B \cdot 10^5 = 5.3 \pm 1.1$) was obtained and published in Dubna in a joint paper by the Joint Institute for Nuclear Research and the Physics Institute, Academy of Sciences of the USSR: M. A. Azimov et al., Phys. Lett., 24B, 349 (1967); and R. G. Astvatsaturov et al., Phys. Lett., 27B, 45 (1968). Table 2 shows that this result has been brilliantly confirmed — Editor's note.

TABLE 1. Values Obtained for Parameters by Two Methods of Approximation of Data in Fig. 6 by Expression (4)

Method of approximation	Results of approximation $\frac{\gamma_\omega^2 \sigma_{\rho N}}{\gamma_\rho^2 \sigma_{\omega N}}$	φ, deg	C
$\delta(m) = 1$; $m_c = 700$ MeV/c^2.	12±4	22±25	1,0±0,2
$\delta(m) = \left(\frac{m_\rho}{m}\right)^4$; $m_c = 650$ MeV/c^2 . .	10,7±3	11±20	1,0±0,2

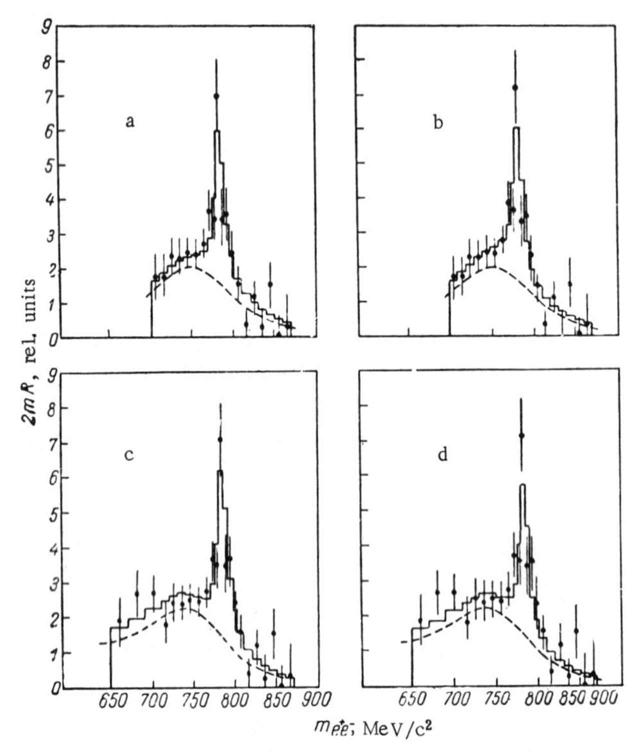

Fig. 7. Comparison of obtained data (points) with calculations (histogram) from Eq. (4) with different assumptions: a) $\delta(m) = 1$, $m_c = 700$ MeV/c^2, $\varphi = 0°$; b) $\delta(m) = 1$, $m_c = 700$ MeV/c^2, $\varphi = 22°$; c) $\delta(m) = (m_\rho/m)^4$, $m_c = 650$ MeV/c^2, $\varphi = 0°$; d) $\delta(m) = (m_\rho/m)^4$, $m_c = 650$ MeV/c^2, $\varphi = 11°$. Changeover from $\delta(m) = 1$ to $\delta(m) = (m_\rho/m)^4$ corresponds to introduction [19] of the Ross–Stodolsky or Kramer–Uretsky correction. The dashed curve corresponds to decays of ρ mesons alone.

For the case where the photon is on the mass shell we use the relationship

$$\frac{\gamma_\rho^2}{4\pi} = \frac{3\Gamma_0}{m_0}\left(1 - \frac{4m^2\pi}{m_c^2}\right)^{-3/2},$$

and from the π meson form factor we obtain

$$\gamma_\rho^2/4\pi = 0{,}53 \pm 0.04.$$

TABLE 2. Results of Measurements of $\rho \to e^+e^-$ Decays

Group	m_ρ, MeV/c²	Γ_ρ, MeV/c²	$B \cdot 10^5$	$\Gamma_{\rho \to e^+ e^-}$, keV/c²
Novosibirsk	754±9	105±20	5,0±1,0	—
Havard	—	97±20	5,8±1,2	—
Orsay	760±4	112±12	6,54±0,72	7,13±0,51*
DESY-MIT.	—	—	6,4+1,5	—
Averages	759±4,0	108±8,5	6,04±0,150	6,52±0,75†

*According to author's approximation.
† From average B.

TABLE 3. Results of Measurements of $\varphi \to e^+e^-$ Decays

Group	Γ_φ, MeV/c²	$B \cdot 10^4$	$\Gamma_{\varphi \to e^+ e^-}$, keV/c²
DESY-MIT	—	2,9±0,8	—
Orsay	4,2±0,3	3,7±0,3	1,58±0,13*

*Author's approximation.

§ 2.3. EXPERIMENTAL RESULTS: BRANCHING RATIO OF $\varphi \to e^+e^-$

An experiment was performed on the DESY synchrotron by the DESY-MIT group using a precision spectrometer and counting techniques.

1. Normalization and Polarization. The photoproduction of φ mesons on carbon (with 10^4 events) and the decay of $\varphi \to e^+e^-$ (with 40 events) were measured with the same apparatus, which reduced the major systematic errors in the normalization. The experimental apparatus had a mass resolution of about ± 5 MeV/c² for $\varphi \to K^+K^-$ decays, and of about ± 20 MeV/c² for $\varphi \to e^+e^-$ decays. It was found with 10^4 events that photoproduction of φ on complex nuclei is dominated by a diffraction mechanism. Thus, the angular distributions of K^+K^- pairs and e^+e^- pairs have the form

$$W_{KK} = \frac{3}{8\pi} \sin^2\theta^*, \quad W_{ee} = \frac{3}{16\pi} (1 + \cos^2\theta^*).$$

2. QED Pair Contamination in Total Yield. The first-order Bethe–Heitler and $\varphi \to e^+e^-$ diagrams contribute to the reaction $\gamma + C \to C + e^+e^-$. However, since the BH cross section varies rapidly with angle ($\sim \theta^{-6}$), while the $\varphi \to e^+e^-$ cross section varies slowly with angle the signal ($\varphi \to e^+e^-$)/background (BH–e^+e^- pairs) will vary in proportion to θ^3.

To reduce BH background the e^+e^- pairs were measured at large angles of 22°–30°. The BH background under the narrow φ peak is then less than half of the total yield.

3. Since the ρ^0 Meson has a Large Width the $\rho \to e^+e^-$ Yield Contaminates the $\varphi \to e^+e^-$ Spectrum. However, since the φ meson has a narrow width a good mass resolution for e^+e^- pairs will enable one to pick out the φ peak with a small ρ background. An estimate obtained by using the $\rho \to e^+e^-$ spectrum measured with the same apparatus and the measured diffraction cross section (pure imaginary amplitude) showed that the overall contamination under the φ peak due to ρ and (ρ, φ) interference did not exceed 10%.

Figure 8 shows the resultant $\varphi \to K^+K^-$ and $\varphi \to e^+e^-$ spectra. Integrating the area under the spectra, we obtain directly

$$\Gamma(\varphi \to e^+e^-)/\Gamma(\varphi \to K^+K^-) = (5.7 \pm 1.7) \cdot 10^{-4}.$$

Using the ratio $\Gamma(\varphi \to K^+K^-)/\Gamma(\varphi \to \text{all modes}) = 0.473$, we obtain $B = (2.90 \pm 0.80) \cdot 10^{-4}$.

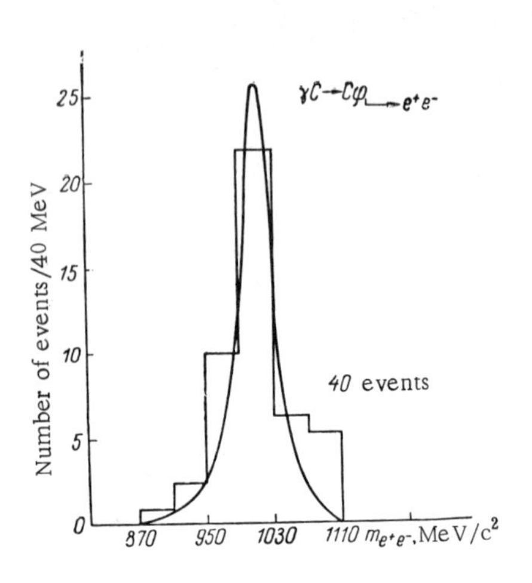

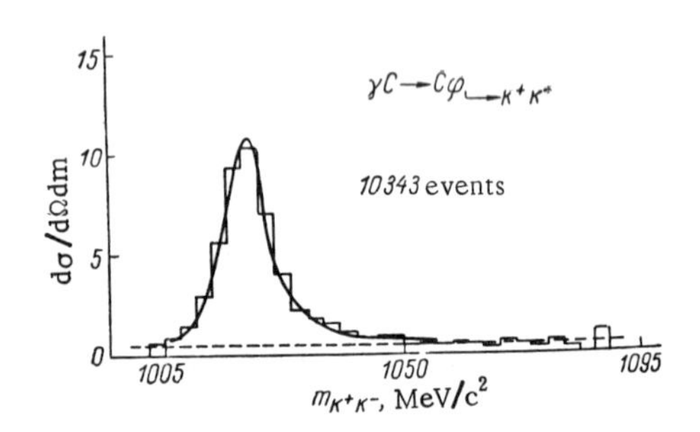

Fig. 8. Mass spectrum of $\varphi \to e^+e^-$ and $\varphi \to K^+K^-$ decays measured by the DESY-MIT group.

TABLE 4. Values of Generalized Mixing Angle θ

Group	θ, deg
CERN	23^{+7}_{-5}
Orsay	35,1±3,5
DESY-MIT	40^{+5}_{-7}*

* The Orsay result was used: $\Gamma_\omega \to e^+e^- =$ 1.04 ± 0.19 keV/c².

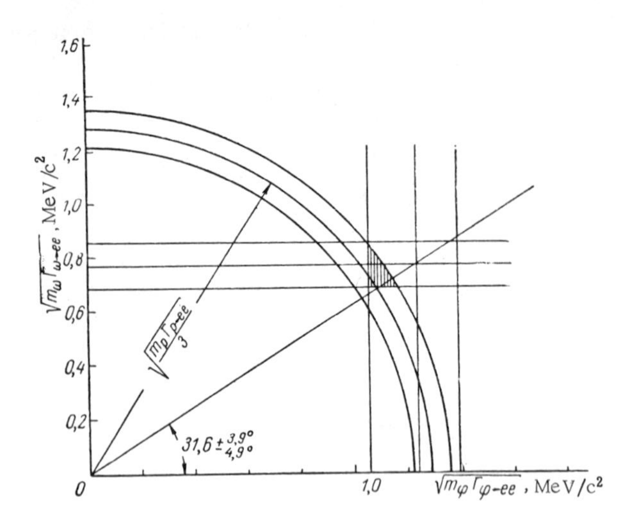

Fig. 9. Comparison of experimental data (average of three investigations) for leptonic decays or vector mesons with Weinberg's first sum rule. Region of data (corrected for errors) hatched.

Table 3 sums up the results of measurements of $\varphi \to e^+e^-$ decays.* Taking into account the new data obtained in Orsay (Γ_{tot} = 4.2 ± 0.3), the results cited above give the first time an accurate value for $\gamma^2_\varphi/4\pi = 3.04^{+1.07}_{-0.66}$.

§ 2.4. SUMMARY

The fair amount of accurate data on leptonic decays of vector mesons (ρ, ω, φ), obtained from production experiments and colliding beam experiments, are in good agreement with each other. The average results of the three independent experiments plotted on the Sakurai circle (based on Weinberg's first sum rule) are shown in Fig. 9. These results give the values of the generalized mixing angle θ shown in Table 4.

A comparison of the average experimental values for $V^0 \to e^+e^-$ with the predictions of various theoretical models is shown in Table 5.

§ 2.5. CONCLUSION

1. Experimental data for $V^0 \to e^+e^-$ agree with the predictions of Weinberg's first sum rule, based on the current mixing model. It should be noted, however, that without any correction for the finite width of the ρ meson the Orsay result differs from the prediction of Weinberg's first sum rule by 1.8 standard deviations.

2. The average values of the partial widths of $V^0 \to e^+e^-$ agree remarkably well with the quark-model calculations of Dar and Weisskopf.

3. The Orsay data and the data obtained by the DESY-MIT group lead to the mixing angle θ = 39° predicted by Das, Mathur, and Okubo [7].

* The first result in the world for $\varphi \to e^+e^-$ ($B \cdot 10^4 = 6.6^{+4.4}_{-2.8}$) was obtained and published in Dubna [with the collaboration of the Joint Institute for Nuclear Research and the Physics Institute, Academy of Sciences of the USSR: R. G. Astvatsaturov et al., Phys. Lett., 27B, 45 (1968)]. The CERN result (6.1 ± 2.6) was published later. These results agree with one another and differ from the results obtained later by DESY-MIT and Orsay on the same standard — Editor's note.

TABLE 5. Comparison of Average Experimental Results for $V^0 \rightarrow e^+e^-$ with Predictions of Various Theoretical Models

Leptonic decays	Average experimental results						Predictions of theoretical models				
							γ_V^{-2}			$\Gamma_{V \rightarrow e^+ e^-}$, keV/c^2	
	B	Γ_{tot}, MeV/c^2	$\Gamma_{V \rightarrow e^+ e^-}$, KeV/c^2	$\frac{\gamma_V^2}{4\pi}$	γ_V^{-2}	Group	SU_3	Sakurai	Das, Mathus, Okubo	Quark model	Mass mixing model
ρ	$(6{,}04 \pm \pm 0{,}50) \times \times 10^{-5}$	$108{,}0 \pm \pm 8{,}5$	$6{,}52 \pm 0{,}75$	$0{,}52 \begin{cases} +0{,}07 \\ -0{,}06 \end{cases}$	9	DESY-MIT Novosibirsk Orsay Harvard	9	9	9	5,7	—
ω	$(6{,}01 \pm \pm 1{,}1) \times \times 10^{-4}$	$12{,}2 \pm 1{,}3$	$0{,}74 \pm 0{,}16$	$4{,}69 \begin{cases} +1{,}24 \\ -0{,}81 \end{cases}$	$1{,}00 \begin{cases} +0{,}21 \\ -0{,}21 \end{cases}$	CERN Orsay	1	0,65	1,21	0,61	0,6
φ	$(3{,}55 \pm \pm 0{,}48) \times \times 10^{-4}$	$4{,}2 \pm 0{,}9$	$1{,}49 \pm 0{,}35$	$3{,}04 \begin{cases} +1{,}07 \\ -0{,}66 \end{cases}$	$1{,}54 \begin{cases} +0{,}43 \\ -0{,}40 \end{cases}$	CERN DESY-MIT Orsay	2	1,33	1,34	0,95	1,2

4. The total widths of vector mesons, determined directly from their leptonic decays, are: Γ_ρ = 108.0 ± 8.5 MeV/c², Γ_ω = 14.0 MeV/c², and Γ_φ = 4.2 ± 0.9 MeV/c². These values are somewhat different from those obtained from the analysis of strong-interaction experiments: Γ_ρ = 90–150 MeV/c², Γ_ω = 12.2 ± 1.3 MeV/c², and Γ_φ = 3.4 ± 0.8 MeV/c².

3. PHOTOPRODUCTION OF VECTOR MESONS

§ 3.1. Purpose of Investigation

Since photons and vector mesons have the same quantum numbers, it is very likely that at high energy and small momentum transfers the reaction $\gamma A \to V^0 A$ has the same features as $\pi A \to \pi A$, viz.:

1) they both conform to the following relationship:

$$\frac{d\sigma}{dt} \quad |J_1(R\sqrt{-t})|^2_{\text{small}\,t} \simeq \exp[a(A, t)t],$$

where a(A, t) is the nuclear density, the value of which depends significantly on the values of t and A;

2) as in the case of πp scattering, the total cross section should either slowly decrease with energy increase or be almost constant;

3) in the case of diffraction scattering the produced vector mesons have the same polarization as the initial photons. Hence, the angular distribution of the K-meson pairs in photoproduction and decay of φ mesons will be proportional to $\sin^2\theta^*$, where θ^* is the angle, in the φ-meson rest system, between the K meson and the recoil nucleus momentum;

4) to study the detailed mechanism by which vector mesons are produced at high-energy and low-momentum transfer on complex nuclei we compare the experimental data with the predictions of the diffraction models of Margolis [13] for the ρ-meson–nucleon scattering cross section $\sigma_{\rho N}$, the cross section for ρ-meson production on hydrogen $|f_0|^2$, and the ρ-meson–photon coupling constant $\gamma_\rho^2/4\pi$.

§ 3.2. Results of Experiment on ρ-Meson Production on Hydrogen

The DESY-MIT group recently [14] published their latest results on the reaction

$$\gamma + p \to p + \pi^+ + \pi^- \tag{8}$$

for forward scattering at collision energy 2.6 to 6.8 GeV and in the two-pion mass range 500–1000 MeV/c².

With a total number of events of about 10^5 the reaction cross sections were measured in energy intervals ΔE_γ = 0.6 GeV and mass intervals Δm = 30 MeV/c².

The aims of this experiment were:

1. To investigate in detail pion-pair production by exact measurements of the dependence of the two-pion spectrum on mass and energy without theoretical assumptions of any kind.

2. To investigate the energy dependence of the ρ-meson production cross section by the simultaneous fitting (to a large number of obtained experimental points) of several parameters: the production cross section, the ρ-meson mass and width, and the total background.

This investigation of ρ-meson production differs appreciably from most earlier ones, which were either limited by statistics or the cross sections were determined from the measured $\pi\pi$ spectra on the basis of arbitrary assumptions regarding the ρ-meson mass and width and on the assumption of zero background [15].

Experiment. The experiment was conducted on the 7.5 GeV DESY electron synchrotron with an average bremsstrahlung intensity of $3 \cdot 10^{10}$ equivalent photons per second. A 60-cm hydrogen target and a pair spectrometer similar to that described above were used. Two threshold Cerenkov counters separated the electrons from the pions; protons were cut off by time of flight. Throughout the experiment the magnetic fields were kept constant to one part in 3×10^4 and the voltages on the counters were constant to within ± 5 V.

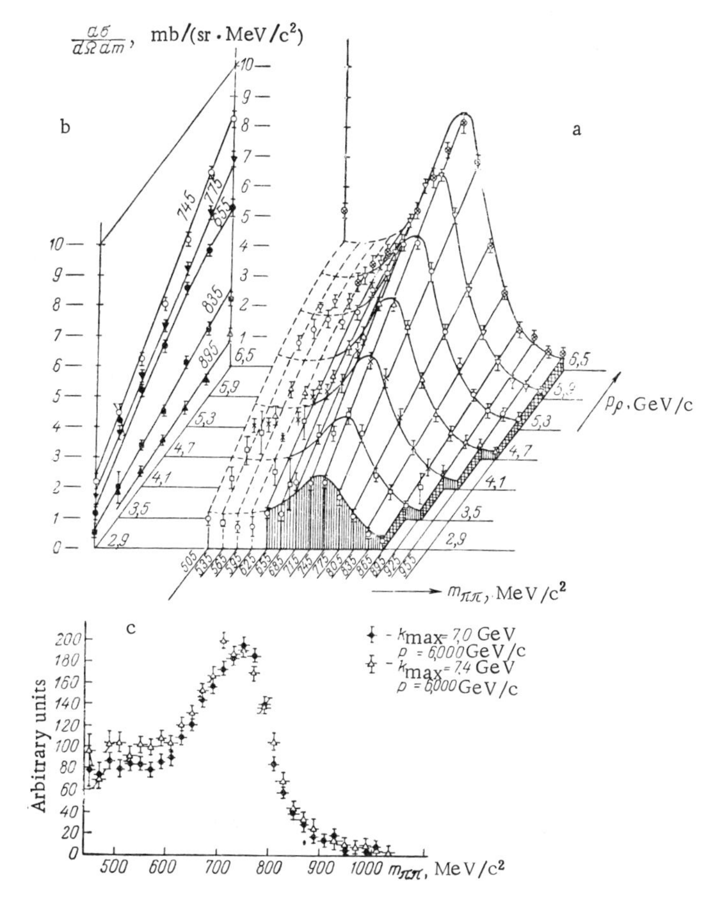

Fig. 10. Experimental results for cross section $d\sigma/d\Omega dm$ of ρ-meson photoproduction of hydrogen (8): a) three-dimensional picture of dependence of cross section on invariant mass of $\pi^+\pi^-$ pair (m) and total momentum of pair in LS (p); curves are result of approximation by expression (9) (background not shown); approximations (10) and (11) gave similar curves (not shown in figure); b) projection of data onto plane (p, $d\sigma/d\Omega dm$) for fixed m; curves are result of approximation by expression $d\sigma/d\Omega dm = p^2 (1 + R/p)^2$; it is quite clear that for a fixed m the data increase more slowly than p^2 ($R > 0$); c) comparison of mass spectra obtained with different maximum energy (k_{max}) of bremsstrahlung spectrum and same average spectrometer momentum (p); the agreement of the spectra above 610 MeV/c² shows that in this region the contributions of inelastic processes are small.

Calibration measurements of ρ-meson production on a carbon target were made every four hours. These showed that the reproducibility was ± 1% throughout the period of measurement. The 22,500 combinations of the hodoscope counters in the magnetic spectrometer gave a mass resolution of $\Delta m/m = \pm 1.7\%$ and a momentum resolution of $\Delta p/p = \pm 3.0\%$. The beam intensity was maintained so that the dead time corrections did not exceed 2%, and accidental coincidences, determined from circuits with different resolving times, were ≤1%. Corrections were introduced for attenuation of the beam in the hydrogen target, for empty-target effects, and for absorption due to nuclear interactions in the hydrogen target, in the target walls, and in the scintillation and Cerenkov counters.

Experimental Results

Figure 10a shows in three-dimensional form the results of measurements of the cross section [16] in relation to the variables m and p. These data allow a direct comparison of the different models. The two main features can be seen in Fig. 10a:

1) The spectra are due mainly to ρ-mesons;

2) In each unit interval of mass from 500 to 1000 MeV/c² the spectrum has a relationship of the type

$$d\sigma/d\Omega\, dm \sim p^2 \left(1 + \frac{R}{p}\right)^2 \text{ for } R > 0,$$

and not a dependence of the form p^2. Such behavior of the spectrum indicates that the $\pi\pi$ forward production spectrum $d\sigma/dtdm$ decreases with increase in energy, like the πN scattering cross section.

Figure 10 b shows typical dependences of the cross section on p for fixed m, obtained by projection of the three-dimensional graph (see Fig. 10a).

Analysis of Results. An analysis of reaction (8) in the region of ρ-meson production is difficult, since there is no theory of broad resonances and the shape of the background is not known. It is for these reasons that Table 6 gives the direct experimental data. To analyze the obtained data we fitted the reaction cross sections, given in Table 6, by the following expressions:

$$\frac{d\sigma}{d\Omega\, dm} = Cg(p)\, 2mR(m) \left(\frac{m_\rho}{m}\right)^4 + BG(p,\ m); \tag{9}$$

$$\frac{d\sigma}{d\Omega\, dm} = Cg(p)\, [2mR(m) + I(m)] + BG(p,\ m); \tag{10}$$

TABLE 6. Data Matrix for ρ-Meson Photoproduction on Hydrogen (8), $d\sigma/d\Omega dm$, mb/(sr · MeV/c^2), as Function of Total Momentum of Pion Pair in LS (p, GeV/c) and Invariant Mass of Pion Pair (m, MeV/c^2)

m \ p	6,5	5,9	5,3	4,7	4,1	3,5	2,9
955	0,51±0,13	0,29±0,06	0,38±0,12	0,25±0,11	—	—	—
925	0,57±0,10	0,42±0,08	0,55±0,10	0,52±0,11	0,17±0,12	—	—
895	1,06±0,11	0,59±0,08	0,64±0,09	0,60±0,09	0,53±0,11	1,02±0,48	—
865	1,32±0,14	1,24±0,09	1,16±0,10	0,87±0,09	0,62±0,09	0,38±0,09	0,26±0,13
835	2,55±0,16	2,20±0,11	2,11±0,12	1,48±0,09	1,39±0,13	0,92±0,12	0,49±0,10
805	4,08±0,19	3,62±0,15	3,20±0,14	2,63±0,11	2,34±0,14	1,43±0,15	1,09±0,20
775	6,86±0,21	6,09±0,17	5,23±0,18	4,33±0,13	3,67±0,16	2,80±0,19	1,70±0,29
745	8,21±0,25	7,42±0,18	6,17±0,20	5,07±0,17	4,24±0,16	3,42±0,23	2,17±0,20
715	7,26±0,22	7,07±0,17	6,08±0,18	5,11±0,19	4,54±0,18	2,79±0,19	2,14±0,15
685	6,29±0,23	5,85±0,17	5,45±0,17	3,96±0,17	3,47±0,20	2,73±0,25	2,17±0,22
655	5,23±0,25	4,84±0,15	4,59±0,18	3,65±0,16	3,19±0,20	3,20±0,51	1,13±0,21
625	4,18±0,32	4,50±0,16	3,84±0,17	2,68±0,15	3,02±0,19	1,08±0,85	1,17±0,14
595	3,13±0,25	4,05±0,23	3,49±0,16	2,34±0,14	2,23±0,18	1,49±0,38	0,73±0,33
565	3,44±0,21	3,21±0,22	3,56±0,20	2,15±0,13	1,67±0,14	2,78±0,69	0,70±0,18
535	2,82±0,26	3,17±0,15	3,24±0,30	2,26±0,15	2,12±0,15	2,21±0,44	0,81±0,09
505	5,17±0,80	2,96±0,17	2,77±0,18	1,33±0,28	2,07±0,16	1,28±0,34	0,95±0,17

$$\frac{d\sigma}{d\Omega\, dm} = Cg(p)\left[2mR(m)\left(\frac{m_\rho}{m}\right)^4 + I(m)\right] + BG(p,\ m); \tag{11}$$

$$R(m) = \frac{1}{\pi}\cdot\frac{m_\rho \Gamma_\rho(m)}{\left(m_\rho^2 - m^2\right)^2 + m_\rho^2\Gamma_\rho^2(m)};$$

$$\Gamma_\rho(m) = \frac{m_\rho}{m}\left[\frac{(m/2)^2 - m_\pi^2}{(m_\rho/2)^2 - m_\pi^2}\right]^{3/2}\Gamma_0;\ g(p) = p^2(1 + R/p)^2;$$

$$I(m) = A\,\frac{m^2 - m_\rho^2}{\left(m_\rho^2 - m^2\right)^2 + m_\rho^2\Gamma_\rho^2(m)};$$

$$BG(p,\ m) = \left(\sum_{i=1}^{4} a_i p^{i-1}\right)\left(\sum_{j=1}^{4} b_j m^{j-1}\right) \geqslant 0.$$

Here g(p) is the energy dependence of the forward production cross section. If $g(p) = p^2$ $(R = 0)$ we have classic diffraction scattering with a constant total cross section. When $g(p) = p^2(1 + R/p)^2$ and $R > 0$ the cross section $d\sigma/dt|_{t=0}$ decreases with increase in ρ-meson energy. The background function BG(m, p) has the form of a product of two polynomials of third order in m and p. The function I(m) (Soeding term [17]) represents an interference of the ρ amplitude with the part of the nonresonance background corresponding to diffractive pair production. R(m) and Γ_ρ (m) are Jackson's relativistic formulas for a p-wave resonance [18].

Equation (9) corresponds to an approximation of the ρ-meson production cross section where the resonance term contains the Ross-Stodolsky coefficient [19], $(m_\rho/m)^4$, and the total background is taken into account. The coefficient $(m_\rho/m)^4$ appears also in the Kramer-Uretsky model [19] and was used at DESY and SLAC to take into account the shifts and distortion of the shape of the mass spectrum in ρ-meson photoproduction. Equation (10) is another common method of fitting the spectrum, when the shift and distortion of the shape are taken into account by the interference term I(m), the value of which (A) is determined by the fit. Equation (11) assumes the existence of both mechanisms.

Fitting was performed in two ways. Firstly, the data contained in Table 6 were approximated by each of the expressions (9)-(11) by using the CERN program MINUIT [20]. The free parameters were the following quantities: a_i, b_j, c, m_ρ, Γ_0, R, and A. The lower limit m_0 used in the approximation of the data was selected in the following way. Figure 10c shows two mass spectra from 460 to 1000 MeV/c^2. One of them was obtained at a maximum bremsstrahlung energy of 7.4 GeV, and the other at 7.0 GeV. The two spectra were obtained with a central spectrometer momentum of 6.0 GeV/c. The graph shows that the spectra coincide above 610 MeV/c^2. From this we can conclude that the inelastic contribution in the region of higher masses is small. The same result was found for the spectra at $K_{max} = 5.55$ and 5.25 GeV. Hence, m_0 was chosen as 610 MeV/c^2. The validity of this choice was checked by varying m_0 in the range $610 < m_0 < 760$ MeV/c^2. The values of m_ρ, Γ_0, C, R, A, and the background function were insensitive to these changes. The background function BG(m, p) was also investigated and it was found that a product of second-order polynomials leads to the same result as the higher-order polynomials described above. The values of m_ρ and Γ_0 obtained in all these fits agreed with the values $m_\rho = 765 \pm$ MeV/c^2 and $\Gamma_0 = 145 \pm 10$ MeV/c^2.

Secondly, for the values of $m_\rho = 765$ MeV/c^2 and $\Gamma_0 = 145$ MeV/c^2, obtained above, the mass spectra in each momentum interval were separately fitted by expressions (9)-(11). In this case, both the cross sections and the background had an arbitrary dependence on the momentum, in contrast to treatment by the first method, where the cross section and background were assumed to be smooth functions of the momentum. The results of these fits were insensitive to the values of m_ρ and Γ_0 in the limits $m_\rho = 765 \pm 10$ MeV/c^2 and $\Gamma_0 = 145 \pm 10$ MeV/c^2, indicated above. Figure 11 shows the dependences of $d\sigma/dt|_{t=0}$ on p, obtained by the described method [21]. It is apparent that the results do not depend on the function (9), (10), or (11), used for fitting. Figure 11 also shows the results of other experiments. It is evident that they are in good agreement with the present experiment. (The data obtained in Cornell by the counter method are not included, since they were analyzed without any subtraction of background and with a fixed width of 120

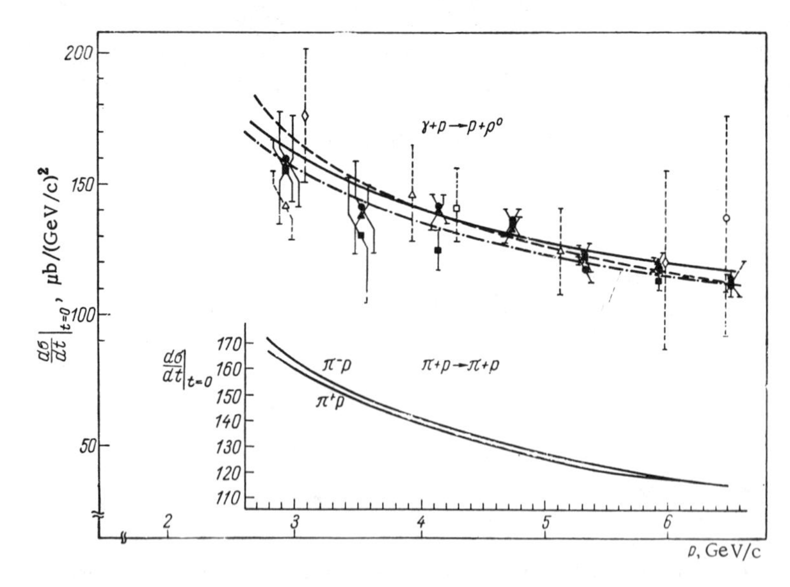

Fig. 11. Results for cross section of ρ-meson photoproduction on hydrogen at t = 0, $[d\sigma/dt]_{t=0}$, obtained by an approximation of the data of Table 6 in three ways: —·—·—, ■) by (9); – – –) ●) by (10); ——) ◆ by (11). The data of other investigations [1] are given for comparison: Δ) Aachen–Bonn, etc., collaboration; ◇) Davier et al.; □) Eisenberg et al.; ○) Jones et al. The data of [15] are not given. The bottom curves are the πp scattering cross sections: $(1.0 + 1.2/p)^2$ for π^-p and $(1.0 + 1.1/p)^2$ for π^+p.

MeV/c², and not of 145 MeV/c², which we and the bubble chamber groups used.) From the presented fits we obtained the following results.

1. Fitting of $d\sigma/dt|_{t=0}$ to a dependence on p of the form $(1 + R/p)^2$ gave:

a) using expression (9) (Ross–Stodolsky coefficient):

$$d\sigma/dt\,|_{t=0} \sim \left(1 + \frac{1.0 \pm 0.3}{p}\right)^2;$$

b) using (10) (inclusion of interference):

$$d\sigma/dt\,|_{t=0} \sim \left(1 + \frac{1.3 \pm 0.3}{p}\right)^2;$$

c) using (11) (inclusion of Ross–Stodolsky coefficient and interference term):

$$d\sigma/dt\,|_{t=0} \sim \left(1 + \frac{1.0 \pm 0.4}{p}\right)^2.$$

The obtained formulas should be compared with the momentum dependence of the π^+p and π^-p scattering cross sections in the energy region (3.0–7.0) GeV [22]:

$$d\sigma/dt\,|_{t=0} \sim \sigma^2_{\pi^+p}(1 + \alpha^2_+) \sim \left(1 + \frac{1.1 \pm 0.1}{p}\right)^2;$$
$$d\sigma/dt\,|_{t=0} \sim \sigma^2_{\pi^-p}(1 + \alpha^2_-) \sim \left(1 + \frac{1.2 \pm 0.1}{p}\right)^2,$$

where p is the momentum, GeV/c, and α_+ and α_- are equal to the ratios of the real part of the π^+p scattering amplitude to the imaginary part. Hence, it has been established that irrespective of our assumptions the energy behavior of the ρ-meson production reaction is very similar to that of πN scattering in the same energy region.

The ρ-meson mass measured from $\rho \to \pi^+\pi^-$ decays is approximately the same as that obtained from $\rho \rightleftharpoons e^+e^-$ reactions, but the ρ-meson width is approximately 30 MeV/c^2 greater [12].

3. Using the value $\gamma_\rho^2/4\pi = 0.45 \pm 0.10$ measured by the same group and the approximation for $d\sigma/dt|_{t=0}$ at 6.0 GeV, we obtain $\sigma_{\rho N} = 24 \pm 3$ mb.

§ 3.3. Results of Experiment on ρ-Meson Production on Complex Nuclei

The DESY-MIT group has also published its data on the reaction

$$\gamma + A \to A + \pi^+ + \pi^-, \tag{12}$$

obtained for 14 elements: hydrogen (A = 1), beryllium (9), carbon (12), aluminum (27), titanium (47.9), copper (63.5), silver (107.9), cadmium (112.4), indium (114.7), tantalum (181), tungsten (183.9), gold (197), lead (207.2), and uranium (238.1).

Measurements were made for 25 intervals of two-pion mass (m) between 400 and 1000 MeV/c^2, for ten intervals of ρ-meson momentum between 3.5 and 7 GeV/c, and for 20 intervals of transverse momentum transfer to the nucleus ($t_\perp$) from 0.001 to 0.04 GeV/c^2.

Thus, the results of the measurements are a four-dimensional matrix $(A, m, p, t_\perp) = (14, 25, 10, 20)$ containing approximately one million registered $\pi^+\pi^-$ events. The good statistics of the experiment (10^2–10^3 times greater than in the previous work) and the variety of investigated elements (twice as great as in previous experiments) [24-26] allowed an accurate investigation of the physics of ρ-meson production. This information, together with the measured dependence of the production cross section on the energy, mass, and momentum transfer, allowed an accurate determination of the total ρN scattering cross section $\sigma_{\rho N}$ and the ρ-meson–photon coupling constant $(\gamma_\rho^2/4\pi)$.

Purpose of Experiment. Accurate measurement of the cross section for ρ-meson photoproduction on complex nuclei as a function of the variables A, m, p, $t_\perp$ sheds light on the following questions.

1. The nuclear density distribution. For a given A the t dependence gives information (to an accuracy of ± 2%) on the nuclear radius R(A) seen by the ρ meson. By comparing this value with radii obtained from electron scattering and from strong interactions we hope to investigate the nature of the interaction of ρ mesons with nuclear matter.

2. The absolute and relative cross section for forward ρ-production $d\sigma/d\Omega dm\,(A, m, p, t_\perp)$. Measurements of the $\pi\pi$ spectra at fixed A, p, and $t_\perp$ provide a unique determination of the shape of the ρ peak and background.

3. Cross section ($\sigma_{\rho N}$) and γ–ρ coupling constant $(\gamma_\rho^2/4\pi)$. Measurements of the nuclear density distributions and the production cross sections on 14 elements can be used to determine the coefficient of reabsorption of ρ mesons by nuclear matter and the effective forward production cross section per nucleon $|f_0|^2$.

Thus, $\sigma_{\rho N}$ and $\gamma_\rho^2/4\pi$ can be determined in a self-consistent manner $[\gamma_\rho^2/4\pi = (\alpha/64\pi)(\sigma_{\rho N}^2/|f_0|^2)]$.

Experiment. The experiment was carried out on the 7.5- GeV electron synchrotron. The bremsstrahlung photon flux interacted with the target and the produced pairs were then detected by a wide-aperture magnetic spectrometer, which has been described separately [27]. The 22,500 hodoscope combinations allowed the description of each event with accuracies of $\Delta m = \pm 15$ MeV/c^2, $\Delta p = \pm 150$ MeV/c, and $\Delta t_\perp = 0.001$ (GeV/c)^2.

Vacuum pipes and helium bags were placed inside the spectrometer to reduce multiple scattering and nuclear absorption of pions.

Numerous checks were made throughout the experiment to ensure that the spectrometer operated as intended and that all the systematic errors were taken into account. We list the following ten examples.

1. To ensure that the results were not affected by second-order effects occurring in the target, such as attenuation of the photon beam and pion absorption, we measured the yield of reaction (12) to within 1% in relation to target thickness in the range 0–5 g/cm^2 of carbon. The corrected counting rate increased linearly with increase in target thickness.

2. The number of accidental coincidences was determined by means of a series of duplicate logic circuits with different resolving times and was kept to less than 2% by controlling the beam intensity.

3. The electronics dead time was monitored by continuous recording of the counting rate of the individual counters. The beam intensity was adjusted so that the dead time was less than 2%.

4. Nuclear absorption of pions by matter in the spectrometer was investigated by the introduction of additional material and variation of the gas pressure in the Cerenkov counters. The measured pion pair loss agreed to within 1% with the results of calculations based on published data.

5. To avoid any possible effects of asymmetry of the spectrometer we obtained half of the data with one polarity of the spectrometer, and the other half with the opposite polarity. The counting rates for the two different polarities of the spectrometer were identical.

6. The voltage on all the counters was kept constant to within ± 5 V and all the magnetic fields were stable at the $3:10^4$ level.

7. To ensure that at large Z the pion pairs with low effective mass were not contaminated with e^+e^- pairs we used Cerenkov counters to count and reject e^+e^- pairs. They showed that the maximum contamination was less than $1 \cdot 10^4$.

8. After every few hours we made calibration measurements to check the reproducibility of the data. Throughout the period of the measurements the reproducibility of the system was constant to within ± 1%.

9. We obtained all our data with p close to the maximum photon energy k_{max} to avoid errors due to inelastic contributions to reaction (12).

10. The purity of the materials used for the targets was more than 99.9%. The thickness of the target was chosen so that the corrections for beam attenuation and pion absorption were the same for all the elements and the counting rate without the target was small.

Results. The results were corrected for small systematic effects, such as beam attenuation, rate without target, nuclear absorption, dead time, accidental coincidences, etc. We checked all these corrections by measurements on the same spectrometer within an accuracy of 1%. We calculated the acceptance of the spectrometer by the Monte Carlo method, using fourth-order equations for beam transportation in the magnetic field. We took into account such effects as multiple scattering, decay angular distribution, decay in flight, etc.

For the Monte Carlo method we used a sufficiently large number of events to ensure that no additional errors would be introduced in the cross section calculations.

To facilitate the analysis and allow an investigation of the dependence of the cross section on all the dynamic variables we assembled the experimental data in a four-dimensional matrix $(A, m, p, t_\perp) = (14, 25, 10, 20)$.

Because of lack of space we cannot show all its 70,000 constituents here. We give only some of the characteristic features of the experimental data and the results of the overall analysis of all the data from [28].

The projection of 2% of the data onto three-dimensional space $(A, m, t_\perp)$ at $p = 6.2 \pm 0.2$ GeV/c is shown in Fig. 12. It is readily apparent that in all the spectra the ρ-meson dominates and that it is diffractively produced off the nucleus. The mass profile changes sharply with change in A and m and, hence, not all the π-meson pairs result from ρ decay and there must be a considerable nonresonant background which depends on A, m, and $t_\perp$.

Figure 13 shows the projection of 5% of the data onto the (A, m) plane [29] for $\langle p \rangle = 6.0$ GeV/c. The shape and width of the spectrum (total width at half height) vary considerably between elements with small A and heavy elements. This again indicates that the nonresonant background depends greatly on A.

To check the experimental method we calculated the hypothetical ρ-meson cross sections from our measured cross sections for π-meson pair production near mass values $m = m_\rho$ at momentum 6.2 GeV/c, assuming 120 MeV/c^2 for the ρ-meson width. The results were found to be in good agreement with the cross sections at $\theta = 0°$ obtained by the same method in Cornell [25] (Fig. 14). The obtained values are not, however, the true cross sections, since the background is ignored, the detailed $\pi\pi$ profile is not taken into account, and an arbitrary assumption regarding the ρ-meson width is made.

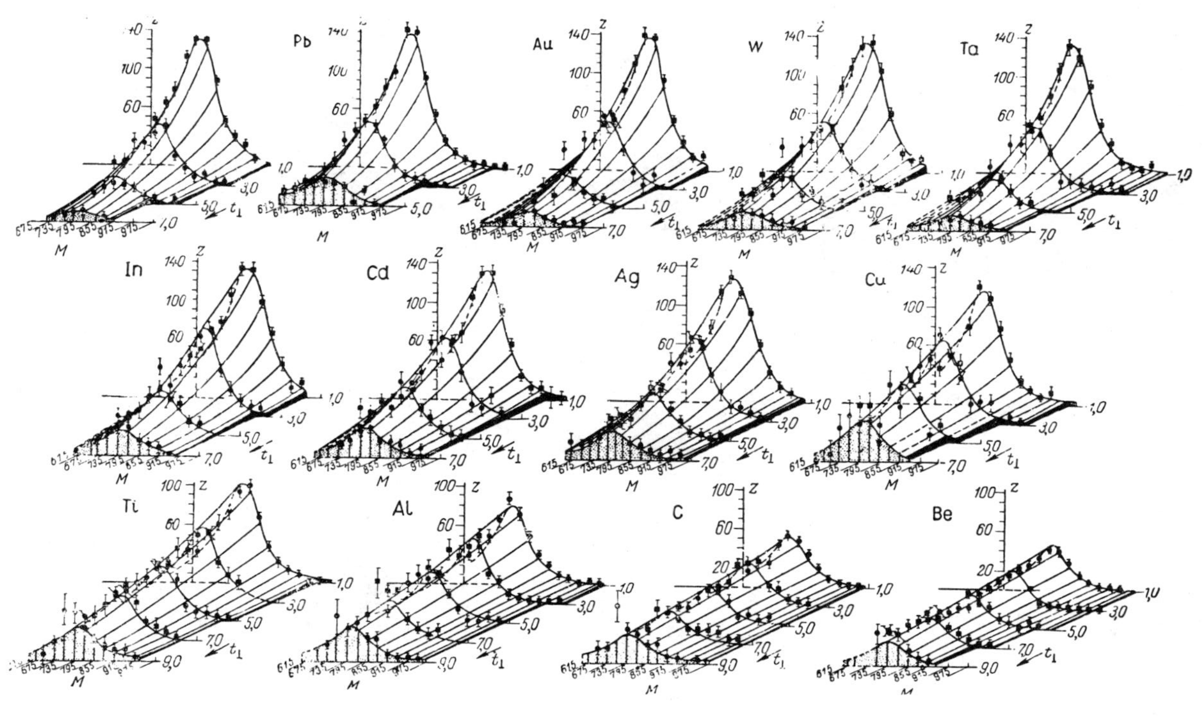

Fig. 12. Cross section for ρ-meson photoproduction on various nuclei, $z = d\sigma/d\Omega dm$, $\mu b/(\text{nucleon} \cdot \text{sr} \cdot \text{MeV}/c^2)$. Data for $p = 6.2$ GeV/c in relation to m and $t_\perp$ are given; this corresponds to only 2% of all the data [28]. The curves represent the best fit.

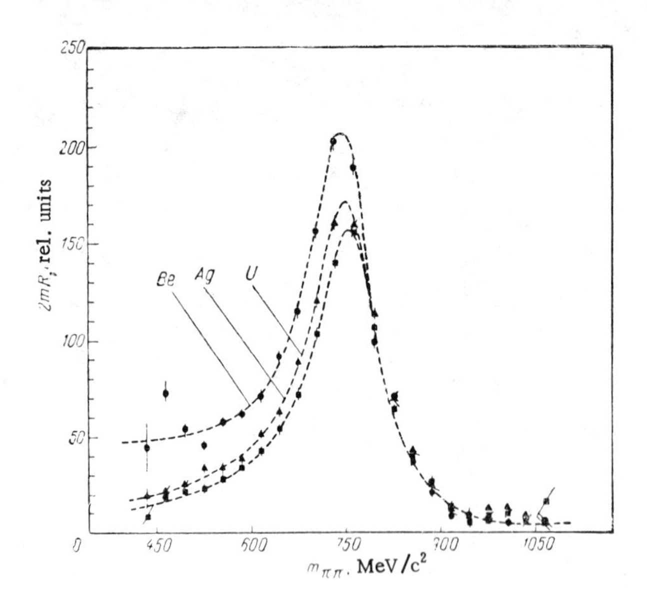

Fig. 13. Spectra of 2mR at p = 6.0 GeV/c in relation to m (5% of all the data). If there was no background all the spectra would be identical.

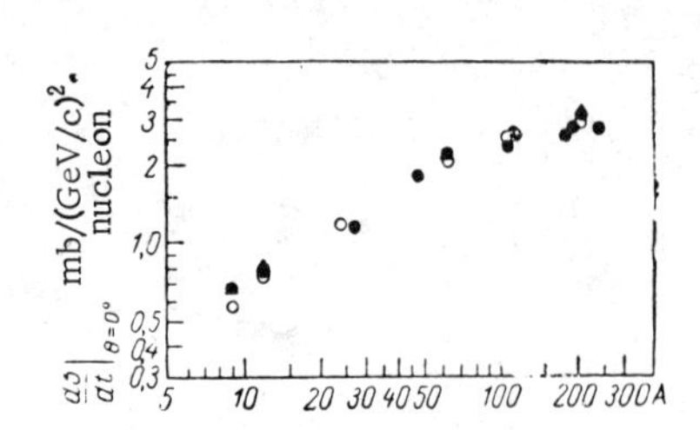

Fig. 14. Comparison with data of [25] (Cornell, U.S.A.) for value of $d\sigma/dt(\theta=0°)$ assuming that there is no background and Γ_ρ = 120 MeV/c². ▲) DESY I; ●) DESY II; ○) Cornell.

The results obtained in SLAC [26] differ greatly from our results and this disagreement cannot be attributed entirely to our inadequate understanding of the physics of nuclear processes. As Fig. 14 shows, the data of earlier experiments at DESY are in good agreement with the results obtained in the present measurements.

Analysis. On the basis of the formulated principles and the conclusion, derived above, that ρ-meson production dominates in $\pi\pi$ spectra, we performed the following analysis.

The four-dimensional data matrix $d\sigma/d\Omega dm(A, m, p, t_\perp)$ was approximated by a theoretical function of the form

$$\frac{d\sigma}{d\Omega dm}(A, m, p, t_\perp) = p^2\, 2mR(m)\,(f_c + f_{mc}) + BG(A, m, p, t_\perp). \tag{13}$$

The first term represents the main contribution from ρ production, and the second term is the contribution from the nonresonant background or any possible interference between ρ production and the p wave part of the background

$$f_c = f_c(A, t_\perp, t_\parallel, \sigma, \beta) = \left(\frac{\sigma'}{\sigma}\right)^2 \left| 2\pi f_0 \int_0^\infty b\,db \int_{-\infty}^{\infty} dz J_0(b\sqrt{t_\perp}) \exp(iz\sqrt{t_\parallel})\, \rho(z, b) \exp\left[-\frac{\sigma'}{2}(1-i\beta)\int_z^\infty \rho(z', b)\,dz'\right]\right|^2;$$

$$\sigma' = \sigma(1-\xi\eta\sigma);\quad \xi = \frac{1}{16\pi a}\int \exp(-b^2/4a)\, g(\mathbf{b}, z)\, d^2b\,dz;$$

$$\eta = \eta(\sigma) = \int \exp[-T(b)\sigma/2]\, Q(b)\, d^2b \Big/ \int \exp[-T(b)\sigma/2]\, T(b)\, d^2b;$$

$$Q(b) = \int_{-\infty}^{\infty} \rho^2(b, z)\,dz,\quad \sigma = \sigma_{\rho N},\quad T(b) = \int_{-\infty}^{\infty} \rho(b, z)\,dz.$$

Here f_c is the coherent production cross section [13]; it is taken into account in the expression that the ρ meson produced on an individual nucleon with effective forward cross section $|f_0|^2$ is then absorbed in nuclear matter in accordance with an $\exp[-\frac{1}{2}\sigma'(1-i\beta)\int_z^\infty \rho dz]$ law. The factor $J_0(b\sqrt{t_\perp})\rho(b, z)$ depends on the shape of the nucleus: $\rho = \rho_0\{1+[\exp(z-R)/\alpha]\}^{-1}$ is the Woods-Saxon potential, where R is the nuclear radius and α = 0.545 F. $\exp(iz\sqrt{t_\parallel})$ takes into account the difference in the initial and final mass, and σ' is the effective ρ-nucleon total cross section, which includes second-order correlation effects between the nucleons inside the nucleus. Here, ξ denotes the correlation length, and $g(r_1, r_2)$ the correlation wave function [30]; β is the ratio of the real part of the nucleon production amplitude to the imaginary part. For β we took the value $\beta = -0.2$ from the analysis of measurements of the total γp cross section at 6.0 GeV [31]; f_{inc} is the incoherent production cross section where the recoil nucleus is left in an excited or fragmented state. The incoherent contribution (~10%) is most appreciable at small A and becomes negligibly small at A = 100. Finally, zmR(m) is the Breit-Wigner relativistic mass distribution.

The background function consists of $BG(a, m, p, t_\perp) = G(A, m, p, t_\perp) + I(m)$, where G is a general polynomial in $(A, m, p, t_\perp)$ space and the function

$$I(m) = \frac{\text{const}\,(m^2 - m_\rho^2)}{(m_\rho^2 - m^2)^2 + m_\rho^2 \Gamma^2}$$

is due to interference between the ρ meson and the p-wave background.

A comparison of the data of the 70,000 fins of the four-dimensional matrix of results with the theoretical function (13) allows a direct determination of the parameters m_ρ, Γ_0, $\sigma_{\rho N}$, R(A), $|f_0|^2$, $d\sigma/dt \cdot (A)|_{\theta=0°}$, $\gamma_\rho^2/4\pi$. The fitting was made by means of the CERN program MINUIT. The procedure in determining the various parameters can readily be understood from the following.

1. Determination of the background function BG(A, m, p, t). The fit was effected with and without inclusion of a background term of the form

$$BG(A, m, p, t_\perp) = \left(\sum_{i=0}^{l} a_i(A)\, m^i\right)\left(\sum_{j=0}^{m} b_j(A)\, p^j\right)\left(\sum_{k=0}^{n} c_k(A)\, t_\perp^k\right) + I(m).$$

In the region $m > m_c$ (where m_c is the cut-off parameter, equal to 600 MeV/c^2) the background depends in a complicated manner on m, p, t, and A. It was found that the fit was improved significantly by the choice of $l = 2$, m = n = 0. After this there were no improvements due to an increase in l, m, or n, or to inclusion of a term of the form I(m). The results of the fit were insensitive to changes in $m_c > 600$ MeV/c^2, and subsequently all the data were analyzed with $m_c > 600$ MeV/c^2. The percentage background content was fairly high at small A and m, but decreased with increase in A and m, and was small for uranium.

2. Determination of R(A) — the nuclear density parameter. For fixed A the t dependence of the cross sections is determined by the nature of the diffraction off the investigated nucleus. A comparison of the diffraction curves with Eq. (13) enables one to find the average radius of the nucleus seen by the ρ meson. The results of measurement of the radius are given in Fig. 15 and Table 7. The data lead to the equation $R(A) = (1.12 \pm 0.02)A^{1/3}$ F, which is the most precise expression today for the nuclear radius in strong interactions [33]. The nuclear radii determined in this way are larger than the corresponding values from experiments on electron scattering on nuclei [$(R(A) \simeq 1.08\ A^{1/3}$ F].

3. Determination of m_ρ and Γ_ρ — mass and width of ρ meson. For fixed A, p, $t_\perp$ the mass and width m_ρ and Γ_ρ are found directly from a comparison of the dependence of $d\sigma/d\Omega dm$ on mass with expression (13). We obtain $m_\rho = 765 \pm 5$ MeV/c^2; $\Gamma_\rho = 140 \pm 10$ MeV/c^2.

4. Determination of coherent cross sections $d\sigma/dt(A)|_{\theta=0°}$. The data matrix was compared with expression (13). Our experimentally determined values were used for R(A), m_ρ, Γ_ρ, and BH. To check the consistency of the data and the analysis, the different subsystems of data were approximated in restricted ranges of m, p, and t. The results agreed well with all the values given in Table 7.

5. Determination of $\sigma_{\rho N}$, $|f_0|^2$, and $\gamma_\rho^2/4\pi$. The cross sections $d\sigma/dt(A)|_{\theta=0°}$ were approximated by expression (13) with $\beta = -0.2$ and the values of R(A) found above.

The approximation (Fig. 16) led to the following values for the parameters:

$$\sigma_{\rho N} = 26 \pm 2 \text{ mb}, \qquad |f_0|^2 = \frac{d\sigma}{dt}(t = 0,\ A = 1) = 117 \pm 5 \text{ mb/(GeV/c)}^2$$

and $\gamma_\rho^2/4\pi = 0.54 \pm 0.10$.

The value found for $\sigma_{\rho N}$ differs by a factor of approximately 1.5 from the result 38 ± 3 mb obtained in Cornell [25], and our $\gamma_\rho^2/4\pi$ differs by a factor of two from the value 1.1 obtained in Cornell and SLAC

TABLE 7. Typical Results for Cross Sections of ρ-Meson Photoproduction on Various Nuclei at 0° and Nuclear Radii R

Nucleus	$d\sigma/dt(\theta = 0°)$, $t_{\parallel} = 0.002\ (GeV/c)^2$, $p = 6.54\ (GeV/c) \cdot mb/[(GeV/c^2)/$ nucleon]		R, F
	$\frac{k_{max}}{p} = 1.23$	$\frac{k_{max}}{p} = 1.12$	
Be	652± 32	615± 30	2,35±0,12
C	804± 50	741± 53	2,50±0,16
Al	1336± 62	1290± 63	3,37±0,16
Ti	1748± 89	1838± 65	3,94±0,10
Cu	2172±125	2073±110	4,55±0,11
Ag	2671± 98	2577± 59	5,35±0,09
Cd	2679±114	2640± 72	5,40±0,14
In	2810±118	2680± 62	5,56±0,25
Ta	2993±181	2940±127	6,50±0,15
W	2964±111	2896±170	6,30±0,12
Au	3039±141	3068±115	6,45±0,27
Pb	3302± 65	3147±122	6,82±0,20
U	3160± 92	3093± 93	6,90±0,14

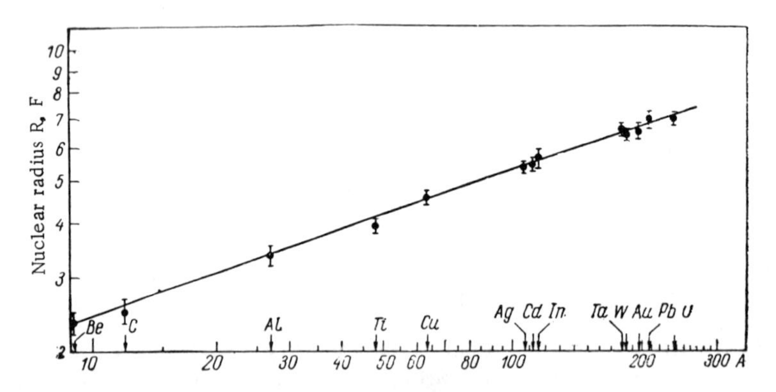

Fig. 15. Nuclear radii obtained from measured t dependence of cross section for ρ-meson photoproduction on nuclei. The straight line is the result of approximation, $R = (1.12 \pm 0.02)A^{1/3}$.

[26]. The sources of the discrepancy are as follows.* Using the published data of SLAC and the parameters R(A), $\sigma' = \sigma_{\rho N}$, and $\beta = \xi = 0$, we were able to reproduce our published results. The discrepancy is due mainly to differences in the experimental data. The reasons for the disagreement with the Cornell results are more complex. The cross sections also fail to agree. For instance, as Fig. 16 shows, our measured cross section on lead [$3302 \pm 65\,\mu b/(GeV/c)^2$] differs by 25% from that obtained in Cornell [$2630 \pm 80\,\mu b/(GeV/c)^2$]. The two values were measured at $\theta = 0°$ in the same kinematic conditions. In addition, we could not reproduce their published results by using their data and fitting parameters.

Our value for $|f_0|^2$ agrees with our previous measurement of the cross section on H_2: $d\sigma/dt|_{t=0} = 119 \pm 7\,\mu b/(GeV/c)^2$ at 6.0 GeV/c. Our result for $\gamma_\rho^2/4\pi$ agrees with the initial result of DESY (0.45± 0.10) and with the value obtained independently at DESY from measurements of the total hadronic cross sections $\sigma_{A\gamma}$ [34].

Comparing our results for $\gamma_\rho^2/4\pi$ with the value $0.52^{+0.07}_{-0.06}$ determined from $\rho \to e^+e^-$ we conclude that the ρ-meson–photon coupling constant is independent (to within ±20%) of the value of m_γ (photon mass) in the range $0 < m_\gamma < m_\rho$.

* The comparison with the data of other investigations and the analysis of the reasons for the discrepancy are rather casually dealt with in the original text. Interested readers will have to refer to the original publications – Editor's note.

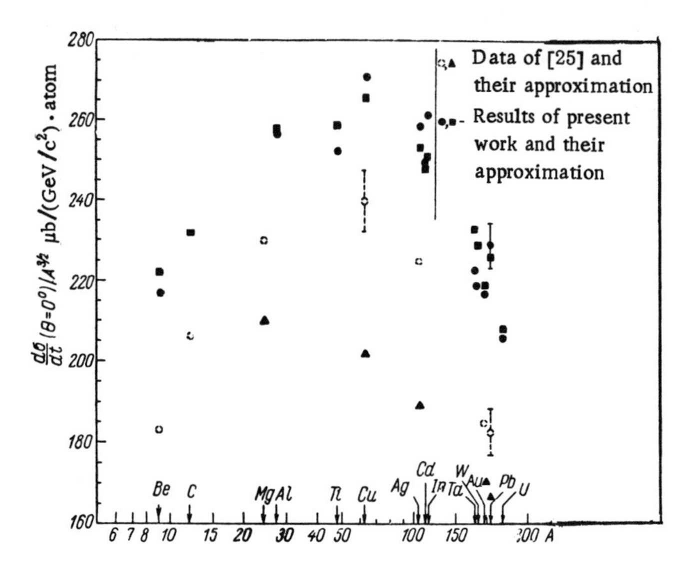

Fig. 16. Results for cross section of ρ-meson photoproduction on nuelci $d\sigma/dt$ ($\theta = 0°$) at $p/k_{max} = 1.23$ and their approximation. For comparison we show the data of [25] (Cornell, U.S.A.) and our approximation of these data with the following values of parameters: $\sigma_{\rho N} = 38$ μb; $d\sigma/dt$ ($t = 0$; $A = 1$) $= 124$ μb/(GeV/c)^2; $\beta = \xi = 0$; $m_\rho = 770$ MeV/c^2; $p = 6.2$ GeV/c; $t_{\parallel} = m_\rho^4/4p^2 = 0.0023$ (GeV/c)^2.

§ 3.4. Total Photon-Hadron Cross Sections

In the vector dominance model the cross section for photoproduction of vector mesons can be related to the nucleon-vector meson cross sections. In particular, the total γp cross section can be expressed as:

$$\sigma_{\text{tot}}(\gamma p) = \sqrt{4\pi\alpha}\left[\left(\frac{\left.\frac{d\sigma}{dt}\right|_{t=0}(\gamma p\mid p\rho)}{\gamma_\rho^2/4\pi}\right)^{1/2} + \left(\frac{\left.\frac{d\sigma}{dt}\right|_{t=0}(\gamma p\mid p\omega)}{\gamma_\omega^2/4\pi}\right)^{1/2} + \left(\frac{\left.\frac{d\sigma}{dt}\right|_{t=0}(\gamma p\mid p\varphi)}{\gamma_\varphi^2/4\pi}\right)^{1/2}\right].$$

The cross sections $d\sigma/dt|_{t=0}(\gamma p/pV^0)$, $V^0 = \rho, \omega, \varphi$ are all measured values, while $\gamma_\rho^2/4\pi$, $\gamma_\omega^2/4\pi$, and $\gamma_\varphi^2/4\pi$ are determined from leptonic decays. Hence, measurement of $\sigma_{tot}(\gamma p)$ allows direct verification of the predictions of the vector dominance model and shows convincingly that the coupling constants $\gamma_\rho^2/4\pi$, etc., have the same value, irrespective of whether the photon or ρ meson is on the mass shell.

§ 3.5. Summary of Results for Coupling Constant $\gamma_\rho^2/4\pi$ from ρ^0 Meson Photoproduction Experiments

1. From the data of the present investigation we have for ρ^0-meson production on nuclei: $\gamma + A \to A + \rho^0$

$$\frac{\gamma_\rho^2}{4\pi} = 0.54 \pm 0.10.$$

2. From the data of Meyer (DESY) [34] and Guiragossian [1] for the total cross section on hydrogen: $\sigma_{tot}(\gamma p) = 110$ μb; $d\sigma/dt|_{t=0}(\gamma p|p\rho) = 120$ μb/(GeV/c^2).

$$\sigma_{\text{tot}}(\gamma p) = \sqrt{4\pi\alpha}\left(\frac{\left.\frac{d\sigma}{dt}\right|_{t=0}(\gamma p\mid p\rho)}{\gamma_\rho^2/4\pi}\right)^{1/2}$$

$$\frac{\gamma_\rho^2}{4\pi} = 0.5 \pm 0.1.$$

3. From Meyer's data [34] for the total cross section on Be, C, Al, and Cu nuclei

$$\sigma_{\text{tot}}\ (\gamma A)|_{E\to\infty} = \sqrt{4\pi\alpha}\left(\frac{\left.\frac{d\sigma}{dt}\right|_{t=0}(\gamma A\mid A\rho)}{\gamma_\rho^2/4\pi}\right)^{1/2}$$

$$\frac{\gamma_\rho^2}{4\pi} = 0.4 \pm 0.1.$$

4. From the data of Margolis [30] $\sigma_{\rho N} \cong 25$ mb

$$\frac{\gamma_\rho^2}{4\pi} = 0.4 \pm 0.1.$$

LITERATURE CITED

1. A complete bibliography can be found in the review: S. C. C. Ting, Proceedings of 1967 International Conference on Photons and Electrons, Stanford, California, and in J. J. Sakurai, Vector Mesons 1960-1968, Preprint EFI 68-59, University of Chicago.
2. A. Dar and V. F. Weisskopf, Phys. Lett., 26B, 670 (1968).
3. M. Gell-Mann, D. Sharp, and W. G. Wagner, Phys. Rev. Lett., 8, 261 (1962).
4. H. Joos, Proceedings of 1967 Conference on High-Energy Physics, Heidelberg, Germany.
5. H. Alvensleben et al., Phys. Rev. Lett., 21, 1501 (1968).
6. R. G. Parsons and R. Weinstein, Phys. Rev. Lett., 20, 1314 (1968); M. Davier, Phys. Lett., 27B, 27 (1968).
7. S. C. C. Ting, Rapporteur's Summary of Proceedings of the 14th International Conference on High-Energy Physics, Vienna (1968), CERN Scientific Information Service, Geneva (1968), p. 43.
8. J. Asbury et al., Phys. Rev. Lett., 19, 869 (1967). Experiments to find $\mu^+\mu^-$ pairs in the region of ρ, ω masses are listed below. No effect from ω has been detected. J. K. de Pagter et al., Phys. Rev. Lett., 16, 35 (1966); P. L. Rothwell et al., Proceedings of Third International Symposium on Electron and Photon Interactions at High Energies, Stanford (1967), Clearing House of Federal Scientific and Technical Information, Washington, D. C. (1968), p. 644; S. Hayes et al., Phys. Rev. Lett., 21, 1134 (1969).
9. H. R. Collard, L. R. B. Elton, and R. Hofstadter, Landolt-Boernstein Tables I. Vol. 2. Nuclear Radii, Springer Verlag, Berlin (1967).
10. S. D. Drell and C. L. Schwartz, Phys. Rev., 112, 568 (1958).
11. H. Alvensleben et al., To be published; E. Lohrmann, SLAC Conference Report (1967).
12. J. E. Augustin et al., Phys. Lett., 28B, 508, 513 (1969).
13. K. S. Koelbig and B. Margolis, Nucl. Phys., B6, 85 (1968).
14. Earlier investigations of this reaction can be found in the paper "Cambridge Bubble Chamber Collaboration," Phys. Rev., 146, 994 (1966); 163, 1510 (1967). J. Ballam et al., Phys. Rev. Lett., 21, 1541, 1544 (1968); L. J. Lanzerotti et al., Phys. Rev., 166, 1365 (1968); H. Blechschmidt et al., Nuovo Cimento, 52A, 1348 (1967). W. G. Jones et al., Phys. Rev. Lett., 21, 586 (1969). ABBHHM Collaboration, 175, 1669 (1968); Phys. Lett., 27B, 54 (1968); Y. Eisenberg et al., Phys. Rev. Lett., 22, 669 (1969); M. Davier et al., Phys. Rev. Lett., 21, 841 (1968); Phys. Lett., 28B, 619 (1969).
15. G. McClellan et al., Phys. Rev. Lett., 22, 374 (1969); J. G. Asbury et al., Phys. Rev., 161, 1344-1355 (1967); M. J. Longo and B. J. Moyer, Phys. Rev., 125, 701 (1962).
16. When the data obtained at angles $<\theta_{\pi\pi}> < 1°$ with an average $t \simeq 0.01$ (GeV/c^2) are collected, the effects of the t dependence and the angular distribution are small. These special features have been measured [14] and were included in the calculation of the cross sections.
17. P. Soeding, Phys. Lett., 19, 702 (1965).
18. J. D. Jackson, Nuovo Cimento, 34, 1644 (1964).
19. For use of term $(m_\rho/m)^4$ in Breit-Wigner formula, see: M. Ross and L. Stodolsky, Phys. Rev., 149, 1172 (1966); G. Kramer and J. L. Uretsky, ANL/HEP (1968).
20. F. James and M. Ross, CERN 6600 Computer Library Long Write-up 67/623/1.
21. In obtaining $d\sigma/dt|_{t=0}$ from $d\sigma/d\Omega dm$ there was an uncertainty of $\pm 10\%$ in the normalization owing to the spectral functions used. This error is not shown in Fig. 2.

22. M. N. Foccacci and G. Giacomelli, CERN Report 66-18.
23. S. C. C. Ting, Rapporteur's Summary of the 14th International Conference on High-Energy Physics, Vienna (1968).
24. J. G. Asbury et al., Phys. Rev. Lett., 19, 867 (1967); L. J. Lanzerotti et al., Phys. Rev., 166, 1305 (1968).
25. G. McClellan et al., Phys. Rev. Lett., 22, 377 (1969).
26. F. Bulos et al., Phys. Rev. Lett., 22, 490 (1969).
27. J. G. Asbury et al., Phys. Rev., 161, 1344 (1967); M. T. Longo and B. T. Moyer, Phys. Rev., 125, 701 (1962).
28. These data are recorded on magnetic tape and can be transmitted.
29. In Fig. 10b the experimental cross section $d\sigma/d\Omega dm$ is divided by $p^2(f_c + f_{inc})$. Thus, the ρ-meson production mechanism (but not the background is separated from the mass spectrum.
30. G. V. Bochmann, B. Margolis, and C. L. Tang, to be published; D. Walecka et al., Contributions to Conference on Nuclear Interaction, Columbia University (1969).
31. Information on the experimental determination of β can be found in: J. Weber, Ph. D. Thesis, DESY (1969). We thank Profs. A. Dar and B. Margolis, who pointed out to us the importance of including this term.
32. J. S. Trefil, Nuclear Physics, B11, 330 (1969).
33. R. J. Glauber and G. Mathiae, ISS 67/16.
34. H. Meyer, Private communication.

SHORT-RANGE REPULSION AND BROKEN CHIRAL SYMMETRY IN LOW-ENERGY SCATTERING*

V. V. Serebryakov and D. V. Shirkov

The introduction of short-range repulsive "potentials" into the low-energy equations for the lower partial waves makes it possible to eliminate the main difficulties of the pure elastic low-energy (Pele) approximation. In principle, it is then possible to obtain solutions with short s-wave scattering lengths and broad resonances. The use of threshold conditions that follow from chiral symmetry enables one (under certain simple additional conditions) to express the main resonance scattering parameters in terms of the pion decay characteristics. Thus, on the basis of the approximation of broken chiral symmetry and unitarity dispersion equations for low-energy $\pi\pi$ and πN scattering, expressions are obtained for the masses, lifetimes, and coupling constants for p-wave resonances, it being only necessary to specify the pion and nucleon masses and lifetimes and the Fermi coupling constant.

1. PHYSICAL INCOMPLETENESS OF THE LOW-ENERGY REGION

1.1. Physical Content of the Pure Elastic Low-Energy Model

The pure elastic low-energy (Pele) model (see Ch. 2 in [1]) is based on the following fundamental assumptions.

1. Strict analyticity for the partial waves.

2. Two-particle unitarity (i.e., one can neglect the three-particle and higher mass states in the unitarity condition).

3. Neglect of the higher partial waves f_l, $l > l_{max}$ (usually, one considers only s and p waves, i.e., $l_{max} = 1$).

4. Approximate crossing symmetry for the lower partial waves f_l ($l \le l_{max}$) obtained by combining the dispersion relations for forward and backward scattering (differential approximation).

Assumptions 1 and 3 and also assumption 4, which is based on 3, are good approximations in the low-energy region, in which the contributions of the inelastic (multiparticle) channels are either absent or numerically small and, as a rule, the higher partial waves are also numerically small.

The physical content of the pure elastic low-energy model is basically determined by assumption 2. Because of this assumption, the integral contributions to the direct channel (s channel) are exhausted by the graphs depicted in Fig. 1 (for the illustration we show only the contributions to the $\pi\pi$ and πN scattering).

---

* This paper is a significantly revised exposition of the Chapter "Short-Range Repulsion" from the English edition [1a] of the book "Dispersion Theories of Strong Interaction at Low Energy" (this chapter was not included in the Russian edition [1]). In contrast to [1a], we introduce the concept of chiral symmetry in the present paper.

Institute of Mathematics, Siberian Branch, Academy of Sciences of the USSR, Novosibirsk. Translated from Problemy Fiziki Élementarnykh Chastits i Atomnogo Yadra, Vol. 1, No. 1, pp. 171–226, 1970.

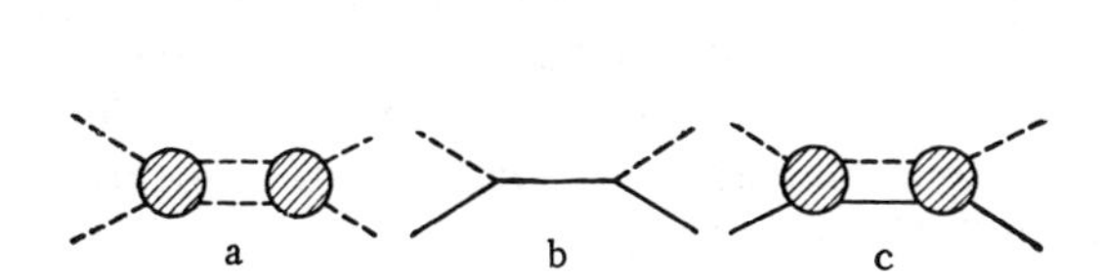

Fig. 1. Contributions from the direct channel to $\pi\pi$ (a) and πN scattering (b, c).

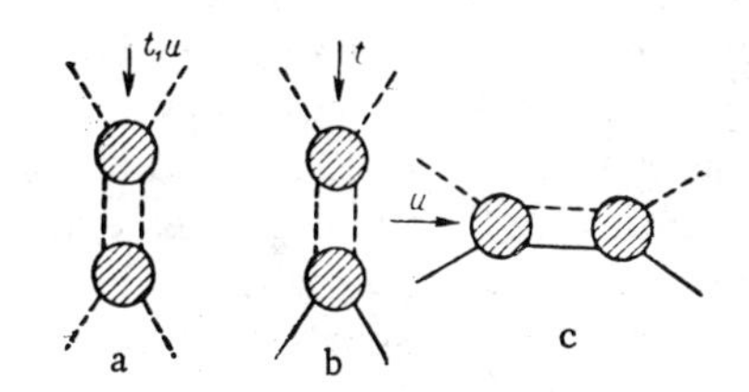

Fig. 2. Integral contributions from the crossed channels to $\pi\pi$ scattering (a) and πN scattering (b, c).

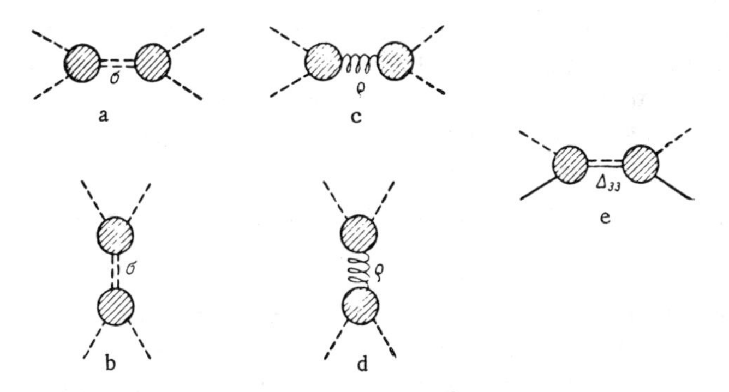

Fig. 3. Resonance contributions to the $\pi\pi$ and πN scattering.

The integral contributions from the crossed channels correspond to the graphs of Fig. 2.

Since the imaginary parts of the amplitudes are approximated by a small number of lower partial waves (by virtue of assumption 3), the integrals are basically saturated by the low-energy resonance contributions from the corresponding two-particle states. For the $\pi\pi$ system these are the contributions from the ρ meson (m = 765 MeV, I = l = 1) and the σ resonance† ($m_\sigma \approx$ 500–1000 MeV, I = l = 0) (Fig. 3a, b); for the πN system they are the Δ_{33} contributions (M_{33} = 1236 MeV, I = l = 3/2) (see Fig. 3c, e).

In other words, in the direct channel the scattering proceeds through the lowest excited resonance states and the contribution of the crossing channels corresponds to exchange of the lowest partial excitations.

The integral contributions from the higher parts of the spectrum are suppressed by conditions 2 and 3 and also by the fact that in the pure elastic low-energy model one considers especially solutions that decrease fairly rapidly with increasing energy (the condition of physical self-consistency).

Thus, in the pure elastic low-energy models, the scattering is almost completely described by an intermediate state with relatively small masses $M_i^*(m_\rho, m_\sigma, M_{33})$ (of order 1 GeV or less). This means that the scattering has a peripheral nature, since the wave functions of these states are concentrated mainly at relatively large distances from the center of mass:

$$R_i^* \sim \frac{\hbar c}{M_i^*} \sim (0.5 - 1)\, 10^{-13} \text{ cm.} \tag{1.1}$$

If, in agreement with the quantum-mechanical description of scattering, the contributions of the crossed channels are associated with exchange forces, the ranges of the forces are determined by the relation (1.1) and are relatively large. The exchange forces have a long-range nature.

The characteristics of the exchanged quasi-particles are associated with the resonances in the direct channel by conditions that ensure the correct threshold behaviors for the considered (i.e., not small) partial

† The experimental data for the σ meson are discussed in more detail in Sec. 5.4.

waves with $l > 1$. Since the equations that describe the low-energy scattering are integral equations with Cauchy kernels, these conditions are equivalent to the requirement that the partial waves decrease sufficiently rapidly, in accordance with the unitarity condition, at high energies. In the case of $\pi\pi$ scattering, these conditions relate the positions and the widths of the $\pi\pi$ resonances. For the more complicated cases of πN and NN scattering these conditions express the constants (the widths and masses) of the interaction of the unstable mesons (ρ, σ, ω) with the baryons and pions through the characteristics (scattering lengths and resonance widths) of the partial waves of the πN and NN scattering.

This kind of relationship can be interpreted as conditions of low-energy self-consistency: the low-energy resonances arise solely because of forces associated with the exchange of such low-energy resonances. The resulting situation closely resembles the "low-energy bootstrap" program (see Sec. 30.2 in [1]).

1.2. Effective Lagrangian

It is convenient to consider separately intermediate states corresponding to the annihilation t channel. These states do not have a baryon charge and correspond to the ρ and σ mesons. The contributions of these states to the scattering amplitudes can be described by means of the Born approximation (i.e., by means of the pole Feynman graphs) based on the "effective Lagrangian"

$$L_{\text{eff}} = g_{1V}\,\bar{\Psi}\tau\gamma_\mu\Psi_\rho\rho_\mu + \frac{g_{2V}}{M}\bar{\Psi}\sigma_{\mu\nu}\tau\Psi(\partial_{\mu\rho}\rho_\nu - \partial_{\nu\rho}\rho_\mu) + g_{\sigma NN}\,\bar{\Psi}\Psi\sigma + g_\rho\rho_\mu\left[\frac{\partial\pi}{\partial x_\mu}\pi\right] + g_\sigma(\pi\pi)\,\sigma. \tag{1.2}$$

Here Ψ is the operator of the nucleon field, γ_μ are the Dirac matrices, ρ_μ is the operator of the ρ-meson field, a Lorentz four-vector and an isotopic three-vector, τ are the isotopic Pauli matrices, $\sigma_{\mu\nu}$ is the spin matrix tensor of the σ mesons, a Lorentz and isotopic scalar, and π is the operator of the pion field. The expression (1.2) gives the part of the effective Lagrangian that is needed to describe only the $\pi\pi$- and πN-scattering processes. To describe the NN-scattering potential it would be necessary, for example, to consider also the interaction of the nucleons with the ω and A_2 mesons.

Of course, the parameters associated with the effective Lagrangian (1.2), i.e., the coupling constants and the masses of the meson states, are not known a priori. However, considering the interaction processes in the sequence corresponding to the hierarchical scheme, according to which it is impossible to describe meson-baryon scattering theoretically without knowledge of the properties of meson-meson scattering (see Ch. 1 of [1]), these parameters can be determined by consistency considerations for all the reactions ($\pi\pi$, πN, and NN scattering) as a whole. Thus, the constants g_ρ and g_σ and the masses m_ρ and m_σ can be determined from $\pi\pi$ scattering. At the same time, for example, the width of the ρ meson can be expressed in terms of g_ρ and m_ρ as follows:

$$\Gamma_\rho^{\text{tot}} = \frac{g_\rho^2}{6\pi}\cdot\frac{q_\rho^3}{m_\rho^2} = \frac{1}{12}\left(\frac{g_\rho^2}{4\pi}\right)\frac{(m_\rho^2 - 4\mu^2)^{3/2}}{m_\rho^2}. \tag{1.3}$$

Similarly, the constants g_{1V} and g_{2V} can be determined from the condition of consistency of the contributions to the πN and NN scattering or consistency of the πN scattering with the electromagnetic nucleon form factor.

The effective Lagrangian is also convenient in that it can be used to take into account the additional symmetry principles that are not imposed directly in the dispersion approach. For example, Sakurai's principle of universality of the interaction of the ρ-meson field with the vector current leads to the following relationships between the constants of the ρ-meson interaction:

$$g_{1V} = \frac{1}{2}g_\rho,\quad g_{2V} : g_{1V} = 2\mu_N : 1. \tag{1.4}$$

Here $\mu_N = 1.9$ is the anomalous nucleon magnetic moment.

We must emphasize particularly that we require the Lagrangian (1.2) to construct the t-channel contributions to the dispersion-type equations. These contributions have the same poles and residues as the single-particle Feynman graphs. However, in contrast to the latter, they decrease at infinity.* One can

* For the case of unsubtracted dispersion relations.

therefore say that the dispersion contributions are the long-range parts of the corresponding Feynman graphs.

It is shown in Ch. 3 of the monograph [1] that the pure elastic low-energy theory of $\pi\pi$ scattering leads to only qualitative agreement with the experimental data. In particular, it cannot explain the ρ-meson width of ~ 100 MeV. We shall now show that the pure elastic low-energy description of πN scattering also possesses a similar shortcoming. In particular, it strongly contradicts Sakurai's universality principle.

1.3. Long-Range Model of πN Scattering

We shall now obtain the system of equations for the s and p waves of πN scattering using the pure elastic low-energy assumptions formulated above. We shall also make the following simplifying assumptions.

5. The static limit, i.e., the ratio of the pion-nucleon mass, will be assumed to be infinitely small: $\mu/M \ll 1$.

6. We shall approximate the contributions of the annihilation channel $\pi\pi \to N\overline{N}$ by pole graphs (see Fig. 3b, d) corresponding to the effective Lagrangian (1.2).

Let us now consider the dispersion relations for forward and backward scattering written down for the following four combinations of the structure functions of πN scattering:

$$\varphi^{\pm}(\nu, \cos\theta) = \frac{A^{\pm} + \frac{s-u}{4M} B^{\pm}}{4\pi}; \quad \beta^{\pm}(\nu, \cos\theta) = \frac{B^{\pm}}{8\pi M}.$$

These combinations are convenient in that the lower partial waves go over into them very simply in the static limit:

$$\left.\begin{aligned} \varphi^{\pm} \simeq \left[s_1 + \binom{2}{-1} s_3\right] + \frac{\cos\theta}{3}\Big[p_{11} + 2p_{13} \\ + \binom{2}{-1}(p_{31} + 2p_{33})\Big] \equiv s^{\pm} + \cos\theta p^{\pm}; \\ \beta^{\pm}(\nu, \theta) \simeq \frac{1}{3}\left[h_{11} - h_{13} + \binom{2}{-1}(h_{31} - h_{33})\right] \equiv \beta^{\pm}(\nu). \end{aligned}\right\} \tag{1.5}$$

Here $\nu = q^2$ is the square of the meson momentum in the center of mass system, θ is the scattering angle in the center of mass system, s_{2I} is the s wave of the scattering with isospin I, and $p_{2I,\,2J}$ and $h_{2I,\,2J}$ are p waves with isospin I and total angular momentum J. At the same time

$$s_k = \frac{e^{i\delta_k} \sin\delta_k}{q}, \quad p_{ik} = \frac{e^{i\delta_{ik}} \sin\delta_{ik}}{q} = q^2 h_{ik}. \tag{1.6}$$

We now write down unsubtracted forward dispersion relations for $\varphi^{\pm}$ and $\beta^{\pm}$:

$$\varphi^{+}(\nu, 1) = \frac{1}{\pi}\int_0^{\infty} \frac{\operatorname{Im}\varphi^{+}(\nu', 1)}{\nu' - \nu} d\nu'; \tag{1.7a}$$

$$\varphi^{-}(\nu, 1) = \frac{2f^2}{\omega} + \frac{\omega}{\pi}\int_0^{\infty} \frac{\operatorname{Im}\varphi^{-}(\nu', 1)}{\omega'(\nu' - \nu)} d\nu'; \tag{1.7b}$$

$$\beta^{\pm}(\nu, 1) = -\frac{2f^2}{\omega} + \frac{\omega}{\pi}\int_0^{\infty} \frac{\operatorname{Im}\beta^{+}(\nu', 1)\, d\nu'}{\omega'(\nu' - \nu)}; \tag{1.7c}$$

$$\beta^{-}(\nu, 1) = \frac{1}{\pi}\int_0^{\infty} \frac{\operatorname{Im}\beta^{-}(\nu', 1)\, d\nu'}{\nu' - \nu}. \tag{1.7d}$$

Equations (1.7) differ from the well-known dispersion relations for forward πN scattering in two respects.

First, they are written down in the static limit, i.e., we have used assumption 5. In this approximation the laboratory energy is $E = \omega = \sqrt{\nu + 1}$.

Secondly, and this is very important, they are written down without subtractions, since we assume that the approximate representations 4 and 5 are valid in the whole range of integration; it follows by virtue of 4 that $\mathrm{Im}\varphi^{\pm}$ and $\mathrm{Im}\beta^{\pm}$ decrease sufficiently rapidly and the unsubtracted integrals converge.

The dispersion relations for backward scattering written down without subtractions in the static limit have the form:

$$\varphi^{\pm}(\nu, -1) = \frac{1}{\pi}\int_0^{\infty} \frac{\mathrm{Im}\,\varphi^{+}(\nu', -1)}{\nu' - \nu}\, d\nu' + \frac{\alpha_{\sigma}}{1 + \nu_{\sigma} + \nu}; \tag{1.8a}$$

$$\varphi^{-}(\nu_1 - 1) = -\frac{2f^2}{\omega} + \frac{\omega\alpha_{\rho 1}}{1 + \nu_{\rho} + \nu} + \frac{\omega}{\pi}\int_0^{\infty} \frac{\mathrm{Im}\,\varphi^{-}(\nu', -1)}{\omega'(\nu' - \nu)}\, d\nu'; \tag{1.8b}$$

$$\beta^{+}(\nu, -1) = -\frac{2f^2}{\omega} + \frac{\omega}{\pi}\int_0^{\infty} \frac{\mathrm{Im}\,\beta^{+}(\nu', -1)}{\omega'(\nu' - \nu)}\, d\nu'; \tag{1.8c}$$

$$\beta^{-}(\nu, -1) = \frac{\alpha_{\rho 1} + \alpha_{\rho 2}}{2M(1 + \nu + \nu_{\rho})} + \frac{1}{\pi}\int_0^{\infty} \frac{\mathrm{Im}\,\beta^{-}(\nu', -1)}{\nu' - \nu}\, d\nu'. \tag{1.8d}$$

The last terms on the right sides of Eqs. (1.8a), (1.8b), and (1.8d) are the contributions from the annihilation channel and are approximated by the σ- and ρ-meson exchange graphs by means of the effective Lagrangian (1.2). In this case

$$\alpha_{\sigma} = \frac{g_{\sigma NN}\, g_{\sigma}}{8\pi}; \quad \alpha_{\rho_1} = \frac{g_{1V}\, g_{\rho}}{8\pi}; \quad \alpha_{\rho 2} = \frac{g_{2V}\, g_{\rho}}{8\pi}, \tag{1.9}$$

and the parameters ν_{σ} and ν_{ρ} are related to the masses m_{σ} and m_{ρ} by the equations

$$m_{\sigma}^2 = 4(1 + \nu_{\sigma}),\quad m_{\rho}^2 = 4(1 + \nu_{\rho}).$$

To go over from the functions $\varphi^{\pm}$ to the partial waves, we must, in accordance with (1.5), consider the half-sums and half-differences of Eqs. (1.7a) and (1.7b) and (1.8a) and (1.8b):

$$s^{\pm}(\nu) = \frac{\varphi^{\pm}(\nu, 1) + \varphi^{\pm}(\nu, -1)}{2},\quad p^{\pm}(\nu) = \frac{\varphi^{\pm}(\nu, 1) - \varphi^{\pm}(\nu, -1)}{2}.$$

As regards the p-wave combinations $\beta^{\pm}(\nu)$, these must be determined by Eqs. (1.7c) and (1.7d) and (1.8c) and (1.8d). Since we have neglected d waves, a certain ambiguity arises. We shall define the $\beta^{\pm}$ combinations as the half-sum

$$\beta^{\pm}_{(\nu)} = \frac{\beta^{\pm}(\nu, 1) + \beta^{\pm}(\nu, -1)}{2},$$

since the half-sums do not contain d waves.

As a result, we obtain a system of equations for the s and p waves:

$$s^{+}(\nu) = \frac{\alpha_{\sigma}}{2(1 + \nu_{\sigma} + \nu)} + \frac{1}{\pi}\int_0^{\infty} \frac{\mathrm{Im}\, s^{+}(\nu')\, d\nu'}{\nu' - \nu}; \tag{1.10a}$$

$$s^{-}(\nu) = \frac{\omega\alpha_{\rho 1}}{2(1 + \nu_{\rho} + \nu)} + \frac{\omega}{\pi}\int_0^{\infty} \frac{\mathrm{Im}\, s^{-}(\nu')\, d\nu'}{(\nu' - \nu)\,\omega'}; \tag{1.10b}$$

$$p^{+}(\nu)=-\frac{\alpha_{\sigma}}{2(1+\nu_{\sigma}+\nu)}+\frac{1}{\pi}\int_{0}^{\infty}\frac{\operatorname{Im} p^{+}(\nu')\,d\nu'}{\nu'-\nu}; \tag{1.10c}$$

$$p^{-}(\nu)=\frac{2f^{2}}{\omega}-\frac{\omega\alpha_{\rho 1}}{2(1+\nu_{\rho}+\nu)}+\frac{\omega}{\pi}\int_{0}^{\infty}\frac{\operatorname{Im} p^{-}(\nu')\,d\nu'}{\omega'(\nu'-\nu)}; \tag{1.10d}$$

$$\beta^{+}(\nu)=-\frac{2f^{2}}{\omega}+\frac{\omega}{\pi}\int_{0}^{\infty}\frac{\operatorname{Im}\beta^{+}(\nu')\,d\nu'}{\omega'(\nu'-\nu)}; \tag{1.10e}$$

$$\beta^{-}(\nu)=\frac{\alpha_{\rho 1}+\alpha_{\rho 2}}{4M(1+\nu_{\rho}+\nu)}+\frac{1}{\pi}\int_{0}^{\infty}\frac{\operatorname{Im}\beta^{-}(\nu')\,d\nu'}{\nu'-\nu}. \tag{1.10f}$$

In conjunction with the elastic unitarity conditions for the partial waves, Eqs. (1.10) form a complete system of equations for the s and p waves of πN scattering derived in the pure elastic low-energy model for the static case.

$$\operatorname{Im} s_i = q\,|s_i|^2,\quad \operatorname{Im} p_{ik} = q\,|p_{ik}|^2,\quad 0\leqslant q<\infty.$$

We shall not solve this system. Our aim is to verify its internal consistency and agreement with the main experimental data of low-energy πN scattering.

Consider first the threshold behaviors of the $p^{\pm}$ combinations of the p waves. In accordance with the quantum-mechanical threshold conditions

$$p_{ik}(q)\simeq a_{ik}\,q^2 \quad \text{as} \quad q^2\to 0,$$

where the constants a_{ik} are the p-wave scattering lengths. Thus, the combinations $p^{\pm}$ must vanish at the threshold. Equations (1.10c) and (1.10d) yield

$$-\frac{\alpha_{\sigma}}{2(1+\nu_{\sigma})}+\frac{1}{\pi}\int_{0}^{\infty}\frac{\operatorname{Im} p^{+}\,d\nu}{\nu}=0; \tag{1.11a}$$

$$2f^{2}-\frac{\alpha_{\rho 1}}{2(1+\nu_{\rho})}+\frac{1}{\pi}\int_{0}^{\infty}\frac{\operatorname{Im} p^{-}\,d\nu}{\omega\nu}=0. \tag{1.11b}$$

Let us now turn to the asymptotic behavior of the functions $\beta^{\pm}$. By virtue of (1.5) and (1.6), they must decrease not slower than q^{-3} as $q\to\infty$.

With allowance for (1.10e) and (1.10f), this yields the two conditions

$$-2f^{2}-\frac{1}{\pi}\int_{0}^{\infty}\frac{\operatorname{Im}\beta^{+}}{\omega}\,d\nu=0; \tag{1.12a}$$

$$\frac{\alpha_{\rho 1}+\alpha_{\rho 2}}{4M}-\frac{1}{\pi}\int_{0}^{\infty}\operatorname{Im}\beta^{-}\,d\nu=0. \tag{1.12b}$$

One can readily show that the asymptotic conditions (1.12) are threshold conditions for the functions $q^2\beta^{\pm}$, which are, in accordance with (1.5), linear combinations of the p waves in the usual normalization (1.6). At the same time, it must be assumed that unsubtracted dispersion relations hold for the functions $q^2\beta^{-}$ and $(q^2/\omega)\beta^{+}$.

We shall now assume that the solutions of the system (1.10) include a solution that agrees closely with the experimental data in the low-energy region, i.e., all the s and p waves with the exception of p_{33} are small and the p_{33} pass through a resonance. Accordingly, we shall neglect all the imaginary parts with the exception of $\mathrm{Im}\, p_{33}$, which we represent in the form

$$\mathrm{Im}\, p_{33} = \pi \nu_{33} \Gamma_{33} \delta(\omega - \omega_{33}). \tag{1.13}$$

Substituting (1.13) into (1.12a), we obtain

$$\Gamma_{33} = \frac{3}{2} f^2. \tag{1.14}$$

Equations (1.11b) and (1.12b) yield

$$\alpha_{\rho 1} = 0, \quad \frac{\alpha_{\rho 2}}{4M} = \frac{2}{3} \Gamma_{33} \omega_{33}. \tag{1.15}$$

Equation (1.11a) leads to the relationship

$$\frac{\alpha_\sigma}{m_\sigma^2} = \frac{4}{3} \Gamma_{33} \omega_{33}. \tag{1.16}$$

The energy dependence of the s waves is described by the exchange terms. With allowance for (1.14)-(1.16), we obtain

$$s^+(\nu) = \frac{4f^2 \omega_{33}}{1 + 4\frac{q^2}{m_\sigma^2}}, \quad s^- = 0. \tag{1.17}$$

Let us discuss the results obtained. The relation (1.14) is known from the Chew-Goldberger-Low-Nambu theory [3]. This is not surprising, since Eqs. (1.11b) and (1.12a) used to derive (1.14) do not contain contributions from the annihilation channel. However, the relationship (1.14) is numerically unsatisfactory. Substituting the generally adopted value $f^2 = 0.08$, we obtain $\Gamma_{33} = 0.12$.

In accordance with the experimental data the phase δ_{33} passes through a resonance at $M_3 = \sqrt{s_{33}} = 1236$ MeV. The width of the resonance in its proper system is 120 MeV. In the laboratory system this corresponds to $E_{33}^{\mathrm{lab.kin}} = 190$ MeV. The total width in the laboratory system calculated from $\sin^2 \delta_{33}$ is $\gamma_{\mathrm{lab}}^{\mathrm{tot}} = 170$ MeV. Going over to the center of mass system by means of the Jacobian $\partial\omega_{\mathrm{sms}}/\partial E_{\mathrm{lab}}$, we obtain

$$\omega_{33,\,\mathrm{exp}} = 265 \text{ MeV} = 1.92\ \mu, \quad \gamma_{\mathrm{sms,\ exp}}^{\mathrm{tot}} = 75 \text{ MeV} \simeq 0.54\ \mu,$$

which, for the "reduced half-width" in (1.13), yields

$$\Gamma_{33,\,\mathrm{exp}} = \frac{\gamma^{\mathrm{tot}}}{2q_{33}^2} = 0.088.$$

At the same time, a calculation of the integrals $\int \mathrm{Im}\, h_{33} d\omega$ and $\int \mathrm{Im}\, h_{33} \omega\, d\omega$ from the experimental phases leads to the effective values

$$\Gamma_{33,\,\mathrm{exp}}^{\mathrm{eff}} = 0.055, \quad \omega_{33,\,\mathrm{exp}}^{\mathrm{eff}} \simeq 1.9. \tag{1.18}$$

The difference between these numbers and those calculated by means of the Jacobian (in the δ approximation) is due to the fairly large width of the 33 resonance.

The relationship (1.15) strongly contradicts the universality principle (1.4), which is satisfied experimentally (see [4]):

$$\alpha_{\rho 1}^{\mathrm{exp}} \simeq 0.7 - 0.5, \quad \alpha_{\rho 2}^{\mathrm{exp}} \simeq 2.65 - 1.90. \tag{1.19}$$

The second of the relations (1.15) enables one to express ω_{33} in terms of Γ_{33} and $\alpha_{\rho 2}$. Setting $\Gamma_{33} = 0.12$ and using (1.19), we find $\omega_{33} \approx 1.25$, which is smaller than the experimental value (1.18) by a factor of 1.5.

Finally, the expressions for the s waves also strongly contradict the experimental data, according to which

$$s^{+}(0) = a^{+} = 0; \quad s^{-}(0) = a^{-} = 0.08 - 0.09. \tag{1.20}$$

We, therefore, conclude that the pure elastic low-energy description of the low-energy πN scattering based on allowance for only the long-range effects fails to give quantitative agreement with the experimental results for the p waves and fails to give even qualitative agreement for the s waves.

As we have already mentioned, a similar situation obtains for $\pi\pi$ scattering.

1.4. Need for Allowance for Short-Range Effects

Let us now analyze the various means by which the pure elastic low-energy approximation could be improved. Of course, assumptions 5 and 6 used to study the πN system introduce definite errors, but it is known that, although they are important, the relativistic corrections to the static limit can change the numerical results by $\leq 50\%$ in the low-energy region. In addition, assumption 5 does not affect the $\pi\pi$ problem.

Equally, assumption 6 can be improved by the introduction of the f_0 resonance in the two-pion system with I = 0, l = 2, (m_{f_0} = 1260 MeV, $\Gamma_{f_0}^{tot}$ = 120 MeV) and also by allowance for the complex structure of the so-called σ resonance (see Sec. 5.4). However, the corresponding corrections are also small.

Let us return to the original approximations of the pure elastic low-energy model, i.e., the assumptions 2-4. One can show (see, for example, Sec. 17.2 of [1]) that allowance for the nearest neglected partial wave yields small numerical effects. On the other hand, an analysis of the experimental data (phase analysis of πN scattering and the approximate equality of the cross sections at the ρ and f_0 maxima and the geometric cross sections) indicates that the nearest inelastic corrections (the contributions of the three- and four-particle states) to the imaginary parts of the partial amplitudes are also numerically small. Thus, the defect of the pure elastic low-energy model is most probably to be found in the neglect of the influence of the high-energy region on the low-energy region, i.e., the assumption that the low-energy region is physically complete, with its consequence that only long-range effects are important.

Let us illustrate this by taking the example of the pure elastic low-energy model of πN scattering that we have just considered. In this approximation, the scattering lengths a_1 and a_3 (as also in $\pi\pi$ scattering) are strongly overestimated; thus, we obtained $a_1 \approx a_3 \approx 5f^2 \approx 0.4$ (the experimental values are $a_1 \approx 0.16$, $a_3 \approx -0.08$). It is obvious that the theory ignores effects that make important negative contributions to the scattering lengths. To understand the source responsible for these effects, let us consider the function $\varphi^{+}(\nu, 1)$.

It follows from Eq. (1.7a) and the definition (1.5) for φ^{+} that $a^{+} > 0$ for arbitrary solutions [since all f and p waves make positive contributions to Im $\varphi^{+}(\nu, 1)$]. Thus, only the trivial solution can be compatible with $a^{+} = 0$. However, such a solution requires that all the inhomogeneous terms in the system (1.10) vanish, and is clearly meaningless.

Equation (1.7a) corresponds to a forward dispersion relation for the crossing-even combination

$$\varphi^{+}(\nu, 1) = \frac{T(\pi^{+}p) + T(\pi^{-}p)}{2} \tag{1.21}$$

and differs from it by the use of the approximations 2-5 and also by the absence of a subtraction. It is precisely because of the subtraction that the rigorous dispersion relation for (1.21) agrees with the experimental data and, in particular, with the first of Eqs. (1.20). It is therefore clear that we could improve the situation by adding to the right side of Eq. (1.7a) the negative subtraction constant

$$\varphi^{+}(\nu, 1) = \frac{1}{\pi} \int \frac{\operatorname{Im} \varphi^{+}(\nu', 1)}{\nu' - \nu} d\nu' - v,$$

where

$$v = \frac{1}{\pi} \int_0^\infty \frac{\operatorname{Im} \varphi^+(\nu, 1)}{\nu} d\nu \simeq \frac{8}{3} \omega_{33} \Gamma_{33} > 0.$$

This constant reflects the fact that we approximate the original dispersion relation, which does not exist unsubtracted, by the unsubtracted equation (1.7a). Thus, the constant v reflects the properties of the original dispersion relation of the high-energy region, i.e., it has a short-range nature.

Let us elucidate what we have said in more detail by considering the model example of the scattering of neutral spinless particles.

1.5. Analysis of the Short-Range Contributions

Let us consider the dispersion relation for the forward scattering of neutral spinless particles:

$$A(x, 1) = A(0, 1) + \frac{x}{\pi} \int_1^\infty \frac{\operatorname{Im} A(x', 1)\, dx'}{x'(x' - x)}. \tag{1.22}$$

Here $A(x, 1)$ is the forward scattering amplitude considered as a function of the variable

$$x = \left(\frac{s - 2m^2}{2m^2}\right)^2 = (2\nu + 1)^2;$$

in accordance with the optical theorem

$$\operatorname{Im} A(x, 1) = \frac{\sqrt{x-1}}{8\pi} \sigma_{\text{tot}}(x),$$

and a subtraction has been made at the point $x = 0$, which is a symmetry point in the original variable

$$z = \frac{s - 2m^2}{2m^2} = -\frac{u - 2m^2}{2m^2}.$$

Let us follow in detail the procedure for obtaining the equation for the lowest partial wave from the dispersion relation (1.22). We split the range of variation $(1, \infty)$ of the energy variable x into two parts: 1) the low-energy region $1 < x < \Lambda$; 2) the high-energy region $\Lambda < x < \infty$.

In the low-energy region we shall assume that the forward scattering amplitude can be well-approximated by the elastic s wave:

$$A(x, 1) \simeq A_0(x), \quad \operatorname{Im} A_0(x) = K(x) | A_0(x) |^2; \quad 1 < x < \Lambda. \tag{1.23}$$

Here $K(x)$ is the ratio of the momentum to the energy:

$$K(x) = \left(\frac{\sqrt{x} - 1}{\sqrt{x} + 1}\right)^{1/2}.$$

In the high-energy region all the partial waves are important. We shall assume that in this region the s-wave partial cross section is negligibly small compared with the total cross section:

$$\operatorname{Im} A_0(x) \ll \operatorname{Im} A(x, 1) = \frac{\sqrt{x-1}\, \sigma_{\text{tot}}}{8\pi}; \quad \Lambda < x < \infty. \tag{1.24}$$

Under real physical conditions the low-energy region defined by (1.23) and the high-energy region defined by (1.24) are separated by a region of intermediate energies in which there is a smooth transition from the low- to the high-energy conditions. The first approximation consists of assuming that this region is infinitely narrow (or rather sufficiently narrow for one to be able to neglect its integral contributions to the low-energy region).

Let us rewrite identically Eq. (1.22):

$$A(x, 1) = \lambda + \frac{x}{\pi}\int_1^\infty \frac{\mathrm{Im}\, A_0(x')\, dx'}{x'(x'-x)} + \frac{x}{\pi}\int_1^\infty \frac{\mathrm{Im}\,[A(x', 1) - A_0(x')]}{x'(x'-x)}\, dx'. \tag{1.25}$$

Here,

$$\lambda = A(0, 1). \tag{1.26}$$

Using (1.23) and (1.24), we now rewrite (1.25) for the region $1 < x < \Lambda$ in the form

$$A_0(x) = \frac{1}{\pi}\int_1^\infty \frac{\mathrm{Im}\, A_0(x')\, dx'}{x'-x} + V(x), \tag{1.27}$$

where

$$V(x) = \lambda - \frac{1}{\pi}\int_0^\infty \frac{\mathrm{Im}\, A_0(x)\, dx}{x} + \frac{x}{8\pi^2}\int_\Lambda^\infty \frac{\sqrt{x'-1}\,\sigma(x')\, dx'}{x'(x'-x)} \tag{1.28}$$

is the short-range contribution to the low-energy region. In the region $x < \Lambda$ this short-range effective potential can be approximated by a polynomial of the first degree:

$$V(x) \simeq -v + xI, \tag{1.29}$$

where

$$v = \frac{1}{\pi}\int_1^\infty \frac{\mathrm{Im}\, A_0(x')}{x'}\, dx' - \lambda = \lambda_0 - \lambda, \quad I = \frac{1}{8\pi^2}\int_\Lambda^\infty \frac{\sqrt{x-1}\,\sigma(x)\, dx}{x}. \tag{1.30}$$

Equation (1.27) differs from the pure elastic low-energy equation for the s wave by the presence of the potential $V(x)$. Equation (1.27) describes low-energy scattering in the field of the potential $V(x)$. Let us investigate this potential in more detail.

The first important property of $V(x)$ is that it is negative ($v > 0$). The potential $V(x)$ is repulsive. This property is common to the different processes ($\pi\pi$, KN, πN, etc.) and can be established by fairly general considerations related to the behavior of the forward scattering amplitude (or rather its imaginary part) in the high-energy region.

To illustrate this, let us consider the inverse function of the forward scattering amplitude $H(x, 1) = A^{-1}(x, 1)$. Its discontinuity across the cut $1 \le x \le \infty$ is

$$2i\,\mathrm{Im}\, H(x, 1) = -\frac{2i\,\mathrm{Im}\, A(x, 1)}{|A(x, 1)|^2}, \tag{1.31}$$

and, as $x \to \infty$,

$$\mathrm{Im}\, H(x, 1) < -\frac{8\pi}{\sqrt{x}\,\sigma(\infty)}.$$

It follows that the spectral integral that represents $H(x, 1)$ converges. The spectral representation for $H(x, 1)$ has the form

$$H(x, 1) = \lambda^{-1} - \frac{x}{\pi}\int_1^\infty \frac{\mathrm{Im}\, A(x', 1)}{|A(x', 1)|^2}\cdot\frac{dx'}{x'(x'-x)} - \frac{x\alpha_0}{x_0(x_0-x)} - \sum_i \frac{x\beta_i}{x_i(x_i-x)}. \tag{1.32}$$

It follows from the representation (1.22) that $A(x, 1)$ is an R function in the complex plane of X. It follows that $H(x, 1)$ is also an R function, i.e., it cannot have poles in the complex plane, with the possible exception of the real axis. Therefore, $\alpha_0 > 0$ and $\beta_i > 0$. In (1.32) x_0 is the possible position of a zero that lies in front of the reaction threshold, $x_0 < 1$, and x_i are possible zeros in the physical region $x_i > 1$. A necessary condition for Im A to increase as $\sqrt{x}$ as $x \to \infty$ is $H(x, 1) \to 0$ as $x \to \infty$. This gives

$$\lambda^{-1} = \frac{-1}{\pi} \int \frac{\operatorname{Im} A}{|A|^2} \cdot \frac{dx}{x} - \frac{\alpha_0}{x_0} - \sum \frac{\beta_i}{x_i}. \tag{1.33}$$

It can be seen from Eq. (1.33) that $\lambda < 0$ when the amplitude $A(x, 1)$ does not have a zero for $x < 0$, i.e.,

$$\lambda < \lambda_0 = \frac{1}{\pi} \int \frac{\operatorname{Im} A_0}{x} dx \text{ and } v = \lambda_0 - \lambda > 0.$$

However, if $\lambda > 0$, then we must have $\alpha_0 \neq 0$ and $x_0 < 0$, i.e., the function Re A passes through zero in the region $x < 0$. Consequently, we must augment the unsubtracted representation for $A_0(x)$ (1.27) with a negative term to ensure that for $x < 0$ the function $\operatorname{Re} A(x, 1) \approx \operatorname{Re} A_0(x)$ passes through zero; hence, in this case, too, $v > 0$.

This argument is based solely on the very general fact that the imaginary part of the forward scattering amplitude (which is positive because of the unitarity condition) increases, albeit weakly, in the high-energy region:

$$\operatorname{Im} A(x, 1) \geqslant c (\ln x)^{1+\varepsilon}, \quad \varepsilon > 0.$$

Numerical estimates for the values of v in a number of cases can be obtained from the experimental values for the s-wave scattering lengths and the low-energy total cross sections that occur in the crossing-even forward scattering amplitude, which requires subtractions. We shall see below that the values of v in both $\pi\pi$ and πN scattering are $\sim \mu^{-1}$.

The range of the forces associated with the potential $V(x)$ can be estimated by means of a numerical estimate of the coefficient of I in the linear term in (1.29). This linear term becomes important for

$$x \sim x_0 = I^{-1}.$$

Setting, for example, for $\pi\pi$ scattering

$$\Lambda = \left(\frac{s - 2\mu^2}{2\mu^2} \right)^2 \Bigg|_{s \sim 1 \, \mathrm{GeV}^2} \simeq (21.5)^2$$

and $\sigma_{tot} = 0.8\mu^{-2}$ (which corresponds to 16 mb), we obtain

$$x_0 = 5\pi^2 \sqrt{\Lambda} \simeq 1060, \tag{1.34}$$

which corresponds to $s_0 \simeq 2\mu^2(\sqrt{x_0} + 1) \simeq 68\mu^2 \simeq (1.15 \text{ GeV})^2$. Therefore, the "range" of the repulsive forces is smaller by a factor of 10 than the pion Compton length.

Thus, the high-energy effects lead to the appearance in the low-energy region of an effective repulsion with an intensity of ~ 1 (in reciprocal pion masses) and a range of ~ 0.1 of the reciprocal pion mass.

Note that the very fact of a small change of the potential in the low-energy region ($x \ll 1000$) is due solely to the fact that at high energies the cross sections of all processes are less than 0.1 b.

Our next aim is to construct a solvable model for the lower partial waves taking into account this short-range repulsion.

2. NEUTRAL MODEL

2.1. Choice of the Model

The simplest such model can be obtained by the approximate replacement of the potential (1.28) by the pole expression

$$V(x) \to v(x) = -\frac{v}{1 + \frac{x}{p}}; \quad p = \frac{v}{I}. \tag{2.1}$$

The "approximate potential" $v(x)$ possesses the following properties.

A. In the physical region of low energies it is numerically very close to the approximate expression (1.28), having two identical terms in the Laurent expansion.

B. It is finite in the whole of the physical region ($1 < x < \infty$).

C. It has a singularity (a pole at the unphysical partial point $x = -p$).

By virtue of property B the equation for the neutral case with the potential (2.1)

$$A_0(x) = -\frac{v}{1+\frac{x}{p}} + \frac{1}{\pi}\int_0^{\infty} \frac{\operatorname{Im} A_0(x')\,dx'}{x'-x} \tag{2.2}$$

is compatible with the unitarity condition

$$\operatorname{Im} A_0(x) = K(x)\,|A_0(x)|^2$$

in the whole of the interval $1 < x < \infty$ and, hence, can be solved analytically. Of course, among the solutions of Eq. (2.2) one must take those for which $A_0(x)$ is small for $x > \Lambda$. It is only subject to this condition that the solution has physical meaning in the low-energy region for $x < \Lambda$.

By virtue of property C a partial amplitude that satisfies Eq. (2.2) has a pole at $x = -p$. This violates strict analyticity (Sec. 1.1).

However, because of the short-range nature of the phenomenological potential we have introduced, the value of p is large [cf. with (1.34)]. It follows that the analytic properties are violated only in the high-energy part of the complex plane.

It follows that proposition 1 in Sec. 1.1 is replaced by: 1a) the correct analyticity properties of the partial waves in the low-energy part of the complex energy plane (in a circle of radius $|x| \approx \Lambda$).

Here it should be noted that, in accordance with assumptions 2-4, we also have departures from strict crossing symmetry and unitarity in the high-energy region.

2.2. Solution of the Equation

The analytic solution of Eq. (2.2) can be represented in the form

$$A_0(x) = \frac{N(x)}{D(x)}, \tag{2.3}$$

where

$$N(x) = \lambda + \frac{x\beta}{x+p} \tag{2.4}$$

and

$$D(x) = 1 - x\sum \frac{\alpha_i}{x_i(x_i - x)} - cx - \lambda I(x) - \beta x\frac{I(x) - I(-p)}{x+p}. \tag{2.5}$$

Here $I(x) = \frac{x}{\pi}\int_1^{\infty} \frac{K(x')\,dx'}{x'(x'-x)}$.

As in the case of the pure elastic low-energy approximation, the solution (2.3)-(2.5) is multiparametric. It depends on the parameters λ, α_i, and x_i. The value of β can be determined from the condition that at the point $-p$ the function A_0 has a residue equal to $-v$:

$$\beta = \frac{v}{1 - vpI'(-p)}\left\{1 + p\sum \frac{\alpha_i}{x_i(x_i+p)} + pc - \lambda I(-p)\right\}. \tag{2.6}$$

Since p is large,

$$\beta \simeq \frac{v}{1-\frac{v}{\pi}}\left\{1+p\sum\frac{\alpha_i}{x_i(x_i+\rho)}+pc+\frac{\lambda\ln p}{\pi}\right\}. \tag{2.7}$$

It is obvious that for $\lambda > 0$ and $v < \pi$ the function N(x) [i.e., $A_0(x)$] has a zero at the point

$$x_0=-\frac{\lambda p}{\lambda+\beta}.$$

For fixed λ, α_i, and x_i the zero at the point x_0 is nearer the reaction threshold the larger is v. As $v \to 0$, $\beta \to 0$, and $x_0 \to -\rho$, the solution (2.3) goes over into the pure elastic low-energy solution. The presence of this zero means that the solution we have obtained is essentially different from the solution of the pure elastic low-energy approximation, its proximity to the subthreshold region being a measure of the influence of the high energy contribution. For a fixed value of v the value of λ is bounded in the interval $\lambda_{min} \le \lambda \le \lambda_{max}$. Here λ_{min} is determined by the equation $\lambda_{min} + \beta = 0$, when $\alpha_i = 0$, reflecting the fact that the zero in the function N need not lie beyond the pole at the point $-p$. This yields

$$\lambda_{min}=-\frac{1}{\frac{1}{v}-\frac{1}{\pi}+\frac{\ln p}{\pi}}<0;$$

λ_{max} (as in the pure elastic low-energy approximation) is bounded by the condition that there are no bound states, $D(1) = 0$, in the interval $0 < x < 1$. For $p \gg 1$

$$\lambda_{max}\simeq\frac{1-\sum\frac{\alpha_i}{x_i(x_i-1)}-c-\frac{v}{\pi-v}\left\{1+p\sum\frac{\alpha_i}{x_i(x_i+p)}+cp\right\}\frac{\ln p}{p}}{I(1)+\frac{v}{\pi p}\cdot\frac{\ln^2 p}{\pi-v}}.$$

It can be seen from Eqs. (2.5) and (2.4) that there exist two types of solution. If the zero x_0 is sufficiently distant, i.e., $\beta \ll p$, then $\lambda > 0$ and the solution in the low-energy region is virtually the same as the corresponding solution in the pure elastic low-energy approximation. However, if the point x_0 is not far from the threshold, i.e., $\beta \sim p$, then $|\lambda| \sim (\ln p)^{-1}$; the scattering length is $|a_0| \sim (\ln p)^{-1}$ and the solution can have a low-energy resonance since $\beta \ln p/\pi \tilde{p} x_r^{-1}$ [see the last term in (2.5)] can be finite. In this case the width of the resonance is fairly small, $\sim(\ln p)^{-1}$, but it is very difficult to approximate the expression for the s wave by a Breit-Wigner type formula. The s wave can be approximated better by the expression

$$A_0\simeq\frac{\lambda\left(1+\frac{x}{x_0}\right)}{1-\frac{x}{x_r}-i\lambda\left(1+\frac{x}{x_0}\right)K(x)}. \tag{2.8}$$

In order to get an idea of the order of the quantities that restrict the constant λ and the scattering length for solutions that depend strongly on the high-energy contribution, we have given the relevant quantities in Table 1.

TABLE 1. Dependence of λ_{min} and a_{min} on the Position of the Pole p in the Repulsive Potential

p	100	1 000	10 000
$W(p)$	660 MeV	1,1 GeV	2 GeV
$-\lambda_{min}$	1,1	0,61	0,41
$-a_{min}$	0,49	0,36	0,28

The quantity W(p) in the second row characterizes the distance which separates the pole p from the threshold in the energy scale $W = 280\left[\frac{\sqrt{p}+1}{2}\right]^{1/2}$ MeV.

Table 1 shows that even if the pole is at an appreciable distance solutions are possible with a negative s-wave scattering length that is not small.

When the conditions $(\nu/\pi)\ln p \ll 1$ and $\lambda \ll 1$ are satisfied, the resonance solution can be represented conveniently in the form

$$\mathrm{Re}\,A_0(x) = \frac{\lambda+v}{1-x/x_r}; \quad \mathrm{Im}\,A_0 = \pi x_r \Gamma\delta(x-x_r); \tag{2.9}$$
$$\Gamma = \lambda + v = \lambda_0.$$

The integral in Eq. (2.2) can be well-approximated by the resonance contributions. In what follows, we shall exploit this fact to make an approximate analysis of the equations that describe the scattering of real particles.

From our investigation we conclude that the scattering amplitude in the low-energy part of the complex plane $|x| \ll \Lambda$ begins to vanish (or is very nearly equal to zero) at certain points because of the influence of the high-energy contributions. Thus, the threshold conditions for the partial waves with $l \geq 1$ cannot be satisfied without correct allowance for the short-range contributions. On the other hand, the properties of the low-energy amplitude, including the distribution of its zeros, can be obtained by a completely different approximation, namely, the massless pion approximation.

3. CURRENT ALGEBRA AND CHIRAL SYMMETRY

3.1. The Algebra of Charges and the Chiral Group

This section is devoted to an extremely condensed exposition of the main features of the method of current algebra. The reader who is unacquainted with this method should consult the works listed in the bibliography and also the review literature [5].

The appearance of the current-algebra method [6] opened up a new field of activity in the physics of elementary particles. The main quantities in this method, the vector j^μ_A and axial j^μ_{5A} currents, are isotopic three-vectors (A = 1, 2, 3) and a Lorentz four-vector and pseudovector, respectively ($\mu = 0, 1, 2, 3$). It is most convenient to represent j^μ_A and j^μ_{5A} as the currents associated with the variation of a Lagrangian under unitary infinitesimal transformations of the fields:

$$\Psi(x) \to \Psi'(x) = U\Psi(x)U^{-1}; \quad U = 1 - i\alpha_A(x)Q^A - i\beta_B(x)Q^B_5. \tag{3.1}$$

The currents and their divergences are defined in terms of the derivatives of the variation of the total Lagrangian [7]:

$$j^\mu_A = -\frac{\partial\delta L(x)}{\partial(\partial_\mu\alpha_A(x))}; \quad j^\mu_{5B} = -\frac{\partial\delta L(x)}{\partial(\partial_\mu\beta_B(x))}; \tag{3.2}$$

$$\partial_\mu j^\mu_A(x) = -\frac{\partial\delta L(x)}{\partial\alpha_A(x)}; \quad \partial_\mu j^\mu_{5B}(x) = -\frac{\partial\delta L(x)}{\partial\beta_B(x)}. \tag{3.3}$$

The transformation properties of the currents are determined by the properties of the operator generators Q_A and Q_B of the transformation (3.1). If the Lagrangian is invariant under (3.1) with parameters α_A and β_B that do not depend on the coordinates $[\partial_\mu\alpha_A(x) = \partial_\gamma\beta_B(x) = 0]$, (3.3) yields continuity equations for the currents:

$$\partial_\mu j^\mu_A(x) = \partial_\nu j^\nu_{5B}(x) = 0.$$

In this case, the "charges"

$$Q_A(x^0) = \int d^3x j^0_A(x), \quad Q_{5B}(x^0) = \int d^3x j^0_{5B}(x) \tag{3.4}$$

do not depend on the time and must be identified with the generators of the transformation (3.1):

$$Q_A(x^0) = Q_A, \quad Q_{5B}(x^0) = Q_{B5}.$$

One of the most important physical hypotheses of the method is that the currents j^{μ}_{A} and j^{μ}_{5B} describe simultaneously both the weak and the electromagnetic interactions of the hadrons. This hypothesis enables one to establish a relationship between the characteristics of the strong interactions (such as the form factors) and the parameters of the weak interactions of the hadrons (the weak decay lifetimes).

At the same time, the currents, which are the main operator quantities in this method, satisfy definite commutation relations. The form of the commutation relations for the current densities j^{μ}_{A} and j^{ν}_{5B} depends essentially on the specific physical properties attributed to the structure of the elementary particles or the structure of the strong interactions. For example, the commutation relations for the vector currents that follow from the quark particle model [6, 8] differ appreciably from the commutation relations obtained on the assumption of vector dominance [9].

However, irrespective of the concrete commutation relations between the current densities, the commutation relations between the corresponding charges have a very simple form:

$$[Q_A, Q_B] = i\,\varepsilon_{ABC} Q_C; \tag{3.5a}$$

$$[Q_A, Q_{5B}] = i\,\varepsilon_{ABC} Q_{5C}. \tag{3.5b}$$

Here ε_{ABC} is the antisymmetric unit tensor of third rank.

The relations (3.5a) reflect the fact that the three quantities Q_1, Q_2, and Q_3 are the generators of the isotopic group SU(2), i.e., they are the operators of the isospin system,*

$$\mathbf{Q} = \mathbf{I}.$$

The relations (3.5b) indicate that the triplet Q_{51}, Q_{52}, and Q_{53} behaves as a vector under isotopic transformations.

It is usually assumed that the commutator of the axial charges has the form [6]

$$[Q_{5A}, Q_{5B}] = i\,\varepsilon_{ABC} Q_C. \tag{3.5c}$$

This relation, which closes the charge algebra, is the main model assumption. It defines the structure of the theory.

To establish this structure it is convenient to introduce left- and right-polarized charges:

$$\mathbf{Q}^{\pm} = \frac{\mathbf{Q} \pm \mathbf{Q}_5}{2}.$$

The commutation relations (3.1) for the charges take the form

$$[Q^{\pm}_A, Q^{\pm}_B] = i\,\varepsilon_{ABC} Q^{\pm}_C;$$
$$[Q^{+}_A, Q^{-}_B] = 0. \tag{3.7}$$

Thus, the components Q^+ and Q^- form individual algebras of isotopic type [SU(2) algebras]. The charges $\mathbf{Q}$ and $\mathbf{Q}_5$ must therefore be regarded as generators of the group SU(2) × SU(2).

Under the influence of the parity operator P the polarized charges go over into one another:

$$P\mathbf{Q}^{\pm} P^{-1} = \mathbf{Q}^{\mp}.$$

Since the properties of the charges $\mathbf{Q}^{\pm}$ have a great resemblance to those of chiral projection operators (and, for example, also contain them explicitly in the quark realization) the algebra that is obtained is also known as the chiral SU(2) × SU(2) algebra.

The invariance of the Lagrangian under the corresponding group of transformations generated by $\mathbf{Q}$ and $\mathbf{Q}_5$ is known as chiral symmetry.

* Here and in what follows, the symbol $\mathbf{a}$ always denotes an isotopic three-vector. Its components satisfy

$$[Q_A, Q_B] = i\,\varepsilon_{ABC} a_C. \tag{3.6}$$

3.2. The PCAC Hypothesis and the Massless Pion Approximation

Another important hypothesis in the treatment of processes in which pions participate is that of the partially conserved axial current (the PCAC hypothesis), which defines explicitly the divergence of the axial current [7, 10]

$$\partial_\mu j_5^\mu = f_\pi \mu^2 \pi. \tag{3.8}$$

Using this operator relationship one can express the amplitudes of processes with n + 1 external pions in terms of the amplitude of processes with n pions and, in particular, one can relate the amplitude for scattering of a pion by a particle H with the pion form factor of this particle. Here an important role is played by the procedure for extrapolation in the square of the external pion mass from the point $m_\pi^2 = 0$ [the PCAC relation (3.8) is used at this point] to the physical point $m_\pi^2 = \mu^2 = 0.02\ \mathrm{GeV}^2$.

It is obvious that in the massless pion approximation ($m_\pi^2 = 0$) both the vector and the axial currents are conserved [see (3.8)], i.e., the charges Q and Q_5 commute with the Hamiltonian and we have exact chiral symmetry. In this limit, as in quantum electrodynamics, there exist low-energy theorems for the scattering amplitudes (considered at the unphysical point $m_\pi^2 = 0$). One goes onto the mass shell by analytic continuation in m_π^2. In the case of scattering of pions by a heavy target (pion-baryon scattering) $\pi + B \rightarrow \pi + B$, when the "natural" unit of the mass scale is large ($M_B^2 \gg \mu^2$) extrapolation from $m_\pi^2 = 0$ to $m_\pi^2 = \mu^2$ is completely justified. However, the situation is not so favorable in the case of pion-meson scattering. Therefore, in particular, the threshold characteristics of $\pi\pi$ scattering can only be explained in the framework of broken chiral symmetry.

3.3. Different Realization of the Chiral Group

In contrast to, for example, isotopic symmetry, chiral symmetry is a dynamic symmetry. This means that the Lagrangian is symmetric only as a whole, i.e., the Lagrangian of the free fields and the interaction Lagrangian are not symmetric individually.

It has been shown in a number of investigations [11] that all the results of current algebra are equivalent to calculations by means of the lowest orders of perturbation theory for a chiral-symmetric Lagrangian. At the same time, it is necessary to calculate only the contributions corresponding to tree diagrams, which do not contain divergences.

The problem now reduces to the construction of this effective Lagrangian, in which two possibilities arise.

In the first variant the Lagrangian is constructed in terms of field operators that are multiplets or linear realizations of the chiral group SU(2) × SU(2) [12]. In this case one must introduce bosons of opposite parity and specify their quantum numbers. The infinitesimal increments of the fields under the action of the group generators are here linear functions of these fields. In the general case they are given by the corresponding infinitesimal commutators:

$$\delta_A \Phi = -i\,[Q_A, \Phi]\,\alpha_A, \quad \bar{\delta}_B \Phi = -i\,[Q_{5B}, \Phi]\,\beta_B.$$

(Here and in what follows, the symbol $\bar{\delta}$ denotes the increment associated with the transformation generated by the axial charge; α_A and β_B are infinitesimally small parameters and there is no summation on the right sides.)

In this case the invariant Lagrangian is a polynomial function of the field operators and their first derivatives.

The second possibility is based on the fact that an invariant Lagrangian can be constructed in the form of a complicated (nonpolynomial!) function of a small number of fields, for example, the pion and nucleon fields. The pion field π and the nucleon field Ψ, being multiplets of the isotopic group, do not belong in this case to any linear representation of the chiral group as a whole, and their infinitesimal increments depend nonlinearly on the π field [see (4.9) and (6.5)]. In this case, we have a nonlinear realization of the chiral group.

We shall show below how one constructs the Lagrangian and which are the most important properties of the Born-Feynman graphs of this Lagrangian for $\pi\pi$ and πN scattering [13].

We shall apply this approximation to obtain the threshold characteristics for the low-energy equations of $\pi\pi$ and πN scattering. We shall show that the equations can be reconciled with the massless pion approximation provided a short-range repulsion is introduced in the equations and that the simplest solutions of these equations without Castillejo-Dalitz-Dyson terms (i.e., that do not contain additional parameters) already possess a strong and even resonance p-wave interaction in a reasonable energy range. This clearly indicates that in the framework of the theory that is nonlinear in the π field we obtain dynamical resonances with higher spins that phenomenologically realize the chiral group SU(2) × SU(2) linearly.

4. MASSLESS PION APPROXIMATION IN THE $\pi\pi$ INTERACTION

4.1. Simplest Linear Realization of Chiral Symmetry (Sigma Model)

Since the chiral transformations (3.1) do not conserve parity, the simplest linear realization that contains a pion must also contain a scalar particle σ. If the isotopic spin of this particle is, like that of the pion, equal to 1, the (π, σ) multiplet can be constructed [14] from the linear representations (0, 1) and (1, 0) of the chiral group. This case is not of interest physically since experiments have not revealed even a trace of the σ particle (I = 1). If one assumes that the isotopic spin of the σ particle vanishes, then a multiplet can be constructed from the linear representation (1/2, 1/2) and the corresponding model is known as the Gell-Mann–Levy sigma model [7].

The importance of this model is also due to the fact that it is intimately related to the PCAC approximation.

In this model the variations of the π and σ fields are given by the equations

$$\left.\begin{aligned}\bar{\delta}\pi_A &\equiv -i\left[\beta_B Q_5^B, \pi_A\right] = \beta_A \sigma;\\ \bar{\delta}\sigma &\equiv -i\left[\beta_B Q_5^B, \sigma\right] = -\beta_A \pi_A;\end{aligned}\right\} \tag{4.1}$$

$$\left.\begin{aligned}\delta\pi_A &\equiv -i\left[\alpha_B Q^B, \pi_A\right] = -\varepsilon_{ABC}\alpha_B \pi_C;\\ \delta\sigma &\equiv 0.\end{aligned}\right\} \tag{4.2}$$

The chiral-symmetric Lagrangian corresponding to (4.1) and (4.2) has the form

$$L = \frac{1}{2}(\partial_\mu \pi)^2 + \frac{1}{2}(\partial_\mu \sigma)^2 - \frac{m^2}{2}(\pi^2 + \sigma^2) + \sum_{n \geqslant 2} \lambda_n (\pi^2 + \sigma^2)^n. \tag{4.3}$$

This Lagrangian leads to conservation of the axial current and equality of the masses of the π and σ mesons. We shall now show that the addition to the Lagrangian of the term

$$L_{\mathrm{br}} = f_\pi \mu^2 \sigma, \tag{4.4}$$

which violates the conservation of the axial current, automatically leads to mass inequality. To see this, we calculate the corresponding variation of the Lagrangian:

$$\partial_\mu j_5^\mu = -\frac{\partial(L + L_{\mathrm{br}})}{\partial\beta} = f_\pi \mu^2 \pi, \tag{4.5}$$

i.e., the PCAC condition (3.4). On the other hand, the interaction (4.4) leads to the appearance of Feynman graphs in which there exists a $\sigma \to$ vacuum transition ("tadpole" graphs) [15]. Such graphs can be eliminated by means of the shift transformation

$$\sigma = \sigma_0 + \sigma'; \quad \sigma_0 = \langle 0|\sigma|0\rangle.$$

Restricting ourselves in the Lagrangian (4.3) to the term n = 2, we obtain the condition for such graphs to compensate each other:

$$-m^2\sigma_0 + 4\lambda\sigma_0^3 + f_\pi \mu^2 = 0.$$

At the same time the interaction term ($\sim \lambda$) leads to mass renormalization:

$$\mu^2 = m^2 - 4\lambda\sigma_0^2;$$
$$m_\sigma^2 = m^2 - 12\lambda\sigma_0^2.$$

The condition of compatibility of these relations is

$$\sigma_0 = f_\pi.$$

The constant λ can be expressed in terms of the physical masses μ and m_σ as follows:

$$\lambda = -\frac{m_\sigma^2 - \mu^2}{8 f_\pi^2}. \tag{4.6}$$

Note that in the limit $\mu^2 \to 0$ the symmetry of the Lagrangian is restored. The $\pi\pi$ scattering lengths in the Born approximation have the form

$$a_0 = \frac{1}{4\pi}\left(5\lambda + \frac{24\lambda^2 f_\pi^2}{m_\sigma^2 - 4\mu^2} + \frac{16\lambda^2 f_\pi^2}{m_\sigma^2}\right);$$
$$a_2 = \frac{1}{4\pi}\left(2\lambda + \frac{16\lambda^2 f_\pi^2}{m_\sigma^2}\right). \tag{4.7}$$

As $\mu^2 \to 0$, (4.7) yields [with allowance for (4.6)]

$$a_0 = a_2 = 0. \tag{4.8}$$

Thus, we have seen that in the limit in which chiral symmetry is satisfied ($\mu^2 \to 0$) the simplest linear model that satisfies the PCAC condition gives nonvanishing scattering lengths.

4.2. Nonlinear Realization of the Chiral Group*

The same result can be obtained without introducing the σ meson explicitly. To do this one must regard the pion field as a nonlinear realization of the chiral group. The nonlinear realization is defined by the functions of the π-meson field that characterize the increment of the pion field under the infinitesimal transformation associated with Q_5. This increment has the form

$$\delta\pi_A = -i[\beta_B Q_5^B, \pi_A] = \beta_A f(\pi) + \pi_A(\beta\pi) g(\pi^2). \tag{4.9}$$

To satisfy the Jacobi identity

$$[Q_5^A, [Q_5^B, \pi^C]] - [Q_5^B, [Q_5^A, \pi^C]] = [[Q_5^A, Q_5^B], \pi^C],$$

where the right side is calculated by means of (3.5c) and (3.6), it is necessary that the functions f and g satisfy

$$1 + 2f'(x) f(x) = g(x)\{f(x) - 2xf'(x)\}, \quad x = \pi^2, \quad f' = df/dx. \tag{4.10}$$

Depending on the choice of f, this equation gives the function g and, consequently, specification of f completely characterizes the transformation law (4.9). One might get the impression that Eq. (4.9) defines a complete class of inequivalent realizations of the chiral group. However, this is not so. One can show [13, 16] that the indeterminacy in the choice of f is entirely transferred to the indeterminacy associated with the redefinition of the pion field:

$$\tilde{\pi}_A = \pi_A \Phi(\pi^2)$$

and, as a result, $\tilde{f}(\pi^2) = f(\pi^2)\Phi(\pi^2)$. It follows that a concrete specification of f is equivalent to a definite specification of the physical π field that describes the real π meson.

---

* Here we follow the exposition of Weinberg [13].

Let us now turn to the construction of the corresponding invariant Lagrangian. It can be seen from (4.9) that the variation of the gradient $\overline{\delta\partial_{\mu}\pi}$ is proportional to the gradient itself. As a result, one introduces the so-called covariant derivative

$$D_{\mu}\pi_{A}=d_{AB}(\pi)\,\partial_{\mu}\pi_{B}$$

for the invariant generalization of the kinetic term $(1/2)(\partial_{\mu}\pi)^2$.

The function d_{AB} must be chosen so as to ensure that the commutation relations have the form:

$$\begin{aligned}\left[Q_5^A, D_{\mu}\pi^C\right] &= -iV^{AB}(\pi)\,\varepsilon_{BCD}D_{\mu}\pi^D;\\ \left[Q^A; D_{\mu}\pi^C\right] &= -i\varepsilon_{ABC}D_{\mu}\pi_{\beta}.\end{aligned} \tag{4.11}$$

The presence on the right sides of these equations of antisymmetric tensors ensures invariance of the generalized kinetic term. Using the Jacobi identities and the commutation relations of the operators Q^A and Q_5^B, we obtain

$$V_{AB}(\pi)=\varepsilon_{ABC}\pi_C W(\pi^2), \tag{4.12a}$$

where

$$W(x)=\left[f(x)+\sqrt{f^2(x)+x}\right]^{-1}, \tag{4.12b}$$

and also

$$D_{\mu}\pi\sim\frac{\partial_{\mu}\pi}{[f^2(\pi^2)+\pi^2]^{1/2}}-\frac{f'(\pi^2)+\frac{1}{2}W(\pi^2)}{f^2(\pi^2)+\pi^2}\,\pi\partial_{\mu}\pi^2. \tag{4.13}$$

The following special cases are most widely known.

A. Weinberg's definition [13] of the π field: $g=\lambda$, $f(\pi^2)=(1-\lambda^2\pi^2)/2\lambda$. From (4.12) and (4.13) we also have $W=\lambda$.

Using (4.9) one can show that it is impossible to construct an invariant expression that depends only on the field π (but not on its gradients). The Lagrangian therefore has the form

$$L=\frac{1}{2}(D_{\mu}\pi)^2=\frac{1}{2}\cdot\frac{(\partial_{\mu}\pi)^2}{(1+\lambda^2\pi^2)^2}\,.$$

In this expression the normalization factor omitted in (4.13) is chosen in such a manner that the usual expression is obtained in the limit $\lambda\to 0$.

B. The definition of the π field corresponding to the sigma model [17], $g=0$ [cf. with (4.1)], $f=\sqrt{f_0^2-\pi^2}$.

$$W=\left(f_0+\sqrt{f_0^2-\pi^2}\right)^{-1}$$

and

$$L=\frac{1}{2}(\partial_{\mu}\pi)^2+\frac{1}{2}\cdot\frac{(\pi\partial_{\mu}\pi)^2}{f_0^2-\pi^2}\,.$$

This Lagrangian corresponds to the Lagrangian (4.3) to within additive c numbers if one makes the identification

$$\sigma=\sqrt{f_0^2-\pi^2}.$$

It is now clear that, since the Lagrangians have a purely kinetic form, their Born terms in $\pi\pi$ scattering are proportional to the scalar derivatives of the four-momenta. Since the mass of the π meson vanishes, the scattering lengths also vanish.

4.3. Broken Symmetry in the Nonlinear Case

Massless pions correspond to exact symmetry; broken symmetry corresponds to pions with a mass. We shall describe the symmetry breaking by adding a term of the form $\Phi(\pi^2)$ to the Lagrangian. The linear term of the expansion of this function yields the mass term:

$$\pi^2 \Phi'(0) = -\frac{\mu^2}{2}\pi^2,$$

and the quadratic term the contact $\pi\pi$ interaction:

$$\frac{1}{2}(\pi^2)^2 \Phi''(0) = \lambda(\pi^2)^2.$$

In the Born approximation the isotopic combinations

$$2A_0(s, t, u) - 5A_2 \text{ and } A_1(s, t, u)$$

do not therefore contain terms of order λ, i.e., they do not depend on the form of the symmetry breaking function $\Phi(\pi^2)$. Moreover, one can show that they depend only on the single parameter f_π, which is determined by the lifetime of the $\pi^\pm$ mesons. To show this, let us consider the contact $\pi\pi$ scattering terms that follow from the Lagrangian

$$L = \frac{1}{2}(D_\mu \pi)^2 + \Phi(\pi^2). \tag{4.14}$$

It is obvious from (4.13) that to do this we must specify the first two terms of the expansion of f:

$$f(x) = f_0 + x f_1.$$

Using Eq. (4.10), one can express f_1 in terms of f_0 and $g_0 = g(0)$:

$$f_1 = \frac{g_0}{2} - \frac{1}{2f_0}.$$

Then the expansion of the Lagrangian (4.14) in the π fields becomes

$$L = \frac{(\partial_\mu \pi)^2}{2} - \frac{\mu^2 \pi^2}{2} - \frac{g_0}{2f_0}\pi^2(\partial_\mu \pi)^2 + \left(\frac{1}{2f_0^2} - \frac{g_0}{f_0}\right)(\pi\,\partial_\mu \pi)^2 + \lambda(\pi^2)^2 + \ldots \tag{4.15}$$

Calculating the Born terms, we obtain

$$2A_0(s, t. u) - 5A_2(s, t, u) = \frac{3(2s-t-u)}{32\pi f_0^2}; \tag{4.16}$$

$$A_1(s, t, u) = \frac{t-u}{32\pi f_0^2}. \tag{4.17}$$

Using this Lagrangian to calculate the axial current,

$$j_{5\mu}^A = -\frac{\delta L}{\delta \partial_\mu \beta_A},$$

we note that the coefficient of $\partial_\mu \pi$ is exactly equal to f_0. It is precisely this term that is responsible for the decay of the charged pion $\pi \to \mu + \nu$. It follows that f_0 can be calculated from the lifetime of the

TABLE 2. Dependence of the s-Wave $\pi\pi$ Scattering Lengths on the Type of Chiral Symmetry Breaking

Type of symmetry breaking	$\xi = a_0/a_2$	a_0	a_2
Equation (4.20), $g_0 f_\pi = 0,5$	$-1/2$	0,057	$-0,105$
The same, $g_0 = 0$	∞	0,315	0
" " $g_0 = -\infty$	5/2	$+\infty$	$+\infty$
Equation (4.21), $N = 1$	$-7/2$	0,184	$-0,0525$
The same, $N = 2$	∞	0,315	0
" " $N = 3$	95/14	0,495	0,73
" " $N \to \infty$	5/2	$+\infty$	$+\infty$

π meson. However, it is usually determined by means of the Goldberger-Treiman relation [18] (see Sec. 6.1):

$$f_0 = f_\pi = -\frac{g_A}{g_V} \cdot \frac{M}{g} = 0.615\ \mu = 86\ \text{MeV}. \tag{4.18}$$

Equations (4.16) and (4.17) yield the scattering lengths*

$$2a_0 - 5a_2 = 0.63; \quad A_1' = 0.035\ q^2. \tag{4.19}$$

The ratio of the scattering lengths a_0 and a_2 can be obtained by fixing a definite form of the chiral symmetry breaking.

Thus, setting $\mu \neq 0$ and $\lambda = 0$ in (4.15), i.e., breaking the symmetry only through the mass term, we obtain

$$\frac{a_0}{a_2} = \frac{5}{2} - \frac{3}{2g_0 f_\pi}. \tag{4.20}$$

Another breaking mechanism, which may be called algebraic, is to attribute definite transformation properties to the term Φ (4.14) under transformations associated with the chiral group. Weinberg [13] shows that if the term $\Phi(\pi^2)$ belongs to the representation (N/2, N/2), the symmetry-breaking contact interaction has the form

$$\frac{N(N+2)+2}{40 f_\pi^2} \mu^2 (\pi^2)^2.$$

For the ratio of the scattering lengths this gives

$$\frac{a_0}{a_2} = \frac{5}{2} + \frac{30}{N(N+2)-8}. \tag{4.21}$$

In Table 2 we give the numerical values of a_0 and a_2 calculated by means of (4.19)-(4.21) for a number of specific cases of symmetry breaking, including those considered above.

The fourth row of the table (N = 1) corresponds to Weinberg's well-known result [19]: $a_0 \approx 0.2$, $a_2 \approx -0.06$, which he first obtained in an investigation of the soft pion approximation in current algebra. It follows that the case N = 1 corresponds to the PCAC formula (4.5). The other scattering lengths given in Table 2 correspond to expressions for $\partial_\mu j_5^\mu$ in which expansion terms of higher orders in the powers of the π field are added on the right side of (4.5).

Let us note a further interesting fact. In the framework of the linear realization of chiral symmetry (sigma model) the Weinberg scattering lengths can be obtained by letting the mass m_σ of the σ meson tend to infinity in Eqs. (4.7) and (4.6) [11].

* In calculating (4.18) we have used the new value $g_A/g_V = -1.23$. In [13] $f_\pi = 0.59\,\mu$ was obtained by means of the old value, viz. -1.18. Note also that the calculation from the pion lifetime yields $0.68\,\mu$ = 95 MeV. As a result, the reliability of the numerical values on the right sides of (4.19) are evidently ~ ±10%.

5. PION SCATTERING

5.1. Equations

The dispersion relations for $\pi\pi$ forward scattering can be considered conveniently for the following combinations of the structure amplitudes:

$$\left.\begin{aligned} T_0 &= B + \frac{A+C}{3}; \\ T_1 &= A - C; \\ T_2 &= A + C. \end{aligned}\right\} \quad (5.1)$$

If s and u are transposed, they transform as follows:

$$T_I(s, u) = (-1)^I T_I(u, s). \quad (5.2)$$

Here the index 1 corresponds to the isospin in the t channel.

To discuss the high-energy behavior of these combinations we shall use the concept of Regge poles. For the leading asymptotic term for t = 0, we obtain

$$T_0 \sim s, \quad T_1 \sim s^{\alpha_\rho(0)} \sim \sqrt{s}. \quad (5.3)$$

At the present time, no single resonance with isospin 2 and mass less than 1.5 GeV has been detected. It is therefore natural to assume that even if there exists a corresponding Regge trajectory it lies much lower, i.e.,

$$s^2 T_2 \to 0 \quad \text{as} \quad s \to \infty. \quad (5.4)$$

Thus, a subtraction is required for only the single dispersion relation for T_0 or, equivalently, the dispersion relation for the crossing-even structure function B. It follows that the vacuum singularities in the low-energy equations can be handled in exactly the same way as the neutral case. As we shall see below, the repulsion parameter is $v \sim 0.5$. The contribution from the high-energy region to T_1 due to the ρ-meson trajectory has the form zv_1 (5.2). The parameter v_1 can be estimated on the basis of the universality hypothesis for the ρ-meson interaction [20]:

$$v_1 = \frac{2}{8\pi^2} \int_{z_H}^{\infty} \frac{\sigma_{+-} - \sigma_{++}}{z^2} \sqrt{z^2 - 1}\, dz \sim 0.02 \quad \text{for} \quad z_H \simeq 25. \quad (5.5)$$

In what follows we shall neglect this contribution.

For the forward scattering dispersion relation, the relationship (5.4) gives the so-called superconvergent sum rule [21]:

$$\frac{2}{\pi} \int \left\{ \frac{\operatorname{Im} A_0}{3} - \frac{\operatorname{Im} A_1}{2} + \frac{\operatorname{Im} A_2}{6} \right\} z\, dz = 0. \quad (5.6)$$

In low-energy physics, this sum rule plays the role of a bootstrap type condition: in the resonance approximation it relates the masses and widths of the resonances. At the same time, the presence of the factor z in the integrand means that this sum rule is very sensitive to resonances with large mass and to the nonresonance background environment. In other words, it depends strongly on the cross sections in the intermediate energy range. Unfortunately, these cross sections are virtually unknown for $\pi\pi$ scattering and we cannot actually use Eq. (5.6) to solve the low-energy system of equations. Nevertheless, using the low-energy characteristics of the scattering amplitude and (5.6), one can estimate the contributions from the intermediate region.

Restricting ourselves to s and p waves and using assumptions 2-4 of Sec. 1.1, we obtain the following system of equations for the two s waves $A_0 \equiv A^{l=0}_{I=0}$ and $A_2 \equiv A^{l=0}_{I=2}$ and the p wave $A_1 \equiv A^{l=1}_{I=1}$:

$$A_i(z) = -\alpha_i v(z) + \frac{1}{\pi}\int_1^\infty \frac{\operatorname{Im} A_i}{z' - z}\, dz' + \frac{b_{ik}}{\pi}\int_1^\infty \frac{\operatorname{Im} A_k\, dz'}{z' + z}, \tag{5.7}$$

where b_{ik} is the crossing symmetry matrix (see Sec. 13.1 in [1])

$$b_{ik} = \begin{pmatrix} 1/3 & -3 & 5/3 \\ -1/9 & 1/2 & 5/18 \\ 1/3 & 3/2 & 1/6 \end{pmatrix},$$

$\alpha = (1, 1/3, 1)$, and the potential of the short-range repulsion $v(z)$ can be represented in the form

$$v(z) = v - z^2 \frac{v}{p^2} \simeq \frac{v}{1 + z^2/p^2}. \tag{5.8}$$

The threshold condition for the p wave takes the form

$$\frac{1}{\pi}\int_1^\infty \frac{\operatorname{Im} A_1}{z-1}\, dz + \frac{1}{\pi}\int \frac{dz}{z+1}\left[-\frac{\operatorname{Im} A_0}{9} + \frac{\operatorname{Im} A_1}{2} + \frac{5 \operatorname{Im} A_2}{18}\right] = \frac{v(1)}{3}. \tag{5.9}$$

Unfortunately, it is very difficult to obtain analytically an exact solution of the system (5.7). Like the neutral model it has many solutions. This is already evident in an investigation of the analogous system of equations without repulsion (see Ch. 3 in [1]). As is well known, the characteristic features of the solution in this case are a small width of the ρ meson and large positive scattering lengths a_0 and a_2. Now, after the introduction of the repulsion, it can be seen from the threshold condition (5.9) that the parameter $v(1)$ increases the ρ-meson width (increases the integral contribution to the threshold condition from the ρ wave). At the same time the scattering lengths decrease.

To obtain a crude estimate of v one can use Eq. (5.8), according to which $v \approx v(1)$ for $p \gg 1$.

Identifying the expansion term in (5.8) that is proportional to $z^2 = x$ with the linear term of a formula analogous to (1.28), we obtain

$$v = \frac{p^2}{4\pi}\int_{z_H}^\infty \frac{dz\sqrt{z^2-1}}{z^3}\, \sigma_B(z) \simeq \frac{p^2 \sigma_\infty^B}{4\pi^2 z_H};$$

now, setting

$$p \sim z_H \gg 25\,(W_H = 1\,\text{GeV} \text{ and } \sigma_B(\infty) \simeq \frac{0.8}{\mu^2} (\sim 16\text{ mb}),$$

we find

$$v \gtrsim 0.5.$$

This estimate shows that the high-energy effects exert an important influence on the s-wave scattering lengths and the ρ-meson width.

We shall now consider in more detail the influence of the short-range repulsion on the threshold characteristics of the s and p waves.

5.2. Threshold Analysis

We recall that the expressions for the combinations $2A_0 - 5A_2$ and A_1 [(4.16) and (4.17)] obtained in the investigation of the phenomenological Lagrangians do not depend on the method chosen to break the chiral symmetry. At the same time, we considered only the phenomenological approximation [16], i.e., the approximation that takes into account only tree graphs (Feynman graphs without internal loops). It is therefore natural to assume that such an approximation works sufficiently well in the analyticity region of the scattering amplitude in the neighborhood of the point z = 0. In other words, Eqs. (4.16) and (4.17) define expansions in z of the solutions of Eqs. (5.7) for small

$$2A_0 - 5A_2 = \frac{3(3z+1)}{16\pi f_\pi^2}; \tag{5.10a}$$

$$A_1 = \frac{z-1}{48\pi f_\pi^2}. \tag{5.10b}$$

Let us now expand Eq. (5.7) in powers of z. Retaining terms of order z, we have

$$A_i(z) \approx -a_i v + \frac{1}{\pi}\int \frac{dz'}{z'} [\operatorname{Im} A_i + b_{ik} \operatorname{Im} A_k] + \frac{z}{\pi}\int \frac{dz'}{z'^2} [\operatorname{Im} A_i - b_{ik} \operatorname{Im} A_k] + \ldots \tag{5.11}$$

The expansions (5.11) have a radius of convergence z = 1. However, if the Im $A_i(z')$ for $z' \sim 1$ are small, the first terms of such expansions can also be used for $z \sim 1$. Therefore, we shall use Eq. (5.11) to analyze solutions with short scattering lengths.

Comparing Eqs. (5.10) and (5.11), we obtain

$$3v + \frac{1}{\pi}\int \frac{dz}{z}\left[\operatorname{Im} A_0 - \frac{27}{2}\operatorname{Im} A_1 - \frac{5}{2}\operatorname{Im} A_2\right] = \frac{3}{16\pi f_\pi^2}. \tag{5.12}$$

Now, defining v in terms of integrals of Im A_1 from the threshold condition for the p wave

$$3v = \frac{27}{2\pi}\int \frac{\operatorname{Im} A_1}{z} dz + \frac{1}{\pi}\int \frac{dz}{z}\left(-\operatorname{Im} A_0 + \frac{5}{2}\operatorname{Im} A_2\right) + \frac{1}{\pi}\int \frac{dz}{z^2}\left[\operatorname{Im} A_0 + \frac{9}{2}\operatorname{Im} A_1 - \frac{5}{2}\operatorname{Im} A_2\right]$$

in the approximation (5.11), we obtain the following expression from (5.12):

$$\frac{1}{\pi}\int \frac{dz}{z^2}\left\{\operatorname{Im} A_0 + \frac{9}{2}\operatorname{Im} A_1 - \frac{5}{2}\operatorname{Im} A_2\right\} = \frac{3}{16\pi f_\pi^2}. \tag{5.13}$$

One can show that (5.13) ensures that both the equations (5.10) are satisfied at the symmetry point z = 0 and also on the threshold z = 1 to within terms $\sim \int dz z^{-3}$. The accuracy with which (5.10) is satisfied can be increased by requiring

$$\frac{1}{\pi}\int \frac{dz}{z^3}\left\{\operatorname{Im} A_0 - \frac{27}{2}\operatorname{Im} A_1 - \frac{5}{2}\operatorname{Im} A_2\right\} = 0. \tag{5.14}$$

The conditions (5.13) and (5.14) ensure that (5.10) is satisfied to within terms $\sim \int dz z^{-4}$.

The scattering lengths have the form

$$a_0 = \frac{5}{2\pi}\int \frac{dz}{z}\left\{\operatorname{Im} A_0 - \frac{9}{2}\operatorname{Im} A_1 + \frac{1}{2}\operatorname{Im} A_2\right\} + \frac{1}{16\pi f_\pi^2}; \tag{5.15}$$

$$a_2 = \frac{2}{3\pi}\int \frac{dz}{z}\left\{\operatorname{Im} A_0 - \frac{9}{2}\operatorname{Im} A_1 + \frac{1}{2}\operatorname{Im} A_2\right\} - \frac{1}{8\pi f_\pi^2}. \tag{5.16}$$

We expect that the system of equations has a solution with a resonance in the p wave whose characteristics are very similar to the experimental data. In addition, we shall assume that the partial wave A_2 is small over a wide energy range (right up to 1 GeV) and we shall therefore neglect its imaginary part. To make a quantitative estimate of (5.15) and (5.16), we shall use the fact that at the present time the position and width of ρ can be assumed to be fairly well established:

$$m_\rho = 765 \text{ MeV}, \quad \Gamma_\rho^{tot} = 120 \text{ MeV}. \tag{5.17}$$

TABLE 3. Dependence of the Parameter of Strong Repulsion v and the s-Wave Contribution to the Threshold Conditions for the p Wave and the Ratio of the $\pi\pi$ Scattering Lengths on the Way in Which the Chiral Symmetry Is Broken

Method of symmetry breaking	Ratio of scattering lengths	$\frac{1}{\pi}\int_1^\infty \frac{dz}{z}\,\mathrm{Im}\,A_0$	v
$g_0 f_0 = 0,5$	$-1/2$	0,747	0,56
1 (PCAC)	$-7/2$	0,796	0,54
2	$\pm\infty$	0,906	0,51

It follows that the ρ meson is fairly narrow and that the imaginary part of the corresponding partial wave can be approximated by a δ function:

$$\mathrm{Im}\,A_1(z) = \pi\gamma_1 z_1 \delta(z - z_1). \tag{5.18}$$

Here,

$$z_1 = \frac{m_\rho^2 - 2\mu^2}{2\mu^2} = 14.0; \quad \gamma_1 = \frac{m_\rho \Gamma_\rho^{\mathrm{tot}}}{m_\rho^2 - 2\mu^2} \simeq 0.17.$$

The integral from the p wave in (5.18) makes a contribution to (5.15) that is numerically equal to 1.28 of the pion Compton wavelength. At the same time $3/16\pi f_\pi^2 = 0.158$ [see (4.18)]. Therefore, short scattering lengths corresponding to weak chiral symmetry breaking [small N in Eq. (4.21)] are possible only when the integral of Im A_0 in (5.15) and (5.16) is compensated by the integral of Im A_1, i.e., one must have approximately

$$\frac{1}{\pi}\int_1^\infty \mathrm{Im}\,A_0 \frac{dz}{z} \simeq \frac{9}{2\pi}\int \mathrm{Im}\,A_1 \frac{dz}{z} = 0.756.$$

Table 3 can serve as an illustration of this result.

Note that if such compensation occurs the parameter of the high-energy repulsion is strongly nonzero. In this case we can express the integral in (5.9), which contains Im A_0, in terms of a_2, obtaining

$$v = 3\gamma_1 - \frac{a_2}{2}.$$

However, if v = 0, such compensation is impossible, and we arrive at long scattering lengths:

$$a_0 \simeq \frac{15}{\pi}\int \mathrm{Im}\,A_1 \frac{dz}{z} \simeq 15\gamma_1 \simeq \frac{5}{2}a_2 = 2.5.$$

Such a situation corresponds to large N (> 10) in Eq. (4.21).

Expressing the integral of Im A_0 in terms of the repulsion parameter, we obtain simple formulas for the scattering lengths:

$$a_0 = -5v + 2.84; \quad a_2 = -2v + 1.02. \tag{5.19}$$

It follows from (5.19) that v ~ 0.5 when a_2 is small, which corresponds to the estimate from the high-energy integrals (see Sec. 5.1). This is an independent indication that the system of equations (5.7) has a solution with the experimental p wave and short scattering lengths.

Thus, the introduction of a short-range repulsion in the low-energy dispersion theory for the low partial waves enables one to reconcile the theory with the approximation based on a small pion mass. It turns out that the strength of the repulsion v is a very convenient parameter for making a quantitative description of $\pi\pi$ scattering in the low-energy region. It enables one to relate the method of symmetry breaking to the high-energy scattering characteristics due to the vacuum contributions.

5.3. Solvable Model $A_2 = 0$

In our above threshold analysis we have shown that short s-wave scattering lengths arise only if there is a definite compensation of the large integral contributions from Im A_1 and Im A_0 (5.17). One can exploit the idea of this compensation and, using Eqs. (5.7), construct a solvable model for the partial waves A_0 and A_1, whose solution in the region up to 1 GeV gives a reasonable description of the energy dependence of the δ_0^0 and δ_1^1 phases and also corresponds to (5.10).

We shall base such a model on the assumption that in the considered physical region both Im A_2 and Re A_2 may be assumed to vanish. Then, using the third of Eqs. (5.7) to express the repulsion potential v(z) in terms of the crossing integrals of Im A_0 and Im A_1, and substituting the resulting expressions into the first two equations of (5.7), we obtain a solvable model for the waves A_0 and A_1:

$$A_0(z) = \frac{1}{\pi}\int_1^\infty \frac{\operatorname{Im} A_0(z')}{z'-z}\,dz' - \frac{9}{2\pi}\int_1^\infty \frac{\operatorname{Im} A_1(z')}{z'+z}\,dz';$$

$$A_1(z) = \frac{1}{\pi}\int_1^\infty \frac{\operatorname{Im} A_1(z')}{z'-z}\,dz' - \frac{2}{9\pi}\int_1^\infty \frac{\operatorname{Im} A_0(z')}{z'+z}\,dz'. \tag{5.20}$$

This model has the simple crossing symmetry

$$A_0(-z) = -\frac{9}{2}A_1(z). \tag{5.21}$$

It follows from (5.21) that the wave A_0 has a zero on the crossing threshold for $z = -1$. The solution obtained by the transition to the inverse function has the form

$$H(z) = A_0^{-1}(z) = -\frac{2}{9}A_1^{-1}(-z) = \frac{1}{\lambda} - \frac{R_0 z}{z+1} - I(z) + \frac{2}{9}I(-z) - \sum_i \frac{z\alpha_i}{z_i(z_i-z)} - cz. \tag{5.22}$$

Here R_0, α_i, $c > 0$, $|z_i| > 1$, and also

$$I(z) = \frac{z}{\pi}\int \sqrt{\frac{z'-1}{z'+1}}\cdot\frac{dz'}{z'(z'-z)}.$$

The absence of bound states, i.e., the absence of poles in the interval $-1 < z < 1$, leads to an upper bound on the possible values of λ:

$$\lambda < \lambda_{\max} = \left|\frac{R_0}{2} + I(1) - \frac{2}{9}I(-1) + \sum \frac{\alpha_i}{z_i(z_i-1)} + c\right|^{-1}. \tag{5.23}$$

The scattering length a_0 of the wave A_0 can take only positive values. This follows from the solution. It also follows directly from Eq. (5.20); for, in accordance with this equation, $dA_0/dz > 0$ for $-1 < z < 1$, and also $A_0(-1) = 0$.

Using similar arguments one can show that the p-wave scattering length is also positive [in the solution (5.22) this corresponds to $R = (4/9)\, a_1 > 0$].

Let us consider the simplest solution without Castillejo-Dalitz-Dyson terms, i.e., for $c = 0$ and $\alpha_i = 0$). For large z we have the asymptotic behavior

$$(A_0)^{-1} \simeq \frac{7}{9\pi}\ln z, \quad (A_1)^{-1} \simeq -\frac{7}{2\pi}\ln z. \tag{5.24}$$

Since a_1 is positive and A_1 has no zeros in the physical region it now follows that the phase δ_1^1 must definitely pass through a resonance. The position and width of this resonance are determined by the two parameters λ and R_0. If we now determine these parameters by means of the expansion (5.10b) for the p wave:

$$R_0 = \frac{1}{\lambda} = \frac{32\pi}{3} f_\pi^2, \tag{5.25}$$

the mass and width of the ρ meson are determined by the f_π and the pion mass μ by equations which, for $m_\rho^2 \gg \mu^2$, have the form

$$\frac{192}{7} f_\pi^2 = m_\rho^2 \left(\frac{1}{\pi} \ln \frac{m_\rho^2}{\mu^2} - \frac{1}{2} \right); \tag{5.26}$$

$$\frac{\Gamma_\rho}{m_\rho} = \left(\frac{7}{2\pi} + \frac{96\pi f_\pi^2}{m_\rho^2} \right)^{-1}. \tag{5.27}$$

In accordance with (4.18), we set $f_\pi = 0.615\,\mu$ ($R_0 = 12.7$), obtaining $m_\rho \sim 930$ MeV and $\Gamma_\rho^{tot} \sim 230$ MeV. It is interesting to note, going over from Γ_ρ to g_ρ by means of (1.3), that (5.27) is replaced by

$$\frac{2 f_\pi^2 g_\rho^2}{m_\rho^2} = \left(1 + \frac{7 m_\rho^2}{192 \pi^2 f_\pi^2} \right)^{-1}. \tag{5.27a}$$

This formula can be regarded as a generalization of the well-known Kawarabayashi-Suzuki-Fayyazuddin-Riazzudin relation [22, 23] which takes into account unitary corrections.

5.4. Scattering Phases in the Region up to 1 GeV

Curve 1 in Fig. 4 gives the energy dependence of the p-wave phase for the case (5.25).

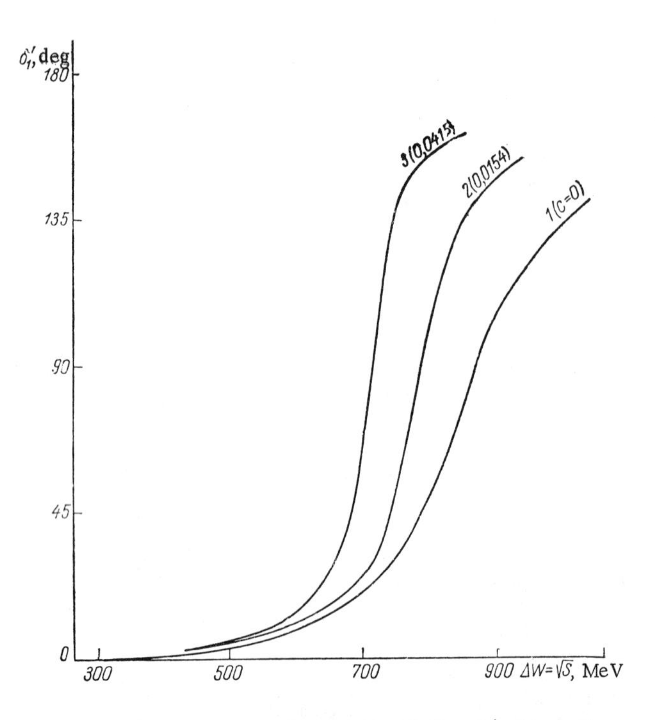

Fig. 4. Energy dependence of the δ_1^1 phase for different c.

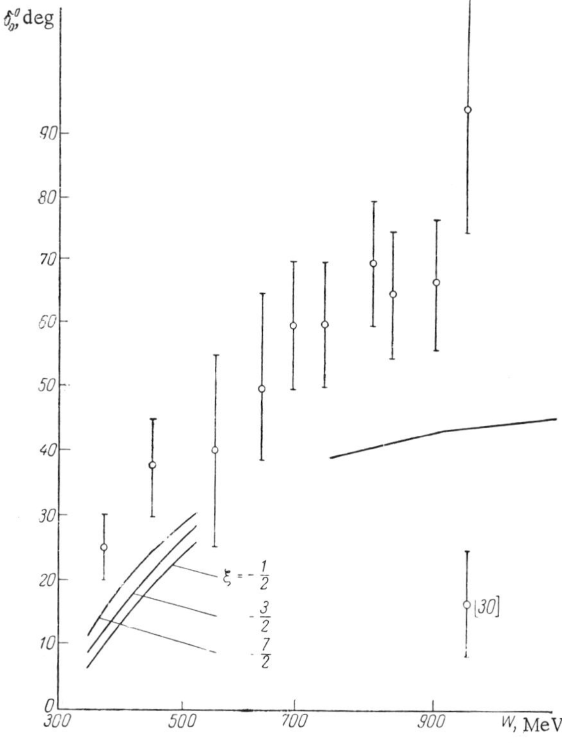

Fig. 5. Energy dependence of the phase δ_0^0 for different ξ.

To calculate the energy dependence of the s phase, we note that the first of Eqs. (5.23) is sufficiently good only in regions far from the threshold. In the region around the threshold it must be corrected by using the expressions (5.10a) and introducing the parameter $\xi = a_0/a_2$, which characterizes the manner in which the symmetry is broken. Then one can readily find that A_0 in the neighborhood of $z = 0$ can be represented in the form

$$A_0 \simeq \frac{z+z_0}{8\pi f_\pi^2},$$

where

$$z_0 = \frac{2\xi+5/2}{\xi-5/2}. \tag{5.28}$$

If there is only a small amount of symmetry breaking (see Table 2), the position of the zero z_0 is a long way from $z = 0$ and it is therefore to be expected that the corrections that take into account unitarity shift the zero by a small amount. Consequently, A_0 in the neighborhood of the threshold in the physical region can be approximated by the expression (cf. [24])

$$A_0 = \frac{1}{\dfrac{3R_0}{4(z+z_0)} - I(z)}. \tag{5.29}$$

For $z \gtrsim 10$ one can neglect A_2 and it is therefore more correct to use Eq. (5.22). The dependence of the s phase δ_0^0 on the energy for different values of ξ is shown in Fig. 5 ($R_0 = 12.7$).

It can be seen from Fig. 4 (curve 1) and Fig. 5 that the values of δ_0^0 lie below the experimental data and that, at the same time, ρ-meson parameters (m_ρ, Γ_ρ) are overestimated. This disagreement with the experimental data was to be expected, since our treatment above is merely a first approximation intended to give only a qualitative picture of the $\pi\pi$ interaction.

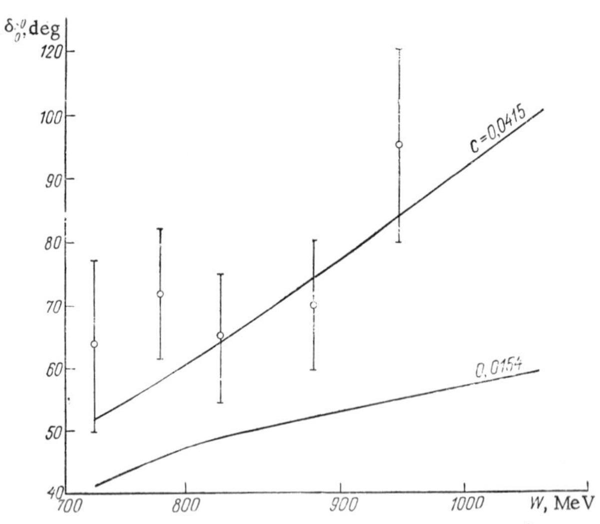

Fig. 6. Energy dependence of the phase δ_0^0 for different c.

We have, for example, ignored a number of important effects: first, we have restricted ourselves to s and p waves, but it is known that the d wave (I = 0) is not small and passes through a resonance in the region of 1.2 GeV (the f_0 meson). A similar situation also obtains for the isotopic amplitude A_1. The result is that the integral contributions from the higher resonances in the region 0.5–1 GeV cannot be neglected. Secondly, we have already estimated the contribution from the ρ-meson trajectory to the real part of the scattering amplitude at the threshold (5.5). At this point it was found to be small. However, in the region of the ρ meson this contribution is of the order of a few tenths of the meson Compton wavelength and it must also be taken into account. It turns out that the s phase δ_0^0 is then increased by 5–10°. Finally, we must not forget that the amplitude A_2 is also nonvanishing.

The most important effect is that due to the contributions from the intermediate region (1–2 GeV). To take these into account we must formulate the problem for the s, p, d, and f partial waves and consider the solution of the resulting equations. However, at this juncture we shall not study this system. We shall only mention that one can also take into account the influence of the higher waves on the s and p solutions by assuming that the parameter c in the expression (5.22) is nonvanishing.

This can be readily seen by considering the example of the scattering of neutral pions (2.2). Representing the unitarity condition in the form

$$\operatorname{Im} A = K\,|\,A(z)\,|^2 + 2\varphi(z)\operatorname{Re} A(z) + \eta(z),$$

where the functions φ and η are determined by the d wave, we solve Eq. (2.2). We can then show that the solution of this equation without Castillejo–Dalitz–Dyson terms corresponds to the solution (2.3)–(2.5) with purely elastic unitarity, but with a Castillejo–Dalitz–Dyson term with c ≠ 0 in the low-energy region if the functions φ and η are essentially nonvanishing at energies greater than 1 GeV.

The solutions (2.22) for two values of c ≠ 0 are shown in Fig. 4 (curves 2 and 3) and Fig. 6. It can be seen that at small c the mass of the resonance in the p wave and its width approach the experimental values and the s phase increases. A resonance in the region of 1 GeV becomes possible. This resonance has a fairly large width. To obtain a crude estimate of this width in the δ approximation one can use the threshold condition for the second of Eqs. (5.20).

This gives $\gamma_\sigma = 9/2\gamma_\rho$ or

$$\Gamma_\sigma = \frac{9m_\sigma}{2m_\rho}\,\Gamma_\rho \simeq 700 \text{ MeV} \quad (\text{for } m_\sigma \sim 1 \text{ GeV}).$$

Of course, such a broad resonance can only be conventionally called a quasiparticle (σ meson). It would be very difficult to distinguish in the bipion mass spectrum (for example, in the region $\pi + N \to \pi' + \pi'' + N$) from the effects of the interaction of the particles in the final state when this interaction is fairly strong but is not necessarily a resonance interaction.

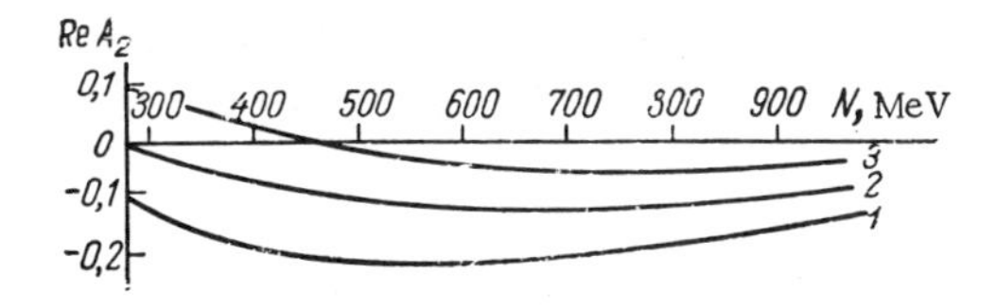

Fig. 7. Energy dependence of the amplitude A_2 for different a_2.

Finally, a few words about the wave A_2. Using the definition (5.8) and the numerical values of A_0 and A_1 obtained from the third equation of (5.7), one can calculate the small wave A_2. Here, the main indeterminacy is associated with the strength of the repulsion v. It can be eliminated by specifying the scattering length a_2 [see the second equation of (5.19)]. The behavior of the phase of this wave for three different values of a_2, namely, +0.10, 0, and −0.10, is shown in Fig. 7 (the details are discussed in [25]).

This figure shows that the phase δ_2^0 is small and negative over a wide energy range. It is therefore clear that the original approximation Im $A_2 = 0$ is consistent with this behavior of the phase.

6. PION-NUCLEON SCATTERING

6.1. Massless Pion Approximation in πN Scattering

The approximation in which symmetry under the chiral group SU(2) × SU(2) is satisfied gives results that are close to the experimental values of the threshold characteristics. Therefore, as in the case of $\pi\pi$ scattering, the consequences of the massless pion approximation can be regarded as boundary conditions for the solutions of the system of equations for the πN scattering partial waves.

As in Sec. 4, we obtain consequences of chiral symmetry for πN scattering from the Born terms of the corresponding effective Lagrangian. As we have already mentioned, this Lagrangian can be constructed in two ways.

The first method is based on the assumption that there exists a certain set of elementary particles that realize linear representations of SU(2) × SU(2). Let us suppose that the nucleon field belongs to the representation (1/2, 1/2). The variations of this field under the action of the generators Q_5^A and Q^A are given by the expressions

$$\delta\Psi = i\,\frac{\tau\alpha}{2}\,\Psi;\quad \tilde{\delta}\Psi = i\,\frac{\tau\beta}{2}\,\gamma_5\,\Psi. \tag{6.1}$$

Of course, the mass term in the Lagrangian is not invariant under such a transformation. However, one can ensure that the total Lagrangian is invariant by introducing an interaction with the π and σ fields, which also belong to the representation (1/2, 1/2) and transform in accordance with Eqs. (4.1).

The simplest invariant Lagrangian (the Gell-Mann–Levy sigma model [7, 12]) has the form

$$L = \overline{\Psi}\,(i\gamma_\mu\partial_\mu - M)\,\Psi + \frac{1}{2}\,\{(\partial\sigma)^2 + (\partial\pi)^2\} - \lambda\left\{\sigma^2 - \frac{2M}{g}\,\sigma + \pi^2\right\} + g\overline{\Psi}\,[\sigma + \gamma_5\pi\tau]\,\Psi. \tag{6.2}$$

Note that the σ field in (6.2) differs from the $\bar{\sigma}$ field in (4.3) by the additive constant $\sigma = \tilde{\sigma} + M/g$ and the coupling constant g is expressed in terms of the previously introduced (Sec. 1) constant f by the well-known relation

$$f^2 = \frac{g^2}{4\pi}\left(\frac{\mu}{2M}\right)^2.$$

The Born terms of the Lagrangian (6.2) describe only the amplitude T^+. The corresponding scattering length defined in terms of the nucleon and σ-meson exchange graphs have the form

$$a^+ = -\frac{g^2}{4\pi M} + \frac{2\lambda M}{\pi m_\sigma^2}.$$

Now it follows from (6.2) that $m_\sigma^2 = 8M^2\lambda/g^2$; therefore

$$a^+ = 0. \tag{6.3}$$

To obtain the amplitude T^- we must augment (6.2) with terms of the interaction with vector particles. The most suitable candidates are the ρ and A_2 mesons. The corresponding fields, which realize the representation (1, 0) + (0.1), transform as follows:

$$\delta\rho^{\mu}_{A} = -\varepsilon_{ABC}\alpha_B\rho^{\mu}_{C}; \quad \bar{\delta}\rho^{\mu}_{A} = -\varepsilon_{ABC}\beta_B A^{\mu}_{2C};$$
$$\delta A^{\mu}_{2A} = -\varepsilon_{ABC}\alpha_B A^{\mu}_{2C}; \quad \bar{\delta} A^{\mu}_{2A} = -\varepsilon_{ABC}\beta_B\rho^{\mu}_{C}.$$

Since G parity is conserved, only the ρ-meson exchange makes a contribution to the Born term of the amplitude T^-:

$$\varphi^- = \frac{2g_{1V}g_{\rho}}{m_{\rho}^2 - t}.$$

It follows from the universality of the interaction of vector mesons that $2g_{1V} = g_{\rho}$. Therefore (in units of μ^{-1}),

$$a^- = \frac{g_{\rho}^2}{4m_{\rho}^2\pi}. \tag{6.4a}$$

Taking the experimental value of the ρ-meson width (5.17) as our starting point and using (1.3), we can obtain $g_{\rho}^2/4\pi = 2.5$, from which it follows that $a^- = 0.084$.

The other method for constructing the effective Lagrangian is based on the assumption that the pion field is a nonlinear representation of the chiral group [13]. To obtain an invariant Lagrangian one does not need to introduce other fields. One need only specify the commutators

$$[Q^A, \Psi] = -\frac{\tau^A}{2}\Psi; \quad [Q_5^A, \Psi] = V_{AB}(\pi)\frac{\tau^B}{2}\Psi, \tag{6.5}$$

which determine the variation of the nucleon field Ψ.

The find the function V_{AB} one must use the Jacobi identities and the group relations (3.5c). This gives the same expression for V_{AB} as Eq. (4.12), from which the transformation law for the covarient derivative $D_{\mu}\pi$ is found. It follows that the nucleon mass term in the Lagrangian is invariant under transformations of the chiral group.

The invariant interaction Lagrangian has the form

$$L_{\text{int}} = \frac{ig}{2M}\bar{\Psi}\gamma_{\mu}\gamma_5\tau\Psi D_{\mu}\pi, \tag{6.6}$$

where $D_{\mu}\pi$ is defined in (4.13).

To construct the total Lagrangian one must also determine the covariant derivative $D_{\mu}\Psi$, which transforms in the same way as the field Ψ (6.5).

Standard manipulation with the Jacobi identities yields

$$D_{\mu}\Psi = \partial_{\mu}\Psi + \frac{iV(\pi^2)}{\sqrt{f^2(\pi^2)+\pi^2}}\frac{\tau}{2}(\pi\,\partial_{\mu}\pi)\Psi. \tag{6.7}$$

The last term yields the πN interaction that augments (6.6). Thus, the total chiral-symmetric Lagrangian of the πN system is

$$L = \bar{\Psi}(i\,\partial_{\mu}\gamma_{\mu} - M)\Psi + \frac{v(\pi^2)}{\sqrt{f^2(\pi^2)+\pi^2}}\bar{\Psi}\gamma_{\mu}\frac{\tau}{2}\Psi(\pi\,\partial_{\mu}\pi) + \frac{ig}{2M}\bar{\Psi}\gamma_{\mu}\gamma_5\tau\Psi D_{\mu}\pi + \frac{1}{2}(D_{\mu}\pi)^2. \tag{6.8}$$

Using (6.8) to determine the expressions for the axial vector current and considering the coefficient in front of $\bar{\Psi}\gamma_\mu\gamma_5\Psi$, we can obtain the Goldberger–Treiman relation:

$$\left(\frac{g_A}{g_V}\right) = -\frac{f_\pi}{M} g. \tag{4.18}$$

Calculating the Born terms, we obtain

$$A^+ = \frac{g^2}{M}, \quad A^- = 0$$

and

$$B^+ = g^2\left\{\frac{1}{M^2-s} - \frac{1}{M^2-u}\right\};$$
$$B^- = -\frac{g^2}{2M^2} + \frac{1}{2f_\pi^2} + g^2\left\{\frac{1}{M^2-s} + \frac{1}{M^2-u}\right\}. \tag{6.9}$$

It follows that

$$a^+ = 0 \tag{6.3}$$

and

$$a^- = \frac{1}{8\pi f_\pi^2}. \tag{6.4b}$$

For $f_\pi = 0.615$ [see (4.18)] we obtain $a^- = 0.10$. Comparing (6.4a) and (6.4b) and using (4.18), we obtain a relationship between the characteristics of the ρ meson and $g_{\pi NN}$ [22, 23]:

$$a_{\rho 1} = \frac{g_{1V} g_\rho}{8\pi} = \left(\frac{g_V}{g_A}\right)^2 \frac{m_\rho^2}{8M^2} \cdot \frac{g^2}{4\pi}. \tag{6.10}$$

Below we shall use Eqs. (6.3), (6.4), and (6.10) as additional conditions imposed on the system of the s- and p-wave equations with repulsion.

6.2. Introduction of Repulsion

As we have already indicated in Secs. 1 and 4, one must introduce a short-range repulsion $V^+(\nu)$ in the low-energy equation for the function $\varphi^+(\nu, 1)$. By analogy with (2.1), we shall approximate this potential by a pole expression of form-factor type:

$$V^+(\nu) \simeq v^+(\nu) = -\frac{v}{1+\frac{\nu}{\rho^2}}. \tag{6.11}$$

The parameter v is related to the subtraction constant and is the strength of the repulsion; the quantity p^2 characterizes the range of the repulsion. If we assume that the nominal boundary of the low-energy region occurs at $E_{lab} \approx 500$ MeV, then $p^2 \gtrsim 50$.

Equation (1.7a) now takes the form

$$\varphi^+(\nu, 1) = \frac{1}{\pi}\int_0^\infty \frac{\mathrm{Im}\,\varphi^+(\nu', 1)}{\nu'-\nu} d\nu' - \frac{v}{1+\frac{\nu}{p^2}}. \tag{6.12}$$

Using the term of the short-range repulsion, which is proportional to v, we can now obtain small (including vanishing) values of the scattering length a^+. It follows that the introduction of the repulsive term is needed to achieve agreement between the dispersion approach and the consequences of chiral symmetry.

We must also modify the third of Eqs. (1.7). The point is that at high-energies the spin-flip scattering amplitude f_2 is less than the amplitude without spin flip, i.e., $f_2 < f_1$. It follows that Im $\nu_L B^+/4\pi$ tends to the same limit as Im φ^+. Thus, the dispersion relation for $\nu_L B^+$ also calls for a subtraction. In the corresponding low-energy equation written down without subtraction, one must introduce the potential of the short-range repulsion:

$$v_B(q^2) = \frac{-v_B}{1+q^2/P_B^2},$$

where [cf. (1.30)]

$$v_B = \frac{1}{2\pi^2}\int_1^\infty \operatorname{Im} B_0^+(E')\,dE' + 4Mf^2, \tag{6.13}$$

and B_0^+ is the low-energy (p wave) component of the function B^+.

Going over to the function $\beta^+(\nu)$ in the static limit, we find that (1.7c) is replaced by

$$\beta^+(\nu) = -\frac{2f^2}{\omega} + \frac{\omega v_B}{2M(p_B^2+\nu^2)} + \frac{\omega}{\pi}\int_0^\infty \frac{\operatorname{Im}\beta^+(\nu')\,d\nu'}{\omega'(\nu'-\nu)}. \tag{6.14}$$

Since we must assume that $p_B^2 \sim p^2 \gtrsim 50$, this equation is virtually identical with Eq. (1.7c) in the low-energy region. The term we have introduced becomes important in the region of higher energies, where it significantly alters the condition (1.12a) and ensures the correct decrease of the function β^+. The system of equations for the s and p waves takes the form:

$$s^+(\nu) = \frac{\alpha_\sigma}{2(1+\nu_\sigma+\nu)} - \frac{v}{2\left(1+\frac{\nu}{p^2}\right)} + \frac{1}{\pi}\int_0^\infty \frac{\operatorname{Im} s^+(\nu')\,d\nu'}{\nu'-\nu}; \tag{6.15a}$$

$$s^-(\nu) = \frac{\omega\alpha_{\rho 1}}{2(1+\nu_\sigma+\nu)} + \frac{\omega}{\pi}\int_0^\infty \frac{\operatorname{Im} s^-(\nu')\,d\nu'}{\omega'(\nu'-\nu)}, \tag{6.15b}$$

$$p^{(+)}(\nu) = -\frac{\alpha_\sigma}{2(1+\nu_\sigma+\nu)} - \frac{v}{2\left(1+\frac{\nu}{p^2}\right)} + \frac{1}{\pi}\int_0^\infty \frac{\operatorname{Im} p^+(\nu')\,d\nu'}{\nu'-\nu}; \tag{6.16a}$$

$$p^{(-)}(\nu) = \frac{2f^2}{\omega} - \frac{\omega\alpha_{\rho 1}}{2(1+\nu_\rho+\nu)} + \frac{\omega}{\pi}\int_0^\infty \frac{\operatorname{Im} p^-(\nu')\,d\nu'}{\omega'(\nu'-\nu)}; \tag{6.16b}$$

$$\beta^+(\nu) = -\frac{2f^2}{\omega} + \frac{\omega v_B}{2M(\nu+p_B^2)} + \frac{\omega}{\pi}\int_0^\infty \frac{\operatorname{Im}\beta^+\,d\nu'}{\omega'(\nu'-\nu)}; \tag{6.16c}$$

$$\beta^-(\nu) = \frac{\alpha_{\rho 1}+\alpha_{\rho 2}}{4M(1+\nu+\nu_\rho)} + \frac{1}{\pi}\int_0^\infty \frac{\operatorname{Im}\beta^-\,d\nu'}{\nu'-\nu}. \tag{6.16d}$$

Note that all these relations except (6.16c) are obtained by combining the forward and backward dispersion relations. Equation (6.16c) is derived entirely from the forward dispersion relation since the long-range contributions from the annihilation channel to the function $\beta^+(\omega, -1)$ are related to the higher resonances ($l \geq 2$) of the $\pi\pi$ scattering, which lie above 1 GeV ($\nu_{f_0}+1 = 20$).

In order to avoid explicitly the introduction of the f_0-meson characteristics, we eschewed the use of the backward dispersion relation to obtain this equation. Explicit allowance for the f_0 meson would lead to a modification of the second term on the right side of (6.16c).

6.3. Threshold Analysis

The threshold and asymptotic conditions for the p waves

$$p^{+}(0)=p^{-}(0)=0,$$
$$q\beta^{+}\rightarrow 0,\; q^{2}\beta^{-}\rightarrow 0 \text{ as } q\rightarrow\infty$$

yield expressions for the πN-scattering characteristics in terms of the parameters of the effective Lagrangian (1.2), which describes the subthreshold region of the reaction $\pi\pi\rightarrow N\bar{N}$:

$$R=\frac{\alpha_{\rho 1}}{M_{\rho}^{2}}=0{,}025;\quad S=\frac{\alpha_{\sigma}}{M_{\sigma}^{2}};\quad \varkappa=2\mu_{N}+1=\frac{\alpha_{\rho 2}}{\alpha_{\rho 1}}+1=4.8. \tag{6.17}$$

Here, μ_N is the anomalous nucleon magnetic moment.

It follows from the asymptotic condition on Eq. (6.16d) that there does not exist a solution in which the partial waves h_{13} and h_{31} are large but the waves h_{33} and h_{11} are small. It follows from the threshold condition for Eq. (6.16b) that there does not exist a solution with large h_{11} and small h_{33}. Thus, the system of threshold and asymptotic conditions imposes rather stringent restrictions on the possible types of solution. However, this system is compatible with a 33-dominant solution and also a solution in which there exists a resonance in the wave h_{11} as well as the 33 resonance.

Let us consider the 33-dominant solution in which we neglect the imaginary parts of all the partial waves with the exception of h_{33}, for which we use the delta approximation (1.13). For this solution, we combine the threshold and asymptotic conditions for the p wave, obtaining

$$\Gamma_{33}=\frac{3}{2}f^{2}-\frac{3}{2}R;\quad \omega_{33}=\frac{RM_{\rho}^{2}\varkappa}{4M(f^{2}-R)}; \tag{6.18}$$

$$v=-4S+2R\varkappa\frac{M_{\rho}^{2}}{M},\quad v_{B}=4Mf^{2}-\frac{8}{3}M\Gamma_{33}. \tag{6.19}$$

Using the numerical values (6.17), we obtain the following numerical values for the parameters of the 33 resonance:

$$\Gamma_{33}=0.09\simeq f^{2},\quad \omega_{33}=1.85. \tag{6.20}$$

For v_B we have $v_B \approx 0.67$.

From Eqs. (6.15) we obtain the following expressions for the s-wave scattering lengths:

$$a^{+}=4S-R\varkappa\frac{M_{\rho}^{2}}{M};\quad a^{-}=2R=\frac{g_{1V}\,g_{\rho}}{4\pi m_{\rho}^{2}}=\frac{g_{\rho}^{2}}{8\pi m_{\rho}^{2}}. \tag{6.21}$$

It is obvious that a^- is smaller by a factor of two than the value obtained from (6.4a). This is a serious difficulty of the whole approximation based on the static limit, which takes into account only forces due to ρ, N, and Δ_{33} exchange and high-energy repulsion.

The discrepancy between Eqs. (6.21) and (6.4a) indicates that we are not justified in neglecting the threshold contributions of the integrals over the high-energy and intermediate regions. However, if these contributions are small, which can be established theoretically by investigating a much more complicated system of equations that includes the higher partial waves and inelastic effects, there exists a further alternative. It is possible that in the calculation of the dispersion integrals from the annihilation channel, the effective mass of the ρ meson must be taken smaller than the physical mass. This is the conclusion reached in [26] on the basis of an investigation of the amplitude in the relativistic approximation. Let us consider two "limiting" cases.

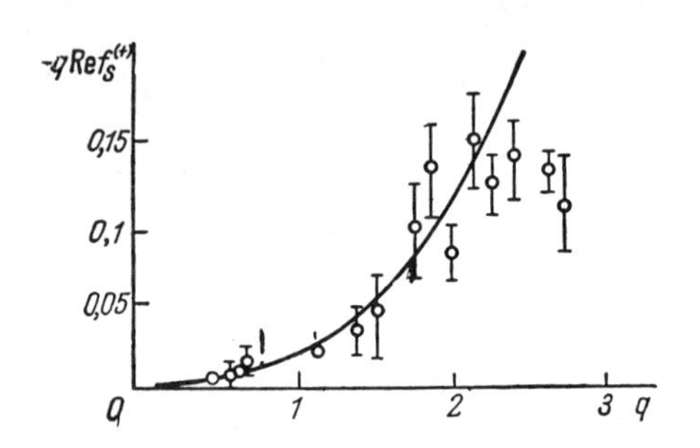

Fig. 8. Energy dependence of $-q\,\mathrm{Re}\,f_S^{(+)}$ on q for $m_\sigma^2 = 50$.

1. Suppose a^- is determined entirely by the effective reductions of the ρ-meson mass. In Eqs. (6.15) and (6.16) we make the substitution $m_\rho^2 \to \tilde{m}^2 < m_\rho^2$. Then, using the relationship between $\alpha_{\rho 1}$ and f^2 from Eq. (6.10), in which m_ρ^2 is the physical ρ-meson mass, we find that (6.18) is replaced by

$$\Gamma_{33} = \frac{3}{2} f^2 \left[1 - \frac{1}{2}\left(\frac{g_V}{g_A} \cdot \frac{m_\rho}{\tilde{m}}\right)^2\right];$$

$$\omega_{33} = \frac{m_\rho^2}{8M}\left(\frac{g_V}{g_A}\right)^2 \frac{\varkappa}{1 - \frac{1}{2}\left(\frac{g_V}{g_A} \cdot \frac{m_\rho}{\tilde{m}}\right)^2}.$$

Instead of the second equation (6.21), we have

$$a^- = \left(\frac{g_V}{g_A}\right)^2 \left(\frac{m_\rho}{\tilde{m}}\right)^2 f^2.$$

For satisfactory values of a^- and Γ_{33} these equations yield an excessively large mass of the 33 resonance. For example, setting $\left(\frac{m_\rho}{\tilde{m}}\right)^2 = 1.5$, and taking the value $\left(\frac{g_V}{g_A}\right) = 1.23$, we obtain

$$a^- = f^2 = 0{,}08, \quad \Gamma_{33} = \frac{3}{4} f^2 \simeq 0{,}06, \quad \omega_{33} = \frac{m_\rho^2 \varkappa}{6M} \simeq 3{,}6. \tag{6.22}$$

To calculate a^+ and v we must know the value of S. Here, it is necessary to remember that the σ meson is a rather complicated formation and, even if it corresponds to a resonance, the latter is broad (see Sec. 5.4); it follows that the Lagrangian description in the form (1.2) is very conditional and the corresponding coupling constants and the "mass" m_σ are certain averaged quantities. Of course, the averaged value of the mass m_σ and the coupling constant g_σ can be extracted from the ππ scattering characteristics. However, to determine the constant $g_{\sigma NN}$ we have no simple theoretical arguments at our disposal.

We shall therefore determine the constant S by means of the first equation (6.21) from the condition $a^+ = 0$ (6.3). With allowance for (6.19), this gives S = 0.125 and v = 0.50.

From Eqs. (6.15) we now obtain expressions that describe the energy dependence of the s waves; we shall assume that the phases are small, i.e., the integrals of the imaginary parts can be neglected:

$$s^+ = -\frac{1}{m_\sigma^2} \cdot \frac{q^2}{1 + 4q^2/m_\sigma^2}; \quad s^- = \frac{a^-}{1 + 4q^2/m_\rho^2}. \tag{6.23}$$

The behavior of the s^+ wave depends on the parameter m_σ^2. Setting $m_\sigma^2 = 50$ (which corresponds to an energy of 1 GeV) and $p^2 = 50$, we obtain the curve shown in Fig. 8. The corresponding curve for the case $m_\sigma^2 = 30$ passes above this at a height that is increased by a factor of ~ 1.5, i.e., along the upper edge of the experimental errors. Thus, for $m_\sigma^2 \approx 30$-50 the first equation (6.23) gives a good description of the experimental data right up to q ~ 2.5.

It should be noted that these rather high values of the averaged mass correspond approximately to the experimental curve of the phase δ_0^0 of the ππ scattering shown in Fig. 6.

Equation (6.23) for s^- leads to a curve that lies below the experimental points (Fig. 9). This means that the effective ρ-meson mass is seriously underestimated.

2. We shall assume that the correction from the annihilation channel can be described sufficiently well by a ρ-meson pole term in which the ρ meson has the physical mass. Then the contributions from the integrals over the intermediate and the high-energy regions are important in a calculation of a^-.

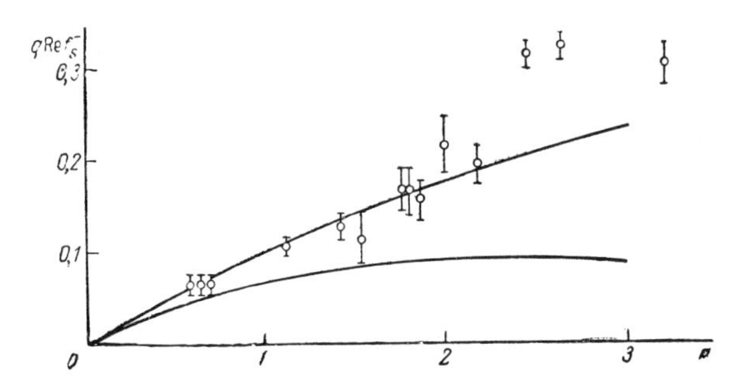

Fig. 9. Energy dependence of $-q\,\mathrm{Re}\,f_s^{(-)}$ on the energy q. The lower curve corresponds to $(m_\rho/m)^2 = 1.5$.

Equations (6.15b) and (6.16b) must be replaced by

$$s^- = \frac{\omega\alpha_{\rho 1}}{2(1+\nu_\rho+\nu)} + \frac{\omega c}{2\left(1+\frac{\nu}{r^2}\right)} + \frac{\omega}{\pi}\int \frac{\mathrm{Im}\, s^-}{\omega'(\nu'-\nu)}\,d\nu';$$
$$p^- = \frac{2f^2}{\omega} - \frac{\omega\alpha_{\rho 1}}{2(1+\nu_\rho+\nu)} + \frac{\omega c}{2(1+\nu/r^2)} + \frac{\omega}{\pi}\int \frac{\mathrm{Im}\, p^-}{\omega'(\nu'-\nu)}\,d\nu'. \quad (6.24)$$

Here, c and r^2 are phenomenological parameters that describe the strength and the range of the forces due to the neglected contributions. One can readily show that this system of equations, too, has a solution with only a large 33 wave. Using the relations obtained in the massless pion approximation (6.4a) and (6.10), we obtain

$$c = R = \frac{1}{2}\left(\frac{g_V}{g_A}\right)^2 f^2.$$

Instead of (4.18) and (4.19), we have

$$\left.\begin{aligned} \Gamma_{33} &= \frac{3}{2} f^2 = 0.12; \\ \omega_{33} &= \frac{m_\rho^2}{8M}\left(\frac{g_V}{g_A}\right)^2 \varkappa = 1.8; \\ a^- &= 2\left(\frac{g_V}{g_A}\right)^2 f^2 = \frac{4}{3} f^2 = 0.10. \end{aligned}\right\} \quad (6.25)$$

In this case (taking satisfactory values of the scattering length a^- and the position of the Δ_{33} resonance) we obtain a slightly large width of the Δ_{33} resonance. The parameter of the high-energy repulsion can be found, as in (6.19), by the normalization $a^+ = 0$ (6.3):

$$v = \frac{f^2 m_\rho^2}{2M}\left(\frac{g_V}{g_A}\right)^2 \varkappa = 0.57, \quad S = 0.14. \quad (6.26)$$

The behavior of the s waves is given by the expressions:

$$s^+ = -\frac{q^2}{m_\sigma^2(1+4q^2/m_\sigma^2)}; \quad s^- = \frac{4}{3} f^2 \frac{m_\rho^2+2q^2}{m_\rho^2+4q^2}. \quad (6.27)$$

The curves corresponding to the second of Eqs. (6.27) are shown in Fig. 9. In the region $q < 1.5$, it passes somewhat higher than the experimental points.

Summing up, we conclude that there is indeed an important effective reduction of the ρ-meson mass, and that one must take into account the effects from the regions of higher energy.

6.4. System of Equations for the p Waves

We now come to the p waves. Combining Eqs. (6.16) and (6.24), we obtain a system of equations for the p waves:

$$h_i(\omega) = \frac{3f^2}{\omega^2}(\Lambda_1)_i + \Phi_i(\omega) + \frac{1}{\pi}\int_1^\infty \left\{\frac{\mathrm{Im}\, h_i(\omega')}{\omega'-\omega} + A_{ik}\frac{\mathrm{Im}\, A_k(\omega')}{\omega'+\omega}\right\} d\omega'. \quad (6.28)$$

Here, Φ_i are the p-wave potentials defined by the expressions:

$$3\Phi_i(\omega)=\frac{2v}{M_p^2+4q^2}+\frac{8S}{m_\sigma^2+4q^2}+\frac{Rm_\rho^2\varkappa}{M\left(m_\rho^2+4q^2\right)}(\Lambda_2)_i+\frac{8\omega R(\Lambda_3)_i}{m_\rho^2+4q^2}+\frac{2\omega c}{r^2+q^2}(\Lambda_3)_i-\frac{2\omega v_B}{M_p^2+4q^2}(\Lambda_1+\Lambda_3)_i; \tag{6.29}$$

$$M_p^2=4p^2,\quad h_i=e^{i\delta_i}\sin\delta_i/q^3,$$
$$i=\{(1.1),\ (1.3),\ (3.1),\ (3.3)\},$$

and also

$$A_{ik}=\frac{1}{9}\begin{pmatrix}1 & -4 & -4 & 16\\ -2 & -1 & 8 & 4\\ -2 & 8 & -1 & 4\\ 4 & 2 & 2 & 1\end{pmatrix};\quad \Delta_1=\begin{pmatrix}-4\\ -1\\ -1\\ 2\end{pmatrix};$$
$$\Lambda_2=\begin{pmatrix}4\\ -2\\ -2\\ 1\end{pmatrix};\quad \Lambda_3=\begin{pmatrix}2\\ 2\\ -1\\ -1\end{pmatrix}. \tag{6.30}$$

The system (6.28) differs from the Chew-Goldberger-Low-Nambu system by the presence of the potentials Φ_i. These potentials, like the corresponding expressions in Eqs. (6.15) for the s waves, contain short-range repulsion terms as well as the σ- and ρ-meson exchange terms. It should be noted that the low-energy properties of the equations are virtually independent of the large parameters M_p^2 and r^2. In the equations for the s waves only the constant terms v and c "work"; these normalize the combinations s^+ and s^- to the correct scattering length $a^+=0$ and $a^-\approx f^2$. In Eqs. (6.28) for the p waves, the corresponding terms in the region of low energies proportional to p^{-2} and r^{-2} are negligibly small. In the high-energy region they make a contribution to the asymptotic behavior of h_{ik} proportional to v and c. Accordingly, these terms play an important role in ensuring the quantum-mechanical threshold conditions for the p waves p_{ik}. As we have already seen, it is only after the introduction of these terms that the system of equations (6.28) has a solution with a sensible 33 wave.

At low energies the potentials Φ_i can be approximated by the expression

$$3\Phi_i(\omega)=f^2\left(\frac{g_V}{g_A}\right)^2\left[\frac{m_\rho^2\varkappa}{M\left(m_\sigma^2+4q^2\right)}+\frac{(\Lambda_2)_i}{2M}\cdot\frac{m_\rho^2\varkappa}{\left(m_\rho^2+4q^2\right)}+\frac{4\omega(\Lambda_3)_i}{m_\rho^2+4q^2}\right]. \tag{6.31}$$

In this approximation a calculation of the p-wave scattering lengths yields

$$a_{33}=0.271;\ a_{13}=-0.033;$$
$$a_{11}=-0{,}099;\ a_{31}=0{,}039.$$

Thus, the use of the relations that follow from the massless pion approximation give a satisfactory qualitative description of N scattering even in the static limit. To achieve this, we had to introduce a short-range repulsion whose strength could be estimated from either the high-energy integrals or their low-energy theorems, the same result being obtained in each case. As in the case of $\pi\pi$ scattering, we have seen that the solution of the equations for the low-energy πN scattering gives a strong p-wave interaction that is even resonant in the 33 wave in a reasonable region. Ultimately, using the massless pion approximation and the unitarity equations for the low-energy $\pi\pi$-and πN-scattering processes, we have obtained the masses of the p-wave resonance, their lifetimes, and the main coupling constants; to do this we needed to specify only the masses of the pion and nucleon and their lifetimes and the Fermi constant.

Of course, our theory can only be regarded as a first approximation. The accuracy with which the main characteristics are obtained is only ~25%. To increase the accuracy we must allow more accurately for the relativistic corrections. In the system of low-energy equations we must introduce equations for the higher partial waves, which take into account inelastic effects and also make a more detailed allowance for the high-energy contributions. We are convinced that in a realistic theory that takes into account all these corrections it will not be necessary to introduce additional parameters.

7.1. Universality of the Repulsion

As we have just seen, the introduction of a short-range repulsion term into the low-energy equations for $\pi\pi$ and πN scattering is needed to obtain quantitative agreement between the solutions of the dispersion equations and the consequences of chiral symmetry in the low-energy region.

As Weinberg has shown [27], a decrease of the combination of the total cross section corresponding to $I = 1$ in the t channel for all scattering processes in which pions participate ensures that the current commutation relations of the chiral group are satisfied. In Secs. 5.2 and 6.2 we have shown that allowance for the short-range "vacuum" contributions corresponding to the nondecreasing combinations of the total cross sections ($I = 0$ in the t channel) are essential to achieve agreement with the requirement of chiral symmetry in the low-energy region. As we have noted in Sec. 5.2, a reduction of the vacuum repulsion is equivalent to an increase in the chiral symmetry breaking and the limit $v \to 0$ corresponds to maximal symmetry breaking [$N \to \infty$ in (4.21)].

The strengths and ranges of the repulsion potentials are parameters that can be used to give a phenomenological description of the influence of the high-energy contributions on the low-energy region. It follows that the existence of any simple dependences between the scattering amplitudes of the various processes in the high-energy region must lead to corresponding relationships between the parameters of the repulsion potentials.

Let us see how such relationships arise in a simple example. We shall assume that in the high-energy region a certain linear combination of the scattering amplitudes of various processes

$$T = \sum_i c_i T_i \tag{7.1}$$

decreases fairly rapidly, and that this combination satisfies an unsubtracted dispersion relation. Of course, the question immediately arises of the arguments of the various terms on the right side of (7.1), i.e., the question of the determination of an energy scale common to the various physical processes.

It is natural to associate this scale with the invariant square of the total energy and choose it in such a way that it explicitly reflects the crossing symmetry properties for forward scattering. The variable

$$x = s - m_i^2 - M_i^2 \tag{7.2}$$

satisfies these conditions; here m_i is the mass of the incident particle and M_i is the mass of the target for the i-th reaction.

We shall assume that each of the T_i is the crossing-even half-sum of the amplitudes for the scattering of a particle and antiparticle by the target [i.e., an amplitude of the type (1.21)]; it then follows from the optical theorem that the imaginary part of T_i is associated with a nondecreasing cross section. The dispersion relation for each of the T_i separately definitely requires a subtraction, and the low-energy model for the corresponding partial waves needs the introduction of a repulsion. The dispersion relation for the function (7.1) written down without a subtraction in the variable x yields

$$T(0) = \frac{2}{\pi} \int \sum_i c_i \operatorname{Im} T_i \left(x + m_i^2 + M_i^2\right) \frac{dx}{x}. \tag{7.3}$$

The integral on the right side of (7.3) includes pole terms and possible subthreshold unphysical regions. Suppose that for $x > x_q$ the value of Im T vanishes in a sufficiently good approximation. Then (7.3) can be rewritten in the form

$$\sum_i c_i v_i (x_q) = 0, \tag{7.4}$$

where

$$v_i (x_q) = \frac{2}{\pi} \int_0^{x_q} \operatorname{Im} T_i \frac{dx}{x} - T_i (0). \tag{7.5}$$

If we now identify x_q with the boundary of the low-energy region $2m_\pi^2\sqrt{\Lambda} = x_\Lambda$, then the quantities $v_i(x_q)$ are none other than the strengths of the short-range repulsion [cf. (1.30)]. Equation (7.4) yields linear relations between the repulsion strengths for the different processes.

Combinations of the form (7.1) that decrease fairly rapidly at infinity can be obtained on the basis of various models of high-energy scattering. At the present time the most widely known are the model of quark additivity [28] and also the model of Regge asymptotic behavior. Both of these models lead to approximately the same combinations of the form (7.1) (in which the coefficients c_i are numbers of order unity) for all the main hadron scattering processes. It follows that the parameters $v_i(x_q)$ for the various processes are also numbers of the same order. In this case the parameter x_q corresponds to the lower boundary of the asymptotic high-energy region, i.e., the region in which the corresponding Regge or quark relations for T_i are satisfied fairly accurately.

Of course, the values of x_q obtained in this manner are appreciably higher than the values x_Λ corresponding to the upper limits of the low-energy regions. However, as we see in the examples of $\pi\pi$ and πN scattering considered earlier, the simple relationships between $v_i(x_q)$ in fact remain in force for the strengths of the short-range repulsion $v_i(x_\Lambda)$. One can show that there is similar agreement for NN and KN scattering [29]. Thus, we see that the short-range repulsion is a general property of all low-energy processes. The repulsion parameters for the various processes may be related to one another. In this sense one may say that the repulsion is universal.

7.2. General Physical Picture

We shall now attempt to sketch the general picture of the various low-energy scattering processes in terms of a single energy variable which can be used to define the low- and high-energy regions simultaneously for all reactions.

The simplest candidate for this universal variable is the quantity x introduced in (7.2). As follows from Fig. 10, it is also a fairly successful candidate. In this figure we have represented schematically the main characteristics of the $\pi\pi$-, πN-, KN-, and NN-scattering processes in the region of low and intermediate energies plotted on the scale of x. We have indicated the positions of the well-defined resonances that basically have a single-channel two-particle decay, i.e., play an important role in the elastic two-particle scattering. It can be seen that the purely elastic resonances [ρ, Δ_{33}, and also Λ(1405)] are situated in the region $x < 1$ GeV^2. These are all p-wave resonances. It is therefore natural to separate the low-energy region $x < 1$ GeV^2 in which the p (and possibly the s) waves are important and in which the inelastic effects can be neglected.

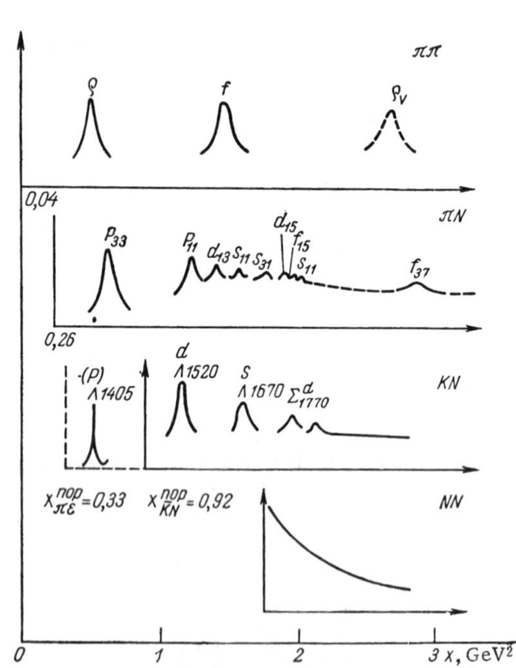

Fig. 10. Picture of the resonances on the scale $x = s - m_1^2 - M_2^2$.

In the region $1 < x < 2$ GeV^2 the d- and f-wave resonances are also manifested. Here, inelasticity becomes important. Above $x = 2$ GeV^2 the total cross sections become smooth functions and many partial waves are manifested; the scattering gradually acquires a diffraction character. It is well known that NN scattering, whose threshold corresponds to $x = 1.76$ GeV^2, does not contain resonances.

Thus, there is a well-defined "elastic low-energy region" $x < 1$ GeV^2 in which only the s and p waves are important. It is precisely in this region that the low-energy assumptions made in Sec. 1 are valid. In this region the scattering can be quantitatively described by a system of equations for the s and p waves, though, of course, the system must include the contributions to the real parts of the partial waves from the region of intermediate and high energies.

The region $1 < x < 2$ GeV^2 may be termed the "inelastic low-energy region." In this region assumption 2 of Sec. 1 is still valid ($l_{max} = 4$), but inelastic processes become important. The system of low-energy equations for this region, which describes the s, p, d, and f waves, must take into account not only the high-energy contributions to the real parts of the partial waves, but also the inelastic contributions to their imaginary parts.

In the region above $x = 2$ GeV^2 the low-energy assumptions clearly cease to have even an approximate validity. It follows that $x = 2$ GeV^2 is the upper limit for the low-energy dispersion schemes of such a kind.

LITERATURE CITED

1. D. V. Shirkov, V. V. Serebryakov, and V. A. Mescheryakov, Dispersion Theories of Strong Interactions at Low Energy [in Russian], Moscow, Nauka (1967).

1a. D. V. Shirkov, V. V. Serebryakov, and V. A. Mescheryakov, Dispersion Theories of Strong Interactions at Low Energy, North-Holland, Amsterdam-London (1969).

2. J. J. Sakurai, Ann. of Phys., 11, 1 (1960).
3. G. Chew, M. Goldberger, F. Low, and V. Nambu, Phys. Rev., 106, 1337 (1957).
4. J. J. Sakurai, Phys. Rev. Lett., 17, 1021 (1966).
5. S. Adler and R. Dashen, Current Algebras and Applications to Particle Physics, Benjamin, New York (1968); B. Renner, Current Algebras and Their Applications, Pergamon Press, Oxford (1968).
6. M. Gell-Mann, Phys. Rev., 125, 1067 (1962); M. Gell-Mann, Physics, 1, 63 (1964).
7. M. Gell-Mann and M. Levy, Nuovo Cimento, 16, 705 (1960).
8. R. P. Feynman, M. Gell-Mann, and G. Zweig, Phys. Rev. Lett., 13, 678 (1964).
9. T. D. Lee, S. Weinberg, and B. Zumino, Phys. Rev. Lett., 18, 1029 (1969).
10. Y. Nambu, Phys. Rev. Lett., 4, 380 (1960).
11. S. Weinberg, Phys. Rev. Lett., 18, 198 (1967); J. Wess and B. Zumino, Phys. Rev., 163, 1727 (1967); J. Cronin, Phys. Rev., 161, 1483 (1967); R. Arnowitt, M. M. Friedmann, and P. Math, Phys. Rev. Lett., 19, 1085 (1967).
12. G. Kramer, Chiral Symmetric Meson-Nucleon Lagrangians, Argonne preprint, ANL/HEP 6817(1968).
13. S. Weinberg, Phys. Rev., 166, 1568 (1968).
14. S. Gasiorowicz and D. A. Geffen, Effective Lagrangians and Field Algebras with Chiral Symmetry, Argonne preprint ANL/HEP 6809 (1968); Rev. Mod. Phys., 41, No. 3, 531-573, July (1969).
15. S. L. Glashow and S. Weinberg, Phys. Rev. Lett., 20, 224 (1968).
16. S. Coleman, J. Wess, and B. Zumino, Phys. Rev., 177, 2239, 2247 (1969).
17. W. A. Bardeen and B. W. Lee, Nuclear and Particle Physics (ed. by B. Margolis and C. Lam), Benjamin, New York (1968).
18. M. L. Goldberger and S. B. Treiman, Phys. Rev., 110, 1178 (1958).
19. S. Weinberg, Phys. Rev. Lett., 17, 336 (1966).
20. V. Barger and M. Olsson, Phys. Rev. Lett., 18, 1428 (1966).
21. A. A. Logunov, L. D. Solovjov, and A. N. Tavkhelidze, Phys. Lett., 24B, 181 (1967).
22. K. Kawarabayashi and M. Suzuki, Phys. Rev. Lett., 16, 255 (1966).
23. Riazzudin and Fayyazuddin, Phys. Rev., 147, 1071 (1966).
24. Y. Fujii, Phys. Lett., 24B, 190 (1967).
25. V. V. Serebryakov and D. V. Shirkov, Yad. Fiz., 6, 170 (1967).
26. J. Engels, G. Höhler, and B. Petersson, "NΔ and ρ Exchange in πN Scattering," Karlsruhe preprint (1969).
27. S. Weinberg, Algebraic Realization of Chiral Symmetry, Massachusetts preprint CTP-39 (1968).
28. E. M. Levin and L. L. Frankfurt, ZhÉTF Pis. Red., 45, 244 (1965).
29. V. I. Lend'el and V. V. Serebryakov, Yad. Fiz., 7, 879 (1968); E. F. Krasnopevtsev and V. L. Chernyak, Yad. Fiz., 7, 1114 (1968).
30. W. D. Walker et al., Phys. Rev. Lett., 18, 630 (1967).

CP VIOLATION IN DECAYS OF NEUTRAL K-MESONS

S. M. Bilen'kii

The problem of CP-invariance violation in the decays of neutral K-mesons is reviewed. An effective equation for the system, described by the superposition of K^0- and $\overline{K}^0$-states, is discussed. The $K_L \to 2\pi$ decays and the lepton decays of K_L-mesons are considered. The model of the superweak interaction is discussed.

1. INTRODUCTION

Soon after it was shown that the weak interaction Hamiltonian is not invariant under space inversion P and charge conjugation C, Landau [1] proposed the hypothesis that all interactions in nature, including weak interactions, are invariant under CP. The natural consequence of this hypothesis was the two-component neutrino theory.

In 1964, Cronin et al. [2] discovered the rare decay of long-lived K^0-meson into two π-mesons. If CP invariance holds for all interactions, such a decay is forbidden.

In the years that followed, the results of Cronin et al. were verified and refined. Dorfan, Bennet, and co-workers [3-4] observed charge asymmetry in $K_L \to \pi^\pm l^\mp \nu$ decays, which is also evidence for CP violation.

Many hypotheses were proposed in connection with the discovery of CP violation in the decays of long-lived K^0-mesons. These led to experiments to check the C and T invariance of strong, electromagnetic, and weak interactions. Up to the present time these experiments have not uncovered any evidence for T and C violation in the strong and electromagnetic interactions. The exception is [5] in which an asymmetry in the η-meson decay was observed.

However, as shown in [6], an asymmetry in η-meson decay can be caused by interference with the background.

This review will discuss the phenomenological analysis of CP violation in long-lived K^0-meson decay. First, the effective equation for the wave function of the system, described as a superposition of K^0- and $\overline{K}^0$-states, is found and the state vectors of the long-lived (K_L)-meson and short-lived (K_S)-mesons are examined in detail. Then the $K_L \to 2\pi$ decays and leptonic decays of K_L-mesons are discussed. In conclusion, Wolfenstein's superweak interaction model [7] is examined.

We will also examine possible methods of checking CPT-invariance based on measurements of parameters characterizing neutral K-meson decays. The questions discussed here are examined in many review articles (see, for example, [8-13]). In reviews [13-14] CP violation models were examined in detail.

2. THE WAVE EQUATION FOR A SYSTEM DESCRIBED BY THE SUPERPOSITION OF K^0- AND $\overline{K}^0$-STATES

Let us obtain the effective equation for the wave function of a system which is described by the superposition of K^0- and $\overline{K}^0$-states [15, 9]. We start from Schrödinger's equation

$$i\frac{\partial|\Psi(t)\rangle}{\partial t} = H|\Psi(t)\rangle. \tag{1}$$

Joint Institute for Nuclear Research, Dubna. Translated from Problemy Fiziki Élementarnykh Chastits i Atomnogo Yadra, Vol. 1, No. 1, pp. 227-253, 1970.

H is the full Hamiltonian and $|\Psi(t)>$ is the state vector. The formal solution of Eq. (1) can be written as

$$|\Psi(t)\rangle = e^{-iHt}|\Psi(0)\rangle, \tag{2}$$

where $|\Psi(0)>$ is the state vector at $t = 0$. Let $|\Psi_n>$ be the complete set of eigenvectors of the Hamiltonian H. We have

$$H|\Psi_n\rangle = E_n|\Psi_n\rangle. \tag{3}$$

Expanding the vector $|\Psi(0)>$ in $|\Psi_n>$ and substituting this expansion into Eq. (2), we have

$$|\Psi(t)\rangle = e^{-iHt}\sum_n|\Psi_n\rangle\langle\Psi_n|\Psi(0)\rangle = \sum_n e^{-iE_n t}|\Psi_n\rangle\langle\Psi_n|\Psi(0)\rangle. \tag{4}$$

Using the relation

$$e^{-iE_n t} = \frac{-1}{2\pi i}\int_{-\infty}^{\infty}\frac{e^{-iEt}dE}{E-E_n+i\varepsilon} \quad (t>0), \tag{5}$$

we find from Eq. (4) that the general solution to Eq. (1) for $t > 0$ can be written

$$|\Psi(t)\rangle = \frac{-1}{2\pi i}\int_{-\infty}^{\infty} G_+(E)\,e^{-iEt}\,dE \quad |\Psi(0)\rangle. \tag{6}$$

Here

$$G_+(E) = \frac{i}{E-H+i\varepsilon}. \tag{7}$$

The full Hamiltonian equals

$$H = H_0 + H_W, \tag{8}$$

where H_W is the weak interaction Hamiltonian and H_0 is the sum of the free Hamiltonian and the strong and electromagnetic interaction Hamiltonians. Let $|K^0>$ and $|\bar{K}^0>$ be the normalized state vectors of K^0 and $\bar{K}^0$ particles in their rest frames. The vectors $|K^0>$ and $|\bar{K}^0>$ are eigenvectors of the strangeness operator. The CP operator anticommutes with the strangeness operator. We can set

$$|\bar{K}^0\rangle = CP|K^0\rangle. \tag{9}$$

Further, the vectors $|K^0$ and $|\bar{K}^0>$ are eigenfunctions of the operator H_0 (they cannot be eigenfunctions of the full Hamiltonian H because the operator H_W does not commute with the strangeness operator). We will assume that the Hamiltonian H_0 is invariant under CPT. It is not hard to show using Eq. (9) that both $|K^0>$ and $|\bar{K}^0>$ belong to the same eigenvalue of the operator H_0, i.e.,

$$H_0|K^0\rangle = m|K^0\rangle, \quad H_0|\bar{K}^0\rangle = m|\bar{K}^0\rangle \tag{10}$$

[m is the mass of the K^0 ($\bar{K}^0$) particle].

Let $|i>$ be the rest of the eigenvectors of the Hamiltonian H_0. We have

$$H_0|i\rangle = E_i|i\rangle. \tag{11}$$

We write the vectors $|K^0\rangle$ and $|\bar{K}^0\rangle$ as $|\alpha\rangle$ ($\alpha = 1$ and 2, $|\bar{K}^0\rangle \equiv |1\rangle$, $|K^0 \equiv |2\rangle$). We expand the vector $|\Psi(t)\rangle$ in the complete set of vectors $|\alpha\rangle$ and $|i\rangle$. We get

$$|\Psi(t)\rangle = \sum_\alpha|\alpha\rangle a_\alpha(t) + \sum_i|i> b_i(t). \tag{12}$$

Let us assume that at the initial moment of time t the state vector of the system equals

$$|\Psi(0)\rangle = \sum_{\alpha} |\alpha\rangle a_{\alpha}(0), \tag{13}$$

where a_{α} are arbitrary.

We are interested in the time development of the system whose state vector at t = 0 is a superposition of $|K^0\rangle$ and $|\overline{K}^0\rangle$ states. It is clear that processes in such a system result from the existence of weak interactions (decays, $|K^0\rangle \rightleftarrows |\overline{K}^0\rangle$ transitions).

We left-multiply Eq. (6) by $\langle\alpha|$. Using Eqs. (12) and (13), we get

$$\langle\alpha|\Psi(t)\rangle = a_{\alpha}(t) = \frac{-1}{2\pi i}\int_{-\infty}^{\infty} dE e^{-iEt} \sum_{\alpha'} \langle\alpha|G_{+}(E)|\alpha'\rangle a_{\alpha'}(0). \tag{14}$$

It is clear that $a_{\alpha}(t)$ is the amplitude of the probability of finding the state $|\alpha\rangle$ at time t. We will see that the amplitude $a_{\alpha}(t)$ is a solution of a Schrödinger equation with a non-Hermitian Hamiltonian.

We discuss the weak interactions according to perturbation theory. It is easy to see from Eqs. (7) and (8) that the operator $G_{+}(E)$ satisfies the following equation:

$$G_{+}(E) = \frac{1}{E-H_0+i\varepsilon} + \frac{1}{E-H_0+i\varepsilon} H_W G_{+}(E). \tag{15}$$

From this we get an equation for the matrix element of interest $\langle\alpha|G_{+}(E)|\alpha'\rangle$. From Eq. (15) we find

$$\langle\alpha|G_{+}(E)|\alpha'\rangle = \frac{\delta_{\alpha'\alpha}}{E-m+i\varepsilon} + \frac{1}{E-m+i\varepsilon}\Big[\sum_{\alpha''}\langle\alpha|H_W|\alpha''\rangle\langle\alpha''|G_{+}(E)|\alpha'\rangle + \sum_{i}\langle\alpha|H_W|i\rangle\langle i|G_{+}(E)|\alpha'\rangle\Big]. \tag{16}$$

We use the operator equation (15) to connect the matrix element $\langle i|G_{+}(E)|\alpha'\rangle$ entering into the right-hand side of Eq. (16) with the matrix element $\langle\alpha''|G_{+}(E)|\alpha'\rangle$. From Eq. (15) we get

$$\langle i|G_{+}(E)|\alpha'\rangle = \frac{1}{E-E_i+i\varepsilon}\Big[\sum_{\alpha''}\langle i|H_W|\alpha''\rangle\langle\alpha''|G_{+}(E)|\alpha'\rangle + \sum_{i'}\langle i|H_W|i'\rangle\langle i'|G_{+}(E)|\alpha'\rangle\Big]. \tag{17}$$

In obtaining this result we used the fact that

$$\langle i|\alpha'\rangle = 0.$$

Iterating Eq. (17), we find

$$\langle i|G_{+}(E)|\alpha'\rangle = \frac{1}{E-E_i+i\varepsilon}\sum_{\alpha''}\Big[\langle i|H_W|\alpha''\rangle + \sum_{i'}\langle i|H_W|i'\rangle\frac{1}{E-E_{i'}+i\varepsilon}\langle i'|H_W|\alpha''\rangle + \dots\Big]\langle\alpha''|G_{+}(E)|\alpha'\rangle. \tag{18}$$

Inside the square brackets of this expression is an expansion in powers of the weak interaction constant G. Substituting Eq. (18) into (16), we get

$$\langle\alpha|G_{+}(E)|\alpha'\rangle = \frac{\delta_{\alpha\alpha'}}{E-m+i\varepsilon} + \frac{1}{E-m+i\varepsilon}\sum_{\alpha''}\langle\alpha|R(E)|\alpha''\rangle\langle\alpha''|G_{+}(E)|\alpha'\rangle, \tag{19}$$

where

$$\langle\alpha|R(E)|\alpha''\rangle = \langle\alpha|H_W|\alpha''\rangle + \sum_{i}\langle\alpha|H_W|i\rangle\frac{1}{E-E_i+i\varepsilon}\langle i|H_W|\alpha''\rangle + \sum_{i,i'}\langle\alpha|H_W|i\rangle\frac{1}{E-E_i+i\varepsilon} \times\langle i|H_W|i'\rangle\frac{1}{E-E_{i'}+i\varepsilon}\langle i'|H_W|\alpha''\rangle + \dots \tag{20}$$

In the future we will drop terms third order and higher in G from this expression. We write R(E) and $G_{+}(E)$ as two-by-two matrices with elements $\langle\alpha|R(E)|\alpha'\rangle$ and $\langle\alpha|G_{+}^{-}(E)|\alpha'^{-}\rangle$. Equation (19) can be written in matrix form as follows:

$$\underline{G}_{+}(E) = \frac{1}{E-m+i\varepsilon} + \frac{1}{E-m+i\varepsilon}\underline{R}(E)\,\underline{G}_{+}(E). \tag{21}$$

This implies

$$\underline{G}_{+}(E)=\frac{1}{E-m-\underline{R}(E)+\mathrm{i}\,\varepsilon}. \tag{22}$$

Substituting this expression into Eq. (14), we get

$$a(t)=\frac{-1}{2\pi\,\mathrm{i}}\int_{-\infty}^{\infty}\frac{\mathrm{e}^{-\mathrm{i}\,Et}\,dE}{E-m-\underline{R}(E)+\mathrm{i}\,\varepsilon}\,a(0), \tag{23}$$

where

$$a(t)=\begin{pmatrix}a_1(t)\\ a_2(t)\end{pmatrix}.$$

It is clear from Eq. (20) that the matrix elements of $\underline{R}(E)$ are much smaller than the mass m. It follows that the pole of the integrand in Eq. (23) is near $\bar{m}$. We therefore make the substitution $R(E)\to R(m)$ in Eq. (23). This approximation is called the Weiskopf-Wigner approximation [16, 15] and depends on the weakness of the interaction causing transitions between basis states compared to the energy of these states.

In the Weiskopf-Wigner approximation, Eq. (23) becomes*

$$a(t)=\mathrm{e}^{-\mathrm{i}\,\mathfrak{X}\,t}\,a(0), \tag{24}$$

where the two-by-two matrix $\mathfrak{X}$ equals

$$\mathfrak{X}=m+\underline{R}\,(m). \tag{25}$$

In this fashion, we get the following equation for the amplitude $a(t)$ in the Weiskopf-Wigner approximation:

$$\mathrm{i}\,\frac{\partial a\,(t)}{\partial t}=\mathfrak{X}\,a\,(t). \tag{26}$$

Let us examine the Hamiltonian $\mathfrak{X}$. Taking into account

$$\frac{1}{m-E_i+\mathrm{i}\varepsilon}=P\,\frac{1}{m-E_i}-\mathrm{i}\pi\delta\,(m-E_i), \tag{27}$$

we find from Eqs. (20) and (25) that

$$\mathfrak{X}=M-\frac{\mathrm{i}}{2}\Gamma. \tag{28}$$

M and Γ are two-by-two matrices whose elements, up to second order in G, equal

$$M_{\alpha\alpha'}=m\delta_{\alpha\alpha'}+\langle\alpha\,|\,H_W\,|\,\alpha'\rangle+P\sum_i\langle\alpha\,|\,H_W\,|\,i\rangle\,\frac{1}{m-E_i}\,\langle i\,|\,H_W\,|\,\alpha'\rangle; \tag{29a}$$

$$\Gamma_{\alpha\alpha'}=2\pi\sum_i\langle\alpha\,|\,H_W\,|\,i\rangle\,\langle i\,|\,H_W\,|\,\alpha'\rangle\,\delta\,(E_i-m). \tag{29b}$$

Using the Hermiticity of the Hamiltonian H_W, it is easy to show that

$$M^{+}=M,\quad \Gamma^{+}=\Gamma. \tag{30}$$

It follows that the effective Hamiltonian $\mathfrak{X}$ is non-Hermitian.

---

* We note that a necessary condition for Eq. (24) to hold is that the eigenvalues of the matrix $\mathfrak{X}$ have negative imaginary parts. We will show that this is the case below.

It is clear from Eq. (20) that Γ_{11} (Γ_{22}) is the total decay probability of a K^0 ($\bar{K}^0$)-particle, while M_{11} (M_{22}) is the mass of the K^0 ($\bar{K}^0$)-particle taking into account corrections due to the weak interactions. The nondiagonal elements of the matrices Γ and M are nonzero, and this, as we will see, is crucial for the physics of neutral K-mesons.

Let us see what limitations invariance principles impose on the matrix elements $\mathfrak{X}$. If the full Hamiltonian H is invariant with respect to CPT-transformations, i.e., if

$$(CPT)^{-1} H\, CPT = H, \tag{31}$$

it follows from Eq. (29), that

$$\mathfrak{X}_{11} = \mathfrak{X}_{22}, \tag{32}$$

i.e., that if Eq. (31) holds, the masses and lifetimes of K^0 and $\bar{K}^0$-particles coincide. It is also clear that CPT-invariance does not put any limitations on the matrix elements $\mathfrak{X}_{12}$ and $\mathfrak{X}_{21}$.

If the Hamiltonian is invariant under DP-transformations, then, besides Eq. (32), we get

$$\mathfrak{X}_{12} = \mathfrak{X}_{21}. \tag{33}$$

We note also that if the Hamiltonian is invariant under the T-transformation, the nondiagonal elements of the $\mathfrak{X}$ are equal.

3. WAVE FUNCTIONS OF K_S- AND K_L-MESONS. UNITARITY RELATIONS

Let us find the eigenfunctions of the Hamiltonian $\mathfrak{X}$. We have

$$\left.\begin{aligned} \mathfrak{X} a_S &= \lambda_S a_S; \\ \mathfrak{X} a_L &= \lambda_L a_L. \end{aligned}\right\} \tag{34}$$

The operator $\mathfrak{X}$ is non-Hermitian. Its eigenvalues are complex numbers. We write them in the form

$$\left.\begin{aligned} \lambda_S &= m_S - \frac{i}{2}\Gamma_S; \\ \lambda_L &= m_L - \frac{i}{2}\Gamma_L, \end{aligned}\right\} \tag{35}$$

where $m_{S,\,L}$ and $\Gamma_{S,\,L}$ are real. Let us clarify the physical meaning of these quantities. We assume that a_S and a_L are normalized. Using Eqs. (28), (30), (34), and (35), we find

$$\left.\begin{aligned} (a_S^{+} \Gamma a_S) &= \Gamma_S; \\ (a_L^{+} \Gamma a_L) &= \Gamma_L. \end{aligned}\right\} \tag{36}$$

Further, from Eq. (29), we get

$$\left.\begin{aligned} \Gamma_S &= 2\pi \sum_i |\langle i | H_W | K_S\rangle|^2 \delta(E_i - m); \\ \Gamma_L &= 2\pi \sum_i |\langle i | H_W | K_L\rangle|^2 \delta(E_i - m), \end{aligned}\right\} \tag{37}$$

where

$$\left.\begin{aligned} |K_S\rangle &= \sum_\alpha |\alpha\rangle a_S(\alpha); \\ |K_L\rangle &= \sum_\alpha |\alpha\rangle a_L(\alpha). \end{aligned}\right\} \tag{38}$$

It follows that Γ_S and Γ_L are the total decay probabilities of states described by the vectors $|K_S\rangle$ and $K_L\rangle$. We find in a similar fashion that $m_S = (a_S^{+} M a_S)$ and $m_L = (a_L^{+} M a_L)$ are the masses of the corresponding states.

If a_S and a_L are chosen as the initial functions in Eq. (24), then it follows from Eqs. (34) and (35) that

$$\left.\begin{aligned} a_S(t) &= e^{-i\mathfrak{X}t} a_S = e^{-i m_S t - \Gamma_S/2\, t} a_S; \\ a_L(t) &= e^{-i\mathfrak{X}t} a_L = e^{-i m_L t - \Gamma_L/2/t} a_L. \end{aligned}\right\} \quad (39)$$

In this fashion, the wave function of such states depends exponentially on time and $\tau_S = 1/\Gamma_S$ and $\tau_L = 1/\Gamma_L$ are lifetimes. Experimentally [17]

$$\tau_S = (0.87 \pm 0.009)\cdot 10^{-10}\ \text{sec}; \quad \tau_L = (5.73 \pm 0.25)\cdot 10^{-8}\ \text{sec}.$$

The particles described by the wave functions a_S and a_L are conventionally called short-lived (K_S) and long-lived (K_L) K-mesons.

Let us now find the functions a_S and a_L. First we assume that the CPT-theorem holds. In that case $\mathfrak{X}_{11} = \mathfrak{X}_{22}$ and the matrix $\mathfrak{X}$ can be written in the form

$$\mathfrak{X} = \mathfrak{X}_{11} I + B, \quad (40)$$

where I is the unit two-by-two matrix, while

$$B = \begin{pmatrix} 0 & \mathfrak{X}_{12} \\ \mathfrak{X}_{21} & 0 \end{pmatrix}. \quad (41)$$

We get from Eqs. (34) and (40)

$$B a_{S,L} = \mu_{S,L} a_{S,L}, \quad (42)$$

where $\mu_{S,L} = \lambda_{S,L} - \mathfrak{X}_{11}$.

Writing

$$a_{S,L} = \begin{pmatrix} a_{S,L}(1) \\ a_{S,L}(2) \end{pmatrix}, \quad (43)$$

we easily get

$$\left.\begin{aligned} \mu_{S,L} &= \pm\sqrt{\mathfrak{X}_{12}\mathfrak{X}_{21}}; \\ a_{S,L}(2) &= \pm\sqrt{\frac{\mathfrak{X}_{21}}{\mathfrak{X}_{12}}}\, a_{S,L}(1). \end{aligned}\right\} \quad (44)$$

We introduce the notation

$$\sqrt{\mathfrak{X}_{12}} = p; \quad \sqrt{\mathfrak{X}_{21}} = q. \quad (45)$$

Using the normalization conditions we find that a_S and a_L equal

$$a_{S,L} = \frac{\alpha_{S,L}}{\left(1 + \left|\frac{q}{p}\right|^2\right)^{1/2}} \begin{pmatrix} 1 \\ \pm\frac{q}{p} \end{pmatrix}, \quad (46)$$

where $\alpha_{S,L}$ are arbitrary phase multipliers. For the eigenvalues λ_S and λ_L we get

$$\lambda_{S,L} = \mathfrak{X}_{11} \pm pq. \quad (47)$$

The phase multipliers can be included in the definition of the functions a_S and a_L.

The functions a_S and a_L are thus characterized by the complex parameter q/p. If CP-invariance holds, then $q = p$ and Eq. (46) yields

$$a_{1,2} = \frac{1}{\sqrt{2}} \begin{pmatrix} 1 \\ \pm 1 \end{pmatrix}. \quad (48)$$

It is clear that the functions a_1 and a_2 are eigenfunctions of the CP operator.

Instead of q/p, we introduce the parameter

$$\varepsilon = \frac{p-q}{p+q}. \tag{49}$$

The parameter ε characterizes CP-invariance violation ($\varepsilon = 0$ if CP is conserved).

It follows from Eqs. (49) and (46) that the functions a_S and a_L can be defined in the following manner:

$$a_{S,L} = \frac{1}{\sqrt{2(1+|\varepsilon|^2)}}\begin{pmatrix} 1+\varepsilon \\ \pm(1-\varepsilon)\end{pmatrix}. \tag{50}$$

The corresponding vectors $|K_S\rangle$ and $|K_L\rangle$ [see Eq. (38)] equal

$$\left.\begin{aligned} |K_S\rangle &= \frac{1}{\sqrt{2(1+|\varepsilon|^2)}}[(1+\varepsilon)|K^0\rangle + (1-\varepsilon)|\overline{K}^0\rangle]; \\ |K_L\rangle &= \frac{1}{\sqrt{2(1+|\varepsilon|^2)}}[(1+\varepsilon)|K^0\rangle - (1-\varepsilon)|\overline{K}^0\rangle]. \end{aligned}\right\} \tag{51}$$

We will show that the parameter ε depends on the difference of matrix elements $\mathfrak{X}_{12} - \mathfrak{X}_{21}$ and experimentally determinable quantities. To do this we multiply both numerator and denominator in Eq. (49) by p + q. Using the obvious relation

$$(p+q)^2 = \frac{4pq}{1-\varepsilon^2}, \tag{52}$$

we get

$$\frac{\varepsilon}{1-\varepsilon^2} = \frac{\mathfrak{X}_{12}-\mathfrak{X}_{21}}{4pq}. \tag{53}$$

We get from Eqs. (47) and (35)

$$2pq = \lambda_S - \lambda_L = (m_S - m_L) - \frac{i}{2}(\Gamma_S - \Gamma_L). \tag{54}$$

We note that experimentally [16]

$$\Delta m = m_L - m_S = (0.46 \pm 0.02)\,\Gamma_S,$$
$$\Gamma_L = 1.63 \cdot 10^{-3}\,\Gamma_S.$$

Finally, using Eqs. (53) and (54), we get

$$\frac{\varepsilon}{1-\varepsilon^2} = \frac{\mathfrak{X}_{12}-\mathfrak{X}_{21}}{2(\lambda_S - \lambda_L)}. \tag{55}$$

The denominator of this expression is the order Γ_S, i.e., it is determined by the square of the weak interaction constant. The numerator has terms linear in the weak interaction constant [see Eq. (29)]. The structure of the expression for the parameter ε gives rise to the interesting possibility of explaining the experimentally observed effects ($|\varepsilon| \sim 10^{-3}$) by introducing a new CP-violating interaction with an extremely small coupling constant ($\sim 10^{-9}$ G).

We will examine this possibility (Wolfenstein's superweak interaction [17]) in detail below.

We now examine the state vectors of the K_S- and K_L-mesons in the general case, without assuming CPT-invariance of the interaction Hamiltonian. Solving Eq. (34), we have

$$\lambda_{S,L} = \frac{1}{2}\left(\mathfrak{X}_{11} + \mathfrak{X}_{22} \pm \sqrt{(\mathfrak{X}_{11} - \mathfrak{X}_{22})^2 + 4\mathfrak{X}_{12}\mathfrak{X}_{21}}\right); \tag{56}$$

$$a_{S,L}(2) = A_{S,L}\, a_{S,L}(1), \tag{57}$$

where

$$A_{S,L}=\frac{\lambda_{S,L}-\mathfrak{X}_{11}}{\mathfrak{X}_{12}}. \tag{58}$$

Thus, the normalized functions describing the short- and long-lived K-meson, in the general case, equal

$$a_{S,L}=\frac{1}{(1+|A_{S,L}|^2)^{1/2}}\begin{pmatrix}1\\ A_{S,L}\end{pmatrix}. \tag{59}$$

We introduce the parameters

$$\varepsilon_S=\frac{1-A_S}{1+A_S};\quad \varepsilon_L=\frac{1+A_L}{1-A_L}. \tag{60}$$

We conclude from Eqs. (59) and (60) that the vectors $|K_S\rangle$ and $|K_L\rangle$ can be expressed in the following form:

$$\left.\begin{aligned}|K_S\rangle&=\frac{1}{\sqrt{2(1+|\varepsilon_S|^2)}}\left[(1+\varepsilon_S)|K^0\rangle+(1-\varepsilon_S)|\bar{K}^0\rangle\right];\\ |K_L\rangle&=\frac{1}{\sqrt{2(1+|\varepsilon_L|^2)}}\left[(1+\varepsilon_L)|K^0\rangle-(1-\varepsilon_L)|\bar{K}^0\rangle\right].\end{aligned}\right\} \tag{61}$$

Thus, in the general case, if CPT-invariance is not assumed, the vectors $|K_S\rangle$ and $|K_L\rangle$ are characterized by two complex parameters. If CP-invariance holds ($\mathfrak{X}_{11}=\mathfrak{X}_{22}$ and $\mathfrak{X}_{12}=\mathfrak{X}_{21}$), then it follows from Eqs. (56) and (58) that $A_S=1$ and $A_L=-1$. In that case $\varepsilon_S=\varepsilon_L=0$. If CPT-invariance holds but CP-invariance does not ($\mathfrak{X}_{11}=\mathfrak{X}_{22}$, $\mathfrak{X}_{12}\neq\mathfrak{X}_{21}$), then

$$A_L=-A_S \text{ and } \varepsilon_S=\varepsilon_L. \tag{62}$$

We also note that if T-invariance holds but CP-invariance does not ($\mathfrak{X}_{12}=\mathfrak{X}_{21}$, $\mathfrak{X}_{11}\neq\mathfrak{X}_{22}$), Eqs. (56), (58), and (60) yield

$$A_L A_S=-1;\quad \varepsilon_S=-\varepsilon_L. \tag{63}$$

We now find the connection between elements of the matrix $\mathfrak{X}$ and the parameters ε_S and ε_L. To do this we introduce

$$\varepsilon_+=\frac{1}{2}(\varepsilon_S+\varepsilon_L);\quad \varepsilon_-=\frac{1}{2}(\varepsilon_S-\varepsilon_L). \tag{64}$$

We have, from Eq. (60), that

$$\varepsilon_+=\frac{1+A_S A_L}{(1+A_S)(1-A_L)};\quad \varepsilon_-=\frac{-(A_S+A_L)}{(1+A_S)(1-A_L)}. \tag{65}$$

It is easy to show, using Eq. (6), that

$$(1+A_S)(1-A_L)=\frac{2(A_S-A_L)}{1-\varepsilon_S\varepsilon_L} \tag{66}$$

holds.

Using Eqs. (56), (58), (65), and (66), we have

$$\frac{\varepsilon_+}{1-\varepsilon_L\varepsilon_S}=\frac{\mathfrak{X}_{12}-\mathfrak{X}_{21}}{2(\lambda_S-\lambda_L)}; \tag{67}$$

$$\frac{\varepsilon_-}{1-\varepsilon_L\varepsilon_S}=\frac{\mathfrak{X}_{11}-\mathfrak{X}_{22}}{2(\lambda_S-\lambda_L)}. \tag{68}$$

We make the following remark with respect to the latter equation. The denominator of the right side of Eq. (68) is $\sim\Gamma_S$ (a quantity that depends on the square of the weak interaction coupling constant). If a CPT-violating interaction with an extremely small coupling constant of the order of the superweak interaction constant exists ($\sim 10^{-9}G$; G is the normal weak interaction constant), then the parameter $|\varepsilon_-|$ could be of the order 10^{-3}. It follows that a comparison of ε_S and ε_L is an extremely sensitive test of CPT-invariance. We also note that if one introduces the parameter

$$\gamma=\frac{\mathfrak{X}_{11}-\mathfrak{X}_{22}}{\mathfrak{X}_{11}+\mathfrak{X}_{22}}, \tag{69}$$

characterizing possible CPT-violation, then, assuming $|\varepsilon_-|\lessapprox 10^{-3}$, one gets from Eqs. (56) and (68) the well-known estimate

$$|\gamma|\lesssim 10^{-3}\frac{\Delta m}{m_K}\sim 10^{-16}. \tag{70}$$

Here $\Delta m = m_L - m_S$, and m_K is the K-meson mass. Let us return to our examination of Eq. (34). It is easy to see if CP is violated, the functions a_S and a_L are nonorthogonal. Using Eq. (34), it is easy to relate the scalar product $(a_S^+ a_L)$ with experimentally measurable quantities. We get

$$\left.\begin{aligned}(a_S^+\mathfrak{X}\,a_L)&=\lambda_L\,(a_S^+a_L);\\ (a_S^+\mathfrak{X}^+a_L)&=\lambda_S^*\,(a_S^+a_L).\end{aligned}\right\} \tag{71}$$

Subtracting the second equation from the first and using Eq. (28) and (30), we get

$$(a_S^+\Gamma a_L)=i\,(\lambda_L-\lambda_S^*)\,(a_S^+a_L). \tag{72}$$

This expression, called the unitarity relation, was first presented in [8]. Important consequences of the unitarity relation will be found below.

4. $K_L\to 2\pi$ DECAYS. LEPTONIC DECAYS OF K_L-MESONS

Let us examine the two π and $K_L\to\pi\mp l\pm\nu$ decays of K_L-mesons. We first assume that CPT-invariance holds. We later discuss methods of verifying the CPT theorem based on measurements of $K_L\to 2\pi$ decay parameters.

The following parameters are experimentally measured:

$$\eta_{+-}=\frac{\langle\pi^+\pi^-|T|K_L\rangle}{\langle\pi^+\pi^-|T|K_S\rangle};\quad \eta_{00}=\frac{\langle\pi^0\pi^0|T|K_L\rangle}{\langle\pi^0\pi^0|T|K_S\rangle}. \tag{73}$$

Here T is the decay matrix while $|\pi^+\pi^-\rangle$ ($|\pi^0\pi^0\rangle$) is the state vector of π^+ and π^--mesons (two π^0-mesons). The parameters η_{+-} and η_{00} characterize CP-violation in $K_L\to 2\pi$ decays (if CP is conserved then $|K_L\rangle$ is an eigenvector of the CP operator with eigenvalue -1 while the vector $|\pi\pi\rangle$, describing two π-mesons in an $l=0$ state, is an eigenvector of the CP operator with eigenvalue $+1$; so in the case of CP-invariance, $\eta_{+-}=\eta_{00}=0$).

Let us examine these parameters. We expand the vectors $|\pi^+\pi^-\rangle$ and $|\pi^0\pi^0\rangle$ into states with definite total isospin. Taking into account the fact that the vector $|\pi^+\pi^-\rangle$ describing the $\pi^+-\pi^-$ system with zero orbital angular momentum is symmetric under exchange of the π^+ and π^- (Bose-Einstein statistics), we get

$$\left.\begin{aligned}|\pi^+\pi^-\rangle&=\sqrt{\tfrac{2}{3}}\,|0\rangle+\sqrt{\tfrac{1}{3}}\,|2\rangle;\\ |\pi^0\pi^0\rangle&=\sqrt{\tfrac{1}{3}}\,|0\rangle-\sqrt{\tfrac{2}{3}}\,|2\rangle.\end{aligned}\right\} \tag{74}$$

Here $|\,0>$ and $|\,2>$ are the state vectors (with total isospin 0 and 2 and I_z of zero) of the two π-meson system arising from the decay of $K_{S,\,L}$-mesons. It is clear from Eqs. (73) and (74) that parameters η_{+-} and η_{00} are determined by the matrix elements $< I|\,T\,|\,K_L >$ and $< I|\,T\,|\,K_S >$ (I = 0.2). The transition to an I = 2 state is suppressed with respect to transitions to I = 0 because of the $\Delta I = 1/2$ rule. It is therefore natural, after substituting Eq. (74) into Eq. (73), to divide the numerator and denominator of Eq. (73) by the allowed $\Delta I = 1/_2$ rule and CP-conserving amplitude $< 0|\,T\,|\,K_S >$. We get

$$\left.\begin{aligned}\left(1+\frac{1}{\sqrt{2}}\omega\right)\eta_{+-}&=\varepsilon_0+\varepsilon';\\(1-\sqrt{2}\,\omega)\,\eta_{00}&=\varepsilon_0-2\varepsilon',\end{aligned}\right\}\tag{75}$$

where

$$\left.\begin{aligned}\varepsilon_0=\frac{\langle 0\,|\,T\,|\,K_L\rangle}{\langle 0\,|\,T\,|\,K_S\rangle};\quad \varepsilon'=\frac{1}{\sqrt{2}}\,\frac{\langle 2\,|\,T\,|\,K_L\rangle}{\langle 0\,|\,T\,|\,K_S\rangle};\\ \omega=\frac{\langle 2\,|\,T\,|\,K_S\rangle}{\langle 0\,|\,T\,|\,K_S\rangle}.\end{aligned}\right\}\tag{76}$$

The parameter ω characterizes the $\Delta I = 1/2$ rule violation in $K_S \rightarrow 2\pi$ decays. Using expansion (74), we have

$$R=\frac{\Gamma\,(K_S\rightarrow\pi^0\,\pi^0)}{\Gamma\,(K_S\rightarrow\pi^0\,\pi^0)+\Gamma\,(K_S\rightarrow\pi^+\,\pi^-)}=\frac{1}{3}\cdot\frac{|1-\sqrt{2}\,\omega\,|^2}{1+|\,\omega\,|^2}.\tag{77}$$

The value of R has been measured in several experiments. The mean value of R equals [13]

$$R_{av}=0.31\pm0.04,\tag{78}$$

in agreement with the $\Delta I=\frac{1}{2}\left(R_{\Delta I=\frac{1}{2}}=\frac{1}{3}\right)$ rule. Comparing Eqs. (77) and (78) and assuming that $|\,\omega|^2 \ll 1$, we find [13]

$$\mathrm{Re}\,\omega=(2\pm4)\cdot10^{-2}.\tag{79}$$

Neglecting terms of the order $1/\sqrt{2}\,\omega$ and $\sqrt{2\omega}$ in Eq. (75), we get [18]

$$\left.\begin{aligned}\eta_{+-}&=\varepsilon_0+\varepsilon';\\ \eta_{00}&=\varepsilon_0-2\varepsilon'.\end{aligned}\right\}\tag{80}$$

Let us now examine the unitarity relation (72). We separate out the $|\,\pi\pi>$ state from the sum over intermediate states $|i>$ in the left side of Eq. (72). We get

$$R\Gamma_S\left(\frac{1-R}{R}\,\eta_{+-}+\eta_{00}\right)+\beta\Gamma_S=i\,(\lambda_L-\lambda_S^*)\,(a_S^+\,a_L).\tag{81}$$

where R is the experimentally measurable ratio of K_S-meson decay probabilities [see Eq. (77)], Γ_S is the total decay probability of a K_S-meson, while $\beta\,\Gamma_S$ is the contribution to the left side of the unitarity relation (72) of the states $|\,3\pi>$, $|\,\pi l\nu\,>$, $\pi\pi\gamma\,>$, etc. We first examine the contribution from $|\pi l\nu\,>$ states. We assume that the $\Delta Q = \Delta S$ rule holds. We denote the part of the decay matrix (see Eq. (29b) which is determined by transitions to $|\,\pi^+l^-\nu>$ and $|\,\pi^-l^+\nu>$ intermediate states (l = e and μ) by $\Gamma_{\alpha'\alpha}$ (lept). It follows from CPT-invariance that

$$\Gamma_{11}\,(\text{lept})=\Gamma_{22}\,(\text{lept}).\tag{82}$$

If the $\Delta Q = \Delta S$ rule holds, it is clear that

$$\Gamma_{12}\,(\text{lept})=\Gamma_{21}\,(\text{lept})=0.\tag{83}$$

Using Eqs. (50), (82), and (83), we find

$$\left.\begin{aligned}&(a_S^+\Gamma(\text{lept})a_L)=\frac{2\,\mathrm{Re}\,\varepsilon}{1+|\varepsilon|^2}\Gamma_{11}(\text{lept});\\&(a_L^+\Gamma(\text{lept})a_L)=(a_S^+\Gamma(\text{lept})a_S)\\&=\Gamma(K_S\to\text{lept})=\Gamma(K_L\to\text{lept})=\Gamma_{11}(\text{lept}).\end{aligned}\right\}\tag{84}$$

Here

$$\Gamma(K_{S,L}\to\text{lept})=\sum_{l=e,\mu}\left[\Gamma(K_{S,L}\to\pi^-l^+\nu)+\Gamma(K_{S,L}\to\pi^+l^-\bar{\nu})\right].\tag{85}$$

From Eqs. (50) and (51), we obtain

$$(a_S^+a_L)=\langle K_S|K_L\rangle=\frac{2\,\mathrm{Re}\,\varepsilon}{1+|\varepsilon|^2}.\tag{86}$$

Finally, the contribution of $|\pi l\nu>$ states to the left side of the unitarity relation equals

$$(a_S^+\Gamma(\text{lept})a_L)=(a_S^+a_L)\Gamma(K_L\to\text{lept}).\tag{87}$$

Experimentally, [17],

$$\Gamma(K_L\to\text{lept})\simeq10^{-3}\Gamma_S.\tag{88}$$

It follows from experiment that $|a_S^+a_L)|\sim10^{-3}$, as will be seen below. Thus

$$|(a_S^+\Gamma(\text{lept})a_L)|\sim10^{-6}\Gamma_S.$$

The $|\pi\pi>$ states of the order $\sim10^{-3}\Gamma_S$ ($|\eta_{+-}|$ and $|\eta_{00}|$ are of the order 10^{-3}) contribute to the unitarity relation. It follows that one can neglect the $|\pi l\nu>$ contribution to the left side of Eq. (81) in comparison to the contribution of $|\pi\pi>$ states.

To estimate the $|3\pi>$ state contribution we use the inequality

$$|(a_S^+\Gamma(G)a_L)|\leqslant\sqrt{\Gamma(K_S\to G)\Gamma(K_L\to G)}.\tag{89}$$

where $\Gamma_{\alpha\alpha'}(G)$ is the part of the decay matrix determined by intermediate state G. Equation (89) can be easily derived from Eq. (29) by using the Schwartz inequality. Inequality (89) implies that

$$|(a_S^+\Gamma(3\pi)a_L)|\leqslant\beta_{3\pi}\Gamma_S,\tag{90}$$

where

$$\beta_{3\pi}=\sqrt{\frac{\Gamma(K_L\to3\pi)\Gamma(K_S\to3\pi)}{\Gamma_S^2}};\tag{91}$$

$$\Gamma(K_{S,L}\to3\pi)=\Gamma(K_{S,L}\to\pi^+\pi^-\pi^0)+\Gamma(K_{S,L}\to\pi^0\pi^0\pi^0).$$

Experimentally [16],

$$\frac{\Gamma(K_L\to3\pi)}{\Gamma_S}=6.1\cdot10^{-4};\quad\frac{\Gamma(K_S\to3\pi)}{\Gamma_S}<10^{-4}.$$

It follows that

$$\beta_{3\pi}<2.5\cdot10^{-4}.\tag{92}$$

We have found the upper limit of $|3\pi>$ contribution to the left side of the unitarity relation (72).

Experiments studying $K_L\to3\pi$ decays have not found CP-violation effects ($\sim0.4\%$ accuracy). This means that the magnitude of $|(a_S^+\Gamma(3\pi)a_L)|$ is significantly less than the upper limit, and the $|3\pi>$ contributions can be neglected.

It is easy to convince oneself by using Eq. (89) that one can neglect the contribution of all other states in comparison to the $|\pi\pi>$ contribution. It follows that only $|i>$ states should be included in the sum over intermediate states $|\pi\pi>$ in the left side of the unitarity relation.

Dropping the term $\beta\Gamma_S$ in Eq. (81), neglecting Γ_L in comparison with Γ_S, and assuming R = 1/3 [see Eq. (78)[, we get

$$\frac{1}{3}\Gamma_S(2\eta_{+-}+\eta_{00})=\left(i\Delta m+\frac{1}{2}\Gamma_S\right)\langle K_S|K_L\rangle. \tag{93}$$

Equation (80) gives

$$(2\eta_{+-}+\eta_{00})=3\varepsilon_0. \tag{94}$$

Using Eqs. (93) and (94), we get

$$\frac{\varepsilon_0\Gamma_S}{i\Delta m+\frac{1}{2}\Gamma_S}=\langle K_S|K_L\rangle. \tag{95}$$

The right side of this relation is a real number. Setting the imaginary part of the left side of Eq. (95) equal to zero we find [18]

$$\frac{\text{Im}\,\varepsilon_0}{\text{Re}\,\varepsilon_0}=\frac{2\Delta m}{\Gamma_S}. \tag{96}$$

Therefore, the tangent of the phase of the parameter ε_0 is completely determined by experimentally measurable quantities. It turns out that the quantity $\text{Re}\varepsilon_0$ is experimentally measurable. We examine the decays

$$K_L\to\pi^{\mp}e^{\pm}\nu. \tag{97}$$

We denote the decay amplitude of K^0 and $\overline{K}^0$-particles by

$$\left.\begin{aligned}\langle\pi^- e^+\nu|T|K^0\rangle&=f;\\ \langle\pi^- e^+\nu|T|\overline{K}^0\rangle&=g.\end{aligned}\right\} \tag{98}$$

Amplitude f describes $\Delta Q = \Delta S$ transitions while g describes $\Delta Q = -\Delta S$ transitions. CPT-invariance and S-matrix unitarity imply that

$$\left.\begin{aligned}\langle\pi^+ e^-\bar{\nu}|T|\overline{K}^0\rangle&=f^*;\\ \langle\pi^+ e^-\bar{\nu}|T|K^0\rangle&=g^*.\end{aligned}\right\} \tag{99}$$

We find further

$$\left.\begin{aligned}\langle\pi^- e^+\nu|T|K_L\rangle&=\frac{1}{\sqrt{2(1+|\varepsilon|^2)}}[(1+\varepsilon)f-(1-\varepsilon)g];\\ \langle\pi^+ e^-\bar{\nu}|T|K_L\rangle&=\frac{1}{\sqrt{2(1+|\varepsilon|^2)}}[(1+\varepsilon)g^*-(1-\varepsilon)f^*].\end{aligned}\right\} \tag{100}$$

Experimentally, one measures the charge asymmetry of decays (97), which is defined in the following way:

$$\delta_L=\frac{\Gamma(K_L\to\pi^- e^+\nu)-\Gamma(K_L\to\pi^+ e^-\nu)}{\Gamma(K_L\to\pi^- e^+\nu)+\Gamma(K_L\to\pi^+ e^-\bar{\nu})}. \tag{101}$$

Let g = 0 (valid if $\Delta Q = \Delta S$). Using Eqs. (100) and (101) we get the following expression for the charge asymmetry:

$$(\delta_L)_{\Delta Q=\Delta S}=\frac{2\,\text{Re}\,\varepsilon}{1+|\varepsilon|^2}=\langle K_S|K_L\rangle. \tag{102}$$

In the general case, we get

$$\delta_L = \langle K_S | K_L \rangle \frac{1-|x|^2}{|1-x|^2}, \tag{103}$$

where $x = g/f$ is a parameter characterizing the $\Delta Q = \Delta S$ rule violation. We note that terms of order ε^2 and εx were dropped in getting Eq. (103). The quantity $1-|x|^2/|1-x|^2$ can be determined from independent experiments. The latest experiments give

$$\frac{1-|x|^2}{|1-x|^2} = 1.06 \pm 0.06.$$

Measuring the charge asymmetry δ_L thus allows the determination of the scalar product $\langle K_S | K_L \rangle$. The charge asymmetry has been measured in two experiments 73, 47. The following values for the asymmetry were found:

$$\delta_L = (2.24 \pm 0.36) \cdot 10^{-3} \text{ for } K_{Le3} \text{ decays},$$
$$\delta_L = (4.05 \pm 1.7) \cdot 10^{-3} \text{ for } K_{L\mu 3} \text{ decays}.$$

We will show that the scalar product $\langle K_S | K_L \rangle$ determining charge asymmetry in $K_L \to \pi \mp l \pm \nu$ decays is related to the parameter ε_0 [see Eq. (76)], characterizing $K_L \to 2\pi$ can be expressed in terms of the parameter ε and the amplitudes $\langle 0 | T | K^0 \rangle$ and $\langle 0 | T | \overline{K}^0 \rangle$. CPT-invariance and S-matrix unitarity imply that the matrix elements $\langle I | T | K^0 \rangle$ and $\langle I | T | \overline{K}^0 \rangle$ ($I = 0.2$) are not independent and are related through the expression

$$\langle I | T | \overline{K}^0 \rangle^* = e^{-2i\delta_I} \langle I | T | K^0 \rangle. \tag{104}$$

Here δ_I is the $\pi\pi$ scattering phase shift in the $| I_I \rangle$ state (orbital angular momentum $l = 0$, total isospin equal to 1, total center of mass energy equal to the K-meson mass). S-matrix unitary ($S+S = 1$) implies

$$S^+(1 + iT) = 1, \tag{105}$$

where

$$S = 1 + iT. \tag{106}$$

Equations (105) and (106) yield

$$S^+ T = T^+. \tag{107}$$

This gives us

$$\langle I | T^+ | K^0 \rangle = \langle I | S^+ T | K^0 \rangle = \sum_n \langle I | S^+ | n \rangle \langle n | T | K^0 \rangle. \tag{108}$$

It is clear that the main contribution to the sum over intermediate states $|n\rangle$ is due to the state $|I\rangle$ ($2\pi \to 3\pi$ transitions are forbidden by G-parity conservation, $|\pi\pi\gamma\rangle$ states make a contribution of order $\lesssim \alpha$ compared to the $|I\rangle$ state, etc.). Thus, up to terms of order $\lesssim \alpha$, we have

$$\langle I | T^+ | K^0 \rangle = \langle I | S^+ | I \rangle \langle I | T | K^0 \rangle. \tag{109}$$

Taking into account

$$\langle I | S | I \rangle = e^{2i\delta_I}$$

(δ_I is the $\pi\pi$ scattering phase shift for the $| I \rangle$ state), we get the following relation from Eq. (109):

$$\langle K^0 | T | I \rangle^* = e^{-2i\delta_I} \langle I | T | K^0 \rangle. \tag{110}$$

Further, CPT-invariances gives

$$\langle K^0 | T | I \rangle = \langle I | T | \overline{K}^0 \rangle. \tag{111}$$

Finally, Eqs. (110) and (111) give Eq. (104).

We note that if CP-invariance holds,

$$\langle I | T | K^0 \rangle = \langle I | T | \bar{K}^0 \rangle$$

while Eq. (104) yields

$$\langle I | T | K^0 \rangle^* = e^{-2i\delta_I} \langle I | T | K^0 \rangle. \quad (112)$$

It follows from Eq. (112) that the phase of matrix element $\langle I | T | K^0 \rangle$ equals the $\pi\pi$-scattering phase shift in the $| I \rangle$ state (the well-known final-state interaction theorem).

We now examine Eq. (76) for the parameter ε_0. Using Eq. (51), we get

$$\varepsilon_0 = \frac{1-b}{1+b}, \quad (113)$$

where

$$b = \frac{(1-\varepsilon) \langle 0 | T | \bar{K}^0 \rangle}{(1+\varepsilon) \langle 0 | T | K^0 \rangle}. \quad (114)$$

It is obvious that

$$b = \frac{1-\varepsilon_0}{1+\varepsilon_0}. \quad (115)$$

The scalar product $\langle K_S | K_L \rangle$ equals

$$\langle K_S | K_L \rangle = \frac{1 - \left| \frac{1-\varepsilon}{1+\varepsilon} \right|^2}{1 + \left| \frac{1-\varepsilon}{1+\varepsilon} \right|^2} = \frac{1 - |b|^2}{1 + |b|^2}. \quad (116)$$

We used the relation

$$|\langle 0 | T | K^0 \rangle = |\langle 0 | T | \bar{K}^0 \rangle|, \quad (117)$$

which is a consequence of Eq. (104) in getting Eq. (116). Using Eqs. (115) and (116), we get

$$\langle K_S | K_L \rangle = \frac{2 \operatorname{Re} \varepsilon_0}{1 + |\varepsilon_0|^2}. \quad (118)$$

The phenomenological analysis of experiments measuring charge asymmetry in $K_L \to \pi^{\pm} l^{\mp} \nu$ decays and experiments studying $K_L \to 2\pi$ decays is based on Eqs. (80), (96), (103), and (118). Measurements of the charge asymmetry δ_L can determine $\operatorname{Re}\varepsilon_0$ [Eqs. (103) and (118)*]. The tangent of the phase of parameter ε_0 is given by Eq. (96). Thus, the parameter ε_0 (magnitude and phase) can be determined if one knows the value of $1 - |x|^2 / |1-x|^2$, the charge asymmetry δ_L, the K_L- and K_S-mass difference Δm, and the total K_S-meson decay probability. Recent experiments [19] give

$$\arg \varepsilon_0 = 42^\circ \pm 1^\circ; \quad |\varepsilon_0| = (1.5 \pm 0.25) \cdot 10^{-3}. \quad (119)$$

The parameter η_{+-} is experimentally measurable. The magnitude of η_{+-} equals [19]

$$|\eta_{+-}| = (1.90 \pm 0.05) \cdot 10^{-3}. \quad (120)$$

The following values for the phase of η_{+-} have been found in recent experiments [19]: $46^\circ \pm 15^\circ$, $51.2^\circ \pm 11^\circ$, and $68^\circ \pm 7.5^\circ$. If η_{+-} and ε_0 are known, then the parameter η_{00} ($\eta_{00} = 3\varepsilon_0 - 2\eta_{+-}$) characterizing $K_L \to \pi^0\pi^0$ decay can be determined from Eq. (80)

The magnitude of η_{00} is presently being measured in various laboratories. The latest experiments yield [19]: $|\eta_{00}| \cdot 10^3 = 3.6 \pm 0.4$, 3.6 ± 0.6, 2.2 ± 0.4, and 2.3 ± 0.3 $|\eta_{00}| \cdot 10^3 < 3.0$.

---

* $K_L \to 2\pi$ experiments give $|\varepsilon_0| = 1/3 | 2\eta_{+-} + \eta_{00} | \sim 10^{-3}$. One can thus neglect $|\varepsilon_0|^2$ compared to unity in the denominator of Eq. (118).

It is clear from the preceding that if the assumptions made in deriving Eqs. (80), (96), (103), and (118) are true (the smallness of ω, the $\Delta I = 1/2$ rule, inclusion of $|\pi\pi\rangle$ states only in the unitarity relation), then a measurement of η_{00} will allow the verification of the basic principles on which these relations are based.

In connection with this remark, let us return to the unitarity relation (72). This equation is based only on the assumption that the Wieskopf-Wigner approximation holds. Assuming that one can neglect the contributions of all but the $|\pi\pi\rangle$ states in the left side of Eq. (72), we write the unitarity relation in the form

$$\frac{\Gamma_S[(1-R)\eta_{+-}+R\eta_{00}]}{\left(i\Delta m+\frac{1}{2}\Gamma_S\right)}=\langle K_S|K_L\rangle. \tag{121}$$

It follows from Eq. (61) that, in the general case, without assuming CPT-invariance, the scalar product $\langle K_S|K_L\rangle$ equals

$$\langle K_S|K_L\rangle=\frac{\varepsilon_S^*+\varepsilon_L}{\sqrt{(1+|\varepsilon_S|^2)(1+|\varepsilon_L|^2)}}. \tag{122}$$

Thus, in the general case, the right side of Eq. (121) is a complex number. If CPT-invariance holds, then $\varepsilon_L = \varepsilon_S$ and the scalar product is real. Setting the imaginary part of the left side of Eq. (121) to zero, we get

$$\sin(\Phi_{+-}-\delta)+\frac{R}{1-R}\frac{|\eta_{00}|}{|\eta_{+-}|}\sin(\Phi_{00}-\delta)=0 \text{ (CPT-invariance)}, \tag{123}$$

where $\frac{1}{2}\Gamma_S+i\Delta m=\sqrt{\frac{1}{4}\Gamma_S^2+(\Delta m)^2}\,e^{i\delta}$; $\eta_{+-}=|\eta_{+-}|e^{i\Phi_{+-}}$; $\eta_{00}=|\eta_{00}|e^{i\Phi_{00}}$.

If CPT-invariance holds, but T-invariance is violated, we have $\varepsilon_S = -\varepsilon_L$ and

$$\langle K_S|K_L\rangle=\frac{2i\,\mathrm{Im}\,\varepsilon_L}{1+|\varepsilon_L|^2}. \tag{124}$$

Setting the real part of Eq. (121) to zero, we get in this case [12, 22]

$$\cos(\Phi_{+-}-\delta)+\frac{R}{1-R}\frac{|\eta_{00}|}{|\eta_{+-}|}\cos(\Phi_{00}-\delta)=0 \text{ (T-invariance)}. \tag{125}$$

All quantities entering into these relations are experimentally determinable. Verification of Eq. (123) [and Eq. (125)] is proof that the interaction causing $K_L \to 2\pi$ decay is CPT-invariant (T-invariant) (if the assumptions used in deriving these equations hold).

Preliminary information of the phase η_{00} has recently become available ($\Phi_{00} = 17°\pm31°$) [24]. Using these data we find [24]

$$\cos(\Phi_{+-}-\delta)+\frac{R}{1-R}\frac{|\eta_{00}|}{|\eta_{+-}|}\cos(\Phi_{00}-\delta)=1.5\pm0.3 \text{ (exptl)}, \tag{125a}$$

which contradicts Eq. (125). At the same time, the data presently available agree with Eq. (123).

In conclusion, we examine Eq. (76) for ε' and ω. Using Eqs. (51) and (104), we find

$$\varepsilon'=\frac{e^{i\delta_2}[e^{-i\delta_2}\langle 2|T|K\rangle(1+\varepsilon)-e^{i\delta_2}\langle 2|T|K\rangle^*(1-\varepsilon)]}{\sqrt{2}\,e^{i\delta_0}[e^{-i\delta_0}\langle 0|T|K\rangle(1+\varepsilon)+e^{i\delta_0}\langle 0|T|K\rangle^*(1-\varepsilon)]}. \tag{126}$$

We introduce amplitudes

$$A_I = e^{-i\delta_I} \langle I|T|K\rangle \quad (I=0 \text{ and } 2). \tag{127}$$

Using Eq. (104), we find that the quantity b [see Eq. (114)] equals

$$b = \frac{(1-\varepsilon) A_0^*}{(1+\varepsilon) A_0}. \tag{128}$$

From Eqs. (126)-(128) we easily get

$$\varepsilon' = \frac{1}{\sqrt{2}} e^{i(\delta_2 - \delta_0)} \frac{\frac{A_2}{A_0} - \frac{A_2^*}{A_0^*} b}{1+b}. \tag{129}$$

Using Eq. (115), which relates b with the parameter ε_0, we get

$$\varepsilon' = \frac{1}{\sqrt{2}} e^{i(\delta_2 - \delta_0)} \left[i \operatorname{Im} \frac{A_2}{A_0} + \varepsilon_0 \operatorname{Re} \frac{A_2}{A_0} \right]. \tag{130}$$

Similarly, we find

$$\omega = e^{i(\delta_2 - \delta_0)} \left[\operatorname{Re} \frac{A_2}{A_0} + i\varepsilon_0 \operatorname{Im} \frac{A_2}{A_0} \right]. \tag{131}$$

Dropping terms of order $|\varepsilon_0|\,|\omega|$ and $|\varepsilon_0|^2$ from Eqs. (130) and (131), we get the following expression for the parameter ε' [18]:

$$\varepsilon' = \frac{1}{\sqrt{2}} e^{i(\delta_2 - \delta_0)} \operatorname{Im} \frac{A_2}{A_0}. \tag{132}$$

Thus, the phase of parameter ε' is determined by the difference of $\pi\pi$ scattering I = 2 and I = 0 S-wave phase shifts at the total center of mass energy equal to the K-meson mass. Dropping terms of order $|\varepsilon_0|^2$ and $|\varepsilon\varepsilon'|$ in Eqs. (130) and (131), we also find that

$$\omega = e^{i(\delta_2 - \delta_0)} \operatorname{Re} \frac{A_2}{A_0}. \tag{133}$$

Finally, it follows from Eq. (104) that

$$|\langle I|T|K^0\rangle| = |\langle I|T|\bar{K}^0\rangle|. \tag{134}$$

Since the phases of the $|K^0\rangle$ and $|\bar{K}^0\rangle$ vectors are arbitrary* this relation means that these phases can be chosen so that matrix elements $\langle 0|T|K^0\rangle$ and $\langle 0|T|\bar{K}^0\rangle$ (or $\langle 2|T|K^0\rangle$ and $\langle 2|T|\bar{K}^0\rangle$) are equal. Since $\langle 2|T|K^0\rangle| \ll |\langle 0|T|K^0\rangle|$ ($\Delta I = 1/2$ rule), it is natural to fix the relative phase of $|K^0\rangle$ and $|\bar{K}^0\rangle$ states in such a way that [18]

$$\langle 0|T|K^0\rangle = \langle 0|T|\bar{K}^0\rangle. \tag{135}$$

It follows from Eqs. (113) and))14) that with such a choice of phase

$$\varepsilon_0 = \varepsilon. \tag{136}$$

Thus, if the phase of $|K^0\rangle$ and $|\bar{K}^0\rangle$ states is chosen so that Eq. (136) holds, then $|\varepsilon| = |\varepsilon_0| \sim 10^{-3}$. We also note that in this case

$$A_0 = A_0^*. \tag{137}$$

---

* The operator H_0 commutes with the strangeness operator S. Multiplying Eq. (10) by $e^{i\chi S}$ (χ an arbitrary real parameter) we find that the vectors $e^{i\chi S}|K^0>$ and $e^{-i\chi}|\bar{K}^0>$ are also eigenvalues of the operator H_0 and can thus be chosen as a basis.

5. SUPERWEAK INTERACTION MODEL

In conclusion, we examine Wolfenstein's superweak interaction model [7, 32]. We first assume that CPT-invariance holds. We will discuss the general case later. We turn to Eq. (55) for ε. The numerator of this expression is a series beginning with a term linear in H_W. The denominator contains a term of order Γ_S, i.e., a term quadratic in the weak interaction constant G. Let us assume that the weak interaction Hamiltonian can be written in the form

$$H_W = H_W^0 + H_W', \tag{138}$$

where H_W^0 is the ordinary CP-invariant weak interaction Hamiltonian, while the Hamiltonian H_W' allows $\Delta S = 2$ transitions and does not commute with the CP operator, i.e.,

$$\langle K^0 | H_W' | \bar{K}^0 \rangle \neq \langle \bar{K}^0 | H_W' | K^0 \rangle. \tag{139}$$

It is clear from Eq. (55) that if $|\varepsilon| \sim 10^{-3}$ then the constant G' characterizing the CP-odd part of the Hamiltonian H_W' must be of order

$$G' \sim |\varepsilon| (G m_K^2)\, G \sim 10^{-9}\, G, \tag{140}$$

Thus, a CP-violating interaction with an extremely small coupling constant ($\sim 10^{-9}$ G) could explain the experimentally observed CP-violation effects.

Let us examine the parameters η_{+-} and η_{00}. Using Eq. (51), we have

$$\eta_{+-} = \frac{\varepsilon + y_{+-}}{1 + \varepsilon y_{+-}}, \tag{141}$$

where

$$y_{+-} = \frac{\langle \pi^+\pi^- | T | K^0 \rangle - \langle \pi^+\pi^- | T | \bar{K}^0 \rangle}{\langle \pi^+\pi^- | T | K^0 \rangle + \langle \pi^+\pi^- | T | \bar{K}^0 \rangle}. \tag{142}$$

If a superweak interaction is responsible for CP-violation then

$$|y_{+-}| \sim \frac{G'}{G} \sim 10^{-9}. \tag{143}$$

Dropping terms of this order we get

$$\eta_{+-} = \varepsilon \tag{144}$$

from Eq. (141) in the superweak interaction model. Similarly, it is easy to show that

$$\eta_{00} = \varepsilon \tag{145}$$

in this model. We examine Eq. (55). It is clear that in the superweak interaction model only first-order perturbation theory terms should be kept in the difference $\mathfrak{X}_{12} - \mathfrak{X}_{21}$. Equations (29) and (55) then yield

$$\varepsilon = \frac{-i \,\mathrm{Im}\, M_{12}}{\Delta m + \frac{i}{2} \Gamma_S}. \tag{146}$$

It follows that

$$\frac{\mathrm{Im}\, \varepsilon'}{\mathrm{Re}\, \varepsilon} = \frac{2\Delta m}{\Gamma_S}. \tag{147}$$

We note that this relation can also be derived from the unitarity relation (72) if $\Gamma_{12} - \Gamma_{21}$ and terms of order $\sim \varepsilon^2$ are dropped.

Thus, if CP-violation in K_L-meson decay is due to a superweak interaction, the parameters characterizing CP-violation satisfy Eqs. (144) and (145). Equation (144) implies that

$$\arg \eta_{+-} = \arg \varepsilon = 42^\circ \pm 1^\circ. \tag{148}$$

Experimental results (see p. 247) do not contradict Eq. (148). Further, Eqs. (144) and (147) yield

$$\mathrm{Re}\,\varepsilon = |\eta_{+-}| \frac{\Gamma_{S/2}}{\left[(\Delta m)^2 + \frac{1}{4}\Gamma_S^2\right]^{1/2}}. \tag{149}$$

From this we have $\mathrm{Re}\varepsilon = (1.44 \pm 0.10) \cdot 10^{-3}$ (superweak interaction). Experiments measuring charge asymmetry in $K_L \to \pi^\pm l^\mp \nu$ decays yield $\mathrm{Re}\varepsilon = (1.09 \quad 0.18) \cdot 10^{-3}$ (experiment). Thus, the experimental results do not contradict Eq. (144).*

We now examine the superweak interaction model in the general case of a CPT-violating Hamiltonian H'_W [23]. In this case the $|K_S\rangle$ and $|K_L\rangle$ states are characterized by two complex parameters ε_S and ε_L. Neglecting terms of order $G'/G \sim 10^{-9}$, we have

$$\eta_{+-} = \eta_{00} = \varepsilon_L \frac{(1+|\varepsilon_S|^2)^{1/2}}{(1+|\varepsilon_L|^2)^{1/2}}. \tag{150}$$

It follows from this that

$$|\varepsilon_L| \leqslant \frac{|\eta_{+-}|}{(1-|\eta_{+-}|^2)^{1/2}} \approx 2 \cdot 10^{-3}. \tag{151}$$

Using the unitarity relation (72) and the inequality (89), we find

$$|\langle K_S | K_L \rangle|^2 \leqslant \frac{\sum_G \Gamma(K_S \to G)\,\Gamma(K_L \to G)}{\left[(\Delta m)^2 + \frac{\Gamma_S^2}{4}\right]}$$

from which we get [7]

$$|\langle K_S | K_L \rangle| \leqslant 10^{-2}. \tag{152}$$

This inequality yields

$$|\varepsilon_S| < 10^{-2}. \tag{153}$$

Neglecting $|\varepsilon_S|^2$ and $|\varepsilon_L|^2$ in comparison with unity, we get

$$\eta_{+-} = \eta_{00} = \varepsilon_L. \tag{154}$$

It is clear that the charge asymmetry in $K_L \to \pi^\pm l^\mp \nu$ decays is given by the following expression:

$$\delta_L = 2\,\mathrm{Re}\,\varepsilon_L \frac{1-|x|^2}{|1-x|^2}. \tag{155}$$

Further, the superweak interaction model being discussed yields

$$\left.\begin{aligned} \mathfrak{X}_{12} - \mathfrak{X}_{21} &= 2i\,\mathrm{Im}\,M_{12}, \\ \mathfrak{X}_{11} - \mathfrak{X}_{22} &= H_{11} - M_{22}. \end{aligned}\right\} \tag{156}$$

Neglecting $\varepsilon_S \varepsilon_L$ in comparison with unity in Eqs. (67) and (68) we get, using Eq. (156),

$$\left.\begin{aligned} \varepsilon_+ &= \frac{-i\,\mathrm{Im}\,H_{12}}{\left(\Delta m + \frac{i}{2}\Gamma_S\right)}, \\ \varepsilon_- &= \frac{-(H_{11} - H_{22})}{\Delta m + \frac{i}{2}\Gamma_S}. \end{aligned}\right\} \tag{157}$$

---

* We note that the fact that Eqs. (144) and (145) hold does not yet prove that Wolfenstein's model is correct. A whole series of other models also yield these relations.

From this we get

$$|\varepsilon_S| = |\varepsilon_L|. \tag{158}$$

Using Eq. (157), it is easy to show that

$$\text{tg}\,\frac{\Phi_S + \Phi_L}{2} = \frac{2\Delta m}{\Gamma_S}, \tag{159}$$

where $\Phi_{S,L} = \arg \varepsilon_{S,L}$.

Thus, in the case of a CPT-violating superweak interaction, the parameters ε_S and ε_L can only differ by a phase. The half-sum of the phases must satisfy Eq. (159).

LITERATURE CITED

1. L. D. Landau, Zh. Éksper. i Teor. Fiz., 32, 405 (1957).
2. J. H. Christenson et al., Phys. Rev. Lett., 13, 138 (1964).
3. D. Dorfan et al., Phys. Rev. Lett., 19, 987 (1967).
4. S. Bennett et al., Phys. Rev. Lett., 19, 993 (1967).
5. M. Gormly et al., Phys. Rev. Lett., 21, 402(1968).
6. H. Yuta and S. Okubo, Phys. Rev. Lett., 21, 781 (1968).
7. L. Wolfenstein, Phys. Rev. Lett., 13, 286 (1964).
8. J. S. Bell and J. Steinberger, Proc. Int. Conf. on Elementary Particles, Oxford (1965), p. 195.
9. N. Byers, S. W. McDowell, and C. N. Yang, High-Energy Physics and Elementary Particles, Vienna (1965), p. 953.
10. M. V. Terent'ev, Uspekhi Fiz. Nauk, 86, 231 (1965).
11. L. B. Okun', Uspekhi Fiz. Nauk, 89, 603 (1966).
12. M. Gourdin and G. Charpak, Preprint CERN 67-18 (1967).
13. L. B. Okun' and C. Rubbia, Proc. Int. Conf. on Elementary Particles, Heidelberg (1967), p. 301.
14. L. Wolfenstein, Proc. Int. School of Physics, "Ettore Majorana" (1968).
15. T. D. Lee, R. Oehme, and C. N. Yang, Phys. Rev., 106, 340 (1957).
16. V. Weisskopf and E. P. Wigner, Z. Physik, 63, 54 (1930); 65, 18 (1930).
17. A. H. Rosenfeld et al., UCRL 8030 (1967).
18. T. T. Wu and C. N. Yang, Phys. Rev. Lett., 13, 380 (1964).
19. J. W. Cronin, Proc. Intern. Conf. on High-Energy Physics, Vienna (1968), p. 28.
20. S. Bennet et al., Phys. Lett., 27B, 239 (1968).
21. L. Wolfenstein, Nuovo Cimento, 62A, 17 (1966).
22. L. I. Lapidus, Preprint of the Joint Institute for Nuclear Research 2-3622, Dubna (1967).
23. T. D. Lee and L. Wolfenstein, Phys. Rev., 138A, 1490 (1965).
24. J. Steinberger, Topical Conf. on Weak Interactions, CERN 69-7 (1969), p. 291.

NONLOCAL QUANTUM SCALAR-FIELD THEORY

G. V. Efimov

The nonlocal theory of the quantized one-component scalar field is considered; the axioms of nonlocal quantum theory are formulated. The class of relativistically invariant generalized functions is considered which can play the role of form factors in constructing perturbation-theory series for the S-matrix. It is shown that for interaction Lagrangians of the $L_I(x) = g\varphi^n(x)$ $(n \geq 3)$ type the perturbation-theory series for the S-matrix is finite and satisfies the unitarity and macrocausality conditions in each order of perturbation theory.

INTRODUCTION

In nonlocal quantum field theory two methods exist for introducing nonlocality into the theory. The first [1, 2] is based on the proposition that at short distances the quantities characterizing the field are already not measurable. This leads to a situation in which the field potentials do not commutate with the coordinates, i.e.,

$$[x_\mu, A(x)]_- \neq 0,$$

where $A(x)$ is a potential which characterizes a certain field. Theories in which it is assumed that the notion of a specific point in space-time has no precise meaning, i.e., the components of the coordinate operator do not commutate,

$$[x_\mu, x_\nu]_- \neq 0 \text{ for } \mu \neq \nu,$$

are also included in this. This leads to the theory of quantum space-time [3-6]. Up to the present such theories still have unclear prospects.

The second approach consists in assuming the free field to be local while introducing nonlocality into the interaction (this approach was proposed for the first time in [7-10]; see [11, 12] for further development of this trend). The physical idea which forms the basis of this trend is the assumption that in small space-time domains it is possible to have other forms of causal coupling than those which are characteristic of large space and time scales.

The present paper applies to the second trend in nonlocal field theory, and therefore hereafter we shall mean precisely this trend by the term "nonlocal quantum field theory."

It should be noted that the original idea of introducing nonlocality consisted in an attempt to eliminate the difficulties with the infinite electron mass in quantum electrodynamics. Then, as the versions of nonlocal theories were investigated, it turned out that the difficulties in the nonlocal theory are even greater than in the local theory: The problems of relativistic invariance, unitarity, macrocausality, and gauge invariance could not be solved simultaneously. The combination of the requirements of relativistic invariance, unitarity, and macrocausality turned out to be an essentially difficult problem. The schemes proposed (see, for example, [11)] cannot yet be treated as a good basis for future theory.

The present paper likewise does not yet solve all of the problems which face practical quantum field theory. We shall compose a nonlocal scalar neutral field theory which satisfies the requirements of relativistic invariance, unitarity, and causality.

Joint Institute for Nuclear Research, Dubna. Translated from Problemy Fiziki Élementarnykh Chastits i Atomnogo Yadra, Vol. 1, No. 1, pp. 255-291, 1970.

The problem of introducing gauge invariance into the theory (i.e., consideration of nonlocal charged fields) shall not be considered here. This is a task of subsequent investigations.

What then are the principal problems facing nonlocal quantum field theory?

In formulating theories with nonlocal interaction of quantized fields two most essential problems develop: first, how to formulate the macrocausality condition which is imposed on the S-matrix of the theory; second, what type of form factors may be used to realize the imposed macrocausality conditions. These two problems are not only closely interrelated, but to a considerable degree the solution of the second problem determines the solution of the first one; namely, knowledge of the properties of specific classes of nonlocal functions determines which nonlocality requirement is imposed on the S-matrix of the theory.

Let us dwell at first on the formulation of the macrocausality condition. Although physically the meaning of this condition is clear — the absence of noncausal influence at macroscopically large distances is required — there is still no precise quantitative formulation of this condition (i.e., no indication of those requirements of the matrix element whose fulfilment guarantees impossibility of experimentally recording a noncausal influence at macroscopically large distances). Therefore, restriction to the qualitative aspect of the problem — clarification of the possibility of making the noncausal influence a short-range one — is the usual practice in investigating the causality problem. Many years (see, for example, [11, 13-18]) have been devoted to an investigation of the macroscopic causality condition, but a complete solution of this problem is still far from completion.

Here a second problem develops immediately: What type of form factor may be used in introducing nonlocality into the theory? In the simplest version of nonlocal coupling it is assumed that the interaction, for example, of an electron field, which is described by the current $J_\mu(x)$, and an electromagnetic field, which is described by the potential $A_\mu(x)$, does not take place at a single point of space-time but is "spread" by means of a relativistically invariant form factor $F(x-x')$ so that the interaction is described by the function

$$W = \iint J_\mu(x) F(x-x') A_\mu(x') \, dx \, dx'.$$

It is assumed from the requirement of relativistic invariance that the form factor $F(x-x')$ must be a function of the interval

$$s^2 = (x_0 - x_0')^2 - (\mathbf{x} - \mathbf{x}')^2$$

and must fall off sufficiently rapidly for $|s^2| \to \infty$, for example, $F(s^2) = \exp\left\{-\left(\frac{s^2}{l^2}\right)^2\right\}$. When form factors of this type are introduced, two kinds of difficulties develop. First, if $F(s^2)$ allows violation of causality in the invariant domain $|s^2| < l^2$, where l^2 is a certain short length, then in directions close to the light cone the space and time extents of the noncausal domain may turn out to be arbitrarily large. Second, difficulties develop in proving the unitarity of the S-matrix of nonlocal theory, since the form factor $F(s^2)$ must be real for real s^2, and this excludes the Feynman rules governing the bypassing of singularities for the Fourier transform of $F(s^2)$, resulting in the impossibility of transition to a Euclidian metric in the amplitudes of the physical processes, to the appearance of angular divergences, [19], and consequently, to violation of unitarity of the S-matrix.

Attempts have been made to get rid of these difficulties. In the first case an additional timelike vector P_μ is introduced into the nonlocal theory, by means of which one can "localize" the violation of causality [20-23]. However, it is completely unclear as to what physical meaning may be attributed to this vector. In the second case one can preserve unitarity either by rejecting analyticity of the amplitudes of the physical processes at sufficiently high energies in momentum space [24] or by introducing an indefinite metric. It seems to us that such an avoidance of the difficulties of nonlocal theory is not completely satisfactory.

How then does one actually construct a nonlocal theory starting from the reasonable physical idea of the possible violation of causality in the small? Roughly speaking, a perturbation-theory series corresponding to local interaction is taken for the S-matrix, and attempts are made to substitute into the matrix elements form factors which would eliminate ultraviolet divergences, preserve unitarity of the S-matrix, and have a form which would permit the appropriate physical interpretation to be made of the form factors.

Under these conditions no attempt at all has been made at giving a definition of the nonlocal nature of the theory from a purely mathematical point of view. In such an approach a mass of varied difficulties develops concerning which we have spoken above, and statements concerning the groundlessness and impossibility of constructing a nonlocal field theory (see, for example, [15]) and the general skeptical attitude towards such a theory are a consequence of all this.

Let us note the principal difference between the modern local and nonlocal theories. In the local theory a rigorous mathematical definition is given of locality, and a space of principle functions is chosen in connection with this definition. In nonlocal theory blind searches are made for form factors, the expansion of the S-matrix in perturbation theory is manipulated, and attempts are made at giving some reasonable physical interpretation of the individual matrix elements. The difference, as is evident, is enormous.

The present paper constitutes an attempt at filling precisely this gap in the formulation of nonlocal quantum field theory. Our solution of the problems of nonlocal quantum field theory consists in the following. It turns out [25] that a large intermediate class of generalized functions exists between the local form factors having a moderate growth

$$F(x-x')=\sum_{n=0}^{N} a_n \Box^n \delta^{(4)}(x-x') \tag{1}$$

and the nonlocal form factors having the form

$$F(x-x')=F\left((x-x')^2\right), \tag{2}$$

where $F(s^2)$ is a certain function of the interval s^2; these generalized functions have properties which are suitable from the point of view of the requirements imposed on the nonlocal theory [25].

The discovery of this intermediate class allowed the gradual transition from rigorous local quantum field theory to the existing construction of nonlocal theory with form factors of the form (2) to be traced. It turned out that the generalized functions of the form

$$F(x-x')=\sum_{n=0}^{\infty} a_n \Box^n \delta^{(4)}(x-x')$$

[here, unlike (1), the coefficients a_n are nonvanishing for any m] are concentrated in a restricted domain in x-space for specific requirements imposed on a_n because they are relativistically invariant functions. Under these conditions the Fourier transform

$$\tilde{F}(p^2)=\sum_{n=0}^{\infty} a_n (p^2)^n,$$

which is an integer function of a certain growth order, is fully capable of playing the role of a cutoff function in momentum space.

Since the integer functions $\tilde{F}(p^2)$ have no singularities at finite complex p^2, it may be expected that the S-matrix of the theory will be unitary and will not contain additional singularities in comparison with the local case.

I. THE AXIOMS OF NONLOCAL QUANTUM FIELD THEORY

1. Introduction

We shall follow the procedure for the axiomatic construction of the theory, which was proposed by N. N. Bogolyubov, D. V. Shirkov, B. V. Medvedev, and M. K. Polivanov [26, 27]. The method of constructing the theory originates from the program advanced by Heisenberg [28] in which only those matrix elements of the S-matrix are considered which correspond to transitions between asymptotically stable states. The aggregate of such matrix elements can be represented in the form of a functional expansion in normal products of asymptotic fields

$$S=\sum_{n=0}^{\infty}\frac{(-i)^n}{n!}\int dx_1 \dots \int dx_n\, F_n(x_1, \dots, x_n) : \varphi(x_1) \dots \varphi(x_n). \tag{1.1}$$

where the $\varphi(x)$-field satisfies the equation

$$(\Box - m^2)\varphi(x) = 0, \tag{1.2}$$

Let us formulate the principal physical assumptions on which the considered version of the theory is based. Under these conditions we do not set ourselves the goal of formulating a noncontradictory, complete, and independent system of axioms, as is done in the exposition of the Wightman approach; we merely enumerate those physical propositions which will be required in order to construct the theory. We shall divide these propositions into two groups: general properties which apply to the scattering matrix on the energy surface, and special local properties which apply to the expansion of the scattering matrix beyond the energy surface.

2. The General Properties of the S-Matrix

The Space Considered. The asymptotic states of the system contain infinitely distant particles and their bound complexes. The interaction between such particles and complexes is equal to zero, and consequently the principal dynamic characteristics of the system (the type of energy, momentum, angular momentum, etc.) are additive. Such states can be described by amplitudes $|\ \dots\ \rangle$ which are elements of Hilbert space.

Relativistic Covariance. There exists a group G of transformations of L which includes the Lorentz group as a subgroup and may likewise include other transformations (for example, isotopic, gauge, etc.). Due to the action of L from G the amplitudes of the states are transformed by means of a certain unitary representation U_L. If in the state $|\, p\,\rangle$ the energy-momentum vector p has a specific value and L_a is the translation $x \to x + a$, then

$$U_{L_a}|p\rangle = e^{-ipa}|p\rangle. \tag{2.1}$$

The Existence of a Vacuum. There exists a unique state for which

$$U_L|0\rangle = |0\rangle \tag{2.2}$$

for all U_L. This is the vacuum state.

Completeness and Spectrology. There exists a system of eigenamplitudes of the four-momentum states $|\, n, k_n\,\rangle$ which correspond to nonnegative energy values; this system together with the amplitudes $|\, 0\,\rangle$ is complete, so that

$$\langle\alpha|AB|\beta\rangle = \langle\alpha|A|0\rangle\langle 0|B|\beta\rangle + \sum_n \int dk_n \langle\alpha|A|n, k_n\rangle\langle k_n, n|B|\beta\rangle. \tag{2.3}$$

Here n is the ensemble of all the remaining discrete and continuous quantum numbers; together with k_n it fully defines the state. If the asymptotic states are exhausted by the states of spinless particles of one kind, then the completeness relationship is written in the form

$$\langle\alpha|AB|\beta\rangle = \sum_{m=0}^{\infty} \int dk_1 \dots \int dk_m \langle\alpha|A|k_1 \dots k_m\rangle\langle k_1 \dots k_m|B|\beta\rangle. \tag{2.4}$$

The Existence of Unitarity of the S-Matrix. The amplitude of the probability of a transition from the state $|\alpha\rangle$ to the state $|\beta\rangle$ is given by the matrix element $\langle\beta\,|\,S\,|\,\alpha\rangle$ of the operator S (the scattering matrix) which satisfies the unitarity condition on the mass shell; i.e.,

$$\langle\alpha|SS^+|\beta\rangle = \langle\alpha|\beta\rangle, \tag{2.5}$$

where $|\,\alpha\,\rangle$ and $|\,\beta\,\rangle$ are arbitrary asymptotic states of the system.

For the transformation L from the group G the scattering matrix S is transformed by means of the unitary representation U_L.

Stability. If $|\alpha\rangle$ is the amplitude of vacuum for a state containing one real particle or one stable complex, then the condition for stability of such states has the form

$$S|\alpha\rangle = |\alpha\rangle. \tag{2.6}$$

Asymptotic states corresponding to the presence of a specific number n of particles of specific kinds α_i having specific momenta p_i can be obtained if production $a^+_{\alpha ipi}$ and annihilation $a^+_{\alpha ipi}$ operators are introduced conventionally for a particle of the $a^+_{\alpha ipi}$ kind having the momentum p_i and are then used to act on the amplitudes of the vacuum state

$$|\alpha_1, p_1; \ldots; \alpha_n, p_n\rangle = \frac{1}{\sqrt{N}} a^+_{\alpha_1 p_1} \ldots a^+_{\alpha_n p_n} |0\rangle, \tag{2.7}$$

where $N = \nu_1! \ldots \nu_n!$; ν_i is the number of particles of the same kind present in the state described by the amplitude (2.7). Since spatially separate particles do not interact, the operators $a^+_{\alpha ipi}$ and $a_{\alpha ipi}$ satisfy the conventional commutation relationships

$$\begin{aligned} \left[a_{\alpha p}, a^+_{\beta q}\right]_- &= \delta_{\alpha\beta}\,\delta(p-q), \\ [a_{\alpha p}, a_{\beta q}]_- &= \left[a^+_{\alpha p}, a^+_{\beta q}\right]_- = 0. \end{aligned} \tag{2.8}$$

The solutions of Eq. (1.2) may be written in the form

$$\varphi(x) = \frac{1}{(2\pi)^{3/2}} \int \frac{dk}{\sqrt{2\omega}} \left(a_k e^{-ikx} + a^+_k e^{ikx}\right), \tag{2.9}$$

where $kx = \omega x_0 - \mathbf{ux}$, $\omega = \sqrt{m^2 - k^2}$.

The commutation rules

$$\begin{aligned} \left[\varphi(x), a^+_p\right]_- &= \frac{1}{(2\pi)^{3/2}} \cdot \frac{e^{-ipx}}{\sqrt{2p_0}}; \\ & \qquad\qquad p_0 = \sqrt{m^2 + p^2} \\ [a_p, \varphi(x)]_- &= \frac{1}{(2\pi)^{3/2}} \cdot \frac{e^{ipx}}{\sqrt{2p_0}} \end{aligned} \tag{2.10}$$

derive from (2.8) and (2.9).

3. The Local Properties of the S-Matrix

Now let us extend the scattering matrix which defines the expansion (1.1) to arbitrary but commutative fields $\varphi(x)$. Such an expansion is nonunique to a higher degree. We shall consider only those expansions for which the property of relativistic invariance of the S-matrix is conserved, and we shall require that the following local properties be fulfilled.

Integrability. The extended operator S has variational derivatives of any order with respect to the asymptotic fields. The radiation operators

$$R^{(n)}(x_1, \ldots, x_n) = \frac{\delta^n S}{\delta\varphi(x_1) \ldots \delta\varphi(x_n)} S^{-1} \tag{3.1}$$

and their products are integrable; i.e., all of the matrix elements

$$\langle\alpha| R^{(m)}(x_1, \ldots, x_m) R^{(n)}(y_1, \ldots, y_n) |\beta\rangle \tag{3.1a}$$

are generalized functions which are integrable on a certain appropriate space of principal functions.

The Macrocausality Condition. A reasonable nonlocal theory must be constructed in such a fashion that the nonlocal interactions of the quantized fields do not lead to violation of causality at macroscopically large distances. As we said in the introduction, we shall consider only those theories in which the noncausal behavior of the S-matrix is due to nonlocal coupling between interacting fields. In this case the elementary length is the characteristic domain in which nonlocal interaction of local fields occurs. Let us go over to the mathematical formulation of our physical conditions. We shall follow the idea developed by N. N. Bogolyubov [26]. We shall introduce the operation of "turn-on" and "turn-off" of interaction. Assume that the function $f(x)$ having values in the interval $(0, 1)$ characterizes the intensity of interaction turn-on. Assume now that $g_1(x)$ is nonvanishing in the domain $G_1 \subset R^4$, while $g_2(x)$ is nonvanishing in the domain $G_2 \subset R^4$. Then the S-matrix of the theory satisfies the macrocausality condition if

$$S(g_1+g_2) = S(g_2)\,S(g_1) \quad \text{for } G_2 \gtrsim G_1. \tag{3.2}$$

The notation $G_2 \sim G_1$ means that all points of the domains G_2 are spatially similar to all points of the domain G_1. The notation $G_2 > G_1$ means that all of the points of the domain $G_2\,(G_1)$ lie in the future (in the past) relative to a certain time t.

We shall assume that the S-matrix satisfies the macrocausality condition if

$$S(g_1+g_2) = S(g_2)\,S(g_1) + M(g_2, g_1) \quad \text{for } G_2 \gtrsim G_1, \tag{3.3}$$

where the operator $M(g_2, g_1)$ is such that any matrix elements $\langle \alpha \mid M(g_1, g_2) \mid \beta \rangle$ fall off fairly rapidly for $\rho(G_1, G_2) \to \infty$, where

$$\rho(G_1, G_2) = \min_{x\in G_1,\, y\in G_2} \sqrt{(x_0-y_0)^2+(\mathbf{x}-\mathbf{y})^2}.$$

Since for x, t $\to \infty$ causal signals attenuate as $1/\mid \mathbf{x} \mid^{3/2}$ (the property of a retarded function $\Delta_{re}(x)$), it is the practice in physics to assume that the function falls off fairly rapidly if it falls off more rapidly than any reciprocal power of the polynomial; i.e.,

$$\lim_{|x|\to\infty} |x|^N\, |f(x)| = 0 \tag{3.4}$$

for any $N > 0$.

Moreover, the idea of introducing nonlocality indicates that we may not separate any two physical events at fairly short distances. Therefore, in checking the macrocausality condition (3.3) one may not use functions with a constrained carrier in x-space as interaction turn-on functions, but rather functions which are of the order of unity in certain domain $G \subset R^4$ and fall off fairly rapidly outside this domain.

Finally, let us formulate the macrocausality condition. If the functions $g_1(x)$ and $g_2(x)$ are of the order of unity in the domains G_1 and G_2, respectively, and, moreover, $G_1 \lesssim G_2$, then*

$$\lim_{\rho(G_1, G_2)\to\infty} [\rho(G_1, G_2)]^N\, |\langle\alpha| \{S[g_2+g_1] - S[g_2]\,S[g_1]\} |\beta\rangle| = 0 \tag{3.5}$$

for any $N > 0$ and for any states $\mid \alpha \rangle$ and $\mid \beta \rangle$. The matrix elements of the scattering matrix can be transformed into vacuum average radiation operators by means of the formal commutation relations (2.10). From this it follows that one can write the equations

$$\begin{aligned} [S, a_p^+]_- &= \frac{1}{(2\pi)^{3/2}} \int d^4x \frac{\delta S}{\delta\varphi(x)} \cdot \frac{e^{-ipx}}{\sqrt{2p_0}}, \\ [a_p, S]_- &= \frac{1}{(2\pi)^{3/2}} \int d^4x \frac{\delta S}{\delta\varphi(x)} \cdot \frac{e^{+ipx}}{\sqrt{2p_0}} \end{aligned} \tag{3.6}$$

for S-matrix commutators having production and annihilation operators.

4. Statement of the Problem

Let us formulate the problem which we shall be solving hereafter. We shall construct the scattering matrix from the stipulated interaction Lagrangian $L_I(x)$ in the form of a formal expansion in powers of the

---

* Note that, as a consequence of the derivation of the causality condition in [26], if $SS^+ \neq 1$ outside the mass shell, then S^{-1} appears in the causality condition instead of S^+.

interaction constant g:

$$S=1+\sum_{n=1}^{\infty}\frac{(ig)^n}{n!}\cdot\int dx_1\ldots\int dx_n\,S_n(x_1,\ldots,x_n). \tag{4.1}$$

The convergence of this formal series is not considered here. Fairly weighty foundations exist for the proposition that this series is divergent [29, 30]. However, we shall assume that the formal expansion (4.1) is the source of reasonable asymptotic approximations if the coupling constant g is fairly small.

In order to relate the formal expansion for the S-matrix (4.1) to the interaction Lagrangian $L_I(x)$ let us use the correspondence principle: for infinitely small g the matrix S(g) must have the form

$$S=1+\mathrm{ig}\int L_I(x)\,dx, \tag{4.2}$$

whence it follows that

$$S_1(x)=L_I(x) \tag{4.3}$$

in the expansion (4.1). According to the procedure developed by Bogolyubov [26], knowing $S_1(x)$ one may construct the remaining $S_n(x_1, \ldots, x_n)$ by requiring fulfillment of the relativistic covariance, unitarity, and causality conditions imposed on the S-matrix in each order of perturbation theory. In the case of local theory it turns out that

$$S_n(x_1.\ldots,x_n)=\mathrm{T}\left(L_I(x_1)\ldots L_T(x_n)\right), \tag{4.4}$$

where the symbol T is understood to represent the procedure of chronological ordering of the operators. However, this expression is not the most general expression which satisfies the conditions indicated. The stipulation of the interaction Lagrangian turns out to be insufficient for the complete determination of the S-matrix, and it is necessary additionally to stipulate an infinite chain of quasi-local operators [26]:

$$\Lambda_2(x_1,\ x_2);\ \Lambda_3(x_1x_2x_2);\ldots;\ \Lambda_n(x_1,\ldots,x_n);\ldots \tag{4.5}$$

since Eq. (4.4) is indefinite for coinciding arguments.

In the case of nonlocal theory the symbol T cannot be treated as the procedure of rigorous time ordering of the field operators. Our problem consists in giving meaning to Eq. (4.4). Let us preserve the symbol T, but we shall assume it to mean a certain operation according to which we shall construct $S_n(x_1, \ldots, x_n)$ from the stipulated Lagrangian $L_I(x)$. Essentially, this operation introduced here is an additional postulate of the theory, which guarantees uniqueness of S-matrix construction from a stipulated interaction Lagrangian.

Naturally, we shall require that the S-matrix (4.1) satisfy all of the requirements enumerated in Secs. 2 and 3 in each order of perturbation theory.

Let us make one essential comment. The S-matrix which we have constructed will satisfy the formulated macrocausality requirement in each order of perturbation theory. However, at present it is the practice to relate the magnitude of the elementary length to the behavior of the amplitude at high energies. Then it turns out that in constructing the expansion of the S-matrix in the coupling constant the noncausal domain expands with increasing order of perturbation theory (i.e., the magnitude of the elementary length, which characterizes the nonlocality scale, rises from order to order of perturbation theory, reaching the value $l_n = (n-1)l$ in the n-th order). This is a direct consequence of unitarity of the S-matrix. Actually, assume that we have a perturbation-theory S-matrix:

$$S=1+igS_1+(ig)^2S_2+(ig)^3S_3+\ldots$$

and assume that the S-matrix is unitary in each order of perturbation theory; i.e.,

$$\begin{gathered}S_1^+-S_1=0,\\ S_2^++S_2-S_1^+S_1=0,\\ \cdots\cdots\cdots\cdots\\ (-)^nS_n^++S_n+(-)^nS_{n-1}^+S_1-S_{n-1}S_1^++\ldots=0.\end{gathered} \tag{4.6}$$

Usually nonlocality is introduced in the second order of perturbation theory and is associated with the behavior of the amplitude at high energies. Let us present a simple example. Assume that at high energies we have $S_2 \sim e^{lE}$ for some amplitudes; then $S_n \sim e^{(n-1)lE}$, i.e., $l_n = (n-1)l$. The same result is obtained for a more complex dependence of the amplitude on energy. This means that the noncausal domain expands from one order of perturbation theory to another. However, at present nothing can be said concerning what occurs in the limit of the complete series. Fairly weighty grounds exist to support the proposition that the perturbation-theory series diverges. Therefore, the problem of determining the complete S-matrix of the theory is connected with the problem of summing the entire perturbation-theory series. In principle, it can be imagined that some method of summation exists which leads to a finite, unitary, macrocausal S-matrix. Then the expansion of the nonlocality domain in perturbation theory has no physical meaning, since as a result of the summation we obtain a completely determinate magnitude of the elementary length, which is associated with the behavior of the complete S-matrix at asymptotically high energies. All of this reasoning is currently in the realm of guesswork and good intentions, but in the future it will form the basis of the requirements imposed on the method of summing perturbation-theory series.

Nevertheless, the problem may be stated physically in such a way that the degree to which causality is violated will be confined within the limits of the conventional requirements imposed on nonlocal series. Actually, we are required to construct an S-matrix in the form of a perturbation-theory series in a small coupling constant from the Lagrangian of the system of quantized fields. We are interested only in perturbation theory, and nothing is known concerning the properties of the overall series within the framework of modern methods. Therefore, it can be required that the causality violation which occurs at macroscopic distances and is described by the higher orders of perturbation theory must be vanishingly small.

Actually, in conventional nonlocal theory it is required that the noncausal signals attenuate sufficiently rapidly with time or distance, for example, as $\exp\{-\Lambda t\}$, where Λ is the cutoff momentum. For example, in the case of weak interactions, in which $\Lambda \sim 100$ GeV is usually chosen, the effect of the noncausal signal is extremely small (namely $\sim \exp\{-10^{11}\}$) for a time of the order of the atomic dimensions (i.e., for $t \sim 10^{-18}$ sec $\sim 1\ \mathrm{eV}^{-1}$). Let us see what occurs in the considered scheme of nonlocal theory with expansion of the S-matrix in a small coupling constant. Once again let us consider weak interactions. The small parameter of the perturbation-theory series is the quantity $G\Lambda^2$ in this case, where G is the weak-interaction constant and $1/\Lambda = l$ is the elementary length. For example, let $G\Lambda^2 = Gl^{-2} < 1$. The violation of causality at distances of the order of the atomic dimensions will be described in the perturbation-theory order $n \sim r/l \sim 10^{11}$ if, as previously, we choose a value of the order of the atomic dimensions:

$$r \sim 10^{-8}\ \mathrm{cm} \sim 1\ \mathrm{eV}^{-1} \text{ and } l \sim 10^{-2}\ \mathrm{GeV}^{-1}.$$

This means that the magnitude of the effects associated with the noncausal behavior at distances of the order of atomic distances will be $\sim (Gl^{-2})^n \sim \exp\{-10^{11}\}$ (i.e., it is just as small as in the usually allowed versions of nonlocal series [11]).

It should be noted that the estimates presented are valid only for fairly low energies. At high energies the actual expansion parameter is a quantity of the type $ge^{a(E)}$, where $a(E)$ increases with an increase in the energy. Therefore, perturbation theory is valid only for those energies for which $g \exp\{a(E)\} \lesssim 1$. If under these conditions $a(E) \sim l_{\mathrm{eff}} E$, then $E \lesssim 1/l_{\mathrm{eff}} \ln 1/g$.

An interesting physical problem is the investigation of those interactions for which the coupling constant g is fairly small, so that it is possible to use perturbation theory at energies exceeding the quantity $1/l_{\mathrm{eff}}$. This allows a realistic approach to a domain in which causality is already violated but reliable calculation methods still exist.

Thus, the suggested scheme for constructing a nonlocal theory for interactions characterized by a small coupling constant is, it seems to us, physically acceptable.

II. NONLOCAL INTERACTION LAGRANGIANS

1. Statement of the Problem

Let us consider a one-component scalar field $\varphi(x)$ which can be described by a Lagrangian of the following form:

$$L(x) = L_0(x) + gL_I(x). \tag{1.1}$$

Here $L_0(x)$ is a conventional free-field Lagrangian, while $L_I(x)$ describes the self-action of the field $\varphi(x)$. In the case of conventional local field theory the interaction Lagrangian is a certain polynomial in the field $\varphi(x)$, for example, $L_I(x) = \varphi^4(x)$.

Let us consider the following problem. We shall assume that the interaction Lagrangian includes not the field $\varphi(x)$ but the field $\Phi(x)$ which is defined as follows:

$$\Phi(x) = \int dy A(x-y)\varphi(y) = A(\Box)\varphi(x), \tag{1.2}$$

$$A(x-y) = A(\Box)\delta^{(4)}(x-y), \tag{1.3}$$

where $A(\Box)$ is a certain operator in $\Box = -\partial^2/\partial x_0^2 + \partial/\partial x^2$.

Let us perform the conventional transformations. Formally, the S-matrix can be written in the form of a T-product

$$S = T\exp\left\{ig\int dx L_I(x)\right\}, \tag{1.4}$$

where $L_I(x)$ is now a polynomial in the field $\Phi(x)$, for example, $L_I(x) = \Phi^4(x)$. We expand the S-matrix into a series in the coupling constant g and go over to the N-products of the field operators $\Phi(x)$ in accordance with the Wick theorem in which the "chronological" convolution of the operators $\Phi(x)$ shall be understood to mean

$$D_c(x-y) = \overline{\Phi(x)\,\Phi}(y) = A(\Box_x)A(\Box_y)\overline{\varphi(x)\,\varphi}(y) = A(\Box_x)A(\Box_y)\Delta_c(x-y) = \frac{1}{(2\pi)^4 i}\int \frac{d^4p\,[\tilde{A}(p^2)]^2}{m^2-p^2-i\varepsilon}e^{ip(x-y)}. \tag{1.5}$$

Such a choice of the convolution of two field operators corresponds to the so-called T-product or T* operation [31]. Our first postulate of the theory consists in this (see Chap. 1, sec. 4).

Further, operating by conventional methods accepted in quantum field theory, we obtain a perturbation-theory series which is conventional in structure with the sole difference that the conventional causal scalar-field functions are replaced by the functions (1.5).

Let us state the following problem. Can we choose the operators $A(\Box)$ in such a way that the function $[\tilde{A}(p^2)]^2$ plays the role of a cutoff function or a form factor in the perturbation-theory series, i.e., so that the integrals corresponding to any Feynman diagram converge and the requirements of the unitarity and causality imposed on the S-matrix of the theory are fulfilled?

Let us study the properties of the generalized functions

$$A(x-y) = \sum_{n=0}^{\infty}\frac{a_n}{(2n)!}\Box^n\delta^{(4)}(x-y). \tag{1.6}$$

We write the Fourier transform of this operator in the form

$$\tilde{A}(z) = \sum_{n=0}^{\infty}\frac{a_n}{(2n)!}z^n; \quad z = p^2. \tag{1.7}$$

What requirements must the function $\tilde{A}(z)$ satisfy? First of all, $\tilde{A}(z)$ must be an integer analytic function in the plane of the complex variable $z = p^2$. Otherwise, any singularities of the functions $\tilde{A}(z)$ lead to the appearance of certain additional nonphysical singularities in the amplitudes of the physical processes for finite z. This means that the S-matrix will not be unitary.

Thus, the function $\tilde{A}(z)$ in (1.7) is an integer function. A detailed investigation of generalized functions of the form (1.6) was carried out in [25]. Here we shall consider three cases:

I. $$\overline{\lim_{n\to\infty}}|a_n|^{\frac{1}{n}} = 0.$$

II. $$\overline{\lim_{n\to\infty}} \frac{|a_n|^{\frac{1}{n}}}{n^{\sigma}} = 0, \quad \sigma < 1.$$

III. $$\overline{\lim_{n\to\infty}} \frac{|a_n|^{\frac{1}{n}}}{n^2} = 0.$$

Functions of class I are integer functions of order less than 1/2; i.e., for any $\varepsilon > 0$ there exists $C_\varepsilon > 0$ such that

$$|\tilde{A}(z)| < C_\varepsilon e^{\varepsilon \sqrt{|z|}}. \tag{1.8}$$

The generalized function $A(x-y)$ is a local generalized function in this case, as shown in [25]. In accordance with the theory of integer functions (see, for example, [42]) not a single direction along which $\tilde{A}(z)$ could fall off exists in the complex plane z for a function $\tilde{A}(z)$ of order less than 1/2. This means that such functions cannot play the role of cutoff functions in the construction of the perturbation theory for the S-matrix (i.e., no local form factors exist).

Functions of the class II are integer functions of order less than unity; i.e., for any $\varepsilon > 0$ there exist C_ε and A_ε which are such that

$$|\tilde{A}(z)| < C_\varepsilon e^{A_\varepsilon |z|^{1-\varepsilon}}. \tag{1.9}$$

The functions of the class III are integer functions having an arbitrary growth order. For functions of the classes II and III there may exist directions in the complex plane z along which the functions decrease. Therefore, these functions may play the role of cutoff functions in the construction of the perturbation theory for the S-matrix.

The generalized functions $A(x-y)$ are nonlocal in cases II and III.

The partitioning of integer functions into the II and III classes is associated with the fact that for generalized class-II functions one can introduce an improper regular limiting-transition procedure which is such that it turns out to be possible to define the product of two generalized functions uniquely [25]. Such a regularization will be considered further on. In the case of generalized class-III functions no such regular procedure exists.

Below we shall consider only generalized class-II functions.

2. The Space-Time Properties of the Generalized Functions

We shall consider the generalized class-II functions (1.6). We find the space of principal functions and denote it by Z_2. If the function $f(x) = f(x_0, \mathbf{x})$ belongs to the principal space Z_2, then the functional

$$(A, f)(x) = \int dx' A(x-x') f(x') = \sum_{n=0}^{\infty} \frac{a_n}{(2n)!} \Box^n f(x = \int dp e^{ipx} \tilde{A}(p^2) \tilde{f}(p) \tag{2.1}$$

must be definite. In accordance with condition II, for any $\tilde{A}(p^2)$ belonging to class II there exists a $\delta > 0$ and constants $C_\delta > 0$ and $A_\delta > 0$ which are such that

$$|\tilde{A}(p^2)| < C_\delta e^{A_\delta |p^2|^{1-\delta}} < C_\delta e^{A_\delta \sum_{j=0}^{3} |p_j|^{2(1-\delta)}}; \tag{2.2}$$

then for convergence of the integral (2.1) it is necessary that for any $\varepsilon > 0$ there exist constants C and b_j (j = 0, 1, 2, 3) which are such that

$$|\tilde{f}(p)| < C e^{-\sum_{j=0}^{3} b_j |p_j|^{2-\varepsilon}}. \tag{2.3}$$

It can easily be shown [25] that the space of principal functions in the case considered will be the space of integer functions having the exact order 2. The space Z_2 may be made countable-normalized if a system of norms is introduced into it, for example, in the following manner:

$$\|f\|_m = \sup_{z_0, \ldots, z_3} |f(z_0, \ldots, z_3)| e^{-\sum_{j=0}^{3} |z_j|^{2+\frac{1}{m}}} \tag{2.4}$$

for any m = 1, 2, 3, ...

The space of principal functions on which not only generalized class-II functions (1.6) are defined but also all the matrix elements (3.1a) (Chap. I) of the radiation operators (3.1) (Chap. I) should be chosen to be the space Z_2^{eff} having the following system of norms:

$$\|f\|_{n,m} = \sup_{z_0, \ldots, z_3} |x_0 x_1 x_2 x_3|^n |f(x_0 + i y_0, \ldots, x_3 + i y_3)| e^{-\sum_{j=0}^{3} |y_j|^{2+\frac{1}{m}}} \tag{2.4a}$$

What is the nature of the space-time properties of the generalized functions $A(x-y)$? Since there are no functions with a constrained carrier in the space of principal functions, it is not simple to define the notion of the carrier of a functional [25]. However, for our purposes it is sufficient to proceed as follows. Let us choose functions $f(x) \in Z_2$ of the principal space which are such that they are of the order of unity in the neighborhood of the point x = 0 and decrease fairly rapidly with distance from it (see Chap. I, Sec. 3).

We shall assume that the generalized function (2.1) is concentrated in a small bounded domain near the point x = y if, for any function $f(x) \in Z_2$ of the principal space, which is such that $f(x)$ is of the order of unity near the point x = 0 and decreases fairly rapidly with distance from it, the functional

$$(A, f)(x) = \int dy\, A(x-y) f(y) \tag{2.5}$$

decreases fairly rapidly for $\|x\| = \sqrt{x_0^2 + \mathbf{x}^2} \to \infty$, i.e.,

$$\lim_{\|x\| \to \infty} \|x\|^N |(A, f)(x)| = 0 \tag{2.6}$$

for any $N > 0$.

In the case considered we shall choose functions from the spaces S_α^β ($\beta \leq 1/2$; $\alpha + \beta \leq 1$) as such trial functions in accordance with the Gel'fand-Shilov terminology. We recall that if $f(x) \in S_\alpha^\beta$, then $f(x + iy)$ is an integer analytic function in the x + iy plane of order $\rho = 1/(1-\beta) \leq 2$ and is such that

$$|f(x + iy)| < C e^{-a|x|^{\frac{1}{\alpha}} + b|y|^{\frac{1}{1-\beta}}} \tag{2.7}$$

for certain constants a, b, and C.

Note that $S_\alpha^\beta \subset Z_2$ for $\beta \leq 1/2$, $\alpha + \beta \leq 1$. Fourier transformation converts the space S_α^β into the space S_β^α; i.e., $\widetilde{S_\alpha^\beta} = S_\beta^\alpha$:

$$|\tilde{f}(p + iq)| < Ce^{-a|p|^{\frac{1}{\beta}} + \beta|q|^{\frac{1}{1-\alpha}}}. \tag{2.8}$$

Then for the functional (2.5), we obtain

$$F(x) = (A, f)(x) = \int dp\, e^{ipx} \tilde{A}(p^2) \tilde{f}(p). \tag{2.9}$$

Since $\tilde{f}(p) \in S_\beta^\alpha$ for $\beta \leq 1/2$, $\alpha + \beta \leq 1$, while $\tilde{A}(p^2)$ is an integer of order rigorously less than 2, it follows that $\tilde{A}(p^2) f(p) \in S_\beta^\alpha$, whence $F(x) \in S_\alpha^\beta$, and consequently decreases rapidly for $\|x\| \to \infty$.

Thus, the generalized functions considered are concentrated in a small bounded domain in accordance with our definition.

3. The Physical Meaning of the Nonlocal Generalized Functions

As was shown in Sec. 1, the introduction of nonlocality into the interaction Lagrangian in accordance with (1.2) effectively leads to exactly the same perturbation-theory series as in the local case, but with a modified causal function:

$$\Delta_c(x-y) \to D_c(x-y) = \Delta_c(x-y) - iK(x-y), \tag{3.1}$$

where

$$D_c(x-y) = \frac{1}{(2\pi)^4 i} \int \frac{d^4k\,[\tilde{A}(k^2)]^2}{m^2-k^2-i\varepsilon} e^{ik(x-y)}; \tag{3.2}$$

$$K(x-y) = \frac{1}{(2\pi)^4 i} \int d^4k \frac{[\tilde{A}(k^2)]^2-1}{m^2-k^2} e^{ik(x-y)} = \frac{1}{i} \sum_{n=0}^{\infty} \frac{c_n}{(2n)!} \Box^n \delta(x-y). \tag{3.3}$$

We assume that the operator $A(\Box)$ is normalized as follows:

$$\tilde{A}(m^2) = 1. \tag{3.4}$$

Note that the generalized function $K(x-y)$ belongs to the same class as does $A(x-y)$. The change in the free propagator in accordance with (3.1) can be interpreted differently. The propagator of a free particle

$$\Delta_c(x_1-x_2) = \langle 0|T(\varphi(x_1)\varphi(x_2))|0\rangle \tag{3.5}$$

is indeterminate for coinciding arguments $x_1 = x_2$. In the case of local theory one can add any quasi-local generalized class-I function in the right side of (3.5). The "nonlocality" hypothesis consists in the fact that the T-product in (3.5) is indeterminate not only at coinciding points $x_1 = x_2$ but also in a certain small bounded domain near $x_1 = x_2$. One of the possible mathematical realizations of this hypothesis resides in the fact that one can add a nonlocal generalized function which is concentrated in a small space-time domain (i.e., a generalized class-II function) in the right side of (3.5); this is just what occurs according to Eq. (3.1).

Let us present still another physical concept which will allow us to impose additional constraints on the class of form factors which are introduced. Assume that there are two infinitely heavy point sources situated at a certain distance from one another; the interaction between them is accomplished via exchange and scalar mesons. Assume the interaction Lagrangian of this system has the form

$$L_I(x) = [\delta(\mathbf{x}-\mathbf{x}_1) + \delta(\mathbf{x}-\mathbf{x}_2)]\,\Phi(x_0, \mathbf{x}), \tag{3.6}$$

where the field $\Phi(x)$ is related to the field $\varphi(x)$ in accordance with Eq. (1.2). The sources are situated at the points $\mathbf{x}_1$ and $\mathbf{x}_2$. The S-matrix describing the interaction of this system is written according to the conventional rules:

$$S = T\exp\left\{ig\int d^4x L_I(x)\right\}. \tag{3.7}$$

The first nonvanishing correction to the energy of each source and to the energy of interaction between them is given by the matrix elements of the S-matrix in the second order of perturbation theory and is proportional to

$$\Delta E \sim ig^2 \int d^4x \langle 0|T(L_I(x)L_I(0))|0\rangle = 2g^2 W(0) + 2g^2 W(\mathbf{x}_1-\mathbf{x}_2), \tag{3.8}$$

where

$$W(\mathbf{x}_1-\mathbf{x}_2) = i\int_{-\infty}^{\infty} dx_0 \langle 0|T(\Phi(x_0, \mathbf{x}_1)\Phi(0, \mathbf{x}_2))|0\rangle. \tag{3.9}$$

The first term in Eq. (3.8) represents the correction to the proper energy of the source, while the second term represents the potential energy between the two sources due to exchange of a scalar meson.

The natural physical requirements are

$$\left.\begin{array}{l}\text{I. } W(0) < \infty, \\ \text{II. } W(\mathbf{x}) \to \dfrac{1}{|\mathbf{x}|} e^{-m|\mathbf{x}|} \quad \text{for} \quad |\mathbf{x}| \to \infty.\end{array}\right\} \tag{3.10}$$

In conventional local quantum field theory one cannot satisfy these requirements while remaining within the limits of the requirements of relativistic invariance. Let us see what occurs in the nonlocal theory considered. In accordance with (3.9) and (1.5), we obtain

$$W(\mathbf{r}) = i \int_{-\infty}^{\infty} dx_0 \frac{1}{(2\pi)^4 i} \int \frac{d^4p\, [\tilde{A}(p^2)]^2}{m^2 - p^2 - i\varepsilon} e^{i p_0 x_0 - i \mathbf{p}\mathbf{r}} = \frac{1}{(2\pi)^3} \int \frac{d\mathbf{p}\, [\tilde{A}(-\mathbf{p}^2)]^2}{m^2 + \mathbf{p}^2} e^{-i\mathbf{p}\mathbf{r}}. \tag{3.11}$$

If the form factor $\tilde{A}(-\mathbf{p}^2)$ is such that

$$\tilde{A}(\mathbf{p}^2) = 0\left(\frac{1}{p^2}\right) \quad \text{for} \quad p^2 \to -\infty, \tag{3.12}$$

i.e.,

$$\int d\mathbf{p}\, [\tilde{A}(\mathbf{p}^2)]^2 < \infty, \tag{3.13}$$

then the function

$$V(\mathbf{r}) = \frac{1}{(2\pi)^3} \int d\mathbf{q}\, e^{i\mathbf{r}\mathbf{q}} [\tilde{A}(-\mathbf{q}^2)]^2 \tag{3.14}$$

has the following properties:

$$\left.\begin{array}{l} V(0) < \infty, \\ V(\mathbf{r}) = O\left(e^{-\left(\frac{r}{l}\right)^{\gamma}}\right) \quad \text{for} \quad \dfrac{r}{l} \gg 1, \end{array}\right\} \tag{3.15}$$

where l is a certain parameter having the dimensionality of a length — the so-called "elementary length"; $\gamma = 2\rho/2\rho - 1$ [here ρ is the growth order of the integer function $A(p^2)$]. Since $1/2 \leq \rho < 1$, it follows that $2 < \gamma < \infty$.

We finally obtain

$$W(\mathbf{r}) = \int d\boldsymbol{\varrho}\, V(|\mathbf{r} - \boldsymbol{\varrho}|) \frac{1}{|\rho|} e^{-m|\rho|} = \frac{1}{r} e^{-mr} - \frac{1}{r} e^{-h(r)}. \tag{3.16}$$

Here $h(r)$ is a continuous function which is such that $h(0) = 0$ and $h(r) = O((r/l)^{\gamma}$, where $\gamma > 2$, for $r \gg l$.

Thus, we see that the introduction of nonlocality into the theory changes the Yukawa potential at short distances. A kind of "spreading" of the Yukawa potential at distances of the order of the elementary length l occurs, while at large distances the portion of the potential $W(\mathbf{r})$ associated with the nonlocality which has been introduced decreases more rapidly than any Gaussian exponential.

Let us focus attention on the supplementary condition (3.12) imposed on the form factor $A(p^2)$. Only in the case in which this condition is fulfilled do we obtain a clear physical picture of the interaction between two point sources having a finite energy of interaction between them.

However, if $A(p^2) \to \infty$ for $p^2 \to -\infty$, which does not contradict the main principles of constructing the S-matrix, the interaction potential $W(\mathbf{r})$ can no longer be described simply by the function (3.16) but is a generalized function of the form

$$W(\mathbf{r}) = \int d\boldsymbol{\varrho}\, V(\mathbf{r} - \boldsymbol{\varrho}) \frac{e^{-m|\rho|}}{|\rho|}, \tag{3.17}$$

where

$$V(\mathbf{r}-\boldsymbol{\varrho})\sim\sum_{n=0}^{\infty}c_n\Delta^n\delta^{(3)}(\mathbf{r}-\boldsymbol{\varrho}). \tag{3.18}$$

Since the interaction energy is a physically observable quantity, the potential W(r) must be a good smooth function. Here, however, $V(r-\rho)$ is a nonlocal generalized function which is defined on a certain space of integer functions. Therefore, the requirement of smoothness of the potential W(r) in this case leads to a situation in which an infinitely heavy source cannot be a finite source, while its distribution in space must be described by an integer function of the principal space. In this case we write the interaction Lagrangian in the form

$$L_I(x)=[g(\mathbf{x}-\mathbf{x}_1)+g(\mathbf{x}-\mathbf{x}_2)]\,\Phi(\mathbf{x},x_0), \tag{3.19}$$

where g(x) is a certain integer function of the principal space.

Consequently, in this case we do not have that clear physical interpretation in potential language which applies in the first case. Moreover, the requirement (3.12) may be treated as the physical substantiation of the introduction of the Euclidian metric – the transition to integration over the space-like domain of the momentum variables ($q_0\to iq_4$). As was noted in [38], only in this case is the S-matrix unitary.

4. The Perturbation-Theory Series for the S-Matrix

As has already been noted, the perturbation-theory series (1.4) is constructed in accordance with the conventional Feynman diagram technique, only instead of the conventional causal function the function (1.5) is used. In x-space the matrix elements of some process in the n-th approximation of perturbation theory may be represented by the sum of expressions of the form

$$F(x_1,\ldots,x_n)=\prod_{i,\,j\in G}D_c(x_i-x_j), \tag{4.1}$$

where i and j take integer values from 1 to n in accordance with the specific choice of the Feynman diagram G. If in (4.1) one goes over formally to the momentum representation, then one obtains

$$\tilde{F}(p_1,\ldots,p_n)=(2\pi)^4\,\delta^{(4)}(p_1+\ldots+p_n)\,T(p_1,\ldots,p_n); \tag{4.2}$$

$$\tilde{F}(p_1,\ldots,p_n)=\int d^4x_1\ldots\int d^4x_n\,e^{i(p_1x_1+\ldots+p_nx_n)}\,F(x_1,\ldots,x_n); \tag{4.3}$$

$$T(p_1,\ldots,p_n)=\int\ldots\int\prod_{i=1}^{L}d^4l_i\prod_{j=1}^{N}\frac{[\tilde{A}(k_j^2)]^2}{m^2-k_j^2-i\varepsilon}, \tag{4.4}$$

where N is the number of internal lines; L is the number of independent integrations. Here k_j is the four-momentum corresponding to a given line in the diagram; l_i are the four-momenta over which the integration is performed.

However, Eq. (4.4) as yet has no mathematical meaning. The introduction of the form factor $A(k^2)$ still does not make the integral (4.4) convergent, since if the function $A(k^2)$ decreases for $k^2\to-\infty$, then it increases for $k^2\to+\infty$ faster than any polynomial. Thus, the introduction of one form factor still does not ensure convergence of the integrals into a perturbation-theory series.

In order to give meaning to Eq. (4.4) we use the regular procedure of an improper transition in the limit. Instead of the causal function $D_c(x)$ we introduce the following regularized function in (4.1):

$$D_c^{\lambda}(x)=\frac{1}{(2\pi)^4\,i}\int\frac{d^4k\,[\tilde{A}(k^2)]^2}{m^2-k^2-i\varepsilon}\,e^{ikx}\,R^{\lambda}(k^2). \tag{4.5}$$

We choose the function $R^{\lambda}(k^2)$ in the following form:

$$R^{\lambda}(z)\doteq\exp\left\{-\lambda(z+iM^2)^{\frac{1}{2}+\nu}\,e^{-i\pi\sigma}\right\}, \tag{4.6}$$

where $0 < \nu < \sigma < 1/2$, while M^2 is a certain parameter. For the regularizing function $R^\lambda(z)$ the following estimates are valid for large $|z|$:

$$|R^\lambda(z)| \sim e^{-\lambda|z|^{\frac{1}{2}+\nu}}; \quad -\pi a_2 < \arg z < \pi(1+a_1);$$
$$|R^\lambda(z)| \sim e^{\lambda|z|^{\frac{1}{2}+\nu}}; \quad \pi(1+a_1) < \arg z < 2\pi\left(1-\frac{a_2}{2}\right), \tag{4.7}$$

where $a_1 = 2(\sigma - \nu/(1+2\nu) > 0$, $a_2 = (1-2\sigma)/(1+2\nu) > 0$.

In other words, in the complex plane $zR^\lambda(z)$ is analytic and decreases as an exponential of order $\rho_1 = 1/2 + \nu < 1$ in $|z|$ in the upper half-plane, including the real axis.

Since $\tilde{A}(k^2)$ is an integer function of $\rho < 1$, it follows that by choosing $\rho < 1/2 + \nu < 1$, i.e.,

$$\rho - \frac{1}{2} < \nu < \frac{1}{2}, \tag{4.8}$$

we find that the integral (4.5) converges well for $\lambda > 0$. Then substituting $D_c^\lambda(x)$ into (4.1) and going over to the momentum representation, we obtain

$$T^\lambda(p_1, \ldots, p_n) = \int \ldots \int \prod_{i=1}^{L} d^4 l_i \prod_{j=1}^{N} \frac{[\tilde{A}(k_j^2)]^2}{m^2 - k_j^2 - i\varepsilon} R^\lambda(k_j^2) \tag{4.9}$$

instead of (4.4). For $\lambda > 0$ this integral converges well.

Hereafter we shall consider only such functions $\tilde{A}(k^2)$ which satisfy the conditions (3.12); i.e., $\tilde{A}(k^2) = O(1/k^2)$ for $k^2 \to -\infty$.

Let us now consider how to go over to the limit $\lambda \to 0$ in the integral (4.9). We proceed as follows. In the integral (4.9) we go over to the Euclidian metric after having turned the contours for integration over l_{j_0} through an angle $\pi/2$; i.e., $l_{j_0} \to il_{j_4}$. Simultaneously we go over to Euclidian external momenta by means of the transition $p_{j_0} \to ip_{j_4}$. If rotation is carried out simultaneously with respect to all of the arguments $l_{\gamma 0}$ ($\nu = 1, \ldots, L$) and p_{j_0} ($j = 1, \ldots, n$), i.e., if we place

$$l_{\nu 0} = r_\nu e^{i\varphi} \quad (\nu = 1, \ldots, L);$$
$$p_{\mu 0} = \rho_\mu e^{i\varphi} \quad (\mu = 1, \ldots, n), \tag{4.10}$$

then each argument k_j^2 ($j = 1, \ldots, N$) will be situated in the upper half-plane of the complex variable k_j^2. Actually, the momentum k_j corresponding to any chosen line in the diagram is the sum of a certain number of momenta $l\nu$ over which the integration is performed, and a certain number of external momenta p_μ:

$$k_j = \sum_{\nu=1}^{L} \Theta_\nu^{(j)} l_\nu + \sum_{\mu=1}^{n} \vartheta_\mu^{(j)} p_\mu. \tag{4.11}$$

The numbers $\Theta_\gamma^{(j)}$ and $\vartheta_\mu^{(j)}$ may take only one of three values: -1.0 or $+1$, depending on the lines chosen in the diagram. In the case of simultaneous rotation (4.10), we obtain

$$\operatorname{Im} k_j^2 = \operatorname{Im}\left\{\left[\sum_{\nu=1}^{L} \Theta_\nu^{(j)} r_\nu e^{i\varphi} + \sum_{\mu=1}^{n} \vartheta_\mu^{(j)} \rho_\mu e^{i\varphi}\right]^2 - \mathbf{k}_j^2\right\} = \sin 2\varphi \left\{\sum_{\nu=1}^{L} \Theta_\nu^{(j)} r_\nu + \sum_{\mu=1}^{n} \vartheta_\nu^{(j)} \rho_\mu\right\}^2 \geqslant 0 \tag{4.12}$$

for all $0 \le \varphi \le \pi/2$. After the rotation (4.10) one can go over to the limit $\lambda = 0$ in the integral (4.9), since the form factor $\tilde{A}(k^2)$ satisfies the condition (3.12), and consequently, the integrals corresponding to any Feynman diagrams converge. Finally, we obtain

$$T((p_1)_E, \ldots, (p_n)_E) = i^L \int \ldots \int \prod_{i=1}^{L} d^4 (l_i)_E \prod_{j=1}^{N} \frac{[\tilde{A}(-(k_j)_E^2)]^2}{m^2 + (k_j)_E^2 - i\varepsilon}. \tag{4.13}$$

Thus, the amplitude $T((p_1)_E, ..., (p_n)_E)$ is determined by means of the convergent integrals (4.13). It depends on scalar products of Euclidian external momenta:

$$(p_{iE}\, p_{jE}) = +(p_{i4}\, p_{j4} + \mathbf{p}_i\, \mathbf{p}_j). \tag{4.14}$$

The real physical amplitude depends on scalar products of pseudo-Euclidian external momenta:

$$(p_i\, p_j) = p_{i0}\, p_{j0} - \mathbf{p}_i\, \mathbf{p}_j. \tag{4.15}$$

Note further that a) all possible scalar products of n pseudo-Euclidian and Euclidian momenta are stipulated by one and the same number of invariant variables, and b) the amplitude $T(p_1, \ldots, p_n)$ is an analytic function of its invariant variables.

The integral (4.15) determines the amplitude T throughout the entire domain of Euclidian external momenta. In order to obtain the physical amplitude corresponding to a given Feynman diagram it is necessary to continue the amplitude (4.15) analytically in invariant momentum variables to physical values of the scalar products $(p_i p_j)$.

It should be noted that the procedure expounded is fully equivalent to the conventional method in the consideration of Feynman diagrams in perturbation theory. If we place $A(\Box) = 1$, then we obtain the conventional expressions for the amplitudes of the scalar theory.

Thus, we have obtained a perturbation-theory series which is free from ultraviolet divergences. Before undertaking a check of the unitarity and causality of the S-matrix, let us investigate what analytic singularities the amplitudes obtained have.

5. The Analytic Properties of Euclidian Amplitudes

Assume that a certain arbitrary Feynman diagram having n external lines is stipulated. Let us construct the amplitudes corresponding to this diagram in four-dimensional Euclidian momentum space. We shall assume that the n external momenta are Euclidian and satisfy the conservation law $q_1 + \ldots + q_n = 0$. Each internal line is juxtaposed with a propagator

$$D(k^2) = \frac{V(k^2)}{m^2 + k^2}. \tag{5.1}$$

Here k is the Euclidian four-momentum. The function $V(\xi)$ is an integer function in the plane of complex ξ and decreases fairly rapidly for $\mathrm{Re}\,\xi \to +\infty$. Then the amplitude corresponding to the stipulated diagram can be described by an integral of the form

$$\tilde{F} = \int \ldots \int \prod_i d^4 l_i \prod_j \frac{V_j(k_j^2)}{m_j^2 + k_j^2}. \tag{5.2}$$

Here k_j is the Euclidian four-momentum corresponding to the given line in the diagram; m_j is the mass of the corresponding particle. The integration in (5.2) is carried out over four-dimensional Euclidian momentum space; l_i are the four-momenta over which the integration is performed.

This integral converges well, since it is assumed that the functions $V_j(k^2)$ decrease fairly rapidly for $k^2 \to \infty$.

As was said above, the Euclidian amplitude $\tilde{F}$ (5.2) coincides (correct to a constant factor) with the real amplitude which corresponds to the process described by the same Feynman diagram in the Euclidian domain of spacelike external momenta p_j on which the real amplitude depends. The transition to the physical domain of external momenta must be accomplished by the analytical continuation of the amplitude in the corresponding invariant momentum variables. Under these conditions one should remember that all the masses have negative imaginary increments $m_j = m_j - i\varepsilon$.

It should be emphasized that in the case when $V_j(k^2) = 1$ are simply polynomials in k^2 the amplitude $\tilde{F}$ in (5.2) coincides with the real pseudo-Euclidian amplitude in the domain of Euclidian external momenta indicated above, since in this case the Euclidian and pseudo-Euclidian formulations are equivalent [32].

Note that L. D. Landau's analysis [33[of the amplitude singularities in perturbation theory begins precisely with the expression for the amplitude in the Euclidian metric.

Thus, our problem consists in studying the analytic singularities of the amplitude $\tilde{F}$ (5.2) in the invariant momentum variables. It can easily be shown that compared with the analogous amplitude in the local theory the amplitude $\tilde{F}$ does not have supplementary singularities associated with the presence of integer functions $V_j(k^2)$ for finite values of the momentum variables. In fact, after Feynman parametrization we obtain

$$\tilde{F}=(N-1)!\int_0^1 d\alpha_1\ldots\int_0^1 d\alpha_N\,\delta\left(1-\sum_{i=1}^N\alpha_i\right)\int\ldots\int\frac{\prod_i d^4 l_i\prod_j V_j(k_j^2)}{\left[\sum_j\alpha_j(m_j^2+k_j^2)\right]^N},\tag{5.3}$$

where N is the number of internal lines.

The transformation of the integration variables from the expression appearing in the denominator can always be used to eliminate terms which are linear in l_j, after which we obtain (we follow the reasoning of L. D. Landau [33] exactly):

$$\sum_{j=1}^N\alpha_j(k_j^2+m_j^2)=\varphi(\alpha,\,q_i,\,q_j,\,m^2)+K(\alpha,\,l').\tag{5.4}$$

Here K is a homogeneous quadratic form of the new integration variables l' having coefficients which depend solely on the parameters α_j; φ is an inhomogeneous quadratic form of the vectors q_j which characterize the free ends of the diagram considered.

Since the quadratic form φ depends on the scalar products q_iq_j, we obtain the original expression for which Landau began the derivation of his well-known equations.

The dependence of the numerator on the external momenta cannot lead to any additional singularities in a finite domain of invariant momentum variables, since the numerator contains an integer function of the scalar products q_iq_j and the parameters α_j.

There remains, however, the very important question of the magnitudes of the breaks of the function $\tilde{F}$ on the corresponding cuts. This problem is directly related to the unitarity of the S-matrix of the theory. We shall prove the following property of the amplitude $\tilde{F}$, which is known as the Cutkosky rule [34] for normal thresholds. Assume that the graph corresponding to the amplitude $\tilde{F}$ can be partitioned into two blocks $\tilde{F}_I$ and $\tilde{F}_{II}$ which are connected by r internal lines (Fig. 1):

$$\tilde{F}=\int\ldots\int d^4k_1\ldots d^4k_r\,\tilde{F}_I(q_j,k_i)\prod_{\nu=1}^r\frac{V_\nu(k_\nu^2)}{m_\nu^2+k_1^2}\tilde{F}_{II}(q_j',k_i)\,\delta^{(4)}(q-k_1-\ldots-k_r).\tag{5.5}$$

Here q_j ($j = 1, \ldots, n_1$) and q'_j ($j = 1, \ldots, n_2$) are the external momenta of blocks I and II, respectively, the equation $q = q'_1 + \ldots + q'_{n_2} = -(q_1 + \ldots + q_{n_1})$ being fulfilled under these conditions ($n = n_1 + n_2$ is the number of external lines). The functions $\tilde{F}_I(q_j, k_i)$ and $\tilde{F}_{II}(q'_j, k_i)$ describe blocks I and II; they depend on the scalar products of the vectors q_i, q'_j, and k_i.

Then the amplitude $\tilde{F}$, considered as a function of the complex variable $z = -q^2$, has a branching line associated with the given partitioning, which begins at the point

$$z=(m_1+\ldots+m_r)^2,\tag{5.6}$$

while the break of the function $\tilde{F}$ on this cut is given by the equation

$$\Delta\tilde{F}(z)=i(2\pi)^r\prod_{\nu=1}^r V_\nu(-m_\nu^2)\int d^4\tilde{k}_1\ldots\int d^4\tilde{k}_r\prod_{\nu=1}^r\Theta(\tilde{k}_{\nu0})$$
$$\times\,\delta(m_\nu^2+\tilde{k}_\nu^2)\,\delta^{(4)}(\tilde{q}-\tilde{k}_1-\ldots-\tilde{k}_r)\,\tilde{F}_I(q_j,\tilde{k}_i)\,\tilde{F}_{II}(q_j',\tilde{k}_i).\tag{5.7}$$

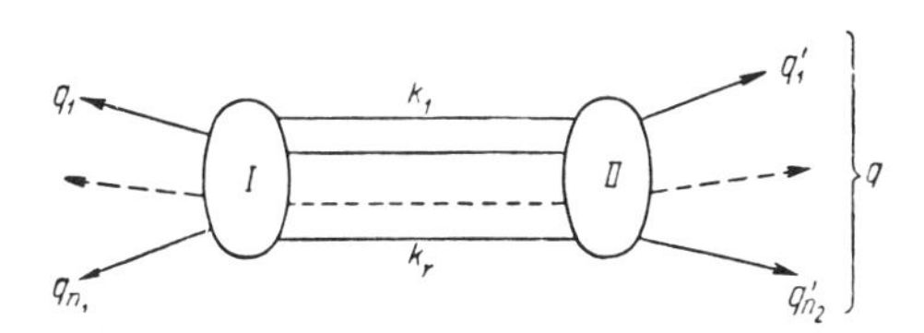

Fig. 1

Here $\tilde{k}_i$ are understood to represent four-dimensional vectors having the components $(i\,k_{i0}, \mathbf{k}_j)$, so that $k_j^2 = -k_{j0}^2 + \mathbf{k}_j^2$ and $(k_j q) = \mathbf{k}_j \tilde{\mathbf{q}} + i\, k_{j0} q_4$, while $d^4k_j = dk_{j0} d\mathbf{k}_j$. The vector $\tilde{q}\,(iq_0, \mathbf{q})$ satisfies the relationship $q^2 = \mathbf{q}^2 - q_0^2 = -z$. The functions $\tilde{F}_I(q_j, \tilde{k}_j)$ and $\tilde{F}_{II}(q'_j, \tilde{k}_j)$ should be understood to represent the analytic continuation to the appropriate values of the scalar arguments $(q_j, \tilde{k}_j)$, etc., of the original functions $\tilde{F}_I(q_j, k_j)$ and $\tilde{F}_{II}(q'_j, k_j)$ describing blocks I and II. The proof of Eq. (5.7) is given in [41].

We proved the Cutkosky rule for normal thresholds in the case of arbitrary Feynman diagrams when the diagrams are written in Euclidian momentum space, while integer functions are chosen as the cutoff functions.

Note once again that in the case $V_j(k^2) = 1$ the proof which has been carried out is the proof of the Cutkosky rule for conventional quantum field theory diagrams, since the pseudo-Euclidian integrals for the amplitudes of the physical processes in the domain of Euclidian external momenta can always be written in Euclidian space, while the transition to the physical domain may always be treated as the analytical continuation in invariant momentum variables.

As far as the anomalous singularities of the diagrams are concerned, they develop conventionally when the analytic properties of blocks I and II are considered. For example, the appearance of an anomalous singularity in a conventional triangular or rectangular diagram can be traced by a method similar to the one used in [35].

Finally, let us note that, as derived from the proof presented, the introduction of an integer cutoff function $V(k^2)$ which violates equivalence in the conventional sense of the Euclidian and pseudo-Euclidian formulations of the theory does not violate the analytic properties of the theory in any finite domain of the momentum variables. The validity of the Cutkosky rule makes the proof of unitarity in the theory considered rather simple.

6. The Unitarity of the S-Matrix

Let us show that the S-matrix constructed is unitary in each order of perturbation theory on the mass shell [i.e., (2.5) from Chap. I is fulfilled]. First of all let us focus attention on the algebraic character of the unitarity condition. We have in mind the following. Assume

$$S = 1 + iT; \tag{6.1}$$

$$T = \sum_{n=1}^{\infty} g^n T_n. \tag{6.2}$$

From the conditions (1.2.5) and (6.1) it follows that

$$-i\langle\alpha|(T - T^+)|\beta\rangle = \langle\alpha|TT^+|\beta\rangle. \tag{6.3}$$

Let us substitute the expansion (6.2) into (6.3) and introduce the expansion in the complete system of functions $|\mathbf{k}_1, \ldots, \mathbf{k}_n\rangle$ into the right side. Then in each order of powers of the coupling constant we obtain

$$2\operatorname{Im}\langle\alpha|T_n|\beta\rangle = \sum_{m_1+m_2=n} \sum_N \int d\mathbf{k}_1 \ldots \int d\mathbf{k}_N \langle\alpha|T_{m_1}|\mathbf{k}_1, \ldots, \mathbf{k}_N\rangle\langle\mathbf{k}_1, \ldots, \mathbf{k}_N|T^+_{m_2}|\beta\rangle. \tag{6.4}$$

Summation over N is not carried out to infinity but is bounded by the total energy of the state $|\alpha\rangle$ or $|\beta\rangle$. Moreover, we make use of the equation

$$\langle\alpha|T_n|\beta\rangle = \langle\beta|T_n|\alpha\rangle, \tag{6.5}$$

which is valid in one-component scalar field theory.

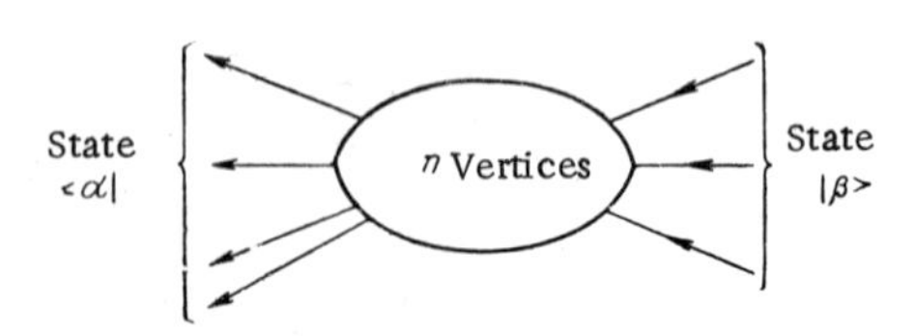

Fig. 2

The amplitude $\langle\alpha|T_n|\beta\rangle$ represents the sum of all possible Feynman diagrams in the n-th order of perturbation theory which are of the form shown in Fig. 2, with n_β entering lines and n_α issuing lines.

From Eq. (6.4) it follows that the imaginary part of the amplitude $\langle\alpha|T_n|\beta\rangle$ must be equal to the set of amplitudes of the lower-order physical processes. If we now formally cut all possible Feynman diagrams represented in the amplitude $\langle \alpha | T_n | \beta \rangle$ in all possible ways, leaving state $| \alpha \rangle$ at the left and state $| \beta \rangle$ at the right (see Fig. 2), then we formally obtain a structure representing the right side of Eq. (6.4) with the sole difference that instead of the causal functions D_c the functions $\Delta^{(-)}$ appear on the cut lines. It is precisely this which is the algebraic property of the perturbation-theory series and is associated with the algebraic property of the Wick theorem for expansion of the T-product for an S-matrix having the form

$$S = \mathrm{T} \exp \left\{ \mathrm{i} g \int d^4 x L_I(x) \right\}, \tag{6.6}$$

where $L_I(x)$ is a certain polynomial in the field $\varphi(x)$. The Cutkosky theorem ensures precisely this property of the perturbation-theory amplitudes. Therefore, if in perturbation theory a) the amplitudes have singularities which are the same as in conventional local theory, and b) the Cutkosky theorem is valid, then the S-matrix of such a theory will be unitary on the mass shell in each order of perturbation theory.

In our case we have precisely this picture: The analytic singularities of the amplitudes are the same as they are for local interactions, and the magnitude of the break at the normal thresholds is determined by the same Cutkosky formula as in the local case. Consequently, the S-matrix which we have constructed is unitary on the mass shell.

Note that the unitarity condition (1.2.5) remains valid if a scalar external field $a(x)$ is introduced into the interaction Lagrangian in the following manner, for example:

$$gL_I(x) = g\Phi^4(x) + \varphi^2(x)\, a(x). \tag{6.7}$$

Actually, on the Feynman diagram the propagator connecting the vertex g with the external field $a(x)$ is equal to the function

$$\overline{\Phi(x)\, \varphi}(y) = \frac{1}{(2\pi)^4 \mathrm{i}} \int \frac{d^4 p \tilde{A}(p^2)}{m^2 - p^2 - \mathrm{i}\varepsilon}\, \mathrm{e}^{\mathrm{i}p(x-y)} \tag{6.8}$$

instead of the propagator (1.5). Since $\tilde{A}(p^2)$ is an integer function of the same type as $[\tilde{A}(p^2)]^2$, it follows that the Cutkosky formula remains in force and the S-matrix is unitary.

7. The Causality of the S-Matrix

In order to check the causality of the S-matrix (the matrix is a functional of the operator $\Phi(x)$) we shall consider the expression

$$C(x, y) = \frac{\delta}{\delta\Phi(x)} \left(\frac{\delta S}{\delta\Phi(y)} S^{-1} \right). \tag{7.1}$$

In the case of local interaction, in which the theory is microcausal, the operator $C(x, y) = 0$ for $x \tilde{<} y$ [26]. Let us determine what the operator C (x, y) in (7.1) is equal to in the case considered.

First of all let us show that the operators $\Phi(x)$ in (1.2) satisfy the local commutative relationships

$$[\Phi(x_1),\ \Phi(x_2)]_- = [\varphi(x_1),\ \varphi(x_2)]_- = \Delta(x_1 - x_2), \tag{7.2}$$

where $\Delta(x_1 - x_2)$ is a conventional scalar-field commutator. For this purpose we make use of the completed condition. We have

$$\begin{aligned} [\Phi(x_1),\ \Phi(x_2)]_- = \sum_{n=0}^{\infty} \int d\mathrm{k}_1 \ldots \int d\mathrm{k}_n \{ \Phi(x_1) | \mathrm{k}_1 \ldots \mathrm{k}_n \rangle \langle \mathrm{k}_1 \ldots \mathrm{k}_n | \Phi(x_2) \\ - \Phi(x_2) | \mathrm{k}_1 \ldots \mathrm{k}_n \rangle \langle \mathrm{k}_1 \ldots \mathrm{k}_n | \Phi(x_1) \}. \end{aligned} \tag{7.3}$$

Let us now consider the expression

$$\begin{aligned} &\Phi(x) | \mathrm{k}_1 \ldots \mathrm{k}_n \rangle = A(\Box)\, \varphi(x) | \mathrm{k}_1 \ldots \mathrm{k}_n \rangle \\ &= A(\Box) \frac{1}{(2\pi)^{3/2}} \int \frac{d\mathrm{k}}{\sqrt{2\omega}} \{ \mathrm{e}^{-\mathrm{i}kx} a_{\mathrm{k}} | \mathrm{k}_1 \ldots \mathrm{k}_n \rangle + \mathrm{e}^{\mathrm{i}kx} a_{\mathrm{k}}^{+} | \mathrm{k}_1 \ldots \mathrm{k}_n \rangle \} \\ &= A(m^2)\, \varphi(x) | \mathrm{k}_1 \ldots \mathrm{k}_n \rangle = \varphi(x) | \mathrm{k}_1 \ldots \mathrm{k}_n \rangle. \end{aligned} \tag{7.4}$$

Substituting (7.4) into (7.3), we obtain (7.2). The same may also be said of the $D^{(\pm)}$-function:

$$D^{(\pm)}(x_1-x_2)=\Delta^{(\pm)}(x_1-x_2), \tag{7.5}$$

where $\Delta^{(\pm)}$ are the corresponding scalar-field functions.

Let us now consider Eq. (7.1). Our S-matrix (1.4) is a functional solely of the field operator $\Phi(x)$. For expansion into a series in normal products of the field operators $\Phi(x)$ we obtain a series of the form

$$S=\sum_{n=0}^{\infty}\frac{1}{n!}\int d^4x_1 \dots \int d^4x_n\, S_n(x_1,\dots,x_n):\Phi(x_1)\dots\Phi(x_n):, \tag{7.6}$$

where the coefficient functions $S_n(x_1, \dots, x_n)$ are constructed from the causal function (1.5) which is equal to the sum of the conventional causal scalar-field function and the generalized function $K(x_1-x_2)$ [see Eq. (3.1)]. We substitute the series (7.6) into (7.1) and go over once again to the N-product. Under these conditions $D^{(-)}$-functions of the field operators $\Phi(x)$ develop. But in accordance with (7.5) a $D^{(-)}$-function is exactly equal to the $\Delta^{(-)}$-function of the conventional scalar field $\varphi(x)$. Therefore, if the generalized functions $K(x_1-x_2)$ are neglected, we obtain the conventional condition for microcausality of the local scalar field (i.e., the operator $C(x, y)$ in (7.1) vanishes for $x \tilde{<} y$). But in the presence of the generalized functions $K(x_1-x_2)$ Eq. (7.1) in (3.1) will be proportional to the nonlocal generalized function of the type we are considering in the domain $x \tilde{<} y$. Therefore, the macrocausality condition (3.5) from Chap. I will be fulfilled in each order of perturbation theory.

8. The Stability of the S-Matrix

In order to satisfy the stability condition of the matrix (1.2.6) it is necessary to perform renormalization of the mass of scalar particles and the wave function. For this purpose one should introduce two counterterms into the interaction Lagrangian. Finally, the interaction Lagrangian must have the form

$$gL_I(x)=g:\Phi^n(x):-\delta m^2:\Phi^2(x):-Z_2:\Phi(x)(\Box-m^2)\Phi(x): \tag{8.1}$$

The renormalization constants δm^2 and Z_2 are chosen in such a way that the proper-energy operator of a meson $\Sigma_r(p^2)$ is normalized as follows:

$$\left.\begin{aligned}\Sigma_r(p^2)\,|_{p^2=m^2}=0;\\ \frac{d}{dp^2}\Sigma_r(p^2)\,|_{p^2=m^2}=0.\end{aligned}\right\} \tag{8.2}$$

For this purpose it is necessary to regularize the operator $\Sigma(p^2)$, which is a sum of all irreducible proper-energy diagrams, as follows:

$$\Sigma_r(p^2)=\Sigma(p^2)-\Sigma(p^2)\,|_{p^2=m^2}-\left.\frac{\partial\Sigma(p^2)}{\partial p^2}\right|_{p^2=m^2}(p^2-m^2)=\Sigma(p^2)-\delta m^2-Z_2(p^2-m^2), \tag{8.3}$$

where

$$\delta m^2=\Sigma(p^2)\,|_{p^2=m^2}=\delta m^2(g,m);$$

$$Z_2=\left.\frac{\partial\Sigma(p^2)}{\partial p^2}\right|_{p^2=m^2}=Z_2(g,m).$$

In the nonlocal theory considered, the constants δm^2 and Z_2 are finite quantities.

The procedure expounded corresponds fully to conventional local quantum field theory methods [26, 36].

9. Renormalization of the Coupling Constant

The parameter g which is included in the interaction Lagrangian (8.1) is not the physical coupling constant. The point is the following [37]. Experimentally we measure certain physical quantities. We arbitrarily call them cross sections and denote them by

$$\sigma_A,\ \sigma_B,\ \sigma_C\dots \tag{9.1}$$

The purpose of the theory consists in predicting the results of experiments B, C, ..., if experiment A has yielded the value σ_A.

On the other hand, knowing the interaction Lagrangian (8.1), one can calculate the quantities (9.1) mathematically. As a result of calculations we obtain

$$\left.\begin{array}{l} \sigma_A = \sigma_A(g, m, \varepsilon) = \sum\limits_{k=n_A}^{\infty} g^k \sigma_k^{(A)}(m, \varepsilon); \\ \sigma_Б = \sigma_Б(g, m, \varepsilon) = \sum\limits_{k=n_B}^{\infty} g^k \sigma_k^{(Б)}(m, \varepsilon), \\ \dots\dots\dots\dots\dots\dots \end{array}\right\} \tag{9.2}$$

where ε are the energy and other variables on which the cross section may depend.

In the nonlocal theory considered all of the quantities σ_A, σ_B, ..., are given in the form of series (9.2) in the parameter g, each term of the series being finite.

The physical coupling constant is found from experiment, but it is utterly unclear as to what experiments it should be determined from. We may by definition assume that the physical coupling constant g_r is such that for specific values of the variables $\varepsilon = \varepsilon_0$ in experiment A the measured quantity σ_A is equal to

$$\sigma_A = g_r^{n_A} = \sum_{k=n_A}^{\infty} g^k \sigma_k^{(A)}(m, \varepsilon_0). \tag{9.3}$$

Consequently, Eq. (9.3) is the definition of the physical coupling constant g_r. From (9.3) one can find

$$g = \sum_{k=1}^{\infty} g_r^k a_k(m, \varepsilon_0) \tag{9.4}$$

using perturbation theory. Then substituting (9.4) into (9.2), we obtained

$$\sigma_A(\varepsilon) = \sum_{k=n_A}^{\infty} g_r^k \tilde{\sigma}_k^{(A)}(m, \varepsilon, \varepsilon_0), \text{ etc.} \tag{9.5}$$

Once again we focus attention on the fact that in the nonlocal theory considered the coupling between g_r and g is given in the form of a series (9.3) having finite functions $\sigma_k^{(A)}(m, \varepsilon_0)$. In conventional local renormalizable theory the quantities $\sigma_k^{(A)}$ are represented in the form of diverging integrals, and, therefore still another counterterm eliminating this divergence [26, 36] is added in the interaction Lagrangian (8.1). In our case there are no divergences. Therefore, we may omit the additional renormalization of the constant g in the interaction Lagrangian and simply assume that the coupling (9.4) exists between the constants g and g_r.

10. The Samples of Possible Form Factors and the Second Order of Perturbation Theory

As shown in Sec. 5, the amplitudes corresponding to any Feynman diagram may be written in the form of a converging integral in Euclidian space in the Euclidian domain of spacelike external momenta. Under these conditions the causal function (1.5) has the following form in Euclidian momentum space:

$$D_c(k^2) = \frac{V(k^2)}{m^2 + k^2}, \tag{10.1}$$

where $V(k^2) = [\tilde{A}(-k^2)]^2$. Since $\tilde{A}(-k^2)$ satisfies the condition (3.12), it follows that $D_c(k^2) = O(k^2)^{-3}$ for $k^2 \to +\infty$.

In Euclidian x-space a change of the propagator (10.1) denotes multiplication of the causal function $\Delta_c(x)$ (which is real in Euclidian space) by a certain positive continuous function $\vartheta(x^2)$; i.e.,

$$D_c(x) = \Delta_c(x)\,\vartheta(x^2), \tag{10.2}$$

$$\text{where } \vartheta(x^2) = \begin{cases} O(x^2) & \text{for } x^2 \to 0 \\ 1 - O\left(e^{-[\sqrt{x^2}]^\gamma}\right) & \text{for } x^2 \to \infty. \end{cases} \tag{10.3}$$

Here $x^2 = x_1^2 + x_2^2 + x_3^2 + x_4^2$, and $\gamma = 2\rho/2\rho-1$, where $\rho < 1$ is the order of the function $V(k^2)$.

For the coefficients C_n which determine the generalized function $K(x-y)$ in (3.3) from (10.2) it is easy to obtain the following result for $n \gg 1$:

$$|c_n| \sim \int_0^\infty du\, u^{2+2n}\left[1 - \vartheta\left(\frac{u^2}{m^2}\right)\right] K_1(u), \tag{10.4}$$

where $K_1(u)$ is a Macdonald function. From this it is evident that the least growth of the coefficient C_n for $n \to \infty$ is achieved from the function $\vartheta(x^2) = 1$, beginning with a certain l^2; i.e.,

$$\vartheta(x^2) = 1 \quad \text{for} \quad x^2 \gg l^2. \tag{10.5}$$

In this case

$$|c_n| \sim (ml)^{2n}. \tag{10.6}$$

If, however,

$$|1 - \vartheta(x^2)| < Ae^{-a(\sqrt{x^2})^\gamma} \quad (m^2 x^2 \gg 1), \tag{10.7}$$

where A, a, and γ are positive constants, then we have

$$|c_n| \sim \Gamma\left(\frac{2n}{\gamma}\right) \tag{10.8}$$

for asymptotically large n. An example of such a function can be found in

$$\vartheta(x^2) = 1 - \exp\left\{-\left(\frac{\sqrt{x^2}}{l}\right)^\gamma\right\}.$$

By making different choices of the functions $\vartheta(x^2)$ one can obtain all of the nonlocal classes of the generalized class-II functions which were mentioned in Sec. 1.

The simplest formulas are obtained from the simplest choice of the function $\vartheta(x^2)$:

$$\vartheta(x^2) = \theta(x^2 - l^2), \tag{10.9}$$

where

$$\theta(u) = \begin{cases} 1 & \text{for } u > 0 \\ 0 & \text{for } u < 0. \end{cases}$$

As an example, let us consider the S-matrix corresponding to the nonrenormalizable interaction Lagrangian

$$L_I(x) = :\varphi^\nu(x):, \tag{10.10}$$

where γ is a certain integer.

The second-order amplitudes of perturbation theory for the interaction (10.10) can be described by Feynman diagrams of the form shown in Fig. 3. Here ν lines converge to each point on the diagram. Assume that there are r internal lines. Let p denote the sum of the external momenta which enter the point x_2.

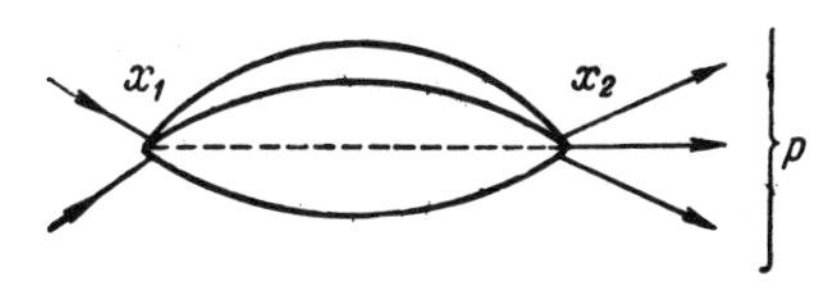

Fig. 3

Then the amplitude corresponding to this diagram can be described by the following integral in the domain $p^2 < 0$:

$$T_2(p^2) = g^2 \int d^4 x e^{iqx} D_c^r(x), \tag{10.11}$$

where $D_c(x)$ is given in (10.2), while q is a Euclidian vector and $q^2 = -p^2$. Let us substitute the propagator $D_c(x)$ from (10.2) into (10.11), and let us make use of the identity

$$[\Delta_c(x)]^r = \int\limits_{(rm)^2}^{\infty} d\varkappa^2 \Omega_r(\varkappa^2) \Delta_c(\varkappa, x), \tag{10.12}$$

where $\Delta_c(\varkappa, x)$ is the causal function of a scalar field having mass $\varkappa$; $\Omega_r(\varkappa^2)$ is the phase volume of r scalar particles having mass m:

$$\Omega_r(\varkappa^2) = \frac{1}{(2\pi)^{3(r-1)}} \int \frac{dk_1}{2\omega_1} \cdots \int \frac{dk_r}{2\omega_r} \delta^{(4)}(k - k_1 - \ldots - k_r), \tag{10.13}$$

where $\omega_j = \sqrt{m^2 + \mathbf{k}_j^2}$, $k^2 = k_0^2 - \mathbf{k}^2 = \varkappa^2$. We represent the arbitrary form factor $[\vartheta(x^2)]^n$ in the form

$$[\vartheta(x^2)]^r = \int\limits_0^{x^2} dl^2 \frac{d}{dl^2} [\vartheta(l^2)]^r = r \int\limits_0^{\infty} dl^2 \vartheta'(l^2) [\vartheta(l^2)]^{r-1} \theta(x^2 - l^2). \tag{10.14}$$

Substituting Eqs. (10.12) and (10.14) into (10.11), we obtain

$$T_2(p^2) = g^2 r \int\limits_0^{\infty} dl^2 \vartheta'(l^2) [\vartheta(l^2)]^{r-1} \int\limits_{(rm)^2}^{\infty} \frac{d\varkappa^2 \Omega_r(\varkappa^2) d(\varkappa^2 l^2, p^2 l^2)}{\varkappa^2 - p^2 - i\varepsilon}, \tag{10.15}$$

where

$$d(\varkappa^2 l^2, p^2 l^2) = \varkappa l \left[I_0(l\sqrt{p^2}) K_1(\varkappa l) + \frac{I_1(l\sqrt{p^2})}{l\sqrt{p^2}} \varkappa l K_0(\varkappa l) \right], \tag{10.16}$$

after simple calculations; here I_k and K_k are Bessel functions. The integral (11.15) converges well, since for $\varkappa \to \infty$, $\Omega_r(x^2) \sim \varkappa^{2r}$, while $K_k(\varkappa l) \sim e^{-\varkappa l}$. Note that

$$d(\varkappa^2 l^2, \varkappa^2 l^2) \equiv 1, \tag{10.17}$$

and therefore for $p^2 > (rm)^2$ we have

$$\operatorname{Im} T_2(p^2) = \pi g^2 \Omega_r(p^2). \tag{10.18}$$

The asymptotic behavior of the real part of the amplitude is determined by the behavior of the Bessel functions $I_0(l\sqrt{p^2})$ and $I_1(l\sqrt{p^2})$ for $|p^2|$:

$$\begin{aligned} \operatorname{Re} T_2(p^2) &\sim g^2 \int\limits_0^{\infty} dl^2 \vartheta'(l^2) e^{l\sqrt{p^2}} \sim g^2 e^{(p^2)^\rho} \quad &\text{for } p^2 \to +\infty; \\ \operatorname{Re} T_2(p^2) &\sim O\left(\frac{1}{(p^2)^2}\right) \quad &\text{for } p^2 \to -\infty. \end{aligned} \tag{10.19}$$

The higher-order matrix elements are constructed just as described in Sec. 4. The perturbation-theory series for the S-matrix will satisfy all of the axioms of nonlocal theory (see Chap. 1).

CONCLUSIONS

The proposed scheme for constructing a nonlocal theory consists of the following principal stages.

1. A nonlocal interaction Lagrangian is introduced, which effectively leads to a change in the free-particle propagator in the perturbation-theory series:

$$\frac{1}{m^2-p^2-i\varepsilon}\to\frac{[\tilde{A}(p^2)]^2}{m^2-p^2-i\varepsilon}$$

where $\tilde{A}(p^2)$ is an integer function of order $\rho < 1$ in the plane of complex p^2 and is such that

$$\tilde{A}(p^2)=O\left(\frac{1}{p^2}\right)\quad \text{for}\ p^2\to-\infty.$$

2. In constructing the perturbation-theory series a regular procedure is introduced for an improper transition in the limit $R^{\lambda}(p^2)$, which allows translation to the Euclidian metric.

3. The Cutkosky theorem is proved for normal thresholds in the Euclidian formulation of the theory.

These principal stages lead to a finite unitary macrocausal perturbation-theory series for the S-matrix in the case of any nonrenormalizable interaction.

It seems to us that the advantage of the scheme developed consists in the following: First, the entire arbitrariness in the choice of the cutoff form and the magnitude of the "elementary length" l has been successfully incorporated in the interaction Lagrangian; second, the amplitudes of physical processes do not have additional singularities in a finite domain of variation of the invariant momentum variables, as compared with local theory.

LITERATURE CITED

1. M. A. Markov, Zh. Éksperim. i Teor. Fiz., 10, 1311 (1940).
2. H. Yukawa, Phys. Rev., 77, 219 (1950); 80, 1047 (1950).
3. H. Snyder, Phys. Rev., 71, 38 (1947).
4. Yu. A. Gol'fand, Zh. Éksperim. i Teor. Fiz., 37, 504 (1959); 43, 256 (1962).
5. V. G. Kadyshevskii, Zh. Éksperim. i Teor. Fiz., 41, 1885 (1961); Dokl. Akad. Nauk SSSR, 147, 588 (1962).
6. M. E. Tamm, in: Report to the 12th International Conference on High-Energy Physics, Vol. II [in Russian], Dubna (1964), p. 229.
7. G. Wataghin, Z. Phys., 88, 92 (1934).
8. D. I. Blokhintsev, Vestnik MGU (Fizika), 3, 77 (1946); 1, 83 (1948); Zh. Éksperim. i Teor. Fiz., 16, 480 (1964); Usp. Fiz. Nauk, 61, 137 (1957).
9. H. McManus, Proc. Roy. Soc., A9, 195, 323 (1948).
10. R. Peierls, Proc. Roy. Soc., A214, 143 (1952).
11. D. A. Kirzhnits, Usp. Fiz. Nauk, 90, 129 (1966).
12. Transactions of the International Conference on Nonlocal Quantum Field Theory, Preprint of the Joint Institute for Nuclear Research R2-3590, Dubna (1968).
13. C. Bloch, Kgl. Dansk. Mat. Fys., 27, No. 8 (1952).
14. M. Chretien and R. Peierls, Nuovo Cimento, 10, 688 (1953).
15. E. Stueckelberg and G. Wanders, Helv. Phys. Acta, 27, 607 (1954).
16. E. M. Ebel, Kgl. Danske Vid. Selskab. Mat. Fys. Medd., 29, No. 2 (1954).
17. D. A. Slavnov and A. D. Sukhanov, Zh. Éksperim. i Teor. Fiz., 36, 1472 (1959).
18. D. I. Blokhintsev and G. I. Kolerov, Preprint of the Joint Institute for Nuclear Research E-250, Dubna (1965).
19. D. A. Kirzhnits, Zh. Éksperim. i Teor. Fiz., 41, 551 (1961); 45, 143 (1963); 45, 2024 (1963).
20. G. Wataghin, Nuovo Cimento, 25, 1383 (1962).
21. D. I. Blokhintsev and G. I. Kolerov, Nuovo Cimento, 34, 163 (1964).
22. R. Ingraham, Nuovo Cimento, 39, 361 (1965).
23. D. I. Blokhintsev, Preprint of the Joint Institute for Nuclear Research R-2422, Dubna (1965).
24. A. N. Leznov and D. A. Kirzhnits, Zh. Éksperim. i Teor. Fiz., 48, 622 (1965).
25. G. V. Efimov, Commun. Math. Phys., 7, 1938 (1968); Preprint of the Institute of Technical Physics 68-52 [in Russian], Kiev (1968).

26. N. N. Bogolyubov and D. V. Shirkov, Introduction to Quantum Field Theory [in Russian], Gostekhizdat, Moscow (1957).
27. N. N. Bogolyubov, V. V. Medvedev, and M. K. Polivanov, Problems in the Theory of Dispersion Relations [in Russian], Fizmatgiz, Moscow (1958).
28. W. Heisenberg, Z. Phys., 120, 513 and 673 (1943); Z. Naturforsch., 1, 608 (1966).
29. F. J. Dayson, Phys. Rev., 85, 631 (1952).
30. W. Thirring, Helv. Phys. Acta, 26, 33 (1953).
31. H. Umedzawa, Quantum Field Theory [Russian translation], Izd. Inostr. Lit., Moscow (1958).
32. J. Schwinger, Phys. Rev., 115, 721(1959).
33. L. D. Landau, Zh. Éksperim. i Teor. Fiz., 37, 62 (1959).
34. R. E. Cutkosky, J. Math. Phys., 1, 429 (1950).
35. V. N. Gribov, G. S. Danilov, and I. T. Dyatlov, Zh. Éksperim. i Teor. Fiz., 41, 924 and 1215 (1961).
36. S. Schweber, H. Bethe, and F. Hofman, Mesons and Fields, Vol. 1 [Russian translation], Izd. Inostr. Lit., Moscow (1957).
37. G. Källen, Nuovo Cimento, 12, 217 (1964).
38. G. V. Efimov, Preprint of the Institute of Technical Physics 68-54 [in Russian], Kiev (1968).

PARTICLES AND NUCLEI

N. N. Bogolyubov

Editor-in-Chief

Director, Laboratory for Theoretical Physics
Joint Institute for Nuclear Research
Dubna, USSR

A Translation of Problemy Fiziki Élementarnykh Chastits i Atomnogo Yadra
(Problems in the Physics of Elementary Particles and the Atomic Nucleus)

Volume 1, Part 2

A SPECIAL RESEARCH REPORT / TRANSLATED FROM RUSSIAN

PARTICLES AND NUCLEI

Volume 1, Part 2

PARTICLES AND NUCLEI

Volume 1, Part 1

Elastic Scattering of Protons by Nucleons in the Energy Range 1-70 GeV
V. A. Nikitin
Probability Description of High-Energy Scattering and the Smooth Quasi-potential
A. A. Logunov and O. A. Khrustalev
Hadron Scattering at High Energies and the Quasi-potential Approach in Quantum Field Theory
V. R. Garsevanishvili, V. A. Matveev, and L. A. Slepchenko
Interaction of Photons with Matter
Samuel C. C. Ting
Short-Range Repulsion and Broken Chiral Symmetry in Low-Energy Scattering
V. V. Serebryakov and D. V. Shirkov
CP Violation in Decays of Neutral K-Mesons
S. M. Bilen'kii
Nonlocal Quantum Scalar-Field Theory
G. V. Efimov

Volume 1, Part 2

The Model Hamiltonian in Superconductivity Theory
N. N. Bogolyubov
The Self-Consistent-Field Method in Nuclear Theory
D. V. Dzholos and V. G. Solov'ev
Collective Acceleration of Ions
I. N. Ivanov, A. B. Kuznetsov, É. A. Perel'shtein, V. A. Preizendorf, K. A. Reshetnikov, N. B. Rubin, S. B. Rubin, and V. P. Sarantsev
Leptonic Hadron Decays
É. I. Mal'tsev and I. V. Chuvilo
Three-Quasiparticle States in Deformed Nuclei with Numbers between 150 and 190 (E/T)
K. Ya. Gromov, Z. A. Usmanova, S. I. Fedotov, and Kh. Shtrusnyi
Fundamental Electromagnetic Properties of the Neutron
Yu. A. Aleksandrov

Volume 2, Part 1

Self-Similarity, Current Commutators, and Vector Dominance in Deep Inelastic Lepton–Hadron Interactions
V. A. Matveev, R. M. Muradyan, and A. N. Tavkhelidze
Theory of Fields with Nonpolynomial Lagrangians
M. K. Volkov
Dispersion Relationships and Form Factors of Elementary Particles
P. S. Isaev
Two-Dimensional Expansions of Relativistic Amplitudes
M. A. Liberman, G. I. Kuznetsov, and Ya. A. Smorodinskii
Meson Spectroscopy
K. Lanius
Elastic and Inelastic Collisions of Nucleons at High Energy
K. D. Tolstov

PARTICLES AND NUCLEI

N. N. Bogolyubov

Editor-in-Chief

Director, Laboratory for Theoretical Physics
Joint Institute for Nuclear Research
Dubna, USSR

A Translation of Problemy Fiziki Élementarnykh Chastits i Atomnogo Yadra
(Problems in the Physics of Elementary Particles and the Atomic Nucleus)

Volume 1, Part 2

CONSULTANTS BUREAU • NEW YORK-LONDON • 1972

The original Russian text, published by Atomizdat in Moscow in 1971 for the Joint Institute for Nuclear Research in Dubna, has been revised and corrected for the present edition. This translation is published under an agreement with Mezhdunarodnaya Kniga, the Soviet book export agency.

PROBLEMS IN THE PHYSICS OF ELEMENTARY PARTICLES AND THE ATOMIC NUCLEUS
PROBLEMY FIZIKI ÉLEMENTARNYKH CHASTITS I ATOMNOGO YADRA
Проблемы физики элементарных частиц и атомного ядра

Library of Congress Catalog Card Number 72-83510
ISBN 0-306-17192-9

A Division of Plenum Publishing Corporation
227 West 17th Street, New York, N. Y. 10011

United Kingdom edition published by Consultants Bureau, London
A Division of Plenum Publishing Company, Ltd.
Davis House (4th Floor), 8 Scrubs Lane, Harlesden,
London NW10 6SE, England

Printed in the United States of America

CONTENTS

Volume 1, Part 2

	Eng.	Russ.
The Model Hamiltonian in Superconductivity Theory—N. N. Bogolyubov	1	301
The Self-consistent-Field Method in Nuclear Theory—R. V. Dzholas and V. G. Solov'ev	53	390
Collective Acceleration of Ions—I. N. Ivanov, A. B. Kuznetsov, É. A. Perel'shtein, V. A. Preizendorf, K. A. Reshetnikov, N. B. Rubin, S. B. Rubin, and V. P. Sarantsev	71	391
Leptonic Hadron Decays—É. I. Mal'tsev and I. V. Chuvilo	105	443
Three-Quasiparticle States in Deformed Nuclei with Mass Numbers between 150 and 190—K. Ya. Gromov, Z. A. Usmanova, S. I. Fedotov, and Kh. Shtrusnyi	159	525
Fundamental Electromagnetic Properties of the Neutron—Yu. A. Aleksandrov	170	547

THE MODEL HAMILTONIAN IN SUPERCONDUCTIVITY THEORY

N. N. Bogolyubov

A system of fermions with attraction described by the model Hamiltonian in superconductivity theory with separable interaction is considered. Asymptotically exact estimates (as $V \to \infty$) for the minimal eigenvalue of the Hamiltonian, correlation functions, and Green's functions are obtained.

§1. Statement of the Problem

The simplest model system considered in superconductivity theory is characterized by a Hamiltonian in which only the interaction between particles having opposite momenta and spins is retained:

$$H=\sum_{f} T(f) a_f^+ a_f - \frac{1}{2V}\sum_{f,f'} \lambda(f)\lambda(f') a_f^+ a_{-f}^+ a_{-f'} a_{f'}, \tag{1.1}$$

where $f = (\mathbf{p}, s)$, $s = \pm 1$; $\mathbf{p}$ is the momentum vector. For fixed $V = L^3$,

$$p_x=\frac{2\pi}{L}n_x,\quad p_y=\frac{2\pi}{L}n_y,\quad p_z=\frac{2\pi}{L}n_z,$$

n_x, n_y, n_z are integers; $-f = (-\mathbf{p}, -s)$.

Finally, $T(f) = (p^2/2m)-\mu$, where $\mu > 0$ is the chemical potential,

$$\lambda(f)=\begin{cases} J\cdot\varepsilon(s) & \text{for } \left|\frac{p^2}{2m}-\mu\right|\leqslant\Delta, \\ 0 & \text{for } \left|\frac{p^2}{2m}-\mu\right|>\Delta; \end{cases}$$

$$\varepsilon(s)=\pm 1,\quad J=\text{const}.$$

The application of the Bardeen–Cooper–Schrieffer method [1] and the method of compensation of dangerous diagrams leads to the identical result in the case given. Moreover, in [2] it was shown that a Hamiltonian of the type (1.1) is of great methodological interest, since here we have one of the very few completely solvable problems in statistical physics.

In the paper mentioned it is established that for this problem we may obtain an asymptotically exact (for $V \to \infty$) expression for the free energy.

This result was found there in the following manner. The Hamiltonian (1.1) was partitioned into two parts H_0 and H_1 in a special manner. The problem with the Hamiltonian H_0 was solved exactly. Perturbation theory was used to consider the effect of H_1. It was shown that any n-th term of the corresponding expansion becomes asymptotically small for $V \to \infty$, in connection with which it was concluded that the effect of H_1 may in general be neglected after the transition in the limit $V \to \infty$. Of course, reasoning of this kind

---

Joint Institute for Nuclear Research, Dubna. Translated from Problemy Fiziki Élementarnykh Chastits i Atomnogo Yadra, Vol. 1, No. 2, pp. 301–364, 1971.

cannot pretend to mathematical rigor, but it should nevertheless be underlined that in statistical physics problems still cruder devices are often used. For example, approximate methods based on selective summation of "principal terms" (in some sense) of the perturbation-theory series are very widely used; here the remaining terms are discarded even though they do not vanish even for $V \to \infty$.

Doubts of the validity of the results of [2] also arise in connection with the fact that various attempts at using conventional Feynman diagram techniques (without allowance for "anomalous pairings" $\overline{a_f\, a_{-f}}$, $\overline{a_{-f}^{+}\, a_f^{+}}$, to which canonical u-, v-transformation leads) did not yield the expected result. Furthermore, based on the summation of a certain class of Feynman diagrams, Prange [3] obtained a solution which differed in principle from the solution obtained in [1, 2] and assumed that the latter papers were wrong.

In [4] a study was made of a chain of linked equations for the Green's function without the use of perturbation theory. It was shown there that the Green's function for the Hamiltonian H_0 satisfied this entire chain of equations for the exact Hamiltonian $H = H_0 + H_1$ with an error of order $1/V$. This substantiates the results of [2] and reveals the "inefficiency" of the correction H_1.

However, one can also dwell on the purely mathematical point of view. As soon as we have fixed the Hamiltonian, say in the form (1.1), we have an already fully defined mathematical problem which should be solved rigorously without any "physical assumptions." In this case, the approximate expressions satisfy the exact equations with an error of order $1/V$, and we should estimate the difference between the most exact and approximate expressions.

Having in mind complete parity in the problem of the behavior of a dynamic system having the Hamiltonian (1.1), we shall take precisely such a purely mathematical viewpoint in this paper.

We shall study the Hamiltonian (1.1) at a temperature $\theta = 0$ and demonstrate rigorously that the relative difference $(E-E_0)/E_0$ between the lowest energy levels H and H_0, and likewise between the corresponding Green's functions, tends to vanish for $V \to \infty$; we shall obtain estimates for the order of decrease.

Based on methodological concepts it is convenient to consider a somewhat more general Hamiltonian containing terms which represent sources of creation and annihilation of pairs:

$$\mathcal{H} = \sum_f T(f)\, a_f^+ a_f - \nu \sum_f \frac{\lambda(f)}{2}\left(a_{-f}\, a_f + a_f^+ a_{-f}^+\right) - \frac{1}{2V} \sum_{f,\, f'} \lambda(f)\, \lambda(f')\, a_f^+ a_{-f}^+ a_{-f'} a_{f'}, \tag{1.2}$$

where ν is a parameter which we shall assume to be greater than or equal to zero.

Let us note that the case $\nu < 0$ need not be considered, since it can be reduced to the case $\nu > 0$ by a trivial change in the gauge of the Fermi operators:

$$a_f \to i\, a_f; \qquad a_f^+ \to -i\, a_f^+.$$

Let us emphasize the fact that the case $\nu > 0$ will be considered exclusive of those notions that it is of interest in understanding the situation in the actual case $\nu = 0$.

For the investigation undertaken we shall not need those specific properties of the functions $\lambda(f)$, $T(f)$ of which we spoke above. It will be quite sufficient if the following general conditions are satisfied:

1) the functions $\lambda(f)$ and $T(f)$ are real, piecewise continuous, and have the symmetry conditions

$$\lambda(-f) = -\lambda(f); \qquad T(-f) = T(f);$$

2) $\lambda(f)$ is uniformly bounded throughout the entire space, and $T(f) \to \infty$ for $|f| \to \infty$;

3) $\frac{1}{V} \sum_f |\lambda(f)| \leqslant \text{const} \quad$ for $V \to \infty$;

4) $\lim_{V\to\infty} \frac{1}{2V} \sum_f \frac{\lambda^2(f)}{\sqrt{\lambda^2(f)x + T^2(f)}} > 1$ for sufficiently small positive x.

Let us represent $\mathcal{H}$ (1.2) in the form

$$\mathcal{H} = \mathcal{H}_0 + \mathcal{H}_1, \tag{1.3}$$

where

$$\mathcal{H}_0 = \sum_f T(f) a_f^+ a_f - \frac{1}{2}\sum_f \lambda(f)\{(\nu+\sigma^*) a_{-f} a_f + (\nu+\sigma) a_f^+ a_{-f}^+\} + \frac{|\sigma|^2 V}{2}, \tag{1.4}$$

$$\mathcal{H}_1 = -\frac{1}{2V}\left(\sum_f \lambda(f) a_f^+ a_{-f}^+ - V\sigma^*\right)\left(\sum_f \lambda(f) a_{-f} a_f - V\sigma\right). \tag{1.5}$$

Here σ is a certain complex number.

Let us note that if σ is determined from the condition for the minimum of the least eigenvalue $\mathcal{H}_0$, while $\mathcal{H}_1$ is discarded, we arrive at the well-known approximate solution which was considered in [1,2,4]. Here our problem will consist in finding the estimates for the deviation of the minimal eigenvalues $\mathcal{H}_0$, $\mathcal{H}$ and for the deviation of the corresponding Green's functions. Let us show that these deviations will vanish in the process of the transition in the limit $V \to \infty$.*

§ 2. The General Properties of the Hamiltonian

1. In this section we shall establish certain general properties of the model Hamiltonian $\mathcal{H}$ (1.2). Let us consider the occupancy numbers $n_f = a_f^+ a_f$ and let us show that the differences $n_f - n_{-f}$ are integrals of motion. Actually,

$$a_{-f} a_f (n_f - n_{-f}) - (n_f - n_{-f}) a_{-f} a_f = 0,$$

and likewise

$$a_f^+ a_{-f}^+ (n_f - n_{-f}) - (n_f - n_{-f}) a_f^+ a_{-f}^+ = 0,$$

therefore,

$$\mathcal{H}(n_f - n_{-f}) - (n_f - n_{-f})\mathcal{H} = 0.$$

Consequently,

$$\frac{d}{dt}(n_f(t) - n_{-f}(t)) = 0. \tag{2.1}$$

2. Let us show that for the wave function $\Phi_{\mathcal{H}}$, corresponding to the least eigenvalue of the Hamiltonian $\mathcal{H}$, we may place

$$(n_f - n_{-f})\Phi_{\mathcal{H}} = 0 \tag{2.2}$$

for any f.

In order to prove this let us assume the opposite. Since $(n_f - n_{-f})$ commutates with $\mathcal{H}$ (and with one another) one can always choose $\Phi_{\mathcal{H}}$ in such a way that it is an eigenfunction for all these operators:

---

*Recently papers have appeared [7-12] in which new methods have been developed for finding asymptotically exact expressions for multitemporal correlation functions (Green's functions) in the case of arbitrary temperatures θ. Estimates were likewise constructed for finding expressions for the free energies in model systems of the BCS type which are exact for $V \to \infty$. Based on an analysis and generalization of the papers, it was possible to formulate a new principle — the minimax principle [12] — for an entire class of model problems in statistical physics.

$$n_f - n_{-f} = \begin{cases} 1 \\ 0 \\ -1 \end{cases}.$$

Let us use K_0, K_-, K_+, respectively to denote the ensemble of subscripts f for which

$$(n_f - n_{-f})\Phi_{\mathcal{H}} = 0, \qquad f \in K_0;$$
$$(n_f - n_{-f} - 1)\Phi_{\mathcal{H}} = 0, \qquad f \in K_+;$$
$$(n_f - n_{-f} + 1)\Phi_{\mathcal{H}} = 0, \qquad f \in K_-.$$

This assumption can be reduced to the proposition that the sets K_+, K_- are not empty and that*

$$\langle \Phi^*_{\mathcal{H}} \mathcal{H} \Phi_{\mathcal{H}} \rangle \leqslant \langle \varphi^* \mathcal{H} \varphi \rangle$$

for any function φ.

Further we shall require that φ satisfy the additional conditions

$$(n_f - n_{-f})\varphi = 0. \tag{2.3}$$

Let us note now that if $f \in K_+$, then $n_f = 1$, $n_{-f} = 0$, while if $f \in K_-$, then $n_f = 0$; $n_{-f} = 1$. Therefore, $\Phi_{\mathcal{H}}$ may be represented in the form of the direct product

$$\Phi_{\mathcal{H}} = \Phi_{K_0} \Phi_{K_+} \Phi_{K_-},$$

where

$$\Phi_{K_+} = \prod_{f \in K_+} \delta(n_f - 1)\,\delta(n_{-f}); \quad \Phi_{K_-} = \prod_{f \in K_-} \delta(n_f)\,\delta(n_{-f} - 1),$$

while ΦK_0 is a function of only those n_f for which $f \in K_0$:

$$\Phi_{K_0} = F(\ldots n_f \ldots); \quad \langle \Phi^+_{K_0} \Phi_{K_0} \rangle = 1, \quad f \in K_0.$$

Let us note further that

$$a_{-f}\, a_f\, \delta(n_f - 1)\,\delta(n_{-f}) = 0; \quad a_{-f}\, a_f\, \delta(n_f)\,\delta(n_{-f} - 1) = 0;$$
$$a^+_f\, a^+_{-f}\, \delta(n_f - 1)\,\delta(n_{-f}) = 0; \quad a^+_f\, a^+_{-f}\, \delta(n_f)\,\delta(n_{-f} - 1) = 0,$$

and therefore that

$$a_{-f}\, a_f\, \Phi_{K_+} \Phi_{K_-} = 0; \quad a^+_f\, a^+_{-f}\, \Phi_{K_+} \Phi_{K_-} = 0,$$

if $f \in K_+$ for K_-. Consequently,

$$\mathcal{H}\Phi_{\mathcal{H}} = \Big\{ \sum_{f \in K_+} T(f) + \sum_{f \in K_-} T(f) + \sum_{f \in K_0} T(f)\, n_f - \frac{\nu}{2} \sum_{f \in K_0} \lambda(f)\,(a_{-f}\, a_f + a^+_f\, a^+_{-f})$$
$$- \frac{1}{2V} \sum_{f \in K_0} \sum_{f' \in K_0} \lambda(f)\,\lambda(f')\, a^+_f\, a^+_{-f}\, a_{-f'}\, a_{f'} \Big\} \Phi_{\mathcal{H}}.$$

And thus,

$$\langle \Phi^*_{\mathcal{H}} \mathcal{H} \Phi_{\mathcal{H}} \rangle = \sum_{f \in K_+} T(f) + \sum_{f \in K_-} T(f) + \Big\langle \Phi^*_{K_0} \Big\{ \sum_{f \in K_0} T(f)\, n_f - \frac{\nu}{2} \sum_{f \in K_0} \lambda(f)\,(a_{-f}\, a_f + a^+_f\, a^+_{-f})$$

*The symbol $\langle \Phi^* \Psi \rangle$ will be used to denote the scalar product of the functions Φ and Ψ.

$$-\frac{1}{2V}\sum_{f\in K_0}\sum_{f'\in K_0}\lambda(f)\lambda(f')a_f^+a_{-f}^+a_{-f'}a_{f'}\Big\}\Phi_{K_0}\Big\rangle.$$

Let us now partition the set $K_+ + K_-$ into two sets

$$K_+ + K_- = Q_+ + Q_-$$

in such a way that Q_+ will include those subscripts f from $K_+ + K_-$ for which $T(f) \geq 0$, while Q_- will include those subscripts from $K_+ + K_-$ for which $T(f) < 0$. In view of the symmetry of $T(f) = T(-f)$ the subscript f will always be included in Q_+ and Q_- simultaneously with $-f$.

We have

$$\langle\Phi_{\mathcal{H}}^* \mathcal{H}\, \Phi_{\mathcal{H}}\rangle = \sum_{f\in Q_+}|T(f)| - \sum_{f\in Q_-}|T(f)|$$

$$+\Big\langle\Phi_{K_0}^*\Big\{\sum_{f\in K_0}T(f)n_f - \frac{\nu}{2}\sum_{f\in K_0}\lambda(f)(a_{-f}a_f + a_f^+a_{-f}^+)$$

$$-\frac{1}{2V}\sum_{f\in K_0}\sum_{f'\in K_0}\lambda(f)\lambda(f')a_f^+a_{-f}^+a_{-f'}a_{f'}\Big\}\Phi_{K_0}\Big\rangle.$$

Let us now construct the function φ likewise in the form of a simple product, having placed

$$\varphi = \Phi_{K_0}\Phi_{Q_+}\Phi_{Q_-},$$

where

$$\Phi_{Q_+} = \prod_{f\in Q_+}\delta(n_f)\,\delta(n_{-f}); \quad \Phi_{Q_-} = \prod_{f\in Q_-}\delta(n_f - 1)\,\delta(n_{-f} - 1).$$

(Here it is precisely essential that f belong to Q_+ or Q_- simultaneously with $-f$.) For such a function

$$\langle\varphi^* \mathcal{H}\varphi\rangle = -2\sum_{f\in Q_-}|T(f)| + \Big\langle\Phi_{K_0}^*\Big\{\sum_{f\in K_0}T(f)n_f - \frac{\nu}{2}\sum_{f\in K_0}\lambda(f)(a_{-f}a_f + a_f^+a_{-f}^+)$$

$$-\frac{1}{2V}\sum_{f\in K_0}\sum_{f'\in K_0}\lambda(f)\lambda(f')a_f^+a_{-f}^+a_{-f'}a_{f'}\Big\}\Phi_{K_0}\Big\rangle - \frac{1}{2V}\sum_{f\in Q_-}\lambda^2(f).$$

As is evident,

$$\langle\Phi_{\mathcal{H}}^* \mathcal{H}\,\Phi_{\mathcal{H}}\rangle > \langle\varphi^* \mathcal{H}\varphi\rangle.$$

On the other hand, the method of construction of φ satisfies all of the additional conditions (3), and we have arrived at a contradiction with Eq. (2). Thus, our statement has been proved. From the statement (2.2) it follows, in particular, that the total momentum for $\Phi_{\mathcal{H}}$ is equal to zero:

$$\sum_{\mathbf{f}}\mathbf{f}n_{\mathbf{f}}\Phi_{\mathcal{H}} = \frac{1}{2}\sum_{\mathbf{f}}\mathbf{f}(n_f - n_{-f})\Phi_{\mathcal{H}} = 0. \tag{2.4}$$

As is evident from what has been said earlier, the eigenfunction Φ for the least eigenvalue $\mathcal{H}$ may always be sought in the class of functions φ which are governed by the additional conditions (2.3). Let us note that for this special class of φ satisfying the conditions (2.3) the Hamiltonian $\mathcal{H}$ may be expressed in terms of Pauli amplitudes.

Let us consider the operators

$$b_f = a_{-f}a_f; \quad b_f^+ = a_f^+a_{-f}^+.$$

Independently of the additional conditions, we have

$$b_f b_{f'} = b_{f'} b_f; \quad b_f^+ b_{f'}^+ = b_{f'}^+ b_f^+; \quad b_f^2 = 0; \quad b_f^{+2} = 0;$$

$$b_f b_{f'}^+ - b_{f'}^+ b_f = 0; \quad f \neq f'.$$

Moreover, with allowance for the additional conditions we have

$$b_f^+ b_f + b_f b_f^+ = n_f n_{-f} + (1 - n_f)(1 - n_{-f}) = 1,$$

since n_f and n_{-f} are simultaneously either both equal to zero or both equal to unity.

Thus, in the class (2.3) investigated the operators b_f, b_f^+ are Pauli amplitudes. In this class of functions the Hamiltonian (1.2) has the form

$$\mathcal{H} = 2\left\{\sum_{f>0} T(f)\, b_f^+ b_f - \frac{\nu}{2}\sum_{f>0}\lambda(f)\{b_f + b_f^+\} - \frac{1}{V}\sum_{\substack{f>0\\ f'>0}}\lambda(f)\lambda(f')\, b_f^+ b_{f'}\right\}. \tag{2.5}$$

We isolated the class of subscripts $f > 0$ so that all operators b_f would be different, since $b_f = -b_{-f}$. A Hamiltonian of this type was considered in our previous paper [5].

§ 3. The Upper Estimate of the Eigenvalue of the Hamiltonian (1.2)

Let us now consider the problem of the upper estimate of the minimal eigenvalue of the Hamiltonian $\mathcal{H}$. We shall start from the representation of the Hamiltonian $\mathcal{H}$ in the form (1.2). Let us use $E_{\mathcal{H}}$ to denote the least eigenvalue of the Hamiltonian $\mathcal{H}$ (1.2) and $E_0(\sigma)$ to denote the least eigenvalue of the Hamiltonian $\mathcal{H}_0$ (1.4). Note that the operator $\mathcal{H}_1 < 0$, and therefore the minimal eigenvalue of the Hamiltonian $\mathcal{H}_0$, is larger than the minimal eigenvalue of the Hamiltonian $\mathcal{H} = \mathcal{H}_0 + \mathcal{H}_1$:

$$E_0(\sigma) \geqslant E_{\mathcal{H}} \tag{3.1}$$

for any σ. Thus, the minimal eigenvalues of the Hamiltonian $\mathcal{H}_0$ majorize the minimal eigenvalue of $\mathcal{H}$. The best estimate is obtained for σ which yields $\min E_0(\sigma)$.

Let us now go over to calculating the eigenvalues of the Hamiltonian $\mathcal{H}_0$. Carrying out the appropriate canonical transformation which diagonalizes the quadratic form of $\mathcal{H}_0$ (1.4), we obtain the identity

$$\mathcal{H}_0 = \sum_f \sqrt{\lambda^2(f)(\nu+\sigma^*)(\nu+\sigma) + T^2(f)}\,(a_f^+ u_f + a_{-f} v_f^*)(u_f a_f + v_f a_{-f}^+)$$

$$+ \frac{1}{2} V\left\{\sigma^*\sigma - \frac{1}{V}\sum_f\left[\sqrt{\lambda^2(f)(\nu+\sigma^*)(\nu+\sigma) + T^2(f)} - T(f)\right]\right\}, \tag{3.2}$$

where

$$\left.\begin{aligned} u_f &= \frac{1}{\sqrt{2}}\sqrt{1 + \frac{T(f)}{\sqrt{\lambda^2(f)(\nu+\sigma^*)(\nu+\sigma) + T^2(f)}}}, \\ v_f &= \frac{-\varepsilon(f)}{\sqrt{2}}\sqrt{1 - \frac{T(f)}{\sqrt{\lambda^2(f)(\nu+\sigma^*)(\nu+\sigma) + T^2(f)}}}\,\frac{\sigma+\nu}{|\sigma+\nu|}. \end{aligned}\right\} \tag{3.3}$$

Here

$$\lambda(f) = \varepsilon(f)\,|\lambda(f)|; \quad \varepsilon(f) = \operatorname{sign}\lambda(f). \tag{3.4}$$

Obviously,

$$u(-f)=u(f);\quad v(-f)=-v(f);\quad u^2+|v|^2=1, \tag{3.5}$$

where u is real, and v is complex.

From this it is evident that the amplitudes

$$\left.\begin{aligned} \alpha_f &= u_f a_f + v_f a^+_{-f};\\ \alpha^+_f &= u_f a^+_f + v^*_f a_{-f} \end{aligned}\right\} \tag{3.6}$$

are fermion amplitudes. Consequently, the expression for $\mathcal{H}_0$ may be rewritten in the form

$$\mathcal{H}_0=\sum\sqrt{\lambda^2(f)(\nu+\sigma^*)(\nu+\sigma)+T^2(f)}\;\alpha^+_f\alpha_f$$
$$+\frac{1}{2}V\left\{\sigma^*\sigma-\frac{1}{V}\sum_f\left[\sqrt{(\nu+\sigma^*)(\nu+\sigma)\lambda^2(f)+T^2(f)}-T(f)\right]\right\}. \tag{3.7}$$

It is obvious that min $\mathcal{H}_0$ will be reached for the occupancy numbers $\overset{+}{\alpha}_f\alpha_f = 0$. Consequently, for the ground-state energy of the Hamiltonian $\mathcal{H}_0$ we obtain

$$E_0(\sigma)=\frac{1}{2}V\left\{\sigma^*\sigma-\frac{1}{V}\sum_f\left[\sqrt{\lambda^2(f)(\nu+\sigma^*)(\nu+\sigma)+T^2(f)}-T(f)\right]\right\}. \tag{3.8}$$

In order to improve the upper estimate of $E_{\mathcal{H}}$ it is necessary to take $E_0(\sigma)$.

Let us consider the following cases separately:

1. The case $\nu=0$. Let us place $x = \sigma^*\sigma > 0$; then $E_0(\sigma) = 1/2\, VF(\sigma^*\sigma)$, where

$$F(x)=x-\frac{1}{V}\sum_f\{\sqrt{\lambda^2(f)x+T^2(f)}-T(f)\}.$$

In this case, as is evident from the minimum condition, one may determine only the modulus of σ but not its phase. We have

$$F'(x)=1-\frac{1}{2V}\sum_f\frac{\lambda^2(f)}{\sqrt{\lambda^2(f)x+T^2(f)}};$$
$$F''(x)=\frac{1}{4V}\sum_f\frac{\lambda^4(f)}{(\sqrt{\lambda^2(f)x+T^2(f)})^3}.$$

As is evident, $F''(x)>0$ in the interval $0\le x\le\infty$, and therefore, $F'(x)$ may have no more than one root in this interval. Taking account of the properties of the functions $\lambda(f)$ and $T(f)$ (see §1), we shall have $F'(0)<0$; $F'(\infty)>0$. And, consequently, in the interval $0<x<\infty$ there exists a single solitary solution of the equation $F'(x) = 0$; it is this solution which realizes the absolute minimum. Thus, we finally have

$$\frac{V}{2}\min F(x)\geqslant E_{\mathcal{H}}\qquad(0<x<\infty). \tag{3.9}$$

2. The case $\nu > 0$. Let us place $(\nu+\sigma^*)(\nu+\sigma)=x$ (it is obvious that $x > 0$), and note that

$$\sigma^*\sigma=x+\nu^2-\nu(\sigma+\nu+\sigma^*+\nu)=(\sqrt{x}-\nu)^2+\nu\{2\sqrt{x}-(\sigma+\nu+\sigma^*+\nu)\}.$$

Here the root, as always, is assigned the sign +. Then

$$\sigma+\nu=\sqrt{x}\,e^{i\varphi};\quad \sigma^*+\nu=\sqrt{x}\,e^{-i\varphi}$$

and

$$\sigma^*\sigma = (\sqrt{x}-\nu)^2 + 2\nu\sqrt{x}\,(1-\cos\varphi).$$

Therefore,

$$E_0(\sigma) = \frac{V}{2}F(x) + V\nu\sqrt{x}\,(1-\cos\varphi), \tag{3.10}$$

where

$$F(x) = (\sqrt{x}-\nu)^2 - \frac{1}{V}\sum_f \{\sqrt{\lambda^2(f)\,x + T^2(f)} - T(f)\}.$$

Further we have

$$F'(x) = 1 - \frac{\nu}{\sqrt{x}} - \frac{1}{2V}\sum_f \frac{\lambda^2(f)}{\sqrt{\lambda^2(f)\,x + T^2(f)}};$$

$$F''(x) = \frac{\nu}{2x^{3/2}} + \frac{1}{4V}\sum_f \frac{\lambda^4(f)}{(\lambda^2(f)\,x + T^2(f))^{3/2}}.$$

Since $F''(x) > 0$, we see that $F'(x)$ may not have more than one root in the interval $[0, \infty]$. But $F'(0) = -\infty$, $F'(\infty) = 1$. Therefore, a x_0 exist in the interval $0 < x_0 < \infty$, for which $F'(x_0) = 0$. It is precisely for this value of x_0 that the function $F(x)$ has an absolute minimum.

From (3.10) it is evident that the only possible choice of σ corresponding to the absolute minimum will be

$$x = x_0, \qquad \varphi = 0. \tag{3.11}$$

Thus, we have

$$\sigma + \nu = \sqrt{x}, \quad \sigma = \sqrt{x} - \nu.$$

Thus, in the case given ($\nu > 0$) the phase of σ can also be determined. As we can see, σ must be real. We likewise have

$$\frac{V}{2}\min F(x) \geqslant E_{\mathcal{H}} \qquad (0 < x < \infty). \tag{3.12}$$

The simple concepts used in [2] show that in Eq. (1.3) the additional term $\mathcal{H} - \mathcal{H}_0 = \mathcal{H}_1$ is ineffective for $V \to \infty$. However, the rigorous establishment of this property is complicated by the fact that we have only the upper estimate for $E_{\mathcal{H}}$ and do not have an analogous lower estimate. In general, it would be desirable to cancel the term

$$\left(\sum_f \lambda(f)\,a_f^+ a_{-f}^+ - V\sigma^*\right)\left(\sum_f \lambda(f)\,a_{-f}\,a_f - V\sigma\right).$$

This could be achieved by making σ the operator

$$L = \frac{1}{V}\sum_f \lambda(f)\,a_{-f}\,a_f$$

rather than a number. But with an operator one cannot perform canonical transformations from a-fermions to α-fermions. However, we shall try to generalize the identity (3.2) for such a case. One need merely establish the order of the operators correctly. It is precisely in this way that we prove the theorem to the effect that using $\mathcal{H}_0$ one may obtain the asymptotically exact solution for $\mathcal{H}$ when $V \to \infty$.

§ 4. The Lower Estimate of the Eigenvalue of the Hamiltonian

In order to obtain the lower estimate of the Hamiltonian (1.2) we first of all generalize the identity (1.3) in such a way that the term $\mathcal{H}_1$ (1.5) vanishes. This may be done by treating σ as a certain operator L rather than as a c-number:

$$L=\frac{1}{V}\sum_f a_{-f}\,a_f\,\lambda(f). \tag{4.1}$$

Instead of the c-number $(\nu+\sigma^*)(\nu+\sigma)$ we introduce the operators

$$K=(L+\nu)(L^{+}+\nu)+\beta^2,\quad \widetilde{K}=(L^{+}+\nu)(L+\nu)+\beta^2, \tag{4.2}$$

where β is a certain constant.

We now introduce the operators

$$p_f=\frac{1}{\sqrt{2}}\sqrt{\sqrt{K\lambda^2(f)+T^2(f)}+T(f)}\,;\quad p_f=p_f^{+};$$

$$q_f=-\frac{\varepsilon(f)}{\sqrt{2}}\sqrt{\sqrt{K\lambda^2(f)+T^2(f)}-T(f)}\cdot\frac{1}{\sqrt{K}}(L+\nu). \tag{4.3}$$

Obviously,

$$p_f q_f=-\frac{\lambda(f)}{2}(L+\nu); \tag{4.4}$$

$$p_f^2=\frac{1}{2}\{\sqrt{K\lambda^2(f)+T^2(f)}+T(f)\}; \tag{4.5}$$

$$q_f^{+}q_f=(L^{+}+\nu)\frac{1}{2K}\{\sqrt{K\lambda^2(f)+T^2(f)}-T(f)\}(L+\nu). \tag{4.6}$$

Taking account of the fact that for any operator ξ the identity

$$\xi^{+}F(\xi\xi^{+})\xi=\xi^{+}\xi F(\xi^{+}\xi) \tag{4.7}$$

is valid, Eq. (4.8) can be written in the form

$$q_f^{+}q_f=(L^{+}+\nu)(L+\nu)\frac{1}{2\widetilde{K}}\{\sqrt{\widetilde{K}\lambda^2(f)+T^2(f)}-T(f)\}$$

$$=\frac{1}{2}\{\sqrt{\widetilde{K}\lambda^2(f)+T^2(f)}-T(f)\}-\frac{\beta^2}{2\widetilde{K}}\{\sqrt{\widetilde{K}\lambda^2(f)+T^2(f)}-T(f)\}. \tag{4.8}$$

Going on to apply Lemma II [see Appendix, Eqs. (A1.9)-(A1.10)], we write

$$q_f^{+}q_f=\frac{1}{2}\left\{\sqrt{\lambda^2(f)\left(K+\frac{2s}{V}\right)+T^2(f)}-T(f)\right\}$$

$$-\frac{1}{2}\left\{\sqrt{\lambda^2(f)\left(K+\frac{2s}{V}\right)+T^2(f)}-\sqrt{\lambda^2(f)\widetilde{K}+T^2(f)}\right\}$$

$$-\frac{\beta^2}{2\widetilde{K}}\{\sqrt{\widetilde{K}\lambda^2(f)+T^2(f)}-T(f)\}. \tag{4.9}$$

Note that the second term on the right is not negative, while s is the upper estimate of the expression $\frac{1}{V}\sum_f |\lambda(f)|^2$:

$$\frac{1}{V}\sum_f |\lambda(f)|^2 \leqslant s. \tag{4.10}$$

Moreover, we have

$$p_f^2 = \frac{1}{2}\left\{\sqrt{\left(K+\frac{2s}{V}\right)\lambda^2(f)+T^2(f)}+T(f)\right\}$$

$$-\frac{1}{2}\left\{\sqrt{\left(K+\frac{2s}{V}\right)\lambda^2(f)+T^2(f)}-\sqrt{K\lambda^2(f)+T^2(f)}\right\}. \tag{4.11}$$

Let us now consider the equation

$$\Omega = \sum_f (a_f^+ p_f + a_f q_f^+)(p_f a_f + q_f a_{-f}^+). \tag{4.12}$$

Making use of the equation $q_f^+ q_f = q_{-f}^+ q_{-f}$ and (4.4), we obtain

$$\Omega = \sum_f a_f^+ p_f^2 a_f + \sum_f a_f q_f^+ q_f a_f^+$$

$$-\sum_f \frac{\lambda(f)}{2}\{(L^+ + \nu) a_{-f} a_f + a_f^+ a_{-f}^+ (L+\nu)\} + R_1, \tag{4.13}$$

where

$$R_1 = \sum_f \frac{\lambda(f)}{2}\{(L^+ a_{-f} - a_{-f} L^+) a_f + a_f^+ (a_{-f}^+ L - L a_{-f}^+)\}. \tag{4.14}$$

Note that

$$\sum_f \frac{\lambda(f)}{2}\{(L^+ + \nu) a_{-f} a_f + a_f^+ a_{-f}^+ (L+\nu)\} = VL^+ L + \frac{V}{2}(\nu L + \nu L^+) \tag{4.15}$$

and, consequently,

$$\Omega + \frac{V}{2} L^+ L - \sum_f a_f^+ p_f^2 a_f - \sum_f a_f q_f^+ q_f a_f^+ = -\frac{V}{2}\{L^+ L + \nu(L + L^+)\} + R_1. \tag{4.16}$$

Or, by virtue of (4.9) and (4.11), we have

$$\sum_f \{a_f^+ p_f + a_{-f} q_f^+\}\{p_f a_f + q_f a_{-f}^+\}$$

$$+\frac{1}{2}\sum_f a_f^+ \left\{\sqrt{\left(K+\frac{2s}{V}\right)\lambda^2(f)+T^2(f)} - \sqrt{K\lambda^2(f)+T^2(f)}\right\} a_f$$

$$+\frac{1}{2}\sum_f a_f \left\{\sqrt{\left(K+\frac{2s}{V}\right)\lambda^2(f)+T^2(f)} - \sqrt{\tilde{K}\lambda^2(f)+T^2(f)}\right\} a_f^+$$

$$+\frac{1}{2}\sum_f a_f \frac{\beta^2}{2\tilde{K}}\left\{\sqrt{\tilde{K}\lambda^2(f)+T^2(f)} - T(f)\right\} a_f^+$$

$$-\frac{1}{2}\sum_f a_f^+\left\{\sqrt{\left(K+\frac{2s}{V}\right)\lambda^2(f)+T^2(f)}+T(f)\right\}a_f$$

$$-\frac{1}{2}\sum_f a_f\left\{\sqrt{\left(K+\frac{2s}{V}\right)\lambda^2(f)+T^2(f)}-T(f)\right\}a_f^+ +\frac{V}{2}L^+L$$

$$=-\frac{V}{2}\{L^+L+\nu(L+L^+)\}+R_1. \tag{4.17}$$

Let us introduce the notation

$$\Delta_1=\frac{1}{2}\sum_f a_f^+\left\{\sqrt{\left(K+\frac{2s}{V}\right)\lambda^2+T^2}-\sqrt{K\lambda^2+T^2}\right\}a_f; \tag{4.18}$$

$$\Delta_2=\frac{1}{2}\sum_f a_f\left\{\sqrt{\left(K+\frac{2s}{V}\right)\lambda^2+T^2}-\sqrt{\tilde{K}\lambda^2+T^2}\right\}a_f^+; \tag{4.19}$$

$$\Delta_3=\frac{1}{2}\sum_f a_f\frac{\beta^2}{2\tilde{K}}\left\{\sqrt{\tilde{K}\lambda^2+T^2}-T\right\}a_f^+. \tag{4.20}$$

Then, by virtue of Lemma II [see Appendix, Eqs. (A1.9), (A1.10)]

$$\Omega\geqslant 0;\quad \Delta_1\geqslant 0;\quad \Delta_2\geqslant 0;\quad \Delta_3\geqslant 0. \tag{4.21}$$

Thus,

$$\Omega+\Delta_1+\Delta_2+\Delta_3-\frac{1}{2}\sum_f a_f^+\left\{\sqrt{\left(K+\frac{2s}{V}\right)\lambda^2+T^2}+T\right\}a_f$$

$$-\frac{1}{2}\sum_f a_f\left\{\sqrt{\left(K+\frac{2s}{V}\right)\lambda^2+T^2}-T\right\}a_f^+ +\frac{V}{2}L^+L$$

$$=-\frac{V}{2}\{L^+L+\nu(L+L^+)\}+R_1. \tag{4.22}$$

Let us set

$$R_2=\frac{1}{2}\sum_f a_f^+\left\{\sqrt{\left(K+\frac{2s}{V}\right)\lambda^2+T^2}\,a_f-a_f\sqrt{\left(K+\frac{2s}{V}\right)\lambda^2+T^2}\right\}; \tag{4.23}$$

$$R_3=\frac{1}{2}\sum_f a_f\left\{\sqrt{\left(K+\frac{2s}{V}\right)\lambda^2+T^2}\,a_f^+-a_f^+\sqrt{\left(K+\frac{2s}{V}\right)\lambda^2+T^2}\right\}. \tag{4.24}$$

Then

$$\Omega+\Delta_1+\Delta_2+\Delta_3-R_2-R_3+\frac{V}{2}L^+L-\frac{1}{2}\sum_f a_f^+a_f\left\{\sqrt{\left(K+\frac{2s}{V}\right)\lambda^2+T^2}+T\right\}$$

$$-\frac{1}{2}\sum_f a_f a_f^+\left\{\sqrt{\left(K+\frac{2s}{V}\right)\lambda^2+T^2}-T\right\}=-\frac{V}{2}\{L^+L+\nu(L+L^+)\}+R_1. \tag{4.25}$$

But

$$\frac{1}{2}\sum_f a_f^+ a_f\left\{\sqrt{\left(K+\frac{2s}{V}\right)\lambda^2+T^2}+T\right\}$$

$$+\frac{1}{2}\sum_f a_f a_f^+\left\{\sqrt{\left(K+\frac{2s}{V}\right)\lambda^2+T^2}-T\right\}$$

$$=\frac{1}{2}\sum_f\left\{\sqrt{\left(K+\frac{2s}{V}\right)\lambda^2+T^2}-T\right\}+\sum_f T(f)a_f^+ a_f. \tag{4.26}$$

Consequently,

$$\Omega+\Delta_1+\Delta_2+\Delta_3-R_1-R_2-R_3+\frac{V}{2}(L^+L-LL^+)$$

$$+\frac{1}{2}V\left[LL^+-\frac{1}{V}\sum_f\left\{\sqrt{\left(K+\frac{2s}{V}\right)\lambda^2+T^2}-T\right\}\right]$$

$$=\sum_f T(f)a_f^+a_f-\frac{V}{2}\{L^+L+\nu(L+L^+)\}$$

$$=\sum_f T(f)a_f^+a_f-\nu\sum_f\frac{\lambda(f)}{2}(a_{-f}a_f+a_f^+a_{-f}^+)$$

$$-\frac{1}{2V}\sum_{ff'}\lambda(f)\lambda(f')a_f^+a_{-f}^+a_{-f'}a_{f'}=\mathscr{H}. \tag{4.27}$$

Thus, we finally will have

$$\mathscr{H}=\frac{1}{2}V\left\{LL^+-\frac{1}{V}\sum_f\left[\sqrt{\left(K+\frac{2s}{V}\right)\lambda^2(f)+T^2(f)}-T(f)\right]\right\}$$

$$+\Omega+\Delta_1+\Delta_2+\Delta_3-R_1-R_2-R_3+\frac{V}{2}(L^+L-LL^+). \tag{4.28}$$

Equation (4.28) represents an identical transformation of the Hamiltonian (1.2). The first term of Eq. (4.28) will be treated as the principal term; as far as the terms R_1, R_2, and R_3 are concerned, we shall show that they are asymptotically small, while the terms Ω, Δ_1, Δ_2, and Δ_3 will be dropped. Since they are positive (4.21), we shall obtain the lower estimate for $\mathscr{H}$.

It can easily be shown that when (2.2) is considered, we have

$$-R_1+\frac{V}{2}(L^+L-LL^+)=-\frac{1}{V}\sum_f\lambda^2(f), \tag{4.29}$$

where in accordance with (4.10), we have $\frac{1}{V}\sum_f \lambda^2(f) \leqslant s$. Further, by virtue of Lemma IV (inequality A1.30)

$$|R_2|+|R_3| \leqslant C, \tag{4.30}$$

where

$$C=\frac{4}{\pi}\cdot\frac{1}{V}\sum_f |\lambda(f)|^2\left[1+\frac{|\lambda(f)|\left(\frac{1}{V}\sum|\lambda(f)|+V\right)}{2\frac{1}{V}\sum|\nu(f)|^2+V\frac{T^2(f)}{\lambda^2(f)}}\right]\int_0^\infty \frac{\sqrt{t}}{(1+t)^2}\,dt. \tag{4.31}$$

Thus, for any normalized function Φ the inequality

$$\langle\Phi^*\mathcal{H}\Phi\rangle \geqslant -(s+C)+\frac{1}{2}V\left\langle\Phi^*\left(LL^+-\frac{1}{V}\right.\right.$$

$$\left.\left.\times\sum_f\left[\sqrt{\left\{(L+\nu)(L^++\nu)+\beta^2+\frac{2s}{V}\right\}\lambda^2+T^2(f)}-T(f)\right]\right)\Phi\right\rangle$$

is valid by virtue of (4.21). But s and C do not depend on β. Therefore, performing a transition in the limit $\beta\to 0$, we find

$$\langle\Phi^*\mathcal{H}\Phi\rangle \geqslant -(2s+C)+\frac{1}{2}V\left\langle\Phi^*\left\{LL^*+\frac{2s}{V}\right.\right.$$

$$\left.\left.-\frac{1}{V}\sum_f\left[\sqrt{\left\{(L+\nu)(L^++\nu)+\frac{2s}{V}\right\}\lambda^2(f)+T^2(f)}-T(f)\right]\right\}\Phi\right\rangle. \tag{4.32}$$

Further, we have

$$\left.\begin{aligned} LL^+&=(L+\nu)(L^++\nu)-\nu\{L+\nu+L^++\nu\}+\nu^2;\\ LL^++\frac{2s}{V}&=\left\{(L+\nu)(L^++\nu)+\frac{2s}{V}\right\}-\nu\{(L+\nu+L^++\nu)\}+\nu^2.\end{aligned}\right\} \tag{4.33}$$

We set

$$(L+\nu)(L^++\nu)+\frac{2s}{V}=X. \tag{4.34}$$

Then

$$LL^++\frac{2s}{V}=(\sqrt{X}-\nu)^2+\nu\{2\sqrt{X}-(L+\nu)-(L^++\nu)\}. \tag{4.35}$$

But according to Lemma I (see A1.1, A1.2), having placed $\xi = L+\nu$; $\xi^+ = L^+ + \nu$ in the inequality, we obtain

$$2\sqrt{(L+\nu)(L^++\nu)+\frac{s}{V}}-(L+\nu)-(L^++\nu)\geqslant 0, \tag{4.36}$$

and the more so since

$$2\sqrt{X}-(L+\nu)+(L^++\nu)\geqslant 0. \tag{4.37}$$

Let us consider the function F(x) (3.11):

$$F(x)=(\sqrt{x}-\nu)^2-\frac{1}{V}\sum_f\left[\sqrt{x\lambda^2(f)+T^2(f)}-T(f)\right].$$

Then Eq. (4.32) may be written in the form

$$\langle \Phi^* \mathcal{H} \Phi \rangle \geqslant -(2s+C) + \frac{1}{2} V \langle \Phi^* F(X) \Phi \rangle$$

$$+ V \frac{\nu}{2} \langle \Phi^* \{ 2\sqrt{X} - (L+\nu+L^+ +\nu) \} \Phi \rangle$$

$$\geqslant -(2s+C) + \frac{1}{2} V \langle \Phi^* F(X) \Phi \rangle, \tag{4.38}$$

where X is an operator which is determined by Eq. (4.34).

Assume $E_{\mathcal{H}}$ is the least eigenvalue of $\mathcal{H}$; $\Phi_{\mathcal{H}}$ is the corresponding eigenfunction. Assume further that $E_{\mathcal{H}_0}$ is the least eigenvalue of $\mathcal{H}_0$; then with allowance for (3.11), we have $E_{\mathcal{H}_0} = \frac{V}{2} \min F(x)$. Assume the absolute minimum can be reached for F(x) for

$$x = x_0 = C^2. \tag{4.39}$$

We have

$$\frac{V}{2} F(C^2) \geqslant E_{\mathcal{H}} = \langle \Phi^*_{\mathcal{H}} \mathcal{H} \Phi_{\mathcal{H}} \rangle \geqslant -(2s+C)$$

$$+ \frac{1}{2} V \langle \Phi^* F(X) \Phi \rangle \geqslant -(2s+C) + \frac{V}{2} F(C^2). \tag{4.40}$$

From this, taking account of the fact that the energy of the system is proportional to the volume of the system, we obtain the final estimate for the eigenvalues of the Hamiltonian $\mathcal{H}$ (1.2):

$$0 \leqslant \frac{E_{\mathcal{H}_0} - E_{\mathcal{H}}}{V} \leqslant \frac{2s+C}{V}. \tag{4.41}$$

Note that C (4.31) and s remain finite for $V \to \infty$ in accordance with the conditions of §1. Therefore, the difference between the eigenvalues of the approximate Hamiltonian $\mathcal{H}_0$ (1.4) and the exact Hamiltonian $\mathcal{H}$ (1.2), normalized to the system volume, decreases as 1/V for $V \to \infty$. Thus, the solution of the approximate Hamiltonian $\mathcal{H}_0$ (1.4) yields an asymptotically exact solution of the Hamiltonian $\mathcal{H}$ (1.2) for $V \to \infty$.

Let us now show that the operator X (4.34) may be treated as a c-number with asymptotic accuracy (i.e., accurate to 1/V). For this purpose we choose any normalized function Φ such that

$$\langle \Phi^* \mathcal{H} \Phi \rangle - E_{\mathcal{H}} \leqslant C_1 = \text{const}. \tag{4.42}$$

Now on the basis of (4.38), (4.40), and (4.42), we have

$$\langle \Phi^* (F(X) - F(C^2)) \Phi \rangle + \nu \langle \Phi^* (2\sqrt{X} - (L+\nu+L^+ +\nu)) \Phi \rangle \leqslant \frac{l}{V}; \tag{4.43}$$

$$l = 2(2s+C+C_1).$$

Note that both terms in the left side of (4.43) are positive. In particular, in view of the positiveness of the second term in the left side of the inequality (4.43) (see Lemma I, Eqs. A1.1 and A1.2), we obtain

$$\langle \Phi^* (F(X) - F(C^2)) \Phi \rangle \leqslant \frac{l}{V}, \tag{4.44}$$

but

$$F(X)-F(C^2)=\frac{1}{2}F''(\xi)(X-C^2)^2; \tag{4.45}$$

$$F''(x)=\frac{\nu}{2x^{3/2}}+\frac{1}{4}\cdot\frac{1}{V}\sum_f\frac{\lambda^4(f)}{(x\lambda^2(f)+T^2(f))^{3/2}},$$

$$\frac{1}{2}F''(\xi)\geqslant\alpha=\text{const}>0. \tag{4.46}$$

From this we obtain

$$\langle\Phi^*|X-C^2|^2\Phi\rangle\leqslant\frac{l}{\alpha V}. \tag{4.47}$$

From (4.46) it follows that the operator X may be treated as a c-number with asymptotic accuracy.

For the case $\nu > 0$ one may obtain more complete information on the mathematical expectation of the operators L, L^+. Specifically, we shall show that the mean-square deviation of the operator L from the quantity C (4.39) is asymptotically small for $V\to\infty$. We have the obvious inequality

$$(\sqrt{X}-C)^2=\frac{(X-C^2)^2}{(\sqrt{X}+C)^2}\leqslant\frac{1}{C^2}(X-C^2)^2. \tag{4.48}$$

From this, making use of (4.47), we obtain

$$\langle\Phi^*(\sqrt{X}-C)^2\Phi\rangle\leqslant\frac{l}{\alpha C^2V}. \tag{4.49}$$

Define

$$\langle\Phi^*\sqrt{X}\Phi\rangle=C_0. \tag{4.50}$$

Then for consideration of (4.49), we have

$$\langle\Phi^*(\sqrt{X}-C_0)^2\Phi\rangle\leqslant\langle\Phi^*(\sqrt{X}-C)^2\Phi\rangle\leqslant\frac{l}{\alpha C^2V}. \tag{4.51}$$

This is actually so, since

$$\langle\Phi^*(\sqrt{X}-C)^2\Phi\rangle=(C-C_0)^2+\langle\Phi^*(\sqrt{X}-C_0)^2\Phi\rangle. \tag{4.52}$$

The bound for the mathematical expectation of the operator X derives from the estimate (4.51) and is given by the relationship

$$\langle\Phi^*X\Phi\rangle-C_0^2\leqslant\frac{l}{\alpha C^2V}. \tag{4.53}$$

Finally, from (4.51) and (4.52), we obtain the estimate for the difference $(C-C_0)^2$:

$$(C-C_0)^2\leqslant\frac{l}{\alpha C^2V}. \tag{4.54}$$

Assume now that

$$\xi=L+\nu;\ \xi^+=L^++\nu; \tag{4.55}$$

then for the mean-square deviation of ξ from C_0 we have the following result when (4.34) is considered:

$$\langle\Phi^*(C_0-\xi)(C_0-\xi^+)\Phi\rangle\leqslant C_0^2+\langle\Phi^*X\Phi\rangle-C_0\langle\Phi^*(\xi+\xi^+)\Phi\rangle. \tag{4.56}$$

Then, using (4.53) and (4.43), we obtain

$$\langle\Phi^*(C_0-\xi)(C_0-\xi^+)\Phi\rangle \leqslant 2C_0^2+\frac{l}{\alpha C^2 V}-C_0\langle\Phi^*(\xi+\xi^+)\Phi\rangle$$

$$=\langle\Phi^*\{2\sqrt{X}-(\xi+\xi^+)\}\Phi\rangle C_0+\frac{l}{\alpha C^2 V}\leqslant\frac{lC_0}{\nu V}+\frac{l}{\alpha C^2 V}. \tag{4.57}$$

Thus, for the quantity $<\Phi^*(C-\xi)\times(C-\xi^+)\Phi>$ of interest to us, we obtain the following estimate:

$$\langle\Phi^*(C-\xi)(C-\xi^+)\Phi\rangle=\langle\Phi^*(C-C_0+C_0-\xi)(C-C_0+C_0-\xi^+)\Phi\rangle$$

$$\leqq 2(C-C_0)^2+2\langle\Phi^*(C_0-\xi)(C_0-\xi^+)\Phi\rangle$$

$$\leqq\frac{2l}{\alpha C^2 V}+\frac{2lC_0}{\nu V}+\frac{2l}{\alpha C^2 V}\leqq\frac{\text{const}}{V}=\frac{I}{V} \tag{4.58}$$

$$(I=\text{const}).$$

Note that the bound obtained is valid only for $\nu > 0$, since ν is included in the denominator of the right side of the inequality (4.58).

Let us discuss the results obtained. Assume $\nu=0$. Then, as we have seen, for states having an average energy which is asymptotically close to the least E_H, the operator L^+L is equal to the c-number C^2 with asymptotic accuracy. These states, however, do not have such properties of the operators L and L^+ proper. Let us consider the state Φ_H having the least energy E_H. In general, a case of degeneration may arise such that we shall have not one state Φ_H but a certain linear manifold $\{\Phi_H\}$ of possible states having the same least energy E_H.

Since the operator $N = \Sigma a_f^+ a_f$, which is the total number of particles in the case $\nu = 0$, commutates exactly with H, one can always find within the manifold $\{\Phi_H\}$ a state Φ_f' which is such that N takes a certain value N_0. Then it is obvious that

$$\langle\Phi_H^{*'}L\Phi_H'\rangle=0;\quad\langle\Phi_H^{*'}L^+\Phi_H'\rangle=0.$$

This means that L cannot take even an approximately definite value in the state Φ_H', since otherwise L^+L for this state would turn out to be approximately equal to 0 rather than to C^2.

Let us now consider a manifold $\{\Phi\}$ of states having an energy asymptotically close to E_H. Since L, L^+ approximately commutates with H, it is natural to expect that in $\{\Phi\}$ one may choose a Φ for which L, L^+ take definite values with asymptotic accuracy. And this is the case in reality. For example, the Φ_{H_0}-state having the least energy for H_0 has the property given. In fact, Φ_{H_0} is determined by the relationship $\alpha_f\Phi = 0$, where $\alpha_f = u_f a_f - v_f a_{-f}^+$;

$$u_f=\frac{1}{\sqrt{2}}\sqrt{1+\frac{T(f)}{\sqrt{\lambda^2(f)C^2+T^2(f)}}};$$

$$v_f=\frac{\varepsilon(f)}{\sqrt{2}}\sqrt{1-\frac{T(f)}{\sqrt{\lambda^2(f)C^2+T^2(f)}}}.$$

Having expressed L in terms of the fermions α, α^+, we find

$$L=\frac{1}{V}\sum_f\lambda(f)\{u_f^2\,\alpha_{-f}a_f-v_f^2\,\alpha_f^+\alpha_{-f}^+-2u_f v_f\alpha_f^+\alpha_f\}+\frac{1}{V}\sum_f u_f v_f\lambda(f).$$

But

$$\frac{1}{V}\sum_f u_f v_f=\frac{1}{2V}\sum_f\frac{|\lambda(f)|^2 C}{\sqrt{\lambda^2(f)C^2+T^2(f)}}=C,$$

and, consequently,

$$\langle \Phi^*_{H_0} (L^+ - C)(L - C) \Phi_{H_0} \rangle \leqslant \frac{\text{const}}{V};$$
$$\langle \Phi^*_{H_0} (L - C)(L^+ - C) \Phi_{H_0} \rangle \leqslant \frac{\text{const}}{V}.$$

For Φ_{H_0}, L and L^+ are approximately equal to C. It is specifically this fact which produced the success of the approximate method in which we replaced the Hamiltonian H with the exact law for conservation of N by H_0 for which N is no longer the exact integral of motion. Now it is likewise evident that the approximate method could be formulated so that the law for conservation of N would not be formally violated. For this purpose it would be necessary to replace the fermions α_f by the amplitude

$$\alpha_f = u_f a_f - v_f \frac{L}{|C|} a^+_{-f},$$

which satisfy the commutation relations for the Fermi amplitudes with asymptotic accuracy. Then α_f decreases N by one, while α^+_f increases N by one. Such amplitudes have analogs with the amplitude

$$b_f = \frac{a_0^+}{\sqrt{N_0}} a_f,$$

which were introduced in superfluidity theory [6] during the isolation of the condensate. In general there is a strong analogy between the amplitudes a_0, a_0^+ for a Bose-condensate and the amplitudes L, L^+ in the case considered.

When we include a term with pair sources ($\nu > 0$) in $\mathcal{H}$, the operators L, L^+ immediately begin to take definite values with asymptotic accuracy for the states having an energy close to $\mathcal{H}$. Here we obtain an analogy with the theory of ferromagnetism in an isotropic medium. In the absence of a magnetic field the direction of the magnetization axis is indefinite. For turn-on of a magnetic field which is arbitrarily weak and acts in a definite direction, the magnetization vector is immediately established precisely along this direction. Finally, from the relationships

$$L^+ L \approx C^2 \quad (\nu = 0);$$
$$\nu + L \approx C; \quad \nu + L^+ \approx C \quad (\nu > 0)$$

one can find that the correlation averages

$$\langle \Phi^*_H \ldots \alpha_{f_i}(t_i) \ldots a^+_{f_j}(t_j) \ldots \Phi_H \rangle$$

for the Hamiltonian $\mathcal{H}$ are also asymptotically equal to the corresponding averages for the Hamiltonian $\mathcal{H}_0$. For $\nu > 0$ this applies to all averages of the given type, while for $\nu = 0$ it applies, of course, only to those for which the c-number is equal to the number a^+ (i.e., the averages of those operators which conserve N).

§ 5. The Green's Function for the Case $\nu > 0$

In this section we shall deal with asymptotic estimates for the Green's functions and correlation averages of the case $\nu > 0$. From these estimates it will follow that the solution of the equations for the Green's functions, constructed on the Hamiltonian $\mathcal{H}_0$ (1.4), will be different to an asymptotically small degree from the corresponding solutions for such a model Hamiltonian $\mathcal{H}$ (1.2) for $V \to \infty$.

Let us consider the equations of motion for the operators a_f, a_f^+. Taking account of (1.2), we obtain

$$\left.\begin{aligned} i\frac{da_f}{dt} &= T(f) a_f - \lambda(f) a^+_{-f} (\nu + L); \\ i\frac{da^+_f}{dt} &= -T(f) a^+_f + \lambda(f)(\nu + L^+) a_{-f}, \end{aligned}\right\} \quad (5.1)$$

whence we likewise obtain

$$\left.\begin{aligned} i\frac{da_{-f}}{dt} &= T(f)\,a_{-f} + \lambda(f)\,a_f^+(\nu+L); \\ i\frac{da_{-f}^+}{dt} &= -T(f)\,a_{-f}^+ - \lambda(f)\,(\nu+L^+)\,a_f. \end{aligned}\right\} \tag{5.2}$$

Let us set (see Eqs. (3.3) and (3.11))

$$\left.\begin{aligned} u_f &= \frac{1}{\sqrt{2}}\sqrt{1+\frac{T(f)}{\sqrt{C^2\lambda^2(f)+T^2(f)}}}; \\ v_f &= -\frac{\varepsilon(f)}{\sqrt{2}}\sqrt{1-\frac{T(f)}{\sqrt{C^2\lambda^2(f)+T^2(f)}}} \end{aligned}\right\} \tag{5.3}$$

and let us introduce a new fermion amplitude

$$\alpha_f^+ = u_f a_f^+ + v_f a_{-f}. \tag{5.4}$$

We have

$$\begin{aligned} i\frac{d\alpha_f^+}{dt} &= u_f\, i\frac{da_f^+}{dt} + v_f\, i\frac{da_{-f}}{dt} \\ &= u_f\{-T(f)\,a_f^+ + \lambda(f)\,(\nu+L^+)\,a_{-f}\} + v_f\{T(f)\,a_{-f} + \lambda(f)\,a_f^+(\nu+L)\} \\ &= -a_f^+\{T(f)\,u_f - \lambda(f)\,v_f(\nu+L)\} + \{\lambda(f)\,(\nu+L^+) + T(f)\,v_f\}\,a_{-f} \\ &= -a_f^+\{T(f)\,u_f - \lambda(f)\,v_f C\} + \{\lambda(f)\,Cu_f + T(f)\,v_f\}\,a_{-f} + R_f, \end{aligned}$$

where

$$\left.\begin{aligned} R_f &= R_f^{(1)} + R_f^{(2)}; \\ R_f^{(1)} &= u_f\lambda(f)\,(L^+ + \nu - C)\,a_{-f}; \\ R_f^{(2)} &= v_f\lambda(f)\,a_f^+(L+\nu-C). \end{aligned}\right\} \tag{5.5}$$

From the identities

$$\left.\begin{aligned} T(f)\,u_f - \lambda(f)\,v_f C &= \sqrt{C^2\lambda^2(f)+T^2(f)}\,u_f, \\ T(f)\,v_f + \lambda(f)\,u_f C &= -\sqrt{C^2\lambda^2(f)+T^2(f)}\,v_f \end{aligned}\right\} \tag{5.6}$$

it follows that

$$i\frac{d\alpha_f^+}{dt} + \sqrt{C^2\lambda^2(f)+T^2(f)}\,\alpha_f^+ = R_f, \tag{5.7}$$

and this means that

$$i\frac{d\alpha_f}{dt} - \sqrt{C^2\lambda^2(f)+T^2(f)}\,\alpha_f = -R_f^+. \tag{5.8}$$

Let us estimate the quantities associated with R, R^+. We have

$$\begin{aligned} \left\langle \Phi_{\mathscr{H}}^* R_f R_f^+ \Phi_{\mathscr{H}} \right\rangle &\leq 2\left\langle \Phi_{\mathscr{H}}^* R_f^{(1)} R_f^{(1)+} \Phi_{\mathscr{H}} \right\rangle \\ +2\left\langle \Phi_{\mathscr{H}}^* R_f^{(2)} R_f^{(2)+} \Phi_{\mathscr{H}} \right\rangle &= 2u_f^2\lambda^2(f)\,\langle \Phi_{\mathscr{H}}^* (L^+ + \nu - C)\,a_{-f}\,a_{-f}^+(L+\nu \\ -C)\Phi_{\mathscr{H}}\rangle &+ 2v_f^2\lambda^2(f)\,\langle \Phi_{\mathscr{H}}^* a_f^+ (L+\nu-C)\,(L^+ + \nu - C)\,a_f \Phi_{\mathscr{H}}\rangle. \end{aligned}$$

But since $|a_{-f} a_{-f}^{+}| \leq 1$, it follows that

$$\langle \Phi_{\mathcal{H}}^{*} (L^{+}+\nu-C)\, a_{-f}\, a_{-f}^{+} (L+\nu-C)\, \Phi_{\mathcal{H}} \rangle$$

$$\leqslant \langle \Phi_{\mathcal{H}}^{*} (L^{+}+\nu-C)(L+\nu-C)\, \Phi_{\mathcal{H}} \rangle,$$

and further, by virtue of (A1.18), we have

$$\langle \Phi_{\mathcal{H}}^{*} a_{f}^{+} (L+\nu-C)(L^{+}+\nu-C)\, a_{f}\, \Phi_{\mathcal{H}} \rangle$$

$$\leqslant \frac{2s}{V} + \langle \Phi_{\mathcal{H}}^{*} a_{f}^{+} (L^{+}+\nu-C)(L+\nu-C)\, a_{f}\, \Phi_{\mathcal{H}} \rangle$$

$$= \frac{2s}{V} + \langle \Phi_{\mathcal{H}}^{*} (L^{+}+\nu-C)\, a_{f}^{+} a_{f} (L+\nu-C)\, \Phi_{\mathcal{H}} \rangle$$

$$\leqslant \frac{2s}{V} + \langle \Phi_{\mathcal{H}}^{*} (L^{+}+\nu-C)(L+\nu-C)\, \Phi_{\mathcal{H}} \rangle.$$

Thus,

$$\langle \Phi_{\mathcal{H}}^{*} R_{f} R_{f}^{+} \Phi_{\mathcal{H}} \rangle \leqslant 2\lambda^{2}(f) \langle \Phi_{\mathcal{H}}^{*} (L^{+}+\nu-C)(L+\nu-C)\, \Phi_{\mathcal{H}} \rangle + 2\lambda^{2}(f)\, v_{f}^{2} \frac{2s}{V}. \tag{5.9}$$

Completely analogously we obtain

$$\langle \Phi_{\mathcal{H}}^{*} R_{f}^{+} R_{f} \Phi_{\mathcal{H}} \rangle \leqslant 2\lambda^{2}(f) \langle \Phi_{\mathcal{H}}^{*} (L+\nu-C)(L^{+}+\nu-C)\, \Phi_{\mathcal{H}} \rangle + 2\lambda^{2}(f)\, u_{f}^{2} \frac{2s}{V}. \tag{5.10}$$

But as was shown earlier (4.58),

$$\langle \Phi_{\mathcal{H}}^{+} (L+\nu-C)(L+\nu-C)\, \Phi_{\mathcal{H}} \rangle \leqslant \frac{I}{V};$$

$$\langle \Phi_{\mathcal{H}}^{*} (L+\nu-C)(L^{+}+\nu-C)\, \Phi_{\mathcal{H}} \rangle \leqslant \frac{I}{V}.$$

Therefore, introducing the constant

$$\gamma = 2(I+2s), \tag{5.11}$$

one may write

$$\begin{aligned} \langle \Phi_{\mathcal{H}}^{*} R_{f} R_{f}^{+} \Phi_{\mathcal{H}} \rangle &\leqslant \frac{\gamma}{V} |\lambda(f)|^{2}; \\ \langle \Phi_{\mathcal{H}}^{*} R_{f}^{+} R_{f} \Phi_{\mathcal{H}} \rangle &\leqslant \frac{\gamma}{V} |\lambda(f)|^{2}. \end{aligned} \tag{5.12}$$

Let us make several additional more general estimates. Let us consider the operators A_f, each of which is a linear combination of the operators a_f and a_{-f}^{+}:

$$A_{f} = p_{f} a_{f} + q_{f} a_{-f}^{+} \tag{5.13}$$

with bounded coefficients

$$|p_{f}|^{2} + |q_{f}|^{2} \leqslant \text{const}. \tag{5.14}$$

Let us show that

$$\left.\begin{aligned} \left| \langle \Phi_{\mathcal{H}}^{*} A_{f_1} \ldots A_{f_l} R_{f} A_{f_{l+1}} \ldots A_{f_m} R_{f}^{+} A_{f_{m+1}} \ldots \Phi_{\mathcal{H}} \rangle \right| &\leqslant \frac{\text{const}}{V}; \\ \left| \langle \Phi_{\mathcal{H}}^{*} A_{f_1} \ldots A_{f_l} R_{f}^{+} A_{f_{l+1}} \ldots A_{f_m} R_{f} A_{f_{m+1}} \ldots \Phi_{\mathcal{H}} \rangle \right| &\leqslant \frac{\text{const}}{V}. \end{aligned}\right\} \tag{5.15}$$

Proof. Note first of all that

$$La_f - a_f L = 0; \quad L^+ a_f^+ - a_f^+ L^+ = 0;$$

$$|La_f^+ - a_f^+ L| \leqslant \frac{2|\lambda(f)|}{V}; \quad |L^+ a_f - a_f L^+| \leqslant \frac{2|\lambda(f)|}{V}.$$

Therefore, for example,

$$\langle \Phi_{\mathcal{H}}^* A_{f_1} \ldots (L+\nu-C) A_{f_j} \ldots (L^+ + \nu - C) A_{f_i} \ldots \Phi_{\mathcal{H}} \rangle$$

$$= Z + \langle \Phi_{\mathcal{H}}^* (L+\nu-C) A_{f_1} \ldots A_{f_n} (L^+ + \nu - C) \Phi_{\mathcal{H}} \rangle,$$

where $|Z| \leq \text{const}/V$. Consequently,

$$|\langle \Phi_{\mathcal{H}}^* A_{f_1} \ldots (L+\nu-C) A_{f_j} \ldots (L^+ + \nu - C) A_{f_i} \ldots \Phi_{\mathcal{H}} \rangle| \leqslant \frac{\text{const}}{V}$$

$$+ |A_{f_1}| \ldots |A_{f_n}| \langle \Phi_{\mathcal{H}}^* (L+\nu-C)(L^+ + \nu - C) \Phi_{\mathcal{H}} \rangle \leqslant \frac{\text{const}}{V}. \tag{5.16}$$

Analogously, it is proved that

$$|\langle \Phi_{\mathcal{H}}^* A_{f_1} \ldots (L^+ + \nu - C) A_{f_j} \ldots (L+\nu-C) A_{f_i} \ldots \Phi_{\mathcal{H}} \rangle| \leqslant \frac{\text{const}}{V}. \tag{5.17}$$

Considering (5.16) and (5.17), we have

$$|\langle \Phi_{\mathcal{H}}^* A_{f_1} \ldots (L+\nu-C) A_{f_j} \ldots (L+\nu-C) A_{f_i} \ldots \Phi_{\mathcal{H}} \rangle|$$

$$\leqslant \frac{\text{const}}{V} + |\langle \Phi_{\mathcal{H}}^* (L+\nu-C) A_{f_1} \ldots A_{f_n} (L+\nu-C) \Phi_{\mathcal{H}} \rangle|$$

$$\leqslant \frac{\text{const}}{V} + \sqrt{\langle \Phi_{\mathcal{H}}^* \{(L+\nu-C) A_{f_1} \ldots A_{f_n} A_{f_n}^+ \ldots A_{f_1}^+ (L^+ + \nu - C)\} \Phi_{\mathcal{H}} \rangle}$$

$$\times \sqrt{\langle \Phi_{\mathcal{H}}^* (L^+ + \nu - C)(L+\nu-C) \Phi_{\mathcal{H}} \rangle} \leqslant \frac{\text{const}}{V} + |A_{f_1}| \ldots |A_{f_n}|$$

$$\times \sqrt{\langle \Phi_{\mathcal{H}}^* (L+\nu-C)(L^+ + \nu - C) \Phi_{\mathcal{H}} \rangle \, \Phi_{\mathcal{H}}^* (L^+ + \nu - C)(L+\nu-C) \Phi_{\mathcal{H}} \rangle} \leqslant \frac{\text{const}}{V}. \tag{5.18}$$

Analogously, it is proved that

$$\langle \Phi_{\mathcal{H}}^* A_{f_1} \ldots (L^+ + \nu - C) A_{f_j} \ldots (L^+ + \nu - C) A_{f_i} \ldots \Phi_{\mathcal{H}} \rangle| \leqslant \frac{\text{const}}{V}. \tag{5.19}$$

The validity of the proved inequalities (5.15) follows from Eqs. (5.16)-(5.19).

Let us now deal with the estimates for the correlation functions. Using Eq. (5.7), we obtain

$$i \frac{d}{dt} \langle \Phi_{\mathcal{H}}^* \alpha_f^+(t) \alpha_f \Phi_{\mathcal{H}} \rangle = -\sqrt{C^2 \lambda^2(f) + T^2(f)}$$

$$\times \langle \Phi_{\mathcal{H}}^* \alpha_f^+(t) \alpha_f \Phi_{\mathcal{H}} \rangle + \langle \Phi_{\mathcal{H}}^* R_f(t) \alpha_f \Phi_{\mathcal{H}} \rangle, \tag{5.20}$$

where, as always, $\alpha_f(0) = \alpha_f$, $\alpha_f^+(0) = \alpha_f^+$.

Recalling that from the equation

$$\mathrm{i}\frac{dJ(t)}{dt}=-\Omega J(t)+R(t)$$

it follows that

$$J(t)=J(0)\,\mathrm{e}^{\mathrm{i}\Omega t}+\mathrm{e}^{\mathrm{i}\Omega t}\int_0^t \mathrm{e}^{-\mathrm{i}\Omega t}R(t)\,dt,$$

we write

$$\langle \Phi^*_{\mathscr{H}}\alpha_f^+(t)\,\alpha_f\Phi_{\mathscr{H}}\rangle=\mathrm{e}^{\mathrm{i}\sqrt{C^2\lambda^2(f)+T^2(f)}\,t}\langle \Phi^*_{\mathscr{H}}\alpha_f^+\alpha_f\Phi_{\mathscr{H}}\rangle$$

$$+\mathrm{e}^{\mathrm{i}\sqrt{C^2\lambda^2(f)+T^2(f)}\,t}\int_0^t \mathrm{e}^{-\mathrm{i}\sqrt{C^2\lambda^2(f)+T^2(f)}\,t}\langle\Phi^*_{\mathscr{H}}R_f(t)\,\alpha_f\Phi_{\mathscr{H}}\rangle\,dt. \tag{5.21}$$

On the other hand, since $\Phi_{\mathscr{H}}$ is an eigenfunction of $\mathscr{H}$, corresponding to the least eigenvalue, its conventional spectral representation yields

$$\langle \Phi^*_{\mathscr{H}}\alpha_f^+(t)\,\alpha_f\Phi_{\mathscr{H}}\rangle=\int_0^\infty J_f(\nu)\,\mathrm{e}^{-\mathrm{i}\nu t}\,d\nu, \tag{5.22}$$

where

$$J_f\geqslant 0 \text{ and } \int_0^\infty J_f(\nu)\,d\nu\leqslant 1. \tag{5.23}$$

Let us place

$$h(t)=\int_0^2 \omega^2(2-\omega)^2\,\mathrm{e}^{-\mathrm{i}\omega t}\,d\omega. \tag{5.24}$$

As is evident, this function is regular on the entire real axis. Using integration by parts, it is not difficult to confirm the fact that for $|t|\to\infty$, $h(t)$ decreases according to the estimate:

$$|h(t)|\leqslant\frac{\text{const}}{|t|^3}. \tag{5.25}$$

Therefore, the integral

$$\int_{-\infty}^{\infty}|th(t)|\,dt \tag{5.26}$$

turns out to be finite. Let us set

$$\sqrt{C^2\lambda^2(f)+T^2(f)}=\Omega \tag{5.27}$$

and let us note that

$$h(\Omega t)=\frac{1}{\Omega^5}\int_0^{2\Omega}\nu^2(2\Omega-\nu)^2\,\mathrm{e}^{-\mathrm{i}\nu t}\,d\nu. \tag{5.28}$$

From such a method of construction it is evident that

$$\int_{-\infty}^{\infty}h(\Omega t)\,\mathrm{e}^{-\mathrm{i}\nu t}\,dt=0 \text{ for } \nu\geqslant 0, \tag{5.29}$$

and by virtue of (5.22)

$$\int_{-\infty}^{\infty} \langle \Phi^*_{\mathcal{H}} \alpha_f^+(t)\, \alpha_f \Phi_{\mathcal{H}} \rangle h(\Omega t)\, dt = 0. \tag{5.30}$$

Therefore, we have

$$\langle \Phi^*_{\mathcal{H}} \alpha_f^+ \alpha_f \Phi_{\mathcal{H}} \rangle \int_{-\infty}^{\infty} e^{i\Omega t} h(\Omega t)\, dt = -\int_{-\infty}^{\infty} h(\Omega t)\, e^{i\Omega t}$$

$$\times \left(\int_0^t e^{-i\Omega t'} \langle \Phi^*_{\mathcal{H}} R_f(t')\, \alpha_f \Phi_{\mathcal{H}} \rangle dt' \right) dt \tag{5.31}$$

from (5.21). But

$$\int_{-\infty}^{\infty} e^{i\Omega t} h(\Omega t)\, dt = \frac{2\pi}{\Omega}. \tag{5.32}$$

This means that

$$\langle \Phi^*_{\mathcal{H}} \alpha_f^+ \alpha_f \Phi_{\mathcal{H}} \rangle \leqslant \frac{\Omega}{2\pi} \int_{-\infty}^{\infty} |h(\Omega t)| \left\{ \int_0^t |\langle \Phi^*_{\mathcal{H}} R_f(t')\, \alpha_f \Phi_{\mathcal{H}} \rangle|\, dt' \right\} dt. \tag{5.33}$$

But in view of (5.12),

$$|\langle \Phi^*_{\mathcal{H}} R_f \alpha_f \Phi_{\mathcal{H}} \rangle| \leqslant \sqrt{|\langle \Phi^*_{\mathcal{H}} R_f R_f^+ \Phi_{\mathcal{H}} \rangle|\, |\langle \Phi^*_{\mathcal{H}} \alpha_f^+ \alpha_f \Phi_{\mathcal{H}} \rangle|}$$

$$\leqslant \left(\frac{\gamma}{V} \right)^{\frac{1}{2}} |\lambda(f)| \left(\langle \Phi^*_{\mathcal{H}} \alpha_f^+ \alpha_f \Phi_{\mathcal{H}} \rangle \right)^{\frac{1}{2}}. \tag{5.34}$$

Consequently,

$$\langle \Phi^*_{\mathcal{H}} \alpha_f^+ \alpha_f \Phi_{\mathcal{H}} \rangle \leqslant \frac{\Omega}{2\pi} \int_{-\infty}^{\infty} |h(\Omega t)|\, |t|\, dt \left(\frac{\gamma}{V} \right)^{\frac{1}{2}} \left(\langle \Phi^*_{\mathcal{H}} \alpha_f^+ \alpha_f \Phi_{\mathcal{H}} \rangle \right)^{\frac{1}{2}}$$

$$= \frac{1}{2\pi\Omega} \int_{-\infty}^{\infty} |h(\tau)\, \tau|\, d\tau \left(\frac{\gamma}{V} \right)^{\frac{1}{2}} \left(\langle \Phi^*_{\mathcal{H}} \alpha_f^+ \alpha_f \Phi_{\mathcal{H}} \rangle \right)^{\frac{1}{2}} |\lambda(f)|.$$

Thus,

$$\langle \Phi^*_{\mathcal{H}} \alpha_f^+ \alpha_f \Phi_{\mathcal{H}} \rangle \leqslant \frac{|\lambda(f)|^2}{2\pi (C^2 |\lambda(f)|^2 + T^2(f))} \cdot \frac{\gamma}{V} \left(\int_{-\infty}^{\infty} |h(\tau)\tau|\, d\tau \right)^2. \tag{5.35}$$

From this we obtain a number of estimates. In view of the Schwartz inequality, the fact that $|a_f^+ a_f| \le 1$, and using (5.35), we obtain

$$|\langle \Phi^*_{\mathcal{H}} \alpha_{f_1}^+ \ldots \alpha_{f_s}^+ \alpha_{g_e} \ldots \alpha_{g_1} \Phi_{\mathcal{H}} \rangle|$$

$$\leqslant \sqrt{\langle \Phi^*_{\mathcal{H}} \alpha_{f_1}^+ \ldots \alpha_{f_s}^+ \alpha_{f_s} \ldots \alpha_{f_1} \Phi_{\mathcal{H}} \rangle \langle \Phi^*_{\mathcal{H}} \alpha_{g_1}^+ \ldots \alpha_{g_i}^+ \alpha_{g_i} \ldots \alpha_{g_1} \Phi_{\mathcal{H}} \rangle}$$

$$\leqslant \sqrt{\langle \Phi^*_{\mathcal{H}} \alpha_{f_1}^+ \alpha_{f_1} \Phi_{\mathcal{H}} \rangle \langle \Phi^*_{\mathcal{H}} \alpha_{g_1}^+ \alpha_{g_1} \Phi_{\mathcal{H}} \rangle} \leqslant \frac{\text{const}}{V}. \tag{5.36}$$

We likewise have

$$\left|\langle \Phi^*_{\mathcal{H}} \alpha_{f_1} \ldots \alpha_{f_s} \Phi_{\mathcal{H}} \rangle\right|$$

$$\leqslant \sqrt{\langle \Phi^*_{\mathcal{H}} \alpha_{f1} \ldots \alpha_{fs-1} \alpha^+_{fs-1} \ldots \alpha^+_{f_1} \Phi_{\mathcal{H}} \rangle \langle \Phi^*_{\mathcal{H}} \alpha^+_{f_s} \alpha_{fs} \Phi_{\mathcal{H}} \rangle} \leqslant$$

$$\leqslant \sqrt{\langle \Phi^*_{\mathcal{H}} \alpha^+_{f_s} \alpha_{fs} \Phi_{\mathcal{H}} \rangle} \leqslant \frac{\text{const}}{\sqrt{V}} \tag{5.37}$$

and

$$\left|\langle \Phi^*_{\mathcal{H}} \alpha^+_{f_1} \ldots \alpha^+_{f_s} \Phi_{\mathcal{H}} \rangle\right| \leqslant \sqrt{\langle \Phi^*_{\mathcal{H}} \alpha^+_{f_1} \alpha_{f1} \Phi_{\mathcal{H}} \rangle} \leqslant \frac{\text{const}}{\sqrt{V}}. \tag{5.38}$$

Let us now compare the averages

$$\langle \Phi^*_{\mathcal{H}} \mathfrak{A}_{f_1} \ldots \mathfrak{A}_{f_s} \Phi_{\mathcal{H}} \rangle$$

(where $\mathfrak{A}_f = a_f$ and a_f^+) with the corresponding averages calculated on the basis of the Hamiltonian $\mathcal{H}_0$, in which $\nu + \sigma = C$. For convenience we shall denote averages of these types by $\langle \mathfrak{A}_{f1} \ldots \mathfrak{A}_{fs} \rangle_{\mathcal{H}}$ and $\langle \mathfrak{A}_{f_1} \ldots \mathfrak{A}_{f_s} \rangle_{\mathcal{H}_0}$, respectively. Let us estimate the magnitude of the difference

$$\langle \mathfrak{A}_{f_1} \ldots \mathfrak{A}_{f_s} \rangle_{\mathcal{H}} - \langle \mathfrak{A}_{f_1} \ldots \mathfrak{A}_{f_s} \rangle_{\mathcal{H}_0}. \tag{5.39}$$

Let us say a few words concerning the way in which $\langle \mathfrak{A}_{f_1} \ldots \mathfrak{A}_{f_s} \rangle_{\mathcal{H}_0}$ is calculated. We make use of the formulas $a_f^+ = u_f \alpha_f^+ - v_f \alpha_{-f}$, $a_f = u_f \alpha_f - v_f \alpha_{-f}^+$ and then reduce the product $\mathfrak{A}_{f1} \cdots \mathfrak{A}_{fs}$ to the sum of products of a normal type in which all α^+ precede α. Since all terms of the type

$$\langle \alpha^+ \ldots \alpha^+ \rangle_{\mathcal{H}_0},\ \langle \alpha \ldots \alpha \rangle_{\mathcal{H}_0},\ \langle \alpha^+ \ldots \alpha \rangle_{\mathcal{H}_0} \tag{5.40}$$

are equal to zero, we obtain the expression for calculating $\langle \mathfrak{A}_{f_1} \ldots \mathfrak{A}_{f_s} \rangle_{\mathcal{H}_0}$. Let us apply this same procedure to the calculation of $\langle \mathfrak{A}_{f_1} \ldots \mathfrak{A}_{f_s} \rangle_{\mathcal{H}}$. As is evident, the difference (5.39) is caused by terms which are proportional to

$$\langle \alpha^+ \ldots \alpha^+ \rangle_{\mathcal{H}},\ \langle \alpha \ldots \alpha \rangle_{\mathcal{H}},\ \langle \alpha^+ \ldots \alpha \rangle_{\mathcal{H}}, \tag{5.41}$$

and which, unlike (5.40), are in general not equal to zero. But for the quantities (5.41) the estimates (5.36)-(5.38) exist. Therefore, we have

$$\left|\langle \mathfrak{A}_{f_1} \ldots \mathfrak{A}_{f_s} \rangle_{\mathcal{H}} - \langle \mathfrak{A}_{f_1} \ldots \mathfrak{A}_{f_s} \rangle_{\mathcal{H}_0}\right| \leqslant \frac{\text{const}}{\sqrt{V}}. \tag{5.42}$$

Let us now consider the bitemporal correlation functions and show that the differences

$$\langle \mathcal{B}_{f_1}(t) \ldots \mathcal{B}_{f_l}(t) \mathfrak{A}_{f_m}(\tau) \ldots \mathfrak{A}_{f_n}(\tau) \rangle_{\mathcal{H}} - \langle \mathcal{B}_{f_1}(t) \ldots \mathcal{B}_{f_l}(t) \mathfrak{A}_{f_m}(\tau) \ldots \mathfrak{A}_{f_n}(\tau) \rangle_{\mathcal{H}_0}, \tag{5.43}$$

where $\mathfrak{A}_f$, $\mathcal{B}_f$ are equal to either a_f or a_f^+, will all be bounded in modulus by quantities of order $1/\sqrt{V}$. Note that on the one hand

$$\langle \alpha^+_{f_1}(t) \ldots \alpha_{f_j}(t) \mathfrak{A}_{f_m}(\tau) \ldots \mathfrak{A}_{f_n}(\tau) \rangle_{\mathcal{H}_0} = 0, \tag{5.44}$$

while on the other hand

$$\left|\langle \alpha^+_{f_1}(t) \ldots \alpha_{f_j}(t) \mathfrak{A}_{f_m}(\tau) \ldots \mathfrak{A}_{f_n}(\tau) \rangle_{\mathcal{H}}\right|$$

$$\leqslant \sqrt{\langle \alpha^+_{f_1}(t) \alpha_{f_1}(t) \rangle_{\mathcal{H}} \langle \omega^+ \omega \rangle_{\mathcal{H}}} = \sqrt{\langle \alpha^+_{f_1} \alpha_{f_1} \rangle_{\mathcal{H}} \langle \omega^+ \omega \rangle_{\mathcal{H}}}, \tag{5.45}$$

where $\omega = \alpha_{f_j}(t)\, \mathfrak{A}_{f_m}(\tau) \ldots \mathfrak{A}_{f_n}(\tau)$, so that

$$|\langle \alpha^+_{f_1}(t) \ldots \alpha_{f_j}(t)\, \mathfrak{A}_{f_m}(\tau) \ldots \mathfrak{A}_{f_n}(\tau) \rangle_{\mathscr{H}}| \leqslant \sqrt{\langle \alpha^+_{f_1} \alpha_{f_1} \rangle} \leqslant \frac{\text{const}}{\sqrt{V}}. \tag{5.46}$$

Therefore, we need only establish the fact that differences of the type

$$\langle \alpha_{f_1}(t) \ldots \alpha_{f_l}(t)\, \mathfrak{A}_{f_m}(\tau) \ldots \mathfrak{A}_{f_n}(\tau) \rangle_{\mathscr{H}} - \langle \alpha_{f_1}(t) \ldots \mathfrak{A}_{f_n}(\tau) \rangle_{\mathscr{H}_0}$$

are bounded in modulus by quantities of order $1/\sqrt{V}$. Let us set

$$\langle \alpha_{f_1}(t) \ldots \alpha_{f_l}(t), \mathfrak{A}_{f_m}(\tau) \ldots \mathfrak{A}_{f_n}(\tau) \rangle_{\mathscr{H}} = \Gamma(t-\tau). \tag{5.47}$$

By virtue of (5.8), we have

$$i\frac{d\Gamma(t-\tau)}{dt} - \{\Omega(f_1) + \ldots + \Omega(f_l)\}\Gamma(t-\tau) = \Delta(t-\tau), \tag{5.48}$$

where $\Omega(f) = \sqrt{C^2\lambda^2(f) + T^2(f)}$ and

$$\Delta(t-\tau) = \Delta_1(t-\tau) + \ldots \Delta_l(t-\tau);$$

$$\Delta_1(t-\tau) = -\langle R^+_{f_1}(t)\, \alpha_{f_2}(t) \ldots \alpha_{f_l}(t)\, \mathfrak{A}_{f_m}(\tau) \ldots \mathfrak{A}_{f_n}(\tau) \rangle_{\mathscr{H}};$$

$$\cdots$$

$$\Delta_l(t-\tau) = -\langle \alpha_{f_1}(t) \ldots \alpha_{f_{l-1}}(t)\, R^+_{f_l}(t)\, \mathfrak{A}_{f_m}(\tau) \ldots \mathfrak{A}_{f_n}(\tau) \rangle_{\mathscr{H}}.$$

But

$$\begin{aligned} &|\Delta_s(t-\tau)| \\ &\leqslant \sqrt{\langle \alpha_{f_1}(t) \ldots \alpha_{f_{s-1}}(t)\, R^+_{f_s}(t) \ldots \alpha_{f_l}(t)\, \alpha^+_{f_l}(t) \ldots R_{f_s}(t) \ldots \alpha^+_{f_1}(t) \rangle_{\mathscr{H}}} \\ &\times \sqrt{\langle \mathfrak{A}^+_{f_n}(\tau) \ldots \mathfrak{A}^+_{f_m}(\tau)\, \mathfrak{A}_{f_n}(\tau) \ldots \mathfrak{A}_{f_n}(\tau) \rangle_{\mathscr{H}}} \\ &= \sqrt{\langle \alpha_{f_1} \ldots \alpha_{f_{s-1}} R^+_{f_s} \ldots \alpha_{f_l} \alpha^+_{f_l} R_{f_s} \ldots \alpha^+_{f_1} \rangle_{\mathscr{H}} \langle \mathfrak{A}^+_{f_n} \ldots \mathfrak{A}_{f_n} \rangle_{\mathscr{H}}} \\ &\leqslant \sqrt{\langle \alpha_{f_1} \ldots \alpha_{f_{s-1}} R^+_{f_s} \ldots \alpha_{f_l} \alpha^+_{f_l} R_{f_s} \ldots \alpha^+_{f_1} \rangle_{\mathscr{H}}}, \end{aligned}$$

and therefore, taking account of (5.15), we have

$$|\Delta_s(t-\tau)| \leqslant \frac{\text{const}}{\sqrt{V}}. \tag{5.49}$$

Consequently,

$$|\Delta(t-\tau)| \leqslant \frac{s}{\sqrt{V}}, \text{ where } s = \text{const}. \tag{5.50}$$

But from (5.48), we have

$$\Gamma(t-\tau) = \Gamma(0)\, e^{-i\{\Omega(f_1)+\ldots+\Omega(f_l)\}(t-\tau)} + \exp\Big[\{-i[\Omega(f_1) + \ldots + \Omega(f_l)](t-\tau)\}$$

$$\times \Big\{ \int_0^{t-\tau} \exp^{i[\Omega(f_1)+\ldots+\Omega(f_l)]\omega}\, \Delta(\omega)\, d\omega \Big\}\Big], \tag{5.51}$$

whence by virtue of (5.50), we have

$$|\Gamma(t-\tau) - \Gamma(0)\, e^{-i\{\Omega(f_1)+\ldots+\Omega(f_l)\}(t-\tau)}| \leqslant \frac{s}{\sqrt{V}}|t-\tau|. \tag{5.52}$$

But, on the other hand,

$$\langle \alpha_{f_1}(t) \dots \alpha_{f_l}(t) \dots \mathfrak{A}_{f_m}(\tau) \dots \mathfrak{A}_{f_n}(\tau) \rangle_{\mathcal{H}_0}$$

$$= e^{-i\{\Omega(f_1)+\dots+\Omega(f_l)\}(t-\tau)} \langle \alpha_{f_1} \dots \alpha_{f_l} \mathfrak{A}_{f_m} \dots \mathfrak{A}_{f_n} \rangle_{\mathcal{H}_0}. \tag{5.53}$$

Thus, according to (5.52), (5.53), we have

$$D \equiv |\langle \alpha_{f_1}(t) \dots \alpha_{f_l}(t) \mathfrak{A}_{f_m}(\tau) \dots \mathfrak{A}_{f_n}(\tau) \rangle_{\mathcal{H}} - \langle \alpha_{f_1}(t) \dots \alpha_{f_l}(t) \mathfrak{A}_{f_m}(\tau) \dots$$

$$\dots \mathfrak{A}_{f_n}(\tau) \rangle_{\mathcal{H}_0}| \leqslant \frac{s}{\sqrt{V}} |t-\tau| + |\langle \alpha_{f_1} \dots \alpha_{f_l} \mathfrak{A}_{f_m} \dots \mathfrak{A}_{f_n} \rangle_{\mathcal{H}}$$

$$- \langle \alpha_{f_1} \dots \alpha_{f_l} \mathfrak{A}_{f_m} \dots \mathfrak{A}_{f_n} \rangle_{\mathcal{H}_0}|.$$

But the difference between the simultaneous averages appears in the second term of the right side, and as was shown previously (see Eq. (5.42)), it is majorized by a term of order $1/\sqrt{V}$. Thus, let us establish the fact that for the bitemporal averages

$$|\langle \mathcal{B}_{f_1}(t) \dots \mathcal{B}_{f_l}(t) \mathfrak{A}_{f_m}(\tau) \dots \mathfrak{A}_{f_n}(\tau) \rangle_{\mathcal{H}}$$

$$- \langle \mathcal{B}_{f_1}(t) \dots \mathcal{B}_{f_l}(t) \mathfrak{A}_{f_m}(\tau) \dots \mathfrak{A}_{f_n}(\tau) \rangle_{\mathcal{H}_0}|$$

$$\leqslant \frac{G_1}{\sqrt{V}} |t-\tau| + \frac{G_2}{\sqrt{V}}, \quad G_1, \ G_2 = \text{const}. \tag{5.54}$$

These bounds may also be generalized for the case of s-time correlation averages

$$\left. \begin{array}{l} \langle \mathfrak{P}_s(t_s) \mathfrak{P}_{s-1}(t_{s-1}) \dots \mathfrak{P}_1(t_1) \rangle; \\ \mathfrak{P}_j(t) = \mathfrak{A}_1^{(j)}(t) \dots \mathfrak{A}_l^{(j)}(t), \end{array} \right\} \tag{5.55}$$

where $\mathfrak{A}_s^{(j)}(t)$ is equal to $a_f(t)$ and $a_f^+(t)$. Let us show that

$$|\langle \mathfrak{P}_s(t_s) \dots \mathfrak{P}_1(t_1) \rangle_{\mathcal{H}} - \langle \mathfrak{P}_s(t_s) \dots \mathfrak{P}_1(t_1) \rangle_{\mathcal{H}_0}| \leqslant \frac{(K_s |t_s - t_{s-1}| + \dots + K_2 |t_2 - t_1| + Q_s)}{\sqrt{V}}, \tag{5.56}$$

where

$$K_j = \text{const}, \ Q_s = \text{const}. \tag{5.57}$$

The proof can easily be carried out by the method of induction. Let us assume that these relationships are true for (s−1)-time averages, and let us prove them for s-time averages. Reasoning as in the two-dimensional case, we see that it is sufficient to prove (5.56) for $\mathfrak{P}_s(t) = \alpha_{f1}(t) \dots \alpha_{fl}(t)$. But then

$$\langle \mathfrak{P}_s(t_s) \mathfrak{P}_{s-1}(t_{s-1}) \dots \mathfrak{P}_1(t_1) \rangle_{\mathcal{H}_0} = \exp\{-i(\Omega_{f_1} + \dots$$

$$+ \Omega_{f_l})(t_s - t_{s-1})\} \langle \mathfrak{P}_s(t_{s-1}) \mathfrak{P}_{s-1}(t_{s-1}) \dots \mathfrak{P}_1(t_1) \rangle_{\mathcal{H}_0}. \tag{5.58}$$

On the other hand, on the basis of (5.8) and (5.15) and the reasoning by means of which the inequality (5.52) was established we verify the fact that

$$|\langle \mathfrak{P}_s(t_s) \mathfrak{P}_{s-1}(t_{s-1}) \dots \mathfrak{P}_1(t_1) \rangle_{\mathcal{H}} - \exp\{-i(\Omega_{f_1} + \dots$$

$$+ \Omega_{f_l})(t_s - t_{s-1})\} \langle \mathfrak{P}_s(t_{s-1}) \mathfrak{P}_{s-1}(t_{s-1}) \dots \mathfrak{P}_1(t_1) \rangle_{\mathcal{H}}|$$

$$\leqslant \frac{K_1^{(s)} |t_s - t_{s-1}|}{\sqrt{V}}, \tag{5.59}$$

where $K_1^{(s)} = \text{const}$.

Thus,

$$|\langle \mathfrak{P}_s(t_s) \ldots \mathfrak{P}_1(t_1)\rangle_{\mathcal{H}} - \langle \mathfrak{P}_s(t_s) \ldots \mathfrak{P}_1(t_1)\rangle_{\mathcal{H}_0}|$$

$$\leqslant \frac{K_1^{(s)} |t_s - t_{s-1}|}{\sqrt{V}} + |\langle \mathfrak{P}_s(t_{s-1}) \mathfrak{P}_{s-1}(t_{s-1}) \ldots \mathfrak{P}_1(t_1)\rangle_{\mathcal{H}}$$

$$- \langle \mathfrak{P}_s(t_{s-1}) \mathfrak{P}_{s-1}(t_{s-1}) \ldots \mathfrak{P}_1(t_1)\rangle_{\mathcal{H}_0}|. \tag{5.60}$$

But the second term in the right side represents the difference between correlation averages with (s−1)-times for which the adopted estimates are established by assumption. For this reason they are also true for the s-time averages. Thus, $\mathcal{H}_0$ yields the asymptotic approximation for all correlation averages of the type $\langle \mathfrak{P}_s(t_s) \ldots \mathfrak{P}_1(t_1)\rangle$. Consequently, the same statement is also valid for the Green's functions constructed on the basis of the operators considered.

Comment. Note that in the estimate of the degree of approximation we would have been able to obtain const/V instead of the const/$\sqrt{V}$ obtained throughout if we had replaced C by C_1 in determining u_f, v_f, $\mathcal{H}_0$:

$$C_1 = \langle L + \nu\rangle_{\mathcal{H}} = \langle L^+ + \nu\rangle_{\mathcal{H}}. \tag{5.61}$$

Since it is obvious (see Eq. (4.58)) that

$$(C - C_1)^2 \leqslant \frac{\text{const}}{V},$$

it follows that all estimates of the type (5.12), (5.15), and (5.35) remain valid. Additional new useful relationships are added:

$$\left.\begin{aligned} |\langle A_{f_1} \ldots R_f \ldots A_{f_n}\rangle_{\mathcal{H}}| &\leqslant \frac{\text{const}}{V}; \\ |\langle A_{f_1} \ldots R_f^+ \ldots A_{f_n}\rangle_{\mathcal{H}}| &\leqslant \frac{\text{const}}{V}. \end{aligned}\right\} \tag{5.62}$$

For their proof it is sufficient to expand the expressions

$$\langle A_{f_1} \ldots R_f \ldots A_{f_n}\rangle_{\mathcal{H}}; \ \langle A_{f_1} \ldots R_f^+ \ldots A_{f_n}\rangle_{\mathcal{H}}, \tag{5.63}$$

having expressed all a and a^+ in terms of α and α^+. Then Eqs. (5.33) may be represented by a sum of terms of the type $\langle \alpha^+ \ldots \alpha\rangle_{\mathcal{H}}$:

$$\left.\begin{aligned} &\langle (L + \nu - C_1) \ldots \alpha\rangle_{\mathcal{H}}; \ \langle \alpha^+ \ldots (L + \nu - C_1)\rangle_{\mathcal{H}}; \\ &\langle (L^+ + \nu - C_1) \ldots \alpha\rangle_{\mathcal{H}}; \ \langle \alpha^+ \ldots (L^+ + \nu - C_1)\rangle_{\mathcal{H}}; \\ &\text{const}\,\langle L + \nu - C_1\rangle_{\mathcal{H}} \equiv 0; \ \text{const}\,\langle L^+ + \nu - C_1\rangle_{\mathcal{H}} \equiv 0 \end{aligned}\right\} \tag{5.64}$$

and commutation terms of order 1/V. (The vanishing of the latter two Eqs. (5.64) derives from (5.61).) Applying the inequality

$$|\langle AB\rangle| \leqslant \sqrt{|\langle AA^+\rangle|}\ \sqrt{|\langle B^+B\rangle|}$$

to (5.64) along with (5.35), we see that all quantities will be of order 1/V, and this proves (5.62).

Let us now make use of these additional relationships. Let us consider the expression $\langle \alpha_{f_1}^+ \ldots \alpha_{f_n}^+\rangle$ which evidently does not depend on t. Therefore,

$$\frac{d}{dt}\langle \alpha_{f_1}^+ \ldots \alpha_{f_n}^+\rangle_{\mathcal{H}} = \left\langle \frac{d\alpha_{f_1}^+}{dt} \ldots \alpha_{f_n}^+\right\rangle_{\mathcal{H}} + \ldots + \left\langle \alpha_{f_1}^+ \ldots \frac{d\alpha_{f_n}^+}{dt}\right\rangle_{\mathcal{H}} = 0. \tag{5.65}$$

Consequently, from (5.7), we obtain

$$(\Omega(f_1)+\ldots+\Omega(f_n))\,\langle \alpha^+_{f_1}\ldots\alpha^+_{f_n}\rangle_{\mathcal{H}}=\langle R_{f_1}\ldots\alpha^+_{f_n}\rangle_{\mathcal{H}}+\ldots+\langle \alpha^+_{f_1}\ldots R_{f_n}\rangle_{\mathcal{H}}. \tag{5.66}$$

But by virtue of (5.62)

$$\left|\langle R_{f_1}\ldots\alpha^+_{f_n}\rangle+\ldots+\langle\alpha^+_{f_1}\ldots R_{f_n}\rangle\right|\leqslant\frac{D}{V},\ D=\text{const}, \tag{5.67}$$

and therefore

$$\left|\langle\alpha^+_{f_1}\ldots\alpha^+_{f_n}\rangle\right|\leqslant\frac{D}{V(\Omega(f_1)+\ldots+\Omega(f_n))}, \tag{5.68}$$

whence, going over to conjugate quantities, we have

$$\left|\langle\alpha_{f_1}\ldots\alpha_{f_n}\rangle\right|\leqslant\frac{D}{V(\Omega(f_1)+\ldots+\Omega(f_n))}. \tag{5.69}$$

Having made use of the new inequalities (5.68) and (5.69) instead of the old ones (5.37) and (5.38), and having retained (5.36), it can be shown that the inequality

$$|\langle\mathfrak{A}_{f_1}\ldots\mathfrak{A}_{f_s}\rangle_{\mathcal{H}}-\langle\mathfrak{A}_{f_1}\ldots\mathfrak{A}_{f_s}\rangle_{\mathcal{H}_0}|\leqslant\frac{\text{const}}{V} \tag{5.70}$$

holds instead of the inequality (5.42). Analogous improvements of the estimates may be fulfilled for all correlation averages of the types considered earlier. We shall not present a general proof here. Let us restrict our analysis to estimating the difference

$$\langle\alpha_{f_1}(t)\ldots\alpha_{f_l}(t)\,\alpha^+_{g_1}(\tau)\ldots\alpha^+_{g_r}(\tau)\rangle_{\mathcal{H}}-\langle\alpha_{f_1}(t)\ldots\alpha^+_{g_r}(\tau)\rangle_{\mathcal{H}_0}. \tag{5.71}$$

We have

$$\Gamma_{\substack{\mathcal{H}\\ \mathcal{H}_0}}(t-\tau)=\langle\alpha_{f_1}(t)\ldots\alpha^+_{g_r}(\tau)\rangle_{\substack{\mathcal{H}\\ \mathcal{H}_0}}. \tag{5.72}$$

We have

$$i\frac{\partial\Gamma_{\mathcal{H}}(t-\tau)}{\partial t}=(\Omega(f_1)+\ldots+\Omega(f_l))\,\Gamma_{\mathcal{H}}(t-\tau)+\Delta_{\mathcal{H}}(t-\tau), \tag{5.73}$$

where

$$\Delta_{\mathcal{H}}=-\sum_j\langle\alpha_{f_1}(t)\ldots R^+_{f_j}(t)\ldots\alpha_{f_l}(t)\,\alpha^+_{g_1}(\tau)\ldots\alpha^+_{g_r}(\tau)\rangle_{\mathcal{H}}. \tag{5.74}$$

Differentiating (5.74) with respect to τ, we find

$$i\frac{\partial\Delta_{\mathcal{H}}(t-\tau)}{\partial\tau}\quad-(\Omega(g_1)+\ldots+\Omega(g_r))\,\Delta_{\mathcal{H}}(t-\tau)+\zeta(t-\tau), \tag{5.75}$$

where

$$\zeta(t-\tau)=-\sum_{j,s}\langle\alpha_{f_1}(t)\ldots R^+_{f_j}(t)\,\alpha^+_{g_1}(\tau)\ldots R_{f_s}(\tau)\ldots\alpha^+_{g_r}(\tau)\rangle_{\mathcal{H}}. \tag{5.76}$$

But in view of (5.15)

$$|\zeta(t-\tau)|\leqslant\frac{Q}{V},\ \text{where } Q=\text{const}. \tag{5.77}$$

Therefore, from (5.75), we obtain the following expression in conventional fashion (see, for example, (5.48)-(5.72)):

$$|\Delta_{\mathcal{H}}(t-\tau)-\Delta_{\mathcal{H}}(0)\exp\{i[\Omega(g_1)+\ldots+\Omega(g_r)](\tau-t)\}|\leqslant\frac{Q}{V}|t-\tau|. \tag{5.78}$$

But by virtue of (5.62) and (5.74), we have

$$|\Delta_{\mathcal{H}}(0)|\leqslant\frac{Q_1}{V}, \quad Q_1=\text{const}. \tag{5.79}$$

This means that

$$|\Delta_{\mathcal{H}}(t-\tau)|\leqslant\frac{Q_1+Q|t-\tau|}{V}. \tag{5.80}$$

Let us substitute this estimate into (5.73). We find that

$$|\Gamma_{\mathcal{H}}(t-\tau)-\Gamma_{\mathcal{H}}(0)\exp\{i[\Omega(f_1)+\ldots+\Omega(f_l)](\tau-t)\}|$$

$$\leqslant\frac{Q_1|t-\tau|+Q|t-\tau|^2\frac{1}{2}}{V}. \tag{5.81}$$

On the other hand,

$$\Gamma_{\mathcal{H}_0}(t-\tau)=\Gamma_{\mathcal{H}_0}(0)\exp\{i[\Omega(f_1)+\ldots+\Omega(f_l)](\tau-t)\}. \tag{5.82}$$

Consequently,

$$|\Gamma_{\mathcal{H}}(t-\tau)-\Gamma_{\mathcal{H}_0}(t-\tau)|\leqslant|\Gamma_{\mathcal{H}}(0)-\Gamma_{\mathcal{H}_0}(0)|+\frac{Q_1|t-\tau|+Q|t-\tau|^2\frac{1}{2}}{V}. \tag{5.83}$$

But from (5.70), we have

$$|\Gamma_{\mathcal{H}}(0)-\Gamma_{\mathcal{H}_0}(0)|=|\langle\alpha_{f_1}\ldots\alpha_{f_l}\alpha^+_{g_1}\ldots\alpha^+_{g_r}\rangle_{\mathcal{H}}$$

$$-\langle\alpha_{f_1}\ldots\alpha_{f_l}\alpha^+_{g_1}\ldots\alpha^+_{g_r}\rangle_{\mathcal{H}_0}|\leqslant\frac{Q_2}{V}, \quad Q_2=\text{const}. \tag{5.84}$$

Thus,

$$|\langle\alpha_{f_1}(t)\ldots\alpha_{f_l}(t)\alpha^+_{g_1}(\tau)\ldots\alpha^+_{g_r}(\tau)\rangle_{\mathcal{H}}-\langle\alpha_{f_1}(t)\ldots\alpha^+_{g_r}(\tau)\rangle_{\mathcal{H}_0}|$$

$$\leqslant\frac{Q_2+Q_1|t-\tau|+Q|t-\tau|^2\frac{1}{2}}{V}.$$

Reasoning further, it is not difficult to raise the order $1/\sqrt{V}$ to $1/V$ throughout in the previously obtained estimate according to this scheme.

§ 6. The Green's Function for the Case $\nu = 0$

In the previous section we obtained all of the necessary asymptotic estimates for the Green's function in the case $\nu > 0$. Since certain estimates [see Eq. (4.58)] which we used in § 5 are meaningless for $\nu = 0$, the results of that section cannot be carried over directly to the case $\nu = 0$. This case requires special consideration.

Since L and L^+ now do not take definite values in the lowest energy state $\Phi_{\mathcal{H}}$, with asymptotic accuracy, we shall, unlike the case considered previously, work with the amplitudes

$$\alpha_f = u_f a_f + v_f a^+_{-f} \frac{L}{C}, \tag{6.1}$$

where

$$\left. \begin{aligned} u_f &= \frac{1}{\sqrt{2}} \sqrt{1 + \frac{T(f)}{\sqrt{C^2 \lambda^2(f) + T^2(f)}}}; \\ v_f &= -\frac{\varepsilon(f)}{\sqrt{2}} \sqrt{1 - \frac{T(f)}{\sqrt{C^2 \lambda^2(f) + T^2(f)}}}. \end{aligned} \right\} \tag{6.2}$$

These amplitudes satisfy the commutation relations for Fermi amplitudes only in an asymptotic approximation rather than exactly. In order to obtain the estimates it will be necessary to establish a number of inequalities.

First of all we consider the expression $\Sigma \Omega(f) \alpha^+_f \alpha_f$ in which

$$\Omega(f) = \sqrt{C^2 \lambda^2(f) + T^2(f)}. \tag{6.2'}$$

With allowance for Eq. (6.1), we have

$$\begin{aligned} \sum \Omega(f) \alpha^+_f \alpha_f &= \sum \Omega(f) \left\{ u_f a^+_f + v_f \frac{L^+}{C} a_{-f} \right\} \left\{ u_f a_f + v_f a^+_{-f} \frac{L}{C} \right\} \\ &= \sum \Omega(f) \left\{ u^2_f a^+_f a_f + v^2_f \frac{L^+}{C} a_{-f} a^+_{-f} \frac{L}{C} + \right. \\ &\quad \left. + u_f v_f \frac{L^+}{C} a_{-f} a_f + u_f v_f a^+_f a^+_{-f} \frac{L}{C} \right\}. \end{aligned}$$

But

$$\begin{aligned} \sum \Omega(f) v^2_f \frac{L^+}{C} a_{-f} a^+_{-f} \frac{L}{C} &= -\sum \Omega(f) v^2_f \frac{L^+}{C} a^+_f a_f \frac{L}{C} \\ + \sum \Omega(f) v^2_f \frac{L^+ L}{C^2} &= -\sum \Omega(f) v^2_f a^+_f \frac{L^+ L}{C^2} a_f + \sum \Omega(f) v^2_f \frac{L^+ L}{C^2}. \end{aligned}$$

Further, since $u_f v_f = C\lambda(f)/2\Omega(f)$ it follows that

$$-\sum \Omega(f) \left\{ u_f v_f \frac{L^+}{C} a_{-f} a_f + u_f v_f a^+_f a^+_{-f} \frac{L}{C} \right\} = VL^+ L.$$

This means that

$$\begin{aligned} \sum \Omega(f) \alpha^+_f \alpha_f &= \sum_f \Omega \left\{ u^2_f a^+_f a_f - v^2_f a^+_f \frac{L^+ L}{C^2} a_f \right\} \\ + \sum \Omega(f) v^2_f \frac{L^+ L}{C^2} - VL^+ L &= \sum_f \Omega(f) (u^2_f - v^2_f) a^+_f a_f \\ - \sum_f \Omega(f) v^2_f a^+_f \frac{L^+ L - C^2}{C^2} a_f &+ \sum \Omega(f) v^2_f \frac{L^+ L}{C^2} - VL^+ L. \end{aligned}$$

But $\Omega(f)\ (u^2_f - v^2_f) = T(f)$, and therefore

$$\begin{aligned} H = \sum T(f) a^+_f a_f - V \frac{L^+ L}{2} &= \sum \Omega(f) \alpha^+_f \alpha_f + \frac{VL^+ L}{2} \\ - \sum \Omega(f) v^2_f \frac{L^+ L}{C^2} &+ \sum \Omega(f) v^2_f a^+_f \frac{L^+ L - C^2}{C^2} a_f, \end{aligned}$$

and, consequently,

$$\begin{aligned} H &= \sum \Omega(f) \alpha^+_f \alpha_f + \frac{V}{2} \left\{ C^2 - \frac{2}{V} \sum \Omega(f) v^2_f \right\} \\ &+ \frac{V(L^+ L - C^2)}{2} - \sum \Omega(f) v^2_f \frac{L^+ L - C^2}{C^2} + \sum \Omega(f) v^4_f \frac{L^+ L - C^2}{C^2} \end{aligned}$$

$$+\sum\Omega(f)v_f^2\left\{a_f^+\frac{L^+L-C^2}{C^2}a_f-v_f^2\frac{L^+L-C^2}{C^2}\right\}.$$

On the other hand, we have

$$C^2\frac{V}{2}-\sum\Omega(f)v_f^2+\sum\Omega(f)v_f^4=C^2\frac{V}{2}-\sum\Omega(f)u_f^2v_f^2$$

$$=\frac{V}{2}\left\{C^2-\frac{1}{2V}C^2\sum\frac{\lambda^2(f)}{\sqrt{C^2\lambda^2(f)+T^2(f)}}\right\}=\frac{V}{2}C^2\mathcal{F}'(C^2),$$

where $\mathcal{F}(C^2)=C^2-\frac{2}{V}\sum\Omega(f)v_f^2$. Therefore,

$$H=\sum\Omega(f)\alpha_f^+\alpha_f-w+\frac{V}{2}(L^+L-C^2)\mathcal{F}'(C^2)+\mathcal{F}(C^2). \tag{6.3}$$

Here

$$w=-\sum\Omega(f)v_f^2\left\{a_f^+\frac{L^+L-C^2}{C^2}a_f-v_f^2\frac{L^+L-C^2}{C^2}\right\}. \tag{6.4}$$

But, by definition, C^2 is the root of the equation (see Eq. (3.8)) $\mathcal{F}'(x)=0$. Furthermore,

$$\langle\Phi_H^*H\Phi_H\rangle\leqslant\mathcal{F}(C^2).$$

Consequently,

$$\langle\Phi_H^*\sum\Omega(f)\alpha_f^+\alpha_f\Phi_H\rangle\leqslant\langle\Phi_H^*w\Phi_H\rangle. \tag{6.5}$$

Let us now undertake to estimate the average value w. We resort to the definition of the amplitudes α (6.1). We have

$$\alpha_f^+=w_f a_f^+ + v_f\frac{L^+}{C}a_{-f};$$

$$\alpha_{-f}=-v_f a_f^+\frac{L}{C}+w_f a_{-f},$$

whence

$$u_f\alpha_f^+-v_f\frac{L^+}{C}\alpha_{-f}=u_f^2a_f^++v_f^2\frac{L^+}{C}a_f^+\frac{L}{C}=a_f^+\left(u_f^2+v_f^2\frac{L^+L}{C^2}\right).$$

Let us set

$$\left.\begin{aligned}\eta_f^+&=a_f^+v_f^2\frac{C^2-L^+L}{C^2}=v_f^2\frac{C^2-L^+L}{C^2}a_f^++\frac{2\lambda(f)L^+}{VC^2}a_{-f};\\ \eta_f&=v_f^2\frac{C^2-L^+L}{C^2}a_f=v_f^2a_f\frac{C^2-L^+L}{C^2}+\frac{2\lambda(f)}{VC^2}a_{-f}^+L.\end{aligned}\right\} \tag{6.6}$$

Then

$$\left.\begin{aligned}a_f^+&=u_f\alpha_f^+-v_f\frac{L^+}{C}\alpha_{-f}+\eta_f^+;\\ a_f&=u_f\alpha_f-v_f\alpha_{-f}^+\frac{L}{C}+\eta_f.\end{aligned}\right\} \tag{6.7}$$

Let us now turn to Eq. (6.4); we write:

$$w=w_1+w_2+w_3;$$

$$w_1 = \sum \Omega(f) v_f^2 u_f \alpha_f^+ \frac{C^2 - L^+ L}{C^2} a_f = \sum \Omega(f) v_f^2 u_f \alpha_f^+ a_f \frac{C^2 - L^+ L}{C^2}$$
$$+ \sum \Omega(f) v_f^2 u_f \alpha_f^+ \frac{2\lambda(f)}{VC^2} a_{-f}^+ L;$$

$$w_2 = \sum \Omega(f) v_f^2 \eta_f^+ \frac{C^2 - L^+ L}{C^2} a_f = \sum \Omega(f) v_f^2 \eta_f^+ a_f \frac{C^2 - L^+ L}{C^2}$$
$$+ \sum \Omega(f) v_f^2 \eta_f^+ \frac{2\lambda(f)}{VC^2} a_{-f}^+ L;$$

$$w_3 = \sum \Omega(f) v_f^2 \left\{ - v_f \frac{L^+}{C} \alpha_{-f} \frac{C^2 - L^+ L}{C^2} a_f - v_f^2 \frac{C^2 - L^+ L}{C^2} \right\}$$
$$= - \sum \Omega(f) v_f^3 \frac{L^+}{C} \left\{ \frac{L^+ L}{C^2} \alpha_{-f} - \alpha_{-f} \frac{L^+ L}{C^2} \right\} a_f$$
$$+ \sum \Omega(f) v_f^2 \left\{ - v_f \frac{L^+}{C} \left(\frac{C^2 - L^+ L}{C^2} \right) (\alpha_{-f} a_f + a_f \alpha_{-f}) \right.$$
$$\left. - v_f^2 \frac{C^2 - L^+ L}{C^2} \right\} + \sum \Omega(f) v_f^3 \frac{L^+}{C} \left(\frac{C^2 - L^+ L}{C^2} \right) a_f \alpha_{-f}.$$

Let us estimate the averages w_1, w_2, and w_3; for w_1 we use the previously proved inequality (4.47) which we write in the form

$$\left. \begin{aligned} &\left\langle \Phi_H^* \left(\frac{C^2 - L^+ L}{C^2} \right) \Phi_H \right\rangle \leqslant \frac{G}{V}, \text{ where } G = \text{const}; \\ &\left\langle \Phi_H^* \left(\frac{C^2 - L L^+}{C^2} \right) \Phi_H \right\rangle \leqslant \frac{G}{V}. \end{aligned} \right\} \tag{6.8}$$

We have

$$|\langle \Phi_H^* w_1 \Phi_H \rangle| \leqslant \sum \Omega(f) v_f^2 u_f \left| \left\langle \Phi_H^* \alpha_f^+ a_f \frac{C^2 - L^+ L}{C^2} \Phi_H \right\rangle \right|$$
$$+ \sum \Omega(f) v_f^2 u_f |\langle \Phi_H^* \alpha_f^+ a_{-f}^+ L \Phi_H \rangle| \frac{2|\lambda(f)|}{VC^2}$$
$$\leqslant \sum \Omega(f) v_f^2 u_f \sqrt{\langle \Phi_H^* \alpha_f^+ a_f a_f^+ \alpha_f \Phi_H \rangle}$$
$$\times \sqrt{\left\langle \Phi_H^* \left(\frac{C^2 - L^+ L}{C^2} \right)^2 \Phi_H \right\rangle} + \sum \Omega(f) v_f^2 u_f$$
$$\times \sqrt{\langle \Phi_H^* \alpha_f^+ \alpha_f \Phi_H \rangle \langle \Phi_H^* L^+ a_{-f} a_{-f}^+ L \Phi_H \rangle} \frac{2|\lambda(f)|}{VC^2}$$
$$\leqslant \sum \Omega(f) v_f^2 u_f \left(\frac{G}{V} \right)^{\frac{1}{2}} \sqrt{\langle \Phi_H^* \alpha_f^+ \alpha_f \Phi_H \rangle}$$

$$+ \frac{1}{V} \sum \Omega(f) v_f^2 u_f \frac{2|\lambda(f)|}{C^2} |L| \sqrt{\langle \Phi_H^* \alpha_f^+ \alpha_f \Phi_H \rangle}$$
$$\leqslant \sqrt{\langle \Phi_H^* \sum \Omega(f) \alpha_f^+ \alpha_f \Phi_H \rangle} \left\{ \sqrt{\frac{G}{V} \sum_f \Omega(f) v_f^4 u_f^2} \right.$$
$$\left. + \frac{2|L|}{C^2 \sqrt{V}} \sqrt{\frac{1}{V} \sum_f \Omega(f) v_f^4 u_f^2 |\lambda(f)|^2} \right\}.$$

Consequently,

$$|\langle \Phi_H^* w_1 \Phi_H \rangle| \leqslant R_1 \sqrt{\langle \Phi_H^* \sum_f \Omega(f) \alpha_f^+ \alpha_f \Phi_H \rangle}, \quad R_1 = \text{const}.$$

We find $|\langle \Phi_H^* w_2 \Phi_H \rangle| \leq R_2$, where $R_2 = \text{const}$, in a completely analogous manner. Let us go over to w_3. Note that

$$\alpha_{-f} a_f + a_f \alpha_{-f} = \left(-v_f a_f^+ \frac{L}{C} + u_f a_{-f}\right) a_f$$
$$+ a_f\left(-v_f a_f^+ \frac{L}{C} - u_f a_{-f}\right) = -\frac{v_f}{C}(a_f^+ L a_f + a_f a_f^+ L)$$
$$= -\frac{v_f}{C}(a_f^+ a_f + a_f a_f^+) L = -\frac{v_f}{C} L.$$

This means (see the expression for w_3) that

$$\Delta \equiv \sum_f \Omega(f) v_f^2 \left\{-v_f \frac{L^+}{C}\left(C^2 - \frac{L^+ L}{C^2}\right)(\alpha_{-f} a_f + a_f \alpha_{-f}) - v_f^2 \frac{C^2 - L^+ L}{C^2}\right\}$$
$$= \sum_f \Omega(f) v_f^2 \left\{v_f^2 \frac{L^+}{C}\left(\frac{C^2 - L^+ L}{C^2}\right)\frac{L}{C} - v_f^2 \frac{C^2 - L^+ L}{C^2}\right\} =$$
$$= \sum_f \Omega(f) v_f^4 \frac{L^+}{C^2}\left(\frac{LL^+ - L^+ L}{C^2}\right) L - \sum_f \Omega(f) v_f^4 \left(\frac{C^2 - L^+ L}{C^2}\right)^2,$$

and therefore (see Eq. II, (1.18))

$$\langle \Phi_H^* \Delta \Phi_H \rangle \leqslant \sum_f \Omega(f) v_f^4 \left\langle \Phi_H^* \frac{L^+}{C}\left(\frac{LL^+ - L^+ L}{C^2}\right)\frac{L}{C} \Phi_H \right\rangle$$
$$= \frac{2}{V^2 C^2} \sum_{f, f'} \Omega(f) v_f^4 \lambda^2(f') \left\langle \Phi_H^* \frac{L^+}{C}(1 - a_{f'}^+ a_{f'} - a_{f'}^+ a_{-f'}) \frac{L}{C} \Phi_H \right\rangle$$
$$\leqslant 2 \frac{|L|^2}{C^4} \cdot \frac{1}{V} \sum_f \Omega(f) v_f^4 \frac{1}{V} \sum_{f'} \lambda^2(f') \leqslant \text{const}.$$

We likewise find

$$\sum_f \Omega(f) |v_f|^3 \left\langle \Phi_H^* \frac{L^+}{C}\left(\frac{L^+ L}{C^2} \alpha_{-f} - \alpha_{-f} \frac{L^+ L}{C^2}\right) a_f \Phi_H \right\rangle \leqslant \text{const};$$

$$\sum_f \Omega(f) |v_f|^3 \left\langle \Phi_H^* \frac{L^+}{C}\left(\frac{C^2 - L^+ L}{C^2}\right) a_f \alpha_{-f} \Phi_H \right\rangle$$
$$\leqslant R_3 \sqrt{\sum_f \langle \Phi_H^* \alpha_f^+ \alpha_f \Phi_H \rangle \Omega(f)}.$$

Thus, combining the expressions for w_1, w_2, and w_3, we have

$$\langle \Phi_H^* w \Phi_H \rangle \leqslant \gamma_1 \sqrt{\left\langle \Phi_H^* \sum_f \Omega(f) \alpha_f^+ \alpha_f \Phi_H \right\rangle} + \gamma_2,$$

where γ_1 = const, γ_2 = const.

Substituting this inequality into (6.5), we obtain

$$\left\langle \Phi_H^* \sum_f \Omega(f) \alpha_f^+ \alpha_f \Phi_H \right\rangle \leqslant \gamma_1 \sqrt{\left\langle \Phi_H^* \sum_f \Omega(f) \alpha_f^+ \alpha_f \Phi_H \right\rangle} + \gamma_2.$$

Let us place $x = \sqrt{\left\langle \Phi_H^* \sum_f \Omega(f) \alpha_f^+ \alpha_f \Phi_H \right\rangle}$. Then $x^2 - \gamma_1 x \le \gamma_2$;

$$\left(x - \frac{\gamma_1}{2}\right)^2 \leqslant \gamma_2 + \frac{\gamma_1^2}{4} \text{ and } x < \frac{\gamma_1}{2} + \sqrt{\gamma_2 + \frac{\gamma_1^2}{4}}.$$

Thus,

$$\left\langle \Phi_H^* \frac{1}{V} \sum_f \Omega(f) \alpha_f^+ \alpha_f \Phi_H \right\rangle \leqslant \frac{R}{V}, \tag{6.9}$$

where

$$R=\left(\frac{\gamma_1}{2}+\sqrt{\gamma_2+\frac{\gamma_1^2}{4}}\right)^2=\text{const.}$$

One can now go over to a consideration of the equations of motion.

For $\nu = 0$, we have

$$\left.\begin{aligned} i\frac{da_f}{dt}&=T(f)a_f-\lambda(f)a_{-f}^{+}L;\\ i\frac{da_{-f}}{dt}&=T(f)a_{-f}+\lambda(f)a_f^{+}L\end{aligned}\right\} \tag{6.10}$$

from Eqs. (5.1) and (5.2). Therefore,

$$\begin{aligned} i\frac{dL}{dt}&=\frac{1}{V}\sum\lambda(f)\{T(f)a_{-f}+\lambda(f)a_f^{+}L\}a_f\\ &+\frac{1}{V}\sum\lambda(f)a_{-f}\{T(f)a_f-\lambda(f)a_{-f}^{+}L\}\\ &=\frac{2}{V}\sum\lambda(f)T(f)a_{-f}a_f+\frac{1}{V}\sum\lambda^2(f)(a_f^{+}a_f-a_{-f}a_{-f}^{+})L\\ &=\frac{2}{V}\sum\lambda(f)T(f)a_{-f}a_f+\frac{1}{V}\sum\lambda^2(f)(a_f^{+}a_f-a_fa_f^{+})L\\ &=\frac{2}{V}\sum\lambda(f)T(f)a_{-f}a_f+\frac{1}{V}\sum\lambda^2(f)(2a_f^{+}a_f-1)L.\end{aligned}$$

Note now that

$$\begin{aligned} &-2\lambda(f)T(f)u_fv_f\frac{L}{C}+\lambda^2(f)(2v_f^2-1)L\\ &=\lambda^2(f)\frac{T(f)}{\Omega(f)}L+\lambda^2(f)\left(1-\frac{T(f)}{\Omega(f)}-1\right)L=0,\end{aligned}$$

and consequently,

$$\left.\begin{aligned} &i\frac{dL}{dt}=D_1+D_2;\\ &D_1=\frac{2}{V}\sum\lambda(f)T(f)\left\{a_{-f}a_f+u_fv_f\frac{L}{C}\right\};\\ &D_2=\frac{2}{V}\sum\lambda^2(f)(a_f^{+}a_f-v_f^2)L.\end{aligned}\right\} \tag{6.11}$$

But with allowance for (6.7),

$$\begin{aligned} a_{-f}a_f+u_fv_f\frac{L}{C}&=\left(u_f\alpha_{-f}+v_f\alpha_f^{+}\frac{L}{C}\right)\left(u_f\alpha_f-v_f\alpha_{-f}^{+}\frac{L}{C}\right)\\ &+\eta_{-f}a_f+a_{-f}\eta_f-\eta_{-f}\eta_f+u_fv_f\frac{L}{C}=u_f^2\alpha_{-f}\alpha_f-\\ &-v_f^2\alpha_f^{+}\frac{L}{C}\alpha_{-f}^{+}\frac{L}{C}-u_fv_f\alpha_{-f}\alpha_{-f}^{+}\frac{L}{C}+u_fv_f\alpha_f^{+}\frac{L}{C}\alpha_f\\ &+\eta_{-f}a_f+a_{-f}\eta_f-\eta_{-f}\eta_f+u_fv_f\frac{L}{C}=u_f^2\alpha_{-f}\alpha_f-\\ &-u_fv_f(\alpha_{-f}\alpha_{-f}^{+}+\alpha_{-f}^{+}\alpha_{-f}-1)\frac{L}{C}-v_f^2\alpha_f^{+}\frac{L}{C}\alpha_{-f}^{+}\frac{L}{C}\\ &+u_fv_f\left(\alpha_f^{+}\frac{L}{C}\alpha_f+\alpha_{-f}^{+}\alpha_{-f}\frac{L}{C}\right)+\eta_{-f}a_f+a_{-f}\eta_f-\eta_{-f}\eta_f.\end{aligned} \tag{6.12}$$

Further, we have

$$\alpha_{-f}\alpha_{-f}^{+}+\alpha_{-f}^{+}\alpha_{-f}-1=\left(-v_fa_f^{+}\frac{L}{C}+u_fa_{-f}\right)$$

$$\times\left(-v_f\frac{L^+}{C}a_f+u_fa_{-f}^+\right)+\left(-v_f\frac{L^+}{C}a_f+u_fa_{-f}^+\right)$$

$$\times\left(-v_fa_f^+\frac{L}{C}+u_fa_{-f}\right)-1=v_f^2a_f^+\frac{LL^+}{C^2}a_f+u_f^2a_{-f}a_{-f}^+$$

$$-u_fv_fa_f^+\frac{L}{C}a_{-f}^+-u_fv_fa_{-f}\frac{L^+}{C}a_f+v_f^2\frac{L^+}{C}a_fa_f^+\frac{L}{C}$$

$$+u_f^2a_{-f}^+a_{-f}-u_fv_fa_{-f}^+a_f^+\frac{L}{C}-u_fv_f\frac{L^+}{C}a_fa_{-f}-1$$

$$=v_f^2a_f^+\frac{LL^+-L^+L}{C^2}a_f+v_f^2\frac{L^+L}{C^2}+u_f^2-1$$

$$-u_fv_fa_f^+\left(\frac{L}{C}a_{-f}^+-a_{-f}^+\frac{L}{C}\right)-u_fv_f\left(a_{-f}\frac{L^+}{C}-\frac{L^+}{C}a_{-f}\right)a_f,$$

and therefore,

$$\alpha_{-f}\alpha_{-f}^++\alpha_{-f}^+\alpha_{-f}-1=\frac{2}{V^2}\sum_{(g)}v_f^2a_f^+\frac{\lambda^2(g)}{C^2}(1-a_g^+a_g-a_{-g}^+a_{-g})a_f$$

$$+v_f^2\frac{L^+L-C^2}{C^2}+u_fv_fa_f^+a_f\frac{2\lambda(f)}{V}+u_fv_fa_f^+a_f\frac{2\lambda(f)}{V}. \tag{6.13}$$

Consequently, we have

$$\langle\Phi_H^*D_1D_1^+\Phi_H\rangle=\frac{2}{V}\sum_f\lambda(f)T(f)u_f^2\langle\Phi_H^*\alpha_{-f}\alpha_fD_1^+\Phi_H\rangle$$

$$-\frac{2}{V}\sum_f\lambda(f)T(f)\left\langle\Phi_H^*\alpha_f^+\left\{v_f^2\frac{L}{C}\alpha_{-f}^+\frac{L}{C}\right.\right.$$

$$\left.\left.-u_fv_f\left(\frac{L}{C}\alpha_f+\alpha_f\frac{L}{C}\right)\right\}D_1^+\Phi_H\right\rangle-\frac{4}{V^3}\sum_{fg}\lambda(f)T(f)u_fv_f^3\frac{\lambda^2(g)}{C^2}$$

$$\times\left\langle\Phi_H^*a_f^+(1-a_g^+a_g-a_{-g}^+a_{-g})a_f\frac{L}{C}D_1^+\Phi_H\right\rangle$$

$$+\frac{2}{V}\sum_f\lambda(f)T(f)\langle\Phi_H^*(\eta_{-f}a_f-\eta_fa_{-f}+[a_{-f}\eta_f+\eta_fa_{-f}]$$

$$-\eta_{-f}\eta_f)D_1^+\Phi_H\rangle-\frac{2}{V}\sum_f\lambda(f)T(f)\left\langle\Phi_H^*\left\{v_f^2\frac{L^+L-C^2}{C^2}\right.\right.$$

$$\left.\left.+2u_fv_fa_f^+a_f\frac{2\lambda(f)}{V}\right\}\frac{L}{C}D_1^+\Phi_H\right\rangle.$$

Taking into account the fact that

$$\langle\Phi_H^*\alpha_{-f}\alpha_fD_1^+\Phi_H\rangle=\langle\Phi_H^*\alpha_{-f}D_1^+\alpha_f\Phi_H\rangle$$
$$+\langle\Phi_H^*\alpha_{-f}(\alpha_fD_1^+-D_1^+\alpha_f)\Phi_H\rangle,$$

one can verify the fact that with allowance for (6.8) and (6.9),

$$\langle\Phi_H^*D_1D_1^+\Phi_H\rangle\leqslant\frac{\Gamma_1}{V},\quad\Gamma_1=\text{const.} \tag{6.14}$$

In the same way one may obtain

$$\langle\Phi_H^*D_1^+D_1\Phi_H\rangle\leqslant\frac{\Gamma_2}{V},\quad\Gamma_2=\text{const.} \tag{6.15}$$

Let us now go over to the expression D_2. We have

$$a_f^+a_f-v_f^2=a_f^+\eta_f+\eta_f^+a_f-\eta_f^+\eta_f+\left(u_f\alpha_f^+-v_f\frac{L^+}{C}\alpha_{-f}\right)$$

$$\times\left(u_f\alpha_f-v_f\alpha^+_{-f}\frac{L}{C}\right)-v_f^2=u_f^2\alpha_f^+\alpha_f+v_f^2\frac{L^+}{C}\alpha_{-f}\alpha^+_{-f}\frac{L}{C}$$

$$-v_f^2-u_fv_f\alpha_f^+\alpha^+_{-f}\frac{L}{C}-u_fv_f\frac{L^+}{C}\alpha_{-f}\alpha_f+a_f^+\eta_f+$$

$$+\eta_f^+a_f-\eta_f^+\eta_f=u_f^2\alpha_f^+\alpha_f+v_f^2\frac{L^+}{C}(\alpha_{-f}\alpha^+_{-f}+\alpha^+_{-f}\alpha_{-f}-1)-\frac{L}{C}$$

$$-v_f^2\frac{C^2-L^+L}{C^2}-v_f^2\frac{L^+}{C}\alpha^+_{-f}\alpha_{-f}\frac{L}{C}-u_fv_f\alpha_f^+\alpha^+_{-f}\frac{L}{C}$$

$$-u_fv_f\frac{L^+}{C}\alpha_{-f}\alpha_f+a_f^+\eta_f+\eta_f^+a_f-\eta_f^+\eta_f. \tag{6.16}$$

Starting from (6.16) and considering the inequalities (6.8) and (6.9), we establish the fact that

$$\left.\begin{aligned}&\langle\Phi_H^*D_2D_2^+\Phi_H\rangle\leqslant\frac{\Gamma_3}{V},\ \Gamma_3=\text{const};\\&\langle\Phi_H^*D_2^+D_2\Phi_H\rangle\leqslant\frac{\Gamma_3}{V}.\end{aligned}\right\} \tag{6.17}$$

From (7.10) we now have

$$\left.\begin{aligned}&\left\langle\Phi_H^*\left(\frac{dL}{dt}\right)^+\frac{dL}{dt}\Phi_H\right\rangle\leqslant\frac{\Gamma}{V},\ \Gamma=\text{const};\\&\left\langle\Phi_H^*\frac{dL}{dt}\left(\frac{dL}{dt}\right)^+\Phi_H\right\rangle\leqslant\frac{\Gamma}{V}.\end{aligned}\right\} \tag{6.18}$$

Let us go back again to the equations of motion (6.11). Considering (6.1), (6.2), we obtain

$$\begin{aligned}\mathrm{i}\frac{d\alpha_f^+}{dt}&=\mathrm{i}\frac{d}{dt}\left(u_fa_f^++v_f\frac{L^+}{C}a_{-f}\right)=u_f\mathrm{i}\frac{da_f^+}{dt}+v_f\frac{L^+}{C}\mathrm{i}\frac{da_{-f}}{dt}\\&+v_f\mathrm{i}\frac{dL^+}{dt}\cdot\frac{a_{-f}}{C}=u_f\{-T(f)a_f^++\lambda(f)L^+a_{-f}\}\\&+v_f\frac{L^+}{C}\{T(f)a_{-f}+\lambda(f)a_f^+L\}+v_f\mathrm{i}\frac{dL^+}{dt}\cdot\frac{a_{-f}}{C}\\&=-a_f^+\left\{T(f)u_f-\lambda(f)v_f\frac{L^+L}{C}\right\}+\frac{L^+}{C}\{u_f\lambda(f)C+T(f)v_f\}a_{-f}\\&+v_f\mathrm{i}\frac{dL^+}{dt}\cdot\frac{a_{-f}}{C}=-a_f^+\{T(f)u_f-\lambda(f)v_fC\}\\&+\frac{L^+}{C}\{u_f\lambda(f)C+T(f)v_f\}a_{-f}-a_f^+\lambda(f)v_f\frac{C^2-L^+L}{C}+v_f\mathrm{i}\frac{dL^+}{dt}\cdot\frac{a_{-f}}{C}.\end{aligned}$$

But (see Eq. (5.6))

$$\left.\begin{aligned}&u_f\lambda(f)C+T(f)v_f=-\Omega(f)v_f;\\&T(f)u_f-\lambda(f)v_fC=\Omega(f)u_f,\end{aligned}\right\} \tag{6.19}$$

and therefore,

$$\mathrm{i}\frac{d\alpha^+}{dt}+\Omega(f)\alpha_f^+=R_f, \tag{6.20}$$

where

$$R_f=-a_f^+\lambda(f)v_f\frac{C^2-L^+L}{C}+v_f(D_1^++D_2^+)\frac{a_{-f}}{C}.$$

Further, we have

$$\langle\Phi_H^* R_f^+ R_f \Phi_H\rangle \leq 2\left\langle\Phi_H^* \frac{C^2 - L^+ L}{C} a_f a_f^+ \frac{C^2 - L^+ L}{C} \Phi_H\right\rangle \lambda^2(f) v_f^2$$
$$+2\left\langle\Phi_H^* \frac{a_{-f}^+}{C}(D_1+D_2)(D_1^+ + D_2^+)\frac{a_{-f}}{C}\Phi_H\right\rangle v_f^2$$
$$\leq 2\lambda^2(f) v_f^2 \left\langle\Phi_H^* \frac{(C^2 - L^+ L)^2}{C^2}\Phi_H\right\rangle$$
$$+2v_f^2\left\langle\Phi_H^*\left\{\frac{a_{-f}^+}{C}(D_1+D_2)(D_1^+ + D_2^+)\frac{a_{-f}}{C}\right.\right.$$
$$\left.\left. -(D_1+D_2)\frac{a_{-f}^+ a_{-f}}{C^2}(D_1^+ + D_2^+)\right\}\Phi_H\right\rangle$$
$$+2\langle\Phi_H^*(D_1+D_2)(D_1^+ + D_2^+)\Phi_H\rangle\frac{v_f^2}{C^2}$$

and likewise

$$\langle\Phi_H^* R_f R_f^+ \Phi_H\rangle \leq 2\left\langle\Phi_H^* a_f^+\left(\frac{C^2 - L^+ L}{C}\right)^2 a_f \Phi_H\right\rangle \lambda^2(f) v_f^2$$
$$+\frac{2v_f^2}{C}\langle\Phi_H^*(D_1^+ + D_2^+)(D_1+D_2)\Phi_{\mathscr{H}}\rangle$$
$$=2\lambda^2 v_f^2\left\langle\Phi_H^*\left\{a_f^+\left(\frac{C^2 - L^+ L}{C}\right)\left(\frac{C^2 - L^+ L}{C}\right)a_f\right.\right.$$
$$\left.\left.-\left(\frac{C^2 - L^+ L}{C}\right)a_f^+ a_f\left(\frac{C^2 - L^+ L}{C}\right)\right\}\Phi_H\right\rangle$$
$$+2\lambda^2 v_f^2\left\langle\Phi_H^*\left(\frac{C^2 - L^+ L}{C}\right)^2\Phi_H\right\rangle$$
$$+2\frac{v_f^2}{C^2}\langle\Phi_H^*(D_1^+ + D_2^+)(D_1+D_2)\Phi_H\rangle.$$

From this it follows that

$$\left.\begin{aligned}&\langle\Phi_H^* R_f R_f^+ \Phi_H\rangle \leq v_f^2\frac{S}{V}, \text{ where } S=\text{const};\\ &\langle\Phi_H^* R_f^+ R_f \Phi_H\rangle \leq v_f^2\frac{S}{V}.\end{aligned}\right\} \tag{6.21}$$

Having Eqs. (6.20) and the inequalities (6.21), one may repeat word-for-word our reasoning from the previous section where we considered the case $\nu > 0$. We now obtain (see Eq. (5.35))

$$\langle\Phi_H^* a_f^+ a_f \Phi_H\rangle \leq \frac{S}{V}\cdot\frac{v_f^2}{2\pi\Omega^2(f)}\left(\int_{-\infty}^{+\infty}|h(\tau)\tau|\,d\tau\right)^2. \tag{6.21'}$$

In comparison with the inequality (6.9) we have substantial progress here.

The inequality (6.9) shows that $<\Phi_H^* a_f^+ a_f \Phi_H>$ is a quantity of order 1/V on the average for f. However, the inequality (6.21') shows that this expression will be of order 1/U for each f.

From (6.21') we may immediately obtain the estimates for the averages which apply to one time. Assume $\mathfrak{A}_f$ is equal to a_f or a_f^+. Let us consider those operators $\mathfrak{A}_{f_1}\mathfrak{A}_{f_2}\cdots\mathfrak{A}_{f_k}$ which conserve the number of particles. Let us show that

$$\left|\langle\mathfrak{A}_{f_1}\mathfrak{A}_{f_2}\ldots\mathfrak{A}_{f_k}\rangle_H - \langle\mathfrak{A}_{f_1}\mathfrak{A}_{f_2}\ldots\mathfrak{A}_{f_k}\rangle_{H_0}\right| \leq \frac{\text{const}}{\sqrt{V}}. \tag{6.22}$$

Now note that Φ_H and Φ_{H_0} satisfy the conditions (2.3)

$$(a_f^+ a_f - a_{-f}^+ a_{-f})\Phi = 0.$$

Therefore,

$$\langle \mathfrak{A}_{f_1} \mathfrak{A}_{f_2} \mathfrak{A}_{f_3} \ldots \mathfrak{A}_{f_k} \rangle$$

may be reduced to the sum of terms of the type

$$\langle \ldots a_f^+ a_f \ldots a_g^+ a_{-g}^+ \ldots a_{-h} a_h \ldots \rangle,$$

where $\pm f$, $\pm g$, $\pm h$ are all different. Of course, the number of subscripts g is equal to the number of subscripts h here. It is likewise obvious that

$$\langle \ldots a_f^+ a_f \ldots a_g^+ a_{-g}^+ \ldots a_{-h} a_h \ldots \rangle_{H_0} = \prod_f v_f^2 \prod_g (-u_g v_g) \prod_h (-u_h v_h).$$

Consequently, we need merely establish the fact that

$$\left| \langle \ldots a_f^+ a_f \ldots a_g^+ a_{-g}^+ \ldots a_{-h} a_h \ldots \rangle_H \right| - \prod_f v_f^2 \prod_g (-u_g v_g) \prod_h (-u_h v_h) \leqslant \frac{\text{const}}{\sqrt{V}}. \tag{6.23}$$

On the basis of what was said previously (see Eqs. (6.12), (6.13), and (6.16)) we note that that

$$a_{-h} a_h + u_h v_h \frac{L}{C} = u_h^2 \alpha_{-h} \alpha_h - u_h v_h \Bigg\{ \frac{2}{V^2} \sum_{(f)} v_h^2 \alpha_h^+ \frac{\lambda^2(f)}{C^2}$$

$$\times (1 - a_f^+ a_f - a_{-f}^+ a_{-f}) \, a_h + v_h^2 \frac{L^+ L - C^2}{C^2}$$

$$+ u_h v_h a_h^+ a_h \frac{4\lambda(h)}{V} \Bigg\} \frac{L}{C} - v_h^2 \alpha_h^+ \frac{L}{C} \alpha_{-h}^+ \frac{L}{C} + u_h v_h$$

$$\times \left(\alpha_h^+ \frac{L}{C} \alpha_h + \alpha_{-h}^+ \alpha_{-h} \frac{L}{C} \right) + \eta_{-h} a_h + a_{-h} \eta_{-h} - \eta_{-h} \eta_h; \tag{6.24}$$

$$a_f^+ a_f - v_f^2 = u_f^2 \alpha_f^+ \alpha_f - v_f^2 \frac{L^+}{C} \alpha_{-f}^+ \alpha_{-f} \frac{L}{C} - u_f v_f \alpha_f^+ \alpha_{-f}^+ \frac{L}{C}$$

$$- u_f v_f \frac{L^+}{C} \alpha_{-f} \alpha_f + v_f^2 \frac{L^+}{C} \Bigg\{ \frac{2}{V^2} \sum_g v_f^2 a_f^+ \frac{\lambda^2(g)}{C^2}$$

$$\times (1 - a_g^+ a_g - a_{-g}^+ a_{-g}) \, a_f + v_f^2 \frac{L^+ L - C^2}{C^2}$$

$$+ u_f v_f \frac{4\lambda}{V} a_f^+ a_f \Bigg\} + a_f^+ \eta_f + \eta_f^+ a_f - \eta_f^+ \eta_f \tag{6.25}$$

We shall now shift α^+ toward the left parenthesis, α to the right parenthesis, and L^+L-C^2/C^2 (for example, those included in η, η^+) to one of them, no matter to which. Since the subscripts $\pm f$, $\pm g$, $\pm L$ are all different, it follows that the commutators which develop in the process of these commutations will be quantities of order 1/V. Considering the constantly used inequality $|\langle AB \rangle| \leq \sqrt{\langle AA^+ \rangle \langle B^+B \rangle}$, we see that as soon as α^+ "touches" the left parenthesis or α touches the right parenthesis, or L^+L-C^2 arrives at either of them, we obtain a quantity having an order of smallness of at least const/$\sqrt{V}$ at that instant. Consequently,

$$|\langle \dots a_f^+ a_f \dots a_g^+ a_{-g}^+ \dots a_{-h} a_h \dots \rangle_H$$

$$\Pi_f v_f^2 \left\langle \dots (-u_g v_g) \frac{L^+}{C} (-u_h v_h) \frac{L}{C} \dots \right\rangle_H \Bigg| \leqslant \frac{\text{const}}{\sqrt{V}} . \tag{6.26}$$

But the number g is equal to the number h, and the commutation of L with L^+ is carried out with an accuracy up to terms of 1/V. Therefore, $\left\langle \dots u_g v_g \frac{L^+}{C} \dots u_h v_h \frac{L}{C} \dots \right\rangle_H$ differs from $\prod_g u_g v_g \prod_h u_h v_h \left\langle \left(\frac{L^+ L}{C^2} \right)^l \right\rangle_H$ by terms of order 1/V. On the other hand,

$$\left\langle \left(\frac{L^+ L}{C^2} \right)^l \right\rangle_H$$

differs from unity by a quantity at least of order $1/\sqrt{V}$.

Thus, the validity of the inequalities (6.23) (and consequently (6.22) as well) has been proved. Let us now go over to the bitemporal correlation averages and show that in general

$$|\langle \mathfrak{B}_{f_1}(t) \dots \mathfrak{B}_{f_l}(t);\ \mathfrak{A}_{g_1}(\tau) \dots \mathfrak{A}_{g_k}(\tau) \rangle_H$$

$$-\langle \mathfrak{B}_{f_1}(t) \dots \mathfrak{B}_{f_l}(t);\ \mathfrak{A}_{g_1}(\tau) \dots \mathfrak{A}_{g_k}(\tau) \rangle_{H_0}| \leqslant \frac{K(t-\tau)+K_1}{\sqrt{V}}, \tag{6.27}$$

$$K = \text{const},\ K_1 = \text{const}.$$

Here $\mathfrak{P}_f$, $\mathfrak{A}_g$ are equal to a or a^+. We assume, as always in such a case, that the operator $\mathfrak{P}_{f_1} \dots \mathfrak{A}_{g_k}$ conserves the number of particles.

By virtue of the additional conditions mentioned earlier, which are satisfied by Φ_H and Φ_{H_0}, the investigated averages may be reduced to a sum of terms of the type

$$\langle \dots a_f^+(t) a_f(t) \dots a_g^+(t) a_{-g}^+(t) \dots a_{-h}(t) a_h(t) \dots$$

$$\dots a_k^+(t) \dots a_g(t) \dots a_{f'}^+(\tau) a_{f'}(\tau) \dots a_{g'}^+(\tau) a_{-g'}^+(\tau) \dots$$

$$\dots a_{-h'}(\tau) a_{h'}(\tau) \dots a_{k'}(\tau) \dots a_{q'}^+(\tau) \dots \rangle, \tag{6.28}$$

by establishing the "proper order" of the operators; here the number of operators a and a^+ is also identical, where the subscripts $\pm f$, $\pm$g, $\pm$h, $\pm$k, $\pm$q and $\pm f'$, $\pm$g', $\pm$h', $\pm$k', $\pm$q' are all different.

In view of the indicated possibility of reduction it is sufficient for us to prove the inequalities (6.27) for averages of the type (6.28). For the "pairs" a^+a, a^+a^+, aa we make use of Eqs. (6.24) and (6.25), while for "singles" a, a^+ we make use of Eqs. (6.7). We shall now shift $\alpha^+(f)$ and $L^+(t)L(t)-C^2$ to the left, while $\alpha(\tau)$ and $L^+(\tau)L(\tau)-C^2$ are shifted to the right. In view of the difference emphasized above between the subscripts, the commutators which appear (all commutations are carried out only between amplitudes which apply to the same time) will yield quantities of order 1/V. Note that as soon as $\alpha^+(t)$ or $L^+(t)L(t)-C^2$ "touch" the right parenthesis, we immediately obtain quantities at least of order $1/\sqrt{V}$. Consequently, it remains merely for us to show that an inequality of the type (6.27) holds for averages of the form

$$\Gamma(t-\tau) = \langle \alpha_{f_1}(t) \dots \alpha_{f_l}(t) L^k(t)\ L^{+q}(t) L^{+q_1}(\tau) L^{k_1}(\tau) \alpha_{g_1}^+(\tau) \dots \alpha_{g_r}^+(\tau) \rangle. \tag{6.29}$$

Let us now make use of the equations of motion (6.19) and the relationships (6.11), (6.18), (6.20), and (6.21) which yield the required estimates. We find

$$\mathrm{i}\,\frac{\partial \Gamma_H(t-\tau)}{\partial t} - \{\Omega(f_1) + \dots + \Omega(f_l)\}\,\Gamma_H(t-\tau) = \Delta(t-\tau),$$

for the condition that $|\Delta(t-\tau)| \leqslant \frac{G}{\sqrt{V}}$, where G = const. From this, since

$$\Gamma_H(t-\tau) = e^{-i\{\Omega(f_1)+\ldots+\Omega(f_l)\}(t-\tau)}\Gamma_H(0)$$

$$+e^{-i\{\Omega(f_1)+\ldots+\Omega(f_l)\}(t-\tau)}\int_0^{t-\tau} e^{-i\{\Omega(f_1)+\Omega(f_2)+\ldots+\Omega(f_l)\}z}\Delta(z)\,dz,$$

we obtain

$$\left|\Gamma_H(t-\tau) - e^{-i\{\Omega(f_1)+\ldots+\Omega(f_l)\}(t-\tau)}\Gamma_H(0)\right| \leqslant \frac{G|t-\tau|}{\sqrt{V}}. \tag{6.30}$$

On the other hand,

$$\Gamma_{H_0}(t-\tau) = e^{i\{\Omega(f_1)+\ldots+\Omega(f_l)\}(t-\tau)}\Gamma_{H_0}(0), \tag{6.31}$$

since

$$\Gamma_{H_0}(t-\tau) = \left\langle \alpha_{f_1}(t)\ldots\alpha_{f_l}(t)\alpha^+_{g_1}(\tau)\ldots\alpha^+_{g_r}(\tau)\right\rangle C^{k+q+q_1+k_1}. \tag{6.32}$$

Thus,

$$|\Gamma_H(t-\tau) - \Gamma_{H_0}(t-\tau)| \leqslant |\Gamma_H(0) - \Gamma_{H_0}(0)|$$

$$+\frac{G|t-\tau|}{\sqrt{V}} = \left|\left\langle \alpha_{f_1}\ldots\alpha_{f_l} L^k (L^+)^{q+q_1} L^{k_1} \alpha^+_{g_1}\ldots\alpha^+_{g_r}\right\rangle_H\right.$$

$$\left. - C^{k+k_1+q+q_1}\left\langle \alpha_{f_1}\ldots\alpha_{f_l}\alpha^+_{g_1}\ldots\alpha^+_{g_r}\right\rangle_{H_0}\right| + \frac{G(t-\tau)}{\sqrt{V}}. \tag{6.33}$$

Assume that among the subscripts $f_1, \ldots, f_l$ there is a pair of identical ones. Then, nothing that

$$\alpha_f^2 = \left(u_f a_f + v_f a^+_{-f}\frac{L}{C}\right)\left(u_f a_f + v_f a^+_{-f}\frac{L}{C}\right)$$

$$= v_f^2 a^+_{-f}\frac{L}{C}a^+_{-f}\frac{L}{C} + u_f v_f\left\{a^+_{-f}\frac{L}{C}a_f + a_f a^+_{-f}\frac{L}{C}\right\}$$

$$= \frac{v_f^2 a^+_{-f}}{C}(La^+_{-f} - a^+_{-f}L)L$$

$$- u_f v_f\{a^+_{-f}a_f + a_f a^+_{-f}\}\frac{L}{C} = -2\frac{v_f^2\lambda(f)}{C^2V}a^+_{-f}a_f L \tag{6.34}$$

will be of order $1/V$, we see that $\langle\ldots\rangle_H$ will also be of the same order. The corresponding $\langle\ldots\rangle_{H_0}$ are simply equal to zero. The same situation naturally develops in the case in which among the subscripts $g_1, \ldots, g_r$ there is just one pair of identical ones.

Assume further that among the subscripts $f_1, \ldots, f_l$ there is at least one subscript f_j which is not included among $g_1, \ldots, g_r$. Then we may shift α_{f_j} to the right parenthesis in $\langle\ldots\rangle_H$, obtaining (along the way) commutators of order $1/V$; thus we verify the results that $\langle\ldots\rangle_H$ in the case given turns out to be a quantity having an order of smallness no lower than $1/\sqrt{V}$. The average $\langle\ldots\rangle_{H_0}$, however, is exactly equal to zero. An analogous situation arises if among $g_1, \ldots, g_r$ there is just one subscript which is not included in $f_1, \ldots, f_l$.

Thus, it remains for us to consider the case in which 1) all $f_1, \ldots, f_l$ are different; 2) the ensemble $g_1, \ldots, g_r$ is the same ensemble $f_1, \ldots, f_l$, but, perhaps, is numbered in a different order.

Now note that in the right side of (6.33) one can establish the "proper order" and replace $\alpha_{g_1}^+ \ldots \alpha_{g_r}^+$ by $\alpha_{f_l}^+ \ldots \alpha_{f_1}^+$. It is natural that in $\langle \ldots \rangle_{H_0}$ we carry out such a substitution exactly, while in $\langle \ldots \rangle_H$ we carry it out with an error in the adopted order which is asymptotically small. Since the operators within $\langle \ldots \rangle$ conserve the number of particles, $k + k_1$ must equal $q + q_1$.

Further, in $\langle \alpha_{f_1} \ldots \alpha_{f_l} L^k (L^+)^{k+k_1} L^{k_1} \alpha_{f_l}^+ \ldots \alpha_{f_1}^+ \rangle$ we carry out the substitution $L^k (L^+)^{k+k_1} L^{k_1} \to (L^+ L)^{k+k_1}$ and transfer it to the right parenthesis. Under these conditions we produce an error of order $1/V$. Note further that

$$\left| \langle \alpha_{f_1} \ldots \alpha_{f_l} \alpha_{f_l}^+ \ldots \alpha_{f_1}^+ (L^+ L)^{k+k_1} \rangle_H - \langle \alpha_{f_1} \ldots \alpha_{f_l} \alpha_{f_l}^+ \ldots \alpha_{f_1}^+ \rangle_{H_0} C^{2(k+k_1)} \right| \leqslant \frac{\text{const}}{\sqrt{V}}. \tag{6.35}$$

Thus, from (6.33), we obtain

$$\left| \Gamma_H (t-\tau) - \Gamma_{H_0} (t-\tau) \leqslant \frac{G |t-\tau|}{\sqrt{V}} + \frac{k}{\sqrt{V}} \right.$$

$$\left. + C^{2(k+k_1)} \left| \left\langle \alpha_{f_1} \ldots \alpha_{f_l} \ldots \alpha_{f_l}^+ \ldots \alpha_{f_1}^+ \right\rangle_H - \left\langle \alpha_{f_1} \ldots \alpha_{f_l} \alpha_{f_l}^+ \ldots \alpha_{f_1}^+ \right\rangle_{H_0} \right| . \right. \tag{6.36}$$

But since all f are different,

$$\left\langle \alpha_{f_1} \ldots \alpha_{f_l} \alpha_{f_l}^+ \ldots \alpha_{f_1}^+ \right\rangle_{H_0} = \left\langle \alpha_{f_1} \alpha_{f_1}^+ \right\rangle_{H_0} \left\langle \alpha_{f_2} \alpha_{f_2}^+ \right\rangle_{H_0} \ldots \left\langle \alpha_{f_l} \ldots \alpha_{f_l}^+ \right\rangle_{H_0} = 1.$$

In $\langle \ldots \rangle_H$ such a distribution may likewise be achieved, but, of course, not exactly but with the allowed asymptotic error.

Thus, our proof has been completed. Just as in the case $\nu > 0$, we could have obtained analogous estimates of the degree of asymptotic approximation for multitemporal correlation functions also. We shall not dwell on this here. The reader may now carry out all of the calculations involved in this himself using the schemes developed above. As in the case of $\nu > 0$, the order of smallness in the case considered may be raised from $\text{const}/\sqrt{V}$ to const/V if in the Hamiltonian H_0 the constant C is replaced by $C_1 = \sqrt{\langle L^+ L \rangle_H}$, which differs from C by a quantity of order $1/\sqrt{V}$.

We shall not prove this remark here.

APPENDIX I

In the present section we present the proofs of certain relationships used in this paper.* All of the operators considered here are assumed to be totally continuous, and we deal only with this kind of operator in the main text.

Lemma I. Assume that the operator ξ satisfies the condition

$$|\xi \xi^+ - \xi^+ \xi| \ll \frac{2s}{V}, \tag{A1.1}$$

where s is a number; $\varepsilon = 1$ or $\varepsilon = -1$. Then the inequality

$$2 \sqrt{\xi^+ \xi + \frac{s}{V}} - \varepsilon (\xi + \xi^+) \geqslant 0 \tag{A1.2}$$

holds.

*Let us arbitrarily denote the norm of the function by $\| \Phi \| = \sqrt{\langle \Phi^* \Phi \rangle}$ and the norm of the operator by $|\mathfrak{A}| = \sup \| \mathfrak{A} \Phi \|$, where $\| \Phi \| = 1$.

Proof. Let us assume the opposite; then one can find a normalized function φ which is such that

$$\left\{2\sqrt{\xi^+\xi+\frac{s}{V}}-\varepsilon(\xi+\xi^+)\right\}\varphi=-\rho\varphi,$$

where $\rho > 0$. From this we have

$$\left(2\sqrt{\xi^+\xi+\frac{s}{V}}+\rho\right)\varphi=\varepsilon(\xi+\xi^+)\varphi. \tag{A1.3}$$

Now let us take into account the fact that $A\varphi = B\varphi$ and A and B are self-conjugate operators, then

$$\langle\varphi^* A^2\varphi\rangle=\langle\varphi^* B^2\varphi\rangle. \tag{A1.4}$$

Considering (A1.4) and (A1.1), we shall have

$$\langle\varphi^*(2\sqrt{\xi^+\xi+s/V}+\rho)^2\varphi\rangle=\langle\varphi^*(\xi+\xi^+)^2\varphi\rangle=2\langle\varphi^*(\xi\xi^++\xi^+\xi)\varphi\rangle$$

$$-\langle\varphi^*(\xi^+-\xi)(\xi-\xi^+)\varphi\rangle\leqslant 2\langle\varphi^*(\xi\xi^++\xi^+\xi)\varphi\rangle$$

$$\leqslant\langle 2\varphi^*(\xi^+\xi+2s/V+\xi^+\xi)\varphi\rangle=4\left\langle\varphi^*\left(\xi^+\xi+\frac{s}{V}\right)\varphi\right\rangle, \tag{A1.5}$$

which is impossible for $\rho > 0$. Thus, the inequality (A1.2) has been proved.

Corollary. We likewise have, having transposed ξ and ξ^+,

$$2\sqrt{\xi\xi^++\frac{s}{V}}-\varepsilon(\xi+\xi^+)\geqslant 0. \tag{A1.6}$$

The following inequalities are also proved analogously:

$$2\sqrt{\xi\xi^++\frac{s}{V}}-\varepsilon\frac{(\xi-\xi^+)}{i}\geqslant 0; \tag{A1.7}$$

$$2\sqrt{\xi^+\xi+\frac{s}{V}}-\varepsilon\frac{(\xi-\xi^+)}{i}\geqslant 0. \tag{A1.8}$$

Lemma II. Assume ξ satisfies the condition

$$|\xi\xi^+-\xi^+\xi|\leqslant 2s/V. \tag{A1.9}$$

Then

$$\sqrt{\xi\xi^++\frac{2s}{V}+A^2}-\sqrt{\xi^+\xi+A^2}\geqslant 0, \tag{A1.10}$$

where A is a real c-number.

Proof. Let us prove the converse. Then one can find a normalized function φ which is such that

$$\left\{\sqrt{\xi\xi^++\frac{2s}{V}+A^2}-\sqrt{\xi^+\xi+A^2}\right\}\varphi=-\rho\varphi. \tag{A1.11}$$

From this we have

$$\left\{\sqrt{\xi\xi^++\frac{2s}{V}+A^2}+\rho\right\}\varphi=\sqrt{\xi^+\xi+A^2}\,\varphi, \tag{A1.12}$$

and using (A1.4), we obtain

$$\left\langle\varphi^*\left(\sqrt{\xi\xi^++\frac{2s}{V}+A^2}+\rho\right)^2\varphi\right\rangle=\langle\varphi^*(\xi^+\xi+A^2)\varphi\rangle\leqslant$$

$$\leqslant \left\langle \varphi^* \left(\xi\xi^+ + \frac{2s}{V} + A^2 \right) \varphi \right\rangle, \tag{A1.13}$$

which is impossible for $\rho > 0$.

Corollary. Changing the role of the operators ξ and ξ^+, we obtain

$$\sqrt{\xi^+\xi + \frac{2s}{V} + A^2} - \sqrt{\xi\xi^+ + A^2} \geqslant 0. \tag{A1.14}$$

If α, λ are real c-numbers, then we have

$$\sqrt{\lambda^2 \left(\xi\xi^+ + \frac{2s}{V} + \alpha^2 \right) + A^2} - \sqrt{\lambda^2 (\xi^+\xi + \alpha^2) + A^2} \geqslant 0; \tag{A1.15}$$

$$\sqrt{\lambda^2 \left(\xi^+\xi + \frac{2s}{V} + \alpha^2 \right) + A^2} - \sqrt{\lambda^2 (\xi\xi^+ + \alpha^2) + A^2} \geqslant 0. \tag{A1.16}$$

Appendix to Lemma II. Let us assume that

$$\xi = \frac{1}{V} \sum_f \lambda(f)\, a_{-f} a_f + \nu \equiv L + \nu. \tag{A1.17}$$

Then

$$\xi\xi^+ - \xi^+\xi = \frac{2}{V^2} \sum_f \lambda^2(f) \left(1 - a_f^+ a_f - a_{-f}^+ a_{-f} \right). \tag{A1.18}$$

Assume $\lambda(f)$ satisfies the condition $\frac{1}{V} \sum_f \lambda^2(f) \leqslant s$; then $|\xi\xi^+ - \xi^+\xi| \leq 2s/V$. Consequently,

$$\sqrt{\lambda^2(f) \{(L+\nu)(L^+ + \nu) + \alpha^2 + 2s/V\} + T^2(f)} - \sqrt{\lambda^2(f) \{(L^+ + \nu)(L + \nu) + \alpha^2\} + T^2(f)} > 0. \tag{A1.19}$$

Lemma III (Generalization of Lemma II). Assume again that $|\xi\xi^+ - \xi^+\xi| \leq 2s/V$. Let us consider the operators $\mathfrak{A}$, $\mathfrak{A}^+$ having the norm $|\mathfrak{A}| < 1$; $|\mathfrak{A}^+| \leq 1$, which are such that

$$|\mathfrak{A}\xi^+ \xi\mathfrak{A}^+ - \xi^+ \mathfrak{A}\mathfrak{A}^+ \xi| \leqslant 2l/V. \tag{A1.20}$$

Then

$$2\sqrt{\xi\xi^+ + \frac{s+l}{V}} - \varepsilon(\xi\mathfrak{A}^+ + \mathfrak{A}\xi^+) \geqslant 0, \tag{A1.21}$$

where $\varepsilon = 1$ or $\varepsilon = -1$.

Proof. Let us assume the converse; then one can find a normalized φ which is such that

$$\left\{ 2\sqrt{\xi\xi^+ + \frac{s+l}{V}} - \varepsilon(\xi\mathfrak{A}^+ + \mathfrak{A}\xi^+) \right\} \varphi = -\rho\varphi, \quad \rho > 0. \tag{A1.22}$$

From this we have

$$\left(2\sqrt{\xi\xi^+ + \frac{s+l}{V}} + \rho \right) \varphi = \varepsilon(\xi\mathfrak{A}^+ + \mathfrak{A}\xi^+)\varphi. \tag{A1.23}$$

Consequently, according to (A1.14),

$$\left\langle \varphi^* \left(2\sqrt{\xi\xi^+ + \frac{s+l}{V}} + \rho \right)^2 \varphi \right\rangle = \langle \varphi^* (\xi\mathfrak{A}^+ + \mathfrak{A}\xi^+)^2 \varphi \rangle =$$

$$=2\langle\varphi^*\{\xi\mathfrak{A}^+\mathfrak{A}\xi^++\mathfrak{A}\xi^+\xi\mathfrak{A}^+\}\varphi\rangle-\langle\varphi^*(\xi\mathfrak{A}^+-\mathfrak{A}\xi^+).$$

$$\times(\mathfrak{A}\xi^+-\xi\mathfrak{A}^+)\varphi\rangle\leqslant 2\langle\varphi^*\{\xi\mathfrak{A}^+\mathfrak{A}\xi^++\mathfrak{A}\xi^+\xi\mathfrak{A}^+\}\varphi\rangle. \quad (A1.24)$$

But since by convention $|\mathfrak{A}|\leq 1$, $|\mathfrak{A}^+|\leq 1$, we have $|\mathfrak{A}^+\mathfrak{A}|\leq 1$, and consequently,

$$\langle\varphi^*\xi\mathfrak{A}^+\mathfrak{A}\xi^+\varphi\rangle\leqslant\langle\varphi^*\xi\xi^+\varphi\rangle. \quad (A1.25)$$

Further, taking account of (A1.20) and (A1.25), we have

$$\langle\varphi^*\mathfrak{A}\xi^+\xi\mathfrak{A}^+\varphi\rangle=\langle\varphi^*\xi^+\mathfrak{A}\mathfrak{A}^+\xi\varphi\rangle$$

$$+\langle\varphi^*\{\mathfrak{A}\xi^+\xi\mathfrak{A}^+-\xi^+\mathfrak{A}\mathfrak{A}^+\xi\}\varphi\rangle\leqslant\langle\varphi^*\xi^+\mathfrak{A}\mathfrak{A}^+\xi\varphi\rangle$$

$$+\frac{2l}{V}\leqslant\langle\varphi^*\xi^+\xi\varphi\rangle+\frac{2l}{V}\leqslant\langle\varphi^*\xi\xi^+\varphi\rangle+\frac{2(l+s)}{V}$$

$$=\left\langle\varphi^*\left(\xi\xi^++\frac{2(l+s)}{V}\right)\varphi\right\rangle. \quad (A1.26)$$

Therefore, taking account of (A1.24), we may write

$$\left\langle\varphi^*\left(2\sqrt{\xi\xi^++\frac{s+l}{V}}+\rho\right)^2\varphi\right\rangle\leqslant 4\left\langle\varphi^*\left(\xi\xi^++\frac{s+l}{V}\right)\varphi\right\rangle. \quad (A1.27)$$

But such an inequality is impossible for $\rho > 0$, which proves the statement (A1.21) of Lemma III.

Appendix to Lemma III. Let us assume $\xi = L + \nu$; $\mathfrak{A} = a_g$. Then

$$|\mathfrak{A}\xi^+\xi\mathfrak{A}^+-\xi^+\mathfrak{A}\mathfrak{A}^+\xi|=|a_g(L^++\nu)(L+\nu)a_g^+-(L^++\nu)a_ga_g^+(L+\nu)|$$
$$=|a_g(L^++\nu)(L+\nu)a_g^+-(L^++\nu)a_g(L+\nu)a_g^+$$
$$+(L^++\nu)a_g(L+\nu)a_g^+-(L^++\nu)a_ga_g^+(L+\nu)|$$
$$\leqslant(|L|+\nu)\{|La_g^+-a_g^+L|+|a_gL^+-L^+a_g|\}\leqslant(|L|+\nu)\frac{4}{V}|\lambda(g)|,$$

where (see the identity (A1.17)) $|L|\leqslant\frac{1}{V}\sum_f|\lambda(f)|$, since $|a_f|\leq 1$. Therefore, in accordance with (A1.21), we have

$$2\sqrt{(L+\nu)(L^++\nu)+\frac{1}{V}\{s+(|L|+\nu)2|\lambda(g)|\}}$$

$$-\varepsilon\{(L+\nu)a_g^++a_g(L^++\nu)\}\geqslant 0. \quad (A1.28)$$

Having placed $\mathfrak{A} = ia_g$, we likewise obtain

$$2\sqrt{(L+\nu)(L^++\nu)+\frac{1}{V}\{s+(|L|+\nu)2|\lambda(g)|\}}$$

$$-\varepsilon\left\{\frac{(L+\nu)a_g^+-a_g(L^++\nu)}{i}\right\}\geqslant 0. \quad (A1.29)$$

Lemma IV. Assume β is a real number $\alpha^2 = \beta^2 + 2s/V$ and $\nu \geq 0$. Then

$$|\sqrt{\{(L+\nu)(L^++\nu)+\alpha^2\}\lambda^2(f)+T^2(f)}\,a_f$$

$$-a_f \sqrt{\{(L+\nu)(L^+ +\nu)+\alpha^2\}\lambda^2(f)+T^2(f)} \leqslant \frac{S_f}{V}, \qquad \text{(A1.30)}$$

where S_f is bounded for $V \to \infty$. (The same inequality holds if in (A1.30) we take a_f^+ instead of a_f.)

Proof. Let us consider an arbitrary normalized function φ and let us formulate the expression

$$\langle \varphi^* \{\sqrt{(Q+\alpha^2)\lambda^2(f)+T^2(f)}\,(a_f+a_f^+)$$

$$-(a_f+a_f^+)\sqrt{(Q+\alpha^2)\lambda^2(f)+T^2(f)}\}\,\varphi\rangle = \mathscr{E}, \qquad \text{(A1.31)}$$

where $Q = (L+\nu)(L^+ + \nu)$. In order to consider the expression (A1.31) let us use the following identical relationship:

$$\sqrt{Z}-\sqrt{Z_0} = \frac{1}{\pi}\int_0^\infty \left\{\frac{1}{Z_0+\omega} - \frac{1}{Z+\omega}\right\}\sqrt{\omega}\,d\omega,$$

where Z_0 is an arbitrary positive number. Note likewise that

$$-\frac{1}{A}B + B\frac{1}{A} = \frac{1}{A}(AB-BA)\frac{1}{A},$$

where A and B are operators. Then we have

$$\mathscr{E} = \frac{1}{\pi}\int_0^\infty \Big\langle \varphi^* \frac{\lambda^2(f)}{(Q+\alpha^2)\lambda^2(f)+T^2(f)+\omega}\{Q(a_f+a_f^+)-(a_f+a_f^+)Q\}$$

$$\times \frac{1}{(Q+\alpha^2)\lambda^2(f)+T(f)+\omega}\varphi\Big\rangle \sqrt{\omega}\,d\omega.$$

But

$$Qa_f - a_f Q = (L+\nu)\{L^+ a_f - a_f L^+\};$$

$$L^+ = \frac{1}{V}\sum_f \lambda(f)\, a_f^+ a_{-f}^+; \quad L^+ a_f - a_f L^+ = -\frac{2}{V}\lambda(f)\, a_{-f}^+,$$

and consequently,

$$Q(a_f+a_f^+)-(a_f+a_f^+)Q = -\frac{2}{V}\lambda(f)(L+\nu)a_{-f}^+ + \frac{2}{V}\lambda(f)\,a_{-f}(L^+ +\nu).$$

Therefore

$$|\mathscr{E}| = \left|\frac{\mathscr{E}}{i}\right| = \frac{2|\lambda(f)|^3}{\pi}\left|\int_0^\infty \Big\langle \varphi^* \frac{1}{(Q+\alpha^2)\lambda^2(f)+T^2(f)+\omega}\right.$$

$$\left.\times \frac{(L+\nu)a_{-f}^+ - a_{-f}(L^+ +\nu)}{i} \times \frac{1}{(Q+\alpha^2)\lambda^2(f)+T^2(f)+\omega}\varphi\Big\rangle \sqrt{\omega}\,d\omega\right|.$$

From this, considering (A1.29) and introducing a new integration variable, we obtain

$$|\mathscr{E}| \leqslant \frac{4|\lambda(f)|^2}{\pi V}\int_0^\infty \Big\langle \varphi^* \frac{\sqrt{Q+\frac{1}{V}(s+2|\lambda(f)|)(|L|+\nu)}}{\left(Q+\alpha^2+\frac{T^2(f)}{\lambda^2(f)}+\tau\right)^2}\varphi\Big\rangle \sqrt{\tau}\,d\tau.$$

But by convention of the lemma we have $\alpha^2 = \beta^2 + 2s/V$, and therefore

$$\sqrt{Q+\frac{1}{V}(s+2|\lambda(f)|)(|L|+\nu)} < \sqrt{Q+\alpha^2+\frac{T^2(f)}{\lambda^2(f)}+\frac{2|\lambda(f)|(|L|+\nu)}{V}} =$$

$$=\sqrt{Q+\alpha^2+\frac{T^2(f)}{\lambda^2(f)}}\cdot\sqrt{1+\frac{2|\lambda(f)|(|L|+\nu)}{VQ+V\alpha^2+V\frac{T^2(f)}{\lambda^2(f)}}}$$

$$<\sqrt{Q+\alpha^2+\frac{T^2(f)}{\lambda^2(f)}}\cdot\sqrt{1+\frac{|\lambda(f)|(|L|+\nu)}{s+\frac{1}{2}V\frac{T^2(f)}{\lambda^2(f)}}}$$

$$<\left(1+\frac{|\lambda(f)|(|L|+\nu)}{2s+V\frac{T^2(f)}{\lambda^2(f)}}\right)\sqrt{Q+\alpha^2+\frac{T^2(f)}{\lambda^2(f)}}\,.$$

Let us place $\Lambda=Q+\alpha^2+\frac{T^2(f)}{\lambda^2(f)}\geqslant\alpha^2$. Then

$$|\mathscr{E}|=\frac{4|\lambda(f)|^2}{\pi V}\left(1+\frac{|\lambda(f)|(|L|+\nu)}{2s+V\frac{T^2(f)}{\lambda^2(f)}}\right)\int\limits_0^\infty\left\langle\varphi^*\frac{\sqrt{\Lambda}}{(\Lambda+\tau)^2}\varphi\right\rangle\sqrt{\tau}\,d\tau.$$

Now let us expand the function φ in eigenfunctions of the operator Λ: $\varphi=\Sigma C_\Lambda\varphi_\Lambda$; $\Sigma|C_\Lambda|^2=1$. We obtain

$$\int\limits_0^\infty\left\langle\varphi^*\frac{\sqrt{\Lambda}}{(\Lambda+\tau)^2}\varphi\right\rangle\sqrt{\tau}\,d\tau=\sum_\Lambda|C_\Lambda|^2\int\limits_0^\infty\frac{\sqrt{\Lambda\tau}\,d\tau}{(\Lambda+\tau)^2}=\sum_\Lambda|C_\Lambda|^2\int\limits_0^\infty\frac{\sqrt{t}\,dt}{(1+t)^2}=\int\limits_0^\infty\frac{\sqrt{t}\,dt}{(1+t)^2}\,.$$

Thus, for an arbitrary normalized function φ we have

$$|\mathscr{E}|=|\left\langle\varphi^*\left[\frac{\sqrt{(Q+\alpha^2)\lambda^2+T^2};a_f+a_f^+}{i}\right]\varphi\right\rangle|\leqslant S_f,$$

where

$$S_f=\frac{4|\lambda(f)|^2}{\pi V}\left(1+\frac{|\lambda(f)|\left(\sum|\lambda(f)|\frac{1}{V}+\nu\right)}{2\frac{1}{V}\sum|\lambda(f)|^2+V\frac{T^2(f)}{\lambda^2(f)}}\right)\int\limits_0^\infty\frac{\sqrt{t}\,dt}{(1+t)^2}\,.$$

But the operator

$$\left[\frac{\sqrt{(Q+\alpha^2)\lambda^2+T^2};\ a_f+a_f^+}{i}\right]$$

is a Hermite operator, and consequently, $\left|\left[\sqrt{(Q+\alpha^2)\lambda^2+T^2};\ a_f+a_f^+\right]\right|\leqslant S_f$. In a completely analogous manner we prove that $\left|\left[\sqrt{(Q+\alpha^2)\lambda^2+T^2};\ a_f-a_f^+\right]\right|\leqslant S_f$. But $|\mathfrak{A}|+|\mathfrak{B}|\geq|\mathfrak{A}+\mathfrak{B}|$, and consequently, $\left|\left[\sqrt{(Q+\alpha^2)\lambda^2+T^2};\ a_f\right]\right|\leqslant S_f$, which is what it was required to prove.

From $|\mathfrak{A}|\leq S_f$ it follows that $|\mathfrak{A}^+|\leq S_f$, whence the validity of the supplementary statement of the lemma is evident.

APPENDIX 2

The principle of weakening the correlations between particles for systems in the state of statistical equilibrium is formulated as follows.

The correlation functions

$$\langle\mathfrak{A}_1(x_1,t_1)\ldots\mathfrak{A}_s(x_s,t_s)\ldots\mathfrak{A}_n(x_n,t_n)\rangle, \tag{A2.1}$$

where $\mathfrak{A}_s(x_s,t_s)$ is the field function $\psi(x_s,t_s)$ or $\psi^+(t_s,x_s)$, can be decomposed into the product of correlation functions

$$\langle \mathfrak{A}_1(x_1, t_1) \ldots \mathfrak{A}_{s-1}(x_{s-1}, t_{s-1})\rangle \langle \mathfrak{A}_{s+1}(x_{s+1}, t_{s+1}) \ldots \mathfrak{A}_n(x_n, t_n)\rangle, \tag{A2.2}$$

if the ensemble of points $x_1, \ldots, x_s$ is placed infinitely far from the ensemble of points $x_{s+1}, \ldots, x_n$ at fixed times $t_1, \ldots, t_s, \ldots, t_n$. Note that for the case in which the numbers of the creation and annihilation operators are not equal in the correlation functions, the averaging $\langle \ldots \rangle$ should be understood in the sense of quasiaverages.

A system having a model Hamiltonian is one of the rare cases in which direct calculations can verify the validity of the principle of correlation weakening. Below, based on the previous asymptotic estimates, we shall show precisely this. Let us consider "vacuum" averages which are formulated from the products of field functions in the spatial representation:

$$\left.\begin{aligned} \Psi_-(t, x) &= \frac{1}{\sqrt{V}} \sum_{(f<0)} a_f(t)\, e^{i(f\cdot x)}; \\ \Psi_+(t, x) &= \frac{1}{\sqrt{V}} \sum_{(f>0)} a_f^+(t)\, e^{-i(f\cdot x)}. \end{aligned}\right\} \tag{A2.3}$$

Here f represents the aggregate of a momentum and a spin $(\mathbf{k}, \sigma)$, the summation $f > 0$, $f < 0$ denoting summation over $\mathbf{k}$ for fixed $\sigma = \pm$, $(f \cdot x) = (\mathbf{k} \cdot \mathbf{r})$. We have, for example,

$$\langle \Psi_{\sigma_1}(t, x)\Psi_{\sigma_2}^+(t, x')\rangle_{H_0} = \frac{1}{V}\sum_{(f>0)} |u_f|^2 e^{if\cdot(x-x')}\,\delta(\sigma_1-\sigma_2)$$

$$= \left\{\frac{1}{V}\sum_{(f>0)} e^{if\cdot(x-x')} - \frac{1}{V}\sum_{(f>0)} |v_f|^2 e^{if\cdot(x-x')}\right\}\delta(\sigma_1-\sigma_2), \tag{A2.4}$$

where u_f and v_f are the coefficients of the canonical transformation. As is evident, the term

$$\frac{1}{V}\sum_{(f>0)} |v_f|^2 e^{if\cdot(x-x')}$$

approaches the following integral for $V \to \infty$:

$$\frac{1}{(2\pi)^3}\int |v_f|^2 e^{if\cdot(x-x')}\, dk.$$

This integral is absolutely convergent, since

$$\int |v_f|^2\, dk = \frac{1}{2}\int \{\sqrt{T^2(f)+\lambda^2(f)C^2} - T(f)\}^2 \frac{dk}{T^2(f)+\lambda^2(f)C^2} < \infty.$$

About the expression $\frac{1}{V}\sum_{(f>0)} e^{if\cdot(x-x')}$ we say that it approaches a "delta-function" $\frac{1}{(2\pi)^3}\int e^{if\cdot(x-x')}$ when $V \to \infty$.

However, at present we, of course, ascribe to the words "limit," "convergence of functions" a different meaning: namely, the meaning adopted in the theory of generalized functions.

Let us recall here what the relationship

$$f_V(x_1, \ldots, x_l) \underset{V\to\infty}{\to} f(x_1, \ldots, x_l), \tag{A2.5}$$

where $f(x_1, \ldots, x_l) = \lim_{V\to\infty} f_V(x_1, \ldots, x_l)$, means in this theory.

Let us consider the class $C(q, r)$ (q, r are positive numbers) of continuous and unboundedly differentiable functions $h(x_1, \ldots, x_l)$ which are such that throughout the space E_l of points $\{x_1, \ldots, x_l\}$ we have

$$\{|x_1|+\ldots+|x_l|\}^{\alpha}|h(x_1,\ldots,x_l)|\leqslant \text{const};\quad \{|x_1|+\ldots+|x_l|\}^{\alpha}\times\left|\frac{\partial^{s_1+\ldots+s_l}h}{\partial x_1^{s_1}\ldots\partial x_l^{s_l}}\right|\leqslant \text{const}.$$
$$\alpha=0,1,\ldots,r \qquad \alpha=0,1,\ldots,r$$
$$s_1+\ldots+s_l=0,1,\ldots,q$$

Then, if the positive numbers q, r can be fixed in such a way that for any function h from the class C(q, r) one may write

$$\int h(x_1,\ldots,x_l)f_V(x,\ldots x_l)\,dx,\ldots,dx_l\to\int h(x_1,\ldots,x_l)f(x_1,\ldots,x_l)\,dx_1\ldots dx_l,$$

we shall say that the generalized limit relation (A2.5) holds. As we have just seen, the averages of the products $\Psi(t, x)$, $\Psi^+(t, x)$ may contain generalized functions. Therefore, the corresponding limit relationships for $V\to\infty$ should be understood in the sense of the theory of generalized functions.

Let us consider the expression

$$\langle\Psi_{\sigma_1}(t_1,x_1)\Psi^+_{\sigma_2}(t_1,x_2)\rangle=\frac{1}{V}\sum_{(f>0)}\langle a_f(t_1)a_f^+(t_1)\rangle e^{if\cdot(x_1-x_2)}\delta(\sigma_1-\sigma_2).$$

We have

$$\int h(x_1-x_2)\langle\Psi_{\sigma_1}(t_1,x_1)\Psi^+_{\sigma_2}(t_2,x_2)\rangle\,dx_1=\frac{1}{V}\sum_{(f>0)}\langle a_f(t_1)a_f^+(t_2)\rangle\tilde{h}(f)\delta(\sigma_1-\sigma_2),$$

where

$$\tilde{h}(f)=\int h(x)\,e^{i(f\cdot x)}\,dx.$$

Having taken the numbers q, r in the class C(q, r) to which h(x) belongs, one may achieve a situation in which $h(f)$ decreases more rapidly than any power of $|f|\to\infty$ for $1/|f|$. It is sufficient merely to ensure that $\frac{1}{V}\sum_f|\tilde{h}(f)|\leqslant K=\text{const}$.

Then, noting that in accordance with (6.36) we have

$$\left|\langle a_f(t_1)a_f^+(t_2)\rangle_H-\langle a_f(t_1)a_f^+(t_2)\rangle_{H_0}\right|\leqslant\frac{s_1|t_1-t_2|+s_2}{\sqrt{V}},$$
$$s_1,s_2=\text{const},$$

we shall have

$$\left|\int h(x_1-x_2)\{\langle\Psi_{\sigma_1}(t_1,x_1)\Psi^+_{\sigma_2}(t_2,x_2)\rangle_H-\langle\Psi_{\sigma_1}(t_1,x_1)\Psi^+_{\sigma_2}(t_2,x_2)\rangle_{H_0}\}\,dx_1\right|$$
$$\leqslant\frac{1}{V}\sum_f\left|\langle a_f(t_1)a_f^+(t_2)\rangle_H-\langle a_f(t_1)a_f^+(t_2)\rangle_{H_0}\right||\tilde{h}(t)|$$
$$\leqslant K\frac{s_1|t_1-t_2|+s_2}{V}\underset{V\to\infty}{\to}0.$$

Consequently, the generalized limit relation

$$\langle\Psi_{\sigma_1}(t_1,x_1)\Psi^+_{\sigma_2}(t_2,x_2)\rangle_H-\langle\Psi_{\sigma_1}(t_1,x_1)\Psi^+_{\sigma_2}(t_2,x_2)\rangle_{H_0}\to 0 \tag{A2.6}$$

holds. But by direct calculation we verify the fact that

$$\langle\Psi_{\sigma_1}(t_1,x)\Psi^+_{\sigma_2}(t_2,x_2)\rangle_{H_0}=\frac{1}{V}\sum_{(f>0)}|u_f|^2e^{-i\Omega(f)(t_1-t_2)+if\cdot(x_1-x_2)}\delta(\sigma_1-\sigma_2),$$

and therefore likewise in the generalized sense we have

$$\langle \Psi_{\sigma_1}(t_1, x_1)\, \Psi^{+}_{\sigma_2}(t_2, x_2)\rangle_H$$

$$-\int |u_f|^2 \exp\left\{-i\Omega(f)(t_1-t_2)+if\cdot(x_1-x_2)\right\} dk\, \delta(\sigma_1-\sigma_2) \underset{V\to\infty}{\to} 0. \tag{A2.7}$$

From (A2.6) and (A2.7) we finally have

$$\lim_{V\to\infty} \langle \Psi_{\sigma_1}(t_1, x_1)\, \Psi^{+}_{\sigma_2}(t_2, x_2)\rangle_H$$

$$=\int |u_f|^2 \exp\{-i\Omega(f)(t_1-t_2)+if\cdot(x_1-x_2)\}\, dk\, \delta(\sigma_1-\sigma_2)$$

$$=\{\Delta(t_1-t_2, x_1-x_2)-F(t_1-t_2, x_1-x_2)\}\, \delta(\sigma_1-\sigma_2); \tag{A2.8}$$

$$\left.\begin{aligned} \Delta(t, x) &= \int \exp\{-i\Omega(f)t+if\cdot x\}\, dk; \\ F(t, x) &= \int |v_f|^2 \exp\{-i\Omega(f)t+if\cdot x\}\, dk. \end{aligned}\right\} \tag{A2.9}$$

In a fully analogous manner we obtain*

$$\lim_{V\to\infty} \langle \Psi^{+}_{\sigma_2}(t_2, x_2)\, \Psi_{\sigma_1}(t_1, x_1)\rangle = F(t_2-t_1, x_1-x_2)\, \delta(\sigma_1-\sigma_2). \tag{A2.10}$$

Let us now consider the binary expressions $\langle\Psi(t_1, x_1)\Psi(t_2, x_2)\Psi^{+}(t_2', x_2') \times \Psi^{+}(t_1', x_1')\rangle$. We have

$$\langle \Psi(t_1, x_1)\, \Psi(t_2, t_2)\, \Psi^{+}(t_2', x_2')\, \Psi^{+}(t_1', x_1')\rangle$$

$$= \frac{1}{V^2} \sum \langle a_{f_1}(t_1)\, a_{f_2}(t_2)\, a^{+}_{g_2}(t_2')\, a^{+}_{g_1}(t_1')\rangle$$

$$\times \{if_1\cdot x_1 + if_2\cdot x_2 - ig_2\cdot x_2' - ig_1\cdot x_1'\}. \tag{A2.11}$$

Since the total momentum is conserved, while for Φ_H and Φ_{H_0} it is equal to zero, it follows that the expressions

$$\langle a_{f_1}(t_1)\, a_{f_2}(t_2)\, a^{+}_{g_2}(t_2')\, a^{+}_{g_1}(t_1')\rangle \tag{A2.12}$$

may be nonvanishing only if

$$f_1+f_2=g_2+g_1. \tag{A2.13}$$

Let us now recall that from (2.1) and (2.2) we have $n_f(t) - n_{-f}(t)$, where $n_f = \overset{+}{a}_f a_f$ is the integral of motion, and Φ_H (and Φ_{H_0}) satisfies the supplementary relations $(n_f - n_{-f})\Phi = 0$. Note finally that $(n_f - n_{-f})a_h = a_h\{(n_f - n_{-f}) - \delta(f-h) + \delta(f+h)\}$. Therefore (for any f)

$$\langle a_{f_1}(t_1)\, a_{f_2}(t_2)\, a^{+}_{g_2}(t_2')\, a^{+}_{g_1}(t_1')\rangle$$

$$\langle\{1+n_f-n_{-f}\}\, a_{f_1}(t_1)\, a_{f_2}(t_2)\, a^{+}_{g_2}(t_2')\, a^{+}_{g_1}(t_1')\rangle$$

$$=\langle\{1+n_f(t_1)+n_{-f}(t_1)\}\, a_{f_1}(t_1)\, a_{f_2}(t_2)\, a^{+}_{g_2}(t_2')\, a^{+}_{g_1}(t_1')\rangle$$

$$=\langle a_{f_1}(t_1)\{1+n_f(t_1)-n_{-f}(t_1)-\delta(f-f_1)+\delta(f+f_1)\}\, a_{f_2}(t_2)$$

$$\times a^{+}_{g_2}(t_2')\, a^{+}_{g_1}(t_1')\rangle = \langle a_{f_1}(t_1)\{1+n_f(t_2)-n_{-f}(t_2)-\delta(f-f_1)+\delta(f+f_1)\}$$

$$\times a_{f_2}(t_2) a^{+}_{g_2}(t_2')\, a^{+}_{g_1}(t_1')\rangle = \langle a_{f_1}(t_1)\, a_{f_2}(t_2)\{1+n_f(t_2)-n-f(t_2)$$

$$-\delta(f-f_1)+\delta(f+f_1)-\delta(f-f_2)+\delta(f+f_2)\}\, a^{+}_{g_2}(t_2')\, a^{+}_{g_1}(t_1')\rangle = \ldots$$

$$=\langle a_{f_1}(t_1)\, a_{f_2}(t_2)\, a^{+}_{g_2}(t_2')\, a^{+}_{g_1}(t_1')$$

* This limit relation likewise holds in the conventional sense in view of the absolute convergence of the integral which defines F(t, x).

$$
\begin{aligned}
&\times \{1+n_f-n_{-f}-\delta(f-f_1)+\delta(f+f_1)+\delta(f+f_2)-\delta(f-f_2)\\
&+\delta(f-g_2)-\delta(f+g_2)+\delta(f-g_1)-\delta(f+g_1)\}\rangle\\
&=\{1-\delta(f-f_1)+\delta(f+f_1)-\delta(f-f_2)+\delta(f+f_2)\\
&+\delta(f-g_2)-\delta(f+g_2)+\delta(f-g_1)-\delta(f+g_1)\}\\
&\times \langle a_{f_1}(t_1)\,a_{f_2}(t_2)\,a^+_{g_2}(t'_2)\,a^+_{g_1}(t'_1)\rangle.
\end{aligned}
$$

This identity shows that the quantities (A2.12) may be nonvanishing only if for any f the relation

$$
\begin{aligned}
&-\delta(f-f_1)+\delta(f+f_1)-\delta(f-f_2)+\delta(f+f_2)\\
&+\delta(f-g_2)-\delta(f+g_2)+\delta(f-g_1)-\delta(f+g_1)=0
\end{aligned}
$$

holds. The latter relation together with (A2.13) is fulfilled only in the following cases:

$$f_1-f_2=0,\quad g_1+g_2=0; \tag{A2.14}$$

$$f_1=g_1,\quad f_2=g_2; \tag{A2.15}$$

$$f_1=g_2,\quad f_2=g_1. \tag{A2.16}$$

Moreover, in (A2.15) and (A2.16) it may always be assumed that $g_1 \neq g_2$, since

$$a^+_g(t'_2)\,a^+_g(t'_1)\,\Phi_H=0. \tag{A2.17}$$

Actually,

$$(n_g-n_{-g})\,a^+_g(t'_2)\,a^+_g(t'_1)\,\Phi_H=a^+_g(t'_2)\,a^+_g(t'_1)\,(n_g-n_{-g}+2)\,\Phi_H=2a^+_g(t'_2)\,a^+_g(t'_1)\,\Phi_H.$$

But the possible eigenvalues n_g-n_{-g} are only ± 1 and 0, and consequently, the latter equation is possible only when (A2.17) is fulfilled. Thus, one can reduce (A2.11) to the form

$$
\begin{aligned}
&\langle \Psi(t_1,x_1)\,\Psi(t_2,x_2)\,\Psi^+(t'_2,x'_2)\,\Psi(t'_1,x'_1)\rangle\\
&=\sum_{f,g}\frac{1}{V^2}\langle a_{-f}(t_1)\,a_f(t_2)\,a^+_g(t'_2)\,a^+_{-g}(t'_1)\rangle \exp\{if\cdot(x_2-x_1)-ig\cdot(x'_2-x'_1)\}\\
&+\sum_{\substack{f,g\\ \left(\substack{f\neq g\\ f+g\neq 0}\right)}}\frac{1}{V^2}\langle a_f(t_1)\,a_g(t_2)\,a^+_g(t'_2)\,a^+_f(t'_1)\rangle \exp\{if\cdot(x_1-x'_1)+ig\cdot(x_2-x'_2)\}\\
&+\sum_{\substack{f,g\\ \left(\substack{f\neq g\\ f+g\neq 0}\right)}}\frac{1}{V^2}\langle a_f(t_1)\,a_g(t_2)\,a^+_f(t'_2)\,a^+_g(t'_1)\rangle \exp\{if\cdot(x_1-x'_2)+ig\cdot(x_2-x'_1)\}.
\end{aligned} \tag{A2.18}
$$

Let us now deal with the transition in the limit $V\to\infty$. Let us consider the class $C(q, r)$ of functions $h(x, y)$, and let us fix q, r in such a way that $\frac{1}{V^2}\sum_{f,g}|\tilde{h}(f,g)| \leqslant \text{const}$, where

$$\tilde{h}(f,g)=\int h(x,y)\,e^{i(fx+gy)}\,dx\,dy.$$

Since (see (5.56)) for fixed t_1, t_2, t'_2, t'_1,

$$\left|\langle a_f(t_1)\,a_g(t_2)\,a^+_f(t'_2)\,a^+_g(t'_1)\rangle_H-\langle a_f(t_1)\,a_g(t_2)\,a^+_f(t'_2)\,a^+_g(t'_1)\rangle_{H_0}\right|<\frac{\text{const}}{\sqrt{V}},$$

we have

$$\left|\int h(x,y)\{\Gamma_H(t_1,t_2,t'_2,t'_1\,|\,x,y)-\Gamma_{H_0}(t_1,t_2,t'_2,t'_1\,|\,x,y)\}\,dx\,dy\right|<\frac{\text{const}}{\sqrt{V}}\xrightarrow[V\to\infty]{}0,$$

where

$$\Gamma(t_1, t_2, t_2', t_1' | x, y) = \frac{1}{V^2} \sum_{\substack{f, g \\ \left(\substack{f \neq g \\ f+g \neq 0}\right)}} \langle a_f(t_1) a_g(t_2) a_f^+(t_2') a_g^+(t_1') \rangle e^{i(f \cdot x + g \cdot y)}.$$

We obtain the generalized limit relations

$$\Gamma_H(t_1, t_2, t_2', t_1', x_1 - x_2', x_2 - x_1') - \Gamma_{H_0}(t_1, t_2, t_2', t_1', x_1 - x_2', x_2 - x_1') \xrightarrow[V \to \infty]{} 0.$$

But direct calculation, just as in the case of (A2.4), yields

$$\Gamma_{H_0}(t_1, t_2, t_2', t_1', x_1 - x_2', x_2 - x_1')$$

$$= -\frac{1}{V^2} \sum_{\substack{f, g \\ \left(\substack{f \neq g \\ f+g \neq 0}\right)}} |u_f|^2 |u_g|^2 e^{-i\Omega(f)(t_1 - t_2') - i\Omega(g)(t_2 - t_1')}$$

$$\times e^{i\left(f(x_1 - x_2') + g(x_2 - x_1')\right)} \to -\{\Delta(t_1 - t_2', x_1 - x_2')$$

$$-F(t_1 - t_2', x_1 - x_2')\} \{\Delta(t_2 - t_1', x_2 - x_1')$$

$$-F(t_2 - t_1', x_2 - x_1')\} \delta(\sigma_1 - \sigma_2') \delta(\sigma_2 - \sigma_1'), \tag{A2.19}$$

where $\Omega(f)$ is determined by the relationship (6.2'), while $\Delta(t, x)$ and $F(t, x)$ are determined by the relation (A2.9).

Consequently,

$$\lim_{V \to \infty} \Gamma_H(t_1, t_2, t_2', t_1', x_1 - x_2', x_2 - x_1')$$
$$= -\{\Delta(t_1 - t_2', x_1 - x_2') - F(t_1 - t_2', x_1 - x_2')\}$$
$$\times \{\Delta(t_2 - t_1', x_2 - x_1') - F(t_2 - t_1', x_2 - x_1')\} \delta(\sigma_1 - \sigma_2') \delta(\sigma_2 - \sigma_1').$$

We also deal in a completely analogous manner with the terms which are included in the right side of Eq. (A2.18). Now let us place

$$\Phi_\sigma(t, x) = -\int u_f v_f e^{-i\Omega(f)t - if \cdot x} d\mathbf{k} = \int \frac{C\lambda(f)}{2\Omega(f)} e^{-i\Omega(f)t - if \cdot \mathbf{k}} d\mathbf{k}. \tag{A2.20}$$

Then we may write the generalized limit relation in the form

$$\lim_{V \to \infty} \langle \Psi_{\sigma_1}(t_1, x_1) \Psi_{\sigma_2}(t_2, x_2) \Psi^+_{\sigma_2'}(t_2', x_2') \Psi^+_{\sigma_1'}(t_1', x_1') \rangle_H$$
$$= \Phi_{\sigma_2}(t_1 - t_2, x_1 - x_2) \Phi_{\sigma_2'}(t_2' - t_1', x_2' - x_1') \delta(\sigma_1 + \sigma_2) \delta(\sigma_1' + \sigma_2')$$
$$+ \delta(\sigma_1 - \sigma_1') \delta(\sigma_2 - \sigma_2') \{\Delta(t_1 - t_1', x_1 - x_1') - F(t_1 - t_1', x_1 - x_1')\}$$
$$\times \{\Delta(t_2 - t_2', x_2 - x_2') - F(t_2 - t_2', x_2 - x_2')\} - \delta(\sigma_1 - \sigma_2') \delta(\sigma_2 - \sigma_1')$$
$$\times \{\Delta(t_1 - t_2', x_1 - x_2') - F(t_1 - t_2', x_1 - x_2')\} \{\Delta(t_2 - t_1', x_2 - x_1')$$
$$- F(t_2 - t_1', x_2 - x_1')\}. \tag{A2.21}$$

Completely analogous formulas are also obtained for other arrangement orders of the operator functions Ψ, Ψ.*

---

* The function $\Delta(t, x)$ itself is a generalized function. This relation likewise holds in the generalized sense.

Using the example of the formula (A2.21) which has been derived, one may illustrate the principle of correlation weakening. One need merely note that for fixed t

$$\left.\begin{aligned} &F(t, x) \to 0 \ |x| \to \infty; \\ &\Phi(t, x) \to 0 \ |x| \to \infty; \end{aligned}\right\} \tag{A2.22}$$

$$\Delta(t, x) \to 0^* \ |x| \to \infty. \tag{A2.23}$$

Let us fix the times t_1, t_2, t_2', t_1' and the spatial differences x_1-x_1', x_2-x_2'. The remaining spatial differences $x_1-x_2, x_1'-x_2', x_1-x_2', x_2-x_1'$ are made to tend to infinity. Then the function considered

$$\lim_{V\to\infty} \langle \Psi_{\sigma_1}(t_1, x_1) \Psi_{\sigma_2}(t_2, x_2) \Psi^+_{\sigma_2'}(t_2', x_2') \Psi^+_{\sigma_2'}(t_1', x_1') \rangle_H \tag{A2.24}$$

will decompose into the product

$$\{\Delta(t_1-t_1', x_1-x_1') - F(t_1-t_1', x_1-x_1')\} \{\Delta(t_2-t_2', x_2-x_2') \\ - F(t_2-t_2', x_2-x_2')\} \delta(\sigma_2-\sigma_1') \delta(\sigma_2-\sigma_2'),$$

which is equal to [see (A2.8)]

$$\lim_{V\to\infty} \langle \Psi_{\sigma_1}(t_1, x_1) \Psi^+_{\sigma_1'}(t_1', x_1') \rangle_H \lim_{V\to\infty} \langle \Psi_{\sigma_2}(t_2, x_2) \Psi^+_{\sigma_2'}(t_2', x_2') \rangle_H. \tag{A2.25}$$

Let us now consider another method of correlation weakening. Let us fix the times t_1, t_2, t_2', t_1 and the spatial differences $x_1-x_2, x_1'-x_2'$ anew. The remaining spatial differences $x_1-x_1', x_2-x_2', x_1-x_2', x_2-x_1'$ are made to tend to infinity. Then the function (A2.24) considered decomposes into the product

$$\Phi(t_1-t_2, x_1-x_2) \Phi(t_2'-t_1', x_2'-x_1') \Phi_{\sigma_2} \Phi_{\sigma_2'} \delta(\sigma_1+\sigma_2) \delta(\sigma_1'+\sigma_2'). \tag{A2.26}$$

For $\nu > 0$,

$$\left.\begin{aligned} &\Phi_\sigma(t_1-t_2, x_1-x_2) = \lim_{V\to\infty} \langle \Psi_{-\sigma}(t_1, x_1) \Psi_\sigma(t_2, x_2) \rangle_H; \\ &\Phi_\sigma(t_2'-t_1', x_2'-x_1') = \lim_{V\to\infty} \langle \Psi^+_\sigma(t_2', x_2') \Psi^+_{-\sigma}(t_1', x_1') \rangle_H, \end{aligned}\right\} \tag{A2.27}$$

so that the function (A2.24) decomposes into the product of averages

$$\lim_{V\to\infty} \langle \Psi_{\sigma_1}(t_1, x_1) \Psi_{\sigma_2}(t_2, x_2) \lim_{V\to\infty} \left\langle \Psi^+_{\sigma_2'}(t_2', x_2') \Psi^+_{\sigma_1'}(t_1', x_1') \right\rangle. \tag{A2.28}$$

In the case given the relationships (A2.25) or (A2.28) which have been found constitute the expression of the principle of correlation weakening (A2.2). For $\nu = 0$ we have $\langle \Psi(t_1, x_1) \times \Psi(t_2, x_2) \rangle_H = 0$, and the relationships (A2.27) have no place. In this case, however, we may introduce the "quasiaverages"

$$\langle \Psi_{\sigma_1}(t_1, x_1) \Psi_{\sigma_2}(t_2, x_2) \rangle_H = \lim_{\substack{\nu>0 \\ \nu\to 0}} \lim_{V\to\infty} \langle \Psi_{\sigma_1}(t_1, x_1) \\ \times \Psi_{\sigma_2}(t_2, x_2) \rangle = \Phi_{\sigma_2}(t_1-t_2, x_1-x_2) \ \delta(\sigma_1+\sigma_2) \tag{A2.29}$$

and take the corresponding product of "quasiaverages" instead of the product of averages (A2.28). The relationships derived above illustrate the general principle of correlation weakening.

I take this opportunity to express sincere thanks to D. N. Zubarev, S. V. Tyablikov,* Yu. A. Tserkovnikov, and E. N. Yakovlev for their discussion.

* Deceased.

LITERATURE CITED

1. J. Bardeen, L. N. Cooper, and J. R. Schrieffer, Phys. Rev., 105, 1175 (1957).
2. N. N. Bogolyubov, D. N. Zubarev, and Yu. A. Tserkovnikov, Dokl. Akad. Nauk SSSR, 177, 788 (1957).
3. R. E. Prange, Bull. Amer. Phys. Soc., 4, 225 (1959).
4. N. N. Bogolyubov, D. N. Zubarev, and Yu. A. Tserkovnikov, Zh. Éksp. Teor. Fiz., 39, 120 (1960).
5. N. N. Bogolyubov, Zh. Éksp. Teor. Fiz., 37, 73 (1958).
6. N. N. Bogolyubov, Izv. Akad. Nauk SSSR, Ser. Fiz., 11, 77 (1947).
7. N. N. Bogolyubov (Jr.), Preprint of the Joint Institute for Nuclear Research, R4-4184 [in Russian], Dubna (1968).
8. N. N. Bogolyubov (Jr.), Preprint of the Joint Institute for Nuclear Research, R2-4175 [in Russian], Dubna (1968).
9. N. N. Bogolyubov, Preprint, ITPh-67-1 [translated from Russian], Kiev (1967).
10. N. N. Bogolyubov, Preprint, ITPh-68-65 [translated from Russian], Kiev (1968).
11. N. N. Bogolyubov, Preprint, ITPh-68-67 [translated from Russian], Kiev (1968).
12. N. N. Bogolyubov (Jr.), Yadernaya Fizika, 10, 425 (1969).

THE SELF-CONSISTENT-FIELD METHOD IN NUCLEAR THEORY

R. V. Dzholos and V. G. Solov'ev

The self-consistent-field method is described in the Bogolyubov formulation. It is shown that this method yields equations for the effective fields in the theory of finite Fermi systems and the secular equations for a model with pairing and multipole forces.

It is difficult to conceive of nuclear physics today without the concept of the self-consistent field of the nucleus. A great amount of experimental information indicates that nucleons in the nucleus behave to a certain extent as independent particles moving in a common potential. For this reason it is natural to construct a nuclear theory based on the concept of a self-consistent field, at least for the low-lying excited states. We describe below the self-consistent-field method in the Bogolyubov formulation [1]. We will use it to derive equations describing the ground and low-lying excited states of the nucleus, and we will show by various examples that the familiar results of the microscopic approach to nuclear structure follows from these equations: secular equations for the case of multipole and spin-multipole forces, equations for pairing-vibration frequencies, and equations for the theory of finite Fermi systems.

We adopt the total Hamiltonian for the system in a quite general form:

$$\left.\begin{aligned} H &= \sum_{ff'} T(f, f') a_f^+ a_{f'} - \frac{1}{4} \sum_{f_1 f_2 f_1' f_2'} G(f_1 f_2; f_2' f_1') a_{f_1}^+ a_{f_2}^+ a_{f_2'} a_{f_1'};\\ T(f, f') &= I(f, f') - \lambda \delta_{ff'}, \end{aligned}\right\} \tag{1}$$

where f is the set of quantum numbers characterizing the one-particle states, a_f^+ and a_f are the Fermion creation and annihilation operators, respectively, λ is the chemical potential, I is the one-particle Hamiltonian, and G is the particle-interaction matrix.

Since the operators a_f^+ and a_f do not commute, and since the Hamiltonian is Hermitian, we see that

$$\left.\begin{aligned} I(f, f') &= I^*(f', f);\\ G(f_1 f_2; f_2' f_1') &= -G(f_1 f_2; f_1' f_2') = -G(f_2 f_1; f_2' f_1')\\ &= G(f_2 f_1; f_2' f_1') = G^*(f_1' f_2'; f_2 f_1). \end{aligned}\right\} \tag{2}$$

We will also use the representation $f = q, \sigma$, where $\sigma = \pm 1$ distinguishes between states which are conjugate with respect to time inversion:

$$\hat{T} a_{q\sigma}^+ \hat{T}^{-1} = s_\sigma a_{q-\sigma}. \tag{3}$$

Here $\hat{T}$ is the time-inversion operator, and the coefficients s_σ have the following properties:

$$s_\sigma s_{-\sigma} = -1; \quad s_\sigma^2 = 1. \tag{4}$$

Joint Institute for Nuclear Research, Dubna. Translated from Problemy Fiziki Élementarnykh Chastits i Atomnogo Yadra, Vol. 1, No. 2, pp. 365-390, 1971.

Using Eq. (3) we can show that the invariance of the Hamiltonian with respect to time inversion yields the properties

$$\left.\begin{aligned} I(q\sigma,\ q'\sigma') &= I^*(q-\sigma,\ q'-\sigma')\, s_\sigma s_{\sigma'}; \\ G(q_1\sigma_1,\ q_2\sigma_2;\ q_2'\sigma_2',\ q_1'\sigma_1') &= G^*(q_1-\sigma_1,\ q_2-\sigma_2; \\ & q_2'-\sigma_2',\ q_1'-\sigma_1')\, s_{\sigma 2} s_{\sigma_1} s_{\sigma_1'} s_{\sigma_2'}. \end{aligned}\right\} \tag{5}$$

Let us consider the function

$$F(f_1,\ f_2) = \langle a_{f_1}^+ a_{f_2} \rangle;\quad \Phi(f_1,\ f_2) = \langle a_{f_1} a_{f_2} \rangle. \tag{6}$$

The averaging is carried over the ground state of the system. We note that the equations of motion yield the following exact relations for the functions F and Φ:

$$\left.\begin{aligned} i\frac{\partial}{\partial t} F(f_1,\ f_2) &= \langle [a_{f_1}^+ a_{f_2},\ H] \rangle \equiv \mathfrak{B}(f_1,\ f_2); \\ i\frac{\partial}{\partial t} \Phi(f_1,\ f_2) &= \langle [a_{f_1} a_{f_2},\ H] \rangle \equiv \mathfrak{A}(f_1,\ f_2). \end{aligned}\right\} \tag{7}$$

In the self-consistent-field method $\mathfrak{A}$ and $\mathfrak{B}$ may be expressed in terms of F and Φ [1, 2]:

$$\begin{aligned} \mathfrak{A}(f_1,\ f_2) &= \sum_f \{\xi(f_1,\ f)\,\Phi(f,\ f_2) + \xi(f_2,\ f)\,\Phi(f_1,\ f)\} \\ &- \frac{1}{2}\sum_{ff_1'f_2'} \Phi(f_2',\ f_1')\,\{G(f_1 f;\ f_2' f_1')\,F(f,\ f_2) + G(f,\ f_2;\ f_2' f_1') \\ &\times F(f,\ f_1)\} + \frac{1}{2}\sum_{f_1' f_2'} \Phi(f_2',\ f_1')\,G(f_1 f_2;\ f_2' f_1'); \end{aligned} \tag{8}$$

$$\begin{aligned} \mathfrak{B}(f_1,\ f_2) &= \sum_f \{\xi(f_2,\ f)\,F(f_1,\ f) - \xi(f,\ f_1)\,F(f,\ f_2)\} \\ &+ \frac{1}{2}\sum_{ff_1'f_2'} \{\Phi^*(f_1,\ f)\,G(f_2 f;\ f_2' f_1')\,\Phi(f_2',\ f_1') \\ &- \Phi(f_2,\ f)\,G(f_1 f;\ f_2' f_1')\,\Phi^*(f_2',\ f_1'), \end{aligned} \tag{9}$$

where

$$\xi(f,\ f') = T(f,\ f') - \sum_{f_1 f_2} G(ff_1;\ f_2 f')\,F(f_1,\ f_2). \tag{10}$$

The functions F and Φ are not independent, being related by

$$\left.\begin{aligned} F(f_1,\ f_2) &= \sum_f \{F(f_1,\ f)\,F(f,\ f_2) + \Phi^*(f,\ f_1)\,\Phi(f,\ f_2)\}; \\ 0 &= \sum_f \{F(f_1,\ f)\,\Phi(f,\ f_2) + F(f_2,\ f)\,\Phi(f,\ f_1)\}. \end{aligned}\right\} \tag{11}$$

If we are not interested in the time-independent ground state, we should solve, instead of Eqs. (3), the following equations [2]:

$$\mathfrak{A}(f_1,\ f_2) = 0;\quad \mathfrak{B}(f_1,\ f_2) = 0. \tag{12}$$

We denote the solutions for Eqs. (12) by $F_0(f_1, f_2)$ and $\Phi_0(f_1, f_2)$. We can obtain the same results as from a solution of Eqs. (12) by assuming that the ground state is the vacuum state for quasiparticles related to ordinary Fermions by a general Bogolyubov transformation:

$$a_f = \sum_\nu \{u(f,\ \nu)\,\alpha_\nu + v(f,\ \nu)\,\alpha_\nu^+\}, \tag{13}$$

whose coefficients satisfy

$$\left.\begin{array}{l}\sum_{\nu}(u(f,\nu)u^*(f',\nu)+v(f,\nu)v^*(f',\nu))=\delta_{ff'};\\ \sum_{\nu}(u(f,\nu)v(f',\nu)+u(f',\nu)v(f,\nu))=0.\end{array}\right\} \tag{14}$$

If we are interested in the spectrum of elementary excitations related to small oscillations about the ground state, we must consider small increments in the functions F_0 and Φ_0:

$$\left.\begin{array}{l}F(f,f')=F_0(f,f')+\delta F(f,f');\\ \Phi(f,f')=\Phi_0(f,f')+\delta\Phi(f,f').\end{array}\right\} \tag{15}$$

Equations can be obtained for δF and $\delta\Phi$ from Eqs. (7):

$$\left.\begin{array}{l}\mathrm{i}\frac{\partial}{\partial t}\delta F(f,f')=\delta\mathfrak{B}(f,f');\\ \mathrm{i}\frac{\partial}{\partial t}\delta\Phi(f,f')=\delta\mathfrak{B}(f,f').\end{array}\right\} \tag{16}$$

Moreover, δF and $\delta\Phi$ are not independent, but are related by auxiliary relations which follow from Eqs. (11):

$$\left.\begin{array}{l}\delta\left\{F(f_1,f_2)-\sum_f F(f_1,f)F(f,f_2)-\sum_f \Phi^*(f,f_1)\Phi(f,f_2)\right\}=0;\\ \delta\left\{\sum_f F(f_1;f)\Phi(f,f_2)+\sum_f F(f_2,f)\Phi(f,f_1)\right\}=0;\end{array}\right\} \tag{17}$$

Since δF and $\delta\Phi$ are related, it is more convenient to represent them in terms of new independent unknowns which automatically satisfy Eqs. (17).

Using canonical transformation (13), we write F, F_0, Φ, and Φ_0 in the form

$$\begin{aligned}F(f_1,f_2)=\langle a^+_{f_1}a_{f_2}\rangle=\sum_g v^*(f_1,g)v(f_2,g)+\sum_{g_1g_2}\{u^*(f_1,g_1)u(f_2,g_2)\langle\alpha^+_{g_1}\alpha_{g_2}\rangle\\ -v^*(f_1,g_1)v(f_2,g_2)\langle\alpha^+_{g_2}\alpha_{g1}\rangle+u^*(f_1,g_1)\\ \times v(f_2,g_2)\langle\alpha^+_{g_1}\alpha^+_{g_2}\rangle+v^*(f_1,g_1)u(f_2,g_2)\langle\alpha_{g_1}\alpha_{g_2}\rangle\};\end{aligned} \tag{18}$$

$$F_0(f_1,f_2)=\sum_g v^*(f_1,g)v(f_2,g); \tag{19}$$

$$\begin{aligned}\Phi(f_1,f_2)=\langle a_{f_1}a_{f_2}\rangle=\sum_g u(f_1,g)v(f_2,g)\\ +\sum_{g_1g_2}\{v(f_1,g_1)u(f_2,g_2)\langle\alpha^+_{g_1}\alpha_{g_2}\rangle-u(f_1,g_1)\\ \times v(f_2,g_2)\langle\alpha^+_{g_2}\alpha_{g_1}\rangle+u(f_1,g_1)u(f_2,g_2)\langle\alpha_{g_1}\alpha_{g_2}\rangle\\ +v(f_1,g_1)v(f_2,g_2)\langle\alpha^+_{g_1}\alpha^+_{g_2}\rangle\};\end{aligned} \tag{20}$$

$$\Phi_0(f_1,f_2)=\sum_g u(f_1,g)v(f_2,g). \tag{21}$$

In the self-consistent-field method, in the nonsteady-state formulation of the problem, the wave function for the nuclear ground state ceases to be a quasiparticle vacuum state. However, the average number of quasiparticles in the ground state is small, so we assume that this number is approximately zero:

$$\langle \alpha_{g_1}^{+} \alpha_{g_2} \rangle = 0. \tag{22}$$

To characterize deviations of the wave function from that of the quasiparticle vacuum state we introduce the coefficients

$$\mu(g_1, g_2) = \langle \alpha_{g_1} \alpha_{g_2} \rangle, \tag{23}$$

which satisfy

$$\mu(g_1, g_2) = -\mu(g_2, g_1). \tag{24}$$

We express δF and ΦF in terms of μ and μ^*:

$$\delta F(f_1, f_2) = \sum_{g_1 g_2} \{v^*(f_1, g_1) u(f_2, g_2) \mu(g_1, g_2) + u^*(f_1, g_1) v(f_2, g_2) \mu^*(g_2, g_1)\}; \tag{25}$$

$$\delta\Phi(f_1, f_2) = \sum_{g_1 g_2} \{u(f_1, g_1) u(f_2, g_2) \mu(g_1, g_2) + v(f_1, g_1) v(f_2, g_2) \mu^*(g_2, g_1)\}. \tag{26}$$

To obtain equations for μ, we write μ in terms of δF and $\delta\Phi$. Multiplying Eq. (25) by $v(f_1, g)$ and Eq. (26) by $u^*(f_1, g)$, combining them, summing over f_1, and using orthonormalization condition (14), we find

$$\begin{aligned}
&\sum_{f_1} \{v(f_1, g) \delta F(f_1, f_2) + u^*(f_1, g) \delta\Phi(f_1, f_2)\} \\
&= \sum_{f_1 g_1 g_2} \{[v(f_1, g) v^*(f_1, g_1) + u^*(f_1, g) u(f_1, g_1)] \\
&\times u(f_2, g_2) \mu(g_1, g_2) + [v(f_1, g) u^*(f_1, g_1) + u^*(f_1, g) v(f_1, g_1)] \\
&\times v(f_2, g_2) \mu^*(g_2, g_1) = \sum_{g_2} u(f_2, g_2) \mu(g, g_2).
\end{aligned} \tag{27}$$

Similarly, we multiply Eq. (25) by $u(f_1, g)$, multiply Eq. (26) by $v^*(f_1, g)$, combine, and sum over f_1 [with account of (14)], finding

$$\sum_{f_1} \{u^*(f_1, g) \delta F^*(f_1, f_2) + v(f_1, g) \delta\Phi^*(f_1, f_2)\} = -\sum_{g_2} v^*(f_2, g_2) \mu(g, g_2). \tag{28}$$

We again use this procedure: we multiply Eq. (27) by $u^*(f_2, g')$, multiply Eq. (28) by $v(f_2, g')$, subtract Eq. (28) from Eq. (27), and sum over f_2; we find

$$\begin{aligned}
\mu(g, g') = \sum_{f_1 f_2} \{&v(f_1, g) u^*(f_2, g') \delta F(f_1, f_2) + u^*(f_1, g) \\
&\times u^*(f_2, g') \delta\Phi(f_1, f_2) - u^*(f_1, g) v(f_2, g') \delta F^*(f_1, f_2) \\
&- v(f_1, g) v(f_2, g') \delta\Phi^*(f_1, f_2)\}.
\end{aligned}$$

Differentiating this expression with respect to t and taking (16) into account, we find an equation for μ:

$$\begin{aligned}
i \frac{\partial}{\partial t} \mu(g_1, g_2) = &\sum_{f_1 f_2} \{u^*(f_1, g_1) u^*(f_2, g_2) \delta\mathfrak{A}(f_1, f_2) \\
&+ v(f_1, g_1) v(f_2, g_2) \delta\mathfrak{A}^*(f_1, f_2) + v(f_1, g_1) u^*(f_2, g_2) \\
&\times \delta\mathfrak{B}(f_1, f_2) + u^*(f_1, g_1) v(f_2, g_2) \delta\mathfrak{B}(f_1, f_2)\}.
\end{aligned} \tag{29}$$

To find an explicit equation for μ, we express $\delta\mathfrak{A}$ and $\delta\mathfrak{B}$ in terms of μ; from Eqs. (8) and (9) we find

$$\begin{aligned}
\delta\mathfrak{A}(f_1, f_2) = &\sum_{f} \{\delta\xi(f_1, f) \Phi_0(f, f_2) + \xi_0(f_1, f) \delta\Phi(f, f_2) \\
&+ \delta\Phi(f_1, f) \xi_0(f_2, f) + \Phi_0(f_1, f) \delta\xi(f_2, f) - \frac{1}{2} \sum_{f f_1' f_2'} \delta\Phi(f_2', f_1') \\
&\times \{G(f_1 f; f_2' f_1') F_0(f, f_2) + G(f f_2; f_2' f_1') F_0(f_1, f)\}
\end{aligned}$$

$$-\frac{1}{2}\sum_{ff'_1f'_2}\Phi_0(f'_2, f'_1)\{G(f_1f; f'_2f'_1)\,\delta F(f, f_2)+G(ff_2; f'_2f'_1)$$

$$\times\,\delta F(f, f_1)+\frac{1}{2}\sum_{f'_1f'_2}\delta\Phi(f'_2, f'_1)\,G(f_1f_2; f'_2f'_1); \tag{30}$$

$$\delta\mathfrak{B}(f_1, f_2)=\sum_{f}\{\delta\xi(f_2, f)\,F_0(f_1, f)+\xi_0(f_2, f)\,\delta F(f_1, f)$$

$$-\delta\xi(f, f_1)\,F_0(f, f_2)-\xi_0(f, f_1)\,\delta F(f, f_2)\}+\frac{1}{2}\sum_{ff'_1f'_2}G(f_2f; f'_2f'_1)$$

$$\times\{\delta\Phi^*(f_1, f)\,\Phi_0(f'_2, f'_1)+\Phi_0^*(f_1, f)\,\delta\Phi(f'_2, f'_1)\}$$

$$-\frac{1}{2}\sum_{ff'_1f'_2}G(f_1f; f'_2f'_1)\{\delta\Phi(f_2, f)\,\Phi_0^*(f'_2, f'_1)+\Phi_0(f_2, f)\,\delta\Phi^*(f'_2, f'_1)\}, \tag{31}$$

where

$$\delta\xi(f, f')=-\sum_{f'_1f'_2}G(ff'_1; f'_2f')\,\delta F(f'_1, f'_2);$$

$$\xi_0(f, f')=T(f, f')-\sum_{f'_1f'_2}G(ff'_1; f'_2f')\,F_0(f'_1, f'_2).$$

We substitute Eqs. (30) and (31) into Eq. (29) and use Eqs. (25) and (26). Grouping similar terms and using the orthonormalization relation, we find, after lengthy calculations, the following equations:

$$i\frac{\partial}{\partial t}\mu(g_1, g_2)=\sum_{g'}(\Omega(g_2, g')\,\mu(g_1, g')-\Omega(g_1, g')$$

$$\times\,\mu(g_2, g'))+\sum_{g'_1g'_2}\{X(g_1g_2; g'_1g'_2)\,\mu(g'_1, g'_2)$$

$$+Y(g_1g_2; g'_1g'_2)\,\mu^*(g'_2, g'_1)\}; \tag{32}$$

$$-i\frac{\partial}{\partial t}\mu^*(g_1, g_2)=\sum_{g'}(\Omega^*(g_2, g')\,\mu^*(g_1, g')-\Omega^*(g_1, g')\,\mu^*(g_2, g'))$$

$$+\sum_{g'_1g'_2}\{X^*(g_1g_2; g'_1g'_2)\,\mu^*(g'_1, g'_2)+Y^*(g_1, g_2; g'_1g'_2)\,\mu(g'_2, g'_1)\}; \tag{33}$$

$$\Omega(g, g')=\sum_{ff'}\xi_0(f, f')\{u^*(f, g)\,u(f', g')-v^*(f, g)\,v(f', g')\}$$

$$-\sum_{f_1f_2}\{c^0_{f_1f_2}u^*(f_1, g)\,v^*(f_2, g')+c^{0*}_{f_1f_2}v(f_2, g)\,u(f_1, g')\}, \tag{34}$$

where

$$c^0_{f_1f_2}=\frac{1}{2}\sum_{f'_1f'_2}G(f_1f_2; f'_2f'_1)\,\Phi_0(f'_2, f'_1);$$

$$X(g_1g_2; g'_1g'_2)=-\frac{1}{2}\sum_{f_1f_2f'_1f'_2}G(f_1f_2; f'_2f'_1)$$

$$\times\{u(f_1, g_2)\,u(f_2, g_1)\,u(f'_1, g'_2)\,u(f'_2, g'_1)+v(f'_1, g_1)$$

$$\times\,v(f'_2, g_2)\,v(f_1, g'_1)\,v(f_2, g'_2)+(v(f'_1, g_1)\,u(f_1, g_2)$$

$$-u(f_1, g_1)\,v(f'_1, g_2))\,(v(f_2, g'_1)\,u(f'_2, g'_2)-v(f_2, g'_2)\,u(f'_2, g'_1))\}; \tag{35}$$

$$Y(g_1 g_2;\ g_1' g_2') = -\frac{1}{2} \sum_{f_1 f_2 f_1' f_2'} G(f_1 f_2;\ f_2' f_1')\{u(f_2, g_1)$$
$$\times u(f_1, g_2) v(f_2', g_1') v(f_1', g_2') + v(f_1', g_1) v(f_2', g_2)$$
$$\times u(f_2, g_2') u(f_1, g_1') + (v(f_1', g_1) u(f_1, g_2) - v(f_1', g_2)$$
$$\times u(f_1, g_1)) (u(f_2, g_1') v(f_2', g_2') - u(f_2, g_2') v(f_1', g_1'))\}. \tag{36}$$

We seek solutions of homogeneous equations (32) and (33) in the form

$$\left.\begin{aligned} \mu(g_1, g_2) &= \sum_\omega e^{-i\omega t} \psi_\omega(g_1, g_2); \\ \mu^*(g_1, g_2) &= \sum_\omega e^{-i\omega t} \varphi_\omega(g_1, g_2), \end{aligned}\right\} \tag{37}$$

where $\varphi_\omega = \psi^*_{-\omega}$.

The functions ψ_ω and φ_ω have the properties

$$\psi_\omega(g_1, g_2) = -\psi_\omega(g_2, g_1), \quad \varphi_\omega(g_1, g_2) = -\varphi_\omega(g_2, g_1). \tag{38}$$

Substituting Eqs. (37) into Eqs. (32) and (33), we find equations for the spectrum of elementary excitations:

$$\omega\psi_\omega(g_1, g_2) = \sum_{g'} \{\Omega(g_2, g')\psi_\omega(g_1, g') - \Omega(g_1, g')$$
$$\times \psi_\omega(g_2, g')\} + \sum_{g_1' g_2'} \{X(g_1 g_2;\ g_1' g_2')\psi_\omega(g_1', g_2')$$
$$- Y(g_1 g_2;\ g_1' g_2')\varphi_\omega(g_1', g_2')\}; \tag{39}$$

$$-\omega\varphi_\omega(g_1, g_2) = \sum_{g'} \{\Omega^*(g_2, g')\varphi_\omega(g_1, g') - \Omega^*(g_1, g')\varphi_\omega(g_2, g')\}$$
$$+ \sum_{g_1' g_2'} \{X^*(g_1 g_2;\ g_1' g_2')\varphi_\omega(g_1', g_2')$$
$$- Y^*(g_1 g_2;\ g_1', g_2')\psi_\omega(g_1', g_2')\}. \tag{40}$$

These equations were first derived in [1]. We note that if ψ_ω, φ_ω, and ω are solutions of Eqs. (39) and (40), the transformations

$$\omega \to -\omega, \quad \psi_\omega \to \varphi^*_\omega, \quad \varphi_\omega \to \psi^*_\omega$$

again yield the solution of this system.

Equations (39) and (40) were obtained without any assumptions whatsoever about the nature of the particle interaction or the structure of the ground state. We assume below that the functions I and G are real and that the functions u, v, and ξ_0 are both real and diagonal:

$$\left.\begin{aligned} u(f, g) &= u_f \delta_{fg};\ u_f \equiv u_{q\sigma} = u_q;\ u_f^* = u_f; \\ v(f, g) &= v_f \delta_{-fg};\ v_f \equiv v_{q\sigma} = s_\sigma v_q;\ v_f^* = v_f; \end{aligned}\right\} \tag{41}$$

$$\xi_0(f, f') = \xi(f)\delta_{ff'}, \quad \xi(f) = \xi^*(f); \tag{42}$$

$$\left.\begin{aligned} c^0_{ff'} &= c^{0*}_{ff'} = c_f \delta_{-f, f'}; \\ c_f &= \frac{1}{2} \sum_{f'} G(f, -f;\ -f', f') u_{f'} v_{f'}. \end{aligned}\right\} \tag{43}$$

Then we have

$$\Omega(g, g') = \sum_{ff'} \xi(f)\delta_{ff'} \{u_f \delta_{fg} u_{f'} \delta_{f'g'} - v_f \delta_{-fg} v_{f'} \delta_{-f'g'}\}$$

$$-\sum_{f_1 f_2} c_{f_1} \delta_{-f_1 f_2} \{u_{f_1} \delta_{f_1 g} v_{f_2} \delta_{-f_2 g'} + u_{f_1} \delta_{f_1 g'} v_{f_2} \delta_{-f_2 g}\}$$
$$= \delta_{gg'} \{\xi(g)(u_g^2 - v_g^2) + 2c_g u_g v_g\}.$$

Using results found from the superfluid model for the nucleus, we find

$$\sqrt{\Omega(g, g')} = \delta_{gg'} \varepsilon(g); \quad \varepsilon(g) = \sqrt{c_g^2 + \xi^2(g)}. \tag{44}$$

In this approximation we have

$$\begin{aligned} X(g_1 g_2; g_1' g_2') = &-\frac{1}{2} G(g_1 g_2; g_2' g_1') u_{g_1} u_{g_2} u_{g_2'} u_{g_1'} \\ &-\frac{1}{2} G(-g_1, -g_2; -g_2', -g_1') v_{g_1} v_{g_2} v_{g_1'} v_{g_2'} \\ &-\frac{1}{2} G(g_1, -g_2'; g_1', -g_2) u_{g_1} v_{g_2} u_{g_1'} v_{g_2'} \\ &-\frac{1}{2} G(-g_1, g_2'; -g_1' g_2) v_{g_1} u_{g_2} v_{g_1'} u_{g_2'} \\ &+\frac{1}{2} G(g_1, -g_1'; g_2', -g_2) u_{g_1} v_{g_2} u_{g_2'} v_{g_1'} \\ &+\frac{1}{2} G(-g_1, g_1'; -g_2', g_2) v_{g_1} u_{g_2} v_{g_2'} u_{g_1'}; \end{aligned} \tag{45}$$

$$\begin{aligned} Y(g_1 g_2; -g_1', -g_2') = &-\frac{1}{2} G(g_1 g_2; g_2' g_1') u_{g_1} u_{g_2} v_{g_1'} v_{g_2'} \\ &-\frac{1}{2} G(-g_1, -g_2; -g_2', -g_1') v_{g_1} v_{g_2} u_{g_1'} u_{g_2'} \\ &-\frac{1}{2} G(g_1, -g_2'; g_1', -g_2) u_{g_1} v_{g_2} u_{g_2'} v_{g_1'} \\ &+\frac{1}{2} G(-g_1, g_2'; -g_1', g_2) v_{g_1} u_{g_2} u_{g_1'} v_{g_2'} \\ &-\frac{1}{2} G(g_1, -g_1'; g_2', -g_2) u_{g_1} v_{g_2} u_{g_1'} v_{g_2'} \\ &-\frac{1}{2} G(-g_1, g_1'; -g_2', g_2) v_{g_1} u_{g_2} v_{g_1'} u_{g_2'}; \end{aligned} \tag{46}$$
$$X^* = X; \quad Y^* = Y.$$

Substituting Eqs. (44) into Eqs. (39) and (40), we find

$$\begin{aligned} \omega\psi_\omega(g_1, g_2) = &(\varepsilon(g_1) + \varepsilon(g_2)) \psi_\omega(g_1, g_2) \\ &+ \sum_{g_1' g_2'} \{X(g_1 g_2; g_1' g_2') \psi_\omega(g_1', g_2') \\ &- Y(g_1, g_2; -g_1', -g_2') \varphi_\omega(-g_1', -g_2')\}; \end{aligned} \tag{47}$$

$$\begin{aligned} -\omega\varphi_\omega(-g_1, -g_2) = &(\varepsilon(g_1) + \varepsilon(g_2)) \varphi_\omega(-g_1, -g_2) \\ &+ \sum_{g_1' g_2'} \{X(-g_1, -g_2; -g_1', -g_2') \varphi_\omega(-g_1', -g_2') \\ &- Y(-g_1, -g_2; g_1' g_2') \psi_\omega(g_1', g_2')\}. \end{aligned} \tag{48}$$

We introduce the new unknowns

$$Z^{(\pm)}(g_1, g_2) = \frac{1}{2} \{\psi_\omega(g_1, g_2) \pm \varphi_\omega(-g_1, -g_2)\}, \tag{49}$$

whose equations are

$$\omega Z^{(\mp)}(g_1, g_2) = (\varepsilon(g_1) + \varepsilon(g_2)) Z^{(\pm)}(g_1, g_2)$$

$$+\frac{1}{2}\sum_{g_1' g_2'}\{[X(g_1 g_2;\ g_1' g_2')+X(-g_1, -g_2;\ -g_1', -g_2')]$$
$$\mp[Y(+g_1, +g_2;\ -g_1', -g_2')+Y(-g_1, -g_2;\ +g_1' g_2')]\}$$
$$\times Z^{(\pm)}(g_1' g_2')+\frac{1}{2}\sum_{g_1' g_2'}\{[X(g_1 g_2;\ g_1' g_2')$$
$$-X(-g_1, -g_2;\ -g_1', -g_2')\pm[Y(g_1 g_2;\ -g_1', -g_2')$$
$$-Y(-g_1, -g_2;\ g_1' g_2')]\}\,Z^{(\mp)}(g_1', g_2'), \tag{50}$$

where

$$[X(g_1 g_2;\ g_1' g_2')+X(-g_1, -g_2, -g_1', -g_2')]\mp[Y(g_1 g_2;\ -g_1', -g_2')$$
$$+Y(-g_1, -g_2;\ g_1' g_2')]=-\frac{1}{2}[G(g_1 g_2;\ g_2' g_1')$$
$$+G(-g_1, -g_2;\ -g_2', -g_1')]\,v^{(\pm)}_{g_1 g_2}v^{(\pm)}_{g_1' g_2'}$$
$$-\frac{1}{2}\{[G(g_1, -g_2';\ g_1', -g_2)\mp G(g_1, -g_1';\ g_2', -g_2)]$$
$$+[G(-g_1, g_2';\ -g_1', g_2)\mp G(-g_1, g_1';\ -g_2', g_2)]\}\,u^{(\pm)}_{g_1 g_2}u^{(\pm)}_{g_1' g_2'}; \tag{51'}$$

$$[X(g_1 g_2;\ g_1' g_2')-X(-g_1, -g_2;\ -g_1', -g_2')]$$
$$\pm[Y(g_1, g_2;\ -g_1', -g_2')-Y(-g_1, -g_2;\ g_1', g_2')]$$
$$=-\frac{1}{2}[G(g_1, g_2;\ g_2', g_1')-G(-g_1, -g_2;\ -g_2', -g_1')]$$
$$\times v^{(\pm)}_{g_1 g_2}v^{(\mp)}_{g_1' g_2'}-\frac{1}{2}\{[G(g_1, -g_2';\ g_1', -g_2)$$
$$\pm G(g_1, -g_1';\ g_2', -g_2)]-[G(-g_1, g_2';\ -g_1', g_2)$$
$$\pm G(-g_1, g_1';\ g_2', g_2)]\}\,u^{(\pm)}_{g_1 g_2}u^{(\mp)}_{g_1' g_2'}; \tag{51''}$$

$$u^{(\pm)}_{gg'}=u_g v_{g'}\pm u_{g'} v_{g'}\quad v^{(\pm)}_{gg'}=u_g u_{g'}\mp v_g v_{g'}. \tag{52}$$

The basic equation may thus be written

$$\omega Z^{(\pm)}(g_1, g_2)=(\varepsilon(g_1)+\varepsilon(g_2))\,Z^{(\mp)}(g_1, g_2)$$
$$-\frac{1}{4}\sum_{g_1' g_2'}[G(g_1, g_2;\ g_2', g_1')+G(-g_1, -g_2;\ -g_2', -g_1')]$$
$$\times v^{(\mp)}_{g_1 g_2}v^{(\mp)}_{g_1' g_2'}Z^{(\mp)}(g_1', g_2')-\frac{1}{4}\sum_{g_1' g_2'}\{[G(g_1, -g_2';\ g_1', -g_2)$$
$$\mp G(g_1, -g_1';\ g_2', -g_2)]+[G(-g_1, g_2';\ -g_1', g_2)$$
$$\mp G(-g_1, g_1';\ -g_2', g_2)]\}\,u^{(\mp)}_{g_1 g_2}u^{(\mp)}_{g_1' g_2'}Z^{(\mp)}(g_1', g_2')$$
$$-\frac{1}{4}\sum_{g_1' g_2'}[G(g_1 g_2;\ g_2', g_1')-G(-g_1, -g_2;\ -g_2', -g_1')]$$
$$\times v^{(\mp)}_{g_1 g_2}v^{(\pm)}_{g_1' g_2'}Z^{(\pm)}(g_1', g_2')-\frac{1}{4}\sum_{g_1' g_2'}\{[G(g_1-g_2';\ g_1'-g_2)$$
$$\pm G(g_1-g_1';\ g_2'-g_2)]-[G(-g_1 g_2';\ -g_1' g_2)$$
$$\pm G(-g_1, g_1';\ -g_2' g_2)]\}\,u^{(\mp)}_{g_1 g_2}u^{(\pm)}_{g_1' g_2'}Z^{(\pm)}(g_1', g_2'). \tag{53}$$

We convert Eq. (53) into the q, σ representation, using relations (5):

$$\omega Z^{(\mp)}(q_1\sigma_1, q_2\sigma_2)=(\varepsilon(q_1)+\varepsilon(q_2))\,Z^{(\pm)}(q_1\sigma_1, q_2\sigma_2)-\frac{1}{4}\times$$

$$\times \sum_{q_1'\sigma_1', q_2'\sigma_2'} G(q_1\sigma_1, q_2\sigma_2; q_2'\sigma_2', q_1'\sigma_1')\left(1+s_{\sigma_1}s_{\sigma_2}s_{\sigma_1'}s_{\sigma_2'}\right)$$

$$\times v^{(\pm)}_{q_1\sigma_1, q_2\sigma_2} v^{(\pm)}_{q_1'\sigma_1', q_2'\sigma_2'} Z^{(\pm)}(q_1'\sigma_1', q_2'\sigma_2') - \frac{1}{4}\sum_{q_1'\sigma_1', q_2'\sigma_2'}$$

$$\times G(q_1\sigma_1, q_2\sigma_2; q_2'\sigma_2', q_1'\sigma_1')\left(1-s_{\sigma_1}s_{\sigma_2}s_{\sigma_1'}s_{\sigma_2'}\right) v^{(\pm)}_{q_1\sigma_1, q_2\sigma_2}$$

$$\times v^{(\mp)}_{q_1'\sigma_1', q_2'\sigma_2'} Z^{(\mp)}(q_1'\sigma_1', q_2'\sigma_2') - \frac{1}{4}$$

$$\times \sum_{q_1'\sigma_1', q_2'\sigma_2'} [G(q_1\sigma_1, q_2'-\sigma_2'; q_1'\sigma_1', q_2-\sigma_2)$$

$$\mp G(q_1\sigma_1, q_1'-\sigma_1'; q_2'\sigma_2', q_2-\sigma_2)]\left(1+s_{\sigma_1}s_{\sigma_2}s_{\sigma_1'}s_{\sigma_2'}\right)$$

$$\times u^{(\pm)}_{q_1\sigma_1, q_2\sigma_2} u^{(\pm)}_{q_1'\sigma_1', q_2'\sigma_2'} Z^{(\pm)}(q_1'\sigma_1', q_2'\sigma_2')$$

$$-\frac{1}{4}\sum_{q_1'\sigma_1', q_2'\sigma_2'} [G(q_1\sigma_1, q_2'-\sigma_2'; q_1'\sigma_1', q_2-\sigma_2)$$

$$\pm G(q_1\sigma_1, q_1'-\sigma_1'; q_2'\sigma_2', q_2-\sigma_2)]\left(1-s_{\sigma_1}s_{\sigma_2}s_{\sigma_1'}s_{\sigma_2'}\right)$$

$$\times u^{(\pm)}_{q_1\sigma_1, q_2\sigma_2} u^{(\mp)}_{q_1'\sigma_1', q_2'\sigma_2'} Z^{(\mp)}(q_1'\sigma_1', q_2'\sigma_2'). \quad (54)$$

To write Eq. (54) for the cases $\sigma_1 = \sigma_2$ and $\sigma_1 = -\sigma_2$, we use relations (41) and the properties of the coefficients s_σ; it follows from (41) that

$$u^{(\pm)}_{q\sigma, q'\sigma} = s_\sigma(u_q v_{q'} \pm u_{q'} v_q) \equiv s_\sigma u^{(\pm)}_{qq'};$$

$$u^{(\pm)}_{q\sigma, q'-\sigma} = s_{-\sigma}(u_q v_{q'} \mp u_{q'} v_q) \equiv s_{-\sigma} u^{(\mp)}_{qq'};$$

$$v^{(\pm)}_{q\sigma, q'\sigma} = u_q u_{q'} \mp v_q v_{q'} \equiv v^{(\pm)}_{qq'};$$

$$v^{(\pm)}_{q\sigma, q'-\sigma} = u_q u_{q'} \pm v_q v_{q'} \equiv v^{(\mp)}_{qq'}.$$

Using these equations, we find

$$\omega Z^{(\mp)}(q_1\sigma, q_2\sigma) = [\varepsilon(q_1) + \varepsilon(q_2)] Z^{(\pm)}(q_1\sigma, q_2\sigma)$$

$$-\frac{1}{2}\sum_{q_1' q_2' \sigma'} G(q_1\sigma, q_2\sigma; q_2'\sigma', q_1'\sigma') v^{(\pm)}_{q_1q_2} v^{(\pm)}_{q_1'q_2'} Z^{(\pm)}(q_1'\sigma', q_2'\sigma')$$

$$-\frac{1}{2}\sum_{g_1' g_2' \sigma'} G(q_1\sigma, q_2\sigma; q_2'-\sigma', q_1'\sigma') v^{(\pm)}_{q_1q_2} v^{(\pm)}_{q_1'q_2'} Z^{(\mp)}(q_1'\sigma', q_2'-\sigma')$$

$$-\frac{1}{2}\sum_{q_1' q_2' \sigma'} [G(q_1\sigma, q_2'-\sigma'; q_1'\sigma', q_2-\sigma)$$

$$\mp G(q_1\sigma, q_1'-\sigma'; q_2'\sigma', q_2-\sigma)]\, s_\sigma s_{-\sigma'} u^{(\pm)}_{q_1q_2} u^{(\pm)}_{q_1'q_2'} Z^{(\pm)}(q_1'\sigma', q_2'\sigma')$$

$$-\frac{1}{2}\sum_{q_1' q_2' \sigma'} [G(q_1\sigma, q_2^{1}-\sigma'; q_1'\sigma', q_2-\sigma)$$

$$\pm G(q_1\sigma, q_1'-\sigma'; q_2'-\sigma', q_2-\sigma)]\, s_\sigma s_{-\sigma'}$$

$$\times u^{(\pm)}_{q_1q_2} u^{(\pm)}_{q_1'q_2'} Z^{(\pm)}(q_1'\sigma', q_2'-\sigma'); \quad (55)$$

$$\omega Z^{(\pm)}(q_1\sigma, q_2-\sigma) = [\varepsilon(q_1) + \varepsilon(q_2)] Z^{(\mp)}(q_1\sigma, q_2-\sigma)$$

$$-\frac{1}{2}\sum_{q_1' q_2' \sigma'} G(q_1\sigma, q_2-\sigma; q_2'-\sigma', q_1'\sigma') v^{(\pm)}_{q_1q_2} v^{(\pm)}_{q_1'q_2'} \times$$

$$\times Z^{(\mp)}(q_1'\sigma', q_2'-\sigma') - \frac{1}{2}\sum_{q_1'q_2'\sigma'} G(q_1\sigma, q_2-\sigma; q_2'\sigma', q_1'\sigma')$$

$$\times v^{(\pm)}_{q_1q_2} v^{(\pm)}_{q_1'q_2'} Z^{(\mp)}(q_1'\sigma', q_2'\sigma') - \frac{1}{2}$$

$$\times \sum_{q_1'q_2'\sigma'} [G(q_1\sigma, q_2'\sigma'; q_1'\sigma', q_2\sigma) \pm G(q_1\sigma, q_1'-\sigma'; q_2'$$

$$-\sigma', q_2\sigma)] \times s_\sigma s_{\sigma'} u^{(\pm)}_{q_1q_2} u^{(\pm)}_{q_1'q_2'} Z^{(\mp)}(q_1'\sigma', q_2'-\sigma') - \frac{1}{2}$$

$$\times \sum_{q_1'q_2'\sigma'} [G(q_1\sigma, q_2'-\sigma'; q_1'\sigma', q_2\sigma) \mp G(q_1\sigma, q_1$$

$$-\sigma'; q_2'\sigma', q_2\sigma)] \times s_\sigma s_{-\sigma'} u^{(\pm)}_{q_1q_2} u^{(\pm)}_{q_1'q_2'} Z^{(\pm)}(q_1'\sigma', q_2'\sigma').$$

In deriving Eqs. (55) and (56), we have made use of the fact that during summation over σ_1' and σ_2' in (54), half of the terms vanish because of the factors $(1 \pm s_{\sigma_2} s_{\sigma_2} s_{\sigma_1'} s_{\sigma_2'})$.

Using relations (4) and (5), we can show that the coefficients

$$\left.\begin{array}{l}
\sum_\sigma G(q_1\sigma, q_2\sigma; q_2'\sigma', q_1'\sigma');\\
\sum_\sigma G(q_1\sigma, q_2\sigma; q_2'-\sigma', q_1'\sigma') s_{\sigma'};\\
\sum_\sigma s_\sigma G(q_1\sigma, q_2-\sigma; q_2'\sigma', q_1'\sigma');\\
\sum_\sigma s_\sigma G(q_1\sigma, q_2-\sigma; q_2'-\sigma', q_1'\sigma') s_{\sigma'};\\
\sum_\sigma G(q_1\sigma, q_2'\sigma'; q_1'\sigma', q_2\sigma);\\
\sum_\sigma G(q_1\sigma, q_2'-\sigma'; q_1'\sigma'; q_2\sigma) s_{-\sigma'};\\
\sum_\sigma s_{-\sigma} G(q_1\sigma, q_2'\sigma'; q_1'\sigma', q_2-\sigma);\\
\sum_\sigma s_{-\sigma} G(q_1\sigma, q_2'-\sigma'; q_1'\sigma', q_2-\sigma) s_{-\sigma'}
\end{array}\right\} \quad (57)$$

do not depend on σ'. For example, we have

$$\sum_\sigma s_{-\sigma} G(q_1\sigma, q_2'\sigma'; q_1'\sigma', q_2-\sigma) = \sum_\sigma s_{-\sigma} s_\sigma s_{-\sigma} s^2_{\sigma'}$$
$$\times G(q_1-\sigma, q_2'-\sigma'; q_1'-\sigma', q_2\sigma)$$
$$= \sum_\sigma s_\sigma G(q_1-\sigma, q_2'-\sigma'; q_1'-\sigma', q_2\sigma)$$
$$= \sum_\sigma s_{-\sigma} G(q_1\sigma, q_2'-\sigma'; q_1'-\sigma', q_2-\sigma).$$

Accordingly, we introduce a new notation for coefficients (57) which reflect this property:

$$\left.\begin{array}{l}
\frac{1}{2}\sum_\sigma G(q_1\sigma, q_2\sigma; q_2'\sigma', q_1'\sigma')\\
\qquad \equiv G^\xi(q_1+, q_2+; q_2'+, q_1'+);\\
\frac{1}{2}\sum_\sigma G(q_1\sigma, q_2\sigma; q_2'-\sigma', q_1'\sigma') s_{\sigma'}\\
\qquad \equiv G^\xi(q_1+, q_2+; q_2'-, q_1'+);\\
\frac{1}{2}\sum_\sigma s_\sigma G(q_1\sigma, q_2-\sigma; q_2'\sigma', q_1'\sigma')\\
\qquad \equiv G^\xi(q_1+, q_2-; q_2'+, q_1'+);
\end{array}\right.$$

$$\left.\begin{aligned}
&\frac{1}{2}\sum_{\sigma} s_{\sigma} G(q_1\sigma, q_2-\sigma;\ q_2'-\sigma',\ q_1'\sigma')\, s_{\sigma'}\\
&\qquad \equiv G^{\xi}(q_1+, q_2-;\ q_2'-,\ q_1'+);\\
&\frac{1}{2}\sum_{\sigma} G(q_1\sigma,\ q_2'\sigma';\ q_1'\sigma',\ q_2\sigma)\\
&\qquad \equiv G^{\omega}(q_1+,\ q_2+;\ q_2'+, q_1'+);\\
&\frac{1}{2}\sum_{\sigma} G(q_1\sigma,\ q_2'-\sigma';\ q_1'\sigma',\ q_2\sigma)\, s_{-\sigma'}\\
&\qquad \equiv G^{\omega}(q_1+,\ q_2+;\ q_2'-,\ q_1'+);\\
&\frac{1}{2}\sum_{\sigma} s_{-\sigma} G(q_1\sigma, q_2'\sigma';\ q_1'\sigma',\ q_2-\sigma)\\
&\qquad \equiv G^{\omega}(q_1+, q_2-;\ q_2'+, q_1'+);\\
&\frac{1}{2}\sum_{\sigma} s_{-\sigma} G(q_1\sigma, q_2'-\sigma;\ q_1'\sigma', q_2-\sigma)\, s_{-\sigma'}\\
&\qquad \equiv G^{\omega}(q_1+, q_2-;\ q_2'-,\ q_1'+).
\end{aligned}\right\} \tag{58}$$

We sum Eq. (55) over σ and multiply Eq. (56) by s_{σ} and sum it over σ; we find

$$\begin{aligned}
&\omega \sum_{\sigma} Z^{(\mp)}(q_1\sigma, q_2\sigma) = (\varepsilon(q_1)+\varepsilon(q_2)) \sum_{\sigma} Z^{(\pm)}(q_1\sigma, q_2\sigma)\\
&- \sum_{q_1' q_2'} G^{\xi}(q_1+, q_2+;\ q_2'+, q_1'+)\, v^{(\pm)}_{q_1 q_2} v^{(\pm)}_{q_1' q_2'} \sum_{\sigma} Z^{(\pm)}(q_1'\sigma, q_2'\sigma)\\
&- \sum_{q_1' q_2'} G^{\xi}(q_1+, q_2+; q_2'-, q_1'+)\, v^{(\pm)}_{q_1 q_2} v^{(\pm)}_{q_1' q_2'} \sum_{\sigma} s_{\sigma} Z^{\mp}(q_1'\sigma, q_2'-\sigma)\\
&- \sum_{q_1' q_2'} [G^{\omega}(q_1+, q_2-; q_2'-, q_1'+) \mp G^{\omega}(q_1+, q_2-;\ q_1'-, q_2'+)]\\
&\qquad \times u^{(\pm)}_{q_1 q_2} u^{(\pm)}_{q_1' q_2'} \sum_{\sigma} Z^{(\pm)}(q_1'\sigma, q_2'\sigma)\\
&- \sum_{q_1' q_2'} [G^{(\omega)}(q_1+, q_2-;\ q_2'+, q_1'+) \pm G^{\omega}(q_1+, q_2-;\\
&q_1'+,\ q_2'+)] \times u^{(\pm)}_{q_1 q_2} u^{(\pm)}_{q_1' q_2'} \sum_{\sigma} s_{\sigma} Z^{\mp}(q_1'\sigma, q_2'-\sigma);
\end{aligned} \tag{59}$$

$$\begin{aligned}
&\omega \sum_{\sigma} s_{\sigma} Z^{(\pm)}(q_1\sigma, q_2-\sigma) = (\varepsilon(q_1)+\varepsilon(q_2)) \sum_{\sigma} s_{\sigma} Z^{\mp}(q_1\sigma, q_2-\sigma)\\
&- \sum_{q_1' q_2'} G^{\xi}(q_1+, q_2-; q_2'-, q_1'+)\, v^{(\pm)}_{q_1 q_2} v^{(\pm)}_{q_1' q_2'} \sum_{\sigma} s_{\sigma} Z^{(\mp)} \times (q_1'\sigma, q_2'-\sigma)\\
&- \sum_{q_1' q_2'} G^{\xi}(q_1+, q_2-;\ q_2'+,\ q_1'+)\, v^{(\pm)}_{q_1 q_2} v^{(\pm)}_{q_1' q_2'}\\
&\times \sum_{\sigma} Z^{(\pm)}(q_1'\sigma, q_2'\sigma) - \sum_{q_1' q_2'} [G^{\omega}(q_1+, q_2+; q_2'+,\ q_1'+)\\
&\pm\ G^{\omega}(q_1+, q_2'+;\ q_1'+, q_2'+)]\, u^{(\pm)}_{q_1 q_2} u^{(\pm)}_{q_1' q_2'} \sum_{\sigma} s_{\sigma} Z^{(\mp)}(q_1'\sigma, q_2'-\sigma)\\
&- \sum_{q_1' q_2'} [G^{\omega}(q_1+, q_2+; q_2'-, q_1'+) \mp G^{\omega}(q_1+, q_2+;\ q_1'-,\ q_2'+)]\\
&\qquad \times u^{(\pm)}_{q_1 q_2} u^{(\pm)}_{q_1' q_2'} \sum_{\sigma} Z^{(\pm)}(q_1'\sigma, q_2'\sigma).
\end{aligned} \tag{60}$$

Since the coefficients $\sum_\sigma Z^{(\pm)}(q\sigma, q'\sigma)$ are antisymmetric with respect to interchange of the indices q and q', while the coefficients $\sum_\sigma s_\sigma Z^{(\pm)}(q\sigma, q'-\sigma)$ are symmetric with respect to this interchange, and since for each type of excitation for fixed q and q' only one of the coefficients $\sum_\sigma Z^{(\pm)}(q\sigma, q'\sigma)$ and $\sum_\sigma s_\sigma Z^{(\pm)}(q\sigma, q'-\sigma)$ is nonvanishing, we see that we can significantly simplify Eqs. (59) and (60). First, we may omit terms containing $\pm G^\omega$ on the basis of the symmetry properties of the coefficients. Second, we can replace the unknowns $\sum_\sigma Z^{(\pm)}(q\sigma, q'\sigma)$ and $\sum_\sigma s_\sigma Z^{(\pm)}(q\sigma, q'-\sigma)$ in the summation over q and q' by their sum, since in each case only one of the terms is nonvanishing. Instead of Eqs. (59) and (60) we then find

$$\omega \sum_\sigma Z^{(\mp)}(q_1\sigma, q_2\sigma) = (\varepsilon(q_1) + \varepsilon(q_2)) \sum_\sigma Z^{(\pm)}(q_1\sigma, q_2\sigma)$$
$$- \sum_{q_1' q_2'} [G^\xi(q_1+, q_2+; q_2'+, q_1'+) + G^\xi(q_1+, q_2+; q_2'-, q_1'+)]$$
$$\times v^{(\pm)}_{q_1 q_2} v^{(\pm)}_{q_1' q_2'} \Big[\sum_\sigma Z^{(\pm)}(q_1'\sigma, q_2'\sigma) + \sum_\sigma s_\sigma Z^{(\mp)}(q_1'\sigma, q_2'-\sigma)\Big]$$
$$-2 \sum_{q_1' q_2'} [G^\omega(q_1+, q_2-; q_2'-, q_1'+) + G^\omega(q_1+, q_2-; q_2'+, q_1'+)]$$
$$\times u^{(\pm)}_{q_1 q_2} u^{(\pm)}_{q_1' q_2'} \Big[\sum_\sigma Z^{(\pm)}(q_1'\sigma, q_2'\sigma) + \sum_\sigma s_\sigma Z^{(\mp)}(q_1'\sigma; q_2'-\sigma)\Big]; \quad (61)$$

$$\omega \sum_\sigma s_\sigma Z^{(\pm)}(q_1\sigma, q_2-\sigma) = [\varepsilon(q_1) + \varepsilon(q_2)] \sum_\sigma s_\sigma Z^{(\mp)}(q_1\sigma, q_2-\sigma)$$
$$- \sum_{q_1' q_2'} [G^\xi(q_1+, q_2-; q_2'-, q_1'+) + G^\xi(q_1+, q_2-; q_2'+, q_1'+)]$$
$$\times v^{(\pm)}_{q_1 q_2} v^{(\pm)}_{q_1' q_2'} \Big[\sum_\sigma Z^{(\pm)}(q_1'\sigma, q_2'\sigma) + \sum_\sigma s_\sigma Z^{(\mp)}(q_1'\sigma, q_2'-\sigma)\Big]$$
$$-2 \sum_{q_1' q_2'} [G^\omega(q_1+, q_2+; q_2'+, q_1'+) + G^\omega(q_1+, q_2+; q_2'-, q_1'+)]$$
$$\times u^{(\pm)}_{q_1 q_2} u^{(\pm)}_{q_1' q_2'} \Big(\sum_\sigma Z^{(\pm)}(q_1'\sigma, q_2'\sigma) + \sum_\sigma s_\sigma Z^{(\mp)}(q_1'\sigma, q_2'-\sigma)\Big). \quad (62)$$

Combining Eqs. (61) and (62), introducing the new unknowns

$$R^{(\pm)}(q, q') = \sum_\sigma Z^{(\pm)}(q\sigma, q'\sigma) + \sum_\sigma s_\sigma Z^{(\mp)}(q\sigma, q'-\sigma)$$

and introducing the notation

$$G^\xi(q_1 q_2; q_2' q_1') \equiv G^\xi(q_1+, q_2+; q_2'+, q_1'+)$$
$$+ G^\xi(q_1+, q_2+; q_2'-, q_1'+) + G^\xi(q_1+, q_2-; q_2'+, q_1'+)$$
$$+ G^\xi(q_1+, q_2-; q_2'-, q_1'+);$$
$$G^\omega(q_1 q_2; q_2' q_1') \equiv G^\omega(q_1+, q_2+; q_2'+, q_1'+)$$
$$+ G^\omega(q_1+, q_2+; q_2'-, q_1'+) + G^\omega(q_1+, q_2-; q_2'+, q_1'+)$$
$$+ G^\omega(q_1+, q_2-; q_2'-, q_1'+),$$

we find equations for the new unknowns:

$$\omega R^{(\mp)}(q_1, q_2) = [\varepsilon(q_1) + \varepsilon(q_2)] R^{(\pm)}(q_1, q_2)$$
$$- \sum_{q_1' q_2'} G^\xi(q_1 q_2; q_2' q_1') v^{(\pm)}_{q_1 q_2} v^{(\pm)}_{q_1' q_2'} R^{(\pm)}(q_1', q_2') -$$

$$-2\sum_{q_1' q_2'} G^{\omega}(q_1 q_2;\ q_2' q_1')\, u^{(\pm)}_{q_1 q_2} u^{(\pm)}_{q_1' q_2'} R^{(\pm)}(q_1', q_2'). \tag{63}$$

The interaction in the particle-particle channel affects the properties of the collective states through the terms of Eq. (63) which are proportional to $v^{(\pm)}_{qq'}$. The contribution of the interaction and the particle-hole channel is contained in terms proportional to $u^{(\pm)}_{qq'}$.

In studying the properties of low-lying nuclear states it should be kept in mind that the $C(g_1g_2;\ g_2'g_1')$ interaction is used for various momenta of the colliding particles. Some of the collective effects associated with quadrupole, octupole, etc., correlations in the particle-hole channel are governed by the interaction with small momentum transfer [in this case, this is $G^0(q_1q_2;\ q_2'q_1')$]. Other effects are related to pairing correlations of the superconducting type. These effects are governed by the interaction with a vanishing net colliding-particle momentum [$G^{\xi}(q_1q_2;\ q_2'q_1')$]. Generally speaking, these two interactions should be considered independent.

In deriving Eqs. (63) we treated the interparticle interaction in general form. We know, however, that the appearance of vibrational levels in nuclei is due primarily to an interaction in the particle-hold channel, which makes a coherent contribution. For this reason we consider the case in which the effect of the interaction in the particle-particle channel on the properties of the vibrational states may be neglected. We write the interaction in the particle-hole channel as a sum of multipole and spin-multipole interactions:

$$G^{\omega}(q_1 q_2;\ q_2' q_1') = \varkappa_f f(q_1, q_2) f(q_1', q_2') + \varkappa_t t(q_1, q_2)\, t(q_1', q_2'), \tag{64}$$

where $f(q, q')$ and $t(q, q')$ are the one-particle matrix elements of the operators corresponding to the multipole and spin-multipole moment, respectively.

In this case Eqs. (63) become, when account is taken of the symmetry properties of the coefficients $R^{(\pm)}(q, q')$,

$$\begin{aligned}\omega R^{(-)}(q_1, q_2) &= [\varepsilon(q_1) + \varepsilon(q_2)]\, R^{(+)}(q_1, q_2)\\ &-2\varkappa_f f(q_1, q_2)\, u^{(+)}_{q_1 q_2} \sum_{q_1' q_2'} f(q_1', q_2')\, u^{(+)}_{q_1' q_2'} R^{(+)}_{(q_1' q_2')};\end{aligned} \tag{65}$$

$$\begin{aligned}\omega R^{(+)}(q_1, q_2) &= [\varepsilon(q_1) + \varepsilon(q_2)]\, R^{(-)}(q_1, q_2)\\ &-2\varkappa_t - t(q_1, q_2)\, u^{(-)}_{q_1 q_2} \sum_{q_1' q_2'} t(q_1', q_2') u^{(-)}_{q_1' q_2'} R^{(-)}(q_1', q_2').\end{aligned} \tag{66}$$

There are no terms in Eqs. (65) and (66) containing

$$\sum_{q_1' q_2'} t(q_1', q_2')\, u^{(+)}_{q_1' q_2'} R^{(+)}(q_1', q_2'),\quad \sum_{q_1' q_2'} f(q_1', q_2')\, u^{(-)}_{q_1' q_2'} R^{(-)}(q_1', q_2'),$$

since these sums vanish due to the symmetry of the coefficients f, t, $R^{(\pm)}$, and $u^{(\pm)}$ with respect to interchange of the indices q_1' and q_2'. The coefficients $R^{(\pm)}(q_1, q_2)$ are antisymmetric with respect to interchange of indices if identical σ_1 and σ_2 correspond to the given q_1 and q_2 for this type of excitation. Otherwise, these coefficients are symmetric with respect to interchange of indices. These symmetry properties are the same as those of the coefficient f and are opposite the symmetry properties of the coefficients t.

To transform Eqs. (65) and (66), we introduce the notation

$$\left.\begin{aligned}V^{(+)} &= 2\varkappa_f \sum_{qq'} f(q, q')\, u^{(+)}_{qq'} R^{(+)}(q, q');\\ V^{(-)} &= 2\varkappa_t \sum_{qq'} t(q, q')\, u^{(-)}_{qq'} R^{(-)}(q, q')\end{aligned}\right\} \tag{67}$$

and we rewrite Eqs. (65) and (66):

$$(\varepsilon(q_1)+\varepsilon(q_2))\,R^{(\pm)}(q_1,q_2)-\omega R^{(\mp)}(q_1,q_2)=\begin{Bmatrix} f(q_1,q_2)\\ t(q_1,q_2)\end{Bmatrix} u^{(\pm)}_{q_1q_2}V^{\pm}. \tag{68}$$

We find

$$\left.\begin{aligned} R^{(+)}(q_1,q_2)&=\frac{f(q_1,q_2)\,u^{(+)}_{q_1q_2}(\varepsilon(q_1)+\varepsilon(q_2))\,V^{(+)}+t(q_1,q_2)\,u^{(-)}_{q_1q_2}\,\omega V^{(-)}}{(\varepsilon(q_1)+\varepsilon(q_2))^2-\omega^2};\\ R^{(-)}(q_1,q_2)&=\frac{t(q_1,q_2)\,u^{(-)}_{q_1q_2}(\varepsilon(q_1)+\varepsilon(q_2))\,V^{(-)}+f(q_1,q_2)\,u^{(+)}_{q_1q_2}\,\omega V^{(+)}}{(\varepsilon(q_1)+\varepsilon(q_2))^2-\omega^2}.\end{aligned}\right\} \tag{69}$$

Substituting Eqs. (69) into Eqs. (67) and setting the determinant of the resulting system of linear equations equal to zero, we find the secular equation, which yields the collective-vibration frequencies:

$$\left(1-2\varkappa_f\sum_{qq'}\frac{f^2(q,q')\,u^{(+)2}_{qq'}(\varepsilon(q)+\varepsilon(q'))}{(\varepsilon(q)+\varepsilon(q'))^2-\omega^2}\right)\left(1-2\varkappa_t\sum_{qq'}\frac{t^2(q,q')\,u^{(-)2}_{qq'}(\varepsilon(q)+\varepsilon(q'))}{(\varepsilon(q)+\varepsilon(q'))^2-\omega^2}\right)$$

$$=4\varkappa_t\varkappa_t\left\{\sum_{qq'}\frac{f(q,q')\,t(q,q')\,u^{(+)}_{qq'}\,u^{(-)}_{qq'}\,\omega}{(\varepsilon(q)+\varepsilon(q'))^2-\omega^2}\right\}^2.$$

This equation was studied in [3] in a treatment of quadrupole states in deformed nuclei. Setting $\varkappa_t = 0$, we find the well-known secular equation for the multipole-multipole interaction:

$$1=2\varkappa_f\sum_{qq'}\frac{f^2(q,q')\,u^{(+)2}_{qq'}(\varepsilon(q)+\varepsilon(q'))}{(\varepsilon(q)+\varepsilon(q'))^2-\omega^2}.$$

The roots of this equation are the energies of the vibrational states. Solutions of equations of this type were found in a study of the vibrational states in spherical [4] and deformed nuclei [5].

In addition to the vibrational levels, whose appearance is due primarily to the interaction in the particle-hole channel, collective states exist in the nuclei whose properties are governed primarily by the interaction in the particle channel. Pairing vibrations are an example of such states. To examine the properties of the pairing vibrational states, we set

$$\left.\begin{aligned} G^{\omega}(q_1q_2;\,q_2'q_1')&=0;\\ G^{\xi}(q_1q_2;\,q_2'q_1')&=G\delta_{q_1q_2}\delta_{q_1'q_2'}.\end{aligned}\right\} \tag{70}$$

Here Eqs. (63) become

$$\omega R^{(\mp)}(q,q)=2\varepsilon(q)\,R^{(\pm)}(q,q)-Gv^{(\pm)}_{qq}\sum_{q'}v^{(\pm)}_{q'q'}R^{(\pm)}(q',q'). \tag{71}$$

We introduce the notation

$$d^{(\pm)}=G\sum_{q}v^{(\pm)}_{qq}R^{(\pm)}(q,q). \tag{72}$$

Then Eqs. (71) may be written

$$2\varepsilon(q)\,R^{(\pm)}(q,q)-\omega R^{(\mp)}(q,q)=v^{(\pm)}_{qq}d^{(\pm)}, \tag{73}$$

from which it follows that

$$R^{(\pm)}(q,q)=\frac{2\varepsilon(q)\,v^{(\pm)}_{qq}d^{(\pm)}+\omega v^{(\mp)}_{qq}d^{(\mp)}}{4\varepsilon^2(q)-\omega^2}. \tag{74}$$

Substituting Eq. (74) into Eq. (72) and setting the determinant of the resulting system of linear equations equal to zero, we find an equation for the pairing-vibration frequencies:

$$\left\{\sum_q \frac{2\varepsilon(q)\, v_{qq}^{(+)2}}{4\varepsilon^2(q)-\omega^2}-\frac{1}{G}\right\}\left\{\sum_q \frac{2\varepsilon(q)}{4\varepsilon^2(q)-\omega^2}-\frac{1}{G}\right\}=\omega^2\left\{\sum_q \frac{v_{qq}^{(+)}}{4\varepsilon^2(q)-\omega^2}\right\}^2. \tag{75}$$

This type of equation was obtained for spherical nuclei in [6] and for deformed nuclei in [7]. These equations form the basis for the theory of pairing vibrations [8, 9].

We consider the more general case in which

$$G^{(\omega)}(q_1 q_2;\, q_2' q_1')=\varkappa f(q_1,\, q_2)\, f(q_1',\, q_2');$$
$$G^{\xi}(q_1 q_2;\, q_2' q_1')=G\delta_{q_1 q_2}\delta_{q_1' q_2'}.$$

Then Eqs. (63) become

$$\left.\begin{aligned}\omega R^{(-)}(q_1,\, q_2)=&\left(\varepsilon(q_1)+\varepsilon(q_2)\right)R^{(+)}(q_1,\, q_2)-\varkappa f(q_1,\, q_2)\, u_{q_1 q_2}^{(+)}\\ &\times\sum_{q_1' q_2'} f(q_1',\, q_2')\, u_{q_1' q_2'}^{(+)} R^{(+)}(q_1',\, q_2')-G v_{q_1 q_1}^{(+)}\delta_{q_1 q_2}\\ &\times\sum_{q'} v_{q' q'}^{(+)} R^{(+)}(q',\, q');\\ \omega R^{(+)}(q_1,\, q_2)=&\left(\varepsilon(q_1)+\varepsilon(q_2)\right)R^{(-)}(q_1,\, q_2)-G\delta_{q_1 q_2}\\ &\sum_{q'} R^{(-)}(q',\, q')\, v_{q' q'}^{(-)}.\end{aligned}\right\} \tag{76}$$

We introduce the notation

$$\left.\begin{aligned}V^{(+)}&=\varkappa\sum_{q,\, q'} f(q,\, q')\, u_{qq'}^{(+)} R^{(+)}(q,\, q');\\ d^{(\pm)}&=G\sum_q v_{qq}^{(\pm)} R^{(\pm)}(q,\, q).\end{aligned}\right\} \tag{77}$$

Then Eqs. (76) become

$$\left.\begin{aligned}&\left(\varepsilon(q_1)+\varepsilon(q_2)\right)R^{(+)}(q_1,\, q_2)-\omega R^{(-)}(q_1,\, q_2)\\ &\quad=f(q_1,\, q_2)\, u_{q_1 q_2}^{(+)} V^{(+)}+v_{q_1 q_1}^{(+)}\delta_{q_1 q_2} d^{(+)};\\ &-\omega R^{(+)}(q_1,\, q_2)+\left(\varepsilon(q_1)+\varepsilon(q_2)\right)R^{(-)}(q_1,\, q_2)=\delta_{q_1 q_2} d^{(-)},\end{aligned}\right\} \tag{78}$$

from which it follows that

$$\left.\begin{aligned}R^{(+)}(q_1,\, q_2)&=\frac{\left(\varepsilon(q_1)+\varepsilon(q_2)\right)f(q_1,\, q_2)\, u_{q_1 q_2}^{(+)} V^{(+)}+2\varepsilon(q_1)\, u_{q_1 q_1}^{(+)}\delta_{q_1 q_2} d^{(+)}+\omega\delta_{q_1 q_2} d^{(-)}}{\left(\varepsilon(q_1)+\varepsilon(q_2)\right)^2-\omega^2};\\ R^{(-)}(q_1,\, q_2)&=\frac{\omega f(q_1,\, q_2)\, u_{q_1 q_2}^{(+)} V^{(+)}+\omega v_{q_1 q_1}^{(+)}\delta_{q_1 q_2} d^{(+)}+2\varepsilon(q_1)\,\delta_{q_1 q_2} d^{(-)}}{\left(\varepsilon(q_1)+\varepsilon(q_2)\right)^2-\omega^2}.\end{aligned}\right\} \tag{79}$$

Substituting Eqs. (79) into Eqs. (77) and setting the determinant of the resulting system of linear equations equal to zero, we find an equation for the collective-excitation energies:

$$\det \begin{vmatrix} \left(\varkappa \sum_{qq'} \frac{\varepsilon(qq') f^2(qq') U_{qq'}^{(+)^2}}{\varepsilon^2(qq') - \omega^2} - 1\right) & \frac{G}{2} \sum_q \frac{f(qq) U_{qq}^{(+)} \omega}{4\varepsilon^2(q) - \omega^2} & \frac{G}{2} \sum_q \frac{2\varepsilon(q) U_{qq}^{(+)} V_{qq}^{(+)} f(qq)}{4\varepsilon^2(q) - \omega^2} \\ \varkappa \sum_q \frac{\omega U_{qq}^{(+)} f(qq)}{4\varepsilon^2(q) - \omega^2} & \left(\frac{G}{2} \sum_q \frac{2\varepsilon(q)}{4\varepsilon^2(q) - \omega^2} - 1\right) & \frac{G}{2} \sum_q \frac{\omega V_{qq}^{(+)}}{4\varepsilon^2(q) - \omega^2} \\ \varkappa \sum_q \frac{2\varepsilon(q) U_{qq}^{(+)} V_{qq}^{(+)} f(qq)}{4\varepsilon^2(q) - \omega^2} & \frac{G}{2} \sum_q \frac{\omega V_{qq}^{(+)}}{4\varepsilon^2(q) - \omega^2} & \left(\frac{G}{2} \sum_q \frac{2\varepsilon(q) V_{qq}^{(+)^2}}{4\varepsilon^2(q) - \omega^2} - 1\right) \end{vmatrix} = 0$$

$$\varepsilon(qq') \equiv \varepsilon(q) + \varepsilon(q') .$$

Similar equations were obtained in [8] for quadrupole states of deformed nuclei.

Having written down equations for the natural vibrations of the system, we turn now to vibrations under the influence of weak external fields; for this purpose we add to Hamiltonian (1) the term

$$\sum_{ff'} \delta I(f, f') a_f^+ a_{f'}, \tag{80}$$

where the function $\delta I(f, f') = \delta I^*(f', f)$ characterizes the external field. The expressions $\delta\mathfrak{A}(f, f')$ and $\delta\mathfrak{B}(f, f')$ should be supplemented by the terms

$$\delta\mathfrak{A}_{ex}(f_1, f_2) = \sum_f \{\delta I(f_1, f) \Phi(f, f_2) + \delta I(f_2, f) \Phi(f_1, f)\}; \tag{81}$$

$$\delta\mathfrak{B}_{ex}(f_1, f_2) = \sum_f \{\delta I(f_2, f) F(f_1, f) + \delta I(f, f_1) F(f, f_2)\}. \tag{81'}$$

Using

$$\delta I(f, f') = \sum_\omega e^{-i\omega t} \delta I_\omega(f, f'); \quad \delta I^*(f, f') = \sum_\omega e^{-i\omega t} \delta I^*_{-\omega}(f, f'), \tag{82}$$

we write Eqs. (39) and (40) (in the presence of an external field) as [1]

$$\begin{aligned} \omega\psi_\omega(g_1, g_2) = & \sum_{g'} \{\Omega(g_2, g') \psi_\omega(g_1, g') - \Omega(g_1, g') \psi_\omega(g_2, g')\} \\ & + \sum_{q_1' q_2'} \{X(g_1, g_2; g_1', g_2') \psi_\omega(g_1', g_2') - Y(g_1, g_2; g_1', g_2') \varphi_\omega(g_1', g_2')\} \\ & + \sum_{ff'} \{v(f', g_1) u^*(f, g_2) - u^*(f, g_1) v(f', g_2)\} \delta I_\omega(f, f'); \end{aligned} \tag{83}$$

$$\begin{aligned} -\omega\varphi_\omega(g_1, g_2) = & \sum_{g'} \{\Omega^*(g_2, g') \varphi_\omega(g_1, g') - \Omega^*(g_1, g') \varphi_\omega(g_2, g')\} \\ & + \sum_{q_1' q_2'} \{X^*(g_1 g_2; g_1' g_2') \varphi_\omega(g_1', g_2') - Y^*(g_1 g_2; g_1' g_2') \psi_\omega(g_1', g_2')\} \\ & + \sum_{ff'} \{v^*(f', g_1) u(f, g_2) - u(f, g_1) v^*(f', g_2)\} \delta I^*_{-\omega}(f, f'). \end{aligned} \tag{84}$$

We rewrite Eqs. (83) and (84) in approximation (41)-(43). We introduce the functions $R^{(\pm)}(q, q')$ and carry out calculations like those involved in the derivations of Eqs. (63); we find

$$\omega R^{(\mp)}(q_1, q_2) = (\varepsilon(q_1) + \varepsilon(q_2)) R^{(\pm)}(q_1, q_2) - \sum_{q_1' q_2'} G^\xi(q_1 q_2; q_2' q_1')$$

$$\times v^{(\pm)}_{q_1 q_2} v^{(\pm)}_{q_1' q_2'} R^{(\pm)}(q_1', q_2') - 2 \sum_{q_1' q_2'} G^{\omega}(q_1, q_2; q_2', q_1')$$
$$\times u^{(\pm)}_{q_1 q_2} u^{(\pm)}_{q_1' q_2'} R^{(\pm)}_{q_1' q_2'} - \frac{1}{2} u^{(\pm)}_{q_1 q_2} \left[\delta I_{\omega}(q_1, q_2) \pm \delta I^{*}_{-\omega}(q_1, q_2)\right], \tag{85}$$

where

$$\delta I_{\omega}(q_1, q_2) = \sum_{\sigma} (\delta I_{\omega}(q_1 \sigma, q_2 \sigma) - s_{\sigma} \delta I_{\omega}(q_1 \sigma, q_2 - \sigma));$$
$$\delta I^{*}_{\omega}(q_1, q_2) = \sum_{\sigma} (\delta I^{*}_{\omega}(q_1 \sigma, q_2 \sigma) - s_{\sigma} \delta I^{*}_{-\omega}(q_1 \sigma, q_2 - \sigma)).$$

To convert Eqs. (85) to a form similar to that in the theory of finite Fermi systems [10], we introduce

$$d^{(\pm)}(q_1, q_2) = \sum_{q_1' q_2'} G^{\xi}(q_1 q_2; q_2' q_1') v^{(\pm)}_{q_1' q_2'} R^{(\pm)}(q_1', q_2'); \tag{86}$$

$$V^{(\pm)}(q_1, q_2) = \sum_{q_1' q_2'} G^{\omega}(q_1 q_2; q_2' q_1') u^{(\pm)}_{q_1' q_2'} R^{(\pm)}(q_1', q_2') + V_0^{(\pm)}(q_1, q_2), \tag{87}$$

$$V_0^{(\pm)}(q_1, q_2) = \frac{1}{2} (\delta I_{\omega}(q_1, q_2) \pm \delta I^{*}_{-\omega}(q_1, q_2)). \tag{88}$$

Then we have

$$(\varepsilon(q_1) + \varepsilon(q_2)) R^{(\pm)}(q_1, q_2) - \omega R^{(\mp)}(q_1, q_2)$$
$$= v^{(\pm)}_{q_1 q_2} d^{(\pm)}(q_1, q_2) + u^{(\pm)}_{q_1 q_2} V^{(\pm)}(q_1, q_2). \tag{89}$$

from which it follows that

$$R^{(\pm)}(q_1, q_2) = \left[(\varepsilon(q_1) + \varepsilon(q_2))^2 - \omega^2\right]^{-1} \{(\varepsilon(q_1) + \varepsilon(q_2))$$
$$\times \left[u^{(\pm)}_{q_1 q_2} V^{(\pm)}(q_1, q_2) + v^{(\pm)}_{q_1 q_2} d^{(\pm)}(q_1, q_2)\right]$$
$$+ \omega \left[u^{(\mp)}_{q_1 q_2} V^{(\mp)}(q_1, q_2) + v^{(\mp)}_{q_1 q_2} d^{(\mp)}(q_1, q_2)\right]\}. \tag{90}$$

Substituting (90) into (86) and (87), we find

$$V^{(\pm)}(q_1, q_2) = V_0^{(\pm)}(q_1, q_2) + 2 \sum_{q_1' q_2'} G^{\omega}(q_1 q_2; q_2' q_1') u^{(\pm)}_{q_1' q_2'}$$
$$\times \left[(\varepsilon(q_1') + \varepsilon(q_2'))^2 - \omega^2\right]^{-1} \left\{ (\varepsilon(q_1') + \varepsilon(q_2')) \left[u^{(\pm)}_{q_1' q_2'} V^{(\pm)}(q_1', q_2')\right.\right.$$
$$\left.\left. + v^{(\pm)}_{q_1' q_2'} d^{(\pm)}(q_1', q_2')\right] + \omega \left[u^{(\mp)}_{q_1' q_2'} V^{(\mp)}(q_1', q_2') + v^{(\mp)}_{q_1' q_2'} d^{(\mp)}_{q_1' q_2'}\right]\right\}; \tag{91}$$

$$d^{(\pm)}(q_1, q_2) = \sum_{q_1', q_2'} G^{\xi}(q_1 q_2; q_2' q_1') \frac{v^{(\pm)}_{q_1' q_2'}}{(\varepsilon(q_1') + \varepsilon(q_2'))^2 - \omega^2}$$
$$\times \left\{(\varepsilon(q_1') + \varepsilon(q_2')) \left[u^{(\pm)}_{q_1' q_2'} V^{(\pm)}(q_1', q_2') + v^{(\pm)}_{q_1' q_2'} d^{(\pm)}(q_1', q_2')\right]\right.$$
$$\left. + \omega \left[u^{(\mp)}_{q_1' q_2'} V^{(\mp)}(q_1', q_2') + v^{(\mp)}_{q_1' q_2'} d^{(\mp)}(q_1', q_2')\right]\right\}. \tag{92}$$

We have obtained a system of equations for the four unknowns $V^{(\pm)}$ and $d^{(\pm)}$, but only two equations are independent, as Eqs. (63) show. For this reason it is more convenient to solve system (63) rather than (91) and (92).

We have thus derived from the equations of the self-consistent-field method the equations of the theory of finite Fermi systems usually obtained by a Green's-function technique. Although the equations of the self-consistent-field method are written for the distribution functions $<\Psi^+(t, r_1)\Psi(t, r_2)>$ and $<\Psi(t, r_1)\Psi(t, r_2)>$, this result cannot be considered unexpected: it is a consequence of the general theorem on the variation of the average value of a dynamic quantity [11]:

$$\delta\langle A(t)\rangle=\langle A(t)\rangle_{H+\delta H}-\langle A(t)\rangle_H$$
$$=2\pi\{e^{-iEt}\langle\langle A, B\rangle\rangle_E\,\delta\xi+e^{iEt}\langle\langle A, B\rangle\rangle^*_{-E}\,\delta\xi^*\},$$

where A(t) is some dynamic quantity in the Heisenberg picture, B is an operator which does not explicitly depend on the time, $\delta\xi$ is an infinitesimal C number, and $\langle\langle A, B\rangle\rangle_E$ is the Green's function in the E picture. This theorem relates variations in the distribution functions to the corresponding Green's function. Using this theorem, introducing weak external fields into the Hamiltonian, and varying with respect to the small parameter, we can always find equations for the Green's functions from equations for the distribution functions.

We have shown that of the mathematical methods available in microscopic nuclear theory the self-consistent-field method is the most general. With certain assumptions, its basic equations yield both equations for the effective field of the theory of finite Fermi systems and secular equations for the model with pairing and multipole forces. Even the self-consistent-field method, however, is not free of limitations. The fact that we use simple rules for splitting up the averages of products of four Fermi operators and the fact that we set the matrix elements $<\alpha_g^+\alpha_{g'}>$ equal to zero mean that we have neglected nonlinear effects. For this reason the self-consistent-field method is actually equivalent to the quasiboson approximation. Moreover, from the purely practical point of view, it is more convenient to use nuclear wave functions as in the method of approximate second quantization, rather than averages of operators as in the self-consistent-field method.

LITERATURE CITED

1. N. N. Bogolyubov, Usp. Fiz. Nauk, 67, 549 (1959).
2. N. N. Bogolyubov, Dokl. Akad. Nauk SSSR, 124, 1011 (1959).
3. N. I. Pyatov, Arkiv Fys., 36, 667 (1967).
4. L. Kisslinger and R. A. Sorensen, Rev. Mod. Phys., 35, 853 (1963).
5. V. G. Solov'ev (Soloviev), Atomic Energy Rev., 3, 117 (1965).
6. J. Hogaassen-Feldman, Nuclear Phys., 28, 258 (1961).
7. V. G. Solov'ev (Soloviev), Nuclear Phys., 69, 1 (1965).
8. A. Bohr, Nuclear Structure. Dubna Symposium, 1968, International Atomic Energy Agency, Vienna (1968).
9. B. Sorensen, Nuclear Phys., A134, 1 (1969).
10. A. B. Migdal, Theory of Finite Fermi Systems and Properties of Atomic Nuclei [in Russian], Nauka, Moscow (1968).
11. N. N. Bogolyubov, JINR Preprint D-781, Dubna (1961).

COLLECTIVE ACCELERATION OF IONS

I. N. Ivanov, A. B. Kuznetsov,
É. A. Perel'shtein, V. A. Preizendorf,
K. A. Reshetnikov, N. B. Rubin,
S. B. Rubin, and V. P. Sarantsev

The collective method in the acceleration of ions is described. Questions relating to the formation of an electron ring charged with ions are considered, together with problems of stability, focusing, and acceleration.

1. Introduction

The rapid development of high-energy physics has already led to a number of discoveries of fundamental significance. Nevertheless, in order to establish the ultimate laws governing the world of elementary particles and the structure of matter, the creation of accelerators producing particles with energies in the range of hundreds and thousands of GeV is absolutely vital.

Work is even now taking place in a number of countries on the manufacture of accelerators with energies of hundreds of gigaelectron volts; preliminary design exercises indicate that enormous amounts of equipment will be required for such accelerators. The weight of the electromagnets and the size and cost of the equipment involved already exceed all reasonable limits, being comparable with the total resources of whole nations. The reasons underlying this rapid rise in the size and cost of accelerators with increasing particle energy are the following.

In linear accelerators the effective field strength acting on the particles is comparatively low, so that a machine with an energy of the order of tens of gigaelectron volts or over must be made extremely long. In cyclical accelerators, which are at present capable of producing the highest energies of all such installations (strong-focusing synchrophasotrons), higher and higher energies can be achieved only by increasing the particle trajectory radius, since the greatest magnetic field now used for retaining particles in orbit is no greater than 12-15 kOe. The size of the electromagnet, its weight, power supply, and the net cost of the whole installation increase accordingly.

It is therefore quite obvious that the creation of accelerators of extremely high energies (order of 1000 GeV) will require the development of fundamentally new methods of acceleration, so that the effective fields acting on the particles (accelerating or retaining) may be far greater than the effective fields employed in present-day automatic-phasing accelerators.

In 1956 V. I. Veksler [1] indicated the possibility of achieving new acceleration mechanisms using collective interactions. The basic idea of these methods is that the field accelerating a particle is created not only by external sources but also by the interaction of the group of particles being accelerated with another group of charges, a stream of electrons, a flow of plasma, or electromagnetic radiation.

Essentially coherent methods of acceleration (all Veksler's new ideas were originally lumped together under this heading) are characterized by the fact that under certain conditions the field acting on an individual particle is proportional to the number of particles actually being aceelerated.

Joint Institute for Nuclear Research, Dubna. Translated from Problemy Fiziki Élementarnykh Chastits i Atomnogo Yadra, Vol. 1, No. 2, pp. 391-442, 1971.

In collective methods [2] the accelerating field is created by another group (assembly, ring, etc.) of charges. The field strength is proportional to the number of charges in the group; the number of accelerated particles may be quite arbitrary. It is not difficult to show that a substantial gain in accelerating field compared with the ordinary linear accelerator can be achieved only if there is a large number of particles (order of 10^{13}–10^{14}) in the accelerating group.

At the present time it is technically feasible to produce electron groups or assemblies containing 10^{14} particles. Such groups may be used for accelerating ions by the collective method. The essence of the method is that a small number of ions captured by the electron group is accelerated, under certain conditions, by its intrinsic field; the electron group itself may in turn be accelerated by external fields of moderate intensity:

$$\mathscr{E} = \mathscr{E}_{\text{к}} \cdot \frac{m_\perp}{M},$$

where $m_\perp$ is the effective mass of the electron,* M is the mass of an ion, $\mathscr{E}_{\text{к}}$ is the field acting on the ion by virtue of the electrons. In view of the great difference between the masses of the ions and electrons, the final energy of the ions is considerably greater than the energy of the electrons (by a factor of $M/m_\perp$ times). The present review is primarily concerned with the physical bases of the collective method of acceleration.

II. Formation of an Electron-Ion Group

1. Principal Requirements Imposed upon a Group and Method of Creating the Latter. One of the main problems arising in the creation of a collective accelerator is the formation of a charged electron-ion group. In this group the number of electrons has to be much greater than the number of ions, and the electron density has to be as high as possible. In any event it is essential that the fields acting on the ions in the group be of the order of 10^6–10^7 V/cm.

The collective method may then constitute an extremely promising technique for producing particles of extremely high energies (hundreds and thousands of gigaelectron volts) and also for making compact accelerators for multiply charged ions with a fair particle density and relatively high energies.

Simple analysis shows that an annular grouping with rotating electrons (an "electron ring") is the most suitable for these purposes. In such a ring the forces of Coulomb repulsion between the electrons are weakened by a factor of $\gamma_\perp^2$ as a result of magnetic attraction ($\gamma_\perp$ is the energy of the transverse motion of the electrons, referred to a unit of mc^2); fairly compact groupings with large numbers of electrons may accordingly be achieved. The more compact the electron ring, the greater will its Coulomb field be, and hence the stronger the forces acting on the ions. When it is the formation of a quiescent grouping which is in question, then these forces will retain the ions in the grouping. If, however, the duly formed ring is moving forward in the course of acceleration (Sec. III), then the Coulomb forces will determine the acceleration of the ions. The ring will in this case be polarized: the center of the ion formation will lie slightly behind the center of the electron formation, and the mean ion-accelerating force will coincide with the Coulomb force acting on the "central" ion by virtue of the electron ring.

At the present time a ring with the required parameters is in fact being created in a time-increasing magnetic field [3]. The electrons are injected from a heavy-current linear induction accelerator, over a period of one or several turns, into the large-radius orbit of a device called an "adhesor." A soft-focusing, barrel-shaped magnetic field is created in this installation by a series of iron-free coils. In the region of the injection radius $n = -(\partial B_z/\partial r \cdot r/B_z) \approx 0.5$; in the center $n = 0$ (B_z is the z component of the magnetic field).

The electron ring so formed is compressed in an adiabatically increasing, azimuthally uniform magnetic field. This compression is accompanied by the azimuthal acceleration of the electrons and an adiabatic reduction in the cross section of the ring. As a result of this, the intensity of the intrinsic electric field rises to the desired value. At the final stage of compression, ions are injected into the ring. The configuration of the coils creating the field enables the ring to be drawn out along the axis.

*The electron may execute a finite transverse motion; then $m_\perp = m\gamma_1$ is the mass of the electron with due allowance for this motion.

2. Equations of Motion of the Particles and Adiabatic Change in the Beam Parameters. The adhesor constitutes a betatron in which the ordinary condition, specifying the constancy of the radius of the equilibrium orbit (the 2 : 1 condition), is not satisfied, so that during the acceleration the electron trajectory takes the form of a deflected spiral. In contrast to the ordinary betatron, in which the accelerated currents are small and the intrinsic fields of the beam may be neglected, in the adhesor the intrinsic fields of the electron ring are comparable with the external fields.

It is convenient to conduct our analysis of particle motion using the approximation of the self-consistent field. Earlier [4, 5] quasistationary adiabatic beam models corresponding to linear intrinsic fields were considered. Models corresponding to two possible cases were chosen:

a) a beam without any energy spread (in the linear approximation), but with radial and axial oscillations ("symmetrical beam");

b) a beam having an energy spread and axial oscillations, but no radial betatron oscillations. Both models correspond to a beam of toroidal shape with a sharp boundary. The cross section of the beam is elliptical.

Apart from the external forces, the equilibrium electron moving along the axis of the ring is acted upon by a constant force due to the mutual influence of the intrinsic fields of different parts of the ring (the electrostatic repulsion of the ring along the major radius). This force introduces a correction into the formula relating the energy of the equilibrium particle to the magnetic field B_s in its orbit (radius R):

$$B_s = -\frac{mc^2 \beta_\theta \gamma_\perp}{eR}(1+\mu P). \tag{II.1}$$

Here e is the charge on the electron, β_θ is the transverse velocity of the electrons relative to the velocity of light c, $\gamma_\perp = 1/\sqrt{1-\beta_\theta^2}$; $\mu = \nu/\gamma_\perp$, where $\nu = r^*(N_e/2\pi R) = (r^*/\beta_\theta c)\cdot(I/e)$; N_e is the number of electrons in the ring, $r^* = 2.8 \times 10^{-13}$ cm is the classical radius of an electron, I is the beam current, $P = 2\ \ln(b+g)]$; b and g are the semiaxes of the elliptical cross section of the torus, referred to the radius of the orbit R (b along the z axis, g along the r axis).

For small beam currents ($\mu P \ll 1$) Eq. (II.1) transforms into the ordinary formula for the equilibrium value of the field in a betatron.

The momentum integral gives the following relation between $B_s(R, t)$ and $R(t)$:

$$B_s(R, t) R^2 [1-\delta(R)] = \text{const}, \tag{II.2}$$

where $\delta(R) = \frac{1}{B_s R^2}\int_0^R n(\xi) B_z(\xi)\, d\xi$ characterizes the nonuniformity of the magnetic field in the adhesor. In the case of a uniform field $\delta(R) \equiv 0$.

The change in the energy of the equilibrium relativistic electron is determined by the equation

$$\gamma_\perp R = \frac{\text{const}}{[1-\delta(R)](1+\mu P)} \approx \gamma_{\perp_0} R_0 \frac{[1-\delta(R_0)]}{[1-\delta(R)]} \approx \text{const}. \tag{II.3}$$

Here, $\mu \approx \mu_0 \frac{[1-\delta(R)]}{[1-\delta(R_0)]} \approx \text{const}$; the zero index relates to the initial values of the quantities.

We see from Eq. (II.3) that a reduction in the radius of the ring is accompanied by an increase in the energy of the electrons, resulting from the accelerating effect of the electric eddy field.

The behavior of the nonequilibrium particles is described by

$$\left.\begin{aligned} &\frac{d}{dt}(\gamma_\perp \dot{\rho}) + \gamma_\perp \omega_s^2 \nu_r^2 \rho = \frac{\omega_s \Delta\widetilde{M}}{Rm}; \\ &\frac{d}{dt}(\gamma_\perp \dot{z}) + \gamma_\perp \omega_s^2 \nu_z^2 z = 0. \end{aligned}\right\} \tag{II.4}$$

Here, $\omega_s = (c\beta_\theta/R)$; $\rho = r-R$; $\Delta\widetilde{M} = \widetilde{M}-\widetilde{M}_s$; $\widetilde{M}$ is the generalized momentum. The index s relates to the equilibrium particle. It follows from Eqs. (II.4) that the nonequilibrium particles execute betatron oscillations around the equilibrium position with dimensionless frequencies

$$\left.\begin{aligned} \nu_r^2 &= (1-n)(1+\mu P) - \left[\frac{4\mu}{g(g+b)\gamma_\perp^2\beta_\theta^2} + \frac{\mu P}{2}\right]; \\ \nu_z^2 &= n(1+\mu P) - \left[\frac{4\mu}{b(g+b)\gamma_\perp^2\beta_\theta^2} + \frac{\mu P}{2}\right]. \end{aligned}\right\} \tag{II.5}$$

The first terms in expressions (II.5) are the ordinary frequencies of betatron oscillations in a weakly-focusing magnetic field, but with a correction due to the repulsion of the ring along the major radius. The terms in square brackets characterize the displacement of the frequency arising from the intrinsic field of the ring.

On slowly changing the magnetic field, the following quantities constitute adiabatic invariants of Eqs. (II.4) in the case of a "symmetrical" beam:

$$J_{r,z} = \gamma_\perp R\beta_\theta \frac{a_{r,z}^2}{R^2}\nu_{r,z}. \tag{II.6a}$$

In the case of a beam with an energy spread, instead of the invariant J_r we have

$$\gamma_\perp R\beta_\theta \frac{a_r}{R}\nu_r^2 = \frac{\Delta\widetilde{M}}{mc} = \text{const}, \tag{II.6b}$$

where $a_{r,z}$ is the amplitude of the betatron oscillations.

Analysis of the adiabatic invariants enables us to find the cross section of the ring in its final state in relation to its initial parameters:

$$b/b_0 = \psi_b; \quad g/g_0 = \psi_g. \tag{II.7}$$

The functions ψ_b and ψ_g characterizing the change in the cross section of the ring are close to unity. It should be remembered that b and g are relative quantities, and hence, the small dimensions of the ring vary approximately as R.

A real beam is more complicated than the foregoing two "extreme" models, and will constitute an intermediate case between them. For one-turn injection the initial state corresponds more to the model of the "symmetrical" beam. For several-turn injection it is useful to vary the injection energy from turn to turn, since the radial betatron oscillations will be small, while the radial dimension associated with the energy spread will fall off more strongly. In this case the beam will be closer to the second model.

3. Intrinsic Field of the Electron Ring. The duly-shaped electron ring is subsequently accelerated by the external fields in the accelerating system, while the ions lying within it are accelerated by the intrinsic Coulomb field of the ring. The greater the intrinsic field of the ring, the greater is the efficiency with which the ions are accelerated in the collective accelerator.

If an ion lies at the edge of the potential well, then the accelerating field acting upon it is

$$\mathscr{E}_\text{K} = \frac{2|e|N_e}{\pi R^2(b+g)}. \tag{II.8}$$

We see from Eq. (II.8) that the greater the current in the ring and the greater its degree of compression the greater will the field accelerating the ions be. If we express all the parameters of the ring in terms of the initial parameters specified on injection, we obtain [3]

$$\mathscr{E}_\text{K} \approx \sqrt{N_e} \approx \sqrt{I_0}. \tag{II.9}$$

The fact that $\mathscr{E}_\text{K}$ depends on the square root of the current is associated with the fact that, although, according to Eq. (II.8), $\mathscr{E}_\text{K} \approx N_e$, the quantities b_0 and g_0 defining $b = b_0\psi_b$ and $g = g_0\psi_g$ in the denominator of this equation are nevertheless proportional to $\sqrt{N_e}$. The calculations carried out in the earlier paper [3] show that for a ring radius of R = 5 cm and an external magnetic field of $B_s = 2 \cdot 10^4$ G.

$$\mathcal{E}_{\text{K}} \approx 2.4 \sqrt{\frac{N_e}{10^{13}}} \text{ [MV/cm]}, \tag{II.10}$$

which for $N_e = 10^{13}$ gives $\mathcal{E}_{\text{K}} = 2.4$ MV/cm, and for $N_e = 10^{14}$ $\mathcal{E}_{\text{K}} = 7.6$ MV/cm.

4. Effect of Screens. In considering the motion of the electrons in a ring we have not so far taken any account of the effect of the image of the ring in the walls of the chamber. This influence is substantial when the ring lies quite close to the walls. The wall limiting the beam radially thus has the greatest effect. The influence of the beam image in this wall is considerable at the initial stage of deflection. Subsequently the beam moves away from the wall, and the effect of the image forces may be neglected.

Allowance for the image forces slightly displaces the initial frequencies of the betatron oscillations and affects the adiabatic change in the parameters of the ring. Consideration shows that the wall focuses the ring in the z direction (increases the frequency of the z oscillations) and defocuses it in the r direction (reduces the frequency of the r oscillations). In the adhesor these changes are not terribly vital. However, in the accelerating system the image forces are used to sustain the z dimensions of the ring (Sec. IV). A certain defocusing along the r direction is not particularly dangerous, since the r dimensions is fully maintained by the external magnetic field.

The effect of the image may also lead to instability of the ring with respect to the major radius, when the ring undergoes a random deviation from the axis of the adhesor and is drawn to the wall. In order to prevent this from happening, the following relation should be satisfied:

$$n < 1 - \frac{\xi(1+\xi)}{2} \cdot \frac{1}{(1-\xi)^2} \mu, \tag{II.11}$$

where $\xi = R/R_w$; R_w is the radius of the wall. For $\xi = 0.6$–0.8; n = 0.5; R = 40 cm, and $\gamma_\perp = 7$, inequality (II.11) is satisfied if $N_e < 2 \cdot 10^{14}$.

5. Part Played by Resonances in the Compression of the Ring. In the course of ring compression the frequencies of the betatron oscillations ν_r and ν_z change and may pass through a series of resonance values.

Passing through resonances in the sense of an increasing frequency is not really dangerous. The increase in the beam cross section arising from the effects of the resonance weakens the space charge and hence the frequency rises, which accelerates passage through the resonance. Calculations show that the tolerances imposed on the magnetic field in respect of tranversing radial and mixed resonances may readily be upheld.

It is also possible, however, to pass through z resonances in the sense of a falling frequency. This extends the resonance interaction and may cause the beam to be drawn into resonance. In order to avoid dangerous passages through resonance it is essential to choose an n(r) law of variation such that

$$0.1 \leqslant \nu_z < 0.25. \tag{II.12}$$

We then avoid the resonances $\nu_z = 1/2$; $1/3$; $1/4$.

6. Admission of the Ions. At the end of the compression process the hydrogen source starts operating. When the relativistic electrons collide with hydrogen molecules, molecular ions are chiefly formed (this is the most probable process, the ionization cross section being $\sigma = 10^{-19}$ cm^2), and in subsequent collisions with electrons these dissociate into a hydrogen atom and a proton. Collisions between ions leading to ionization and charge exchange may be neglected. Estimates show this to be valid for $(N_i/N_e) \ll 10^{-1}$ [6].

The admission of hydrogen into the adhesor takes place along the z axis so as to ensure uniform filling of the ring with neutral molecules. With this form of admission, the time for filling the volume of the ring with hydrogen molecules is determined by the dimensions of the small cross section of the ring, being equal to $t_v \approx 1$–0.1 μsec. This time is much smaller than $t_g \approx t_u$, the characteristic times of dissociation and ionization. For $N_e = 10^{13}$–10^{14}, $t_u \approx (50-5)$ μsec; we may therefore neglect the ring-filling time and consider that the ring is filled with hydrogen instantaneously. For times greater than t_u, the accumulation of protons in the ring takes place in accordance with a linear law. If $N_i = 10^{-2} N_e$, then for a hydrogen pressure of $p = 10^{-6}$ mm Hg the ion-injection time is $t = 50$ μsec. In the case of a greater hydrogen pressure this process may be accelerated.

One further point should be noted. In the presence of ions the betatron-oscillation frequencies will be slightly higher. It may well be that the corresponding shift in the radial frequency on injecting the ions will lead to a passage through the resonance $\nu_r = 1$. However, we must also remember the effect of the metal extraction tube (Section IV), which imparts a defocusing effect in the radial sense. The tube may prevent the frequency from passing through $\nu_r = 1$.

7. Other Possible Means of Creating an Electron Ring. The foregoing method of creating rings in a magnetic field that increases with time is not the only one possible. It is attractive to consider using an adhesor with a static magnetic field in order to create electron rings. It is simpler and cheaper to establish a static magnetic field than a varying one. Furthermore, in this case the number of accelerating cycles per unit time may be substantially increased, being limited solely by the potentialities of the injector.

Several methods of creating rings in a static magnetic field have been proposed [7-9]. The simplest of these [7] is as follows: an electron beam is injected at a certain angle in the weak fringing field of a solenoid. In a magnetic field increasing along the axis of the solenoid, the initial longitudinal momentum of the electrons P_z is transformed into a transverse momentum P_θ, and as a result of this a compressed electron ring is formed in the strong-field region in the center of the solenoid.

However, on using this method, the final transverse dimensions of the ring depend very substantially on the angular and energy spreads of the injected beam. It is accordingly quite impossible to achieve a ring of the desired parameters in this manner, and we shall pay no further attention to it.

There are two further methods by means of which an electron beam of the required parameters might (at least in principle) be obtained.

The first of these [8] lies in the fact that an electron ring formed by injection perpendicularly to the axis of the magnetic field is first accelerated in a magnetic field which diminishes along the z direction. This creates a field configuration satisfying the 2:1 condition:

$$\frac{\partial B_z}{\partial z} = \frac{1}{2} \cdot \frac{\partial \bar{B}_z}{\partial z}, \tag{II.13}$$

where $\bar{B}_z = \frac{2}{r^2} \int_0^r \xi B_z d\xi$ is the mean field in a circle of radius r. On satisfying condition (II.13) the radius of the ring remains constant. Then the accelerated ring is retarded to a complete standstill in a magnetic field adiabatically increasing along z. If at each z the magnetic field is almost uniform with respect to radius, we have $R \sim 1/Bz^{-/2}$. As a result of the asymmetry of this process, the ring finds itself in a stronger magnetic field, and acquires dimensions smaller than at the point of injection.

In view of the axial symmetry of the system, the azimuthal component of the generalized momentum should be conserved, this being written as

$$\tilde{M} = -R^2 \left(B_z - \frac{\bar{B}_z}{2}\right). \tag{II.14}$$

In order to create the required $\bar{B}_z$ in the accelerating section of constant radius, internal coils are required. In order to preserve the wholeness of the ring, $\bar{B}_z$ should be varied adiabatically, and the internal coils must therefore not break off sharply. At the end of the adhesor, the condition $\bar{B}_z = B_z$ should be observed in order to enable the ring to be extracted into the accelerating section. (Satisfaction of the condition $\bar{B}_z = B_z$ is not entirely obligatory if the ring is subsequently accelerated in a falling magnetic field.) Using Eq. (II.14) and the law of conservation of momentum we then have:

$$\left.\begin{aligned} \bar{B}_{z_0} &= 2B_{z_0} - B_{z_K} \frac{1}{s^2}; \\ B_{z_K} &= s \cdot B_{z_0}, \end{aligned}\right\} \tag{II.15}$$

where $s = R_0/R_k$ is the compression factor of the adhesor (the indices 0 and K refer to the beginning and end of the adhesor, respectively).

Maintenance of a small r dimension of the ring is ensured by the external magnetic field. In the z direction there is no focusing by the external field, since in the falling field there is no specially distinguished particle around which betatron oscillations are executed. (It was erroneously stated in one case [8] that the z dimension might also be kept intact by the external field.) Hence, the z dimension has to be preserved by some other means. In principle, focusing by a metal tube may be employed for this purpose (Sec. IV), by making use of the fact that the image forces have a focusing effect in the z direction. The tube profile then has to reproduce the varying radius of the ring. The question of focusing in this case demands special consideration. The injection of the ions in such an adhesor takes place at the final stage of compression, when the longitudinal velocity of the ring is low.

In order to extract the beam it is sufficient to make the field at the end of the adhesor a little smaller than the limiting value of B_{zk}, so that the ring may retain a slight velocity in the axial direction.

The second method of producing an electron beam of the required parameters [9] depends on the use of specially shaped internal and external solenoids; these create a magnetic-field configuration such that on a certain curve $r = r(z)$ $B_r \equiv 0$ [$r(z)$ diminishes monotonically]. Then the equilibrium electrons injected on the curve $r(z)$ in a field increasing along z will experience neither acceleration nor retardation.

The nonequilibrium electrons will execute oscillations around the curve $r(z)$, the frequencies being $\nu_{1,2}^2 = 1$ and 0. Since one of the frequencies is zero, the ring will spread. In order to avoid this, the static magnetic field is augmented by a magnetic pit, depression, or well traveling along the z direction; this ensures focusing of the beam and directs it into the strong-field region. The traveling magnetic well is created by means of a spiral slowing-down system, a current pulse being fed into this. Since the frequency $\nu^2 = 0$ corresponds to a position of neutral equilibrium, only a small traveling field is required in order to ensure focusing.

The foregoing methods of creating electron rings in a static magnetic field have a common failing. The energy of the electrons is constant in a static field, and since, in order to compensate the Coulomb repulsion in the final state, it is essential to have a fairly large value of $\gamma_{\perp k}$, the electrons must be injected with an energy $\gamma_{\perp k}$, which makes the injector expensive and more complicated. Substantial difficulties associated with maintaining a small z dimension of the ring also arise. Furthermore, the creation of complex field configurations by means of internal coils and solenoids is quite a difficult problem.

8. Extraction of the Beam from the Adhesor. After the formation of the electron-ion ring has been completed, it has to be drawn out along the axis of the adhesor for further acceleration. The extraction of the ring from the adhesors considered in Paragraph 7 is not a particularly difficult matter, since in this case there is no magnetic field barrier on passing into the accelerating system. It is therefore of greatest interest to consider the extraction of the beam from an adhesor with a magnetic field increasing in time.

The process of compressing the electron ring takes place in a potential well formed by current-carrying coils or turns symmetrically arranged with respect to the plane in which the electron ring lies. The currents in each pair of turns are identical. The resultant magnetic field increases on both sides of the symmetry axis of the system.

Thus, in order to extract the beam from the adhesor it is essential to overcome the barrier created by the increasing magnetic field. At the same time it is required to preserve the dimensions of the ring achieved by virtue of the compression and also to restrain the ions.

Several ways of extracting the beam from the compression chamber have been suggested [2, 10-12]. These may be divided into two groups. In the first group we have methods of extracting the ring by displacing the potential well together with the ring to a position outside the chamber [10]. In the second group we have extraction methods based on removing the magnetic barrier within the chamber and creating a magnetic field diminishing along the axial direction [2, 11, 12]. In one case [10] the displacement of the potential well is achieved by using additional coils which are switched on after completing the compression process. The magnetic field of these coils changes the original field distribution in the chamber so that the B_z field minimum at which the ring is situated moves slowly in a longitudinal direction in the desired sense. This method is convenient because it creates no disruption in the ring-focusing conditions ($n \approx$ const) and provides the necessary field gradients for the retention of the ions. However, the magnetic barrier is simply moved outside the chamber, not lowered. Furthermore, the longitudinal velocity of the electron-ion beam here equals the velocity of the potential well and is very low ($\beta_z \approx 10^{-5}$–10^{-4}, where β_z is the longitudinal velocity of the ring referred to c). For this reason the second group of methods is preferable, the symmetry of the magnetic field distribution being disturbed in these so as to exert a repulsive force on the ring in the desired direction.

Let us consider the general requirements on any extraction system that distorts the magnetic barrier.

The equation of motion along the z axis for a constant, uniform field* will be:

$$\frac{d}{dt} m\gamma \dot{z}_s = -\mu \frac{dB_z}{dz}. \tag{II.16}$$

The index s signifies the central (equilibrium) particle of the ring on which no Coulomb forces are acting. The quantity μ is the magnetic moment of the particles, constituting an adiabatic invariant:

$$\mu = \frac{m\gamma\beta_{\theta_0}^2 c^2}{2B_{z_0}} = \text{inv.}$$

Since γ = const, Eq. (II.16) may be written

$$\ddot{z}_s = -\frac{\beta_{\theta_0}^2 c^2}{2B_{z_0}} \cdot \frac{dB_z}{dz}. \tag{II.17}$$

For the initial conditions $t = 0$; $z = 0$; $\dot{z} = 0$; $B_z = B_{z_0}$ the first integral of Eq. (II.17) will be

$$\beta_z = \beta_{\theta_0} \sqrt{1 - \frac{B_z}{B_{z_0}}}. \tag{II.18}$$

We thus see that in order to extract the ring from the adhesor a magnetic field falling in the z direction must be created.

Let us estimate the permissible field gradients. In order to ensure that the ions should not be detached from the electron ring it is essential to prevent the acceleration of the electrons from exceeding the accelerations experienced by the ions. For an ion situated at the edge of the electron ring we may, according to Eq. (II.8), write down the following equation of motion:

$$\ddot{z}_i = \frac{2e^2 N_e}{\pi R^2 (b+g) M\gamma_z^3}, \tag{II.19}$$

where $\gamma_z^2 = (1/1-\beta_z^2)$. Equating the left-hand sides of relations (II.17) and (II.19), we obtain the limiting magnetic-field gradient

$$\left|\frac{dB_z}{dz}\right|_{\lim} = \frac{4e^2 N_e B_{z_0}}{\pi R^2 (g+b) M\gamma_z^3 \beta_{\theta_0}^2 c^2}. \tag{II.20}$$

The gradient of the magnetic field falling along the z direction (created in order to extract the ring from the compression chamber) must never exceed the limiting value specified by (II.20).

Focusing of the electron ring during compression is achieved by means of a weak barrel-shaped focusing magnetic field with a specified value of n. For the deviation of the particles from the equilibrium particle we have the linearized oscillation equations (II.4). The focusing conditions are ν_r^2 and $\nu_z^2 > 0$.

When the symmetry of the field distribution in the adhesor starts breaking down and a field diminishing in the z direction is adiabatically created, n falls and may pass through zero. Hence, starting from a certain instant of time, the focusing conditions will cease to be satisfied, and focusing will now have to be achieved by external methods (for example, by virtue of the image forces in a metal tube).

*More details are given in Sec. III, Paragraph 2.

All the methods of extracting the ring by creating a falling magnetic field are restricted by the foregoing common requirements as to field gradients and focusing conditions.

In two cases [2, 11] it was proposed that, in order to extract the ring from the adhesor by lifting the magnetic barrier, two open turns should be placed close to the maximum of the principal field and short-circuited at a specific instant of time. This instant would occur during the rise in the field created by the main turns. The induced field, directed in opposition to the field of the principal turns, would lift part of the magnetic barrier, and, since the current in the main turns would continue rising, at a specific instant of time the conditions required for the extraction of the ring (namely, the creation of a magnetic field falling along the z axis) would be created. This instant of time would be the instant at which the rise in current ceased in the main turns.

In order to smooth the field distribution, additional coils should be placed on the line of motion of the ring, the number of these coils and the currents passing through them being determined by the permissible field gradients.

The change in the magnetic field distribution in the adhesor amounts to a gradual reduction in the depth of the potential well and a simultaneous displacement of the latter, and subsequently to the creation of a diminishing magnetic field. While the depth of the potential well remains fairly large, it secures the focusing of the ring; as it diminishes, however, other means of focusing have to be used, unless one relies on the attainment of a self-focusing ring.

In another case [12] it was proposed that, in order to eliminate the magnetic barrier, the current in one of the main coils should be reduced after the end of the compression process. At the same time the current would remain constant in the symmetrical coil, or even increased. The symmetry of the field distribution would thus be disrupted, the field distorted, and ultimately a magnetic field falling in the direction of extraction of the ring would be created. This method assumes the possibility of creating pulsed currents greater than are required for compression at the specified ultimate dimensions of the ring.

The requirements imposed upon the gradients are satisfied by using a solenoid. The problem of focusing remains the same as in the previous method.

Thus, as a result of one means of distorting the shape of the magnetic field or another, the electron-ion ring falls into a region of diminishing magnetic field, in which the azimuthal velocity of the electrons is converted into longitudinal velocity and the ions in the ring are accelerated. The longitudinal velocity at the exit from the adhesor may be $\beta_i \approx (0.1–0.2)$.

III. Acceleration of an Electron-Ion Ring

1. Preliminary Comments. The problem of accelerating an electron ring charged with ions has a number of special features. The ring constitutes a compact formation with a large charge, and the intrinsic current created by this charge (of the order of tens of kiloamperes) loads the accelerating system severely.

The existence of rotational motion of the electrons leads to a substantial effective "weighting" of the ring. Hence, its acceleration to relativistic velocities takes place much more slowly than the acceleration of a simple group of electrons with the same charge. During acceleration the ring is polarized, and the ions are accelerated by their intrinsic Coulomb forces. These forces are determined by the parameters of the ring and are limited. Hence, the forces accelerating the ring must be equally limited.

These considerations taken together with the technical feasibilities and economic requirements of the accelerator determine the choice of structure for the accelerating system.

The principles to be followed in making this structure may clearly be the same as in the case of an ordinary charged-particle linear accelerator; however, in addition to this, the special characteristics of the problem in hand, involving the acceleration of a group of electrons in the form of a ring, admits a different kind of solution, for example, the use of a falling magnetic field and the combination of such a field with a system of resonators.

We shall now give some further consideration to the determination of the permissible accelerating fields and the use of a falling magnetic field together with a system of resonators for acceleration purposes (as we shall shortly show, this arrangement has great advantages over ordinary methods of acceleration), and finally, we shall discuss the important question of the power supply of the system.

2. Interaction between the Electron and Ion Components of the Ring, and the Permissible Accelerating Fields. The forces acting on the ions by virtue of the accelerated electron ring are determined from the following considerations. If the longitudinal dimensions of the electron ring remain constant in a system of coordinates accompanying the ring under the action of any particular external forces, then on accelerating the ring in a constant longitudinal electric field $\mathscr{E}$ stable acceleration of the ions by a constant force may readily be achieved [2]. The laws of longitudinal motion of the central electron and the central ion then take the following form:

$$m_{\perp}\gamma_{\|}^{3}\ddot{z}=e\left(\mathscr{E}-\frac{N_i}{N_e}\mathscr{E}_{\text{K}}\right); \tag{III.1}$$

$$M\gamma_{\|}^{3}\ddot{z}=e\left(\mathscr{E}_{\text{K}}-\mathscr{E}\right) \tag{III.2}$$

where

$$\mathscr{E}_{\text{K}}=\frac{2eN_e}{\pi r_0^2\left(b_c+g\right)}\Delta; \tag{III.3}$$

$m_{\perp} = m\gamma_{\perp}$ is the "weighted" mass of the electron $\gamma_{\|} = (1-\beta_z^2)^{-1/2}$; $\gamma_{\perp} = (1-\gamma_{\|}^2\beta_\theta^2)^{-1/2}$; β_z is the longitudinal velocity of the electron (along the z axis) referred to the velocity of light, r_0 is the ring radius, b_c is the longitudinal semidimension of the ring cross section in an accompanying system of coordinates, referred to r_0, Δ is the distance between the central particles referred to b_c ($\Delta \ll 1$), which characterizes the degree of polarization.

We see from Eqs. (III.1) and (III.2) that in this case the acceleration of the ring in the accompanying system is constant.

The condition that the laws of motion of the central electron and ion should coincide determines the permissible electric accelerating field:

$$\mathscr{E}_g=\mathscr{E}_{\text{K}}\frac{m_{\perp}}{M}\cdot\frac{1+\dfrac{M}{m_{\perp}}\cdot\dfrac{N_i}{N_e}}{1+\dfrac{m_{\perp}}{M}}, \tag{III.4}$$

The force accelerating the central electron is then equal to

$$e\mathscr{E}_{\text{K}}\frac{m_{\perp}}{M}\cdot\frac{1-\dfrac{N_i}{N_e}}{1+\dfrac{m_{\perp}}{M}},$$

while the force accelerating the central ion is $M/m_{\perp}$ times greater. We thus see that a considerable loading of the electron ring with ions (provided always that $N_i/N_e \ll 1$) has little effect on the efficiency of this method of acceleration.

For the parameters indicated in Paragraph 3 of Sec. II, if $\Delta = 0.5$, $\mathscr{E}_g$ varies between 25 kV/cm ($N_e = 10^{13}$) to 80 kV/cm ($N_e = 10^{14}$).

3. Acceleration of a Ring in a Falling Magnetic Field. Let us now consider the acceleration of a ring in a falling magnetic field (longitudinal and axially symmetric) [2]. In this case the energy accumulated in the rotational motion of the ring is transformed into translational motion. The equation of motion of the electron in such a field takes the form

$$\ddot{r}=r\dot{\theta}\left(\dot{\theta}+\frac{eB_z}{m\gamma c}\right); \tag{III.5}$$

$$\frac{d}{dt}\left(m\gamma r^2\dot{\theta}+\frac{e}{c}\int_0^r B_z\,\xi d\xi\right)=0; \tag{III.6}$$

$$\ddot{z} = -\frac{eB_z}{m\gamma c} r\dot{\theta}\frac{B_r}{B_z}. \quad \text{(III.7)}$$

On satisfying the following conditions:

$$\left|\gamma_{\|}\frac{r}{B_z}\cdot\frac{\partial B_z}{\partial z}\right| = \varepsilon \ll 1; \quad \left|\gamma_{\|}^2\frac{r^2}{B_z}\cdot\frac{\partial^2 B_z}{\partial z^2}\right| \leqslant \varepsilon^2, \quad \text{(III.8)}$$

which correspond to the drift approximation, the equations of motion in the linear approximation with respect to ε admit solutions of the form:

$$\dot{\theta} = -\frac{eB_z}{m\gamma c}; \quad \text{(III.9)}$$

$$r = r_0\sqrt{\frac{B_{z_0} - \bar{B}_{z_0}/2}{B_z - \bar{B}_z/2}}; \quad \text{(III.10)}$$

$$\ddot{z} = r^2\dot{\theta}^2\frac{1}{2B_z}\cdot\frac{\partial\bar{B}_z}{\partial z}. \quad \text{(III.11)}$$

In the case of an almost uniform field we have

$$r = r_0\sqrt{\frac{B_{z_0}}{B_z}}; \quad P_\theta = P_{\theta_0}\sqrt{\frac{B_z}{B_{z_0}}} \qquad (P_\theta = m\gamma r\dot{\theta}). \quad \text{(III.12)}$$

On satisfying the betatron condition

$$B_z = \frac{\bar{B}_z}{2} + \text{const}; \quad \text{(III.13)}$$

$$r = r_0; \quad P_\theta = P_{\theta_0}\frac{B_z}{B_{z_0}}. \quad \text{(III.14)}$$

If we now allow for the fact that the longitudinal motion of the ring should satisfy the condition of constant acceleration in the accompanying coordinate system, it is not hard to obtain the following expressions for the law of variation of the longitudinal magnetic field, which are valid if

$$\frac{N_i}{N_e} \ll 1 \text{ and } \frac{N_i\beta_{\theta_i}}{N_e\beta_{\theta_e}} \ll 1.$$

In the case of an almost uniform field

$$B_z = B_{z_0}\frac{1}{\gamma^2\beta_{\theta_0}^2}\left\{\frac{\gamma_{\perp 0}^2}{\left[1 + \frac{e\mathscr{E}_0(z - z_0)}{mc^2\gamma}\right]^2} - 1\right\}$$

$$\approx B_{z_0}\left[1 - \frac{2}{\gamma_{\|_0}^2\beta_{\theta_0}^2}\cdot\frac{e\mathscr{E}_0(z - z_0)}{mc^2\gamma}\right], \quad \text{(III.15)}$$

the latter if $\frac{e\mathscr{E}_0(z - z_0)}{mc^2\gamma} \ll 1$.

Under the 2:1 condition we have

$$B_z = B_{z_0}\sqrt{\frac{1}{\gamma^2\beta_{\theta_0}^2}\left\{\frac{\gamma_{\perp 0}^2}{\left[1 + \frac{e\mathscr{E}_0(z - z_0)}{mc^2\gamma}\right]^2} - 1\right\}}$$

$$\approx B_{z_0}\left[1-\frac{1}{\gamma_{\|_0}^2\beta_{\theta_0}^2}\cdot\frac{e\mathscr{E}_0(z-z_0)}{mc^2\gamma}\right]. \quad \text{(III.16)}$$

Here the parameter $\mathscr{E}_0$ is chosen from the condition that the ions should be contained, and is equal to $\mathscr{E}_g$ determined from Eq. (III.4) with $\gamma_\perp = \gamma_{\perp 0}$, i.e., equal to the initial value. Equations (III.15) and (III.16) are only valid in a region in which the field is not changing too rapidly (not more than a factor of 2–3). The motion of the equilibrium particles in such fields constitutes a spiral. The angle of the spiral

$$\alpha = \operatorname{arctg}\frac{\beta_\theta}{\beta_z}\approx\frac{1}{\gamma_\|}\quad \text{for}\quad 1\ll\gamma_\|\ll\gamma. \quad \text{(III.17)}$$

The fact that for fairly large $\gamma_\|$ the angle of the spiral is small may clearly be used in order to establish a central solenoid ensuring the satisfaction of the 2:1 condition.

The motion of the deflected particles relative to the equilibrium orbit is approximately described by the usual equation

$$\ddot{\rho}+\dot{\theta}^2\rho=0, \quad \text{(III.18)}$$

whence we readily see that the amplitude of the free radial oscillations varies in accordance with the law

$$a_r = a_{r_0}\sqrt{\frac{B_{z_0}}{B_z}}. \quad \text{(III.19)}$$

It is clear that, if an initially accelerated ring falls into a growing magnetic field, it will start being retarded, and the energy of the forward motion will be converted into rotational energy.

4. Combined Accelerating System. Let us now consider a system using the earlier-mentioned properties of a falling and rising magnetic field in conjunction with a system of accelerating, appropriately phased resonators [2, 13]. This system has the following structure. In the region between the resonators the longitudinal field falls linearly and the ring is accelerated in it as a result of the energy of the rotational motion. Inside the resonator we find a rising longitudinal magnetic field with a configuration such that the energy communicated to the ring in the resonator is mainly converted into rotational motion, and only a part of it, corresponding to the permissible acceleration, passes into forward motion. The longitudinal magnetic field at the exit from the resonators is the same, and hence so is the azimuthal momentum. In the resonators the adiabatic parameter

$$\varepsilon_p = \left|\gamma_\|\frac{r}{B_z}\cdot\frac{\partial B_z}{\partial z}\right|\approx\left|\frac{\mathscr{E}_A}{B_z}\right|\ll 1, \quad \text{(III.20)}$$

where $\mathscr{E}_A$ is the amplitude of the field in the resonator.

In this system the longitudinal magnetic field should satisfy the condition

$$\mathscr{E}_g = -\frac{1}{2}r_0\beta_{\theta_0}\gamma_{\|_0}\gamma_\|\frac{\partial B_z}{\partial z}+\mathscr{E}_A f\cos\Omega_r t, \quad \text{(III.21)}$$

where f is a function reflecting the configuration of the z component of the electric field in the resonator, Ω_r is the frequency of the resonator field. It follows that

$$r_0[B_z(z)-B_z(z_0)] = \frac{2}{\gamma_\|}\left[-\mathscr{E}_g\cdot z-\mathscr{E}_A\int_{z_0}^{z} f\cos\Omega_r t\,d\xi\right] \quad \text{(III.22)}$$

for

$$B_z(z_2) = B_z(z_0). \quad \text{(III.23)}$$

Here z_0 corresponds to the beginning of the gap and z_2 to the beginning of the next gap, i.e., the exit from the resonator.

In the linear approximation the longitudinal motion in such a system will be described by Eq. (III.1), with $\mathscr{E} = \mathscr{E}_g$ determined by Eq. (III.21), and the radial motion will be described by

$$\rho'' + \frac{1}{\gamma_\|^2 \beta_z^2} \rho = \frac{1}{\gamma_\|^2 \beta_z^2} \left[\frac{1}{\gamma_\|} \left(\varepsilon_0 z - \varepsilon_p \int_{z_0}^{z} f \cos \Omega_r t d\xi \right) \right.$$

$$\left. + \frac{\gamma_\| r_0}{2} \varepsilon_p \left(\beta_z \varkappa \bar{f} \sin \Omega_r t - r_0 \frac{\partial \bar{f}}{\partial t} \cos \Omega_r t \right) \right], \qquad \text{(III.24)}$$

where $\varkappa = (2\pi r_0/\lambda)$; λ is the wavelength of the accelerating field, $\bar{f} = \frac{2}{r_0^2} \int_0^{r_0} f r dr$; $\varepsilon_0 = \frac{\mathscr{E}_g}{B_{z_0}}$; $\varepsilon_p = \frac{\mathscr{E}_A}{B_{z_0}}$; the prime denotes differentiation with respect to θ. The solution of this equation may be expressed in the form of the sum of a particular solution of the equation with the right-hand side satisfying the condition

$$\rho(z_0) = \rho(z_2) \text{ and } \rho'(z_0) = \rho'(z_2), \qquad \text{(III.25)}$$

which describes the orbit, and the general solution of the homogeneous equation describing free oscillations around this orbit.

The conditions of adiabatic motion are only not satisfied at the entrance and exit of the resonator.

Let us express f in the following form: $f = \sigma(z - z_1) + \sigma(z_2 - z) - 1$, and hence, $\partial f/\partial z = \delta(z - z_1) - \delta(z - z_2)$, where z_1 corresponds to the entrance into the resonator. Then it is not difficult to obtain a solution describing the orbit. This takes a fairly cumbersome form; however, its main relationships may be expressed in the following way:

$$\rho_1 = \frac{r_0}{\gamma_\|} \varepsilon_p \frac{\frac{z_n}{2r_0 \gamma_\| \beta_z}}{\sin \frac{z_n}{2r_0 \gamma_\| \beta_z}} F, \qquad \text{(III.26)}$$

where $F \sim 1$ is a function depending on z and the parameters of the accelerating system; $z_n = z_2 - z_0$ is the period of the system in question. We see from this equation that the accelerating structure should satisfy the requirement $\sin(z_n/2r_0\gamma_\|\beta_z) \neq 0$, i.e., that the period of the structure should not accommodate a whole number of free oscillations. For

$$\frac{z_n}{2r_0 \gamma_\| \beta_z} \ll 1; \quad \varkappa \frac{z_n}{z_p} \operatorname{tg} \varphi \ll 1; \quad \varkappa^2 \frac{z_n}{z_p} \ll 1, \qquad \text{(III.27)}$$

where $z_p = z_2 - z_1$ is the gap in the resonator, and φ is the phase of the field in the resonator at the instant at which the ring passes through its center, the orbit is proportional to the square bracket on the right-hand side of Eq. (III.24) with a coefficient 0.5. The amplitude of the oscillations of the orbit attenuates as $1/\gamma_\|$, while the amplitude of the free oscillations around this orbit will increase as $\sqrt{\gamma_\|}$.

The period of the oscillations of the ions in the intrinsic system, determined by the parameters of the ring, is $T_c = 6 \cdot 10^{-10}$ sec for $N_e = 10^{14}$. In the laboratory system $T = T_c \gamma_\|$. In the initial part of the accelerating system, T is of the order of the time of flight of the ring through the resonator, and hence, the condition (III.21) must be satisfied in this case to a high degree of accuracy.

For $\gamma_\| \sim 5$–10, T is of the order of the time required to pass through the whole period of the system, and here the requirements need only be imposed upon the integral condition (III.22)–(III.23). For $\gamma_\| \geq 50$–100, T is much greater than the time required to pass through the period of the system. In this case, clearly, modulation of the leading field is not required; discrete acceleration by a field $\mathscr{E}_A$ is equivalent to continuous acceleration by the mean field $\mathscr{E}_D$. The noise generation of oscillations as a result of the discrete nature of the system is clearly slight.

A system with a modulated field enables us to shorten the initial part of the accelerator considerably (for $\gamma_{\parallel} \leq 100$); the possibility of creating a larger field in the resonators in this case improves the energy supply of the beam and eases adjustment.

In the final section of the accelerator, acceleration in a falling magnetic field with a constant or increasing value of r_0 may be employed (Paragraph 2 of this Section).

5. Questions of Energy Supply in the Acceleration of the Ring and Radiation in the Accelerating Structure. Let us now consider the general question of the collection of energy by a relativistic grouping (ring) with a high charge. The fact that the intrinsic current loads the accelerating system severely means, in electrodynamic language, that the energy which has to be communicated to the ring per unit path constitutes a substantial proportion of the energy of the external field, stored in a region from which it can be accepted by the ring. In addition to this, the energy radiated by the ring on passing through the accelerating structure may be comparable with the energy acquired by the ring.

Let us first make an estimate of the power losses J associated with the intrinsic radiation of the electron ring during acceleration, expressed as a ratio of the energy acquired per unit time dE/dt. This ratio [14] is proportional to the square of the charge to mass ratio and is also proportional to the change in the energy per unit path (here the ring is regarded as a single charge with a common mass):

$$\frac{J}{\frac{dE}{dt}} = \frac{2}{3}\left(\frac{e}{m\gamma_{\perp}}\right)^2 \frac{eN_e \mathscr{E}_g}{c^4}. \tag{III.28}$$

Even for $N_e = 10^{14}$, $\mathscr{E}_d = 100$ kV/cm and $\gamma_{\perp} = 50$, the quantity (III.28) is of the order of 10^{-3}, i.e., the intrinsic radiation during acceleration does not constitute a decisive factor in the present problem.

In order to obtain fields capable of accelerating the ring to relativistic velocities, it is fundamentally essential to introduce spatial inhomogeneities into the accelerating tract.* Hence, in collecting energy in the accelerating tract a relativistic ring containing a high charge loses some of its energy in the coherent emission of external charges, which are excited by its electric field on passing along these spatial inhomogeneities, i.e., it loses energy in transitional and Cerenkov radiation.

The intensity of this radiation for technically permissible fields and numbers of particles in the ring may greatly exceed the losses given by Eq. (III.28) [13].

It was suggested earlier [13] that this radiation constituted a considerable proportion of the load on the accelerating system and determined the whole possibility of energy being taken from the external accelerating field by the accelerated ring. Thus, for example, in the case of acceleration in a periodic structure, substantial interaction with the accelerating wave is only possible if the phase velocity of the wave coincides with the velocity of the ring. However, this means that at the same time the condition for the development of Cerenkov radiation, excited by the charge of the moving ring at this frequency, will also be satisfied.

It was also shown earlier [15] that, when a ring passed into a cylindrical resonator in which an external (accelerating) field was excited, the "range of propagation" of the fields excited by the ring at the instant of emerging from the resonator was determined by the following inequalities†:

$$0 \leqslant r \leqslant r_0 + \frac{h}{\beta_z}; \quad 0 \leqslant z \leqslant h, \tag{III.29}$$

where h is the longitudinal dimension of the resonator, r_0 is the radius of the ring, while an unperturbed external field remains in the rest of the volume. Thus, in the course of acceleration only the store of energy of the external field existing in this "region of interaction" acquires a vital significance, not the total store of energy in the whole volume of the resonator.

It is convenient to follow the interaction of the ring with the accelerating field by way of the following example [15, 16]. In a specific interval of time, we consider a closed physical system, consisting of a

*A waveguide with an inner dielectric coating used for slowing the accelerating wave, in particular, also constitutes a spatially inhomogeneous system.

†It is assumed that β_z varies little during the flight of the ring through the gap in the resonator. This condition is naturally satisfied if $\beta_z \approx 1$.

transverse electromagnetic wave excited in the closed volume of a cylindrical resonator, an electron ring traveling through the resonator, and the field excited by its charge both as a result of the acceleration and as a result of the spatial inhomogeneity of the system. In order to describe this system we use the Hamiltonian method [17] and allow for the initial conditions governing the field oscillators and the motion of the charge.

From the condition of the conservation of the total Hamiltonian of the system (using the Coulomb calibration for the potentials)

$$\mathcal{H} = \mathcal{H}_1 + \mathcal{H}_2 - \mathcal{H}_3 = \text{const}, \tag{III.30}$$

where

$$\mathcal{H}_1 = \sqrt{M_0^2 c^4 + c^2 \left\{ \mathbf{P} - \frac{1}{c} \int\limits_V \rho\,[\mathbf{r} - \mathbf{Q}(t)]\,\mathbf{A}(\mathbf{r}, t)\, dV \right\}^2} \tag{III.31}$$

is the Hamiltonian corresponding to the charge considered as a whole [18],

$$\mathcal{H}_2 = \frac{1}{2} \sum_\lambda (p_\lambda^2 + \omega_\lambda^2 q_\lambda^2) \tag{III.32}$$

is the Hamiltonian of the transverse field (λ is the complete set of indices determining the eigenfunctions of the resonator $A_\lambda(\mathbf{r}, t)$); $\mathcal{H}_3$ corresponds to the static Coulomb interaction, $M_0 = mN$ is the total rest mass, $\mathbf{Q}$ is the radius vector of the center of gravity, $\mathbf{P}$ is the canonical momentum, $\rho\,[\mathbf{r} - \mathbf{Q}(t)]$ is the charge density in the ring. The quantities q_λ and $\mathbf{Q}$ are determined from the equations of motion:

$$q_\lambda + \omega_\lambda^2 q_\lambda = \frac{1}{c} \int\limits_V \rho \mathbf{Q} \mathbf{A}_\lambda\, dV; \tag{III.33}$$

$$\frac{d\mathbf{P}}{dt} = -\frac{\partial \mathcal{H}}{\partial \mathbf{Q}}; \quad \frac{d\mathbf{Q}}{dt} = \frac{\partial \mathcal{H}}{\partial \mathbf{P}}. \tag{III.34}$$

We see from Eq. (III.33) that the motion of the charge perturbs all the oscillators of the resonator, i.e., the collection of energy by the charge from the transverse field of the resonator, in accordance with Eqs. (III.30)-(III.34), can only take place in the presence of radiation and through radiation. As the ring moves, there is a redistribution of the energy between the various components of $\mathcal{H}$. Furthermore, the amplitude of the oscillations, and hence, the energy of the earlier-excited harmonic oscillator, changes in accordance with Eq. (III.33), and furthermore the remaining oscillators are excited. The static field of the charges induced on the inner walls of the resonator and entering into $\mathcal{H}_3$ creates an additional nonuniformity in the motion of the ring, in accordance with Eq. (III.34), although ultimately, in view of its potential nature, it does not make any contribution to the redistribution of the energy.

The change in the energy of the ring during its flight is, according to (III.30), equal to

$$\Delta \mathcal{H}_1 = -\Delta \mathcal{H}_2, \tag{III.35}$$

where on allowing for the foregoing discussion $\Delta\mathcal{H}_2$ may be written as

$$\Delta\mathcal{H}_2 = 2\,(\mathbf{A}_{1_0}\, \delta\mathbf{A}_1) + \delta\mathbf{A}_1^2 + \sum_{\lambda \neq 1} \delta\mathbf{A}_\lambda^2. \tag{III.36}$$

Here we have used the notation $A_\lambda = \mathbf{i}P_\lambda + \mathbf{j}\omega_\lambda q_\lambda$ (i and j are unit vectors); $\mathbf{A}_{1_0}$ is the initial excitation of the oscillator of the first harmonic; δA_λ is the perturbation determined by the equations of motion (III.33) and (III.34). The third term on the right-hand side of (III.36) determines the radiation in all the harmonics excluding the first. The second term determines the radiation in the first harmonic, while the first term determines the collection of energy by the charge from the initially excited oscillator.

The first term in Eq. (III.36) may conveniently be written as

$$2\,(\mathbf{A}_{1_0}\, \delta\mathbf{A}_1) = 2\sqrt{\mathbf{A}_{1_0}^2\, \delta\mathbf{A}_1^2}\, \cos\varphi, \tag{III.37}$$

where q is the phase shift between the initial and radiated fields, i.e., it is proportional to the square root of the product of the energy of the initial excitation and the energy of the radiation in the harmonic itself.

Equation (III.36) may be expressed in the form

$$\Delta\mathcal{H}_2 = Aq + Bq^2, \tag{III.38}$$

where g is the total charge of the ring, A and B are coefficients, which, in the case of the specific idealization implied by distinguishing the system as being a closed one, depend on the following parameters: A on the initial external field and the geometry of the system, B on the geometry of the system alone.

The extent to which the system is a closed one in the case of the arguments here presented may be roughly estimated if, for example, we compare the charge arising as a result of the external field on the wall of the resonator with the charge induced in the wall by the ring. If the time of flight of the ring is short in comparison with the period of the external field [we consider that the radius of the resonator $R \gg r_0 + (h/\beta)$, see Eq. (3.29)], the density of the external charge on the end wall of the resonator may be expressed in terms of the amplitude of the external field by the formula $\sigma = \mathcal{E}_A/4\pi$. Hence, the external charge concentrated in the region of interaction equals

$$q_c \approx \left(r_0 + \frac{h}{\beta}\right)^2 \frac{\mathcal{E}_A}{2}.$$

Naturally a necessary condition for the system to be a closed one is that the inequality $q_c \geq q$ should be satisfied. This is equivalent to saying that the energy of the external field stored in the region of interaction is greater than $qh\mathcal{E}_A$.

Equation (III.38) shows that we are concerned with the theory of perturbations, and if the charge in the ring is very great then we are dealing with the most difficult aspect of perturbation theory in which the energy of the perturbation is comparable with the initial energy of the system.

Thus, in approaching the problem it is primarily essential to estimate this perturbation energy, namely, the term $Bq^2 = W_b$. This estimation is usually carried out by calculating the energy of the radiation induced by the ring for a specified law of motion of the latter. Since the transverse components of the electrostatic field of the ring in the laboratory system of coordinates increase as v approaches c, a second important characteristic determining the energy W_b is the relativistic factor γ. The chief and most difficult problem is that of obtaining the $W_b = W_b(\gamma)$ relationship. The difficulty in solving this problem lies in the necessity of considering the large frequency range within which the radiation is excited. With increasing γ, higher and higher frequencies have to be taken into account. This may readily be seen, for example, if we make use of a Fourier expansion of the field components of the moving ring (for simplicity we shall consider that $v = v_0 = \text{const}$), i.e., the field exciting the radiation. The Fourier components are proportional to

$$\left(\frac{\sin \frac{l\omega}{c\beta}}{\frac{l\omega}{c\beta}}\right)^2 \exp\left[-\frac{\omega}{c\beta\gamma}(r - r_0)\right], \tag{III.39}$$

where l and r_0 are, respectively, the longitudinal and transverse characteristic dimensions of the ring, $(r - r_0)$ is the target parameter (distance from the trajectory to the obstacle). The quantity (III.39) constitutes a frequency "cut-off factor," and its effect diminishes with increasing γ. It may well be said that all the frequencies up to $\omega_{max} \approx c\gamma(r - r_0)$ contribute to W_b.

The factor $\left(\sin \frac{\omega l}{c\beta} \Big/ \frac{\omega l}{c\beta}\right)^2$ appearing in the expression (III.39) allows for the incoherence of the exciting effect of the field of the ring for wavelengths smaller than its longitudinal dimension l. It should nevertheless be noted that with increasing γ the value of l becomes smaller: $l = l_c/\gamma$, where l_c is the longitudinal dimension of the ring in its own system.

In a fair number of papers (a list is given in the review articles [19, 20]) the ring-induced radiation has been calculated for various cases of spatial inhomogeneity.* The problems thus solved as yet present no very clear picture as to the dependence of the excited radiation on γ, since in many cases they constitute upper asymptotic estimates or numerical calculations of particular cases.

At the present time, at least in the case of the motion of a ring under the condition $v = v_0 = \text{const}$, we may clearly divide the whole problem into two parts: the flight of the ring past a "single obstacle," and the motion of the ring in a periodic structure.

For the first part of the problem there are two characteristic aspects: the flight of the ring through an aperture in an ideally conducting screen [21], and its flight through a single resonator [15, 16, 20, 22-29].

In the problem relating to the flight of a charge q at a constant velocity v_0 through a circular aperture of radius a in an infinite, ideally conducting screen, we have four parameters: q, a, v_0, and the velocity of light c. From these quantities we can set up only one combination having the dimensions of energy, q^2/a. Hence, the total energy losses may be written in the form

$$\Delta W = \frac{q^2}{a} f\left(\frac{v_0}{c}\right). \tag{III.40}$$

It was shown earlier [21] that f increased in proportion to γ. Analogous results are obtained in the problem relating to the radiation of the charge on flying past an ideally conducting wedge [30], and in that of a charge into or out of a semiinfinite, ideally conducting tube [31]. The results of a calculation carried out for the case of an infinite filament flying past an ideally conducting cylinder [32] show that the linear rise in ΔW with γ is not a consequence of the fact that the obstacles have a sharp edge. The relation giving the energy loss when a charge flies past any single obstacle apparently takes the form (III.40) with a linear dependence of f on γ, some characteristic dimension (e.g., target or aiming parameter) playing the part of a. The only difference lies in the numerical coefficients involved.

The question as to the radiation of a charge passing through a single cylindrical resonator has been considered by a number of authors [15, 16, 20, 22-29], starting with Kotov and Kolpakov [22]. It is very natural that this should be so, since resonators constitute vital elements of accelerating installations. Owing to the complexity involved in considering the true conditions when the resonator contains entrance and exit apertures or feeding waveguides, instead of a resonator certain authors have considered a closed cylindrical volume, in which the charge (either an assembly of charges or a current-carrying ring) penetrates through a wall. Kotov and Kolpakov [22] expanded the fields in terms of the eigenfunctions of the resonator, taking account of the zero initial values for the field oscillators, and obtained expressions for the energy of the radiation left in the resonator by the ring after escaping through the second wall. For a thin charged ring this expression takes the form:

$$\Delta W = \frac{8\pi^2 q^2}{R_r} \sum_{n=1}^{\infty} \sum_{m=0}^{\infty} \left(\frac{2}{1+\delta_{m_0}}\right) \left[\frac{\sin\left(\frac{\pi l}{\beta h}\sqrt{\left(\frac{\nu_n h}{\pi R_r}\right)^2 + m^2}\right)}{\frac{\pi l}{\beta h}\sqrt{\left(\frac{\nu_n h}{\pi R_r}\right)^2 + m^2}} \right]^2$$

$$\times \left[\frac{r_2 J_1\left(\nu_n \frac{r_2}{R_r}\right) - r_1 J_1\left(\nu_n \frac{r_1}{R_r}\right)}{\frac{\nu_n}{R_r}\left(r_2^2 - r_1^2\right)} \right]^2 \frac{1}{\frac{\pi \nu_n}{2} J_1^2(\nu_n)}$$

$$\times \frac{\left(\frac{\nu_n h}{\pi R_r}\right)^3}{\left[\sqrt{\left(\frac{\nu_n h}{\pi R_r}\right)^2 + m^2} + \beta m\right]^2} \cdot \frac{\sin^2\left[\frac{\pi}{2\beta}\left(\sqrt{\left(\frac{\nu_n h}{\pi R_r}\right)^2 + m^2} - \beta m\right)\right]}{\left[\frac{\pi}{2\beta}\left(\sqrt{\left(\frac{\nu_n h}{\pi R_r}\right)^2 + m^2} - \beta m\right)\right]^2}, \tag{III.41}$$

*In what follows we shall not consider the radiation excited by the azimuthal current in the ring, since the retarding action of this radiation differs in no fundamental manner from that of the radiation induced by the charge. Only different modes are excited.

where R_r is the radius of the resonator, h is its length, r_2 and r_1 are the outer and inner radii of the annular group of electrons (the ring), l is its thickness along the z axis, ν_n are the roots of the equation $J_0(\xi) = 0$; δ_{mk} is the Kronecker delta, and $\beta = v/c$, where c is the velocity of the ring along the z axis.

For the limiting cases of an infinitely thin ring and a point charge, the expression for ΔW is logarithmically divergent. For other cases, such an infinitely thin disc or any other two- or three-dimensional charge distribution, the double sum is convergent, although it is certainly very difficult to calculate. However, a complete calculation is not really necessary, since in the limiting case, in which apertures occur in the resonator, according to Eq. (III.39) the very high frequencies make only a small contribution. Kotov and Kolpakov [22] suggested allowing for the entrance and exit apertures phenomenologically, by truncating the spectrum of the radiated waves, assuming that (for purposes of calculation) it was reasonable to confine attention to the terms corresponding to waves having a spatial inhomogeneity of the fields of the order of the diameter of the apertures. On this assumption, the formula obtained for ΔW (in the case of a point charge) was

$$\Delta W = \frac{q^2 h}{2a^2}, \tag{III.42}$$

where a is the radius of the aperture.

In the ultrarelativistic case this formula requires refinement, since the contribution of the high harmonics to Eq. (III.41) becomes greater for large γ.

In another paper [26] the complete spectrum in (III.41) was truncated in respect of the radial harmonics giving a field inhomogeneity in the radial consideration, while all the longitudinal harmonics were taken into account. According to (III.39), however, this type of truncation neglects a number of waves which are in fact excited, but instead of remaining within the resonator pass out through the outer apertures, so that for large γ the result is incomplete.

In order to discover the maximum possible value of ΔW for very large γ with this type of approach, asymptotic estimates of the total sum in (III.41) were subsequently attempted [27]. The resultant upper limit for a group in the form of an infinitesimally thin disc of radius r_0 takes the form

$$\Delta W < \frac{8q^2 h}{r_0^2} \tag{III.43}$$

and is independent of the value of γ. In order to deduce the asymptotic behavior of ΔW it is extremely important that the factor $\sin^2\left[\frac{\pi}{2\beta}\left(\sqrt{\left(\frac{\nu_n h}{\pi R_r}\right)^2 + m^2} - \beta m\right)\right]$ in Eq. (III.41) should be properly taken into account, since for a considerable range of n and m values (expanding with increasing γ) this is much smaller than unity. If due allowance for this fact is not made, and if it is assumed (as has frequently been done) that the factor in question is approximately equal to unity, then ΔW is apparently proportional to γ, and the estimate is far too high. Numerical calculations of the sum (III.41) were made in [28]. In order to allow for the coherence factor in (III.39), rings of different longitudinal dimensions were considered. The dependence of the sum (III.41) on γ was also considered with due allowance for the Lorentz contraction of the longitudinal dimension of the ring, and it was found that over a certain range of γ ($\gamma < 200$) values the rise in ΔW was approximately proportional to $\gamma^{1/2}$. It is nevertheless essential to note that the calculated values of ΔW corresponding to $\gamma = 200$ are very much smaller than the upper estimate (III.43). Hence, on further increasing γ the dependence of ΔW on γ should vanish.

The foregoing approach, with or without phenomenological allowance for the apertures, lies a long way from the true situation, since in this case only the radiation excited during the flight and left in the resonator is taken into account. Actually a certain proportion of the energy is lost by the ring as it enters and leaves. Certain authors accordingly considered another model. In some cases [24, 25] the losses were estimated as the amount of energy of the radiation diffracted within the resonator cavity during the entrance of the charge through the feeding waveguide. The resultant formula, which for an infinitesimally thin ring of radius r_0 takes the form

$$\Delta W \approx \frac{0.44q^2 h^{1/2}}{a(a-r_0)^{1/2}} \gamma^{1/2}, \quad \text{(III.44)}$$

and for a point charge

$$\Delta W \approx 0.6 \frac{q^2 h^{1/2}}{a^{3/2}} \gamma^{1/2}, \quad \text{(III.45)}$$

indicates that the energy rises as $\gamma^{1/2}$.

An analogous relationship with a numerical factor roughly twice as large was also obtained by Sessler [20], who analyzed the same model by partly matching the fields in the resonator cavity and the fields in the waveguide.

Numerical calculations of the energy lost by a point charge in a resonator with semiinfinite entrance and exit waveguides were carried out for $\gamma \leq 30$ in [29]. The relationship so derived may be expressed in the following way for $\gamma > 10$ and $h/R_r = 1$:

$$[\Delta W \approx \frac{4q^2}{\pi a}(0.5+0.09\gamma). \quad \text{(III.46)}$$

For $10 < \gamma < 30$ there are no serious quantitative differences between (III.45) and (III.46).

It is not entirely clear from the foregoing discussion whether any specific conclusion may be drawn regarding the variations in losses with increasing γ.

In connection with problems regarding the radiative energy losses of groups of electrons in the case of open and closed resonators, the question arises as to how the energy losses are distributed along the path of the electron group. The energy lost by a charge on passing through any particular structure may be expressed in the form of the work done by the retarding force of the radiation over a specific path. When the charge passes through a single resonator with apertures or with feeding waveguides, the total energy loss may be written in the form

$$W = q \int_{-\infty}^{\infty} \mathscr{E}_T \, dz. \quad \text{(III.47)}$$

Assuming that $\mathscr{E}_T$ has only one main maximum, this expression may be written in the form

$$W = q\mathscr{E}_{T\max} L, \quad \text{(III.48)}$$

where $\mathscr{E}_{T\max}$ is the maximum value of $\mathscr{E}_T$. Since, from the point of view of the acceleration of the charge in a specific accelerating system, we are interested in the retarding force averaged over a section of path comparable with the period of the system (or with the gap in the resonator), it is extremely important to know what determines the rise in the losses with increasing γ: is it due to a rise in $\mathscr{E}_{T\max}$, a rise in L, or to a rise in both of these, and, if the latter, then in what proportion? It is obvious that, if the rise in the losses with increasing γ is determined mainly by the increases in the path length L within which even the slightest retarding force occurs, then this relationship will not constitute any fundamental obstacle to the creation of an accelerating system.

We may further note that, as a result of the relativistic flattening of the intrinsic field of the group of electrons in the direction of motion (in the laboratory coordinate system), the maximum perturbation of the external charges occurs at the actual instant at which the group flies past an inhomogeneity, and hence the work done by the retarding force depends mainly on how long the group moves in phase with the excited retarding wave.

A good example (and one yielding an exact solution) for tracing the dependence of the retarding force on the position of the group in the course of its motion is that concerned with the excitation of an ideally conducting semiplane by a charged filament with a linear charge density $\varkappa$, flying past it at a distance a [19]. For simplicity, it is convenient to take the trajectory of the filament as being perpendicular to the semiplane. Using the expressions of [19] for the fields and integrating over the whole spectrum of frequencies, we obtain the following exact formula for the retarding force in relation to the position of the filament [33]:

$$F_T = -\frac{\varkappa^2}{2\gamma a\left[1+\left(\frac{y}{\gamma a}\right)^2\right]^2}\left\{\frac{1-\beta\frac{y}{a}}{\frac{y}{a}}\left(\frac{1}{\sqrt{1+\left(\frac{y}{a}\right)^2}}-1\right)\right.$$

$$+\frac{(2-\beta^2)\frac{y}{a}}{\sqrt{1+\left(\frac{y}{a}\right)^2}}-\frac{y}{a}\left(\frac{(2+\beta^2)\beta\frac{y}{a}}{\sqrt{1+\left(\frac{y}{a}\right)^2}}+3\right)+\left(\frac{y}{\gamma a}\right)^2$$

$$\left.\times\left(\frac{\frac{y}{a}}{\sqrt{1+\left(\frac{y}{a}\right)^2}}+\beta\right)-\left(\frac{y}{\gamma a}\right)^2\left(\frac{y}{a}\right)\left[\frac{\beta\frac{y}{a}}{\sqrt{1+\left(\frac{y}{a}\right)^2}}+(2-\beta^2)\right]\right\}. \quad \text{(III.49)}$$

In Eq. (III.49) it is considered that the filament is moving at a velocity v = const from y = $-\infty$ toward y = $+\infty$. It is easy to show from the foregoing equation that, in the ultrarelativistic case, starting from the point y $\approx$ 0 (y = 0 is the coordinate of the half-plane), the retarding force increases almost linearly, reaches a maximum (not depending on the value of γ) at approximately y = 0.7 $a\gamma$, and then falls hyperbolically. Thus, the retarding field arising at the instant of passing the half-plane, which has the form of a pulse with a length of the order of $a/\gamma c$, is first propagated in such a way to overtake the source of excitation, after which dephasing gradually takes place, owing to the slight difference in velocities, and the field outstrips the filament. This retardation process occupies a time t $\approx a\gamma/c$. It is reasonable to suppose that this picture will remain qualitatively the same in any other case in which a single obstacle is excited, for example, in that of a resonator with feeding waveguides, and so on.

If a second obstacle is encountered at a certain distance L_1 after the first, then if the inequality $L_1 \ll 0.7\, a\gamma$ is satisfied it would appear reasonable to treat both obstacles as a single one, and in practice the foregoing picture of the retardation process will change very little. However, in the computing respect, the question as to the excitation of a complex system is made particularly difficult by the necessity of allowing for the screening of one element by another. For $L_1 \gg 0.7\, a\gamma$ the retarding effect of the second obstacle will in practice be independent of that of the first.

We have already pointed out that, for our own purpose of accelerating a ring to ultrarelativistic velocities, the accelerating system should comprise a series of accelerating elements, i.e., it should have a periodic or quasiperiodic structure. Hence, an extremely important problem is that of determining the radiative losses in such structures.

It was shown a long time ago [34] that the energy losses in such cases were of a resonance nature. However, up to the present time there has been no convincing success in elucidating the asymptotic dependence of the losses on the parameter γ.

So far as we know, the only exact solution which has been obtained relates to the excitation of an infinite "comb" of half-planes by a "charged filament" [35]. However, even in this case the analysis of the solution presents substantial difficulties. The asymptotic formula for the losses associated with one period of the structure [35] takes the form ($\gamma \gg 1$)

$$W = \frac{2\varkappa^2 D}{a}\cdot\frac{\beta}{\gamma}, \quad \text{(III.50)}$$

where D is the period of the structure, a is the target (aiming) parameter, $\varkappa$ is the charge per unit length of the filament. However, this formula only allows for the contribution of the high frequencies $ka \geq \gamma$. Numerical calculations were carried out elsewhere for this type of structure [35, 20]. The calculations show that the losses diminish with increasing γ.

Calculations based on the exact formulas in the nonrelativistic and slightly relativistic cases [35] agree closely with those based on (III.50). In one case [29] the relativistic problem ($5 \leq \gamma \leq 200$) yielded a fall in losses varying as $\gamma^{-1/2}$.

The calculation carried out by Sessler [20] also shows a close agreement between the total losses per structure period calculated by integrating the retarding force created by the secondary excited field and the estimated amount of energy radiated by this field in the gap between the corresponding two planes of the structure.

It should be noted that a comb constitutes an open structure, and thus some of the excited waves pass out of the structure into space, and the main contribution to the force retarding the group of electrons arises from the delayed surface waves propagating at a phase velocity equal to the velocity of the particles.

An analogous picture of the retardation of a group by a surface wave is obtained for other open structures. Thus, an approximate solution of the integral equations determining the currents flowing in various elements of the structure [36] yielded asymptotic estimates of the losses for fairly large γ in a system comprising an infinite number of plane screens with apertures of radius a in the center, and also in a system comprising an infinite series of drift tubes. For the first structure, the total loss of energy by the group of particles at a frequency ω on passing through a distance equal to one period D is given by the equation

$$W_{\omega}^{\mathrm{I}}=\frac{2qk}{c\beta\gamma}\,\mathrm{Re}\left\{\int_{a}^{\infty} j_{r\omega}^{0}(r)\,K_{1}\left(\frac{kr}{\beta\gamma}\right) r\,dr\right\}, \tag{III.51}$$

and for the second structure by

$$W_{\omega}^{\mathrm{II}}=-\frac{2qak}{c\beta^{2}\gamma^{2}}\,K_{0}\left(\frac{ka}{\beta\gamma}\right)\mathrm{Im}\left\{\int_{0}^{d} j_{z\omega}^{0}(\xi)\,e^{-i\frac{\omega}{c\beta}\xi}\,d\xi\right\}, \tag{III.52}$$

where d is the length of the tube and a is its radius.

Equations (III.51) and (III.52) are exact, and may be used to obtain the corresponding numerical values of W_{ω}^{I} and W_{ω}^{II} if we know the values of the Fourier components of the currents $j_{r\omega}^{0}(r)$ and $j_{z\omega}^{0}(\xi)$ flowing in any arbitrarily-chosen element of the structure. For $\gamma \gg 1$ estimates of the currents may be obtained by the stationary-phase method, and the corresponding formulas for the losses, integrated over the frequency range within the limits of $ka \geq 1$, are as follows:

$$W^{\mathrm{I}}=\frac{5}{24\pi}\cdot\frac{q^{2}D}{a^{2}}; \tag{III.53}$$

$$W^{\mathrm{II}}\approx A\frac{q^{2}d}{a^{2}\gamma}, \tag{III.54}$$

where A is a certain numerical factor (not derived in [36]).

If the source of excitation is an infinitesimally thin ring of radius r_0, then instead of (III.53) we have

$$W^{\mathrm{I}}\approx 10^{-2}\,\frac{q^{2}D}{a\,(a-r_{0})}. \tag{III.55}$$

The same computing method gives the following expression for the loss of energy by a charged filament in a comb

$$W=\frac{\varkappa^{2}D}{16a}. \tag{III.56}$$

For $(a/\mathrm{D}) < 1$ the latter formula in no way contradicts the earlier numerical results [35, 20], since the quantitative estimates based on Eq. (III.56) lie below these figures.* Equation (III.56) evidently corresponds to a more strongly relativistic situation, and the difference between this and Eq. (III.50) may be explained by the fact that (III.56) allows for the range $\gamma \geq ka \geq 1$, which was neglected in (III.50).

*We note, however, that the results of the earlier paper [20] imply an approximately quadratic dependence on the ratio a/D (for $0.5 \leq (a/\mathrm{D}) \leq 2$).

By basing our considerations on the foregoing discussion as to the retardation of a charge by a single obstacle, we may (at any rate for open structures) qualitatively represent the retardation of charge in a structure as follows. If the period of the structure $D \gg \gamma a$ [here a is the target parameter, or in the case of a ring $(a-r_0)$], the retarding effect of each structural element will manifest itself independently, and the losses per structural period will be proportional to γ. When we increase γ far enough to satisfy the opposite relationship ($\gamma a \gg D$), the interaction between the structural elements during radiation becomes very substantial, and the curve relating the losses to γ develops a plateau. Of course any real accelerating system is closed.

The qualitative picture of the losses (the dependence of the losses on γ and the fact that the charge is retarded by the accompanying wave) in closed structures is plainly the same as in the open structures just considered. Only quantitative changes are to be expected, although these are of course important for designing any specific accelerating system.

However, any calculation of the energy losses associated with the motion of a charge through closed systems (such as diaphragmed waveguides or a system of resonators linked by drift tubes) is extremely difficult, and can certainly only be carried out numerically on an electronic computer, occupying a very considerable amount of machine time.

Nevertheless, for a structure consisting of narrow resonators connected by tubes, an analytical formula was obtained in [37] with due allowance for the low-frequency part of the losses experienced by a point charge ($ka \leq 1$, a is the radius of the apertures):

$$W = \frac{q^2}{2a}\left(\frac{d}{a}\right)\left(\frac{d}{D}\right), \qquad \text{(III.57)}$$

where d is the length of a resonator, D is the period of the structure ($d/a < 1$); this formula may be regarded as a reasonable approximation for the slightly-relativistic case.

Numerical computer calculations [38] were employed in order to analyze the more general case of a system of coupled resonators excited by a ring. Allowance was made for a larger range of wavelengths than in the earlier case [37] ($ka \leq 30$), and the results were extended to the region of $\gamma \leq 50$. The resultant numerical values of the losses were almost independent of γ. It should nevertheless be noted that the calculation made no allowance for a certain range of high frequencies which might well make a finite contribution, since, according to (III.39), the truncating parameter $k(a-r_0)/\gamma$ (r_0 = radius of the ring) is much smaller than unity in the case in question, and its effect is not particularly obvious. This may very well explain why the results obtained [38] vary little with the distance of the ring from the walls of the waveguide.

A comparison between the results obtained for a closed system and the earlier results obtained for an open system of the comb type [20] shows that the losses are several times greater in the closed case. Thus, for example, the ratio of the energy lost per structural period by unit length of ring in a heavily diaphragmed waveguide to the square of the linear charge density of the ring equals two in [38], while in the case of a comb [20] the same ratio equals 0.5. In both cases the ratio of the target parameter (for a ring this is $a-r_0$) to the structural period equals 0.5.

All the foregoing arguments lead to the following conclusions: firstly, the energy lost by a group of electrons (e.g., a ring) as radiation in each period of the structure does not, at any rate, actually increase with increasing relativistic factor γ. Secondly, apart from this qualitative type of conclusion, we cannot at the present time specify any particular general law for determining the extent of these losses, so that careful loss calculations must be carried out on a computer for each specific case, although this will demand a great deal of machine time.

IV. Focusing of an Electron Ring

1. General Comments. In a collective accelerator the ions are captured and contained by the potential well of an electron group, which is itself accelerated by external fields. It is therefore quite clear how important it is to keep the electron density in the group or ring sufficiently high to ensure the required depth of the potential well. The efficiency of the collective method of acceleration largely depends on the solution of this problem.

At the present time the most promising arrangement is that in which the group of electrons constitutes a ring of toroidal shape, formed by relativistically rotating electrons. The repulsion between the electrons in such a ring associated with the Coulomb charge is considerably weakened by the magnetic attraction. In principle no external focusing is really needed if the self-focusing condition $N_i = N_e/\gamma_\perp^2$ is satisfied [2]; however, focusing is certainly necessary until the condition of self-focusing has been established, and even after this condition has been satisfied it is very desirable, since the ring in the potential well is stable with respect to a large number of types of perturbation. Calculations (Sec. V) show that in certain cases a potential well eliminates hydrodynamical instabilities. Subsequently, we shall only consider focusing by external fields and image forces.

For a ring of toroidal shape the requirement that the density should be preserved is equivalent to the necessity of keeping the cross-sectional dimensions small. The major radius of the ring and the radial dimension of the cross section may be kept constant fairly efficiently by means of a magnetic field. We shall therefore devote our principal attention to focusing in the axial direction. Let us make a brief analysis of possible methods of focusing. The simplest method of focusing would appear to be that of using the traveling-wave automatic-phasing effect. However, this method is not very effective for accelerating rings, since it requires the use of powerful generators, the creation of a retarding system, and also a change in the velocity of the wave during the acceleration process. Furthermore, since the gradients of the focusing field in the intrinsic coordinate system fall off as $1/\gamma_\parallel^2$, for a focusing-field amplitude of $\mathcal{E}_0 \approx 100$ kV/cm and $\lambda \leq 10$ cm focusing is only possible up to $\gamma_\parallel = 4$.

Different methods of focusing may find a practical use in different versions of the collective method of acceleration (the acceleration of heavy ions or the acceleration of protons to extremely high energies). We shall consider focusing by means of the azimuthal component of the magnetic field (H_φ focusing), focusing based on the use of the high-frequency Miller potential well involving oppositely directed waves, and focusing based on the use of image forces [39].

2. H_φ Focusing. The mechanism of H_φ focusing may readily be understood by considering a straight filament in a longitudinal magnetic field. Under the influence of the intrinsic repulsive forces the particles in the filament acquire velocities perpendicular to the direction of the magnetic field; the latter turns the particles, not allowing them to leave the filament. In the case of a ring the azimuthal component of the magnetic field (H_φ), created, for example by a current-carrying conductor with the current directed along the z axis, may provide this longitudinal field.

The focusing conditions for a ring moving along the axis are given by two inequalities:

$$\left.\begin{aligned} \omega_{H_\varphi}^2 &> \Omega_\Lambda^2/\gamma_\perp^2\gamma_\parallel^2, \\ \Omega_\Lambda^2/\gamma_\parallel^2 &> \omega_{H_z}^2 - 2\omega_{H_z}\omega_{H_\varphi}\gamma_\parallel, \end{aligned}\right\} \qquad \text{(IV.1)}$$

where Ω_Λ is the Langmuir frequency of the ring, $\omega_{H_\varphi} = eH_\varphi/m\gamma_\perp c$; $\omega_{H_z} = eH_z/m\gamma_\perp c$, Ω_Λ, and $\gamma_\perp$ are taken in the intrinsic coordinate system of the ring.

An analysis of Eq. (IV.1) shows that, on increasing the relativistic factor $\gamma_\parallel$ with a constant field H_z and H_φ, the major radius of the ring becomes greater. This limits the length of the acceleration path of the ring for the type of focusing in question.

The limitation may be removed if as $\gamma_\parallel$ is increased the value of the H_φ field component (in the laboratory coordinate system) is correspondingly lowered.

3. Focusing Based on Oppositely Directed Waves. There is also the so-called Miller mechanism of focusing, which is capable of taking place in a system of two waves moving in opposite directions. It is easy to show that in this case a unique value of the particle velocity may be defined, i.e., we may find an equilibrium particle around which all the other particles execute stable oscillations. A system of this kind has a number of advantages. Let us suppose that in the laboratory coordinate system we have an external "corrugated" magnetic field with a phase $\psi_0 = k_0'z'$, while a wave with a phase $\psi = k'z' - \omega't'$ travels to meet the ring. Let us transform to a system in which the frequencies of these waves coincide; then

$$\left.\begin{aligned} &\omega = \gamma_\parallel(\omega' - vk'); \; k = \gamma_\parallel(k' - v'\omega); \\ &\omega_0 = \gamma_\parallel vk_0'; \; k_0 = \gamma_\parallel k_0' \end{aligned}\right\} \qquad \text{(IV.2)}$$

and $\omega = \omega_0$; hence

$$c\beta_z = \frac{\omega'}{k' + k_0'} < 1. \quad \text{(IV.3)}$$

We accordingly see that such a system requires no retardation; the value of $\omega'/k' + k_0'$ may be varied as acceleration proceeds by using k_0', which is very convenient. If we take $k' \gg k_0'$, $\omega' \approx k'c'$, then it is easy to find $\dot{\gamma}_\parallel \approx \sqrt{\frac{k'}{2k_0'}}$; hence any error in k_0' affects $\gamma_\parallel$, while the error in the velocity (or the condition of exact synchronism between the potential well and the particle) equals $\delta\beta_z \approx 1/\gamma_\parallel^2 \cdot \delta\gamma_\parallel/\gamma_\parallel$, i.e., it diminishes with increasing $\gamma_\parallel$. This demonstrates the usefulness of the method in question for focusing at ultrarelativistic velocities.

Let us write down the equation of the z oscillations in the intrinsic system (as an example, we consider a wave of the H_{01} type):

$$\ddot{z} + \frac{e\beta_0}{m\gamma_\perp}\left(\frac{k}{k'} h_1 \cos\psi - \gamma_\parallel h_0 \cos\psi_0\right) = 0, \quad \text{(IV.4)}$$

where $h_1 = H_1 J_1(kr_0)$ is the amplitude of the traveling wave; $h_0 = H_0 I_0(kr_0)$ is the amplitude of the corrugated field.

We assume that the time required for changes in the external parameters is large compared with the period of the averaged oscillations. This equation may be solved by the Bogolyubov method. We find a well with the following characteristics:

$$\omega^2 \gg \frac{e^2 h_1 h_0}{m^2 c^2 \gamma_\perp^2} > \frac{2}{\gamma_\perp^3} \cdot \frac{c}{a_0^2} \cdot \frac{e^2 N_e}{mc^2 2\pi r_0}; \ \omega' \approx k', \ k' \gg k_0'. \quad \text{(IV.5)}$$

Here a_0 is the small dimension of the ring (cross-sectional radius). The left-hand equation is the condition for the applicability of the method; the right-hand equation constitutes the condition for the compensation of the forces of Coulomb repulsion between the electrons.

The use of this method will be quite feasible if we can obtain powerful sources of radiation in the short-wave range (in order to hold a ring with $N_e = 10^{13}$, we require a traveling wave with an amplitude of about 1 kOe).

4. Focusing by Image Forces. We shall pay rather more attention to the details of this method, since it constitutes the most practical method of focusing in all known installations of the kind in question.

The motion of the electron ring during acceleration takes place under conditions of close screening: in all systems capable of practical use the ring is surrounded by a metal tube or a screen of more complicated configuration.

The interaction of a charged ring with a screen is of a "coherent" nature, i.e., the force acting on each particle is proportional to the number of particles within the ring. This emphasizes the necessity of allowing for screening when considering the motion of the particles in the ring and the behavior of the ring as a whole. One particularly interesting fact is that the screening may be used in order to focus the beam. Let us consider a straight charged filament formed by a flux of electrons and screened by a conducting metal plane. If the distance from the center of the filament to the plane is much greater than its small dimension, then the image field may be replaced by the field of an infinitesimally thin filament formed by charged particles of the opposite sign.

An elementary construction of the forces of interaction between the "image" filament and the extreme particles of the real one gives the focusing force. The value of this force would be much smaller than the forces of Coulomb repulsion between the particles if there were no longitudinal motion of the particles. As a result of this motion, the intrinsic repulsive forces in the beam are weakened by a factor of $\gamma_\perp^2$ and there is a clear possibility of compensating the Coulomb repulsion by the image forces. A filament moving as a whole in a direction perpendicular to the direction of motion of the particles within it may be considered as a "straight" model for a ring traveling in a screening tube. In this case, however, the electric field of the filament becomes time dependent, and this leads to the appearance of a magnetic screening field, which weakens the focusing image forces by a factor of $\gamma_{\perp}^2$ and makes them smaller than the intrinsic defocusing forces. The use of the image forces for focusing is nevertheless very attractive in view of the fact that focusing by this method requires no supplementary power sources and is independent of the velocity of the ring up to fairly large values of $\gamma_{\|}$. This is why attempts have been made at finding systems which will retain the focusing effect of the electric screening while reducing the defocusing effect of the magnetic screening.

If we consider a ring screened by a cylinder, we find that the geometrical curvature itself leads to the desired result. In the intrinsic fields, the curvature may be neglected only if $1/\gamma_0^2 \gg (a_0^2/r_0^2)\ln(8r_0/a_0)$, whereas in the induced fields the corresponding condition is $1/\gamma_{\perp}^2 \gg (a-r_0/r_0)\ln(r_0/a-r_0)$; we may thus find ourselves able to create conditions under which the intrinsic field may be regarded as "straight" (weakened by a factor of $\gamma_{\perp}^2$) while the induced field is "curved" (and thus not weakened at all). Here a is the radius of the screening tube. The force gradient of present interest takes the following form in this case:

$$\frac{1}{\omega_0^2 m\gamma_{\perp}} \cdot \frac{\partial F_z}{\partial z} = -\frac{4e^2}{mc^2} \cdot \frac{1}{\gamma_{\perp}} \cdot \frac{N_e}{2\pi r_0} \int_0^{\infty} dt \cdot t^2 \left[I_0^2(t\xi) \frac{K_0(t)}{I_0(t)} - \beta^2 I_1^2(t\xi) \frac{K_1(t)}{I_1(t)} \right] = -\frac{4e^2}{mc^2} \cdot \frac{1}{\gamma_{\perp}} \cdot \frac{N_e}{2\pi r_0} T_z, \tag{IV.6}$$

where I_n and K_n are modified Bessel functions. In the range $0.8 < \xi = r_0/a < 0.95$, $0.4 < T_z < 0.8$. The force gradient of the Coulomb repulsion is $\approx (4e^2/mc^2) \cdot (1/\gamma_{\perp}) \cdot (N_e/2\pi r_0) \cdot 1/[b_c(g + b_c)] \cdot (1/\gamma_{\perp}^2)$. For $N_e = 10^{13}$, $\gamma_{\perp} = 30$, $g = 0.02$, $r_0 = 5$ cm, and $\xi = 0.8$ may be achieved with a focusing parameter $b_c \approx 0.2$, while the ratio of the square of the frequency of the betatron oscillations in the z direction to the square of the frequency of rotation of a particle in the ring will then be of the order of $3 \cdot 10^{-3}$. It thus follows that, in order to be able to use this type of focusing in practice, it is essential to increase its efficiency.

It is clear that the focusing efficiency increases as the defocusing effect of the screening magnetic field diminishes. Of the various possible ways of reducing the magnetic screening we select the following [40].

We may suppose that, if the metal tube is coated on the inside with a thin layer of dielectric with a fairly high dielectric constant, then the reflection of the electric field will occur at the radius of the dielectric, while the reflection of the magnetic field will occur at the radius of the metal. Since the layer of dielectric is situated closer to the ring than the metal, the contribution of the electric field to the force acting on the particles in the ring will be greater than that of the magnetic field.

This proposition appears very interesting; however, a system incorporating a dielectric may lead to serious losses by way of Cerenkov radiation, and this constitutes its weak point.

Dolbilov [41] proposed a system very effectively reducing the effects of the magnetic screening. This system proposes using a metal cylinder cut into strips along the generators ("squirrel cage"). The ring as a whole moves coaxially with the cylinder; hence, this system does not form a retarding configuration, and no Cerenkov radiation is produced.

A screen of the "squirrel-cage" type gives the following expression for the gradient of the axial force component:

$$\left. \begin{aligned} &\frac{1}{\omega_0^2 m\gamma_{\perp}} \cdot \frac{\partial F_z}{\partial z} = -\frac{4e^2}{mc^2} \cdot \frac{1}{\gamma_{\perp}} \cdot \frac{N_e}{2\pi r_0} \left[\widetilde{T}_0^z + \sum_{n \neq 0}^{\infty} \widetilde{T}_n^z e^{ikn\varphi} \right]; \\ &\widetilde{T}_0^z = \xi^3 \int_0^{\infty} dt \cdot t^2 \left[I_0^2(t\xi) \frac{K_0(t)}{I_0(t)} (x_0^э - 1) \right] \\ &-\beta^2 \xi^3 \int_0^{\infty} dt \cdot t^2 I_1^2(t\xi) \cdot \frac{K_1(t)}{I_1(t)} x_0^{M} \equiv T_{0.e}^z + T_{0.m}^z \end{aligned} \right\} \tag{IV.7}$$

Here k is the number of cuts in the screen.

TABLE 1

ξ	$T^z_{0.e}$	$T^z_{0.m}$	$\tilde{T}^z_0$
0,8	1,84	—0,16	1,68
0,6	0,28	—0,009	0,27

The coefficients T^z_n are determined by the corresponding integrals of $x^{e.m}_n$. Analogous expressions are obtained for the force gradient in the radial direction. The determination of $x^{e.m}_n$ involves the substitution of the boundary conditions at the screen. These conditions lead to the following functional equations:

$$\left.\begin{aligned}
&\sum_n e^{i\,nk\varphi}\, x_n^{e.m} = 0 \quad \text{for} \quad \frac{\pi q^{e.m}}{l} < |\varphi| < \pi;\\
&\sum_{n\neq 0} e^{i\,nk\varphi}\,|n|\,(1-\varepsilon_n^{e.m})\, x_n^{e.m}\\
&= -\varkappa^{e.m}(1 - x_0^{e.m}) \quad \text{for} \quad |\varphi| < \frac{\pi q^{e.m}}{l};\\
&\varepsilon_n^{e} = 1 - \frac{1}{2|n|kI_s(t)K_s(t)};\quad \varepsilon_n^{m} = 1 + \frac{2tI_s'(t)K_s'(t)}{|n|k};\\
&\varkappa^{e} = \frac{l}{4\pi a K_0(t) I_0(t)};\quad \varkappa^{m} = -2t^2 I_1(t) K_1(t),
\end{aligned}\right\} \qquad \text{(IV.8)}$$

where q^e is the width of a strip, q^m is the width of the slit, $l = q^e + q^m$; $s = nk$. This system may be solved by means of an electronic computer.

The possibility of finding a solution to a high accuracy depends on having a small parameter, $\varepsilon_n^{e.m}$, which possesses the following property: $\varepsilon_n^{e.m} \to 0$ as $|n| \to \infty$.

In order to realize this focusing method experimentally we must choose optimum values for ξ, k, and $\zeta = \pi q^e/l$. Let us take $\xi = 0.8$. It is clear that with increasing distance between the wall and the ring the focusing force diminishes.

Table 1 shows how the quantities $T^z_{0.e}$, $T^z_{0.m}$, and $\tilde{T}^z_0$ vary with ξ for $k = 30$ and $\zeta = \pi/2$.

On approaching the screen, instability of the ring as a whole may develop; this increases the demands made upon the accuracy with which the ring is set relative to the tube axis. If the ring originally deviated from the axis, then on satisfying the inequality

$$\frac{8e^2}{mc^2}\cdot\frac{1}{\gamma_\perp}\cdot\frac{N_e}{2\pi r_0}\cdot\frac{1}{\beta^2}\cdot\xi^2\Phi(\xi) \leqslant 1$$

it will not be drawn to the wall, but will execute an angular drift around the tube axis. The function $\Phi(\xi)$ is nonlinear. For $\xi = 0.5$, $\Phi(\xi) \approx 1$, for $\xi = 0.8$, $\Phi(\xi) = 4.5$, and for $\xi = 0.95$, $\Phi(\xi) \approx 140$. However, even on satisfying the foregoing inequality, the deviation of the center of the ring from the axis of the tube should never exceed distances of the order of the smaller radius, in order to prevent the latter from "swelling" as a result of the drift of the ring.

It is convenient to make the lengths of the gaps and strips equal ($\zeta = \pi/2$). For $\zeta = 0$ the "squirrel cage" transforms into a continuous cylinder and the focusing force is simply determined by the difference between the electric and magnetic forces associated with the curvature of the system. The value of $\zeta = \pi$ corresponds to the ring in free space. Calculations show that the focusing force depends very little on ζ when ζ is close to $\pi/2$. The number of divisions is determined by the dependence of the amplitudes of the harmonics of the focusing force on this number. Estimates indicate that

$$T^z_{(n+1)e}/T^z_{ne} \sim (\xi)^k.$$

Numerical calculations confirm this estimate (see Table 2, which represents the values of T^z_{ne} and T^z_{nm} for different n with $k = 30$, $\xi = 0.8$, and $\zeta = \pi/2$).

With increasing number of divisions the focusing force becomes greater and the defocusing force smaller. Table 3 gives the values of $T^z_{0.e}$, $T^z_{0.m}$, and $\tilde{T}^z_0$ for various k and $\xi = 0.8$; $\zeta = \pi/2$.

TABLE 2

n	T^{z}_{ne}	T^{z}_{nm}	n	T^{z}_{ne}	T^{z}_{nm}
0	1,844	—0,1657	2	$0{,}36\cdot 10^{-6}$	$-0{,}39\cdot 10^{-5}$
1	$0{,}53\cdot 10^{-3}$	$-0{,}30\cdot 10^{-2}$	3	$0{,}49\cdot 10^{-9}$	$-0{,}76\cdot 10^{-8}$

TABLE 3

k	$T^{z}_{0.e}$	$T^{z}_{0.m}$	$\tilde{T}^{z}_{0}$
5	1,25	—0,38	0,87
10	1,61	—0,36	1,25
30	1,84	—0,16	1,68

Thus, the "squirrel cage" constitutes an anisotropic screen which holds back the axial electric field and passes the normal magnetic field. The improvement in the anisotropy with increasing number of divisions may be understood in the following manner. For $\zeta = \pi/2$ and a small number of divisions, the distance between the strips is fairly large, and some of the axial electric field penetrates between them and passes outside. On increasing k, the number of lines of force characterizing the electric field increases, the field being defined by the charges on the strips (charges at the edges); this means that the field inside the system increases.

The defocusing magnetic force in F_z is associated with the component H_r, for which the condition $H_r = 0$ should be satisfied in the metal strip. This means that the H_r component of the magnetic field passes freely into the gap between the strips, curving considerably around them. The field associated with the curvature of the magnetic lines of force around the middle of the strip in general remains inside the system (if k = 0, the whole magnetic field is reflected). On increasing k and keeping ζ constant, the part played by the edges of the strips increases more and more, and so does the associated protrusion of the magnetic field.

For these reasons the parameters chosen for the focusing system of the Joint Institute for Nuclear Research model of a collective linear ion accelerator are as follows: $\xi = 0.8$; $\zeta = \pi/2$; and k = 30. This gives $T^{z}_{0} = 1.68$ and $T^{r}_{0} = 1.33$ and ensures the same dimensions of the ring on acceleration as occur in the adhesor at the end of the compression process.

The use of image-force focusing is clearly limited in the case of large $\gamma_{\|}$, for which the effect of the real conductivity of the chamber walls on the motion of the ring becomes appreciable. An estimate gives a maximum value of $\gamma_{\|} \sim 40$.

The foregoing focusing methods are of course not exhaustive. Many laboratories are even now looking for different methods. Among these investigations the complicated and independent problem of obtaining a stabilized ring (Budker) occupies a leading place.

V. Stability of an Electron-Ion Ring

In order to achieve the collective method of acceleration it is essential to hold the parameters of the electron-ion ring within a specified range of values. The characteristic times required for these parameters to undergo changes as a result of various perturbations should be much greater than the acceleration time. It is practically impossible to determine this characteristic time; hence, in analyzing the stability we are compelled to take various model representations and study those cases which are deemed to be the most dangerous.

In considering the foregoing means of creating and accelerating an electron ring, it is natural to divide the problem of stability into two. The first of these is the problem of the stability of the electron ring (without ions) in a magnetic field with weak focusing; the second is a study of the stability of an electron-ion ring (charged plasma).

In this section we shall set out the results (by no means completely) of an analysis of these problems based on earlier review articles [42, 43].

In considering the instabilities of a one-component system (the electron ring), we classify the instabilities arising as, respectively, single-particle and coherent. By "single-particle" we mean instabilities associated with the motion of an individual particle in an external field and the field of all the other particles, which is regarded as being given. Single-particle instabilities are associated with a frequency shift of the betatron oscillations, arising from space charge and resonance perturbations of the external field. We have already given a description of these instabilities. Here we shall consider certain "coherent" instabilities. These instabilities are characterized by a change in both the motion of the particles and the intrinsic fields during the development of the instability. The method of studying such instabilities theoretically lies in solving the Vlasov system of self-consistent equations.

Coherent instabilities of the electron ring are divided into longitudinal (associated with the azimuthal motion of the particles) and transverse. We distinguish three types of longitudinal instability:

1) instability of the negative-mass type (NMI) [44, 45];

2) radiative instability RI (induced cyclotron radiation) [46, 47];

3) resistive longitudinal instability [48].

The main physical characteristic of the first of these instabilities is associated with the specific dependence of the particle rotation frequency in a magnetic field on the energy. If the rotation frequency falls with increasing particle energy, as occurs, for example, in a weakly focusing magnetic field, then the particle moves in the opposite direction to the force under the action of the azimuthal forces, just as if its mass were negative. In association with this, there may be a self-bunching of the particles and an increase in density fluctuations. The mean velocity of the beam remains constant, and the instability is not accompanied by the emission of propagating electromagnetic waves.

An analysis of the negative-mass type of instability in the collective linear method of acceleration is especially important at the initial stage of ring formation, when the ring is close to the cylindrical walls of the chamber. This kind of geometry differs greatly from that traditionally employed in ordinary accelerators, and, since the increment of the negative-mass instability depends on the geometrical factor, special calculations allowing for the effects of specific screening are essential. Calculations of the increments were carried out for a monoenergetic, infinitely thin cylindrical electron beam (an E layer) rotating with relativistic velocity in a constant magnetic field. The beam is surrounded by an infinitely conducting cylindrical screen situated coaxially with respect to the beam [49, 50]. The following expression was obtained for the negative-mass-instability increment (gain) [49]:

$$\operatorname{Im}\omega = \Omega \frac{n}{\gamma_\perp} \sqrt{\frac{2(a-r_0)}{r_0}}. \tag{V.1}$$

Here

$$\Omega = \sqrt{\frac{2\pi r^* \sigma_0 r_0 \omega_0^2}{\gamma_\perp e}};$$

σ_0 is the surface electron density, a is the radius of the cylindrical screen; n is the number of the harmonic of the azimuthal perturbation of the electron beam; ω_0 is the electron rotation frequency. In calculating the increment of the negative-mass instability it was assumed that the following inequality held:

$$\frac{a-r_0}{r_0} \ll \frac{1}{\gamma_\perp^2}, \tag{V.2}$$

this occurring when the beam is immediately next to the screen and $\gamma_\perp$ is not too high. We see from Eq. (V.1) that the geometrical factor $\Lambda = \sqrt{\frac{2(a-r_0)}{r_0}}$ may be very small owing to the influence of the screen.

It was shown elsewhere [50] that on satisfying the condition opposite to (V.2) no negative-mass instabilities arose at all. For beams involving a finite energy spread, on satisfying (V.2) the negative-mass increment falls in relation to (V.1), and an instability threshold appears [51]. The threshold spread may be found by using existing results [44, 45] and the geometrical factor indicated above.

In another paper [51] the negative-mass increment was also calculated for a very thin ring (the ring dimension l along the generator of the cylindrical screen being considered much smaller than the distance from the ring to the screen). In contrast to the case of the E layer, for this kind of ring (without allowing

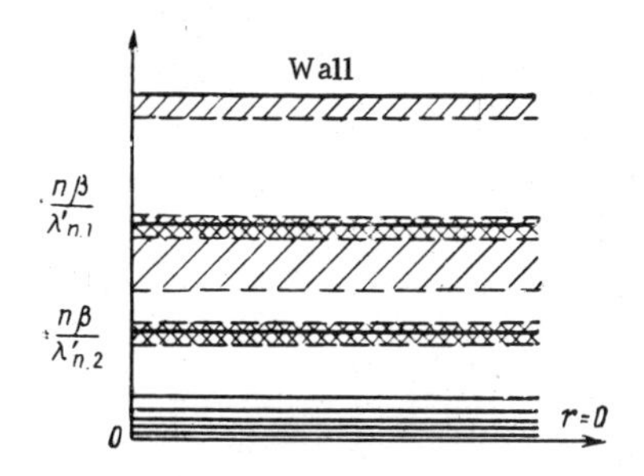

Fig. 1. Region of negative-mass and radiation instability. Values of r/b are shown along the vertical axis.

for the energy spread) no instability cutoff occurs in the relativistic beam. If we take equal values of $2\pi r_0 \sigma_0$ and N/l, the increment calculated for the ring is a factor of $\sqrt{\frac{l}{a-r_0}}$ smaller than (V.1).

Thus, quite a small energy spread is needed in order to suppress negative-mass instability in a ring lying close to the wall. This conclusion is qualitatively supported by experiment [52]. An experiment with a beam close to the screen shows that the initial azimuthal density fluctuations are smoothed out after a few rotations of the particles.

Subsequent consideration of negative-mass instability when the ring moves away from the screen involves computing difficulties. The regions in which this instability appears occur at ring-screen distances such that the harmonic of the mean particle rotation frequency approximately coinciding with the frequency of the perturbation wave differs fairly sharply from the arbitrary eigenfrequency of the chamber corresponding to this harmonic. In this case the condition for the radiation of transverse waves is not satisfied, and we find ourselves concerned with negative-mass instabilities in pure form [51]. It is convenient to describe the situation in terms of beam and chamber impedance, so as to identify the longitudinal instabilities [50, 51, 53]; thus, the regions in which negative-mass instability appears correspond to a capacitive impedance.

These regions are shown shaded in Fig. 1 for the harmonic with number n. The unshaded bands correspond to regions of longitudinal instability of the ring (inductive impedance).

One method of stabilizing negative-mass instability [50] is in fact based on increasing the inductive part of the impedance by means of inductive screens [50]. We also note the possibility of stabilizing this effect in the nonlinear stage of development. If the negative-mass increment is small in comparison with the energy spread in the beam, and the nonlinear interaction of the harmonics of the electromagnetic perturbation field may be neglected, we may use the quasilinear theory in order to analyze the development of negative-mass instability [54]. We find that the energy spread in this case increases with time and reaches a threshold value at which an instability cut-off occurs.

The cross-hatched regions (Fig. 1) represent the region of radiation instability (active impedance). Here we find resonance between the oscillations in the beam and the intrinsic modes of the chamber, and also electromagnetic radiation. The radiation instability was considered for the collective linear method of acceleration in earlier papers [55-60]. We may consider two cases:

1. The radiation practically occurs in only one mode of the chamber oscillations, when the beam is fairly close to the wall, and the time required for the development of the instability is much smaller than the time for the signal to pass from the ring to the wall.

2. In the condition opposite to the foregoing, the resonances merge into a continuous band and radiation occurs as in free space.

In the first case, for a monoenergetic beam, we have the increment [58]

$$\operatorname{Im}\omega = \left(\frac{cr^* N\omega_0 \sqrt{n\omega_0}}{\gamma_\perp a^2} f(n)\right)^{2/5} \tag{V.3}$$

and the real part of the frequency

$$\operatorname{Re}\omega = n\omega_0 + \operatorname{Im}\omega, \tag{V.4}$$

where $f(n)$ is a function increasing as $n^{4/3}$ up to $n \sim \gamma_\perp^3$ and then falling exponentially. We note that for real beams of finite thickness the results presented here are only valid for the first few harmonics, since it was assumed that the length of the perturbation wave was much greater than the thickness of the beam. For a large spread the increments diminish substantially; however, the instability does not have any threshold.

In the second case, the radiation-instability increment for a monoenergetic beam is smaller than (V.3), and furthermore a threshold exists [60]. The threshold number of particles is determined from the following inequality:

$$\frac{r^* N}{2\pi r_0 \gamma_\perp} < n^{2/3} \left(\frac{\Delta\omega_0}{\omega_0}\right)^2. \qquad \text{(V.5)}$$

Here $\Delta\omega_0$ is the scatter in the particle rotation frequencies. On making allowance for the nonlinear interaction of the harmonics in the development of radiation instability in a monoenergetic beam a long way from the screen we find that the instability passes very rapidly into the nonlinear stage with increasing energy spread [60].

All the results which we have just been mentioning were obtained on the assumption of infinite conductivity in the walls of the chamber. The effect of a finite Q in a chamber of the same geometry was discussed earlier [59]. The question as to the transverse instability of the electron beam was discussed by a number of authors [43, 61, 62].

It was shown in [61] that the existence of an infinitely conducting cylindrical screen eliminated the transverse instabilities of the E layer. The condition imposed on the $\Delta\nu_{r,z}$ spread in the betatron frequencies, leading to the suppression of the resistive transverse instability [63] in the case in which the side walls of the chamber exert a substantial influence, takes the form

$$\Delta\nu_{r,z} > \frac{r^* r_0 N}{2\pi\nu_{r,z} a_0^2 \gamma_\perp^3} + \frac{r^* r_0 N}{\pi\nu_{r,z}\gamma_\perp (\Delta h)^2} \left[\frac{r_0 c}{8\pi\sigma(n-\Delta\nu)_{r,z}(\Delta h)^2}\right]^{1/2}, \qquad \text{(V.6)}$$

where a_0 is the small dimension of the ring, Δh is the distance from the wall to the middle plane of the ring, and σ is the wall conductivity.

The case in which the electron ring lies close to the cylindrical surface of the chamber was considered in [62]. The maximum increment in the resistive transverse instability for a monoenergetic beam is

$$\operatorname{Im}\omega = \frac{r^* N}{2\pi r_0 \gamma_\perp} \left(\frac{r_0}{2a}\right)^2 \frac{c}{a} \left[\frac{\omega_0}{2\pi\sigma(1-\nu_{r,z})}\right]^{1/2}. \qquad \text{(V.7)}$$

For beams with an energy spread there is an instability threshold associated with Landau damping. The spread with respect to the longitudinal energy w required to suppress the instability has to satisfy the inequality

$$\Delta w > \frac{1}{2} \cdot \frac{\nu}{\gamma_\perp} \left(\frac{c}{a}\right)^2 \frac{1}{\frac{d}{dw}\left[\omega_0^2(w)(1-\nu_{r,z}(w))\right]\Big|_{w=w(\omega_0)}}. \qquad \text{(V.8)}$$

Now let us turn to instabilities in the electron-ion ring. Here we confine attention to beam instability [64], the stability of the ring with respect to transverse bending [65], and the stability of a slightly inhomogeneous cylindrical layer [66].

Analysis of the beam instability is carried out by using the model of a cylindrical quasicentral plasma filament (pinch) through which a compensated beam of charged particles is moving. In this model no allowance is made for the intrinsic fields of the ring in the stationary state, nor for its curvature. The density of the beam is regarded as being much smaller than the density of the plasma, which in the case of a ring corresponds to the consideration of instability in a coordinate system linked to the electrons.

There are two mechanisms of exciting waves by a beam traveling through the plasma: the Vavilov-Cerenkov effect, and the excitation of waves due to the anisotropy of the particle velocity-distribution function. For three-dimensional waves with a wavelength much smaller than the radius of the filament, the increments in beam instability are the same as in an unlimited medium. In the cases of a relativistic monoenergetic beam of low density currently concerning us, the increment in the instability associated with Cerenkov radiation was given earlier [67]. It should be noted that in the case of a ring this instability may not arise at all if the condition $\omega_0\gamma\ \ /\Omega_e$ is satisfied, where Ω_e is the Langmuir frequency of the electrons. The excitation of an axially symmetric surface wave of the same type in a cylindrical beam is greatly impeded; the corresponding increments are exponentially small [68]. An aperiodic instability of the beam was studied elsewhere [68-70]. In the hydrodynamic approximation, using the linear theory, the instability increment for three-dimensional waves in the laboratory system of coordinates equals

$$\operatorname{Im} \omega = \Omega_i \gamma_\perp, \tag{V.9}$$

where Ω_i is the Langmuir frequency of the ions.

The increment of surface waves is only half as much as (V.9). The quasilinear theory of aperiodic instability shows that in the nonlinear stage the instability breaks off, the relative energy losses of the beam being of the order of $mn_i/Mn_e \ll 1$.

The stability of an electron-ion ring with a partly compensated charge and external focusing with respect to transverse bending was studied in [71, 72]. In these papers the model of two charged cylindrical filaments with a constant density over the cross section was taken, and the instability of the motion of the centers of mass of the filaments during the formation of "snakes" (characteristic configurations) was examined.

In the case currently concerning us, in which the number of ions is small and the condition $\eta = m\gamma/M \cdot n_0/n_i \gg 1$ [71], for the continuous spectrum of wave vectors k there is a region of instability

$$kR = \sqrt{n_{\text{eff}}} + \frac{\omega_i}{\omega_0} + s\sqrt{\frac{\omega_i^3}{\sqrt{n_{\text{eff}}}\,\omega_0^3\,\eta}}, \tag{V.10}$$

where $\omega_i = \frac{c}{a_0}\sqrt{\frac{2m}{M}}\,\nu_e$; n_{eff} is the effective magnetic-field fall-off index (allowing for the intrinsic field of the ring), while s varies over the range $-1 \le s \le 1$. The instability increment is

$$\operatorname{Im} \omega = \frac{\omega_i}{2}\sqrt{\frac{1-s^2}{\sqrt{n_{\text{eff}}}\,\eta}}. \tag{V.11}$$

In a ring the spectrum of wave vectors is discrete ($k = l/r_0$). Here l is an arbitrary whole number; hence, there will be no instability if no wave vector falls in the range (V.10). The conditions of ring stability take the form

$$2\sqrt{\frac{\omega_i^3}{\omega_0^3}} < 1 - \left\{\sqrt{n_{\text{eff}}} + \frac{\omega_i}{\omega_0} - \sqrt{\frac{\omega_i^3}{\sqrt{n_{\text{eff}}}\,\eta\omega_0^3}}\right\} \tag{V.12}$$

or

$$1 > \sqrt{n_{\text{eff}}} + \frac{\omega_i}{\omega_0} + \sqrt{\frac{\omega_i^3}{\sqrt{n_{\text{eff}}}\,\eta\omega_0^3}}, \tag{V.13}$$

where the symbol $\{A\}$ denotes the fractional part of the number A. Consideration of the Landau damping and radiative friction [71] showed that these effects did not lead to any stabilization of the instability.

Low-frequency potential oscillations in an electron-ion ring were considered for the case of a cylindrical layer of relativistic particles rotating in a uniform magnetic field in [66]. Certain simplifying assumptions were made: the frequency of the oscillations was considered to be much smaller than the cyclotron frequency of the particles, the phase velocity of the waves much smaller than the velocity of light, the momentum distribution of the particles along the magnetic field nonrelativistic, and the longitudinal wavelength much smaller than the radius of the layer. In this case, the methods of geometrical optics suffice to find the spectrum of oscillations for the layer, and the increment may be obtained from perturbation theory. The oscillations of the mode with $|n| = 1$ corresponding to ionic sound in a homogeneous plasma are unstable, the growth increment of the oscillations being

$$\operatorname{Im} \omega = \sqrt{\pi} \cdot \frac{n_e}{n_i} \frac{(\operatorname{Re} \omega)^2}{c\,|k_3|\,u_e} \cdot \frac{\int_0^\infty ds f \sqrt{1 + u_{pe}^2 s}\left|\frac{d\Phi}{ds}\right|^2}{\int_0^\infty ds f s \left|\frac{d\Phi}{ds}\right|^2}, \tag{V.14}$$

where k_3 is the wave number in the direction of the magnetic field, the function f describes the particle distribution across the magnetic field, u_{pe}^2 and u_e^2 are, respectively, expressed in terms of the mean squares of the transverse and longitudinal momenta of the electron, and Φ is the potential (in the zero approximation) with respect to a small parameter, namely, the ratio of the increment to the oscillation frequency ω. This kind of kinetic instability corresponds to drift instability in a weakly inhomogeneous nonrelativistic plasma. It was later shown [66] that, for a group of particles limited in the magnetic-field direction, the length of the group being much smaller than the ratio of the thermal velocity of the particles to the wave frequency, this instability could not develop at all. An analysis of the instabilities carried out for specific dimensions of the annular groupings (rings) used in experiments on the collective method of acceleration shows that no serious changes in the ring parameters take place during the period of acceleration.

CONCLUSION

We have thus shown that it is perfectly possible to employ the collective method in order to create charged-particle accelerators. We have considered only one particular system of acceleration, and only from the point of view of creating very-high-energy accelerators. This does not mean, however, that the collective method of acceleration is limited to this. The principle may clearly be used in order to create accelerators of very different types, starting with accelerators capable of being used for the heaviest ions and ending with accelerators yielding extremely high energies. Estimates show that accelerators with beams moving in opposite directions may also be based on the collective principle.

In order to illustrate the prospects of the method under consideration, we here present some general estimates relating to the use of a relativistically stabilized beam in a collective accelerator. For a strong electron beam compensated with ions in such a way that $N_i = N_e/\gamma_\perp^2$, magnetic compression gives rise to a dimunition in the beam cross section as a result of the conversion of some of the energy of the transverse motion of the particles into radiation. This reduction in cross section continue until the scattering of electrons by ions starts to play an appreciable part. A stationary (steady) state then sets in. The cross-sectional dimensions of the ring then amount to 10^{-3}–10^{-4} cm. In such a ring the maximum field strength acting on an ion may reach 10^9 V/cm. This means that if all the potentialities of the annular grouping of particles (the ring) are exploited it should be possible to make accelerators with dimensions of only 1 m for each 100 GeV.

LITERATURE CITED

1. V. I. Veksler, "Coherent principle of the acceleration of charged particles," Symposium, CERN, Vol. 1, 80 (1956); V. I. Veksler, At. Énerg., 5, 427 (1957).
2. V. I. Veksler et al., Preprint of Joint Institute for Nuclear Research (JINR) R9-3440-2, Dubna (1968); V. I. Veksler et al., "Collective linear acceleration of ions," Proc. Sixth International Conf. on High-Energy Accelerators, Cambridge (1967), p. 289.
3. I. N. Ivanov et al., JINR Preprint R9-4132, Dubna (1968).
4. O. I. Yarkovoi, JINR Preprint 2183, Dubna (1965).
5. N. B. Rubin, JINR Preprint 2882-2, Dubna (1966).
6. M. L. Iovnovich and M. M. Fiks, JINR Preprint R9-4849, Dubna (1969).
7. R. E. Berg et al., Phys. Rev. Lett., 22, No. 9, 419 (1969).
8. N. C. Christofilos, Preprint UCRL-71414, Lawrence Radiation Laboratory, Livermore (1969); "Static compression of relativistic electron rings," Phys. Rev. Lett., 22, No. 16, 830 (1969).
9. L. J. Laslett and A. M. Sessler, Preprint UCRL-1858, Lawrence Radiation Laboratory, Berkeley (1969).
10. E. Keil, "Extraction of electron rings by a sequence of axially shifted coil pairs," Symposium on Electron Accelerators, Berkeley, USA (1968).
11. K. A. Reshetnikova and V. P. Sarantsev, JINR Preprint R9-4678, Dubna (1969).
12. D. Keefe, "ERA development at Berkeley, Contribution to the Seventh International Conference on Accelerators, Erevan (1969), (in press).
13. A. G. Bonch-Osmolovskii et al., JINR Preprint R-9-4171, Dubna (1968).
14. J. D. Jackson, Classical Electrodynamics, Wiley (1962).
15. S. B. Rubin and V. N. Mamonov, JINR Preprint 9-3346-2, Dubna (1967).
16. P. L. Morton and V. K. Neil, "The interaction of a ring of charge passing through a cylindrical cavity," Symposium on Electron Ring Accelerators, Berkeley, USA (1968).

17. W. Heitler, Quantum Theory of Radiation, 3rd ed., Oxford Univ. Press (1954).
18. V. L. Ginzburg and V. Ya. Éidman, Zh. Éksp. Teor. Fiz., 36, 1823 (1959).
19. B. M. Bolotovskii and G. V. Voskresenskii, Usp. Fiz. Nauk, 88, 209 (1966).
20. A. M. Sessler, Preprint Lawrence Radiation Laboratory, Berkeley (1968).
21. Yu. N. Dnestrovskii and D. P. Kostomarov, Dokl. Akad. Nauk SSSR, 124, 1026 (1959).
22. O. A. Kolpakov and V. I. Kotov, Zh. Tekh. Fiz., 34, 1387 (1964).
23. A. Faltens, "Radiation by a relativistic particle passing through a cavity," Symposium on Electron Ring Accelerators, Berkeley, USA (1968).
24. T. D. Lawson, Preprint RHEL/M144 (1968).
25. T. D. Lawson, "How can we calculate the radiation loss in the electron ring accelerator," Contribution to the Seventh International Conference on Accelerators, Erevan (1969) (in press).
26. L. K. Orlov and A. V. Ryabtsov, in: Electrophysical Apparatus [in Russian], No. 6, Atomizdat, Moscow (1967).
27. A. B. Kuznetsov and S. B. Rubin, JINR Preprint R9-4909, Dubna (1970).
28. B. S. Levin and A. M. Sessler, Preprint UCRL-18595 (1969).
29. G. V. Voskresenskii and V. N. Kurdyumov, "Energy losses of electron rings in passing through one resonator" [in Russian], Contribution to the Seventh International Conference on Accelerators, Erevan (1969) (in press).
30. I. A. Gilinskii, Dokl. Akad. Nauk SSSR, 150, 767 (1963).
31. B. M. Bolotovskii and G. V. Voskresenskii, Zh. Tekh. Fiz., 34, 704 (1964).
32. A. B. Kuznetsov and S. B. Rubin, JINR Preprint R9-5087, Dubna (1970).
33. A. B. Kuznetsov and S. B. Rubin, JINR Preprint R9-5247, Dubna (1970).
34. A. I. Akhiezer et al., Zh. Tekh. Fiz., 25, 2526 (1955).
35. B. M. Bolotovskii and G. V. Voskresenskii, Usp. Fiz. Nauk, 94, 377 (1968).
36. A. B. Kuznetsov and S. B. Rubin, JINR Preprint R-9-4728, Dubna (1969).
37. O. A. Kolpakov et al., Zh. Tekh. Fiz., 34, 26 (1965).
38. E. Keil, "On the energy loss of a charged ring passing through a corrugated cylindrical waveguide," CERN-IST-TH/69-49.
39. A. G. Bonch-Osmolovskii et al., JINR Preprint R9-4135, Dubna (1968).
40. L. J. Laslett, "Image focusing through use of dielectric," Symposium on Electron Ring Accelerators, Berkeley, Preprint UCRL-18103 (1968).
41. G. V. Dolbilov, JINR Preprint R9-4737, Dubna (1969).
42. A. G. Bonch-Osmolovskii et al., JINR Preprint R9-4138, Dubna (1968).
43. A. M. Sessler, "The electron ring accelerator: general concepts and present state of theoretical understanding," Symposium on Electron-Ring Accelerators, Lawrence Radiation Laboratory (1968), p. 11.
44. A. A. Kolomenskii and A. N. Lebedev, At. Énerg., 7, No. 6, 549 (1969).
45. C. Nielsen et al., CERN Symposium (1959), p. 115; in: Storage of Relativistic Particles [Russian translation], Gosatomizdat, Moscow (1963), p. 133.
46. A. V. Gaponov, Izv. Vuzov. SSSR, Ser. Radiofiz., 11, No. 5, 836 (1959).
47. J. Schneider, Phys. Rev. Lett., 2, 504 (1959).
48. V. K. Neil and A. M. Sessler, Rev. Sci. Instrum., 36, 429 (1965).
49. I. N. Ivanov, JINR Preprint R9-3476-2, Dubna (1967).
50. R. Briggs and V. Neil, Plasma Physics, 9, 209 (1967).
51. A. G. Bonch-Osmolovskii and É. A. Perel'shtein, JINR Preprint R9-4424, Dubna (1969).
52. P. I. Ryl'tsev et al., JINR Preprint R9-4620, Dubna (1969).
53. A. N. Lebedev, Dissertation [in Russian], Physical Institute, Academy of Sciences of the USSR (1968).
54. É. A. Perel'shtein, Zh. Tekh. Fiz., 37, 1177 (1967).
55. I. N. Ivanov, JINR Preprint R9-3474-2, Dubna (1967).
56. R. Briggs, "Coherent radiative instabilities of the compressed electron rings," Symposium on Electron Ring Accelerators, Lawrence Radiation Laboratory (1968), p. 434.
57. C. Pellegrini and A. M. Sessler, "Effect of coherent radiation during ring compression," Symposium on ERA, Lawrence Radiation Laboratory (1968), p. 442.
58. A. G. Bonch-Osmolovskii and É. A. Perel'shtein, JINR Preprint R9-4425, Dubna (1969).
59. V. P. Grigor'ev and A. N. Didenko, Contribution to the Seventh International Conference on Accelerators [in Russian], Erevan (1969) (in press).

60. A. G. Bonch-Osmolovskii et al., Contribution to the Seventh International Conference on Accelerators [in Russian], Erevan (1969) (in press).
61. I. N. Ivanov and V. G. Makhan'kov, JINR Preprint R9-3475-2, Dubna (1967).
62. I. L. Korneev and L. A. Yudin, "On the stability of an electron ring beam in an infinite cylindrical tube with conducting walls," Contribution to the Seventh International Conference on Accelerators [in Russian], Erevan (1969) (in press).
63. L. J. Lasslett et al., Rev. Sci. Instrum., 36, 436 (1965).
64. A. I. Akhiezer and Ya. B. Fainberg, Dokl. Akad. Nauk SSSR, 69, 555 (1949).
65. G. I. Budker, At. Énerg., 5, 9 (1956).
66. M. L. Iovnovich, JINR Preprint 9-3395-2, Dubna (1967).
67. V. P. Silin and A. A. Rukhadze, Electromagnetic Properties of Plasma and Plasma-Like Media [in Russian], Gosatomizdat, Moscow (1961).
68. V. G. Makhan'kov, Zh. Tekh. Fiz., 36, 1752 (1966).
69. V. G. Makah'kov and A. A. Rukhadze, JINR Preprint 1005, Dubna (1962); Yadernyi Sintez, 2, 177 (1962).
70. O. I. Yarkovoi, JINR Preprint 1053, Dubna (1962).
71. B. V. Chirikov, Preprint, Inst. Nuclear Physics, Siberian Branch, Academy of Sciences of the USSR, Novosibirsk (1964).
72. E. E. Mills, "Mutual oscillations of ions and electrons in the electron-ring accelerator," Symposium on ERA, Lawrence Radiation Laboratory, Berkeley (1968), p. 448.

LEPTONIC HADRON DECAYS

É. I. Mal'tsev and I. V. Chuvilo

The main properties of leptonic hadron decays are reviewed from the point of view of the modern theory of weak interactions. The first chapter is devoted to a description of the decay properties of Π and K mesons and the experimental data obtained prior to the middle of 1969 are analyzed. The second chapter is devoted to discussion of leptonic baryon decays in the framework of $SU_{(3)}$ symmetry.

1. Introduction

Leptonic and semileptonic hadron decays form a large class of reactions in elementary particle and nuclear physics; in such reaction a pair of leptons participates in the transition of a hadron A into another hadron B or to the vacuum. The most typical decays are

$$A \to B + l^-(l^+) + \tilde{\nu}_l(\nu_l); \tag{1a}$$

$$A \to l^-(l^+) + \tilde{\nu}_l(\nu_l). \tag{1b}$$

The reactions of this class include the β decay of atomic nuclei (in this case A is a nucleon bound in a nucleus) and the β (or μ) decays of baryons and mesons. The lifetimes of such processes are characterized by a gigantic spread: from 10^{-10} sec (hyperon decay) to 10^{21} years (double β decay of nuclei). These reactions also include the processes of μ and K capture:

$$l^- + A \to B + \nu_l \tag{2}$$

and reactions with a neutrino:

$$\nu_l(\tilde{\nu}_l) + A \to B + l^-(l^+). \tag{3}$$

All these reactions as well as a large class of hadron decay processes in which leptons do not participate and a whole series of other known or conjectured reactions are united in the framework of the present theory of the so-called weak interactions of elementary particles.

Turning to the theoretical aspects of the descriptions of these reactions in terms of modern field-theoretical notions [1], the weak-interaction Lagrangian must satisfy the following requirements: Lorentz invariance, Hermiticity, locality.

In the theory it is also assumed that there is linearity in the four-component Dirac spinor fields (without derivatives) and that the following quantum numbers satisfy conservation laws: the electric charge, the baryon number, the lepton number. Within the framework of these requirements, the weak-interaction Lagrangian can be expressed as a sum of two terms with opposite spatial parities:

$$L_W = \sum_n c_n L_n(B, A, l, \nu_l) + \sum_n c'_n L'_n(B, A, l, \nu_l), \tag{4}$$

Joint Institute for Nuclear Research, Dubna. Translated from Problemy Fiziki Elementarnykh Chastits i Atomnogo Yadra, Vol. 1, No. 2, pp. 443-524, 1971.

TABLE 1

Type of interaction	Invariant	Limit
Scalar S	I	1
Pseudoscalar P	γ_5	0
Axial vector A	$\gamma_5\gamma_\alpha$	$\gamma_5\gamma_i = \frac{1}{i}\ \sigma_i\gamma_4 = \sigma_i$ $\gamma_5\gamma_4 \approx 0$
Vector V	γ_α	$\gamma_4 \approx 1$ $\gamma \approx 0$
Tensor T	$\sigma_{\alpha\beta}$	$\sigma_{il} \simeq \varepsilon_{ikl}\sigma_l$ $\sigma_{i4} \simeq 0$

where for processes in which leptons participate

$$\left.\begin{array}{l} L_n(B, A, l, \nu_l) = [\bar{u}_B O_n u_A]\,[\bar{u}_l O_n u_{\nu_l}]; \\ L'_n(B, A, l, \nu_l) = [\bar{u}_B O_n u_A]\,[\bar{u}_l O_n \gamma_5 u_{\nu_l}]; \end{array}\right\} \tag{5}$$

in this case, (4) is replaced by the more compact Lagrangian

$$L_W = \sum_n [\bar{u}_B O_n u_A[\,[\bar{u}_l O_n (c_n + c'_n \gamma_5) u_{\nu_l}], \tag{6}$$

and in the theory with a two-component neutrino ($m_\nu = 0$, which means 100% violation of spatial parity) this Lagrangian takes the form

$$L_W = \sum_n c_n\, [\bar{u}_B O_n u_A]\,[\bar{u}_l O_n (1 + \varepsilon\gamma_5)\, u_{\nu_l}], \tag{7}$$

where ε may have the values $\varepsilon = \pm 1$. It follows from the data for $l\nu$ correlations in nuclear β decays that $\varepsilon = +1$.

In the above equations, u_i are the Dirac spinor functions of the corresponding particles and O_n are operators characterized by definite transformation properties under spatial rotations and reflections. Using the invariant quantities at our disposal, viz, the four-dimensional momenta P_i and the Dirac matrices γ_α, we can construct five relativistic invariants. These are given in Table 1 together with their nonrelativistic limits.

Assuming in the nonrelativistic limit that the baryons are at rest before and after the decay and ignoring the energy liberated in the decay, we obtain an approximation that describes the so-called allowed transitions, when the Lagrangian can be represented in the form

$$\begin{array}{l} L_n = \bar{u}_B u_A\,[\bar{u}_l (c_S + c_V \gamma_4)\,(1 + \varepsilon\gamma_5)\, u_{\nu_l}] \\ \quad + \bar{u}_B \sigma u_A\,[\bar{u}_l (c_T + c_A \gamma_4)\,\sigma\,(1 + \varepsilon\gamma_5)\, u_{\nu_l}]. \end{array} \tag{8}$$

Here, the first term describes nuclear β transitions of the Fermi type with $\Delta J = 0$; the second, transitions of the Gamow-Teller type with $\Delta J = 0, \pm 1$.

It follows from the relations (4) or (6) that the total Lagrangian in the general case contains ten complex constants c_n and c'_n or 19 real constants, since one can choose one common phase arbitrarily. If the process is invariant under time reversal, c_n and c'_n are real. If conservation of spatial parity is assumed, either all the c_n or all the c'_n vanish.

The formulation of the modern theory of weak interactions has been based in the first place on the experimental data on nuclear β decay and the decay of the $\mu+$ meson:

$$\mu^+ \to e^+ + \nu_e + \tilde{\nu}_\mu. \tag{9}$$

It is not our task to make a detailed analysis of these processes; such an analysis has led to our present understanding of these processes and, to a considerable extent, understanding of the general problem of weak interactions. We shall only mention some of the main results that follow from the currently available experimental data on this subject. It should be noted first that, in accordance with Eq. (8), pseudoscalar couplings do not participate in allowed β transitions. No traces of the existence of couplings of this kind have been found in the forbidden β transitions either. Secondly, an analysis of the data on the various correlations in nuclear β decays indicates that they do not contain tensor or scalar couplings, i.e.,

$$c_P = c'_P = c_T = c'_T = c_S = c'_S = 0. \qquad (10)$$

It follows that the nuclear β decay is due entirely to a mixture of the vector and axial-vector variants of the weak interaction. It turns out that

$$c_A = c'_A; \quad c_V = c'_V; \qquad (11)$$

these conditions mean that there is 100% violation of spatial parity in these processes. Finally, the known correlation properties of the β decay of polarized nuclei shows that the relative sign of the constants c_V and c_A is negative: $\alpha = c_A/c_V < 0$.

Thus, nuclear β decay is described by the V-A variant of the weak interaction [2].

All the available information on processes governed by the weak interaction indicate that this is a universal interaction. In its most general formulation, universality of the weak interactions must mean the equality of the coupling constants for all weak processes and identical fundamental space-time properties of the interaction. It follows that all weak interactions can be described by a single Lagrangian with one and the same constant G and that this Lagrangian contains only vector and axial-vector interactions if one ignores effects due to the presence of strongly interacting particles. The constant G is taken equal to the constant of $\mu+$ decay:

$$G_\mu = (1.4350 \pm 0.0011) \cdot 10^{-49}\ \text{erg} \cdot \text{cm}^3 \qquad (12)$$

It follows that the weak-interaction Lagrangian can be constructed by analogy with electrodynamics. This means that in the case in which we are interested it can be expressed as a product of the leptonic $J_\alpha^{(l)}(x)$ and hadronic $J_\alpha(x)$ currents:

$$L_W(x) = \frac{G}{\sqrt{2}} J_\alpha^{(l)} \cdot J_\alpha + \text{h. c.} \qquad (13)$$

in which G is the μ-decay constant G_μ (12). At the same time each of the currents that occurs in (13) must contain only vector and axial-vector parts, i.e., they must have the form

$$J_\alpha^{(l)}(x) = \bar{u}_l(x)\gamma_\alpha(1+\gamma_5)u_{\nu_l}(x) + \bar{u}_\mu(x)\gamma_\alpha(1+\gamma_5)u_{\nu_\mu}(x); \qquad (14)$$

$$J_\alpha(x) = \bar{u}_B(x)\gamma_\alpha(c_V + c_A\gamma_5)u_A(x) = J_\alpha^{(V)}(x) + J_\alpha^{(A)}(x). \qquad (15)$$

In the weak hadronic current $J_\alpha(x)$ there remain the constants of the vector and axial-vector couplings; in contrast to the leptonic current, for which one assumes $\varepsilon = +1$ in accordance with (7), it is a priori clear that $c_A/c_V \neq 1$ because of the renormalization effects due to the strong interactions. These effects are determined by the structure of the hadrons. At present they cannot be studied theoretically with sufficient accuracy and our only information about them is obtained experimentally.

If we now turn to the experimental facts and compare, for example, two pairs of hadron decays:

$$\left.\begin{matrix} \pi \to l\nu_l, & \text{and} & n \to pe\nu, \\ K \to l\nu_l & & \Lambda \to pe\nu, \end{matrix}\right\} \qquad (16)$$

and take into account the phase-space for these decays, we find that in each of the pairs the matrix elements of the first decays, in which the strangeness quantum number S does not change ($\Delta S = 0$), are much greater than the matrix elements of the second decays, in which the strangeness changes. The same situation is encountered for the pair of Σ^--hyperon decays:

$$\left.\begin{array}{l}\Sigma^- \rightarrow \Lambda e^- \nu; \\ \Sigma^- \rightarrow n e^- \nu,\end{array}\right\} \tag{17}$$

and this condition is satisfied quite generally for all processes that proceed through weak interactions. Thus, if we represent the hadronic current (15) as a sum of two currents, of which one, $J_\alpha^{(0)}(x)$, is due to transitions with $\Delta S = 0$ and the other, $J_\alpha^{(1)}(x)$, to transitions with $|\Delta S| = 1$, i.e.,

$$J'_\alpha(x) = J_\alpha^{(0)}(x) + J_\alpha^{(1)}(x), \tag{18}$$

the constants for these two types of transition must be different. It follows that the V-A theory in its original form is not universal; even for the different processes of the same nature one must introduce separate appropriate interaction constants. Cabibbo [3] made the decisive step in restoring the universality. This is done as follows. We write down two Lagrangians: for the strangeness-conserving decay

$$L_W^{(0)} = \frac{G}{\sqrt{2}} J_\alpha^{(l)} \cdot J_\alpha^{(0)} \tag{19}$$

and for the strangeness-violating decay

$$L_W^{(1)} = \frac{G}{\sqrt{2}} J_\alpha^{(l)} \cdot J_\alpha^{(1)}. \tag{20}$$

The gist of Cabibbo's assumption is that both the hadronic currents $J_\alpha^{(0)}(x)$ and $J_\alpha^{(1)}(x)$ belong to the same octet representation of $SU_{(3)}$ symmetry and that the total hadronic current is turned through an angle θ with respect to the spin axis u_2 in the corresponding eight-dimensional unitary spin space. The upshot is that the hadronic current, which we previously wrote as the sum (18), now takes the form

$$J_\alpha(x) = \cos\theta \cdot J_\alpha^{(0)}(x) + \sin\theta\, J_\alpha^{(1)}(x), \tag{21}$$

while the leptonic current $J_\alpha^{(l)}(x)$ and the total Lagrangian of the weak interaction retain their structure.

We see that the introduction of the angle θ extends the original concept of universality in a distinctive manner: the leptonic current remains unchanged, whereas the hadronic currents acquire the additional factors $\sin\theta$ and $\cos\theta$, which maybe included among the weak-interaction constants, and the universality concept acquires a wider meaning. Now that the universality of the weak interactions has been restored by the introduction of the Cabibbo angle, these interactions are described by two parameters: the common coupling constant G and the angle θ. One could, of course, attempt to go further and make the theory more precise by introducing, for example, different angles for the vector and axial-vector parts of the current and for the components of the current that conserve and change strangeness. It is natural that an appropriate choice of such angles would lead to better agreement between the experiments and theory but the introduction of many new constants detracts from the elegance of the theory of weak interactions.

Finally, it should be emphasized that Cabibbo's assumption does not in fact violate the conception of a universal V-A interaction in its original form but merely extends and makes it more precise. Although the physical significance of the angle θ is not yet completely clear, Cabibbo's conjecture is undoubtedly a step forward in the theory of weak interactions.

We shall now proceed to a systematic exposition of the various aspects of leptonic meson and baryon decays, which, as we have seen, have a common nature and are characterized by the same constants of the theory.

1. Leptonic Meson Decays

In the meson decay processes the leptonic current in the Lagrangian (7) corresponds to the creation of two leptons from the vacuum:

$$\langle l, \nu_l | J_\alpha^{(l)} | 0 \rangle.$$

The hadronic current J_α must either annihilate the original meson A:

$$\langle 0 | J_\alpha | A \rangle,$$

or transform it into a different hadron state B:

$$\langle B | J_\alpha | A \rangle.$$

Both these currents must be constructed from the wave functions ψ_i of the particles participating in the process. The leptonic covariants which must be used to construct $J_\alpha^{(l)}$ are well known:

$$S^l = \bar{\psi}_\nu \psi_l;$$
$$V^l = \bar{\psi}_\nu \gamma_\alpha \psi_l;$$
$$T^l = \bar{\psi}_\nu \sigma_{\alpha\beta} \psi_l;$$
$$A^l = \bar{\psi}_\nu \gamma_5 \gamma_\alpha \psi_l;$$
$$P^l = \bar{\psi}_\nu \gamma_5 \psi_l.$$

To obtain a coupling with the neutrino in the form $\psi_\nu^-(1-\gamma_5)$ one must construct from the existing covariants the definite linear combinations

$$S^l - P^l = \bar{\psi}_\nu (1-\gamma_5) \psi_l;$$
$$V^l - A^l = \bar{\psi}_\nu (1-\gamma_5) \gamma_\alpha \psi_l = \bar{\psi}_\nu \gamma_\alpha (1+\gamma_5) \psi_l;$$
$$T^l - PT^l = \bar{\psi}_\nu (1-\gamma_5) \sigma_{\alpha\beta} \psi_l.$$

On the basis of the Lagrangian (7) we obtain the general form of the Hamiltonian:

$$H_W = S(S^l - P^l) + V(V^l - A^l) + TT^l + A(V^l - A^l) + P(S^l - P^l),$$

where S, V, T, A, and P refer to the hadronic current.

Whereas the leptonic covariants are the same for meson decays of the type $A \to 0$ and transitions of the type $A \to B$, the hadronic part of the Hamiltonian must take into account the specific form of the decay. In this connection let us consider the different transitions of π and K mesons in which leptons participate, i.e., the decays of pseudoscalar mesons.

1.1. Meson Decays that Proceed through the Channel $A \to l + \nu_l$

The leptonic decays of the charged π and K mesons

$$A^\pm \to e^\pm (\mu^\pm) + \nu_l (\tilde{\nu}_l) \tag{1.1}$$

are distinguished from all the other hadron decays by the absence of strongly interacting particles in the final state.

In principle, processes that are free from the effects associated with the strong interaction between particles in the final state enable one to obtain information about the decaying particle in the most unadulterated form. The properties of the decaying hadron, i.e., the π or K meson in the present case, must determine the behavior of certain functions of the kinematic variables corresponding to the given decay. Unfortunately, there are no kinematic variables in two-particle decays with fixed momenta of the secondary particles. The only parameters that characterize the hadrons in the reaction (1.1) are the coupling constants f_π and f_K, which can be compared with each other.

Let us consider this hadron decay mode for the reaction

$$K^{\pm} \to l^{\pm} + \nu_l\,(\tilde{\nu}_l).$$

The leptonic part of the Hamiltonian H_W is known; let us analyze separately each term of the hadronic part of H_W.

With a negative parity of the K meson $\langle 0\,|S|\,K\rangle = 0$, whereas the matrix element $\langle 0\,|P|\,K\rangle$ may be nonvanishing. The only invariant kinematic variable related to the hadronic part of the process is the square of the four-momentum of the K meson, $p_K^2 = m_K^2 = \text{const}$; thus,

$$\langle 0|P|K\rangle = f_p = \text{const}.$$

Arguing in the same way for the vector and axial-vector parts, we obtain

$$\langle 0|V|K\rangle = 0;$$

$$\langle 0|A|K\rangle = f_A \cdot p_K^\alpha,$$

where f_A does not, of course, depend on p_K. For the tensor part we must write

$$\langle 0|T|K\rangle = f_T \cdot p_K^\alpha \cdot p_K^\beta.$$

The total matrix element of the decay obtained from the relation for the total Hamiltonian H_W is

$$\langle l, \nu_l|H_W|K\rangle = f_p \langle l, \nu_l|\bar{\psi}_\nu (1-\gamma_5)\psi_l|0\rangle$$
$$+ f_A \sum_\alpha p_K^\alpha \langle l, \nu_l|\bar{\psi}_\nu (1-\gamma_5)\gamma_\alpha \psi_l|0\rangle + f_T \sum_\alpha \sum_\beta p_K^\alpha p_K^\beta \langle l, \nu_l|\bar{\psi}_\nu$$
$$\times (1-\gamma_5)\,\delta_{\alpha\beta}\,\psi_l\,|0\rangle. \quad (1.2)$$

The last term, which describes the tensor contribution to the matrix element, must vanish, since $\sigma_{\alpha\beta}$ is antisymmetric with respect to its subscripts and the product $p_K^\alpha \cdot p_K^\beta$ is symmetric with respect to the α and β functions. It follows that this kaon decay mode cannot give any information about tensor couplings. Going over to spinors and using the Dirac equation, we obtain the final form of the decay matrix element:

$$\langle l, \nu_l|H_W|K\rangle = (f_p - m_l f_A)\,[\bar{u}_\nu(p_\nu)(1-\gamma_5)\,u_l(p_l)]. \quad (1.3)$$

For decays that proceed through the modes $K \to e\nu_e$ and $K \to \mu\nu_\mu$ there must be an important difference due to the large mass difference $m_e - m_\mu$ and the neutrino helicity. This situation arises because the helicity is determined by the ratio of the particle velocity to the velocity of light (v/c) and is therefore always equal to unity for the neutrino. Conservation of the total angular momentum necessitates an induced helicity of the charged lepton and the phase volume of the interaction is therefore proportional to $1 - v/c$. Since the electron has a small mass, its velocity is nearly equal to c and the decay $K \to e\nu_e$ is suppressed compared with the transition $K \to \mu\nu_\mu$ since we obviously have $v_\mu/c \ll v_e/c$. After summation over the lepton polarizations, we immediately obtain the branching ratio of the electron and muon modes:

$$R_0 = \frac{\Gamma(K \to e\nu_e)}{\Gamma(K \to \mu\nu_\mu)} = \left[\frac{1-(m_e/m_K)^2}{1-(m_\mu/m_K)^2}\right]^2 \cdot \left[\frac{f_P - m_e f_A}{f_P - m_\mu f_A}\right]^2, \quad (1.4)$$

where the first factor is the ratio of the phase volumes.

Let us consider the limiting cases in which one of the two possible couplings predominates. If $f_A = 0$, we have pure S–P coupling and R_0 (the ratio of the phase volumes in this case) is ≈ 1.1. The other limiting case $f_P = 0$ leads to V–A coupling and in this case $R_0 \approx 1.1 \times m_e^2/m_\mu^2 \approx 2.6 \times 10^{-5}$. The values of R_0 for these

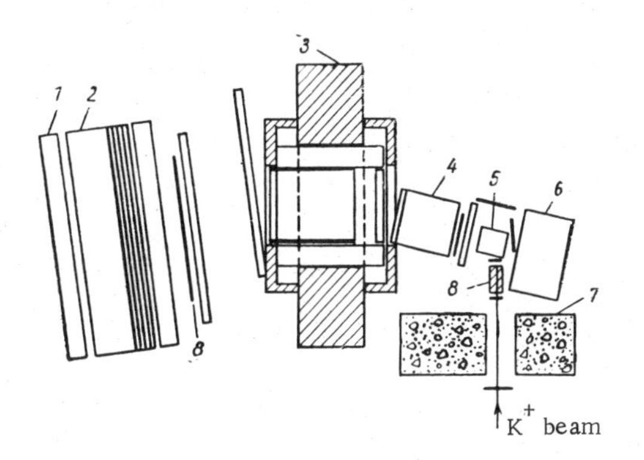

Fig. 1. Spectrometer for the investigation of the decay $K^+ \to e^+\nu e$: 1) range spark chambers; 2) aluminum plates; 3) magnet; 4) Cerenkov counter; 5) beryllium chamber; 6) lead glass; 7) concrete shield; 8) copper moderator.

two possibilities differ from each other by a factor of almost 10^5; it follows that the branching ratio of these decays is extremely sensitive to the nature of the pseudoscalar coupling.

In deriving the decay matrix element we ignored the graphs that describe the radiation corrections that arise because of the emission of a virtual photon and internal bremsstrahlung. When these corrections are taken into account, we obtain $R_1 = 0.815 \times R_0 \approx 2.1 \cdot 10^{-5}$ for the case of pure axial-vector coupling.

Experimental difficulties entailed in investigations of decays in which an electron participates have rendered it impossible until very recently to make any sort of accurate estimate of R_1. These difficulties are due to the low value of the relative rate (less than for the muon), the low accuracy in the determination of the energy of the emitting electron, and the appreciable background from the semileptonic decay $K \to \pi e \nu_e$. All these factors have meant that it has been virtually impossible to investigate the $K \to e\nu_e$ decay by means of emulsions and bubble chambers. Prior to 1967 all one knew was an upper limit, $R_1 \leq 2.6 \cdot 10^{-3}$, and it was only with the use of complicated spectrometers (with spark chambers, Cherenkov and scintillation counters) with a good momentum resolution that it proved possible [4] to measure R_1.

At the present time the mean experimental value of the $K^+ \to e^+\nu_e$ to $K^+ \to \mu^+ \nu_\mu$ branching ratio is $\langle R_1 \rangle = (2.15 \pm 0.35) \cdot 10^{-5}$, in fairly good agreement with the estimate for pure axial-vector coupling [for the processes $\pi \to e\nu_e$ and $\pi \to \mu\nu_\mu$ we have $R_1^\pi = (1.24 \pm 0.03) \cdot 10^{-4}$]. The upper limit for the pseudoscalar form factor f_p obtained from (1.4) is $|f_p| \lessgtr 1.5 \cdot 10^{-3}$. One may therefore conclude that axial-vector coupling predominates in the leptonic decays of K mesons, there being no appreciable admixture of pseudoscalar coupling; however, the experimental accuracy is not yet sufficient to make a rigorous quantitative estimate of $|f_p|$. (The arrangement of the Oxford group's spectrometer and the position momentum spectrum obtained in an investigation of the K_{e2} decay are shown in Figs. 1 and 2.) Considering the parameter R_1 we have assumed that $\mu - e$ universality holds, i.e., that $f_A^e = f_A^\mu$ and $f_p^e = f_p^\mu$. A few words on $\mu - e$ universality are appropriate. In the form adopted the weak-interaction Lagrangian $\mu - e$ universality is assumed, i.e., it is assumed that the interactions are invariant under a simultaneous substitution of the form

$$\left.\begin{array}{l} \mu \rightleftarrows e; \\ \nu_\mu \rightleftarrows \nu_e. \end{array}\right\} \tag{1.5}$$

The differences in the masses m and m_e despite the otherwise complete identity of the properties of these leptons and their quantum numbers is one of the currently most important problems of high-energy physics [5]. The point is that it is quite natural to explain the differences in the masses of other particles, for example, the electron and the pion, by the fact that these particles participate in different interactions. However, this explanation breaks down for the electron and the muon, since both leptons take part in the same hitherto known interactions, i.e., the weak and the electromagnetic interactions, and, what is more, participate universally in the same manner. The experimental data on the different processes in which the electron and muon participate have hitherto confirmed this assumption.

Anticipating our later discussion, let us compare the probability of processes in which the substitution (1.5) holds. Such decays have the same absolute matrix elements and the difference in the probabilities is due solely to the difference in the phase factors, which can usually be calculated easily. For example,

$\Gamma(\pi^+ \to e^+\nu_e)/\Gamma(\pi^+ \to \mu^+\nu_\mu) = \left(\frac{m_e}{m_\mu}\right)^2 \left[\frac{m_\pi^2 - m_e^2}{m_\pi^2 - m_\mu^2}\right] 0.965 \approx 1.23 \cdot 10^{-4}$. In Table 2 we give the data for various decays that are symmetric under (1.5).

Thus, to within the experimental errors, the measured values agree with those predicted by $\mu - e$ universality. Unless specially stipulated, we shall henceforth assume that $\mu - e$ universality is valid. Let us

TABLE 2. Comparison of Decays with Electrons and Muons

Decays	Theory	Experiment
$\Gamma(\pi^+\to e^+\nu_e)/\Gamma(\pi^+\to\mu^+\nu_\mu)$	$1{,}23\cdot10^{-4}$	$(1{,}24\pm0{,}03)\cdot10^{-4}$
$\Gamma(K^+\to e^+\nu_e)/\Gamma(K^+\to\mu^+\nu_\mu)$	$2{,}1\cdot10^{-5}$	$(2{,}15\pm0{,}35)\cdot10^{-5}$
$\Gamma(K^+\to\pi^0 e^+\nu_e)/\Gamma(K^+\to\pi^0\mu^+\nu_\mu)$	0,69	$0{,}703\pm0{,}056$
$\Gamma(\Lambda\to pe\tilde{\nu}_e)/\Gamma(\Lambda\to p\mu\tilde{\nu}_\mu)$	5,88	$5{,}87\pm0{,}75$
$\Gamma(\Sigma\to n\mu\tilde{\nu}_\mu)/\Gamma(\Sigma\to ne\tilde{\nu}_e)$	0,45	$0{,}40\pm0{,}06$

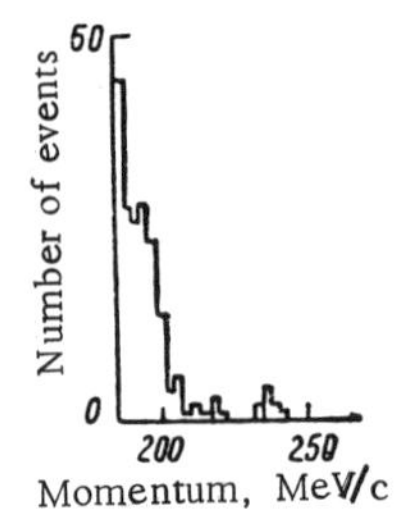

Fig. 2. Momentum spectrum of positrons from the K_{e2} decay.

now turn to the decay $K^+\to e^+\nu_e$. It is interesting to compare the relative strength of the interactions $K\to l\nu_l$ and $\pi\to l\nu_l$. If, as now seems reasonable, $f_P\approx 0$, if follows from the experiments that $f_A(K\to\mu\nu_\mu)/f_A(\pi\to\mu\nu_\mu)\approx 0.52$. There are numerous indications [6] that leptonic decays in which the strangeness changes must be at least three times less in amplitude than decays of nonstrange particles. With allowance for the mass factors in the dimensionless amplitudes, we have $f_A^{K,\pi}=Gm^{K,\pi}\cdot f_A'^{K,\pi}$ and $|f_A'^K|/|f_A'^\pi|\approx 0.14$, which indeed yields the desired suppression.

We now ask what information these decays yield about the value of the Cabibbo angle θ. If (21) is true, the ratio of the amplitudes of the decays $K\to l_\nu l$ and $\pi\to l_\nu l$, which is proportional to the ratio of the matrix elements of the currents $J_\alpha^{(1)}$ and $J_\alpha^{(0)}$, is

$$\frac{A(K\to l\nu_l)}{A(\pi\to l\nu_l)}\sim\frac{\langle 0|J_\alpha^{(1)}|K\rangle}{\langle 0|J_\alpha^{(0)}|\pi\rangle}=\mathrm{tg}\,\theta\,\frac{\langle 0|j_\alpha^{(1)}|K\rangle}{\langle 0|j_\alpha^{(0)}|\pi\rangle}. \tag{1.6}$$

For the decays $K^+\to\mu^+\nu_\mu$ and $\pi^+\to\mu^+\nu_\mu$ this means that $\theta_A=0.257$ rad. From a comparison of the decays $K(\pi)\to\pi l\nu_l$ one obtains $\theta_V\approx 0.26$ rad. We see that there is very good agreement between the two values of θ for the decays $K(\pi)\to l\nu_l$ due to the axial-vector part of the interaction and the decay $K\to\pi l\nu_l$ and $\pi^0 l\nu_l$, in which, as we shall see below, the vector part predominates. Of course, universality of the weak interactions is not restricted to universality of the pion and kaon decays; one must also make a full analysis of all the transitions that proceed through weak interactions. We shall do this later. Here, we shall only point out that the values obtained for the angle θ from the other decays do not contradict the quantities given in the present paper. The experiments yield the following values:

$\sin\theta_A=0.2655\pm0.0005$ from a comparison of $K^+_{\mu2}$ and $\pi^+_{\mu2}$;

$\sin\theta_V=0.220\pm0.003$ " " " K^+_{e3} and π^+_{e3};

$\sin\theta_V=0.207\pm0.004$ " " " K^0_{e3} and $\pi^+_{\mu3}$.

1.2. Selection Rules in Pion and Kaon Decays

Before we consider the other decay modes of pseudoscalar mesons in which leptons participate, we must obtain some general rules for all transitions, i.e., selection rules for a number of quantum numbers. It has been established that V−A coupling is present in leptonic decays in which such a meson is annihilated. However, besides a matrix element of, for example, the form $\langle 0|J_\alpha^{(1)}|K\rangle$, there exist many other matrix elements with the current $J_\alpha^{(1)}$ for which one must establish the type of the coupling and the other characteristics.

As we have seen in the previous section, the leptonic current operator $J_\alpha^{(l)}$ can lead, for example, to the creation of a neutrino and a positive lepton from the vacuum or the disappearance of a negative lepton and the appearance of a neutrino. All such transitions under the influence of the operator $J_\alpha^{(l)}$ have a common feature: the electric charge of the final state is always greater by unity than the charge of the initial state, i.e., there is a change of the charge under the influence of the current $J_\alpha^{(l)}$, $\Delta Q=+1$. Since the charge is conserved in all known interactions, this must lead to the existence of only those matrix elements with the currents $J_\alpha^{(0)+}$ and $J_\alpha^{(1)+}$ which yield a change of the charge $\Delta Q=-1$. Thus, a necessary condition for the matrix element $\langle B|J_\alpha^{(i)}|A\rangle$ to be nonvanishing is $\Delta Q=Q_B-Q_A=-1$. However, since $\langle B|J_\alpha^{(i)+}|A\rangle=\langle A|J_\alpha^{(i)}|B\rangle^*$, this rule means that transitions under the action of $J_\alpha^{(i)}$ lead to $\Delta Q=+1$. It now remains to assume that this rule is valid for all transitions under the influence of $J_\alpha^{(0)}$ and $J_\alpha^{(1)}$. Let us consider how the other hadron quantum numbers behave in different processes. In all interactions known at present the baryon number is strictly conserved. It follows that there is a further selection rule that we can introduce for the current $J_\alpha^{(i)}$, namely $\Delta N = 0$.

Let us consider a specific matrix element with $\Delta Q = 1$ which is certainly nonvanishing, $\langle 0|J_\alpha^{(1)}|K^-\rangle$. Since the strangeness of the K^- meson is $S = -1$ and $S = 0$ for the vacuum, $\Delta S = 1$ for this transition. Let us therefore assume that $\Delta S = 1$ for all transitions under the influence of $J_\alpha^{(1)}$; we immediately obtain the rule $\Delta Q = \Delta S$ for all such processes. In transitions with $\Delta S = 0$, we have the obvious consequence that $\Delta Q = 1$.

Let us now turn to the isotopic structure of our matrix elements. The K meson has isospin 1/2; the final state for a transition to the vacuum does not contain hadrons and the isospin of the vacuum is, of course, zero. If we now assume that $J_\alpha^{(1)}$ is a ΔI vector in the isotopic space, then, from the known nonvanishing matrix elements $\langle 0|J_\alpha^{(1)}|K\rangle$ one can assert that at least some of the matrix elements are such that they satisfy $|\Delta I| = 1/2$. As in the case of the strangeness, we shall again make the simplest assumption that only transitions in which the isospin changes by 1/2 are realized under the influence of the current $J_\alpha^{(1)}$. A similar treatment for transitions with $\Delta S = 0$ leads to $|\Delta I| = 1$.

1.3. Semileptonic Decays

The leptonic decay experiments discussed above made it possible to obtain information about the axial-vector and pseudoscalar coupling and the value of the Cabibbo angle θ_A. However, they do not yield any information about the scalar, vector, and tensor couplings and the corresponding angle θ_V. Such couplings can be studied by investigating the decays

$$\left.\begin{array}{l} K \to \pi + e + \nu_e \quad (K_{e3}); \\ K \to \pi + \mu + \nu_\mu \quad (K_{\mu 3}). \end{array}\right\} \tag{1.7}$$

Apart from the type of the interaction, semileptonic transitions make it possible to verify effects associated with the violation of T and CP invariance, to investigate the structures of the form factors that describe the contribution from the strong interactions, and also such consequences of the theory as, for example, the assumption of the local creation of a lepton pair.

Comparison of the form factors and the decay rates of neutral and charged kaons makes it possible to verify the selection rule $|\Delta I| = 1/2$, whereas equality of the corresponding form factors in the K_{e3} and $K_{\mu 3}$ decays would be confirmation of $\mu - e$ universality. The nature of the time dependence of the decays of neutral kaons depends on the extent to which the rule $\Delta Q = \Delta S$ is satisifed; in an investigation of this dependence one can estimate the value of x, the ratio of the amplitudes of decays with $\Delta Q = -\Delta S$ to the decay amplitudes in which this rule is not violated. The Hamiltonian for the semileptonic decays has the same general form as for the transitions $K(\pi) \to l\nu_l$ but, in contrast to the purely leptonic decays, the current $J_\alpha^{(i)}$ does not annihilate a pion or kaon but transforms it into a pion: $M^{(i)} = \langle \pi|J_\alpha^{(i)}|K(\pi)\rangle$. Arguing as previously and noting that the p parity of the initial and final hadrons is the same in this case, we find that three possibilities can be realized for the decays $K \to \pi l \nu_l$:

scalar interaction $M \sim f_S$;

vector interaction $M \sim 1/2 f_+ (p_K + p_\pi)_\alpha + 1/2 f_- (p_K - p_\pi)_\alpha$;

tensor interaction $M \sim f_T p_K^\alpha p_\pi^\beta$.

(Here, we have at our disposal two kinematic variables, namely, the four-momenta of the kaon p_K and the pion p_π.)

The functions f_i are dimensionless form factors that depend only on the square of the four-momentum transferred to the lepton pair; this follows from the assumption of a local creation of leptons. The form factors are relatively real if the interaction is invariant under time reversal. For the pure vector variant of the coupling, the majority of theoretical estimates indicate that $f_\pm$ must be slowly varying functions of $q^2 = (p_K - p_\pi)^2$, i.e., one is naturally led to expand them in powers of q^2:

$$f_\pm(q^2) = f_\pm(0)\left[1 + \lambda_\pm q^2/m_\pi^2\right],$$

On the other hand, the form factors $f_\pm$ can be represented in the form

$$f_\pm(q^2)=f_\pm(0)\left[X/(X-q^2/m_\pi^2)\right],$$

where $m \equiv m_\pi X^{1/2}$ is the mass of the intermediate K* state ($J^P = 1^-$, $I = 1/2$). For $|\lambda_\pm| \gtrsim 0.1$ both representations of $f_\pm$ are equivalent for $\lambda = 1/X$.

Nature of the Coupling Responsible for the Decays $K \to \pi + l + \nu_l$. Correct interpretation of the experimental data on the $K \to \pi l \nu_l$ processes due to some particular form of interaction are possible only if one allows correctly for the distortions due to the strong interactions and if the assumption of local creation of leptons is correct. Of course, the form of any energy or angular distribution of secondary particles depends on the nature of the coupling responsible for the decay but the different spectra have different sensitivities to the assumptions of the theory and the effect of the strong interactions. As we already know, all the effects from the strong interactions are included in the form factors f_i. In the simplest case of constant form factors, all the distributions are equally suited to an analysis of the nature of the interaction. However, in the general case, when the form factors depend on the energy of the strongly interacting particles, one must always investigate the parameters that are least sensitive in the given case to a change in the pion energy (or rather, to $g_2 = m_K^2 + m_\pi^2 - 2m_K E_\pi$).

Which distributions are most suited to an investigation of the nature of the interaction? The electron momentum spectrum is relatively insensitive to a change in the pion energy, being obtained by integration over the energy E_π, and it follows that the information about the nature of the interaction is not too strongly affected by an inaccuracy in the knowledge of the value and structure of the form factors. The distribution over the angle α between the momenta of the neutrino and the pion in the rest system of the dilepton (lepton plus neutrino) is independent of the change of the form factors; this is also true, of course, for all other spectra obtained for a fixed energy of the pion. (All these arguments hold if there exists pure coupling of a particular kind.) As regards the assumption of local creation of the lepton pair, this also can be most conveniently verified by analyzing the angular correlation in α. A difference in the behavior of the electron (for $m_e \approx 0$) and the neutrino would indicate either violation of the locality principle or the presence of a large contribution from the scalar-tensor term (or both). It should be noted that the asymmetry which arises if there is nonlocal creation of the leptons is manifested only in the structure of the form factor f_+ in the decay $K \to \pi e \nu_e$, whose expansion in powers of q^2 contains a term that depends on the momentum transferred to the lepton:

$$f_+(q^2)=f_+(0)\left[1+\lambda_+ q^2/m_\pi^2+\lambda_e q_e^2/m_\pi^2\right],$$

where $q_e^2 = m_K^2 + m_e^2 - 2m_K E_e$.

Attempts to distinguish a term proportional to λ_e in the structure of $f_+(q^2)$ clearly smack of sophistry but, approaching the Milan group's result [7] purely formally, we may remark that it does not contradict the principle of local ($\lambda_e = 0$) lepton creation $\lambda_e = 0.11^{+0.013}_{-0.011}$. For more than a decade unceasing investigations of semileptonic decays have been made. The accuracy of the measurements and the estimate of the background have been improved and an ever greater body of statistical material has been accumulated. The recent experiments [8, 9] of the groups at Saclay (K_{e3}^0) and Princeton (K_{e3}^+), which were carried out by means of spark chambers and counters (Fig. 3), were based on more events than all the foregoing experiments together. As in the earlier investigations, it was again found that the vector coupling is predominant and that the contribution to the amplitude from the scalar and tensor interactions satisfies the following inequalities: for K_{e3}^0: $A_S/A_V^0 \le 0.11$, $A_T/A_V \le 0.06$ with a 68% confidence limit and for K_{e3}^+: $A_S/A_V \le 0.15$, $A_T/A_V \le 0.04$ with a 90% confidence limit.

The Form Factors $f_\pm$ and ξ in Kaon Decays. If vector coupling is predominant in semileptonic decays, the total matrix element must have the form

$$M=\frac{G}{\sqrt{2}}\sum_\alpha\left[f_+(q^2)(p_K+p_\pi)_\alpha+f_-(q^2)(p_K-p_\pi)_\alpha\right]\bar{u}_e\gamma_\alpha\times(1+\gamma_5)u_\nu, \tag{1.8}$$

where, as we have indicated above, the form factors $f_\pm(q^2)$ can be represented by expansions in q^2. These form factors are directly related to virtual strong interaction in which kaons and pions participate. To

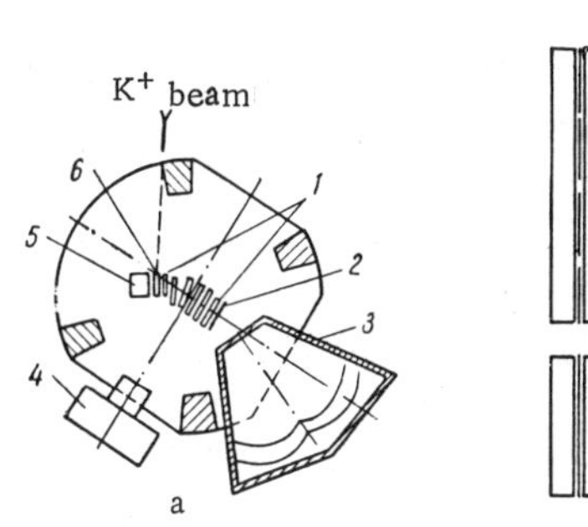

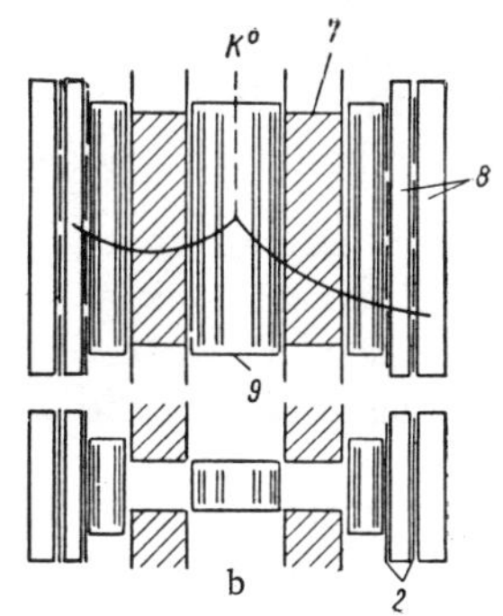

Fig. 3. Apparatus of the groups at Princeton (a) and Saclay (b) for the investigation of the K^+_{e3} and K^0_{e3} decays: 1) spark chambers; 2) scintillation counters; 3) Cerenkov counters; 4) photographic system; 5) detector of π^0 mesons; 6) kaon detector; 7) magnet; 8) range spark chambers; 9) momentum spark chambers.

analyze their energy structure one can apply dispersion relations under the most varied hypotheses, for example, under the assumption [10] that the virtual strong interactions dominate with an intermediate K* state present. In addition, numerous predictions based on current algebra [11] have recently been made concerning the structure of the form factors and their mutual relationships for different decay modes. Of particular interest from the point of view of comparison with the predictions is the form factor $\xi(q^2)$, the ratio of the form factors f_- and f_+:

$$\xi(q^2)=f_-(q^2)/f_+(q^2).$$

The solution of several problems depends on the investigation of ξ. The violation of T invariance in semileptonic decays leads to the appearance of a phase of ξ that is not equal to 0 or 180°. The selection rule $\Delta I=1/2$ gives a quantitative relationship between the form factors ξ in decays of neutral and charged kaons. If only the amplitude with $\Delta I = 1/2$ makes a contribution to the K_{l3} decays, we must have $\xi_0 = \xi_\pm$. On the other hand, if ξ is known, a comparison of the form factors f_+^e and f_+^μ from the K_{e3} and $K_{\mu 3}$ decays (by analogy with the leptonic K_{l2} decays) can be used to verify μ–e universality, which predicts $f_+^e = f_+^\mu$.

Some methods of determining ξ are based on measurements that can only be interpreted correctly when the answers to the questions listed above are known. For example, in many experiments ξ has been determined by measuring the ratio of the partial rates of K-meson decay through the $K_{\mu 3}$ and K_{e3} modes:

$$\eta=\frac{\Gamma(K\to\pi\mu\nu_\mu)}{\Gamma(K\to\pi e\,\nu_e)}\approx 0.65+0.13\,\xi+0.02\xi^2.$$

The branching ratio η has this form because the terms of the matrix element with ξ are proportional to the mass of the leptons and the denominator of the fraction does not contain ξ for $m_e = 0$. In this method of estimating ξ one must assume, first, that f_+ and f_- are constants and secondly, that μ–e universality holds. For example, if the first assumption is false, η takes the form

$$\begin{aligned}\eta &\approx 0.649+0.127\mathrm{Re}\,\xi+0.019\,|\xi|^2+1.34\lambda_+ +0.008\lambda_+\mathrm{Re}\,\xi\\ &+0.459\lambda_-\mathrm{Re}\,\xi+0.163\lambda_-|\xi|^2-0.068\lambda_+|\xi|^2,\end{aligned} \tag{1.9}$$

and to estimate ξ correctly one must know the energy structure of the form factors $f_\pm(q^2)$, i.e., the parameters λ_+ and λ_-. A further possibility of estimating ξ arises in an investigation of the energy spectra of pions and muons and also from a measurement of the muon polarization in the decay $K\to\pi\mu\nu_\mu$. The interpretation of such experiments does not depend on the extent to which μ–e universality holds but additional assumptions must be used. The only method that does not depend on any of these assumptions, i.e., μ–e universality, invariance under time reversal, and the nature of the energy structure of $f_\pm$, is to make measurements for fixed energy of the muons and pions, for then necessarily ξ = const. This can be done by measuring the polarization direction of the muon at each point of the Dalitz plot (E_π, E_μ). The polarization direction of the muon at a given point (100% polarization) depends in this case only on the corresponding value of ξ: $\hat{p} = f(\xi)$. The angular distribution over the directions of the momenta of the electrons $\hat{e}$ from the muon decay relative to the direction of the magnetic field B employed is a simple function of ξ:

$$dN/d(\hat{l}\,\hat{B})\sim\left[1+\alpha\hat{p}_\mu(\xi)\cdot\hat{B}\right](\hat{l}\,\hat{B}),$$

where α is the asymmetry parameter of the muon decay.

Many experiments have now been performed in which ξ has been determined by different methods [12]. The weighted mean of the branching ratio for the $K_{\mu 3}$ and K_{e3} modes are $\eta^+ = 0.73 \pm 0.03$ for K^+ mesons and

$\eta_0 = 0.78 \pm 0.05$ for K^0 mesons. If we assume that $f_\pm$ are constant, i.e., $\lambda_\pm = 0$, and that there is invariance under time reversal, Im $\xi = 0$, then for $\langle\eta\rangle = 0.75$ we obtain $\xi = 0.6 \pm 0.2$. On the other hand, measurements of the muon polarization in decays of charged and neutral kaons yield Re $\xi = -1.0 \pm 0.2$. Thus, $\xi_{br \cdot rat} - \xi_{polar} \approx 1.6 \pm 0.3$, which differs from zero by more than five standard deviations.

It is true that results have recently been obtained which indicate that $\eta_+ = 0.596 \pm 0.025$ and $\xi = -0.72 \pm 0.21$. The last result agrees with the data found from the polarization experiments. However, if such a difference is present, how could one to attempt to explain it? First of all, it would be natural to try an energy dependence of the form factors, i.e., to introduce nonvanishing values of the parameters λ_+ into the expresion for the branching ratio. We must now explain why we take λ_+ and not both λ_+ and λ_-, which would enable one to allow for an energy structure of both f_+ and f_-. The point is that many theoretical estimates show that $|f_-| < |f_+|$ [13] or $\xi \approx 0$, and in the limit of strict $SU_{(3)}$ symmetry we would exact vanishing of ξ. One could introduce a nonvanishing λ_+ into (1.9) without regard to the experimental polarization data since the latter are much less sensitive to λ_+ than the $K_{\mu 3}$ and K_{e3} branching ratio. However, to change the value of ξ from the experimental value $\xi = +0.6$ to $\xi = -1.0$ without changing the branching ratio $\eta = 0.75$ one would have to take $\lambda_+ \approx 0.15$. Such a value of λ_+ contradicts not only many theoretical predictions ($|\lambda_+| \leq 0.02$) but also the majority of experimental data obtained from the study of K_{e3} decays. At the present time nothing is known definitely about the values of λ_+ in the K_{e3}^+ and K_{e3}^0 decays. It follow from the previous experimental data that $\langle\lambda_+\rangle_{K^+} = 0.024 \pm 0.009$ and $\langle\lambda_+\rangle_{K^0} = 0.018 \pm 0.009$. However, there have been recent indications that the values of λ_+ may lie in the interval 0.06-0.09. If $\xi = -1$ and we assume $\lambda_- = 0$, then for $\lambda_+ = 0.02$ the branching ratio η must be equal to 0.6, which differs from the previously found mean value (0.75 ± 0.03) by more than five standard deviations but corresponds to the value of η_+ obtained in recent experiments with K^+ mesons. On the other hand, if $\lambda_+ = 0.09$, then $\eta \approx 0.75$; this agrees with the conclusions drawn from an analysis of the results of the earlier experiments but contradicts the more recent data.

There remains the possibility of reconciling the data by taking $\lambda_- \neq 0$. Although theoretical estimates indicate $|f_-| < |f_+|$, the majority of them do, on the other hand, predict [14] a stronger energy dependence of f_- than that of f_+, i.e., $|\lambda_-| > |\lambda_+|$. Let us consider what values of λ_- could reconcile the experimental data. For $\xi = -1.0$, $\lambda_+ = 0.02$, and $\eta = 0.75$ one must take $|\lambda_-| = 0.6$. Some theoretical models indicate $|\lambda_-| \geq 0.2$ if $|\lambda_+| \leq 0.02$, but if we have such a large value, $\lambda_- = -0.6$, the very expansion of f_- in powers of q^2 in the form $1 + \lambda_- \frac{q^2}{m_\pi^2}$ may be incorrect. We still have $\mu - e$ universality at our disposal; suppose we suddenly find $f_+^e \neq f_+^\mu$; then, for example, for $\lambda_\pm = 0$ and Im $\xi = 0$ we have $\eta = (f_+^\mu / f_+^e)^2 \times (0.065 + 0.13\xi + 0.02 \quad \xi^2)$ and for $|f_+^\mu| \; |f_+^e| > 1$ we then have $\xi_{br.rat} > \xi_{polar}$. To reconcile the existing experimental values of ξ_{polar} and $\xi_{br.rat}$ it is necessary to assume that $|f_+^\mu / f_+^e| \approx 1.16$ for $\xi = -1.0$ and $\eta = 0.73$. However, the violation of $\mu - e$ universality goes against the grain more than, for example, the introduction of a large λ to reconcile the results for ξ obtained by different methods.

The Selection Rules $|\Delta I| = 1/2$ and $\Delta Q = \Delta S$ for Semileptonic Kaon Decays. In Sec. 1.3 we have adduced simple arguments to derive the selection rules $|\Delta I| = 1/2$ and $\Delta Q = \Delta S$. These arguments were based on the assumption that the strangeness-changing part of the weak-interaction Lagrangian transforms as a isospinor.

A more fundamental basis for the introduction of these selection rules was proposed by Cabibbo and others; it reduces to the assumption that the strangeness-changing current $J_\alpha^{(1)}$ is composed of the components of an $SU_{(3)}$-symmetry octet. Within the framework of such an assumption the selection rules $|\Delta I| = 1/2$ and $\Delta Q = \Delta S$ are obtained for all processes with $\Delta S = \pm 1$. For semileptonic decays the rule $|\Delta I| = 1/2$ predicts the branching ratio:

$$R = \Gamma(K^0 \to \pi l \nu_l / \Gamma(K^+ \to \pi l \nu_l) = 2.024,$$

where the differences of the pair arises because of the mass difference of the particles. The experimental value is $R_{exp} = 1.9 \pm 0.08$ and therefore $R_{exp}/R_{theor} = 0.94 \pm 0.04$, which does not differ from unity by more than one standard deviation [15].

One can investigate a possible violation of $\Delta Q = \Delta S$ by considering the four decay amplitudes of the neutral kaons:

$$f = A(K^0 \to \pi^- + l^+ + \nu_l); \; q = A(K^0 \to \pi^+ + l^- + \tilde{\nu}_l); \; f^* = A(\bar{K}^0 \to \pi^+ + l^- + \tilde{\nu}_l); \; q^* = A(\bar{K}^0 \to \pi^- + l^+ + \nu_l),$$

where CPT invariance is assumed. Here, $f\,(f^*)$ is the amplitude of the process that satisfies $\Delta Q = \Delta S$ and g (g*) is the amplitude that violates this rule ($\Delta Q = -\Delta S$). The ratio of these two amplitudes, $x = g/f$, must vanish if the weak interaction satisfies $\Delta Q = \Delta S$ exactly. Violation of this rule must affect the nature of the time distribution of decays with positively and negatively charged leptons and also, if there is a CP-invariance violating amplitude in semileptonic decays, the charge asymmetry

$$\Delta = (\Gamma_+ - \Gamma_-)/(\Gamma_+ - \Gamma_-),$$

where $\Gamma_\pm = \Gamma(K^0 \to \pi l \pm \nu_l)$ are the partial rates of the processes.

Let us consider what possibilities there are for obtaining information about the value of x from an experimental study of the time dependence for the decays $K_L^0 \to \pi l \nu_l$. If at the initial instant there is, for example, a pure $\overline{K}^0$ state, then after a certain τ_S (τ_S is the lifetime of the short-lived K_S^0 state) there will be a mixture of the $\overline{K}^0$ and K^0 states. For small t, negatively charged leptons will arise because of the amplitude f^* that satisfies $\Delta Q = \Delta S$ and positively charged leptons will arise from the amplitude which satisfies $\Delta Q = -\Delta S$. If both f and g have approximately the same energy dependence in the kinematically allowed region, the time distribution of the electronic K_{e3} from the initial $\overline{K}^0$ state is

$$\widetilde{N}^{\pm}(t) \sim |1+x|^2 e^{-\lambda_S t} + |1-x|^2 e^{-\lambda_L t} - 2e^{\frac{1}{2}(\lambda_S+\lambda_L)t}\,[2\mathrm{Im}x \sin \delta t \pm (1-|x|^2)\cos \delta t], \tag{1.10}$$

where the sign $\pm$ refers to the electron charge; λ_S and λ_L are the total rates of the short-lived K_S^0 and the long-lived K_L^0 states and δ is the mass difference $m(K_S) - m(K_L)$. The necessary condition for CP invariance of the semileptonic decays is $\mathrm{Im}\,x = 0$, whereas the simultaneous fulfillment of the conditions $\mathrm{Im}\,x = \mathrm{Re}\,x = 0$ is equivalent to $\Delta Q = \Delta S$. Violation of CP invariance in this case necessarily entails nonfulfilment of $\Delta Q = \Delta S$ but the condition $\Delta Q = \Delta S$ and hence $\mathrm{Im}\,x = 0$ is not sufficient to deduce CP invariance since the latter may be violated in the allowed channel with $\Delta Q = \Delta S$. If CP invariance holds, the sum $N^+(t) + N^-(t)$ is independent of δ. However, this obvious advantage is in reality lost because of the indeterminacy of x due to the quadratic dependence and the large indeterminacy in x resulting from the loss of information about the sign of the lepton charge. If one assumes that the form factor f_- (g_-) does not dominate in the hadron currents, the time distribution of the $K_{\mu 3}$ decays can also be expressed by Eq. (1.10) with the same value of x. We recall that violation of $\Delta Q = \Delta S$ leads to a transition with $|\Delta I| = 3/2$ (the amplitudes g and g*), i.e., in this case the selection rule $|\Delta I| = 1/2$ must necessarily be violated for semileptonic decays.

The time dependence for the K_{e3}^0 decays has been investigated [16] for both the $K_{\mu 3}$ and K_{e3} modes and for the initial states K^0 and $\overline{K}^0$. The combined data from all the experiments yield

$$\mathrm{Re}\,x = 0.021 \pm 0.036;$$
$$\mathrm{Im}\,x = -0.10 \pm 0.005.$$

We see that whereas $\mathrm{Im}\,x$ vanishes to within the experimental errors, $\mathrm{Re}\,x > 0$ by more than four standard deviations. However, the value of the error $\Delta \mathrm{Re}\,x$ should not be taken too seriously. The combined data of the various experiments do not take into account the almost unavoidable systematic errors that could greatly change the result. In addition, the experiments discussed above do not yet indicate a violation of $|\Delta I| = 1/2$, which would be necessary if there is a contribution to the amplitude with $\Delta Q = -\Delta S$.

Summing up, our very tentative interpretation of the measurements of the time dependence of semileptonic decays is that the selection rule $\Delta Q = \Delta S$ may be violated with CP invariance nevertheless holding.

Let us consider the possible effects that could arise if the selection rule $\Delta Q = \Delta S$ is violated in the charge asymmetry of the decays $K_L^0 \to \pi l \pm \nu_l$. It is difficult to determine the contribution to the K_{l3} decays from the $\Delta Q = \Delta S$ violating amplitude by measuring the charge asymmetry $\Delta = (\Gamma_+ - \Gamma_-)/(\Gamma_+ + \Gamma_-)$ because x in this case is a function of many experimental variables:

$$\mathrm{Re}\,x = f(|\eta_{+-}|, |\eta_{00}|, \theta_{+-}, \theta_{00}, \Delta).$$

Here, η_{+-} and η_{00} are the ratios of the amplitudes of two-pion decays: $\eta_{+-} = \langle \pi^+\pi^- | J_\alpha^{(1)} | K_L^{(0)} \rangle | \langle \pi^+\pi^- | J_\alpha^{(1)} | K_S^0 \rangle$ and $\eta_{00} = \langle \pi^0\pi^0 | J_\alpha^{(1)} | K_L^0 \rangle / \langle \pi^0\pi^0 | J_\alpha^{(1)} | K_S^0 \rangle$ with corresponding phases θ_{+-} and θ_{00}. A phenomenological analysis of the majority of the experimental data leads to the value $\theta_{00} = (51 \pm 30)°$. In accordance with many experiments* we have $\theta_{+-} = (42.5 \pm 3)°$. For $|\eta_{+-}|$ and $|\eta_{00}|$ the weighted mean values are

$$\langle |\eta_{+-}| \rangle = (1.92 \pm 0.03)\, 10^{-3};$$

$$\langle |\eta_{00}| \rangle = (2.06 \pm 0.17) \cdot 10^{-3};$$

however, different values were found for $|\eta_{00}|$ in the different experiments. The recent measurements [17] of the charge asymmetry Δ in the K^0_{l3} decays gave

$$\Delta_e = (2.24 \pm 0.36) \cdot 10^{-3};$$

$$\Delta_\mu = (4.00 \pm 1.40) \cdot 10^{-3}.$$

Using all these values and the relationship $2\mathrm{Re}\,x = 3\Delta/(2|\eta_{+-}| \times \cos\theta_{+-} + |\eta_{00}| \cos\theta_{00})$, we obtain $\mathrm{Re}\,x = +0.11 \pm 0.09$ for K^0_{Le} and $\mathrm{Re}\,x = -0.20 \pm 0.26$ for $K^0_{L\mu}$. Within the limits of the errors these values agree and their weighted mean is

$$\mathrm{Re}\,x_\Delta = +0.080 + 0.100.$$

In addition, data on the K^0_{e3} decay yield

$$\mathrm{Im}\,x = -0.12 \pm 0.15;$$

$$\frac{1-|x|^2}{|1+x|^2} = 1.06 \pm 0.06.$$

For the relative probability of the transitions with $\Delta Q = -\Delta S$ we therefore obtain a value less than 2×10^{-2}.

The Problem of T Invariance Violation. The need to verify T invariance arose with the discovery of CP-noninvariant transitions in decays of K^0 mesons. If CP invariance is violated, it immediately follows from the fundamental CPT theorem that T invariance must necessarily be violated.

Experiments with weak decays enable one to verify directly the extent to which invariance under time reversal holds since they are sensitive to terms of the form $\sigma \cdot (p_1 \times p_2)$, where σ is the spin of one of the particles and p_1 and p_2 are the momenta of the two secondary particles that participate in the decay. Under time reversal the signs of both the momentum and the angular momentum are changed; it follows that invariance under time reversal requires the vanishing of $\sigma \cdot (p_1 \times p_2)$ if one can neglect the CP-violating interaction in the final state.

Such an effect can be sought in the decays of neutral and charged kaons through the $K_{\mu 3}$ mode. The parity violating decay of the muon depends on its polarization; with allowance for this fact one can measure the normal component of the polarization vector, which is proportional to $\sigma_\mu \cdot (\bar{p}_\pi \times \bar{p}_\mu)$.

At the present time the most complete data have been obtained in the investigation of neutral kaons [12]. The coefficient of the term $\sigma_\mu \cdot (\bar{p}_\mu \times \bar{p}_\pi)$ is proportional to $\mathrm{Im}\,\xi$, the value of the form-factor ratio taking into account the possible presence of the $\Delta Q = \Delta S$ violating amplitude g: $\xi = (f_- - g_-)/(f_+ - g_+)$. The experimental value is $\mathrm{Im}\,\xi = -0.014 \pm 0.066$. (Similar investigations with charged kaons lead to $\mathrm{Im}\,\xi \approx 0$.) Since both $\mathrm{Im}\,\xi$ and $\mathrm{Re}\,\xi$ have hitherto been determined with a very poor accuracy, the individual measurements are characterized by large fluctuations. Assuming the value $\mathrm{Re}\,\xi = -1.0 \pm 0.2$, which is the weighted mean of the polarization measurements, we obtain $\varphi = 0.8 \pm 3.0°$ for the phase φ ($\xi = |\xi| e^{i\varphi}$). The electromagnetic interaction in the final state may give rise to a phase $\varphi \approx 0.3°$; it follows that the result obtained for $\mathrm{Im}\,\xi$ is in complete agreement with the assumption of T invariance.

Verification of CP Invariance. There has recently been an experimental discovery of charge asymmetry in the semileptonic decays of neutral kaons [17]. The magnitude of this asymmetry is an important parameter in the phenomenological study of the violation of CP invariance. The Californian group, who studied the decay $K^0_L \to \pi\mu\nu$, obtained

*The individual measurements are characterized by a very large spread of the θ_{+-} values.

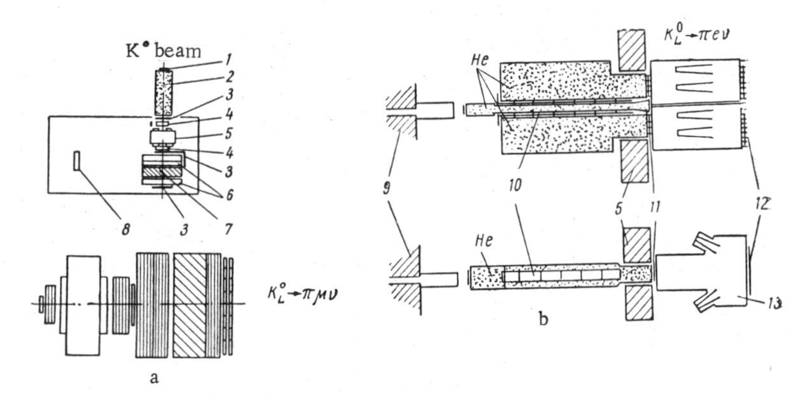

Fig. 4. Experimental arrangement for the investigation of the charge asymmetry in the decays $K_L^0 \to \pi\mu\nu$ (a) and $K_L^0 \to \pi e\nu$ (b): 1) anticoincidence counters; 2) helium bag; 3) scintillation counters; 4) thin-plate spark chambers; 5) magnet; 6) thick-plate spark chambers; 7) pion stopper; 8) photographic system; 9) collimator; 10) hodoscope A; 11) hodoscope B; 12) hodoscope C; 13) Cerenkov counter.

$$R \equiv R_{\mu^+}/R_{\mu^-} = 1.0081 \pm 0.0027, \tag{1.11}$$

and the group at Columbia University found

$$\Delta = (\Gamma_{e^+} - \Gamma_{e^-})/(\Gamma_{e^+} + \Gamma_{e^-}) = (2.24 \pm 0.36) \cdot 10^{-3} \tag{1.12}$$

for the $K_L^0 \to \pi e\nu$ decay.

Both these results gave the first indications that CP invariance may be violated in not only the $K^0 \to 2\pi$ transitions but also in other processes that take place because of the weak interaction; this is because one must have $R = 1$ and $\Delta = 0$ if CP invariance is satisfied (the arrangements of the experiments are shown in Fig. 4). The charge asymmetry data can also be attributed to other properties of the neutral kaon decays. If we denote by α the transition $\alpha = |\langle L|S\rangle|$, where L and S are the long- and short-lived states, then

$$\Delta = \frac{\Gamma_+ - \Gamma_-}{\Gamma_+ + \Gamma_-} = \alpha \frac{(1 - |x|^2)}{|1 + x|^2}.$$

Here, the parameter x is again the ratio of the amplitudes g/f; on the other hand, the parameter Δ can be expressed in terms of parameters known from phenomenological analysis of the decays $K_L^0 \to 2\pi^0$ and $K_L^0 \to \pi^+\pi^-$, that proceed with CP violation, namely, η_{+-}, η_{00}, ε, and ε', which are related to one another by the equations $\eta_{+-} = \varepsilon + \varepsilon'$ and $\eta_{00} = \varepsilon - 2\varepsilon'$. From the definition of ε and α (with allowance for the fact that they are small) we obtain $\mathrm{Re}\,\varepsilon \approx \alpha/2$. Hence, $x = 0$ implies $\Delta = 2\mathrm{Re}\varepsilon$ or $R \approx 1 + 4\mathrm{Re}\varepsilon$ and it follows from (1.11) and (1.12) that $\mathrm{Re}\varepsilon_\mu = 0.0020 \pm 0.0007$ and $\mathrm{Re}\varepsilon_e = 0.0011 \pm 0.0002$.

As can be seen, the results obtained from measurements of the charge asymmetry in the decays K_L^0 $\pi\mu\nu$ and $K_L^0 \to \pi e\nu$ agree well. We must emphasize once more that the observed charge asymmetry is associated with the CP-violating amplitudes of the 2π decays of the K^0 mesons and not with the CP-noninvariant amplitudes of the semileptonic decays, which we have neglected in the analysis. It should be noted that values found for the charge asymmetry in the semileptonic decays agree with the predictions [18] obtained for violation of CP invariance in the two-pion decays, $\Delta \leqslant 4 \cdot 10^{-3}$.

1.4. The Decays $K \to \pi + \pi + l + \nu_l$

Among all the kaon decay processes in which leptons are included among the secondary particles, the K_{l4} decays are distinguished from all decays for which one can expect the accumulation of extensive experimental data by the saturation of their kinematic structure. In these decays one can investigate almost all the consequences of the theory of weak interactions, for example, the validity of the selection rules $|\Delta I| = 1/2$ and $\Delta Q = \Delta S$, and T invariance; one can also verify the conclusions obtained from current algebra, etc. Let us consider the reactions:

$$K^+ \to \pi^+ \pi^- e^+ \nu_e; \tag{1.13a}$$

$$K^+ \to \pi^+ \pi^+ e^- \tilde{\nu}_e. \tag{1.13b}$$

Let us compare the possibility of investigating the degree of violation of the selection rule $\Delta Q = \Delta S$ from a study of the reactions (1.13) and in the K_{e3} decay. The latter is due to the vector current, whereas the reaction (1.13b), which violates $\Delta Q = \Delta S$, is evidently described by an axial-vector current. The decay (1.13a) can contain an admixture of axial-vector and vector currents although it is expected that the axial-vector current predominates. Thus, the decays K_{e4} and K_{e3}, which violate $\Delta Q = \Delta S$, if this violation exists at all, are due to different currents and a comparison can test the notion of the universality of the currents. However, the most important aspect of the $K \to \pi\pi l \nu_l$ transitions is the possibility of obtaining direct information about the $\pi\pi$ interaction in the final state. The fact that two strongly interacting particles appear together with two leptons that participate in weak interactions enables one to study the interaction in the final state of these two pions in a pure form without the influence of other strong interactions. The $\pi\pi$ interaction in the final state must affect the following characteristics of the K_{l4} decay: 1) the decay rate; 2) the form of the dipion mass distribution; 3) the decay symmetry with respect to π^+ and π^- in the dipion rest system, which, in turn, must lead to differences in the π^+ and π^- spectra in the laboratory system; 4) the angular correlation between the dipion and dilepton planes.

The following feature of the decays (1.13) should also be noted. These processes differ from one another not only with regard to the rule $\Delta Q = \Delta S$ but also in the isospin states of the two pions [19, 20]. Whereas a final state with $I = 2$ is realized in (1.13b), states with $I = 0$, 1, or 2 may be present in the reaction (1.13a) providing appropriate angular momenta are present. If both states with $I = 0$ and $I = 1$ are present, the s and p angular momentum states may interfere if they have amplitudes of comparable magnitude. This interference could lead to a forward-backward asymmetry in the emission of one of the pions relative to the dipion direction.

The intensity and form of the K_{l4} decay spectra are functions of five variables. If these variables are chosen appropriately and only the single assumption of an effective local coupling of the lepton pair to the hadron currents is made, the structure of the decay can be expressed in terms of two of the five variables and will be independent of the hadron interactions. All the dynamic effects will be contained in the form factors, which, in the general case, are functions of the three remaining variables. This 2 + 3 separation appreciably simplifies the kinematic situation and can be used to decompose the general structure of the decays into parts that can be internally even more interesting. One such set of variables is: the angles θ_π and θ_l, which describe the "decay" of the dipion and dilepton in their rest systems; the angle φ between the normals to the planes containing the dilepton and dipion; and the invariant masses $m_{2\pi}$ and m_{2l} of the dipion and dilepton. The angles θ_l and φ are "simple" variables and the dependence on them can be expressed exactly. In general, all the form factors depend only on the remaining variables θ_π, $m_{2\pi}$, and m_{2l}.

We shall consider the decay

$$K^+ \to \pi^+ + \pi^- + e^+ + \nu_e, \tag{1.14}$$

since it is only for only this process that more or less sufficient experimental material has so far been accumulated [21-23]. We already know how to construct the matrix elements, so we shall not go into this question in detail. For the decay (1.14)

$$M \sim \frac{G}{\sqrt{2}} \bar{u}_\nu (p_\nu) \gamma_\alpha (1 + \gamma_5) u_e (p_e) \langle \pi^+ \pi^- | J_\alpha^V + J_\alpha^A | K^+ \rangle, \tag{1.15}$$

where J_α^V and J_α^A are the weak vector and axial hadron currents, whose matrix elements can be written in the form [20]

$$\langle \pi^+ \pi^- | J_\alpha^V | K^+ \rangle = \frac{ih}{m_K^3} \varepsilon_{\alpha\mu\nu\gamma} p_K^\mu (p_+ + p_-)^\nu (p_+ - p_-)^\gamma; \tag{1.16}$$

$$\langle \pi^+\pi^- | J_\alpha^A | K^+ \rangle = \frac{f}{m_K}(p_+ + p_-)_\alpha + \frac{g}{m_K}(p_+ - p_-)_\alpha, \quad (1.17)$$

where we have omitted the term in (1.17) that makes a contribution to the matrix element proportional to the square of the lepton mass. The matrix element (1.15) includes the three form factors h, g, and f, which, in the general case, are functions of $(p_K + p_{\pi+})^2$, $(p_K + p_{\pi-})^2$ and $(p_{\pi+} + p_{\pi-})^2$. If we make the assumption that the dependence on $(p_K + p_{\pi+})^2$ and $(p_K + p_{\pi-})^2$ can be neglected, the calculation of the decay amplitudes can be greatly simplified. In the case when the form factors depend only on $(p_{\pi+} + p_{\pi-})$, one can readily see that the f term in Eq. (1.17) is symmetric under transposition of π^+ and π^-, whereas g in (1.17) and h in (1.16) are antisymmetric under this transposition. This fact, together with the selection rule $|\Delta I| = 1/2$, indicates that f corresponds to the amplitude for the emission of two pions in a state with I = 0 and that the g and h terms are the amplitudes for the emission of pions in a state with I = 1.

If invariance under time reversal is assumed, f, g, and h can be represented in the form

$$f = f_0 e^{i\delta_0}, \quad g = g_0 e^{i\delta_1}, \quad h = h_0 e^{i\delta_1},$$

where δ_0 is the s-wave phase shift of $\pi\pi$ scattering for I = 0 and δ_1 is the p-wave phase shift of $\pi\pi$ scattering for I = 1, and g_0, f_0, and h_0 are real quantities.

At the present time international statistics provides about 300 decays for which a sufficiently detailed analysis has been made and individual cases of decay through the mode $K^+ \to \pi^+\pi^-\mu^+\nu_\mu$. The data obtained from $K^+ \to \pi^+\pi^- e^+\nu_e$ have been considered under various assumptions about the behavior of the s-wave of $\pi\pi$ scattering [24]. In the analysis all the form factors were assumed constant with the exception of the possible presence of a factor leading to a monotonic increase of the form factors. The s-wave phase shift δ_0 can be expressed in terms of definite parameters by two methods:

1) By Chew-Mandelstam parametrization with variable scattering length a_0:

$$\operatorname{ctg} \delta_0 = \frac{1}{\beta a_0} + \frac{2}{\pi} \ln \left[\frac{m_K}{2m_\pi} x (1 + \beta) \right], \quad (1.18)$$

where $\beta = (1 - 4m_\pi^2/x^2 m_K^2)^{1/2}$ and $x^2 = (p_{\pi+} + p_{\pi-})^2/m_K^2$.

2) By Breit-Wigner resonance with variable width γ and energy E_R:

$$\operatorname{ctg} \delta_0 (x^2) = 2\,(x^2 m_K^2 - E_R^2)/(\gamma m_K x \beta). \quad (1.19)$$

In the analysis the p-wave scattering phase δ_1 was ignored, an assumption that is perfectly justified in the energy range of the dipion in the given decay. The analysis was made for three cases, which are as follows.

A. All form factors constant; this is the simplest model.

B. The s-wave form factor f depends on the energy through a possible strong $\pi\pi$ and s-wave interaction at low energies. This can be taken into account by introducing, for example, a relativistic increasing Watson factor $f_0 = f_0 \sin \delta_0(x^2)/a_0\beta$.

C. All the form factors depend on the energy. In this case one introduces a so-called increasing ρ factor for g_0 and h_0:

$$g_0 = g_0' (m_\rho^2 - 4m_\pi^2)/(m_\rho^2 - x^2 m_K^2);$$

$$h_0 = h_0' (m_\rho^2 - 4m_\pi^2)/(m_\rho^2 - x^2 m_K^2).$$

The best agreement with the experimental data is obtained in all three cases if the Chew-Mandelstam $\pi\pi$ scattering phase is used in the calculations; this is especially true for the first case. Parametrization in the form of a Breit-Wigner resonance does not greatly affect the values of the form factors. The values of the form factors and the scattering length a_0 for the above cases are

$$a_0 = 1.04 \pm 0.50; \; a_0 = 0.89 \pm 0.44; \; a_0 = 0.84 \pm 0.43;$$
$$f_0 = 1.19 \pm 0.13; \; f_0' = 1.45 \pm 0.16; \; f_0' = 1.42 \pm 0.15;$$
$$g_0 = 1.34 \pm 0.30; \; g_0 = 1.36 \pm 0.30; \; g_0' = 1.25 \pm 0.28;$$
$$h_0 = -4.84 \pm 1.77; \; h_0 = -4.89 \pm 1.73; \; h_0' = -4.57 \pm 1.63.$$

Note that the main contribution to the error of the form factors arises because of the large error in the determination of the decay rate, which was found for the minority of the cases [19]:

$$\Gamma(K^+ \to \pi^+\pi^- e^+ \nu_e) = (2.9 \pm 0.6) \cdot 10^3 \text{ sec}^{-1}.$$

The values found for f_0 and g_0 are in good agreement with the result $|f_0| = |g_0| = 0.97 \pm 0.03$ obtained with the help of current algebra under the assumption that the form factors in the K_{l3} and K_{l4} decays are constant. The value $a_0 = 0.2$, which is also found in [19], does not appear too seriously different from the experimentally determined value, firstly, because of the large error and, secondly, because of the possible effects of unitarity restrictions on the $\pi\pi$ scattering amplitude (which may increase the value of a_0 deduced from current algebra requirements).

The most interesting result of the investigation is the large value of the vector form factor h, which differs from zero by almost three standard deviations. Unfortunately, the statistical reliability is so low and the errors in the values of the form factors are so large that one cannot draw any definite conclusions on this subject. We shall only mention that the form factor h_0 obtained in the framework of $SU_{(3)}$ symmetry (with the assumption of vector dominance) by using the relationship between the K_{l4} decay and the transition $\eta \to \pi\pi\gamma$ is much less (by approximately a factor of four) than the value obtained experimentally.

The accumulation of data on the K_{l4} decay is very desirable. If the conclusions given here are confirmed, it will mean that the K_{l4} decay is even more interesting than is at present assumed.

The Decays $K \to \pi\pi l \nu_l$ and the Selection Rules $\Delta Q = \Delta S$ and $|\Delta I| = 1/2$. The selection rule $\Delta Q = \Delta S$ allows the decays $K^+ \to \pi^+\pi^- l^\pm \nu_l$ and forbids the transitions $K^\pm \to \pi^\pm \pi^\pm l^\mp \nu_l$ with similarly charged pions. Up to now approximately 300 allowed decays $K^\pm \to \pi^+\pi^- l^+ \nu_l$ and some cases of $K^+ \to \pi^+\pi^-\mu^+\nu_\mu$ have been found but not a single transition that is forbbiden by $\Delta Q = \Delta S$. Thus, the upper limit for K_{l4} decays with $\Delta Q = -\Delta S$ is at present $\Gamma(K^+ \to \pi^+\pi^+ l^- \nu)/(\Gamma(K^+ \to \text{all})) \leq 0.5 \cdot 10^{-6}$.

Of course, to be able to say anything about the ratio of the currents corresponding to $\Delta Q = \Delta S$ and $\Delta Q = -\Delta S$ one must take into account fully the interaction effects in final states possessing different isospins. If these effects are negligibly small, the existing experimental data indicate that

$$R = \frac{A(K_{l4}^+, \Delta Q = -\Delta S)}{A(K_{l4}^+, \Delta Q = \Delta S)} \lesssim 0.13.$$

As we have seen in the foregoing section, the experimental results on the verification of $\Delta Q = \Delta S$ for vector currents in neutral kaon decays are compatible with the above value of R. Thus, as far as this question is concerned there has not yet been observed any difference in the behavior of transitions under the influence of the vector and axial-vector currents. Unfortunately, there is still no experimental data on the decay $K^+ \to \pi^0\pi^0 l^+ \nu_l$ and on the K_{l4} decays of neutral kaons; this means that we cannot verify the consequence of the selection rule $|\Delta I| = 1/2$ which relates the relative decay rates of charged and neutral kaons.

1.5. Weakly Electromagnetic Leptonic Meson Decays

The description "weakly electromagnetic" has now been accepted for the meson decays through the modes

$$K \to \pi + \pi + \gamma;$$
$$K \to \pi + \pi + \pi + \gamma;$$
$$K \to l + \nu + \gamma;$$
$$\pi \to l + \nu + \gamma.$$

In this review we shall consider only the last two decays.

The emission of a photon in any hadron decay

$$A \rightarrow a + b + \ldots + \gamma$$

can occur in the general case in two ways: by means of internal bremsstrahlung which necessarily accompanies all nonradiative processes and the possible mechanism of direct emission of a photon on the transition from the initial A state to the final state (a +b +. . .).

The process of internal bremsstrahlung is well known; its amplitude is proportional to the product eG of the constants of the electromagnetic and weak interactions and the bremsstrahlung γ ray is always emitted in a C- and P-invariant manner. The so-called direct (or structural) emission of a photon has not yet been detected experimentally but it happens that a number of interesting problems are related to this mechanism γ-ray creation. Indeed, as follows from the very name of the mechanism, the simple confirmation of the existence of direct processes would enable one to draw certain conclusions about the structure of the decaying meson. For this purpose it is clearly most expedient to investigate the weakly electromagnetic leptonic decays of kaons, for which there is no interfering influence of strongly interacting particles in the final state and the possible intermediate states (see below) have a mass that is closer to the kaon mass than to m_π in the decay $\pi \rightarrow l\nu_l\gamma$. If the direct emission of a photon exists and its amplitude is at least comparable in magnitude with the amplitude of the internal bremsstrahlung, interest in the weakly electromagnetic decay of kaons would greatly increase. In contrast to the CP-invariant amplitude of the internal bremsstrahlung process (for which the parent decay without γ ray, $A \rightarrow a + b + \ldots$, is CP invariant) the amplitude of the direct processes can, in general, be complex, i.e., noninvariant under time reversal. In such a case, violation of T orCP invariance could lead to experimentally observable effects.

The possibility of detecting T-noninvariant effects in weakly electromagnetic kaon decays appears in an investigation of the leptonic reaction

$$K^{\pm} \rightarrow \mu^{\pm} + \nu_\mu (\tilde{\nu}_\mu) + \gamma,$$

in which, in contrast to transitions of the type $K \rightarrow \pi\pi\gamma$, there is a new variable, the nonvanishing spin of the muon. This enables one to investigate a correlation of the form $\sigma_\mu \cdot (\overline{p}_K \times \overline{p}_\mu)$, which changes its sign under time reversal. In the absence of strong interactions in the final state the appearance of a nonvanishing normal component of the muon polarization $\hat{p}_\perp \sim \sigma_\mu \cdot (\overline{p}_K \times \overline{p}_\pi)$ would be a direct indication of noninvariance under time reversal in the given process. As we have seen in Sec. 1.3, a similar effect must obtain in the decay $K^{\pm} \rightarrow \mu^{\pm} \pi^0 \nu_\mu$, which is due to the weak interaction. Careful searches for a normal component of the muon polarization have yielded a negative result, whereas this component of the polarization must attain values of ~20% if CP invariance were violated in the weak interactions.

The decay $K \rightarrow \mu\nu\gamma$ differs quantitatively from the $K_{\mu 3}$ process in that it takes place because of both the weak and electromagnetic interactions. If CP invariance is violated in the electromagnetic interactions, the expected effect must, of course, be much larger in the $K_{\mu\nu\nu}$ decay than in the $K_{\mu\pi\nu}$ decay, in which the electromagnetic interaction appears as a correction to the weak interaction. As in the case of the decay $K \rightarrow \pi\pi\gamma$, such violation can occur only for the amplitude for the direct emission of a γ ray.

If one assumes maximal violation, i.e., $\mathrm{Im} f_i = \mathrm{Re} f_i$ for the form factors corresponding to the direct transitions, the maximal value of the normal component of the muon polarization may attain values of ~ 50% of the total polarization for certain points of the Dalitz plot [25, 26]. However, the integral effect leads to an experimentally observable asymmetry in the muon decay electron distribution of only 1-2% so the search for T-invariance violation for this decay is neither a simple nor a reliable undertaking.

However, as we have already mentioned, these effects can be appreciable only if the direct emission amplitude is at least comparable in magnitude with the internal bremsstrahlung amplitude. Hitherto the weakly electromagnetic $K_{e\nu\gamma}$ leptonic decays have not been investigated experimentally and the search for the direct emission of a photon in the nonleptonic processes $K \rightarrow \pi\pi\gamma$ has shown that the relative rate of transitions with direct emission, if the latter occur, does not exceed $\sim 10^{-4}$.

In this connection it would seem worthwhile to investigate in detail the decay $K \rightarrow l\nu_l\gamma$ since, as we have already mentioned in Sec. 1.1, the leptonic transition $K \rightarrow e\nu_e$ is suppressed because of the neutrino and the small mass of the charged lepton. Thus, in the decay $K \rightarrow e\nu_e\gamma$ the mode that competes with internal bremsstrahlung must be suppressed. For the putative direct-emission decay mode the exclusion rule is lifted by the presence of a third particle and we can reasonably expect that the rate of the process $K \to e\nu_e\gamma$ may be even greater than that of the decay without γ ray provided, of course, there is no special suppression of direct transitions generally.

We may also mention another characteristic feature of the decay $K \to e\nu_e\gamma$ that distinguishes it from the other weakly electromagnetic processes: strongly interacting particles are not present in the final state and the energy and angular correlations between the decay products depend on the properties of a single hadron – the kaon; in principle, this enables one to obtain information about the latter in the purest form. In the decay $K \to \mu\nu_\mu\gamma$ there are also no strongly interacting particles in the final state but this is offset by the internal bremsstrahlung which cannot be suppressed in any manner and is an undesirable background that interferes with the search for direct processes.

Despite the undoubted advantages of the decay $K \to e\nu_e\gamma$ for an investigation of the direct emission of a photon, experiments of this kind are beyond our possibilities at the present time. The trouble is that one must simultaneously and with sufficient accuracy measure the electron and photon energies in order to be able to distinguish the background process $K \to e\nu_e\pi^0$ when one of the γ rays from the decay of the π^0 meson is not detected by the instrument. These factors and the low probability of the process $K \to e + \nu_e$ have meant that hitherto experiments with the decay $K \to \mu\nu\gamma$ have seemed more realistic. In this connection it is natural to ask whether one can separate direct processes if their amplitude is less than or only comparable in magnitude with the bremsstrahlung amplitude. Such a possibility was investigated in [27], which was devoted to an analysis of the angular and energy correlations between the decay products in this process $K \to \mu\nu\gamma$, and in [28], in which an investigation was made of the possibility of polarization measurements in the decay $K \to \mu\nu\gamma \to \mu\nu e^+e^-$. These investigations indicate that the direct-transition effects could (for a certain choice of the kinematic regions of variation of the variables) be appreciable even if the direct-emission amplitude is only ~ 0.5 of the bremsstrahlung amplitude.

Now a few words about the possible structure of the decay $K \to l\nu\gamma$, which proceeds through the mode with the direct emission of a photon. Since two leptons can only arise in the final state as a result of a strangeness-changing weak interaction, the intermediate state required for the emission of the photon must have the same strangeness as the kaon. In the general case the direct transitions may have vector and (or) axial-vector nature. The main contribution to the vector transitions, which are characterized by the form factor f_V, must arise from two-meson $K\pi$ states with threshold $m_K + m_\pi$ and the resonance K^*. For axial-vector direct transitions with the corresponding form factor f_A the intermediate states $K\pi\pi$, $K\rho$, $K\omega$, and $K^*\pi$ are allowed. They are all more distant in the mass than in the case of the vector transition. Therefore, in general, $|f_A| < |f_V|$. In addition, the possible intermediate states have a mass that is greater than the kaon mass in both cases; one must therefore expect that the dependence of the form factors on the momentum transferred to the leptons must be smooth in the physical region and that it can be completely ignored in a first approximation. The total matrix element of the decay $K = l\nu_l\gamma$ in this case has the form

$$\langle l\nu\gamma | J | K \rangle = ie\frac{G}{2}\cdot\frac{1}{(2\pi)^4}\cdot\frac{1}{2\sqrt{E_K E_\gamma}}\delta^4(p_K - p_l - p_\nu - p_\gamma)$$

$$\times u_l\left\{f_K m_l\left[\frac{p_K\varepsilon}{p_K p_\gamma} - \frac{p_l\varepsilon}{p_l p_\gamma} - \frac{\hat{\varepsilon}\hat{p}_\gamma}{2p_l p_\gamma}\right] + \frac{1}{m_K^2}\gamma^\alpha p_K^\beta \varepsilon^\rho\right.$$

$$\left.\times [f_A(\delta_{\alpha\rho}\delta_{\beta\delta} - \delta_{\delta\alpha}\delta_{\beta\rho}) + if_V\varepsilon^{\alpha\beta\delta\rho}]\right\}(1+\gamma_5)u_\nu.$$

This expression shows that the internal bremsstrahlung contribution (the term proportional to the lepton mass) depends on the kaon structure only to the extent that the two-particle decay $K \to l\nu_l$ depends on this structure; this is because the same factor f_K, which depends on the kaon properties, is present in both channels. Finally, we must emphasize once more that at the present time only this range of leptonic transitions remains uninvestigated among the ensemble of weakly electromagnetic decays $K \to a + b + \ldots + \gamma$. The nonleptonic decays have hitherto failed to reveal any appreciable direct transition effects [29] and although one can hardly hope for any unexpected properties of the $K \to l\nu_l\gamma$ processes, their analysis would nevertheless enable one to complete the initial stage of the search for direct transitions.

1.6. Rare Leptonic Kaon Decays

One of the groups of rare leptonic kaon decays consists of the processes which can exist if the so-called neutral currents are present. Before we discuss this problem we should like to make a small digression.

We should first like to emphasize how important it is to make systematic investigations in weak interactions of effects of higher orders in small quantities. Such investigations are very important since they indicate whether the existing weak-interaction Lagrangian plays a purely phenomological role, i.e., is an effective Lagrangian, or whether it can be interpreted as the primary Lagrangian needed to construct a field theory of weak interactions. All the hitherto observed weak processes for the range of momentum tranfers $\lesssim$2 GeV can be well described by an effective local Lagrangian L_{eff} which can be represented as a sum of three terms [30]:

$$L_{eff} = L_{ll} + L_{Ll} + L_{LL}, \tag{1.20}$$

where L_{ll} describes processes in which only leptons participate; L_{Ll}, the semileptonic decays; and L_{LL}, the processes in which leptons do not participate. The matrix elements of lowest order in the interaction described by this Lagrangian at once determine the amplitudes of the observable processes. In the general case the matrix elements of the higher orders are divergent and are rejected. In this sense the Lagrangian (1) is a purely effective Lagrangian.*

If the effects of higher orders are investigated comprehensively, one of the first problems that must be faced is undoubtedly the striking absence of neutral leptonic currents in the first-order of the weak interactions. Indeed, as we have already pointed out in the introduction, the interaction responsible for the majority of reactions can be expressed phenomenologically in the form of the coupling of two currents; now such a formalism of the theory of weak interactions allows the presence of both charged and neutral currents and the absence of the latter is rather mysterious.

In the current-current representation one can rewrite (1.20) in the form

$$L_{eff} = \frac{G}{\sqrt{2}} (l_\alpha l_\alpha^* + L_\alpha l_\alpha^* + L_\alpha L_\alpha^* + \ldots), \tag{1.21}$$

where l_α is the current that includes only leptons and L_α is the corresponding current containing hadrons. The interaction $l_\alpha l_\alpha^* \equiv L_{ll}$ describes, for example, a purely leptonic process, the decay of the muon. In the framework of the V-A interaction, we must choose the leptonic current for this decay in the form

$$l_\alpha = \bar{u}_l \gamma_\alpha (1 + \gamma_5) u_\nu, \tag{1.22}$$

which leads to the existence of at least two components of this current:

$$l_\alpha (e\nu_e) + l_\alpha (\bar{\mu}\nu_\mu).$$

We know only a single purely leptonic decay and we therefore know nothing about the other possible components of the current l_α or, conversely, about other leptonic processes which could be described by such a current. Our knowledge of the form of the hadron currents is even more meager. All the processes in which hadrons participate can be split into three groups: 1) semileptonic; 2) nonleptonic, containing only bosons in the initial and final states; 3) nonleptonic, with fermions in the initial and final states.

The existing experimental data indicate that the semileptonic processes are indeed described by an interaction of the current-current form. As regards the nonleptonic processes, the question remains

*It should be noted that at high energies the Lagrangian (1) ceases to be correct even as an effective Lagrangian. It must be modified at high momentum transfers, i.e., at short distances, by the introduction of a certain "effective" nonlocality to eliminate the divergences that appear, for example, in the cross sections for the scattering $\nu_\mu + e^- \rightarrow \nu_e + \mu^-$.

open since conclusions of this nature are prevented in these processes by the interfering influence of the strong interactions in the final state.

Thus, we have at our disposal at least one group of reactions for which the proposed form of the corresponding interaction is of the current-current nature; in principle, this enables one to use the reactions to establish the nature of the currents and, in particular, to look for the neutral currents allowed by such a structure.

Form of the Leptonic and Hadronic Neutral Currents. The most general form of the leptonic current that satisfies lepton conservation (see below) is

$$l_\alpha = l_\alpha(\bar{e}\nu_e) + l_\alpha(\bar{\mu}\nu_\mu) + l_\alpha(\bar{\mu}\mu) + l_\alpha(\bar{\nu}_e\nu_e) + l_\alpha(\bar{\nu}_\mu\nu_\mu) + l_\alpha(\bar{e}e) + \ldots \tag{1.23}$$

The general form is the hadronic currents that contains fermions or bosons is

$$L_\alpha^F = L_\alpha(\bar{p}n) + L_\alpha(\bar{p}\Lambda) + L_\alpha(\bar{p}p) + L_\alpha(\bar{n}n) + L_\alpha(\bar{n}\Lambda) + \ldots; \tag{1.24}$$

$$L_\alpha^B = L_\alpha(\bar{K}^+\pi^0) + L_\alpha(\bar{K}^+\pi^-) + L_\alpha(\bar{K}^0K^0) + L_\alpha(\bar{\pi}^+\pi^0) + L_\alpha(\bar{K}^+, \pi^+\pi^-) + \ldots \tag{1.25}$$

If the all the components of l_α and L_α^l existed, a very large number of decays and scattering processes would be allowed. In Table 3 we give all such possible kaon decay processes with the exception of those that could be due to currents with $\Delta Q = -\Delta S$. We mention here that in the framework of the V−A theory of weak interactions the components of the current (1.24) must have vector and axial-vector parts.

At present we have no experimental proof of the presence of neutral currents in the purely leptonic interactions. However, their absence cannot be regarded as established either. Whereas earlier the absence of decays of the type

$$\mu \to e + \gamma;$$
$$\mu \to e + e + e$$

could be adduced to prove the absence of neutral leptonic currents, this absence can now be attributed to the separate conservation of the electron L_e and muon L_μ numbers ($L_e = +1$ for e^- and ν_e; -1 for e^+ and $\bar{\nu}_e$; 0 for all other particles; $L_\mu = +1$ for μ^- and ν_μ; -1 for μ^+ and $\bar{\nu}_\mu$; 0 for all other particles).

The most complete experimental data have been obtained in the study of semileptonic processes in which kaons participate. As regards the processes with $\Delta S = 0$, there are no experimental proofs applicable to neutral currents since the predominant electromagnetic interaction competes with such transitions if $\Delta S = 0$. (We shall discuss the question of the electromagnetic "competition" in more detail below.) Since the interfering (from this point of view) electromagnetic interaction conserves strangeness, processes with $\Delta S \neq 0$ should be more sensitive in a search for neutral currents than the processes with $\Delta S = 0$ provided, of course, the corresponding coupling constant is not infinitesimally small. If both the leptonic and the hadronic current contain vector and axial-vector parts, we have at our disposal ten coupling constants that completely characterize these currents:

$$g^{A,V}(\mu\mu),\ g^{A,V}(ee),\ g^{A,V}(\nu\nu),\ g^{A,V}(\mu\nu_\mu),\ g^{A,V}(e\nu_e).$$

These constants are related to the processes

$$\begin{aligned}
&g^V(\mu\nu_\mu) - K \to \pi\mu\nu_\mu; && g^A(\mu\nu_\mu) - K \to \mu\nu_\mu;\\
&g^V(e\nu_e) - K \to \pi e\nu_e; && g^A(e\nu_e) - K \to e\nu_e;\\
&g^V(ee) - K \to \pi ee; && g^A(\mu\mu) - K_2^0 \to \mu\mu;\\
&g^V(\mu\mu) - K \to \pi\mu\mu; && g^A(ee) - K_2^0 \to ee;\\
&g^V(\nu\nu) - K \to \pi\nu\nu; && g^A(\nu\nu) - K_2^0 \to \nu\nu.
\end{aligned}$$

Apart from the above processes, conservation of the lepton number allows the decays [31]

$$K_1^0 \to \mu^+\mu^-,\ K_1^0 \to \pi^0 e^+e^-.$$

These two decays violate CP invariance if the weak interaction is strictly local. The transition $K_2^0 \to \bar{\nu}_\mu \nu_\mu$ is absolutely forbidden because of helicity conservation for $m_\nu = 0$ (provided, of course, ν and $\tilde{\nu}$ are not identical particles). Similarly, the decay $K_2^0 \to e^+e^-$ is also suppressed because of the small electron mass even if the corresponding constant $g^A(ee)$ is very large. In addition, the transitions $K \to \pi\nu_e\bar{\nu}_e$ and $K \to \pi\nu_\mu\bar{\nu}_\mu$ cannot be identified separately; we therefore have eight possible semileptonic transitions in which kaons participate.

Hitherto only the processes

$$K \to \pi + l + \nu_l \text{ and } K \to l + \nu_l$$

have been observed experimentally. Numerous searches have been made for the other decays but no single sufficiently reliable case has yet been reported. All experiments with decays of K^+ mesons have been made in bubble chambers and a search for neutral currents in the decays of K^0 mesons has been made as an additional task in the investigation of the decays $K_L^0 \to 2\pi$. The most complete data on the $K^0 \to ll$ decays have been obtained recently at CERN [32]:

$$K_L^0 \to \mu^+\mu^-; \quad K_L^0 \to \mu^\pm e^\mp;$$
$$K_L^0 \to e^+e^-; \quad K_S^0 \to \mu^+\mu^-$$

and at Princeton [33]:

$$K_L^0 \to \mu^+\mu^-;$$
$$K_L^0 \to \mu^\pm e^\mp.$$

In both cases the experimental set-up was designed to investigate interference effects in the decays $K^0 \to 2\pi$. This naturally meant that the experimental conditions were not optimal for the detection of the decays $K \to ll$ although some necessary additions to the apparatus (for example, pion stoppers that allow the passage of muons etc.) were made. For the upper limit of the branching ratio the Princeton group obtained

$$\Gamma(K_L^0 \to \mu^+\mu^-)/\Gamma(K_L^0 \to \text{all}) \leqslant 3.5 \cdot 10^{-5}$$

with a 90% confidence level.

An estimate of the upper limit of the relative rate of the decay $K_L^0 \to \mu^\pm e^\mp$ (with identification of the $e^\pm$ track) gave

$$\Gamma(K_L^0 \to \mu^\pm e^\mp)/\Gamma(K_L^0 \to \text{all}) \leqslant 6 \cdot 10^{-6}.$$

The group at CERN found

$$\Gamma(K_L^0 \to \mu^+\mu^-)/\Gamma(K_L^0 \to \text{all}) \leqslant 1.6 \cdot 10^{-6};$$
$$\Gamma(K_L^0 \to e^+e^-)/\Gamma(K_L^0 \to \text{all}) \leqslant 1.8 \cdot 10^{-5};$$
$$\Gamma(K_L^0 \to \mu^\pm e^\mp)/\Gamma(K_L^0 \to \text{all}) \leqslant 9 \cdot 10^{-6}.$$

(We emphasize once more that the transition $K_L^0 \to \mu^\pm e^\mp$ is forbidden by the law of separate conservation of the lepton numbers.) In the CERN investigation the exact number of decays of the short-lived K^0 mesons was known; this made it possible to estimate the upper limit of the branching ratio of the CP-violating decay $K_L^0 \to \mu^+\mu^-$:

$$\Gamma(K_S^0 \to \mu^+\mu^-)/\Gamma(K_S^0 \to \text{all}) \leqslant 7.3 \cdot 10^{-5}.$$

As yet no experimental data are available for the decays

$$K_L^0 \to \pi^0\mu^+\mu^- \text{ and } K_L^0 \to \pi^0 e^+e^-,$$

TABLE 3. Allowed Kaon Decays

ΔQ	$\Delta S=0$	$\Delta S=1$
1	$K^0 \to K^+ + e^- + \bar{\nu}_e$	$K^+ \to \pi^0 + e^+ + \nu_e$ $K^0 \to \pi^\pm + e^\mp + \nu_e$ $K^+ \to \pi + \pi + e^+ + \nu_e$
0	—	$K^+ \to \pi^+ + l + \bar{l}$ $K_L^0 \to \mu^+ + \mu^-$ $K_L^0 \to \pi^0 + \mu^+ + \mu^-$ $K^+ \to \pi^+ + \pi^0 + e^+ + e^-$ $K^0 \to \pi^+ + \pi^- + e^+ + e^-$

and for the transitions of charged kaons to $\pi l \bar{l}$ we have the following data:

$$\Gamma(K^+ \to \pi^+ e^+ e^-)/\Gamma(K^+ \to \text{all}) \leqslant 8.8 \cdot 10^{-7} \; [34,\ 35];$$

$$\Gamma(K^+ \to \pi^+ \mu^+ \mu^-)/\Gamma(K^+ \to \text{all}) \leqslant 3 \cdot 10^{-6} \; [35];$$

$$\Gamma(K^+ \to \pi^+ \nu \bar{\nu})/\Gamma(K^+ \to \text{all}) \leqslant 1.1 \cdot 10^{-4} \; [36]\, *;$$

$$\Gamma(K^+ \to \pi^+ \mu^+ e^-)/\Gamma(K^+ \to \text{all}) \leqslant 3 \cdot 10^{-5} \; [37].$$

The last process violates the conservation of the lepton numbers.

One can relate the rates of the decays due to the neutral and charged currents and the corresponding coupling constants.

For the charged K-meson decays [37]

$$\Gamma(K^+ \to \pi^+ \mu^+ \mu^-) = 5.5\, [g^V(\mu^+\mu^-)/g^V(\mu^+\nu_\mu)]^2 \cdot 10^7 \text{ sec}^{-1};$$

$$\Gamma(K^+ \to \pi^+ e^+ e^-) = 1.2\, [g^V(e^+e^-)/g^V(e^+\nu_e)]^2 \cdot 10^8 \text{ sec}^{-1}.$$

and for the K^0-meson decays [38]

$$\frac{\Gamma(\bar{K}_L^0 \to \mu^+\mu^-)}{\Gamma(K^+ \to \mu^+ \nu_\mu)} = 4\, [g^A(\mu^+\mu^-)/g^A(\mu^+\nu_\mu)]^2 \frac{[m_K^3 (m_K^2 - 4m_\mu^2)]^{1/2}}{(m_K^2 - m_\mu^2)^2}.$$

If the above estimates for the upper limit of the branching ratios are used, the following limits for the ratios of the corresponding coupling constants are obtained†:

$$g^V(e^+e^-)/g^V(e^+\nu_e) \approx 7 \cdot 10^{-4};$$

$$g^V(\mu^+\mu^-)/g^V(\mu^+\nu_\mu) \approx 1.5 \cdot 10^{-2};$$

$$g^V(\nu\bar{\nu})/g^V(e^+\nu_e) \approx 6 \cdot 10^{-2};$$

$$g^A(\mu^+\mu^-)/g^A(\mu^+\nu_\mu) \approx 7 \cdot 10^{-4}.$$

Since the processes due to neutral currents have been sought in different reactions including vector and axial-vector strongly interacting currents and for all lepton combinations, the absence of such processes (or, at least, the small values of the corresponding coupling constants) is one of the fundamental properties

*In the most recent experiments the limit was lowered to $1.4 \cdot 10^{-6}$ for this ratio.
†In a recent experiment at Berkeley new data have been obtained on the decays $K_L^0 \to l^+ l^-$, from which it follows that $\Gamma(K_L^0 \to \mu^+\mu^-)/(\Gamma(K_L^0 \to \text{all}) < 8 \cdot 10^{-9}$ and there are similar estimates for the other transitions of this type; this considerably changes the estimates given here for the ratios of the coupling constants.

of the weak interactions. The existing data indicate that if the primary neutral currents do exist, the strength of their coupling is at least three-orders of magnitude less than that of the charged currents. Reaction with four particles in the final state such as

$$K^0 \to \pi^+\pi^-e^+e^- \text{ and } K^0 \to \pi^0\pi^0e^+e^-$$

differ slightly from the decays already considered. Whereas decays into two or three particles are due to vector or axial-vector transitions, processes with four particles are characterized by an interference term between the vector and axial-vector transitions. In addition, such processes cannot proceed through electromagnetic transitions with a Dalitz pair for the 0^+ state of the $\pi\pi$ system (the 0-0 transitions), i.e., the $\pi\pi ee$ final state must be a good subject for the search for primary neutral currents.

Electromagnetic Competition (Induced Neutral Currents). Even if primary neutral leptonic currents are absent in the first-order in the weak interaction, the presence of certain strangeness-conserving neutral hadron currents in the primary Lagrangian is sufficient to generate neutral leptonic currents (the so-called induced neutral leptonic currents). This process may proceed through the intermediate electromagnetic field; for example, a current of the form $\bar{p}\gamma_\mu p$ will generate the current $\bar{l}\gamma_\mu l$ through the process $\bar{p}+p \to \gamma \to \bar{l}l$.

Thus, the combined action of the weak and electromagnetic interactions may lead to processes in which the final state contains leptons with a vanishing total charge. Since usual conservation of the electric charge of the particles always leads to $\Delta Q = 0$ for the hadrons for such processes, one may conjecture that the existence of neutral hadron currents is due to the presence of such processes. The converse assertion is not always valid because of the possible contribution from strong interactions, whose effects require a detailed theoretical analysis. Such induced neutral leptonic currents are quite capable of competing with the primary currents and are an interfering background in this sense.

Numerical estimates of the branching ratios of the different induced processes obtained by different methods are given in Table 4. Comparing the experimental and theoretical estimates for the upper limits of the branching ratios, we see that they do not contradict one another, at least not in their order of magnitude. It should be noted that the smallest calculated branching ratio ($\sim 10^{-8}$) corresponds to the process $K_L^0 \to \mu^+\mu^-$, whereas the experimental limit is also $\sim 10^{-8}$. Thus, for all the remaining processes the estimates of the limits of the branching ratios lie in practice at the level of the theoretical estimates for the induced processes.

Possible Violation of CP Invariance in Processes with Neutral Currents. Suppose that the observed violation of CP invariance can be attributed to the existence of neutral leptonic currents and that the smallness of CP-violating effects is due to the weakness of the coupling of the neutral currents. Such an assumption can be immediately verified by investigating the decay $K^+ \to \pi^+\mu^+\mu^-$, since it must reveal T-noninvariant correlations of the form

$$\sigma_{\mu^+}(\bar{p}_\mu \times \bar{p}_\pi) \quad \sigma_{\mu^-}(\bar{\sigma}_{\mu^+} \times \bar{p}_\pi).$$

In addition, one could observe a number of CP-noninvariant decay modes such as $K_S^0 \to \mu^+\mu^-$.

It is shown in [31] that in the case of strict locality of the weak interaction this decay, if it occurs in the first-order in the weak interaction and proceeds without interference from the electromagnetic field (i.e., the $\mu^+\mu^-$ system is formed in the $^1s_0$ state) is CP violating; for the system of two fermions ll we have $CP = (-1)^{S+1}$. A similar situation obtains for the decay modes

$$K_S^0 \to \begin{matrix} \pi^0 \nu\bar{\nu} \\ \pi^0 \mu^+ \mu^- \\ \pi^0 e^+ e^-, \end{matrix}$$

which are vector transitions, i.e., the $\bar{l}l$ state must be in the ground $^3s_1$ state and the system must have $CP = -1$; it follows that such modes also violate CP invariance.

TABLE 4. Predictions for Induced Neutral Currents

Decay	Branching ratio	Literature	Decay	Branching ratio	Literature
$K^+ \to \pi^+ + e^+ + e^-$	10^{-7}	[39[	—	$(1,8-4)10^{-7}$	[43]
			$K_L^0 \to \pi^0 + e^+ + e^-$	10^{-8}	[40]
—	10^{-6}	[40]	—	$4 \cdot 10^{-8}$	[41]
—	10^{-6}	[41]	$K_L^0 \to \mu^+ + \mu^-$	$10^{-8} - 10^{-9}$	[44]
—	10^{-6}	[42]			

2. Leptonic Baryon Decays

Virtually all the known baryons that are stable against strong interactions (apart from the Ω^- hyperon) are unified in accordance with the $SU_{(3)}$-symmetry classification into a baryon octet with the quantum numbers of the spin J and the parity P satisfying $J^P = 1/2^+$. Therefore, in the semileptonic decay of a baryon A into a baryon B in accordance with the scheme

$$A \to B + l^- + \tilde{\nu}_l \tag{2.1}$$

both baryons are regarded as having the set of quantum numbers $J^P = 1/2^+$.

In the introduction we formulated the general form of the Lagrangian in the universal four-fermion V−A theory for the process (2.1) [the expressions (13)-(15)].

It should be noted that one can, in principle, extend the analogy with electrodynamics and introduce a vector particle (the intermediate boson W, the analog of the photon in electrodynamics) and go over from the formalism of a contact interaction to a nonlocal formulation of the theory of the weak interactions. The nonlocality effects depend on the mass of the intermediate W boson and decrease with increasing mass of this particle. The present-day experimental data indicate that if an intermediate vector W meson exists, its mass is at least greater than 3.5 GeV. If this is its mass, its influence on the phenomena considered below is very small. In what follows we shall therefore content ourselves with a Lagrange formalism of the form (13).

2.1. Structure of Matrix Elements

The semileptonic hadron decays can be described by a matrix element of the form

$$M = l^\alpha X_\alpha, \tag{2.2}$$

where l_α and X_α are the parts of the matrix element due to the leptonic and hadronic currents, respectively. In the presence of strong interactions the form of the hadronic currents [15] changes. In a general form, we therefore have

$$X_\alpha = \langle f | J_\alpha(x) | i \rangle. \tag{2.3}$$

Thus, the expression of the matrix element in the form (2.2) and (2.3) presupposes that the weak interaction is taken into account in the first order of perturbation theory but that the strong interaction has been taken into account fully.

The part X_α can be represented as the sum of a vector V_α and an axial-vector A_α part:

$$X_\alpha = V_\alpha + A_\alpha; \tag{2.4}$$

if the baryons A and B have positive relative parity ($P_{AB} = +1$), each of these parts can be expressed in the most complete form as follows:

$$V_\alpha = \bar{u}(p_B)\left[f_1(q^2)\gamma_\alpha + \frac{f_2(q^2)}{m_A + m_B}\sigma_{\alpha\beta} q_\beta + \frac{f_3(q^2)}{m_A + m_B} q_\alpha\right] u(p_A);$$

$$A_\alpha = \bar{u}(p_B)\left[g_1(q^2)\gamma_\alpha + \frac{g_2(q^2)}{m_A+m_B}\sigma_{\alpha\beta}q_\beta + \frac{g_3(q^2)}{m_A+m_B}q_\alpha\right]\gamma_5 u(p_A). \tag{2.5}$$

Here, p_A and p_B are the four-momenta of the baryons A and B and $q = p_A - p_B$ is the momentum transferred to the leptons. In the case of a negative relative AB parity ($P_{AB} = -1$) these expressions swap their positions. In the general case we therefore have six interactions; because of the renormalization effects from the strong interactions the two original V and A interactions with the form factors $f_1(q^2)$ and $g_1(q^2)$ are augmented by induced interactions which have received the following names: weak magnetism [the form factor $f_2(q^2)$], weak electricity [the form factor $g_2(q^2)$]; the scalar interaction [the form factor $f_3(q^2)$], and the pseudoscalar interaction [the form factor $g_3(q^2)$]. The hadronic part X_α of the matrix element (2.2) is determined by six form factors, which are functions of the square of the four-momentum q^2 transferred to the leptons. One can replace (2.4) and (2.5) by the compact expression

$$X_\alpha = u(p_B)\left[\sum_{k=1}^{6} f_k(q^2)\,O_\alpha^{(k)}\right]u(p_A),$$

where summation is extended over k for all possible variants of the interaction. Here, we have the same five invariants as are listed in Table 1 and these augmented by a pseudotensor (weak electricity) of the form $\sigma_{\alpha\beta}q_\beta\gamma_5$.

In semileptonic baryon decays a small amount of energy is usually liberated. It follows that the range of variation of q^2 is small and one can assume that the form factors change little. As a rule, their q^2 dependence is represented by the approximation

$$f_i(q^2) = f_i(0)\left[1 + \lambda_i q^2/m_\pi^2\right], \tag{2.7}$$

2.2. Expressions for Observables

Thus, the semileptonic baryon decays are described by expressions that depend on 12 real constants in the complex form factors $f_i(0)$ and $g_i(0)$ and also on the six parameters λ_i in the q^2 dependence of the form factors $f_i(q^2)$ and $g_i(q^2)$. If T invariance holds, there remain only six form-factor constants. Estimates show that, generally speaking, the dominant form factors are still the form factors of the vector and axial-vector interactions. The remainder lead to corrections to the main terms. These corrections are $\sim(m_A - m_B)/m_A$ or 1% and less. For example, if one neglects the recoil terms and the lepton mass, the differential probability of the decay $A \to Bl\nu_l$ with the emission of a lepton with energy in the interval from $\eta = E_l/E_l^{max}$ to $\eta + d\eta$ and acccordingly a baryon with kinetic energy in the interval from $\xi = \frac{T(\eta)_{max} - T}{T(\eta)_{max} - T(\eta)_{min}}$ to $\xi + d\xi$ takes the form

$$\frac{d^2W}{d\eta\cdot d\xi} = \eta^2(1-\eta)^2\left[|f_1|^2(1-\xi) + |g_1|^2(1+\xi)\right]. \tag{2.8}$$

This expression in the same approximation also determines the $l\nu$ correlations since $\xi \approx 1/2\,(1 - \cos\theta_{l\nu})$, which yields

$$\frac{dW}{d(\cos\theta_{l\nu})} \sim 1 - \frac{1 - |g_1/f_1|^2}{1 + 3|g_1/f_1|^2}\cdot\cos\theta_{l\nu}. \tag{2.9}$$

It can be seen from (2.9) that the electron energy spectrum in such an approximation is quite independent of the form factors and that the spectrum of the secondary baryons or the $l\nu$ correlations are determined solely by the modulus of the form-factor ratio g_1/f_1. The electron spectrum depends on the form factors in the order $(m_A - m_B)/m_A$. This term is determined by the interference effect of the form factors f_1 and g_1 in the presence of also the interference term of the form factors f_1 and f_2. It has the form

$$(f_1 + 2f_2)\,g_1\,\frac{m_A - m_B}{m_A}\cdot\xi(2\eta - 1) \tag{2.10}$$

and is due to the influence of three form factors at once or two of their ratios, for example, f_2/f_1 and g_1/f_1. Thus, the ensemble of decays $A \to B + l + \nu_l$ can be represented on a Dalitz plot in the variables of the kinetic energies of the charged lepton and the secondary baryon T_B. One can then analyze such diagrams in the

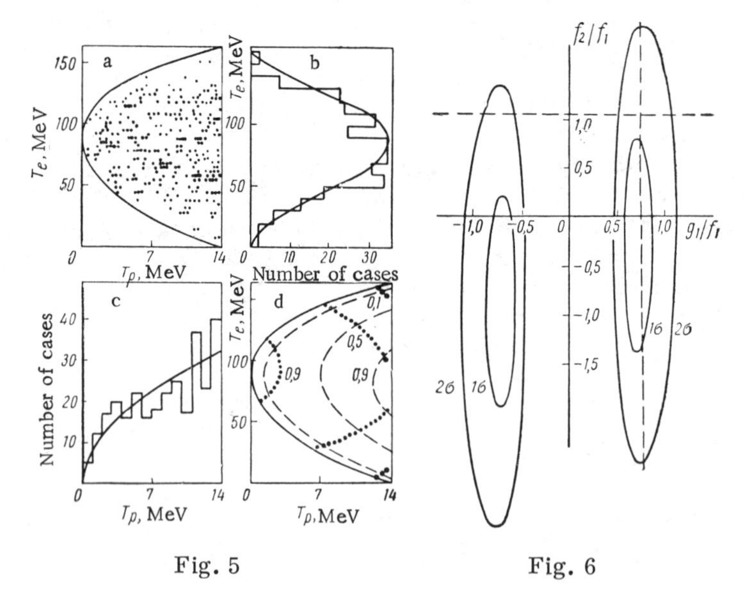

Fig. 5 Fig. 6

Fig. 5. Investigation of the properties of the decay $\Lambda \to pe^- \overline{\nu}_e$: a) Dalitz plot; b) spectrum of the electron kinetic energies; c) spectrum of the proton kinetic energies in the rest system of the Λ hyperon; d) contours of the relative probability in the decay for the purely vector (dashed curves) and purely axial-vector (dotted curves) interaction variants and different values of the ratio g_1/f_1.

Fig. 6. Function of maximal likelihood for the form-factor ratios f_2/f_1 and q_1/f_1 in the decay $\Lambda \to pe\nu$ represented by the contours for one (1σ) and two (2σ) standard deviations. The horizontal dashed line is the ratio f_2/f_1 predicted by hypothesis of a conserved vector current (CVC), and the vertical dashed line is the ratio q_1/f_1 predicted by the Cabibbo theory (see Table 6).

variables g_1/f_1 and f_2/f_1. An example of the distribution of decay events $\Lambda \to pe\nu_e$ on the Dalitz plot is shown in Fig. 5a. An analysis by the method of the function of maximal likelihood in the variables g_1/f_1 and f_2/f_1 for this decay is shown in Fig. 6. This figure shows that the accuracy of the present-day data on the decay $\Lambda \to pe\nu_e$ is not yet sufficient to draw an unambiguous conclusion about the parameters of such a description of the semileptonic decay [45]. A similar conclusion is reached in an analysis of the corresponding data for the decay $\Sigma^- \to ne^-\widetilde{\nu}_e$ [46] (Fig. 7).

Let us now consider in more detail the expressions for the observable quantities. In deriving expressions for observable quantities (spectra, correlations etc.) onedoes not usually use the expressions (2.5) for the hadronic part of the matrix element but more convenient expressions. To obtain these we replace the dimensionless form factors $f_i(q_i^2)$ and $g_i(q^2)$ by the linear combinations [47]

$$\left.\begin{aligned} F_1 &= f_1 + \left(1 + \frac{m_A}{m_B}\right) f_2; \\ F_2 &= -2f_2; \\ F_3 &= f_2 + f_3; \\ G_1 &= g_1 - \left(1 - \frac{m_A}{m_B}\right) g_2; \\ G_2 &= -2g_2; \\ G_3 &= g_2 + g_3. \end{aligned}\right\} \tag{2.11}$$

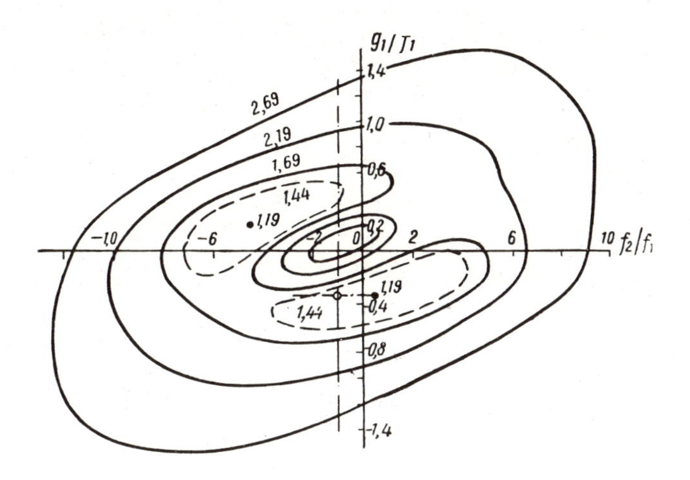

Fig. 7. Function of maximal likelihood for the form-factor ratios f_2/f_1 and g_1/f_1 in the decay $\Sigma^- \to ne\nu$ represented by the contours for one, two, and three standard deviations. The figure also includes lines corresponding to the values of these ratios for the hypothesis of a conserved vector current (CVC) (the vertical dash-dot-dash line) and the Cabibbo theory (the horizontal dash-dot-dash line) with the parameters from Table 6.

Then Eqs. (2.5) are replaced by the following expressions for the calculation of the matrix elements:

$$V_\alpha = \bar{u}(p_B)\left[F_1(q^2)\gamma_\alpha + \frac{F_2(q^2)}{m_A+m_B}(p_B)_\alpha + \frac{F_3(q^2)}{m_A+m_B}q_\alpha\right]u(p_A); \tag{2.12}$$

$$A_\alpha = \bar{u}(p_B)\left[G_1(q^2)\gamma_\alpha\gamma_5 + \frac{G_2(q^2)}{m_A+m_B}(p_B)_\alpha\gamma_5 + \frac{G_3(q^2)}{m_A+m_B}q_\alpha\right]u(p_A). \tag{2.13}$$

Let us determine the q^2 dependence of the new form factors $F_i(q^2)$ and $G_i(q^2)$, again using expressions of the form (2.7). For the products of the form factors we use the approximations

$$F_i(q^2)F_j^*(q^2) = F_i(0)F_j^*(0)\left[1+(\lambda_i+\lambda_j)q^2/m_\pi^2\right]; \tag{2.14}$$

$$G_i(q^2)G_j^*(q^2) = G_i(0)G_j^*(0)\left[1+(\mu_i+\mu_j)q^2/m_\pi^2\right]. \tag{2.15}$$

We now determine the different quadratic expressions in $F_i(q^2)$ and $G_i(q^2)$ as follows:

$$\left.\begin{aligned}
a_{ij} &= \mathrm{Re}\left[F_i(0)F_j^*(0)\right];\\
c_{ij} &= \mathrm{Re}\left[G_i(0)G_j^*(0)\right];\\
b_{ij} &= \mathrm{Re}\left[F_i(0)F_j^*(0)\frac{\lambda_i+\lambda_j}{m_\pi^2}\right];\\
d_{ij} &= \mathrm{Re}\left[G_i(0)G_j^*(0)\frac{\mu_i+\mu_j}{m_\pi^2}\right];\\
e_{ij} &= \mathrm{Re}\left[F_i(0)G_j^*(0)\right];\\
h_{ij} &= \mathrm{Re}\left[F_i(0)G_j(0)\frac{\lambda_i+\mu_j}{m_\pi^2}\right].
\end{aligned}\right\} \tag{2.16}$$

In this terminology the expressions for the energy spectra of the baryons (W_B), leptons (W_l), and also for the angular correlation of the charged lepton and the neutrino ($W_{l\nu}$) have the same form [48]:

$$W(R) = \frac{dW}{dR} = \left[\sum_{i,j=1}^{3} \left(a_{ij} A_k^{ij} + b_{ij} B_k^{ij} + c_{ij} C_k^{ij} + d_{ij} D_k^{ij} + l_{ij} E_k^{ij} + h_{ij} H_k^{ij} \right) \right] \Phi_k(x), \tag{2.17}$$

where the functions $\Phi_k(x)$ in the formulas for the energy spectra of the baryons and leptons have the simple form

$$\Phi_k(x) = \sqrt{x^2 + bx + a} \left(\frac{x+c}{x^2} \right)^2 x^k \frac{G^2}{(4\pi)^3} \cdot 2 \left(\frac{m_B}{m_A} \right)^2;$$
$$x_{B(l)} = -\frac{m_A^2 - m_{B(l)}^2}{m_B^2} + 2 \frac{m_A}{m_B^2} E_{B(l)}, \tag{2.18}$$

where $k = 1, 2, 3, \ldots, 6$; and a, b, and c are constants.

For the $l\nu$ correlations, $\Phi(x)$ has a more complicated form. However, one can use an approximation in which one ignores the terms from the recoil baryon proportional to $(m_A - m_B)/m_A$ and the lepton mass. For the $l\nu$ correlations we then obtain the simple formula

$$W(\cos\theta_{l\nu}) = \frac{1}{2} \left(1 + \frac{|z|^2 - 1}{|z|^2 + 3} \cos\theta_{l\nu} \right), \tag{2.19}$$

where now

$$z = f_1/g_1. \tag{2.20}$$

It is also helpful to give the expression for the angular distribution of the polarization of the baryon B from the decay $A \to B l \nu_l$. In the frame of reference defined by the vectors

$$\alpha = \frac{q_l + q_\nu}{|q_l + q_\nu|}; \quad \beta = \frac{q_l - q_\nu}{|q_l + q_\nu|}; \quad \gamma = \frac{q_l \cdot q_\nu}{|q_l + q_\nu|}, \tag{2.21}$$

this distribution becomes

$$W(S_B) = 1 + \frac{8}{3} S_B \left[\frac{\operatorname{Re} z}{|z|^2 + 3} \alpha + \frac{1}{|z|^2 + 3} \beta + \frac{3\pi}{16} \cdot \frac{\operatorname{Im} z}{|z|^2 + 3} \gamma \right]. \tag{2.22}$$

If we now consider the following decay of the baryon B in accordance with a scheme of the nonleptonic type

$$B \to N\pi, \tag{2.23}$$

then the angular distribution for the decay nucleon in the same frame of reference is

$$W(\mathbf{N} \cdot \alpha) = 1 + \alpha_B \frac{8}{3} \cdot \frac{\operatorname{Re} z}{|z|^2 + 3} \mathbf{N} \cdot \alpha; \tag{2.24}$$

$$W(\mathbf{N} \cdot \beta) = 1 + \alpha_B \frac{8}{3} \cdot \frac{1}{|z|^2 + 3} \mathbf{N} \cdot \beta; \tag{2.25}$$

$$W(\mathbf{N} \cdot \gamma) = 1 + \alpha_B \frac{\pi}{2} \cdot \frac{\operatorname{Im} z}{|z|^2 + 3} \mathbf{N} \cdot \gamma. \tag{2.26}$$

Here, N is a unit vector in the direction of the nucleon momentum and α_B is the asymmetry coefficient in the corresponding decay of the baryon B in accordance with (2.23). Note that since the total lepton momentum is eqüal to the momentum of the baryon B, (2.24) is obviously the angular distribution of the decay nucleons in formula (2.23) with respect to the direction of the momentum of the baryon B.

Let us now consider the decay of a polarized baryon A at rest with polarization vector $\mathbf{p}_A$. The probability of its decay into the state $dE_l d\Omega_l d\Omega_\nu$ is

$$W(E_l, \Omega_l, \Omega_\nu, \mathbf{p}_A)\, dE_l\, d\Omega_l\, d\Omega_\nu = [c_1 + c_2 \mathbf{p}_l \mathbf{p}_A + c_3 \mathbf{p}_\nu \mathbf{p}_A + c_4 \mathbf{p}_A (\mathbf{p}_l \times \mathbf{p}_\nu)] \times \frac{|\mathbf{p}_l||\mathbf{p}_\nu|\, dE_l\, d\Omega_l\, d\Omega_\nu}{m_A - E_l + |\mathbf{p}_l| \cos\theta_{l\nu}}, \tag{2.27}$$

where the coefficients c_i depend on the form factors $F_i(q^2)$ and $G_i(q^2)$ and the kinematic characteristics of the decay (for example, E_l and $\cos\theta_{l\nu}$). For the angular distributions of the recoil baryons $W(\cos\theta_B)$ and the charged leptons $W(\cos\theta_l)$ with respect to the polarization vector of the original baryon $\mathbf{p}_A$ we hence obtain the expressions [48]:

$$W(\cos\theta_{B\,(l)}) = \frac{dW}{d\cos\theta_{B\,(l)}} = \frac{W_0}{2}\left[1 + |\mathbf{p}_A| \frac{W_a^{B\,(l)}}{W_0} \cos\theta_{B\,(l)}\right]. \tag{2.28}$$

Here, W_a^l again has the form (2.17) with $\Phi(x)$ = const and W_a^B is given by

$$W_a^B = \frac{G^2}{2\pi^3}\left[e_{ij} J_e^{ij} + h_{ij} J_h^{ij}\right],$$

i.e., the sum contains terms that depend only on the interference of the form factors of the vector and axial-vector parts of the interaction.

In the formulas, W_0 is the total probability of the given type of semileptonic decay of the baryon A:

$$W_0 = \frac{G^2}{(2\pi)^3} \sum_{i,j=1}^{3} \left(a_{ij} J_a^{ij} + b_{ij} J_b^{ij} + c_{ij} J_c^{ij} + d_{ij} J_d^{ij}\right). \tag{2.29}$$

The structure of the coefficients a_{ij}, b_{ij}, c_{ij}, and d_{ij} [see (2.16)] shows that in this expression the summation is extended only over terms that contain form factors of one parity type, i.e., the VV, AA, A, and PT types but not the types VA, TA, etc.

In all these expressions the coefficients A^{ij} and J^{ij} are determined by the dynamic characteristics of the decay and, ultimately, are functions of only the masses of the particles that participate in the process. The values of these coefficients have been calculated by different authors and they can be found in the corresponding tables.

The expression (2.28) for the angular distribution of the leptons relative to the polarization vector of the original baryon in the above approximation takes the form

$$W(\cos\theta_l) \sim 1 + \alpha_l \cos\theta_l. \tag{2.30}$$

The asymmetry coefficient α_l in this expression is given by

$$\alpha_l = -2\beta_l \frac{g_1/f_1 + (g_1/f_1)^2}{1 + 3(g_1/f_1)^2}, \tag{2.31}$$

where β_l is the lepton velocity.

The expression (2.29) for the total probability can be written down by separating the principal part that depends on the kinematics:

$$W_0 = \frac{G^2}{60\pi^3} \cdot \frac{\Delta^5}{(1+\xi)^3} H_l, \tag{2.32}$$

where $\Delta = m_A - m_B$; $\xi = (m_A - m_B)/(m_A + m_B)$; and H_l is a bilinear function of the form factors:

$$H_l = \sum_{M,N} f_M f_N S^{MN}. \tag{2.33}$$

In this representation we obtain the following relationships for decays with the emission of an electron [49]:

$$\left.\begin{aligned} S^{VV}&=1;\\ S^{AA}&=3;\\ S^{A,PT}&=S^{PT,A}=-\varepsilon;\\ S^{VT}&=S^{TV}=\varepsilon^2;\\ S^{VA}&=S^{AV}=S^{TA}=S^{AT}\approx\frac{1}{2}\varepsilon^2. \end{aligned}\right\} \qquad (2.34)$$

Here, $\varepsilon = W/m_A$; $W = (m_A^2 - m_B^2 + m_C^2)/2m_A$.

Thus, if the form factors f_i and g_i have similar orders of magnitude

$$f_1 \approx f_2 \approx f_3 = g_1 \approx g_2 \approx g_3, \qquad (2.35)$$

then the total probablity of an electronic decay of a baryon (and also of a decay with the emission of a muon) is determined primarily, as one would expect, by the form factors of the vector f_1 and axial-vector g_1 interactions. The contribution of the induced interactions is determined by the interference terms with the form factors g_1 and g_2 and also f_1 and f_2, whose numerical coefficients have the values $\varepsilon \approx 10^{-2}$ and $\varepsilon^2 \approx 10^{-4}$. One can write down the relationship

$$H_l = f_1^2 + 3g_1^2 - \varepsilon g_1 g_2 + \varepsilon^2 f_1 f_2. \qquad (2.36)$$

In the order ε^3 one would obtain interference terms from the form factors with different parities, namely, terms of the form $f_1 g_1$ and $f_2 g_1$.

In the case of a decay with the emission of a muon, calculations show [5] that one may have terms with different form factors as well as terms that arise from the q^2 dependence of the form factors. If one writes down the ratio of the probabilities of μ decay and β decay of a baryon in the form

$$\frac{\Gamma_\mu}{\Gamma_l} = \sigma_1\left(1 + \frac{\Delta}{H_l}\right), \qquad (2.37)$$

where

$$\Delta = (H_\mu - \sigma_1 H_l)/\sigma_1; \qquad (2.38)$$

σ_1 is a constant, and H_l is already defined by (2.36), a comparison of the decays $\Sigma^- \to n\mu^- \tilde{\nu}_\mu$ and $\Sigma^- \to ne^- \tilde{\nu}_e$ yields

$$\Gamma_\mu/\Gamma_e = 0.45\,(1 + 1.33\,\gamma), \qquad (2.39)$$

where, if one does not assume the V−A variant of the theory, the quantity γ is given by

$$\gamma = \frac{\operatorname{Re} f_3 f_1^* + 6 \operatorname{Re} g_1 f_2^* + 6\delta \operatorname{Re} f_2 f_2^* - \delta \operatorname{Re} g_3 g_1^*}{|f_1|^2 + 3|g_1|^2 + 12|f_2|^2 + |f_3|^2}, \qquad (2.40)$$

in which

$$\delta = \frac{m_{\Sigma^-} + m_n}{m_{\Sigma^-} + m_n}. \qquad (2.41)$$

It follows from the presently available experimental data on the semileptonic Σ^- decays [51] that

$$\Gamma_\mu/\Gamma_e = 0.42 \pm 0.06. \qquad (2.42)$$

In the framework of $\mu - e$ universality this corresponds to the following interval for γ:

$$-0.18 < \gamma < 0.02. \tag{2.43}$$

In a comparison of the corresponding Λ-hyperon decays it is found that $(\Gamma_\mu/\Gamma_e)_{theor} = 0.17$ and $(\Gamma_\mu/\Gamma_e)_{exp} = 0.17 \pm 0.02$.. This again suggests that a vanishing value of γ would be compatible.

The influence of the induced-electricity form factor g_2 can be observed from the interference effect in a comparison of the decays $\Sigma^- \to \Lambda e^- \tilde{\nu}_e$ and $\Sigma^+ \to \Lambda e^+ \nu_e$. Here, Eq. (2.32) of [50] gives

$$\frac{\Gamma_-}{\Gamma_+} = \left(\frac{m_{\Sigma^-} - m_\Lambda}{m_{\Sigma^+} - m_\Lambda}\right)^5 \left(\frac{1+\xi_+}{1+\xi_-}\right)^3 \frac{H_-}{H_+} = 1.64\left(1 + \frac{\Delta}{H_+}\right), \tag{2.44}$$

where

$$\Delta = H_- - H_+ \approx \frac{m_- - m_+}{4m_\pi} g_1 g_2 = 0.015\, g_1 g_2. \tag{2.45}$$

At the same time, since $f_1 = 0$ (see below) in these decays,

$$H_+ = 3g_1^2 + 0.52\, g_1 g_2. \tag{2.46}$$

The experimental value of branching ratio (2.44) is at present

$$\frac{\Gamma_-}{\Gamma_+} = 1.60 \pm 0.27, \tag{2.47}$$

which corresponds to the following interval of Δ/H_+:

$$-0.19 < \frac{\Delta}{H_+} < 0.05. \tag{2.48}$$

Using Eq. (2.45) and (2.46), we obtain

$$\frac{\Delta}{H_+} = \frac{0.005\, g_2/g_1}{1 + 0.17\, g_2/g_1}, \tag{2.49}$$

which amounts to $\sim 0.5\%$ for $g_2/q_1 \approx 1$. Thus, to resolve the question of the possible contribution of the form factor g_2 to the decays $\Sigma^\pm \to \Lambda e^\pm \nu_e$ it will also be necessary to increase the accuracy of the experimental data considerably.

The experimental situation can be illustrated most clearly for the problem of semileptonic baryon decays by the example of the now relatively well studied β decay of the neutron:

$$n \to pe^- \tilde{\nu}_e. \tag{2.50}$$

The most recent experiments have given the following result for its half-decay period:

$$\tau_{n \to pe\tilde{\nu}_e} = (10.80 \pm 0.16) \text{ min}, \tag{2.51}$$

i.e., the value of τ for the neutron is known with an accuracy of $\sim 1.5\%$. Since $\varepsilon \approx 10^{-3}$ in this decay, the present-day accuracy in the determination of τ is clearly insufficient to reveal the influence of the induced form factors on the value of τ. The accuracy would have to be increased by at least a further order of magnitude. Recalling that the conservation of the G parity of the nucleon current implies

$$g_2 = 0, \tag{2.52}$$

we see from (2.36) that the detection of a contribution from the weak magnetism form factor f_2 necessitates an experimental value of τ for the neutron with an accuracy not worse than 10^{-6}.

Thus, using the information about the neutron lifetime, one can obtain only the modulus of the ratio of the axial-vector to the vector form factors $|g_1/f_1|$. It is found to be

$$|g_1/f_1| = 1.23 \pm 0{,}01 \text{ (from } n \to pe\nu). \tag{2.53}$$

Measurements of the asymmetry in the emission of electrons in the β decay of polarized neutrons in accordance with Eqs. (2.30) and (2.31) enables one to determine both the value and the sign of the ratio g_1/f_1. Recent experiments gave the following result [52]*:

$$g_1/f_1 = -1{,}25 \pm 0.05 \text{ (from } n \to pe\nu). \tag{2.54}$$

Here, the corrections from the induced weak magnetism form factor again have the order of magnitude

$$\frac{m_n - m_p}{3m_n}, \tag{2.55}$$

i.e., they are $\sim 10^{-3}$.

Finally, measurements of the electron-neutrino correlation in the β decay of the neutron [67] yield [from (2.9)]

$$\alpha_{l\nu} = \frac{1 - |\alpha|^2}{1 + 3|\alpha|^2} = -0.091 \pm 0.039,$$

from which

$$|\alpha| = |g_1/f_1| = 1.33 \pm 0.15. \tag{2.56}$$

Thus, these three experiments give the same results (to within the errors of the measurements) for the ratio g_1/f_1 for β decay. The ratio is negative and is at present assumed to be 1.23 ± 0.01. At the present stage there is no point in considering a possible contribution of the induced form factors to the experimentally observed characteristics of neutron β decay.

The formalism expounded above for the description of the weak interactions is in excellent agreement with the vast experimental material on the β decay of nuclei. However, the β (μ) decay probabilities calculated for hyperons by this theory are in serious contradiction with the experimentally found values. This was a serious problem of the universal theory of weak interactions. Its solution was found in the framework of the idea of $SU_{(3)}$ symmetry in the world of elementary particles.

2.3. Relevant Aspects of $SU_{(3)}$ Symmetry

There are now available a number of excellent reviews of $SU_{(3)}$ symmetry and its development. We shall therefore only discuss the aspects that have a bearing on our problem. The concept of $SU_{(3)}$ symmetry developed out of isospin symmetry, the nucleon spinor

$$N = \begin{pmatrix} p \\ n \end{pmatrix}; \quad \overline{N} = (p, n)$$

being replaced by the three-component spinor

$$b = \begin{pmatrix} p \\ n \\ \Lambda \end{pmatrix}; \quad \bar{b} = (p, n, \Lambda),$$

by means of which the isospin I is augmented by the hypercharge Y as a group characteristic.

*The most recent results give -1.26 ± 0.02 for this ratio.

The states p, n, and Λ are the states of a single particle with the same mass, spin, and parity. They are distinguished by the quantum numbers of the isospin and hypercharge. Then any interaction in which they participate must be invariant under rotations in a certain formal space. This is equivalent to the assertion that under such rotations under the influence of a transformation Ω a bilinear form remains invariant:

$$\bar{b}\Omega^{+}\Omega b=\bar{b}b.$$

It follows that Ω is represented by a 3×3 matrix such that $\Omega^{+}=\Omega^{-1}$, i.e., the Hermitian-conjugate matrix Ω^{+} is equal to the inverse matrix Ω^{-1} and the transformation Ω is therefore unitary. There are nine linearly independent 3×3 matrices which can be taken as the basis matrices in an investigation of a unitary transformation $U_{(3)}$ in three dimensions. One of them is the identity matrix I and the remaining eight matrices $\lambda_1, \lambda_2, \ldots, \lambda_8$ are Hermitian with vanishing traces. A general transformation Ω can be expressed as a product $\Omega = e^{iI\alpha}\cdot e^{i\lambda\theta}$. This corresponds to a representation of the unitary transformation $U_{(3)}$ as a product of a unitary transformation $U_{(1)}$ of dimension 1 and a unitary unimodular $SU_{(3)}$ transformation of rank 2 and dimension 3 since there are three basis vectors p, n, and Λ in the latter case and two of the eight generators can be reduced to diagonal form. This last fact corresponds to the presence in the group of two conserved quantities, namely, the isospin I and the hypercharge Y. Thus, instead of a three-dimensional isotopic space we now deal with an eight-dimensional unitary space. Invariance under $U_{(1)}$ corresponds to conservation of a baryon current of the form $n_\alpha = i\bar{b}\gamma_\alpha b$. Invariance under $SU_{(3)}$ corresponds to conservation of the eight component current of the so-called unitary spin:

$$\vec{\mathcal{F}}_a=i\bar{b}\gamma_a\frac{\lambda}{2}b. \tag{2.57}$$

Like the isospin current, the unitary spin current is conserved in the approximation of strict $SU_{(3)}$ symmetry. The baryon number n and the eight components of the unitary spin are given by the expressions

$$n=-i\int n_4 d^3x;$$
$$F_k=-i\int \vec{\mathcal{F}}_{k4}\,d^3x.$$

The λ_i can be taken as the following eight matrices [63]:

$$\lambda_1=\begin{pmatrix}0&1&0\\1&0&0\\0&0&0\end{pmatrix};\quad \lambda_2=\begin{pmatrix}0&-i&0\\i&0&0\\0&0&0\end{pmatrix};\quad \lambda_3=\begin{pmatrix}1&0&0\\0&-1&0\\0&0&0\end{pmatrix};$$

$$\lambda_4=\begin{pmatrix}0&0&1\\0&0&0\\1&0&0\end{pmatrix};\quad \lambda_5=\begin{pmatrix}0&0&-i\\0&0&0\\i&0&0\end{pmatrix};\quad \lambda_6=\begin{pmatrix}0&0&0\\0&0&1\\0&1&0\end{pmatrix};$$

$$\lambda_7=\begin{pmatrix}0&0&0\\0&0&-i\\0&i&0\end{pmatrix};\quad \lambda_8=\frac{1}{\sqrt{3}}\begin{pmatrix}1&0&0\\0&1&0\\0&0&-2\end{pmatrix}.$$

These matrices satisfy the relations

$$\mathrm{Sp}\,\lambda_k\lambda_l=2\delta_{kl};$$
$$[\lambda_k, \tilde{\lambda}_l]=2if_{klm}\lambda_m;$$
$$[\lambda_k, \lambda_l]=2d_{klm}\lambda_m,$$

where f_{klm} is a completely antisymmetric tensor and d_{klm} is a completely symmetric tensor (under permutations of the subscripts). We have the following nonvanishing values of these tensors:

$$f_{123}=1;$$

$$f_{147}=f_{246}=f_{257}=f_{345}=-f_{156}=-f_{367}=\frac{1}{2};$$

$$f_{458}=f_{678}=\frac{\sqrt{3}}{2};$$

$$d_{118}=d_{228}=d_{338}=-d_{888}=\frac{1}{\sqrt{3}};$$

$$d_{448}=d_{558}=d_{668}=d_{778}=-\frac{1}{2\sqrt{3}};$$

$$d_{146}=d_{157}=d_{256}=d_{344}=d_{355}=-d_{247}=-d_{366}=-d_{377}=\frac{1}{2}.$$

It follows from the structure of the λ_i matrices that the first three matrices λ_1, λ_2, and λ_3 determine the isospin subgroup $SU_{(2)}$. The operators of its projections are the matrices $I_1={}^1\!/_2\lambda_1$, $I_2={}^1\!/_2\lambda_2$, $I_3={}^1\!/_2\lambda_3$. The diagonality of λ_3 corresponds to conservation of the third component of the isospin. As we have already mentioned, there is a further diagonal matrix λ_8. It is associated with the hypercharge, whose operator can be expressed in the form

$$Y=\frac{1}{\sqrt{3}}\lambda_8=\frac{1}{3}\begin{pmatrix}1 & 0 & 0\\ 0 & 1 & 0\\ 0 & 0 & -2\end{pmatrix}.$$

It commutes with the matrices of the projection operators of the isospin. Using the Gell-Mann–Nishijima formula, we obtain the following expression for the electric charge operator Q:

$$Q=e\left(I_3+\frac{1}{2}Y\right)=\frac{1}{2}e\left(\lambda_3+\frac{1}{\sqrt{3}}\lambda_8\right)$$

or

$$Q=\frac{1}{3}e\begin{pmatrix}2 & 0 & 0\\ 0 & -1 & 0\\ 0 & 0 & -1\end{pmatrix}.$$

In the same normalization we have an expression for λ_0, the matrix of the baryon number operator (it corresponds to $U_{(3)}$ and not to $SU_{(3)}$):

$$\lambda_0=\sqrt{\frac{2}{3}}\begin{pmatrix}1 & 0 & 0\\ 0 & 1 & 0\\ 0 & 0 & 1\end{pmatrix}.$$

This yields an expression for the baryon number operator in the form

$$B=\frac{1}{\sqrt{6}}\lambda_0=\frac{1}{3}\begin{pmatrix}1 & 0 & 0\\ 0 & 1 & 0\\ 0 & 0 & 1\end{pmatrix}.$$

The group $SU_{(3)}$ is characterized by a set of irreducible representation, which are taken as the basis for the classification of particles, sets of particles being associated with irreducible representations of the group. The tensors of such representations $\varphi^{\alpha\beta\gamma\cdots}_{\lambda\mu\nu\cdots}$ with q superscripts and p subscripts that take the values from one to three are characterized by definite sets of quantum numbers of the hypercharge and isospin. The number of dimensions of the representation is given by

$$N(p,q)=\frac{1}{2}(p+1)(q+1)(p+q+2).$$

The hypercharge of the tensor component is given by

$$Y=p(3)-q(3)-\frac{p-q}{3}, \qquad (2.58)$$

and I_3, the isospin component, by

$$I_3=\frac{1}{2}\,[p(2)-q(2)-p(1)+q(1)].$$

In these expressions q(i) and p(i) are the superscripts and subscripts, i taking the values 1, 2, and 3. If real particles are characterized by only integral values of Y, it follows from (2.58) that the families of particles can only be associated with representations for which the numbers of the superscripts and subscripts are such that $(p-q)/3$ is an integer.

The lowest of these irreducible representations of $SU_{(3)}$ have the following dimensions: singlet D(0,0), octet D(1,1), decuplet D(3,0), 27-plet D(2,2) etc. In what follows, we shall be interested in the octet representation D(1,1). We write the tensor of this representation in the form of a 3×3 matrix $\varphi_\alpha^\beta(I_3, Y)$, indicating the quantum numbers I_3, the isospin projection, and the hypercharge Y. This matrix has the form

$$\varphi_\alpha^\beta=\begin{pmatrix} \varphi_1^1(0,0) & \varphi_1^2(1,0) & \varphi_1^3\left(\frac{1}{2},+1\right) \\ \varphi_2^1(-1,0) & \varphi_2^2(0,0) & \varphi_2^3\left(-\frac{1}{2},1\right) \\ \varphi_3^1\left(-\frac{1}{2},-1\right) & \varphi_3^2\left(\frac{1}{2},-1\right) & \varphi_3^3(0,0) \end{pmatrix} \tag{2.59}$$

All the presently known baryons that are stable against strong interactions and have spin-parity quantum numbers $J^P=1/2^+$ are united in the unitary octet. We write the matrix for these baryons in the form

$$B=\begin{pmatrix} \frac{\Lambda^0}{\sqrt{6}}+\frac{\Sigma^0}{\sqrt{2}} & \Sigma^+ & p \\ \Sigma^- & \frac{\Lambda^0}{\sqrt{6}}-\frac{\Sigma^0}{\sqrt{2}} & n \\ \Xi^- & \Xi^0 & -\frac{2}{\sqrt{6}}\Lambda^0 \end{pmatrix} \tag{2.60}$$

Similarly, one can write the matrix for the pseudoscalar mesons with $J^P=0^-$:

$$P=\begin{pmatrix} \frac{\eta^0}{\sqrt{6}}+\frac{\pi^0}{\sqrt{2}} & \pi^+ & K^+ \\ \pi^- & \frac{\eta^0}{\sqrt{6}}-\frac{\pi^0}{\sqrt{2}} & K^0 \\ K^- & \bar{K}^0 & -\frac{2}{\sqrt{6}}\eta^0 \end{pmatrix} \tag{2.61}$$

and the other particles that are united in $SU_{(3)}$-symmetry octets.

Finally, forming the scalar products of the eight-dimensional vectors F and λ, we obtain a representation of the unitary spin in the form of a mixed tensor of second rank, which again can be written as a 3×3 matrix with vanishing trace:

$$U=\begin{pmatrix} F_3+\frac{1}{\sqrt{3}}F_8 & F_1+iF_2 & F_4+iF_5 \\ F_1-iF_2 & -F_3+\frac{1}{\sqrt{3}}F_8 & F_6+iF_7 \\ F_4-iF_5 & F_6-iF_7 & -\frac{2}{\sqrt{3}}F_8 \end{pmatrix}$$

We have already discussed the main properties of the unitary spin components F_1, F_2, and F_3. They are always conserved. Here, we should like to mention that, in accordance with (2.59), the components F_4 and F_5 and also F_6 and F_7 are characterized by a strangeness equal to unity and isospin equal to 1/2.

2.4. Semileptonic Baryon Decays in the Framework of $SU_{(3)}$ Symmetry

In the framework of this formalism the vector hadronic current in the weak-interaction Lagrangian (18) has the form

$$J_\alpha^{(V)} = F_{1\alpha} + iF_{2\alpha} + F_{4\alpha} + iF_{5\alpha}. \tag{2.62}$$

Here, the first two terms correspond to transitions with $\Delta S = 0$ and the third and fourth terms to transitions with $|\Delta S| = 1$. A comparison of the matrices (2.59) and (2.61) indicates that these two pairs of terms correspond to the π^+ and K^+ components in the matrix for the pseudoscalar mesons. The vector hadronic current is therefore frequently written as the sum of the two corresponding terms:

$$J_\alpha^{(V)} = V_\alpha^{\pi+} + V_\alpha^{K+}. \tag{2.63}$$

By analogy with (2.57), we shall now define an eight-component axial-vector quantity:

$$A_\alpha = i\bar{b}\frac{\lambda}{2}\gamma_\alpha\gamma_5 b.$$

Then the axial-vector part of the hadronic current in the Lagrangian (18) has the form

$$J_\alpha^{(A)} = A_{1\alpha} + iA_{2\alpha} + A_{4\alpha} + iA_{5\alpha}, \tag{2.64}$$

the terms of this expression having the same properties as those of Eq. (2.62) with respect to the strangeness and isospin quantum numbers. By analogy with (2.63), one can also write

$$J_\alpha^A = A_\alpha^{\pi+} + A_\alpha^{K+}.$$

The total hadronic current is the sum of the currents (2.62) and (2.64). It contains terms corresponding to transitions with $\Delta S = 0$ and $|\Delta S| = 1$:

$$J_\alpha = J_\alpha^{(0)} + J_\alpha^{(1)}, \tag{2.65}$$

where

$$J_\alpha^{(0)} = V_\alpha^{\pi+} + A_\alpha^{\pi+};$$

$$J_\alpha^{(1)} = V_\alpha^{K+} + A_\alpha^{K+}.$$

We now make the decisive step proposed by Cabibbo [3], namely, the hypothesis that the hadronic weak current J_α is characterized like the vector current (2.65) by a unit length but is obtained from the latter by means of a certain rotation in the unitary space through an angle θ. Then the weak hadronic current J_α is not represented by (2.65) but by

$$J_\alpha = \cos\theta J_\alpha^{(0)} + \sin\theta J_\alpha^{(1)}. \tag{2.66}$$

In principle the parameter θ may be different for the vector and axial-vector parts in the currents $J_\alpha^{(0)}$ and $J_\alpha^{(1)}$. We introduce the angles θ_V and θ_A for the vector and axial-vector parts, respectively. We then obtain

$$J_\alpha = \cos\theta_V V_\alpha^{\pi+} + \cos\theta_A A_\alpha^{\pi+} + \sin\theta_V V_\alpha^{K+} + \sin\theta_A A_\alpha^{K+}.$$

Later we shall return to this question but, as a rule, we shall generally use the variant of the theory with a single Cabibbo angle θ in which the hadronic current is described by the expression (2.66).

Thus, we have a set of baryons with $J^P = 1/2^+$ whose wave functions transform in accordance with an octet representation of $SU_{(3)}$ and make up the matrix (2.60). We also have two sets of currents $J_\alpha^{(V)}$ and

$J_{\alpha}^{(A)}$ defined by (2.62) and (2.64), respectively. The components of these currents also transform in accordance with octet representations of $SU_{(3)}$. For the weak-interaction Lagrangian we obtain

$$L = \frac{G}{\sqrt{2}} \left\{ \bar{u}_e \gamma_\alpha (1+\gamma_5) u_{\nu_e} + \bar{u}_\mu \gamma_5 (1+\gamma_5) u_{\nu_\mu} \right\}$$
$$\times \left\{ \cos\theta_V V_\alpha^{\pi^+} + \cos\theta_A A_\alpha^{\pi^+} + \sin\theta_V V_\alpha^{K^+} + \sin\theta_A A_\alpha^{K^+} \right\}.$$

For the hadronic part of the matrix element due to the action of the n-th component of the current $F_{m\alpha} + A_{m\alpha}$, we have an expression of the form

$$\langle \bar{B} | F_{m\alpha} + A_{m\alpha} | A \rangle = i f_{ABm} \Phi_\alpha + d_{ABm} \Delta_\alpha, \qquad (2.67)$$

This is a generalization of the well-known Wigner-Echart theorem in the theory of spatial rotations. The quantities Φ_α and Δ_α play the role of reduced matrix elements and the tensors f_{ABm} and d_{ABm} the role of the Clebsch-Gordan coefficients in the $SU_{(3)}$ formalism. The presence of two reduced matrix elements corresponds to the existence of f and d type couplings in the interactions of the baryon octet with $J^P = 1/2^+$ and the octet of pseudoscalar mesons.

On the other hand, one can write down matrix elements for the vector and axial-vector parts of the hadronic current in the form

$$\begin{aligned} \langle \bar{B} | F_{m\alpha} | A \rangle &= \bar{u}_B \left[i f_{ABm} V^\Phi + d_{ABm} V^\Delta \right] \gamma_\alpha u_A; \\ \langle \bar{B} | A_{m\alpha} | A \rangle &= \bar{u}_B \left[i f_{ABm} A^\Phi + d_{ABm} A^\Delta \right] \gamma_\alpha \gamma_5 u_A, \end{aligned} \qquad (2.68)$$

where V^Φ, V^Δ, A^Φ, and A^Δ determine the contributions of the f and d type couplings to the vector and axial-vector interactions. Comparing Eqs. (2.67) and (2.68), we obtain the structure of the reduced matrix elements in the form

$$\left.\begin{aligned} \Phi_\alpha &= \bar{u}_B (V^\Phi \gamma_\alpha + A^\Phi \gamma_\alpha \gamma_5) u_A; \\ \Delta_\alpha &= \bar{u}_B (V^\Delta \gamma_\alpha + A^\Delta \gamma_\alpha \gamma_5) u_A. \end{aligned}\right\} \qquad (2.69)$$

In this terminology the hadronic part X_α of the matrix element for the semileptonic baryon decay process has the form

$$X_\alpha = T(\theta, \Delta S)\, \bar{u}_B (f_m^V \gamma_\alpha + g_m^A \gamma_\alpha \gamma_5) u_A.$$

Here, the vector f_m^V and axial-vector g_m^A form factors are given by the expressions

$$\begin{aligned} f_m^V &= (i f_{ABm} - f_{ABm+1}) V^\Phi + (d_{ABm} + i d_{ABm+1}) V^\Delta; \\ g_m^A &= (i f_{ABm} - f_{ABm+1}) A^\Phi + (d_{ABm} + i d_{ABm+1}) A^\Delta \end{aligned}$$

and

$$T(\theta, \Delta S) = \begin{cases} \cos\theta & \text{for transitions with } \Delta S = 0, \\ \sin\theta & \text{for transitions with } |\Delta S| = 1, \end{cases}$$

and the subscript m is equal to unity in the case $\Delta S = 0$ and 4 in the case $|\Delta S| = 1$.

Allowance for the induced interactions naturally complicates the reduced matrix elements. Instead of (2.69), we shall now have expressions for the reduced matrix elements obtained by the following substitutions:

$$V^{\Phi(\Delta)} \gamma_\alpha \to f_1^{\Phi(\Delta)}(q^2) \gamma_\alpha + \frac{f_2^{\Phi(\Delta)}(q^2)}{m_A + m_B} \sigma_{\alpha\beta} q_\beta + \frac{f_3^{\Phi(\Delta)}(q^2)}{m_A + m_B} q_\alpha; \qquad (2.70)$$

$$A^{\Phi(\Delta)}\gamma_\alpha\gamma_5 \to g_1^{\Phi(\Delta)}(q^2)\gamma_\alpha\gamma_5 + \frac{g_2^{\Phi(\Delta)}(q^2)}{m_A+m_B}\sigma_{\alpha\beta}q_\beta\gamma_5 + \frac{g_3^{\Phi(\Delta)}(q^2)}{m_A+m_B}q_\alpha\gamma_5. \quad (2.71)$$

In these expressions, as in (2.69), $f_i^{\Phi(\Delta)}$ and $g_i^{\Phi(\Delta)}$ denote the contributions from the forces of f and d type to the corresponding form factors. Arguing as in the derivation of (2.7), we obtain an expression for the q^2 dependence of the form factors $f_i^{\Phi(\Delta)}(q^2)$:

$$f_i^{\Phi(\Delta)}(q^2) = f_i^{\Phi(\Delta)}(0)\left[1+\lambda_i^{\Phi(\Delta)}q^2/m_\pi^2\right]$$

and similar expressions for the form factors $g_i^{\Phi(\Delta)}(q^2)$.

2.5. Selection Rules in Leptonic Baryon Decays

In accordance with this formalism for the description of the weak interactions the following selection rules must be satisfied.

For the strangeness $|\Delta S| \leq 1$ and if $|\Delta S| = 1$ then $\Delta Q = \Delta S$. For the isospin $|\Delta I| = 1$ for transitions with $\Delta S = 0$ and $|\Delta I| = 1/2$ for transitions with $|\Delta S| = 1$.

The selection rule $|\Delta S| \leq 1$ can be verified by attempting to discover direct decays of Ξ^0 hyperons into nucleons:

$$\Xi \to Ne\nu_e;$$
$$\Xi \to N\pi.$$

As yet the upper limit for the branching ratios of these decays is small, the combined results giving [64]

$$\frac{\Gamma_\Xi(\Delta S=2)}{\Gamma_\Xi(\text{all})} \lesssim 10^{-3}.$$

The selection rule $\Delta Q = \Delta S$ can be verified by searching for the forbidden β and μ decays of the Σ^+ hyperon. The complete data indicate that

$$\frac{\Gamma(\Sigma^+ \to ne^+\nu_e)}{\Gamma(\Sigma^+ \to \text{all})} \lesssim 10^{-4}$$

and

$$\frac{\Gamma(\Sigma^+ \to n\mu^+\nu_\mu)}{\Gamma(\Sigma^+ \to \text{all})} \lesssim 5\cdot 10^{-2}.$$

These estimates take into account the following data. Altogether in experiments with bubble chambers and photoemulsions a total of about $2.3\cdot 10^6$ Σ^+ hyperons have been detected. Three events have been regarded as possible candidates for the decay $\Sigma^+ \to ne^+\nu_e$ and one event as a candidate for the decay $\Sigma^+ \to ne^+\nu_e$. For Σ^- hyperons there have been detected 174 of the decays $\Sigma^- \to n\mu^-\tilde{\nu}_\mu$ and 881 of the decays $\Sigma^- \to ne^-\tilde{\nu}_e$.

Finally, these decays should not exhibit effects due to neutral leptonic currents. To verify this assertion searches have been made for decays of the form

$$\Sigma^+ \to pe^+e^-.$$

Three events corresponding to these decays have been found with small effective masses of the e^+e^- system. It follows that

$$\frac{\Gamma(\Sigma^+ \to pe^+e^-)}{\Gamma(\Sigma^+ \to \text{all})} \approx 10^{-5}.$$

However, it is known that there is a radiative Σ^+ decay with branching ratio

$$\frac{\Gamma(\Sigma^+ \to p\gamma)}{\Gamma(\Sigma^+ \to \text{all})} = (1.6 \pm 0.3) \cdot 10^{-3}.$$

Then with allowance for the internal conversion coefficient, which is approximately equal to 1/130, and eliminating the contribution from this effect, we have

$$\frac{\Gamma(\Sigma^+ \to pe^+e^-)}{\Gamma(\Sigma^+ \to \text{all})} \lesssim 10^{-5}.$$

We can thus assume with reasonable accuracy that the above selection rules hold and are satisfied.

2.6. Conserved Vector Current Hypothesis and Semileptonic Baryon Decays

The expressions (2.70) and (2.71) show that in the most general case a semileptonic baryon decay is now described by 24 real constants in the form factors $f_i^{\Phi}(q^2)$, $f_i^{\Delta}(q^2)$, $g_i^{\Phi}(q^2)$ and $g_i^{\Delta}(q^2)$ and also the 12 λ parameters in their q^2 dependence and the two values of the Cabibbo angle θ_V and θ_A for the vector the axial-vector parts of the weak interaction. The universal constant of the weak interaction is taken from the muon decay data. If the semileptonic hadron processes are T invariant, there remain 12 form-factor constants. This is, of course, very many for a comparison with experimental data. It is therefore necessary to invoke additional hypotheses to obtain equations between these constants and thus reduce the number of constants to be determined experimentally.

In the understanding of effects due to the weak interaction great importance also attaches to the consequences of the hypothesis of a conserved vector current (CVC). This hypothesis was first formulated by Zel'dovich and Gershtein [53] and also Marshak and Sudarshan [54] and was then reformulated by Feynman and Gell-Mann [55]. Its original content reflects attempts to explain the rather good agreement between the constant of the vector coupling g_β^V measured in nuclear Fermi transitions with the Fermi coupling constant G obtained from the muon lifetime. Since it is known that

$$\delta G^2 = \frac{G^2 - (g_\beta^V)^2}{G^2} \approx 0.05,$$

this coincidence can be explained by assuming that: 1) the constant of the vector coupling g^V in the weak hadronic interactions is not renormalized by the strong interactions; 2) the value of this unrenormalized constant g^V is equal to the value of the muon decay constant G^μ.

The constant of the axial-vector coupling g^A is renormalized by strong interactions and for the neutron β decay the ratio of the two constants is $g^A/g^V = -1.23 \pm 0.01$.

The universality of the weak interactions follows from the second assertion. The first assertion yields an important consequence for the properties of the vector part of the hadronic current $J_\alpha^V(x)$ as an operator acting in the world of strong interactions. Here, one can employ an analogy with electrodynamics, in which, under the assumption that the unrenormalized electric charges of the electron and proton are equal, the equality of the renormalized charges is a consequence of the conservation of the electric current. By analogy with electrodynamics it also follows that g^V is not renormalized because of the conservation of the vector part of the weak hadronic current, $\partial_\alpha V_\alpha(x) = 0$. In the most general form we have the hypothesis of the isovector nature of the hadronic current, according to which the three operators V_α, $\left(\frac{g_V^\beta}{e}\right)(J_\alpha^{\text{e.m.}})_{I=1}$, and V_α^* form three components of a single isotopic vector. These ideas enable one to relate a number of electromagnetic effects and weak-interaction effects. Important conclusions are drawn about the form factors for $q^2 = 0$ and also about their q^2 dependence. We shall mention only the well-known consequences for the neutron β decay, according to which the form factor of the effective scalar $f_3(0)$ must vanish and $f_1(0) = 1$, $f_2(0) = \mu_p - \mu_n \approx 3.7$.

In addition

$$f_1(q^2) = f_1(0)\, G_Q^V(q^2);$$

$$f_2(q^2) = f_2(0)\, G_M^V(q^2),$$

where $G_Q^V(q^2)$ and $G_M^V(q^2)$ are the isovector parts, respectively, of the charge and magnetic form factors of the nucleons with the normalization $G_Q^V(0) = G_M^V(0) = 1$. We obtain an estimate of the q^2 dependence from the known data on this dependence for the Sachs nucleon form factors found from experiments on elastic ep scattering. It is known [55] that

$$\frac{G_{MP}}{1+\mu_p} = \frac{G_{Mn}}{\mu_n} = G_{EP} = \left[\frac{1}{1+1.25\,q^2/m_p^2}\right]^2,$$

$$\left(\frac{dG_{En}}{dq^2}\right)_{q^2=0} = \frac{(0.563 \pm 0.001)}{m_p^2}.$$

Hence, we find that the form factors of the vector and weak magnetism satisfy

$$\lambda_1 \approx 2; \quad \lambda_2 \approx 3. \tag{2.72}$$

The CVC hypothesis yields important information about other decays of baryons with $\Delta S = 0$. Consider the decays

$$\Sigma^{\pm} \to \Lambda e^{\pm} \nu_e. \tag{2.73}$$

For the matrix elements of the isovector part of the hadronic current we have the expressions

$$\left.\begin{aligned}\langle \Lambda | \bar{J}_\alpha^{+} | \Sigma^{+} \rangle &= -\bar{u}_\Lambda \sum_k g_k O_\alpha^{(k)} u_{\Sigma^+};\\ \langle \Lambda | J_\alpha^{+} | \Sigma^{-} \rangle &= \bar{u}_\Lambda \sum_k g_k O_\alpha^{(k)} u_{\Sigma^-}.\end{aligned}\right\} \tag{2.74}$$

These decays are due to two different isotopic components of one and the same weak baryon current and one can therefore immediately conclude that their matrix elements are the same and that the branching ratio of these decays is determined solely by the difference of the phase volumes due to difference of the masses of the Σ^- and Σ^+ hyperons. Consequently,

$$\Gamma_+/\Gamma_- = 0.61. \tag{2.75}$$

The experimentally found relative probabilities for the decays (2.73) R_+ and R_- and the hyperon lifetime τ_+ and τ_- yield the following value for this ratio:

$$\frac{\Gamma_+}{\Gamma_-} = \frac{R_+ \tau_-}{R_- \tau_+} = 0.62 \pm 0.15, \tag{2.76}$$

which does not contradict the calculated value (2.75) and confirms this general assertion of the hypothesis of a conserved vector current.

The form factors g_k in (2.74) are again related to the form factors of the electromagnetic transition $\Sigma^0 \to \Lambda^0\gamma$:

$$\langle \Lambda | J_\alpha^{e.m} | \Sigma^0 \rangle = \bar{u}_\Lambda \sum_{k=V,T,S} F_k O_\alpha^{(k)} u_{\Sigma^0}. \tag{2.77}$$

This relationship has the form

$$\left.\begin{aligned} g_V(q^2) &= \sqrt{2}\,F_V(q^2) = q^2 \sqrt{2}\,F_V' \to 0;\\ g_S(q^2) &= \sqrt{2}\,F_S(q^2) = -\sqrt{2}\,(m_\Sigma^2 - m_\Lambda^2)\,F_V';\\ g_T(q^2) &= \sqrt{2}\,F_T(q^2).\end{aligned}\right\} \tag{2.78}$$

Here, F_V' is the analog of R_e^2/G in the nucleon electromagnetic form factors when R_e is regarded as the electromagnetic radius of the particle. The tensor form factor $F_T(0)$ determines the magnetic moment of the radiative transition $\Sigma^0 \to \Lambda^0\gamma$ and can be found from the probability of this transition:

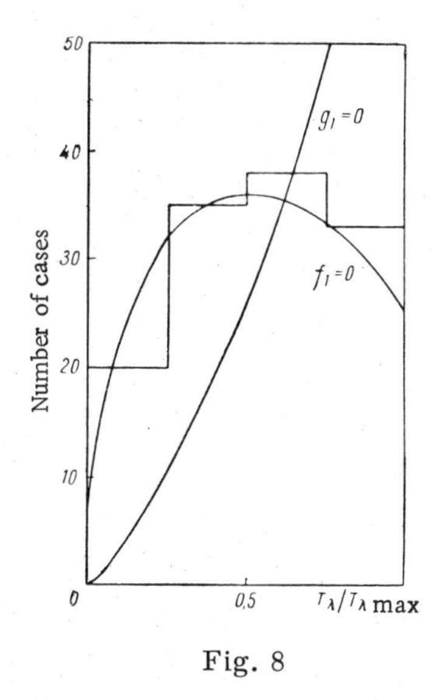

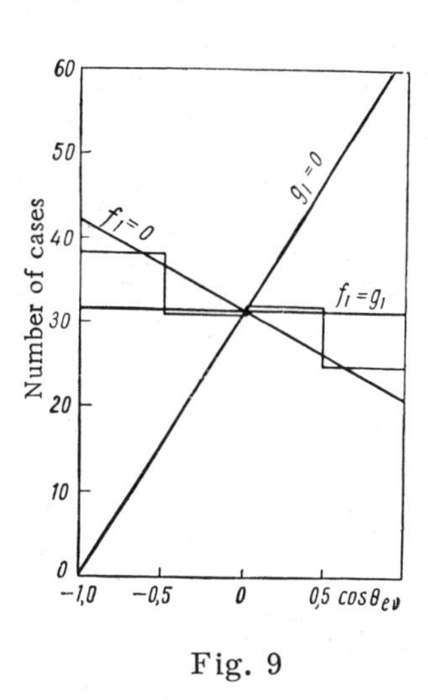

Fig. 8 Fig. 9

Fig. 8. Energy spectrum of the Λ hyperons in the decay $\Sigma\to\Lambda e\nu$. The predictions for the case of purely vector ($q_1=0$) and purely axial-vector ($f_1=0$) interactions are indicated.

Fig. 9. The $e\nu$ correlation in the decay $\Sigma\to\Lambda e\nu$. The predictions for the purely vector ($q_1=0$) and purely axial-vector ($f_1=0$) variants and their equal-probability mixture ($q_1=f_1$) are also shown.

$$\Gamma(\Sigma^0\to\Lambda^0\cdot\gamma)=\frac{e^2}{4\pi}\cdot\frac{4E_\gamma^3}{(m_\Sigma+m_\Lambda)^2}F_T^2(0). \tag{2.79}$$

It follows from the relations (2.78) that, within the framework of the CVC hypothesis, the following limiting relation is satisfied in the decays $\Sigma^\pm\to\Lambda e^\pm\nu_e$:

$$g_V(0)=f_1(0)=0. \tag{2.80}$$

Thus, the form factor of the vector interaction here vanishes and the properties of this decay are primarily determined by the axial-vector interaction. This consequence is confirmed by the experimentally measured [57] energy spectra of the Λ hyperons (Figs. 8 and 9) and the $e\nu$ correlation in this decay. From the combined data obtained from the experiments one can conclude that in the decays (2.73) $|f_1/g_1|=0.26\pm0.20$.

Above [see (2.44)-(2.49)] we have discussed the possible contribution to the Σ^+ decay from the form factor g_2. It follows from theoretical considerations that this form factor vanishes. It is therefore necessary to consider what one can expect here from the weak magnetism effect [58]. In fact, the value of the form factor $f_2(0)$ can, in accordance with (2.79), be obtained from measurements of the lifetime of the Σ^0 hyperon. However, such results are not yet available and one can therefore use the theoretical estimates of τ_{Σ^0}, which give values of $\sim 7\cdot10^{-20}$ sec. Then, using (2.79), we obtain $f_2|_{\Sigma\to\Lambda}=2.98\cos\theta_V$.

From a combined analysis of the data on the semileptonic baryon decays it follows that $g_1|_{\Sigma\to\Lambda}\approx0.618\cos\theta_A$. Setting $\theta_A=\theta_V$, we find $|f_2/g_1|_{\Sigma\to\Lambda}\approx4.83$. This means there is a contribution of $\sim0.5\%$ to the total probability of the decay $\Sigma\to\Lambda e\nu$ from the weak magnetism effect. Another way to estimate the value of this form factor is to attempt to detect a longitudinal polarization of the Λ^0 hyperons. If $f_2(0)\neq0$, then, taking into account the recoil terms, we find that (2.24) is replaced by the following expression for the angular distribution of the protons from the decay $\Lambda^0\to pe^-\tilde{\nu}_e$:

$$W(\mathbf{p}\cdot\boldsymbol{\alpha})=1+\alpha_\Lambda\frac{2.68\,\mathrm{Re}\,z+0.08\,\mathrm{Re}\,z'}{|z|^2+3}\,\mathbf{p}\cdot\boldsymbol{\alpha},$$

where $z'=f_2/g_1$ and $\alpha_\Lambda=0.62$.

Since we assume z = 0, this distribution will be slightly anisotropic because of the term 0.08 Re z'. It can be seen that the asymmetry coefficient in this distribution is very small: for Re z' ≈ 1 we have

$$\alpha_\Lambda \frac{0.08 \operatorname{Re} z'}{3} = 6.2 \cdot 10^{-3}.$$

The distribution of the longitudinal polarization of the Λ hyperon in the decay $\Sigma \to \Lambda e\nu$ is shown in Fig. 10.

2.7. Parameters of the Cabibbo Theory from the Experimental Data on Semileptonic Baryon Decays

First a few words on the T invariance of β-decay processes. If this invariance holds, the form factor constants must be real. Restricting ourselves now to the form factors of the vector and axial-vector interactions, we can assert that in this case too the constants c_V and c_A must be real, i.e., if one writes $\alpha = c_A/c_V = |\alpha| e^{i\varphi}$ in the V−A variant of the theory, the phase must be $\varphi = 180°$. One can verify this consequence experimentally in several ways. For example, there are T-odd correlations of the form $\bar{\sigma}_R \cdot (\bar{p}_e \times \bar{p}_\nu)$ with coefficient equal to $\frac{2 \operatorname{Im} \alpha}{1+3|\alpha|^2}$. Measurements of this coefficient in the decay of polarized neutrons yielded the result [59]: $\varphi - \pi = (1.3 \pm 1.3)°$. It also follows from the data [60] of the β decay of Ne^{19} that $\varphi - \pi = (0.2 \pm 1.6)°$.

One can also attempt to detect T-odd effects by making precise measurements of the energy dependence of the electron polarization,in, for example, the β decay of the nucleus RaE (Bi^{210}). This dependence is sensitive to the value of $\operatorname{Re} \alpha/\alpha$. It follows from the measurements [61] that $\varphi - \pi = (2.0 \pm 2.5)°$. Thus if an admixture of a T-odd weak interaction is present in β decay processes, it is small. In what follows we shall, as a rule, assume that the semileptonic baryon decay processes are described by T-invariant interactions.

In the situation that arises with contributions from induced interactions it is natural to see what the $SU_{(3)}$ and $SU_{(6)}$ symmetries and the hypothesis of a conserved isovector current yield. In this case only f type forces make a contribution to the form factor of the vector interaction. It follows that $V^\Phi = 1$ and $V^\Delta = 0$ in the expressions (2.69). At the same time we find that the weak-magnetism form factor f_m^M is independent of q^2 and satisfies the formula

$$f_m^M = (\mu_A - \mu_B)\left[i f_{ABm}(1 - \alpha_m) + d_{ABm}\alpha_m\right] \frac{2m_A}{m_A + m_B},$$

where $\alpha = \frac{3}{2} \cdot \frac{\mu_A}{\mu_A + \mu_B}$ are the magnetic moments of the corresponding baryons that participate in the process $A \to B l \nu_l$. Their values are known for the Λ^0 and Σ^+ hyperons as well as for the proton and neutron. They do not differ strongly form the values predicted by $SU_{(6)}$ symmetry. We can now therefore use their theoretical expression [62] in terms of μ_p and μ_n. As a result we obtain the form factors f_1, f_2, and g_1, which are given in Table 5 for different semileptonic decays. They have a particularly simple form for the decays (2.73). Using them in the expression (2.32) for the probability, we find that

$$R^- = 1.82 \cdot 10^{-4} \cos^2\theta_A \frac{2}{3} (A^\Delta)^2;$$

$$R^+ = 0.54 \cdot 10^{-4} \cos^2\theta_A \frac{2}{3} (A^\Delta)^2.$$

Using the tabulated data for R^- and R^+, we obtain

$$\cos\theta_A A^\Delta \approx 0.75.$$

It follows form the neutron β decay that

$$\cos\theta_A (A^\Phi + A^\Delta) = 1.23,$$

and then

$$\cos\theta_A \cdot A^\Phi \approx 0.5,$$

TABLE 5. Form Factors of the Vector f_V, Weak Magnetism f_M, and Axial-Vector f_A Interactions in $SU_{(3)}$ Symmetry*

Reaction	ΔS	f_V	f_A	f_M
$n \to pe\nu$	0	1	$A^{\Phi} + A^{\Delta}$	$\frac{1}{2} k_{np} (\mu_p - \mu_n)$
$\Sigma^- \to \Lambda e\nu$	0	0	$-\sqrt{2/3}\, A^{\Delta}$	—
$\Sigma^+ \to \Lambda e\nu$	0	0	$-\sqrt{2/3}\, A^{\Delta}$	$-\frac{1}{2} k_{\Sigma\Lambda} \mu_n$
$\Sigma^- \to \Sigma^0 e\nu$	0	$\sqrt{2}$	$\sqrt{2}\, A^{\Phi}$	—
$\Xi^- \to \Xi^0 e\nu$	0	-1	$A^{\Phi} - A^{\Delta}$	—
$\Lambda \to pe\nu$	1	$\sqrt{3/2}$	$\sqrt{3/2}\left(A^{\Phi} + \frac{1}{3} A^{\Delta}\right)$	$\frac{1}{2}\sqrt{\frac{3}{2}}\, k_{\Lambda p} \mu_p$
$\Sigma^- \to ne\nu$	1	-1	$A^{\Phi} - A^{\Delta}$	$\frac{1}{2} k_{\Sigma n} (\mu_p + 2\mu_n)$
$\Xi^- \to \Lambda e\nu$	1	$-\sqrt{3/2}$	$\sqrt{3/2}\left(A^{\Phi} - \frac{1}{3} A^{\Delta}\right)$	$\frac{1}{2} k_{\Xi\Lambda} (\mu_p + \mu_n)$
$\Xi^- \to \Sigma^0 e\nu$	1	$\sqrt{1/2}$	$\sqrt{1/2}(A^{\Phi} + A^{\Delta})$	—
$\Xi^0 \to \Sigma^+ e\nu$	1	1	$A^{\Phi} + A^{\Delta}$	—

*
$$k_{AB} = \begin{cases} 1 & \text{for exact } SU_{(3)} \text{ symmetry,} \\ \frac{2m_A}{m_A + m_B} & \text{for broken } SU_{(3)} \text{ symmetry,} \end{cases}$$

and hence

$$\alpha = \frac{A^{\Delta}}{A^{\Phi} + A^{\Delta}} \approx 0.61.$$

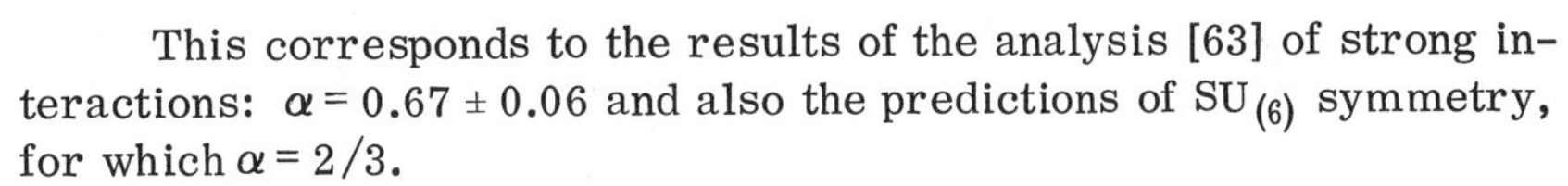

This corresponds to the results of the analysis [63] of strong interactions: $\alpha = 0.67 \pm 0.06$ and also the predictions of $SU_{(6)}$ symmetry, for which $\alpha = 2/3$.

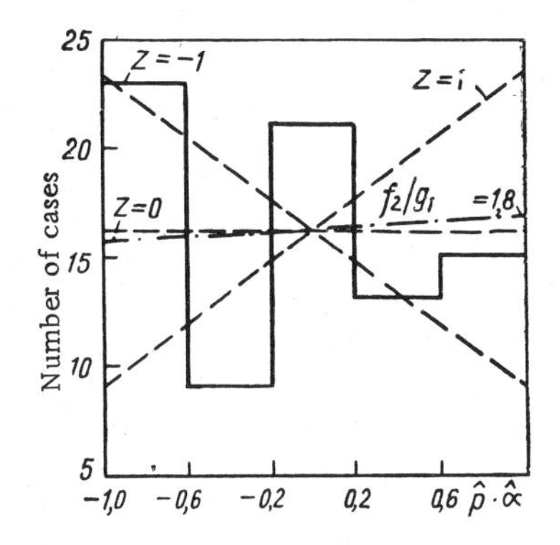

Fig. 10. Distribution of the longitudinal polarization of the Λ hyperons in the decay $\Sigma \to \Lambda e\nu$. The influence of the weak magnetism form factor is shown (the straight line $f_2/q_1 = 1.8$).

A comprehensive statistical analysis [64] of the data on the probabilities of semileptonic baryon decays confirms these estimates. Fitting of data with two Cabibbo angles, the consequences of the CVC hypothesis, and the ideas of current algebra on the relationship between the axial-vector and pseudoscalar form factors yield the results

$$\sin\theta_V = 0.190 \pm 0.035;$$
$$\sin\theta_A = 0.280 \pm 0.030;$$
$$\alpha = 0.66 \pm 0.03.$$

Hence, we obtain the following estimates for the limiting values of the renormalized effects in the axial-vector form factors ($\beta = g_1/f_1$):

$$\beta_{\Sigma\Lambda} \approx \beta_{np} = 1.23;$$
$$(1.38 \pm 0.13) \leqslant \beta_{\Lambda p} \leqslant (1.64 \pm 0.16);$$
$$(1.1 \pm 0.2) \leqslant \beta_{\Sigma^- n} \leqslant (4.4 \pm 0.7);$$
$$(2.3 \pm 1.1) \leqslant \beta_{\Xi^- \Lambda} \leqslant (9.1 \pm 4.2).$$

We have already mentioned on several occasions that the present-day data on the quantities $\Gamma(A \to B l \nu_l)$ are characterized by large experimental errors. Attempts are therefore made to estimate the parameters of the theory on the basis of the complete set of experimental data using the maximum of resonable theoretical arguments. A typical approach of this kind has been developed by the group at Heidelberg University [65]. In this approach the strongly interacting part of the matrix element is used in the form (2.70) and one assumes a q^2 dependence of the form factors:

$$\left. \begin{array}{l} f_i(q^2) = f_i(0)\left[1 + \lambda_1 q^2/m_\pi^2\right]; \\ f_4 = g_1;\ f_5 = g_2;\ f_6 = g_3, \end{array} \right\} \tag{2.81}$$

where $f_i(0)$ is represented in accordance with (2.70) as a sum of reduced matrix elements $f_i^{\Phi(\Delta)}(0)$ with corresponding Clebsch-Gordan coefficients of the group $SU_{(3)}$:

$$f_i^{(k)}(0) = \left[C_F^{(k)} f_i^{(F)}(0) + C_D^{(k)} f_i^{(D)}(0)\right] T_i^{(k)}. \tag{2.82}$$

Here, we again have

$$T_i^{(k)} = \begin{cases} \cos\theta_i & \text{for } \Delta S = 0, \\ \sin\theta_i & \text{for } |\Delta S| = 1, \end{cases} \tag{2.83}$$

and the Cabibbo angle

$$\theta_i = \begin{cases} \theta_V & \text{for } i = 1, 2, 3, \\ \theta_A & \text{for } i = 4, 5, 6. \end{cases} \tag{2.84}$$

The factor $\eta_i^{(k)}$ takes into account the corrections for the difference between the baryon masses of the $SU_{(3)}$ octet with $J^P = 1/2^+$:

$$\eta_i^{(k)} = \begin{cases} 1 & \text{for } i = 1,4 \\ \dfrac{2m_A^{(k)}}{m_A^{(k)} + m_B^{(k)}} & \text{for } i = 2, 3, 5, 6. \end{cases} \tag{2.85}$$

Exact $SU_{(3)}$ symmetry yields

$$f_3^{\Phi}(0) = f_5^{\Delta}(0) = 0, \tag{2.86}$$

and the CVC hypothesis yields the following restrictions on the vector form factor and the weak-magnetism form factor:

$$\begin{array}{c} f_1^{\Phi}(0) = 1;\ f_1^{\Delta}(0) = 0; \\ f_2^{\Phi}(0) = \dfrac{1}{2}\mu_p + \dfrac{1}{2}\mu_n;\ f_2^{\Delta}(0) = -\dfrac{3}{4}\mu_n; \\ \lambda_1 = 2{,}0\dfrac{m_A^2}{m_p^2};\ \lambda_2 = 2{,}6\dfrac{m_A^2}{m_p^2}. \end{array} \tag{2.87}$$

Use of the PCAC hypothesis relates the form factors $f_6^{(k)}$ and $f_4^{(k)}$:

$$\left. \begin{array}{l} f_6(0) = \dfrac{m_A(m_A + m_B)}{m_{\Delta S}^2} f_4(0); \\ \lambda_6 = \lambda_4 - \dfrac{m_p^2}{m_{\Delta S}^2}, \end{array} \right\} \tag{2.88}$$

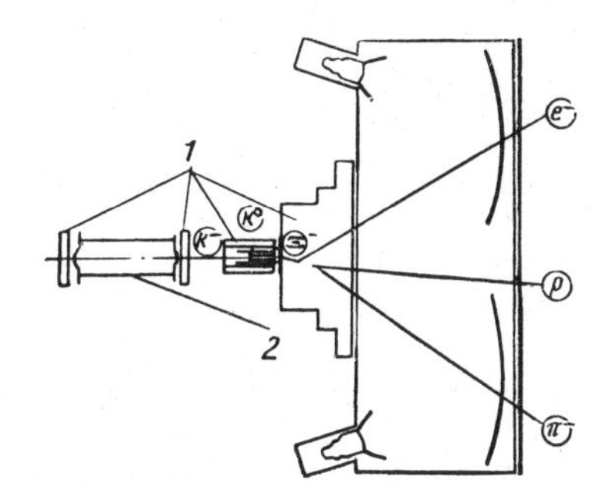

Fig. 11. Arrangement of an experimental electronic apparatus for studying the semileptonic decays of Ξ^- hyperons. The apparatus includes spark chambers (1), a large Cerenkov counter (2), and a large system of scintillation counters.

where

$$m_{\Delta S}=\begin{cases} m_\pi \text{ for transitions with c } \Delta S=0, \\ m_K \text{ for transitions with c } |\Delta S|=1; \end{cases} \tag{2.89}$$

it is also assumed that $\lambda_4 = \lambda_1$. Thus, only the following parameters are free:

$$\theta_V;\ \theta_A;\ f_4^\Phi(0);\ f_4^\Delta(0);\ f_5^\Phi(0);\ f_5^\Delta(0). \tag{2.90}$$

If we lift the restrictions (2.86), there are a further two parameters $f_3^\Delta(0)$ and $f_3^\Delta(0)$, and one can assume that $\lambda_5 = \frac{1}{2}\lambda_2$.

The statistical analysis is based on the requirement of a minimum of the function $\chi^2(h_i)$ of the free parameters h_i defined by the formula

$$\chi^2(h_i)=\sum_k \left[\frac{x^{(k)}_{\text{theor}}(h_i)-x^{(k)}_{\text{exp}}}{\Delta x^{(k)}_{\text{exp}}}\right]^2, \tag{2.91}$$

where $x^{(k)}$ is either $\Gamma^{(k)}$ or $f_4^{(k)}/f_1^{(k)}$.

Establishment of these experimental quantities entails a very substantial amount of experimental work. Because of the small branching ratio $\Gamma(A\to Bl\nu_l)/\Gamma\ (A\to \text{all})$ of the semileptonic processes the experimental groups must investigate ensembles of Λ^0 or $\Sigma^\pm$ decays with $\sim 10^5-10^6$ events with basically nonleptonic modes of the type $N\pi$ recorded in experiments with bubble chambers. About 1×10^2 events of the requisite type with the creation of leptons are detected. The analysis of the selected events yields information about the branching ratios and also about the spectra and correlations in these processes. A recent innovation is the use of Cerenkov and scintillation counters in conjunction with spark chambers in experiments on the semileptonic decays of the Λ^0 and Ξ^- hyperons. The arrangement of one such experiment is shown in Fig. 11.

More detailed information is now available on the characteristics of the decays $\Lambda^0\to pe\nu_e$ and $\Sigma^-\to ne\nu_e$. A typical electron spectrum obtained for the decay $\Sigma^-\to ne\nu_e$ is shown in Fig. 12. An analysis of the Dalitz plot for the decay $A\to Bl\nu_l$ in the variables of the electron and proton kinetic energies T_e and T_p or the spectra of these quantities does not at present enable one to estimate the contributions of the induced form factors. It is therefore assumed that the weak-magnetism form factor can be taken from $SU_{(3)}$ symmetry and the CVC hypothesis, as is done in Table 5. From these data we then find [51] that in the decay $\Lambda\to pe\nu_e$

$$|g_1/f_1|=0.72^{+0.19}_{-0.14}.$$

Measurements of the asymmetry in the emission of the electrons on the decay of polarized Λ^0 hyperons yield, in accordance with (2.30) and (2.31), not only the value but also the sign of the ratio g_1/f_1. The combined data of different experiments yield the following result:

$$g_1/f_1=-0.63\pm 0.08.$$

Finally, an analysis of the νl correlations yields [57]

$$|g_1/f_1|_{\Lambda\to pe\nu}=0.77^{+0.25}_{-0.17}.$$

Thus, in the decay $\Lambda\to pe\nu_e$ these measurements indicate that the ratio g_1/f_1 is negative and has the value

$$\left.\frac{g_1}{f_1}\right|_{\Lambda\to pe\nu}=-0.61\pm 0.065. \tag{2.92}$$

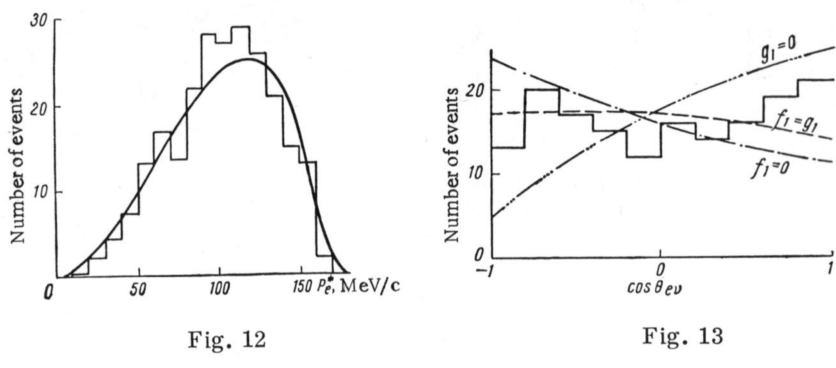

Fig. 12. Electron energy spectrum in the decay $\Sigma^- \to ne\nu$.

Fig. 13. Electron-neutrino correlations in the decay $\Sigma^- \to ne\nu$. The solid line gives the expected spectrum.

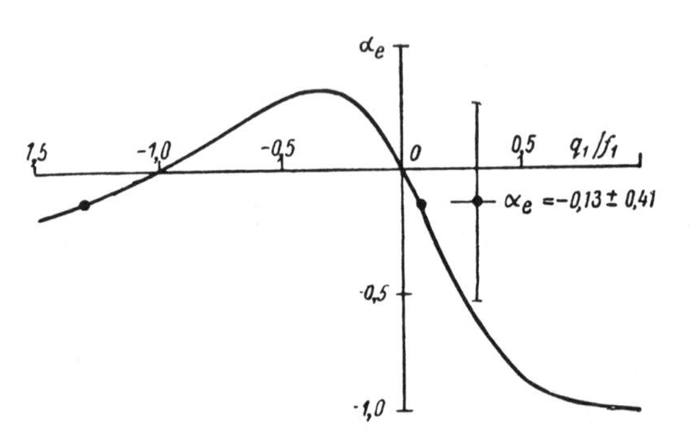

Fig. 14. Asymmetry coefficient for the emission of electrons in the decay of polarized Σ^- hyperons as a function of q_1/f_1. The measured value of α_e is shown.

Now let us consider this ratio in the decay $\Sigma^- \to ne^-\tilde{\nu}_e$. Here the experimental situation is much more complicated since one of the three decay products is charged and the remaining two are neutral. It follows that the direction of emission of the neutron and its energy can be measured only by detecting the recoil proton from np scattering. Approximately 11% of the neutrons from Σ^- decays generate recoil protons in a hydrogen bubble chamber with a track length greater than 2 mm at distances of 20 cm from the decay point. It is therefore very difficult to allow for the background, which account for up to 30% of the orginally chosen candidates for the decay $\Sigma^- \to ne^-\tilde{\nu}_e$.

The combined data at present available on the $e\nu$ correlation are given in Fig. 13. These data indicate that

$$|g_1/f_1|_{\Sigma^-} = 0.36 {+0.18 \atop -0.15}. \tag{2.93}$$

Results have been published of one attempt to measure the asymmetry parameter in the emission of electrons in the decay of polarized Σ^- hyperons. From a total of 49 events the value for this parameter α_e in (2.30) was found to be $\alpha_e = -0.13 \pm 0.41$. Since α_e is a quadratic function of g_1/f_1, two solutions are obtained (this is well shown in Fig. 14):

$$g_1/f_1|_{\Sigma^-} = -0.05 {+0.23 \atop -0.32}$$

or

$$g_1/f_1|_{\Sigma^-} = -1.3 {+0.9 \atop -1.0}.$$

Thus, for $|g_1/f_1|$ in the decay $\Sigma^- \to ne^-\tilde{\nu}_e$ one should now take the value (2.93), but nothing can be said about the sign of the ratio.

The examples we have given characterize the present state of the study of semileptonic baryon decays. The experimental data at present known on this subject are given in Table 6.

It follows from [65] that at the present level of accuracy of the experimental data the contribution from the form factors f_3 and f_5 cannot be determined. It was therefore assumed that $f_3 = f_5 = 0$ and the set of parameters was determined from Eq. (2.90). For $\theta_V = \theta_A = \theta$ [57] the following results were obtained [64]: $\theta = 0.235 \pm 0.06$; $f_4^\Phi(0) = 0.49 \pm 0.02$; $f_4^\Delta = 0.74 \pm 0.02$; $\alpha = 0.60 \pm 0.02$. Fitting for $\theta_V \neq \theta_A$ gives the same result for $f_4^\Phi(0)$ and $f_4^\Delta(0)$ and also $\theta_V = 0.233 \pm 0.012$ and $\theta_A = 0.238 \pm 0.018$.

TABLE 6. Comparison of Experimental Data on Leptonic Baryon Decays and Predictions of the Cabibbo Theory

Decay	Branching ratio $[\Gamma(A\to B+l+\nu_e)/\Gamma(A\to \text{all})]\times 10^4$		Form-factor ratio g_1/f_1	
	Experiment	Cabibbo theory*	Experiment	Cabibbo theory*
$n\to pe\nu$	—	—	$-1,23\pm 0,01$	$-1,227$
$\Sigma^-\Lambda e\nu$	$0,604\pm 0,06$	$0,62$	$f_1/g_1=+0,35\pm\pm 0,18$	$f_1/g_1=0$
$\Sigma^+\to\Lambda e\nu$	$0,202\pm 0,047$	$0,19$	—	$f_1/g_1=0$
$\Lambda\to pe\nu$	$8,60\pm 0,45$	$8,74$	$-0,77^{+0,13}_{-0,09}$	$0,72$
$\Lambda\to p\mu\nu$	$1,35\pm 0,60$	$1,44$	—	$0,72$
$\Sigma^-\to ne\nu$	$10,92\pm 0,43$	$10,6$	$\lvert g_1/f_1\rvert=0,21\pm\pm 0,21$	$-0,31$
$\Sigma^-\to n\mu\nu$	$4,5\pm 0,5$	$5,0$	—	$-0,31$
$\Xi^-\to\Lambda e\nu$	$15,0^{+9,0}_{-6,0}$	$5,4$	—	$0,20$
$\Xi^-\to\Sigma^0 e\nu$	$6,2^{+2,0}_{-3,0}$	$0,09$	—	$1,23$

*The constants of the Cabibbo theory are: $\theta = 0.242 \pm 0.004$; $g_1^{\Phi}(0) = 0.46 \pm 0.02$; $g_1^{\Delta}(0) = 0.77 \pm 0.02$.

The theoretical predictions that follow from this analysis are given in Table 6 and compared with the experimentally measured parameters. The results of a graphical analysis are given in Fig. 15.

It can be seen that there is basically good agreement between the experimental data and the predictions of the Cabibbo theory. However, a great improvement is required in the experimental data on semileptonic baryon decays.

To analyze this question one can also invoke the more far reaching additional theoretical argument based on the ideas of current algebra. An example of such an investigation can be found in [66].

We define in the following manner the constants

$$(G_A)_{pn} = g_{pn}(0)\cos\theta_A;$$
$$(G_V)_{pn} = f_{pn}(0)\cos\theta_V;$$
$$(G_A)_{p\Lambda} = g_{p\Lambda}(0)\sin\theta_A;$$
$$(G_V)_{p\Lambda} = f_{p\Lambda}(0)\sin\theta_V$$

and in an obvious manner all the remaining constants $(G_{A(V)})_{BA}$. Here, the form factors $g_{BA}(0)$ and $f_{BA}(0)$ are taken from the observable constants, for example,

$$\frac{G_\mu}{\sqrt{2}} g_{p\Lambda}(0)\sin\theta_A,$$

$$\frac{G_\mu}{\sqrt{2}} f_{p\Lambda}(0)\sin\theta_V.$$

Using the CVC hypothesis, we obtain $f_{pn}(0)=1$, $f_{p\Lambda}(0)=-\sqrt{\frac{3}{2}}$, etc. (see Table 5). Using current algebra arguments one obtains the following sum rules for the axial-vector constants:

$$g_{\Lambda\Sigma}=\sqrt{\frac{3}{2}}g_{pn}+g_{p\Lambda};\quad g_{\Lambda\Xi}=g_{\Lambda\Sigma}-\sqrt{\frac{3}{2}}g_{\Xi^0\Xi^-};$$

$$g_{\Xi^0\Xi^-}=\sqrt{\frac{3}{2}}g_{\Lambda\Sigma}+\sqrt{\frac{1}{2}}g_{\Sigma^+\Sigma^0};\quad g_{n\Sigma^-}=\sqrt{6}g_{\Lambda\Sigma}-g_{pn};$$

$$g_{\Sigma^+\Sigma^0}=\sqrt{6}g_{\Lambda\Sigma}-g_{\Xi^0\Xi^-};\quad g_{\Sigma^\pm\Sigma^0}=\pm(\sqrt{3}g_{\Lambda\Sigma}-\sqrt{2}g_{pn});$$

where $g_{\Lambda\Sigma}=g_{\Sigma^+\Lambda}=g_{\Sigma^-\Lambda}$.

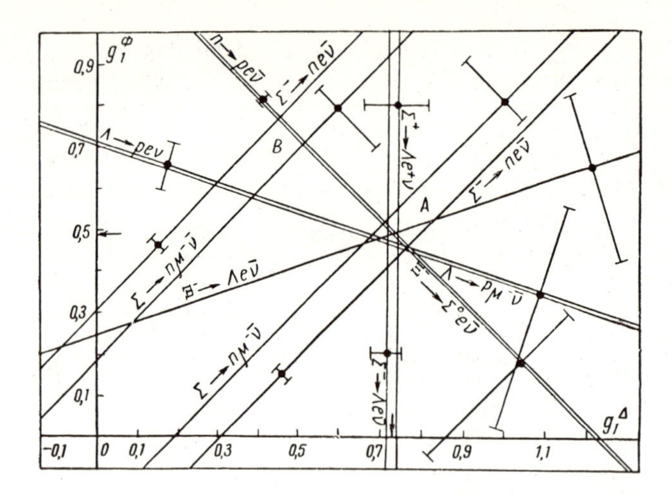

Fig. 15. Graphical representation of the data on the probabilities of semileptonic baryon decays $A \to B + l + \nu_l$ in the plane of the parameters $q_1^{\Phi}(0)$ and $q_1^{\Delta}(0)$ for the Cabibbo angle $\theta = 0.235$ (the arrows indicate their most probable values, which were used to obtain the predictions given in Table 6).

We shall now write down an expression for the partial probability of the process $A \to Bl\nu_l$ as follows:

$$\Gamma_{BA} = \frac{G_\mu^2 f \Delta}{60\pi^3} 3 (G_A)_{BA}^2 \left[\left(1 - \frac{3}{2}\beta + \frac{4}{7}\beta^2\right)(1+a) + \frac{1}{3} x_{BA}^2 \left(1 - \frac{3}{2}\beta + \frac{6}{7}\beta^2\right)(1+b)\right],$$

where

$$\beta = \frac{m_A - m_B}{m_A}; \quad \Delta = m_A - m_B; \quad x_{BA} = \frac{(G_V)_{AB}}{(G_A)_{BA}}$$

and

$$f = \begin{cases} 0.47 & \text{for the decay} \quad n \to pev, \\ 1 & \text{for decays with} \quad |\Delta S| = 1. \end{cases}$$

The factors a and b take into the account the finite mass of the charged lepton in m_l. In what follows we shall assume for the case of electronic decays that $a = b = 0$. The sum rule

$$g_{p\Lambda} = g_{\Sigma^+\Lambda} - \sqrt{\frac{3}{2}}\, g_{pn}$$

yields

$$\operatorname{tg} \theta = \frac{(G_A)_{p\Lambda}}{(G_A)_{\Lambda\Sigma^+} - \sqrt{\frac{3}{2}}\,(G_A)_{pn}}.$$

On the other hand,

$$\operatorname{tg} \theta_V = \frac{x_{p\Lambda} (G_A)_{p\Lambda}}{f_{p\Lambda}},$$

where

$$f_{p\Lambda}=-\sqrt{\frac{3}{2}}.$$

Thus, knowledge of the values of Γ_{fi} and x_{fi} enables one to determine the angles θ_V and θ_A. For example, using the experimental data on the branching ratios of the decays $\Lambda \to pe\nu_e$ and $\Sigma^- \to \Lambda e\nu_e$, and also the value of g_1/f_1 for the decay $\Lambda \to pe\nu_e$, we obtain $\theta_V = 0.225 \pm 0.02$.

From the neutron lifetime (2.51) we now have $|x_{pn}|^{-1} = 1.23 \pm 0.02$, which also corresponds to the data from the asymmetry measurement, from which it follows that $|x_{pn}|^{-1} = 1.25 \pm 0.04$. We then obtain

$$\sin\theta_A = 0.225 \begin{matrix} +0.055 \\ -0.05 \end{matrix}.$$

Thus, in this analysis the values of the angles θ_V and θ_A are equal and for the corresponding constants we obtain the values $g_{pn} = 1.27$, $g_{p\Lambda} = 0.945$, and $g_{\Lambda\Sigma} = 0.618$. In addition, we find $\alpha = (A^\Delta)/A^\Delta + A^\Phi = 0.58$. There are also predictions for the other axial-vector constants:

$$g_{\Sigma^+\Xi^0} = 1.27;\quad g_{n\Sigma^-} = 0.244;\quad g_{\Lambda\Xi^-} = 0.319;\quad g_{\Sigma^\pm\Sigma^0} = \mp 0.726;$$

$$g_{\Xi^0\Xi^-} = 0.244.$$

Allowance for the electromagnetic corrections slightly changes the values of the above constants. For example, if one takes into account only $\pi^0\eta^0$ and $\Sigma^0\Lambda^0$ mixing, the above constants are replaced by modified sum rules from which it then follows that

$$\sin\theta_V = 0.229 \pm 0.02,$$

$$\sin\theta_A = 0.230 \begin{matrix} +0.055 \\ -0.05, \end{matrix}$$

$$|x_{n\Sigma^-}|^{-1} = \begin{cases} \text{either} & -0.38 \pm 0.2, \\ \text{or} & -0.28 \pm 0.2. \end{cases}$$

Similar but less reliable results can be obtained by using other sum rules and corresponding experimental data whose accuracy is less than that of those used above. In this approach the values of the Cabibbo angles θ_V and θ_A are also found to be almost equal.

Since a comparison of the data on mesonic decays with $|\Delta S| = 1$ and $\Delta S = 0$ indicate that the values of θ_A and θ_V may differ, this question must continue to be investigated. Here we should only like to draw attention to the fact that a comparison of mesonic decays yields information about the quantity

$$\mathrm{tg}^2\,\theta_i \frac{f_k^{(i)}(q^2)}{f_\pi^{(i)}(q^2)}$$

and not $\tan^2\theta_i$ and that the value of the Cabibbo angle can be affected both by possible q^2 dependences of the different form factors and, in general, a departure from unity of ratios of the type $f_k^i(q^2)/f_\pi^i(q^2)$. As yet this question remains open.

As regards the Cabibbo angle θ itself, it was introduced into the theory as a parameter needed to save the hypothesis that the weak hadronic current has unit length. Its numerical value, found experimentally, is very close to the ratio of the π- and K-meson masses:

$$\mathrm{tg}\,\theta \approx \frac{m_\pi}{m_K} = 0.28.$$

It is therefore very attractive to try and explain this fact by the idea that the splitting of the particle masses in the unitary multiplets and the orientation of the weak hadronic current in the unitary spin space have a common origin, i.e., that they are due to the common $SU_{(3)}$-symmetry breaking interaction.

Thus, the further development of the universal four-fermion theory of weak interactions based on the inclusion in the theory of the unitary symmetry of the strong interactions has made it possible to achieve further progress in the problem of the weak interactions of elementary particles. No other theories leading to equivalent consequence have as yet been proposed. The predictions obtained on the basis of $SU_{(3)}$-symmetry agree well with the experimental facts. In addition, the numerical values of a number of general parameters of the $SU_{(3)}$-symmetry and its developments, which determine the properties of the strong, electromagnetic, and weak processes, are found to be similar when they are extracted from data on the various processes. This confirms the generality of the effects in the world of elementary particles. More precise experimental data and new developments of the theoretical ideas will enable us to judge the correctness of these tendencies in our construction of a theory of the microscopic world.

2.8. Achievements of the Investigation of Leptonic Hadron Decays

At the present level of our knowledge of the properties of semileptonic hadron decays we have a characteristic accuracy of $\sim 10\%$ for the values of the most important parameters. The main conclusions obtained after almost a decade of investigations of these processes can be briefly summarized as follows.

1. The general properties of semileptonic hadron decays agree well with predictions of the universal four-fermion theory of the weak interaction in the variant proposed by Cabibbo. Possible admixtures from other couplings have hitherto remained undetected.

2. Effects indicating a violation of T invariance of the weak interactions have not been found; agreement has been obtained with the predictions for the semileptonic decays of K^0 mesons induced by violation of CP invariance in the two-pion decays of K_2^0 mesons.

3. The principle of muon-electron universality is satisfied with an accuracy not worse than 15%.

4. There are no appreciable departures from the consequences of the isotopic and also $SU_{(3)}$-symmetry properties of the theory; the selection rule $|\Delta I| = 1/2$ is satisfied with an accuracy of $\sim 5\%$ and the selection rule $\Delta Q = \Delta S$ with an accuracy of $\sim 10\%$.

5. In the range of momentum transfers from 0 to 0.3 GeV^2 the interaction is local; this excludes the possibility of a mass of the hypothetical intermediate W meson less than 1 GeV.

6. The effects associated with the strong interactions do not depend very strongly on the energy.

7. The values of the Cabibbo angle obtained from an analysis of meson and baryon decays are almost equal. The existing indications that the angles θ_V and θ_A may be different must be made more precise both experimentally and theoretically.

8. The relative influence of f and d type forces in the axial-vector form factor are found to be the same as in the strong interactions of the octet of pseudoscalar mesons and baryons.

Of the most important problems that await solution we would mention the following.

1. Verification of the consequences of CPT and CP invariance for processes due to weak interactions. In this connection, the most promising experiments are those in which the properties of the K^+ and K^- mesons are compared.

2. Improvement in the accuracy of data that cast light on $\mu - e$ universality in the weak interaction.

3. Advances in the question of neutral currents.

4. Searches for contributions from couplings distinct from the vector and axial-vector couplings. This would test the consequences of the CVC hypothesis and some deductions of other models at present popular.

5. Improvement in the accuracy of the determination of the parameters of the Cabibbo theory, the angles θ_V and θ_A, and also the mixing parameter of the f and d type forces and the discovery of ways to explain the nature of this phenomenon.

LITERATURE CITED

1. L. B. Okun', The Weak Interaction of Elementary Particles [in Russian], Moscow, Gosatomizdat (1963).
2. M. Gell-Mann, Phys. Rev., 125, 1067 (1962).

3. N. Cabibbo, Phys. Rev. Lett., 10, 531 (1963).
4. D. R. Bowen et al., Phys. Rev., 154, 1314 (1967); D. R. Boterill et al., Phys. Rev., 171, 1402 (1968); R. Macek et al., Phys. Rev. Lett., 22, 32 (1969).
5. M. A. Markov, International School on High-Energy Physics [in Russian], Popradske Pleso, Czechoslovakia. Published in Czechoslavakia (1967), p. 173.
6. T. D. Lee and C. S. Wu, Ann. Rev. Nucl. Sci., 16, 471 (1966).
7. E. Belliotti et al., Nuovo Cimento, A52, 1287 (1967).
8. P. Basile et al., Phys. Lett., B26, 542 (1968).
9. P. T. Eschstruth et al., Phys. Rev., 165, 1487 (1968).
10. L. J. Clavelli, Phys. Rev., 154, 1509 (1967).
11. M. Fitelson and E. Kazes, Phys. Rev. Lett., 20, 304 (1968); J. I. Iliopoulos and R. P. Van Royen, Phys. Lett., B25, 146 (1967).
12. I. R. Botterill et al., Phys. Rev. Lett., 21, 766 (1968); J. A. Helland et al., Phys. Rev. Lett., 21, 257 (1968); D. Gutts et al., Phys. Rev. Lett., 20, 955 (1968); J. Bettels et al., Nuovo Cimento, 56, 1106 (1968).
13. S. Matsudas and S. Oneda, Phys. Rev., 169, 1172 (1968); D. R. Majumdar, Phys. Rev. Lett., 20, 971 (1968); Lar Nam Chang and Y. C. Leung, Phys. Rev. Lett., 21, 122 (1968); N. H. Funchs, Phys. Rev., 172, 1532 (1968).
14. B. D'Espagnat and M. K. Gaillard, Phys. Lett., B25, 346 (1967).
15. N. Barash-Schmidt et al., Rev. Mod. Phys., 41, 109 (1969).
16. D. G. Hill et al., Phys. Rev. Lett., 19, 668 (1967); B. R. Webber et al., Phys. Rev. Lett., 21, 498 (1968).
17. S. Bennet et al., Phys. Rev. Lett., 19, 993 (1967); D. Dorfan et al., Phys. Rev. Lett., 19, 987 (1967).
18. V. L. Fitch, Comm. Nucl. Part. Phys., 11, 6 (1968).
19. E. P. Shabalin, Zh. Éksp. Teor. Fiz., 44, 765 (1963).
20. N. Cabibbo and A. Maksymowicz, Phys. Rev., 137, B438 (1965).
21. R. W. Birge et al., Phys. Rev. Lett., 11, 35 (1963).
22. R. W. Birge et al., Phys. Rev., 139, B1600 (1965).
23. R. W. Birge et al., UCRL-17088 (1966).
24. F. A. Berends et al., Phys. Lett., B26, 109 (1967).
25. S. W. Mac-Dowell, Phys. Rev. Lett., 17, 1116 (1967).
26. J-L. Gervais et al., Phys. Lett., 20, 432 (1966).
27. V. C. Vanashin et al., Preprint JINR R1-3594 [in Russian], Dubna (1967).
28. W. T. Chu et al., Phys. Rev. Lett., 19, 719 (1967).
29. É. I. Mal'tsev, Preprint JINR R1-4557 [in Russian], Dubna (1969).
30. R. P. Feynman and M. Gell-Mann, Phys. Rev., 109, 193 (1958).
31. M. L. Good et al., Phys. Rev., 151, 1195 (1966).
32. M. Bott-Bodenhausen et al., Phys. Lett., B24, 194 (1967).
33. V. L. Fitch et al., Phys. Rev., 164, 1711 (1967).
34. D. Cline et al., Heidelberg Internat. Conf. on High Energy Phys., (1967).
35. U. Camerini et al., Phys. Rev. Lett., 13, 318 (1964).
36. U. Camerini et al., Nuovo Cimento, 37, 1795 (1965).
37. E. De Rafael, Phys. Rev., 157, 1486 (1967).
38. T. O. Lee and C. N. Yang, Phys. Rev., 119, 1410 (1960).
39. N. Cabibbo and E. Ferrary, Nuovo Cimento, 18, 928 (1960).
40. M. Baker and C. Glasgow, Nuovo Cimento, 25, 857 (1962).
41. M. A. Beg, Phys. Rev., 132, 426 (1963).
42. K. Tanaka, Phys. Rev., 140, B463 (1965).
43. V. K. Ignatovich and B. V. Struminsky, Phys. Lett., B24, 69 (1967).
44. L. M. Sehgal, Phys. Rev., 183, 1511 (1969).
45. J. E. Maloney and B. Sechi-Zorn, Phys. Rev. Lett., 23, 425 (1969).
46. F. Eisele et al., Z. Phys., 221, 401 (1969).
47. D. R. Harrington, Phys. Rev., 120, 1482 (1960).
48. O. G. Bokov et al., Preprint JINR R-2278 [in Russian], Dubna (1965).
49. P. Hertel, Z. Phys. 202, 383 (1967).
50. V. M. Shekhter, Zh. Eksp. Teor. Fiz., 47, 262 (1964).
51. N. V. Bagget el al., Phys. Rev. Lett., 23, 249 (1969).
52. M. T. Burgy et al., Phys. Rev., 120, 1827 (1960).

53. S. S. Gershtein and Ya. B. Zel'dovich, Zh. Éksp. Teor. Fiz., 29, 698 (1955).
54. R. E. Marshak and E. C. Sudarshan, Phys. Rev., 109, 1860 (1958).
55. R. Feynman and M. Gell-Mann, Phys. Rev., 109, 193 (1958).
56. T. C. Griffith and L. I. Schiff, Electromagnetic Interactions and the Structure of Elementary Particles [Russian translation], Mir, Moscow (1969), p. 137.
57. H. Filthuth, Topical Conf. on Weak Interactions, CERN, Geneva (1969).
58. P. Desai, Phys. Rev., 179, 327 (1969).
59. B. G. Erozolimskii et al., Yad. Fiz., 11, 1049 (1970).
60. F. P. Calaprice et al., Phys. Rev. Lett., 18, 918 (1967).
61. A. I. Alikhanov et al., Zh. Éksp. Teor. Fiz., 35, 1061 (1959).
62. I. Bender et al., Z. Phys., 212, 190 (1968).
63. J. K. Kim, Phys. Rev. Lett., 19, 1079 (1967).
64. N. Brene et al., Nucl. Phys., B6, 255 (1968).
65. F. Eisele et al., Z. Phys., 225, 383 (1969).
66. S. Matsuda et al., Phys. Rev., 178, 2129 (1969).
67. V. K. Grigor'ev et al., Yad. Fiz., 6, 329 (1967).

THREE-QUASIPARTICLE STATES IN DEFORMED NUCLEI WITH MASS NUMBERS BETWEEN 150 AND 190

K. Ya. Gromov, Z. A. Usmanova,
S. I. Fedotov, and Kh. Shtrusnyi

We have studied the splitting and de-excitation of three-particle states and the possibility of observing these effects in deformed nuclei which have mass numbers between 150 and 190. We have computed the splitting energies for the observed three-particle states, the limiting values of W and α, which determine the interaction force, and the fraction of spin forces in the total pairing forces. An analysis of the experimental date indicates that the de-excitation of three-quasiparticle levels into single-quasiparticle levels, which is an F-forbidden process in the independent quasiparticle model, can be explained by a simple model.

INTRODUCTION

Modern nuclear models [1-4] describe the excited states of odd mass-number deformed nuclei in terms of single-quasiparticle states, three-quasiparticle states, and so on, and collective excitation states such as quadrupole β and γ oscillations, octupole oscillations, and others. Both the theoretical calculations and the experimental data show that the lowest states of the deformed nuclei are well-described by states of single-particle excitation (the level schemes of Nilsson or Saxon and Woods) and their associated rotational levels. However, noticeable amounts of admixed collection states appear in the single-particle states even at excitation energies of 200-300 keV [5]. At energies of 500 keV or higher, levels of a basically collective nature have been observed in addition to those levels which are primarily single-quasiparticle in character.

In deformed, odd mass-number nuclei three-quasiparticle states can appear in the breaking up of neutron or proton pairs. This means that the energy of a three-quasiparticle state must be greater than or near one MeV. Three-quasiparticle states can be one of two kinds: the first is a state in which all three quasiparticles are the same – three protons (3p) or three neutrons (3n); the second state has different quasiparticles such as (2p, n) or (2n,p). In the superfluid model of the nucleus each three-quasiparticle level is four-fold degenerate. Therefore the projections of the total angular momenta of the three quasiparticles onto the symmetry axis of the nucleus can be combined in four different ways*:

$$K=|-\Omega_1+\Omega_2+\Omega_3|;\ K=|\Omega_1-\Omega_2+\Omega_3|;$$
$$K=|\Omega_1+\Omega_2-\Omega_3|;\ K=|\Omega_1+\Omega_2+\Omega_3|.$$

*To describe the single-quasiparticle states we shall use the asymptotic quantum numbers $\Omega^{\pi}[Nn_Z\ \Lambda]\uparrow$, which are applied in both the collective and superfluid models. Here Ω is the projection of the angular momentum of a single-quasiparticle state onto the nuclear axis of symmetry, π is the parity of the state, N is the principal quantum number which determines the number of the main shell in the oscillator potential, n_Z is the quantum number of the oscillator along the symmetry axis of the nucleus, and Λ is the projection of the orbital angular momentum on the symmetry axis. The arrow on the right indicates the spin quantum number Σ. If the arrow points up, $\Sigma=+1/2(\Omega=\Lambda+1/2)$, if it points down, $\Sigma=-1/2(\Omega=\Lambda-1/2)$. K is the projection of the angular momentum of the three-quasiparticle state onto the symmetry axis.

Joint Institute for Nuclear Research, Dubna. Translated from Problemy Fiziki Élementarnykh Chastits i Atomnogo Yadra, Vol. 1, No. 2, pp. 525-546, 1971.

The levels of a three-quasiparticle multiplet can be split by the interaction of the quasiparticles, which was not taken into account in the independent quasiparticle model. This means that experimentally one can observe three-quasiparticle multiplets consisting of four levels.

In 1962 V. G. Solov'ev demonstrated the possibility of observing states having three-quasiparticle character [6]. They computed the energy of the center of gravity of a number of three-quasiparticle multiplets in deformed nuclei. N. I. Pyatov and A. S. Chernyshev [7] have studied the splitting of levels for (2n, p) and (2p, n)-type three-quasiparticle multiplets.

The experimental data on three-quasiparticle states was studied by one of the present authors in 1965 [8]. In this paper we shall analyze the more complete information gathered in recent years concerning three-quasiparticle states.

1. Experimental Identification of Three-Quasiparticle States

The primary source of experimental difficulty in observing and studying three-quasiparticle states is that the splitting energies are 1 MeV or greater. Here one should keep in mind the fact that in this energy range the density of levels in an odd-A nucleus increases, and that near a possible three-quasiparticle state there can be levels of a different kind but having the same spin and parity. Then states of different kinds can be strongly mixed. The amount of such mixing will of course depend on the properties of a given nucleus, such as the nature of the single-particle states in the nucleus, the locations of the collective excitation levels, and so on. But one can state immediately that the low-lying three-quasiparticle states with very high spin values (such as 21/2 or greater) will mix only slightly with other types of states. The reason is, of course, that for energies of 1-2 MeV only three-quasiparticle states can have such high spin values. Other types of states have much lower spins.

Two cases can be cited in which the experimental observation of three-quasiparticle states is simplified. V. G. Solov'ev et al. [6] have shown it to be possible to see three-quasiparticle states in β decay. As mentioned earlier, three-quasiparticle states can be divided into two groups: states like (3n) or (3p) where all three particles are the same, and states like (2n, p) and (2p, n). β decay from the ground state of an odd nucleus (a single-quasiparticle state) to either a (3n) or (3p) three-quasiparticle excited state of the daughter nucleus will be strongly forbidden, because a transition like (p)$\rightarrow$(3n) or (n)$\rightarrow$(3p) requires a number of quasiparticles to change state (F-forbidden [9]). But β decay from either a (2n, p) or (2p, n) three-quasiparticle state is quite another thing. Such β decays can be depicted in this manner:

$$p_1 \rightarrow p_2 p_3 n_4 \text{ (states of the type } 2p, n)$$

or

$$n_1 \rightarrow n_2 n_3 p_4 \text{ (states of the type } 2n, p).$$

Here the indices 1, 2, 3, 4 are a set of asymptotic quantum numbers which characterize the single-quasiparticle states. If the state 1 is identical to state 2 (or 3) then only one other particle must appear in such a transition in the proton and neutron schemes, and the transition will not be F-forbidden. But if states 1, 2, 3 are different, then in one of the systems two quasiparticles appear, and the transition is F-forbidden. Therefore, in the case where state 1 is identical to state 2 β decay to three-quasiparticle states can be observed. The probability of β decay in this case is determined by the probability of a β transition between the single-quasiparticle proton and neutron states labeled 3 and 4. Of special interest then are the neutron and proton states between which unhindered β transitions are allowed:

$$p\,7/2^-\,[523] \rightleftarrows n\,5/2^-\,[523];$$

$$p\,9/2^-\,[514] \rightleftarrows n\,7/2^-\,[514].$$

It is known that rather fast β transitions take place between these states [8], with log ft ranging between 4.6 and 4.8. It seems likely that β transitions to three-quasiparticle states of the type indicated will also be allowed and unhindered, and one would expect that log ft for these transitions will lie in the same range. The systematics of the β decay matrix elements for deformed odd-A nuclei indicate that the allowed unhindered β transitions are rather clearly isolated from the rest of the group; log ft for the allowed hindred β transitions are greater than 5.5. Thus, if β decay to an excited state of an odd-A nucleus is observed, and log ft for the transition is smaller than 5.2, and one can eliminate the possibility of a β transition to a single-quasiparticle state, it can then be concluded that we are dealing with a β transition to a three-quasiparticle state.

It was determined in studies of ^{165}Tm [10] that the 1428-keV level of the daughter nucleus ^{165}Er is populated in 12% of the cases. Preibisz et al [11] measured the ^{165}Tm-^{165}Er decay energy and found the value of log ft for K-capture in the 1428 keV level to be 5.1±0.2. The spin of the ^{165}Tm ground state has been measured to be 1/2 [12]. Thus the ground state of ^{165}Tm is characterized by the asymptotic quantum numbers $1/2^+$ [411]. One cannot have an allowed unhindered β transition from this state to the single-quasiparticle state of ^{165}Er. Other kinds of β transitions must have larger values of log ft. It was therefore shown that the only remaining possibility was to interpret the 1428-keV level as a three-quasiparticle state of the type $\rho_1\ 1/2^+$ [411] $+\rho_2\ 7/2^-$ [523] $-$n $5/2^-$ [523]. Measurements of the multipolarities of the γ transitions from the 1428-keV level showed that this level has spin and parity $3/2^+$, confirming the conclusion reached relative to the type of state [10].

Three-quasiparticle states were found in additional nuclei, based on the identification of allowed unhindered β decay. The data for these states are collected in Table 1. The first column of the table gives the nucleus in which the three-quasiparticle state was observed. The second column contains the configuration of the three-quasiparticle state. The third gives the experimental spin and parity, and the fourth presents the experimental energy for the level. Column five contains the theoretical estimate of the energy of the center of gravity for the three-quasiparticle multiplet. Also shown are the basic experimental data which enabled us to identify the level as a three-quasiparticle state. Column six contains the value of log ft for this particular case (three-quasiparticle states excited by an allowed undelayed β decay of the nucleus).

Another possibility for observing three-quasiparticle states are the three-quasiparticle isomer states. High-spin three-quasiparticle states are excited in various nuclear reactions, among them being the β decay of three-quasiparticle isomer states. As an example, let us consider the isomer state in ^{177}Lu. By irradiating the natural mixture of lutecium isotopes with thermal neutrons, Jorgensen et al [13] observed that in addition to the already well-known radioactive ^{177}Lu nucleus ($T_{1/2}$=6.8 days) there was a new activity having a half-life of 155 days. Studying the gamma-ray and conversion-electron spectra they showed that this new activity is connected with the 969 keV level of ^{177}Lu. It was also determined that the spin and parity of this level is $23/2^-$. It is impossible to explain a level with such high spin on the basis of a single-quasiparticle; in Nilsson's level scheme there is no level with spin greater than 13/2 at this energy. The possibility that this is a collective level is also excluded. The only possible interpretation is that the level is a three-quasiparticle state:

$$n_1\, 9/2^+ [624] + n_2\, 7/2^- [514] + p\, 7/2^+ [404].$$

It was shown in the same paper [13] that a three-quasiparticle state of the type β is excited in ^{177}Hf at 1315 keV during the $23/2^+$ $\{p_1\ 7/2^+$ [404], $p_2\ 9/2^-$ [514], n $7/2^-$ [514], n $7/2^-$ [514]$\}$ decay of the isomer state in ^{177}Lu.

Peker has shown [14] that three-quasiparticle isomer states should be sought in odd-A nuclei for which the neighboring even-even nuclei reveal two-quasiparticle isomers. Thus, in ^{178}Hf, which is a neighbor of ^{177}Lu, an isomer state is observed at 1148 keV ($T_{1/2}$=4.8 sec) of type 8^-. Gallagher and Solov'ev explained it as a two-quasiparticle state $n_1\ 9/2^+$ [624]$+n_2\ 7/2^-$ [514]. One would expect that the lowest-energy three-quasiparticle state in ^{177}Lu is obtained by adding to this state a third particle whose state is the same as that of the proton in the ground state of ^{177}Lu; i.e., p $7/2^+$ [404].

The experimental data which are now available on three-quasiparticle isomer states are also shown in Table 1. The seventh column gives the measured values of the half-lives for the isomer states.

It is clear from Table 1 that all the more or less reliable levels showing a three-quasiparticle nature belong to the (2p, n) or (2n, p) types. The excitation of a three-proton or three-neutron state during β decay of the ground state of a nucleus is highly unlikely. (3p) and (3n) states can be observed in nuclear reactions (such as $n\gamma$ reactions) or during β decay of three-quasiparticle isomer states.

2. Splitting of Three-Quasiparticle States

The energy difference between levels of the (2n, p) or (2p, n) three-quasiparticle types can be computed using the equations obtained in [7]:

$$E_{(K=|-\Omega_1+\Omega_2+\Omega_3|)} - E_{(K=\Omega_1+\Omega_2+\Omega_3)} = (1-4\alpha)\, w\, [A_{12}+B_{12}] - 2\alpha w\, [A_{13}+B_{13}];$$

$$E_{(K=|\Omega_1-\Omega_2+\Omega_3|)} - E_{(K=\Omega_1+\Omega_2+\Omega_3)} = (1-4\alpha)\, w\, [A_{12}+B_{12}] - 2\alpha w\, [A_{23}+B_{23}];$$

$$E_{(K=|\Omega_1+\Omega_2-\Omega_3|)} - E_{(K=\Omega_1+\Omega_2+\Omega_3)} = -2\alpha w\, [A_{23}+B_{23}] - 2\alpha w\, [A_{13}+B_{13}].$$

TABLE 1. Three-Quasiparticle States in Deformed Nuclei with Mass Number A between 150 and 190

Nucleus	Configuration	K^{π}	E_{exp}, keV	E^{*}_{theo}, keV	lg**ft	$T_{1/2}$	Reference
^{163}Dy	p_1 7/2$^-$ [523] p_2 3/2$^+$ [411] n 5/2$^-$ [523]	(5/2$^+$) 1/2$^+$	935 884	≈1200	5,2 4,9	— —	[16]
^{161}Er	p_1 7/2$^-$ [523] p_2 7/2$^+$ [404] n 5/2$^-$ [523]	(9/2$^+$, 5/2$^+$)	1838	≈3200	5,0	— —	[21]
^{163}Er	p_1 7/2$^-$ [523] p_2 1/2$^+$ [411] n 5/2$^-$ [523]	1/2$^+$ 3/2$^+$	1802,0 1538,4	≈1400	4,9 5,3	— —	[19][22]
^{165}Er	p_1 7/2$^-$ [523] p_2 1/2$^+$ [411] n 5/2$^-$ [523]	3/2$^+$	1428	≈1400	5,1	—	[10] [23] [24]
^{175}Yb	p_1 7/2$^-$ [523] p_2 1/2$^+$ [411]*** n 5/2$^-$ [523]	1/2$^+$ 3/2$^+$	2113 1792	≈3000	5,2 5,5	— —	[25]
^{175}Yb	p_1 9/2$^-$ [514] p_2 1/2$^+$ [411] n 7/2$^-$]514]	(1/2$^+$) 3/2$^+$	(1891) 1497	≈1600	4,8 5,0	— —	[25]
^{177}Hf	p_1 9/2$^-$ [514] p_2 7/2$^+$ [404] n 7/2$^-$ [514[	23/2$^+$	1315	≈1400	—	1,12 sec	[26] [27]
^{179}W	p_1 9/2$^-$ [514] p_2 5/2$^+$ [402] n 7/2$^-$ [514]	7/2$^+$ 3/2$^+$	1680,1 720,5	≈1400	5,2 5,2	— —	[18]
^{161}Ho	n_1 5/2$^-$ [523] n_2 3/2$^-$ [521] p 7/2$^-$ [523]	5/2$^-$ 1/2$^-$	1943 1897	≈1800	5,1 4,8	— —	[28]
^{177}Lu	n_1 9/2$^+$ [624] n_2 7/2$^-$ [514] p 7/2$^+$ [404]	23/2$^-$	970	≈1000	—	155 days	[13] [29]

TABLE 1, continued

Nucleus	Configuration	K^{π}	E_{exp}, keV	E^{*}_{theor}, keV	lg ft†	$T_{1/2}$	Reference
^{177}Lu	$n_1\ 7/2^-$ [514] $n_2\ 9/2^+$ [624] $p\ 9/2^-$ [514]	$7/2^+$ $11/2^+$	1241 1230	≈1200	4,4 4,4	— —	[30]
	$n_1\ 1/2^-$ [510] $n_2\ 7/2^-$ [514] $p\ 7/2^+$ [404]	$13/2^+$ $15/2^+$	1503 1357	≈1900	— —	— —	[31]
^{183}Re	$n_1\ 9/2^+$ [624] $n_2\ 11/2^+$[615]‡ $p\ 5/2^+$ [402]	$25/2^+$	1907	≈2300	—	1,02 μsec	[32]

*As in [6], E_{theor} was calculated roughly using the superfluid model without including the interactions of quasiparticles.
†Only those levels whose β decay corresponds to log ft < 5.5 were considered.
‡The interpretation is not unique.

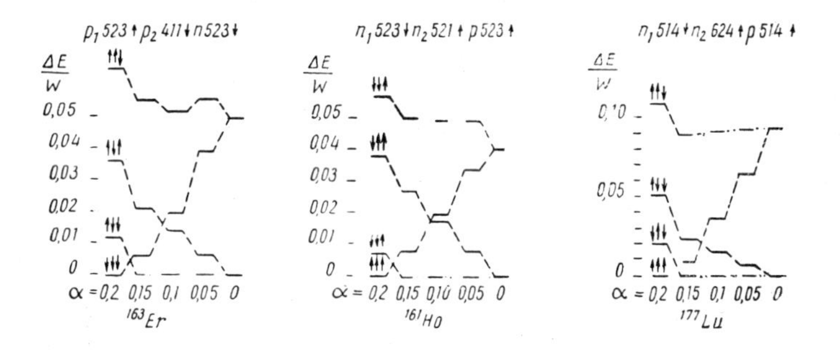

Fig. 1. Splittings of three-quasiparticle multiplets for various states.

The parameter ω determines the nucleon-nucleon interaction force, and is spin dependent. The parameter α determines the contribution of spin forces to the total pairing forces ($0 \leq \alpha \leq 1$). A_{ij} and B_{ij} are coefficients which depend on the asymptotic quantum numbers of the quasiparticles. Both A_{ij} and B_{ij} have been calculated in [7] for a number of single-quasiparticle pairs states (p, n), (p, p) and (n, n).

It is clear from the equation that the character of the splitting – the order of the levels in the multiplet – is determined by α. The splitting energy depends on ω. Figure 1 shows the splitting as a function of α for certain three-quasiparticle states. By examining all possible combinations of asymptotic spin quantum numbers Σ_1, Σ_2 and Σ_3 Pyatov and Chernyshev[7] have concluded that in states like (2p, n) or (2n, p) the highest level in the multiplet will always be the level in which the aymptotic spins of the nucleons in the disrupted pair (p_1, p_2) or (n_1, n_2) are parallel, while the spin of the third particle is antiparallel to them (↑↑, ↓). The three other possible levels are always lower (see Fig. 1). We point out here that we will arrive at the same conclusion if the Gallagher-Moshkovskii rule is extended to three-quasiparticle states. For according to their rule a state of the type (↑↑, ↓) is the most undesirable: the asymptotic spins of the identical particles are parallel while the spin of the third particle is antiparallel to the first two.

The experimental data on the splitting of three-quasiparticle states is meager at present. All available data are collected in Table 2. The first column gives the nucleus in which the three-quasiparticle levels are observed, the second contains the asymptotic quantum numbers of the single-quasiparticle states

TABLE 2. Splittings of Three-Quasiparticle States in Deformed Nuclei with A between 150 and 190

Nucleus	Configuration	K^{π}	$(\Sigma_1, \Sigma_2, \Sigma_3)$	E_{exp}, kev	ΔE_{exp}, kev	ΔE_{theo}, kev	α	w, MeV
^{163}Dy	p_1 7/2$^-$ [523]	5/2$^+$	(↓↓, ↓)	935				
	p_2 3/2$^+$ [411]				51	—	—	—
	n 5/2$^-$ [523]	1/2$^+$	(↓↑, ↓)	884				
^{163}Er	p_1 7/2$^-$ [523]	1/2$^+$	(↓↓, ↓)	1802,0				
	p_2 1/2$^+$ [411]				264	380	less than 0,17	greater than 5,4
	n 5/2$^-$ [523]	3/2$^+$	(↓↑, ↓)	1538,4				
^{175}Yb	p_1 7/2$^-$ [523]	1/2$^+$	(↓↓, ↓)	2113				
	p_2 1/2$^+$ [411]				321	380	less than 0,17	greater than 6,4
	n 5/2$^-$ [523]	3/2$^+$	(↓↑, ↓)	1792				
	p_1 9/2$^-$ [514]	1/2$^+$	(↓↓, ↓)	1891				
	p_2 1/2$^+$ [411]				394	—	—	—
	n 7/2$^-$ [514]	3/2$^+$	(↓↑, ↓)	1497				
^{179}W	p_1 9/2$^-$ [514]	7/2$^+$	(↓↓, ↓)	1680,1				
	p_2 5/2$^+$ [402]				960	about 600	—	—
	n 7/2$^-$ [514]	3/2$^+$	(↓↑, ↓)	720,5				
^{161}Ho	n_1 5/2$^-$ [523]	5/2$^-$	(↓↓, ↓)	1943				
	n_2 3/2$^-$ [521]				46	330	less than 0,19	greater than 1,1
	p 7/2$^-$ [523]	1/2$^-$	(↓↑, ↓)	1897				
^{177}Lu	n_1 7/2$^-$ [514]	7/2$^+$	(↓↑, ↓)	241				
	n_2 9/2$^+$ [624]				—11	740	less than 0,17	greater than 0,1
	p 9/2$^-$ [514]	11/2$^+$	(↓↓, ↓)	1230				
	n_1 1/2$^-$ [510]	13/2$^+$	(↓↓, ↓)	1503				
	n_2 7/2$^-$ [514]				146	—	—	—
	p 7/2$^+$ [404]	15/2$^+$	(↓↑, ↓)	1357				

from which the observed three-quasiparticle state arises, and column three shows the spins and parities of the three-quasiparticle levels. The fourth column gives the relative orientations of the asymptotic spin quantum numbers and column five cites the experimental values for the three-quasiparticle levels. The table indicates a few cases where two levels of the multiplet have been observed, but there are no cases in which a greater number of multiplet levels have been observed experimentally. This makes it impossible to calculate α and w simultaneously from the data.

The parameters α and w, necessary in order to compute the splitting energies of the three-quasiparticle states, were determined in [7] from the experimental energy splittings of two-quasiparticle states in even-even and odd-odd nuclei. By using the values picked in [7] for $(1-4\alpha)$ w = 8.45 MeV and αw = 0.314 MeV, we have computed the energy differences between the corresponding three-quasiparticle levels. These

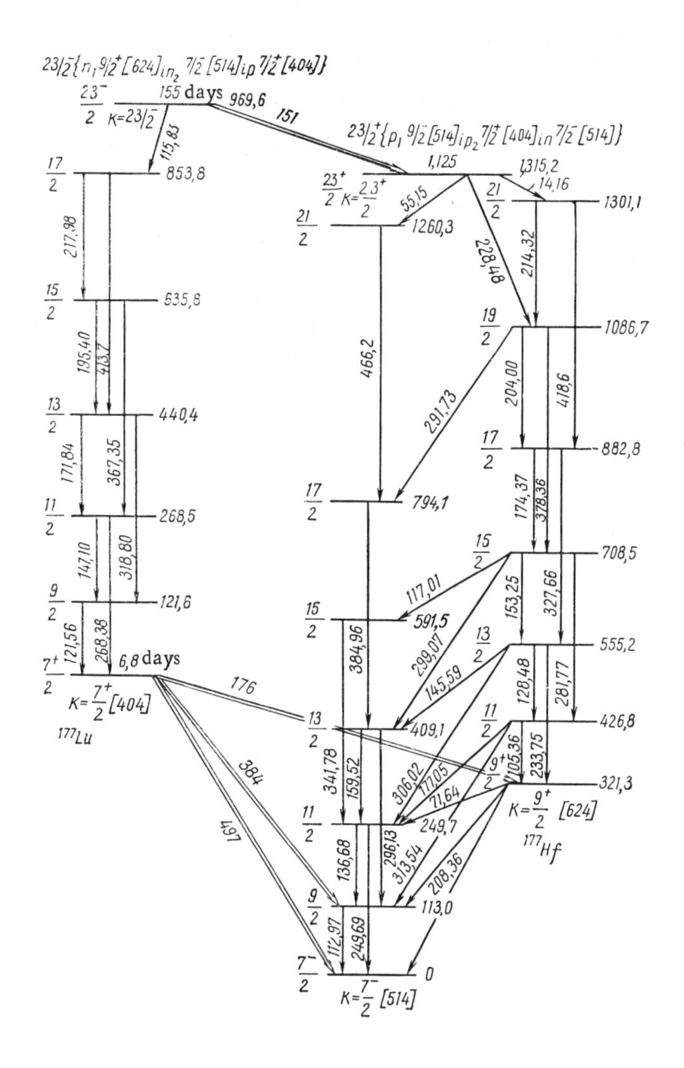

Fig. 2. Discharge of three-quasiparticle states in ^{177}Lu and ^{177}Hf.

are presented in colum seven of Table 2. The experimental values for these splittings are shown in the sixth column. The last two columns give estimates for the limits on α and w obtained on the basis of the experimental energy differences for the three-quasiparticle levels. These estimates were obtained in the following fashion. Using the equation for the corresponding energy difference, we found α when $\Delta E = 0$. The experimental value of ΔE (its sign) enabled us to determine if the value of α (for $\Delta E = 0$) is an upper or lower limit. The values of w shown in the last column were obtained from the experimental values of ΔE under the assumption that $\alpha = 0$. The forms of the expressions for ΔE and the values of $(A_{ij} + B_{ij})$ let us determine how w changes when α is increased and when ΔE = const; we were again able to determine whether we were dealing with an upper or a lower limit. The blanks in the last three columns indicate that A_{ij} and B_{ij} where not computed in [7] for the appropriate pairs of single-quasiparticle states.

By comparing the experimental and theoretical energy splittings (ΔE_{exp} and ΔE_{theo}) we see that in all cases except one we have obtained the right sign for the energy splittings: in a number of cases the agreement in magnitudes is not bad, but for ^{161}Ho ΔE_{exp} and ΔE_{theo} differ by a factor of six or seven. The limiting values for α and w in Table 2 are not inconsistent with each other.

3. De-Excitation of Three-Quasiparticle States

When treating γ transitions from three-quasiparticle states one must keep in mind that for these transitions there is a limitation on the number of particles which can participate in the transition (F-limitation), which is a restriction added to the selection rules on the asymptotic quantum numbers K^{π} for $[Nn_z\lambda]$

usually used in the extended nuclear model. In the de-excitation of (2p, n) or (2n, p) three-quasiparticle states there is no F-limitation if, in the γ transition, one of the identical particles goes into the state occupied by the other particle, forming a pair. The γ transition takes place to the single-quasiparticle state occupied by the unpaired particle. In the independent-quasiparticle model γ transitions to the other single-quasiparticle states are forbidden because they take place with a change in the state of more than one particle.

From analysis of the de-excitation of states showing three-quasiparticle character, it is known that three-quasiparticle states can be classified into two groups. The first group contains three-quasiparticle isomer states; the second group is for states showing three-quasiparticle character and having small spins.

The isomerism of low-lying three-quasiparticle states with high angular momentum is explained very satisfactorily by the selection rules on the angular momentum and its projection on the nuclear symmetry axis (the asymptotic quantum number K). For example, consider the three-quasiparticle isomer state with spin $23/2^-$ in ^{177}Lu (Fig. 2). This state is discharged primarily (78%) by a β transition to the $23/2^+$ three-quasiparticle state in ^{177}Hf. The structure of these three-quasiparticle states in ^{177}Lu and ^{177}Hf is shown in Fig. 2. During the β transition the following conversion takes place n$9/2^+$ [624] $\xrightarrow{\beta}$ p $9/2^-$ [514]; the two other quasiparticles do not alter their state. This β transition belongs to the class of first forbidden unhindered β transitions. The measured matrix element for this transition (log ft=6.1) agrees with the commonly observed values of log ft for first forbidden unhindered β transitions [8]. A type E3 γ transition with energy 115.8 keV from the 970.2 keV three-quasiparticle level of ^{177}Lu to the 17/2 $7/2^+$ [404] – 854.2 keV rotational level of the ground state band is forbidden by the asymptotic quantum number K. The degree of forbiddenness is $\nu = \Delta K - L = 8 - 3 = 5$. The Weisskopf hindrance factor for the 115.8 keV γ transition is $F_W = T_{exp}/T_W \approx 2 \cdot 10^8$.

Another situation is observed in the discharge of states showing three-quasiparticle character (of the type (2p, n) or (2n, p)) and having small angular momenta. In the independent quasiparticle model γ transitions from these levels to all lower single-quasiparticle levels except the basis state* are forbidden, for all such γ transitions involve changes in the states of more than one particle (F-forbidden). Therefore one might expect that the life times for these levels ought to be determined by the probability of γ transitions to the basis state. But it turns out that γ transitions to single-quasiparticle states other than the basis state are observed and the γ transitions to the basis states are relatively weak and generally not observed. This indicates that these states are not pure three-quasiparticle states, but instead are mixtures with other states. It appears that the study of the nature of the de-excitation of levels showing three-quasiparticle character and having small angular momenta permits one to analyze the structure of these states in more detail.

By analyzing the experimental material we have concluded that the discharge of three-quasiparticle states to single-quasiparticle states can be explained by a simple model in which these three-quasiparticle states are mixed with single-quasiparticle states and states of the "quasiparticle+phonon" type. States of the latter kind appear as the result of the interaction of phonons in the even-even nuclear core with the quasiparticles of the odd nucleus. We shall investigate a number of examples.

γ Transitions from the 884-keV level of ^{163}Dy (Fig. 3), interpreted in Table 1 as three-quasiparticle states of the type $1/2^+$ {p_1 $7/2^-$ [523], p_2 $3/2^+$ [411], n $5/2^-$ [523]}, to the rotational band of the ground state have not been observed [16]. But there have been observations of intense E1 transitions to the levels 351 keV – 1/2 $1/2^-$ [521], 390 keV – 3/2 $1/2^-$ [521] and 422 keV 3/2 $3/2^-$ [521]. Calculations [5] show that the states $1/2^-$ [521] and $3/2^-$ [521] in ^{163}Dy are, to a high degree, purely single-quasiparticle states, and that the possible mixtures to these states cannot remove the F-forbiddenness of the γ transitions to them from a three-quasiparticle state. Thus, the observed γ transitions can be explained only by assuming that the 884 keV level has a more complex structure. According to Solov'ev et al. [17] the octupole state $Q_1(32)$ with $K^\pi = 2^-$ in ^{162}Dy at 1148 keV contains these basic two-quasiparticle components:

$$p[523]\uparrow\ p[411]\uparrow\ 51.4\%;$$

$$n[651]\uparrow\ n[521]\downarrow\ 0.9\%;$$

$$n[642]\uparrow\ n[521]\downarrow\ 1.7\%;$$

$$n[633]\uparrow\ n[521]\uparrow\ 24.2\%.$$

*A single-quasiparticle state is called a basis state if it has the same characteristics as the state of the unpaired quasiparticle in a three-quasiparticle state.

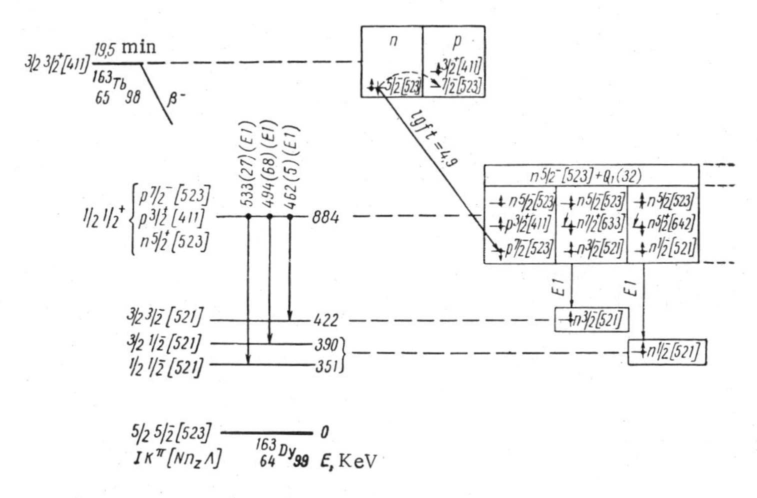

Fig. 3. Discharge of a three-quasiparticle-like state in ^{163}Dy.

Since ^{162}Dy is the even-even core of ^{163}Dy, it can be assumed that the 884 keV level, spin 1/2, is related to the octupole excitation of the even-even core; i.e., this level is n [523] ↓ + Q_1(32). In fact, according to [5], this component makes the major contribution to the 884 keV state with K^π=1/2$^+$. The odd particle in ^{163}Dy will only slightly alter the contribution of the phonon components with K^π=2$^-$ from ^{162}Dy nucleus [5]. One might therefore expect that the largest phonon components with K^π=2$^-$ in ^{162}Dy will also be the largest in ^{163}Dy. The strongest three-quasiparticle component of this state will be n [523] − p_1 [523] ↑ − p_2 [411] ↑ (≈50%). This component explains the allowed unhindered β transition from the ground state of ^{163}Tb. Other components explain the γ transitions from the 884 keV level (Fig. 3). Similar conclusions are reached when considering the de-excitation of the 720.5 keV level in ^{179}W [18].

Now let us study the discharge of the 1538.4 and 1802.0 keV levels in ^{163}Er (Fig. 4). These levels were interpreted in Table 1 as being 3/2$^+$ {$p_1$7/2$^-$[523], $p_2$1/2$^+$[411], n 5/2$^-$ [523]} and 1/2$^+${$p_1$7/2$^-$[523], $p_2$1/2$^+$ [411], n 5/2$^-$[523]} three-quasiparticle states, respectively.

In order to provide an example, we shall discuss just the de-excitation of the 1538.4 keV level because the discharge mechanism for the 1802.0 keV level is of the same character as that which de-excites the 1538.4 keV level (Fig. 4). γ transitions from the 1538.4 keV level to a level of the rotational band of the ground state are not observed [19]. The 1538.4 keV level is discharged through strong γ transitions to the 3/2$^-$ [521] level of the rotational band and the 5/2 5/2$^+$[642] level (see Fig. 4), for which good agreement is found between the experimental and theoretical values of the reduced transition probabilities. In the independent quasiparticle model these γ transitions are F-forbidden. The presence of these γ transitions can be explained by taking into account the interaction phonons in the even-even core of the ^{162}Er nucleus with the quasiparticles of the odd ^{163}Er nucleus. The result of this interaction between the ^{162}Er phonons and the ^{163}Er quasiparticles is the formation of quasiparticle + phonon states. Additions of these states to the 1538.4 keV level of the ^{163}Er nucleus can allow it to discharge to the single-quasiparticle states. Thus, in our case the experimental data indicate that only a mixture of the type 3/2$^-$ [521] + Q_1 (30) can explain the discharge of the 1538.4 keV level to the 3/2$^-$ [521] level of the rotational band (see Fig. 4). In this case we have γ transitions of type E1 with ΔK=0. For transitions of this kind the coupling of particle and rotational motions has little effect on the transition probabilities. We therefore observe good agreement between the experimental and theoretical values of the reduced probabilities for these transitions [20].

The presence of a 1469.0-keV γ transition between the 1538.4-keV level and the 69.21 keV level, (5/2$^+$ [642]), allowing for the theoretical and experimental energies of the single-quasiparticle states, can be explained only by an addition of the 3/2$^+$[651] single-quasiparticle state and a state of the 3/2$^+$ [651] + Q (20) type to the 1538.4-keV level.

According to the calculations of [5], at 1200 keV the ^{163}Er nucleus ought to have a state of spin 3/2$^+$ and configuration 3/2$^+$ [651] 2%, 3/2$^-$ [521] + Q_1 (30) 72%, 3/2$^+$[651] + Q_1 (20) 3%, ..., which can mix with the three-quasiparticle state 3/2$^+${p_1 7/2 [523], p_2 1/2$^+$ [411], n 5/2$^-$ [523]} [5]. Thus, rather good agreement is obtained between our interpretation, which follows from the experimental data and the above-

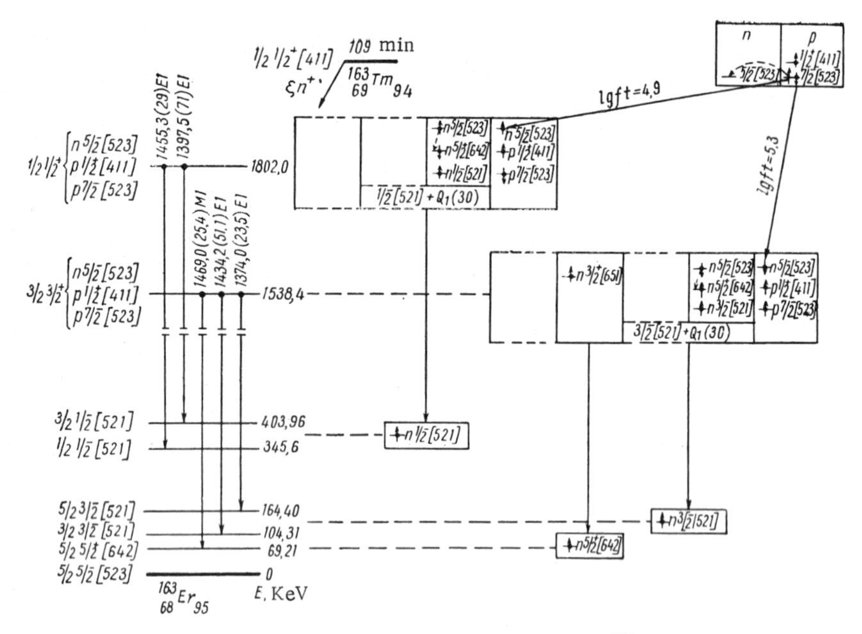

Fig. 4. Discharge of a three-quasiparticle state in ^{163}Er.

described simple model, and the assumption of Solov'ev et al. [5] about the nature of the 1538.4 keV level in ^{163}Er. An analysis of the experimental data on the properties of the levels considered here indicates that use of this simple model will make it possible to explain the discharge of all presently known three-quasiparticle states.

CONCLUSIONS

Among the excited states of odd-A deformed nuclei with mass numbers between 150 and 190 are observed a number of levels having a three-quasiparticle character. All the three-quasiparticle type levels observed until now are of the type (2n, p) or (2p, n). There are as yet no three-quasiparticle levels of the type (3n) or (3p). The energies of the three-quasiparticle levels show satisfactory agreement with the energies calculated using the superfluid model for the center of gravity of the three-quasiparticle states.

The available experimental data on the splitting of three-quasiparticle states is satisfactorily explained qualitatively by the calculations of [7] which include the interaction between quasiparticles. However, quantitative agreement has not yet been achieved.

The experimental data on the discharge of three-quasiparticle-like states enables one to make a more detailed analysis of the structure of these states. To a large extent the large-spin isomer three-quasiparticle states are pure three-quasiparticle states. The state with smaller spins contain significant mixtures of vibrational states.

It is of considerable interest to obtain new experimental data on the three-quasiparticle states. There would be great value in detecting new three-quasiparticle states in nuclear reaction studies, especially in observing (3p) and (3n) type states in (γ, n) reactions. In β decay studies of nuclei far from the stability line one might observe a large number of three-quasiparticle states populated by allowed unhindered decay.

The study of three-quasiparticle states is important in that it can help to explain the degree to which quasiparticle character is preserved in nuclear levels at high excitation energies.

LITERATURE CITED

1. V. G. Solov'ev, The Effect of Superconducting Pair Correlation on Nuclear Properties [in Russian], Gosatomizdat, Moscow (1963).
2. O. Nathan and S. G. Nilsson in: Alpha-, Beta- and Gamma-Ray Spectroscopy, Edited by K. Siegbahn.
3. V. G. Solov'ev, Structure of Complex Nuclei [in Russian], Atomizdat, Moscow (1966), p. 38.
4. V. G. Solov'ev and P. Vogel, Nucl. Phys., A92, 449 (1967).
5. V. G. Solov'ev et al., Izv. AN SSSR, Seriya Fiz., 31, 518 (1967).

6. V. G. Solov'ev, Zh. Éksp. Teor. Fiz., 43, 246 (1962).
7. N. I. Pyatov and A. S. Chernyshev, Izv. AN SSR, Seriya Fiz., 28, 1173 (1964).
8. K. Ya. Gromov, in: Structure of Complex Nuclei [in Russian], Atomizdat, Moscow (1966), p. 299.
9. V. G. Solov'ev, Mat.-fys. Skr. dan. vid. selskab, 1, No. 11 (1961); Izv. AN SSSR, Seriya Fiz., 25, 1198 (1961).
10. N. A. Bonch-Osmolovskaya et al., Nucl. Phys., 81, 225 (1966).
11. Z. Preibisz et al., Phys. Lett., 14, 206 (1965).
12. C. Ekstrom et al., Phys. Lett., 26 B 146 (1968); Erratum Phys. Lett., 26, B 387 (1968).
13. M. Jorgensen et al., Phys. Lett., 1, 321 (1962).
14. L. K. Peker, Izv. AN SSSR, Seriya Fiz., 28, 306 (1964).
15. C. J. Gallagher and V. G. Solov'ev, Mat-fys. Skr. dan. vid. selskab, 2 (1962).
16. L. Funke et al., Nucl. Phys., 84, 424 (1966).
17. V. G. Solov'ev et al., Dokl. Akad. Nauk SSSR, 189, 987 (1969).
18. R. Arl't et al., Preprint JINR R6-4635, Dubna (1969).
19. A. A. Abdurazakov et al., Program and Theses of Reports to the XX Annual Conference on Nuclear Spectroscopy and Nuclear Structure, Leningrad [in Russian], Nauka, Leningrad (1970).
20. L. Funke et al., Dissertation, Dresden (1966).
21. A. A. Abdumalikov et al., Preprint OIyaI 6-4393; Dubna (1969).
22. V. Gnatovich et al., Izv. AN SSSR, Seriya Fiz., 31, 587 (1967).
23. A. A. Abdurazakov et al., Program and Abstracts of Reports to the XX Annual Conference on Nuclear Spectroscopy and Nuclear Structure, Leningrad [in Russian], Nauka, Leningrad (1970).
24. W. Kurcewicz et al., Nucl. Phys., A108, 434 (1968).
25. L. Funke et al., Nucl. Phys., A130, 333 (1969).
26. L. Kristensen et al., Phys. Lett., 8, 57 (1964).
27. E. Bodenstedt et al., Z. Phys., 190, 60 (1966).
28. K. Ya. Gromov et al., Nuclear Physics, 2, 783 (1965).
29. P. Alexander et al., Phys. Rev., 133, B 284 (1964).
30. H. S. Johansen et al., Phys. Lett., 8, 61 (1964).
31a. R. K. Sheline, Proceed. Dubna Symp., (1968), p. 71.
31b. M. R. Beitin', Program and Abstracts of Reports to the XIX Annual Conference on Nuclear Spectroscopy and Nuclear Structure, Erevan, Nauka, Leningrad (1969).
32. M. J. Emmot et al., Phys. Lett., 20, 56 (1966).

FUNDAMENTAL ELECTROMAGNETIC PROPERTIES OF THE NEUTRON

Yu. A. Aleksandrov

This review is devoted mainly to experimental investigations of the fundamental electromagnetic properties of the neutron: electric charge, electric dipole moment, electromagnetic form factors, polarizability, and n-e interaction.

INTRODUCTION

The neutron was discovered in 1932 and probably had more effect on the development of physics than any other elementary particle. The discovery of the neutron initiated the age of nuclear physics, the rapid development of which in the middle of the 20th century led to the birth of nuclear technology and elementary particle physics.

We know a considerable amount about the neutron. Its interaction with nuclei has been fairly well investigated – without this knowledge there would be no nuclear reactors. We know its mass, spin, magnetic moment, and electric charge – the knowledge of these characteristics enables us to use the neutron as a very convenient tool for many investigations, such as the study of the structure of condensed media. In recent years, thanks to the efforts of many physicists, we have managed to build up at least a qualitative picture of the structure of the neutron. We have found that the neutron is a particle with a fairly complicated structure. At any rate, its behavior when it interacts with matter reveals properties which are characteristic of an extended particle. The neutron does not act directly on our sense organs and, hence, all that we know about it has been learned through its interaction with other particles and fields. There are four types of interaction: gravitational, weak, electromagnetic, and strong. The weakest of these is gravitational interaction. The neutron is the only particle with a rest mass which has been experimentally determined from its fall in the earth's gravitational field. There have been no surprises so far – the neutron behaves in the gravitational field like an ordinary macroscopic body.

Weak interaction is responsible for the decay of the neutron into a proton, electron, and antineutrino. Although neutron decay has been studied for a fairly long time, this field is still exceedingly attractive to experimenters. Even such a characteristic as the neutron half-life requires more accurate determination – the difference between different measurements is more than three times the measurement error.

Strong interactions are responsible for the reactions of neutrons with nuclei, and nuclei owe their very existence to strong interactions. Here there is a wealth of experimental material and probably even greater problems, which are unlikely to be settled without an understanding of strong interactions between elementary particles.

Finally, electromagnetic interaction, with the aid of which we have learned about such fundamental properties of the neutron as its electric charge, electric dipole moment, and electromagnetic structure, belongs to a region which we think we understand. So far, at least, we have discovered no effect which contradicts quantum electrodynamics. It is possible that the discovery of such an effect would be the starting point of a new theory, the development of which would lead to an understanding of strong interactions.

Joint Institute for Nuclear Research, Dubna. Translated from Problemy Fiziki Élementarnykh Chastits i Atomnogo Yadra, Vol. 1, No. 2, pp. 547-583, 1971.

TABLE 1

$Q=0$	$Q=Q_p$	$Q=Q_n$	$Q=Q_e$
$\gamma\ (p+p\to p+p+\gamma)$	p	n	e^-
$\pi^0\ (p+p\to p+p+\pi^0)$	$\Sigma^+\ (\Sigma^+\to p+\pi^0)$	$\Lambda^0\ (\Lambda^0\to n+\pi^0)$	$\mu^-\ (\mu^-\to e^- + \nu + \tilde{\nu})$
$K_1^0\ (K_1^0\to 2\pi^0)$	—	$\Sigma^0\ (\Sigma^0\to\Lambda^0+\gamma)$	—
$K_2^0\ (K_2^0\to\pi^+ + \pi^- + \pi^0)$	—	$\Xi^0\ (\Xi^0\to\Lambda^0+\pi^0)$	—

1. Electric Charge of Neutron

The question of the possession of an electric charge by the neutron is closely linked with a more general question: Why are the charges of all elementary particles equal to either $\pm Q_e$, or zero? The neutron is usually regarded as an electrically neutral particle, but there is no theoretical barrier to its having a small charge, and the experimental detection of such a charge would be extremely important for theory.

In modern elementary particle theory the laws of conservation of electric, baryonic, and leptonic charges play a very important role. They are absolute laws, i.e., they are valid for all three types of interaction of elementary particles (strong, electromagnetic, weak). They have been experimentally confirmed with very high accuracy and, together with the laws of conservation of energy, momentum, and angular momentum, govern processes among elementary particles. The essence of these three conservation laws is that particles have three independent sets of numbers Q_i, B_i, L_i (electric, baryonic, and leptonic charge) which satisfy the following relationships:

$$\left.\begin{aligned}\sum_i Q_i &= \text{const in time,}\\ \sum_i B_i &= \text{const in time,}\\ \sum_i L_i &= \text{const in time.}\end{aligned}\right\} \qquad (1)$$

It can be shown [1, 2] that the application of the law of conservation of electric charge in conjunction with the laws of conservation of baryonic and leptonic charges to known elementary-particle reactions does not enable us to determine the relationship between the electric charges of all elementary particles. For instance, the absence of the reaction $p\to e^+ + \pi^0$ and other similar reactions owing to conservation of baryonic charge means that the charge Q_p of the proton is indeterminate relative to the positron charge Q_e. It does not necessarily follow from the law of conservation of electric charge in conjunction with the laws of conservation of baryonic and leptonic charges that the electric charges of the proton and electron are equal in absolute magnitude, and that the neutron charge is zero.

Table 1 gives four sets of particles with electric charges which are equal owing to the existence of the corresponding reactions.

It is clear that even if the scale of electric charge is fixed by measurement of Q_e, the values of Q_p and Q_n are still indeterminate. Thus, the question of the electric charge of the neutron will have to be settled experimentally.

All the experiments aimed at detecting the neutron charge can be divided into direct and indirect experiments. The direct experiments include experiments on the ionization of gases by neutrons and electrostatic deflection of a beam of very slow neutrons. The earliest experiments (Dee [3]) to detect ionization of gases by neutrons led to the estimate $Q_n < |Q_e|/700$. Shapiro and Éstulin [4] attempted to detect deflection of a beam of thermal neutrons in a homogeneous electrostatic field and obtained an estimate $Q_n < 6\cdot 10^{-12}|Q_e|$. The indirect experiments include attempts to detect the charges of un-ionized atoms and molecules [5-7]. The most accurate investigation, by King [5], gave upper limits for the charges of H2, D2, and SF_6 molecules. The author used a method in which the charge of a macroscopic volume of gas was measured after it had been freed from ions and electrons.

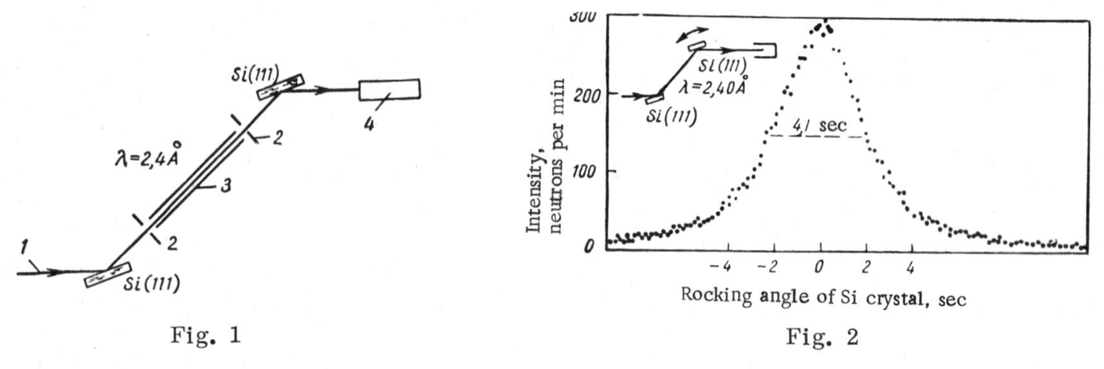

Fig. 1 Fig. 2

Fig. 1. Schematic diagram of double-crystal spectrometer and electrostatic deflecting system: 1) Neutron beam; 2) 1.5-mm slit; 3) electrostatic deflecting plates; 4) detector.

Fig. 2. Neutron intensity as function of angular orientation of second crystal.

If one assumes that $Q_{atom} = Z\Delta Q + NQ_n$, where $\Delta Q = |Q_e| - |Q_p|$, Q_n can be estimated independently of ΔQ from the estimates of $Q(H_2)$ and $Q(D_2)$. This procedure gives $Q_n < 3 \cdot 10^{-20} |Q_e|$. If it is assumed that $\Delta Q = Q_n$, the estimate of $Q(SF_6)$ gives $Q_n < 2 \cdot 10^{-22} |Q_e|$.

The indirect estimates are much more accurate than the direct data. Nevertheless, the latter must be regarded as more reliable, since it is quite possible that when particles combine to form an atom they may have their charge altered in some way. A recent accurate direct experiment is that of Shull, Billman, and Wedgewood [8]. The authors managed to attain extremely high angular sensitivity of the instrument by using two successive Bragg reflections from two perfect silicon crystals. A diagram of the apparatus is shown in Fig. 1, and the curve illustrating its angular sensitivity is shown in Fig. 2. The results of the measurement are given in Fig. 3. Treatment by the least squares method gave $Q_n = (-1.9 \pm 3.7) \cdot 10^{-18} |Q_e|$

Estimates of $Q_n = \Delta Q$ can also be obtained from cosmological considerations. Such estimates were made in [9], in particular. A simple consideration of the balance of gravitational attraction and electrostatic repulsion, which ensures the existence of macromatter, leads to $Q_n < 10^{-18} |Q_e|$, and a charge of $2 \cdot 10^{-18}\ Q_e$ is sufficient to account for the observed expansion of the universe within the framework of Newtonian mechanics.

It was suggested in [10] that the magnetic fields of stars and planets might be due to rotation of weakly charged atoms around a polar axis. It follows from these considerations that Q_n must be of the order of $2 \cdot 10^{-19}\ Q_e$.

A cosmological approach to elementary particle theory has recently been developed [11]. Elementary particles can be regarded in principle as almost closed universes. It is known that a closed world must be electrically neutral, and its total mass must be zero [12]. If an electric charge ε is introduced into such a world, this world will no longer be closed. Its mass will differ from zero and its minimum value will be $m = \varepsilon/\sqrt{\varkappa}$, where $\varkappa$ is the gravitational constant.* If the neutron is considered from this viewpoint, its possible electric charge is $Q_n \lesssim m\sqrt{\varkappa} \sim 10^{-18} |Q_e|$, which is very close to the estimates given above. Thus, it would be very desirable to have direct experiments aimed at detecting an electric charge of the neutron at the $10^{-19} |Q_e|$ level.

We note also that the detection of a small neutron charge would mean that the conservation of baryons would follow from the conservation of electric charge [2].

2. Electric Dipole Moment (EDM) of Neutron†

The discovery of CP violation‡ in K°-meson decays removed the theoretical barrier to the possession of an EDM by elementary particles, particularly the neutron. In its transformation properties the EDM of a particle is a polar vector directed along the spin (axial vector) of the particle and hence, as a result of

*A similar formula for any body of mass m can be derived from five-dimensional field theory [13].
†In this section we consider the question of the natural electric dipole moment of the neutron. For the induced EDM see below.
‡Or T violation in view of the well-known CPT theorem.

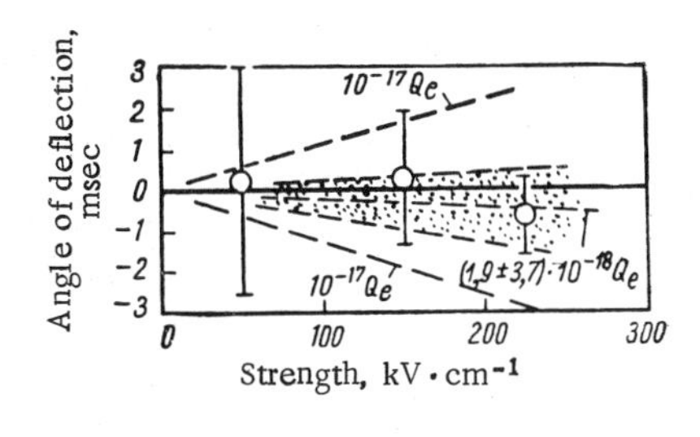

Fig. 3

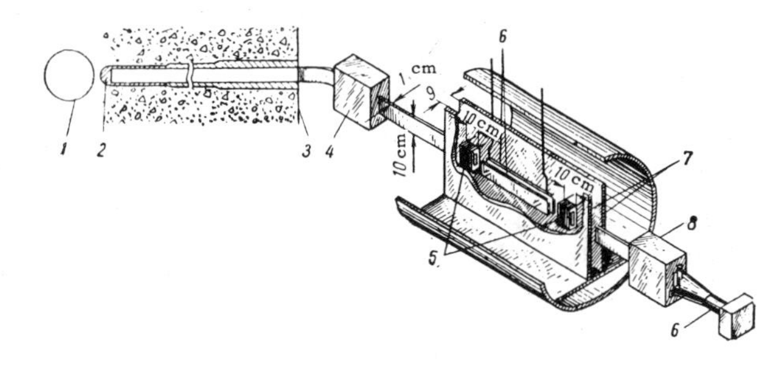

Fig. 4

Fig. 3. Angular deflection of neutron beam by electrostatic field.

Fig. 4. Schematic diagram of Oak Ridge National Laboratory instrument: 1) reactor core; 2) heavy-water moderator; 3) reactor shield; 4) polarizer (Co-Fe mirror); 5) rf coil; 6) electrostatic plates; 7) field magnet pole pieces; 8) analyzer (Co-Fe mirror).

operation of time reversal (reversal of direction of all velocities and replacement of initial state by final state), the relative directions of the EDM and the spin are altered (if the spin and EDM are parallel they become antiparallel). If T invariance exists, the direct and time-reversed states are equally likely and, hence, the observed value of the EDM will be zero [14].

Since numerous searches for violation of T invariance in other processes besides K°-meson decay have been unsuccessful, the detection of an EDM in the neutron, or in any other elementary particle, would be direct evidence of the nonuniversality of the T-invariance principle.

Two experimental methods have been used so far in the search for the neutron EDM. One of these methods — Rabi's magnetic resonance method — was used in 1957 by Smith, Purcell, and Ramsey [15]. They found that the neutron EDM divided by the electron charge $(d_n/e) < 5\cdot10^{-20}$ cm. In 1965–1967, as a result of the discovery of the decay of the K_2° meson into two charged pions, attempts to detect a neutron EDM were renewed [16, 17]. In this new experiment a beam of very slow neutrons (mean velocity 60 m/sec) obtained by means of a bent neutron-conducting tube was used. The neutron beam was subjected to the action of a steady magnetic field H_0 and a strong electrostatic field E which could be either parallel ($E\uparrow\uparrow$) or antiparallel ($E\uparrow\downarrow$), to the field H_0. A weak alternating field H_1 with frequency ω was applied perpendicular to the field H_0. The resonance frequency in this case is

$$\omega_0 = \frac{1}{\hbar I}(\mu_n H_0 \pm d_n E), \tag{2}$$

where I is the neutron spin and μ_n is its magnetic moment. The change in resonance frequency on reversal of the electrostatic field is

$$\Delta\omega_0 = \frac{2d_n E}{\hbar I}. \tag{3}$$

When the frequency ω approaches ω_0 there is a resonance change in the spin orientation, accompanied by a sharp change in the intensity of the detected particle beam. The neutron EDM can be determined from the following relationship:

$$d_n = \frac{\hbar}{2}\cdot\frac{\Delta N}{(E_{\uparrow\uparrow}+E_{\uparrow\downarrow})\dfrac{dN}{d\nu_{res}}}, \tag{4}$$

where ΔN is the change in the count rate due to the change in the electric field from $E_{\uparrow\uparrow}$ to $E_{\uparrow\downarrow}$; $dN/d\nu_{res}$ is the derivative of the count rate with respect to the alternating field frequency.

The sensitivity of the instrument can be increased by increasing the electric field, or by increasing the steepness of the resonance curve. Since the steepness depends on the resonance line width, and the latter, in turn, is limited by the uncertainty relationship, then it is better to increase the transit time of the neutron through the instrument.

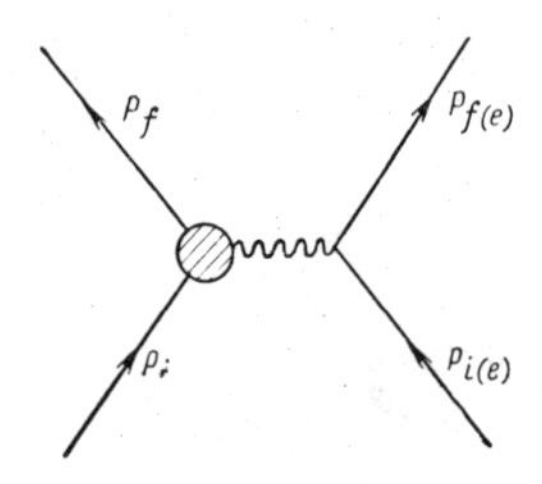

Fig. 5. Feynman diagram for electron scattering by proton in e^2 approximation.

Figure 4 shows a schematic diagram of the Oak Ridge National Laboratory instrument. With E=120 kV/cm and neutron transit time $\tau = 8 \cdot 10^{-3}$ sec, the value obtained in [17] for the neutron EDM was

$$(d_n/e) = (0.02 \pm 0.85) \cdot 10^{-22} \text{ cm}.$$

At the Vienna conference in September, 1968 [18] a new value was reported for the EDM:

$$(d_n/e) = (0.2 \pm 0.2) \cdot 10^{-22} \text{ cm}. \tag{5}$$

The second method was used in the Brookhaven Laboratory [19]. In the presence of an EDM the neutron undergoes an additional interaction with the intra-atomic Coulomb field. The corresponding amplitude in the Born approximation has the form

$$b'' = i\, d_n \frac{Ze\,(1-f)}{\hbar} \cdot \frac{\operatorname{cosec}\theta}{v}\,(\mathbf{pn}), \tag{6}$$

where Ze is the nuclear charge; $1-f$ is an electronic screening factor; f is the charge distribution form factor; ν is the neutron velocity; 2θ is the scattering angle; p is the unit neutron polarization vector, n is the unit scattering vector;

$$\mathbf{n} = \frac{1}{2k \sin\theta}\,(\mathbf{k} - \mathbf{k}_0), \tag{7}$$

where k and k_0 are the wave vectors before and after collision. The amplitude of (6) is imaginary and maximum when vector p is parallel to n and has its sign changed when the sign of the neutron polarization is reversed. The neutron scattering intensity is

$$\mathcal{I} \sim b^2 + (b' + b'')^2, \tag{8}$$

where $b + ib'$ is the nuclear scattering amplitude.

Shull and Nathans in their experiment tried to detect a change in intensity when the neutron polarization was reversed. Since the relative change in intensity is

$$\frac{\Delta\mathcal{I}}{\mathcal{I}} = \frac{4b'b''}{b^2 + b'^2}, \tag{9}$$

it is better to make the measurements on nuclei with small b. They used in their experiment a CdS single crystal, the intensity of reflection of neutrons from the (004) plane of which depends on the difference in the coherent scattering amplitudes of Cd and S:

$$\mathcal{I} \sim |a_{Cd} - a_S|^2, \tag{10}$$

$a_{Cd} = (3.8 + i\,1,\ 2)$ F and $a_S = 3.1$ F. About $4 \cdot 10^8$ neutrons were counted in the experiment over three months. Special measures were taken to compensate the effects of Schwinger scattering, i.e., scattering due to interaction of the magnetic moment of the moving neutron with the Coulomb field of the nucleus. The amplitude of this scattering is also imaginary, but it becomes zero if the polarization vector is exactly parallel to the scattering vector. The final result of the experiment was

$$\left(\frac{d_n}{e}\right) = (+2.4 \pm 3.9) \cdot 10^{-22} \text{ cm}.$$

Theoretical estimates of the neutron EDM are extremely diverse and, according to them, the neutron EDM lies in the range $10^{-19} > d_n/e > 10^{-31}$ cm. The estimates in the region of 10^{-22} cm are based on the assumption of T violation in weak interactions. From this viewpoint it is very desirable to increase the sensitivity of EDM measurements by one or two orders. Work in this direction is being carried out at Oak Ridge and Brookhaven. A very promising way of increasing the sensitivity of the first method is to use ultracold neutrons, which increases the residence time of particles in the instrument [21]. The second method can be improved by the use of single crystals with a very small coherent scattering amplitude, particularly a mixture of tungsten isotopes enriched with tungsten-186. With an appropriate concentration of W186 the coherent scattering amplitude of such a mixture may even be zero [20].

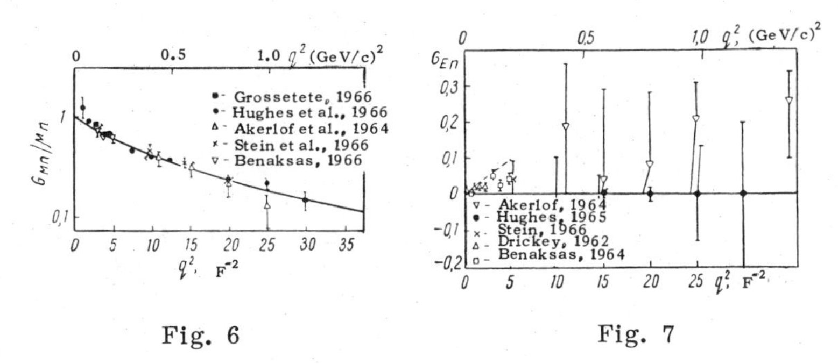

Fig. 6 Fig. 7

Fig. 6. Experimental relationship between magnetic form factor G_{Mn}/μ_n and q^2. Curve calculated from formula (23).

Fig. 7. Experimental relationship between G_{En} and q^2.

3. Electromagnetic Structure of Neutron

Electromagnetic Form Factors of Neutron. The properties and structure of any elementary particle are closely related to the properties of other particles. This is due to the existence, for instance, of a cloud of virtual particles around the real neutron. The particles are responsible for the interaction between the neutron and other particles. Although the virtual particles themselves are unobservable in principle, all the effects which they produce are real, like the nuclear forces engendered by the exchange of virtual mesons.

The characteristic dimensions of the cloud of virtual particles depend on their mass. Thus, in order of magnitude the radius of the π-meson cloud is $r_\pi \sim \hbar/m_\pi c$, where m_π is the π-meson mass; that of the K-meson cloud is $r_k \sim \hbar/m_k c$, where m_k is the K-meson mass; and so on. Our present knowledge of the K-mesonic and other deeper shells of the neutron is very slight. We know a little only about the pion cloud.

The electromagnetic structure of the neutron (or of a nucleon in general) is revealed by the interaction of any charged particle with it (e.g., by the scattering of electrons or muons on the nucleon, etc.). This question has been dealt with most fully for the case of electron scattering on nucleons [22]. The laws of electromagnetic interaction are fairly well known and, hence, information regarding nucleon structure can be derived from scattering data. The scattering cross section is first calculated for a point nucleus; electrostatic interaction and the interaction with the normal and anomalous magnetic moments of the particle are neglected. This cross section is then used for calculation of scattering on an extended nucleon consisting of many particles by summation of the waves scattered by each particle. The method employed here is well known from x-ray scattering theory, but in the case of x-ray scattering one is dealing with ordered scattering centers in a crystal lattice, whereas in the case of the nucleon the density distribution is regarded as irregular. Hence, the sum over the different partial scattering amplitudes is replaced by an integral, which is called the form factor.

A rigorous theoretical examination of this question, satisfying the requirements of Lorentz and gradient invariance in the approximation corresponding to the Feynman diagram (Fig. 5), indicates that the representation of the electromagnetic structure of a nucleon requires two form factors – F_1 and F_2 – which are functions of the square of the transmitted four-momentum q*:

$$q^2 = (p_f - p_i)^2 = (p_{f(e)} - p_{i(e)})^2, \tag{11}$$

where p_i, p_f, $p_{i(e)}$, and $p_{f(e)}$ are the initial and final four momenta of the proton and electron.

It can be shown [23] that when $E \gg m$,† where E is the energy and m the mass of the electron,

$$q^2 = -\frac{(2E \sin \theta/2)^2}{1 + 2\dfrac{E}{M}\sin^2\dfrac{\theta}{2}}. \tag{12}$$

where M is the nucleon mass, and θ the electron scattering angle. For the case described by the Feynman diagram (Fig. 5), $q^2 < 0$ (spacelike values of q).

*We recall that the spatial components of the four-momentum vector are the same as the particle momentum p, while the time component is i E/cj, where E is the particle energy.

†A system of units in which $\hbar = c = 1$ is used.

The Dirac form factor $F_1(q^2)$ describes the spatial distribution of nucleon charge and the associated Dirac magnetic moment. The Pauli form factor $F_2(q^2)$ is related to the spatial distribution of the anomalous magnetic moment. The form factors represent the cumulative result of all the effects produced by any number of virtual particles. This means in diagram terms (Fig. 5) that the interactions occurring at the nucleon-photon vertex are not detailed. The hatched region represents only the ultimate result. In the limiting case of low energies (small momentum transfers), when the nonrelativistic approximation is valid, the form factors represent the Fourier transform of the spatial distribution of electric charge $\rho(r)$ and magnetic moment $m(r)$. For instance, in the case of a spherically symmetrical charged particle $F_1(q)$ is given by:

$$F_1(q)=\frac{4\pi}{qe}\int_0^\infty \rho(r)\sin(qr)\,r\,dr, \tag{13}$$

where $\rho(r)$ is the electric charge distribution density. We find from formula (13) that

$$\rho(r)=\frac{e}{2\pi^2 r}\int_0^\infty F(q)\sin(qr)\,q\,dq. \tag{14}$$

It follows from (13) that $F_1(0)=1$. Similarly, $F_2(0)=1$. A point particle with charge e and total magnetic moment $e\hbar/2Mc+\mu_0$, where $\mu_0=\varkappa e\hbar/2Mc$, is a particle for which F_1 and F_2 are independent of q and equal to 1. The particle has an electromagnetic structure only in the case where $F_1(q^2)$ and $F_2(q^2)$ are not constant. This statement is an exact definition of the electromagnetic structure of a particle for all q^2 [17].

If $\rho(r)$ and $m(r)$ are known, the rms electric and magnetic radii of the particle can be determined:

$$\langle r_e^2\rangle=\frac{\int r^2\rho(\mathbf{r})\,d^3\mathbf{r}}{\int \rho(\mathbf{r})\,d^3\mathbf{r}}=\frac{6F_1'(q^2)_{q^2=0}}{F_1(0)}; \tag{15}$$

$$\langle r_m^2\rangle=\frac{\int r^2 m(\mathbf{r})\,d^3\mathbf{r}}{\int \rho(\mathbf{r})\,d^3\mathbf{r}}=\frac{6F_2'(q^2)_{q^2=0}}{F_2(0)}. \tag{16}$$

In the case of the neutron $F_{1,n}=0$ and the rms radius is given by

$$\langle r_e^2\rangle_n=\frac{6F_{1,n}'(q^2)_{q^2=0}}{F_{1,p}(0)}=6F_{1,n}'(q^2)_{q^2=0}. \tag{17}$$

At small q^2 functions F can be expanded in a series:

$$F_{1,2}(q^2)\approx F_{1,2}(0)+q^2F_{1,2}'(q^2)_{q^2=0}+\ldots=F_{1,2}(0)\left(1+\frac{1}{6}q^2\langle r^2\rangle_{1,2}+\ldots\right). \tag{18}$$

It is clear from (18) that when $q^2\to 0$ the rms radius determines the slope of the form factor and in this sense can serve as a characteristic of the form factor, if the latter is not a rapidly varying function of q^2. Otherwise the value $\langle r^2\rangle$ will not be related to the general behavior of the form factor. From (17) and (18) and for the case of a neutron we obtain

$$F_{1,n}(q^2)\approx q^2\,\frac{1}{6}\langle r_e^2\rangle_n+\ldots; \tag{19}$$

$$F_{2,n}(q^2)\approx F_{2,n}(0)\left(1+\frac{q^2}{6}\langle r_m^2\rangle_n+\ldots\right)=1+\frac{q^2}{6}\langle r_m^2\rangle_n. \tag{20}$$

As already mentioned, each of the form factors F_1 and F_2 does not describe merely the distribution of charge or magnetic moment. There is a coordinate system, however, in which the combinations of form factors F_1 and F_2 correspond very accurately with the distribution of electric charge and magnetic moment. The new form factors are expressed in terms of F_1 and F_2 as follows [24]:

the charge form factor

$$G_E(q^2)=F_1(q^2)+\frac{q^2}{4M^2}\varkappa F_2(q^2), \tag{21}$$

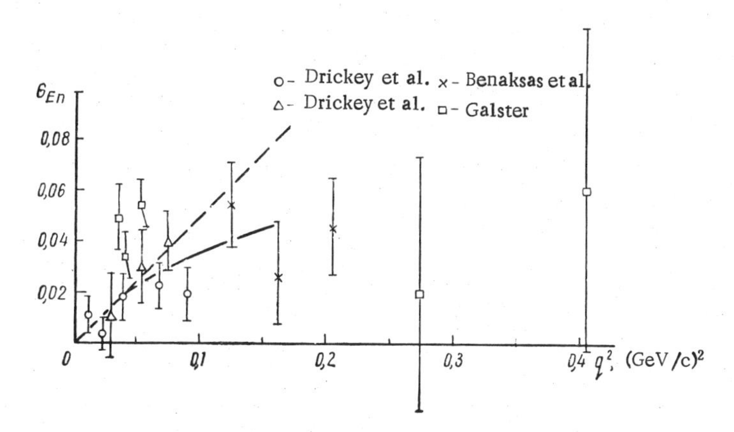

Fig. 8. Experimental relationship between neutron charge for form factor G_{En} and q^2 for $q^2 \leq 0.4$ $(GeV/c)^2$.

the magnetic form factor

$$G_M(q^2) = F_1(q^2) + \varkappa F_2(q^2). \tag{22}$$

The corresponding coordinate system is specified by the requirement that the spatial components of the vector $p_i + p_f$ are equal to zero. In such a system the three momenta of the initial and final protons are equal and opposite in direction, and the energies corresponding to this case are equal. The introduction of G_E and G_M also makes the treatment of the experimental data easier.

The bulk of information on G_E and G_M at present has been obtained from data for electron scattering on protons and deuterons. Figures 6 and 7 show the experimental values of G_{Mn}/μ_n and G_{En} reported at the Stanford Conference in 1967 [25]. It was experimentally established that G_{Ep}, G_{Mp}, and G_{Mn} are connected by the following relationship:

$$G_{Ep}(q^2) = \frac{G_{Mp}(q^2)}{\mu_p} = \frac{G_{Mn}(q^2)}{\mu_n} \approx \frac{1}{(1+q^2/0.71)^2}. \tag{23}$$

The results of recent, very accurate measurements made in Bonn University [26] reveal slight deviations from this relationship.

Figure 8 shows the experimental G_{En} data for $q^2 \leq 0.4$ $(GeV/c)^2$. The broken line – slope at $q^2 = 0$ – was obtained from ne-interaction data. The solid curve is the result of dispersion calculations [27]. Since the neutron form factor values were obtained from an analysis of experiments on electron scattering on deuterons, and there is no strict relativistic theory for the deuteron, these values are not reliable enough. This applies particularly to G_{En} at low q^2. The information about $(\partial G_{En}/\partial q^2)_{q^2=0}$ obtained from ne-interaction experiments is more reliable (see section "Neutron-Electron Interaction").

Polarizability of Neutron. The concept of polarizability of nucleons was introduced in connection with photon scattering and pion photoproduction on nucleons [28, 29], and with neutron scattering on heavy nuclei [30, 31]. Along with the charge, magnetic, and other moments, the polarizability (electric and magnetic) is a characteristic of the particle which has to be introduced for a complete description of the interaction of elementary particles. The polarizability is an index of the deformation of the meson cloud of the particle due to electric and magnetic fields. It is zero if the particle has a rigid, nondeformable structure, or is a point particle. It is rather difficult to determine the polarizability of a nucleon experimentally. Some measurements have been made in the case of the proton. Its electric polarizability has been determined [32]. In the case of the neutron all we can say at present is that there are experimental estimates of quantities of interest. These estimates have been obtained from experiments with photons and neutrons.

Photon Scattering. Scattering of photons by particles with spin 1/2 and an anomalous magnetic moment was examined in [28, 29, 33-38]. In the low-frequency region, i.e,, when $\omega \ll m$, where ω is the photon frequency, and m is the pion mass, the obvious approach is to expand the scattering amplitude in terms of the photon frequency. In [28, 33, 34] linear frequency terms were considered. It was shown that in this case the scattering can be described if the charge, mass, and anomalous magnetic moment are

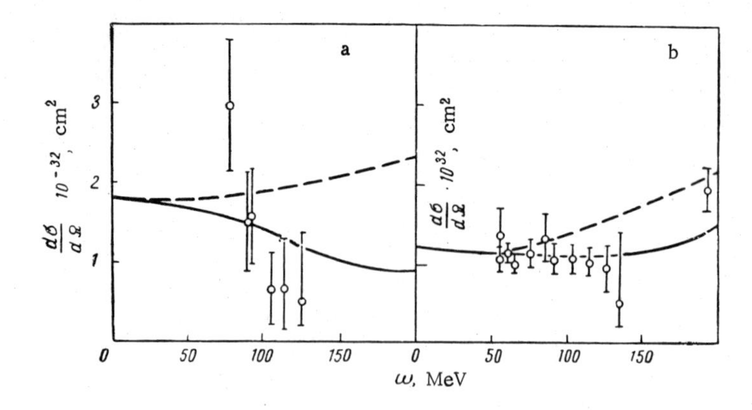

Fig. 9. Cross section for elastic scattering of photons on protons at angles of 45° (a) and 90° (b) as functions of energy. The continuous curve represents the results of calculation with $\alpha_p=0.9\cdot10^{-42}$ cm³ $\beta_p=0.2\cdot10^{-42}$ cm³. The dashed curve represents the results of calculation with $\alpha_p=\beta_p=0$.

known. This description, however, is not exhaustive. In [28] the possible implication of intermediate states corresponding to high excitation of the nucleon in low-energy processes was considered. The absorption of a photon by a nucleon leads to the induction of electric and magnetic moments in the nucleon and their subsequent emission. A similar effect is known in optics under the name of Rayleigh scattering. The Rayleigh scattering amplitude is proportional to the square of the γ-ray frequency. It was shown that at energies below the pion production threshold a similar consideration necessitates the introduction of two new parameters – the α and β coefficients of the electric and magnetic polarizability of a nucleon.

There have been many theoretical investigations devoted to the definition of the concept of polarizability in photon scattering and the theory of the Compton effect. It has been shown that in a consideration of low-energy photon scattering on a system with spin 1/2, accurate to terms containing the cube of the photon frequency, the formula for the scattering amplitude contains another three terms in addition to the charge, mass, and anomalous magnetic moment. These are α, β, and $\langle r_e^2\rangle$, the rms radius of the charge distribution. A fuller review of these investigations can be found in [39].

The values of α_p and β_p can be obtained from experiments on the elastic scattering of photons on protons at energies below the pion production threshold. These experiments were analyzed in [40]. The results of the analysis give the following values for the proton (Fig. 9): $\alpha_p=(0.9\pm0.2)\ 10^{-42}$ cm³, $\beta_p=(0.2\pm0.2)\ 10^{-42}$ cm³.

It should be noted that inclusion of terms higher than the third order in the expansion of the scattering amplitude in terms of the frequency can slightly alter the value of α_p. If their effect is to be excluded,the measurements must be made at γ-ray energies of the order of 10-20 MeV. This involves great difficulties, however, owing to the great reduction of the effect with energy reduction.

It is much more difficult to evaluate the polarizability of the neutron, since it is impossible to obtain direct experimental data on photon scattering by neutrons. It will be possible in the future to carry out a direct experiment in which neutrons will collide with a powerful laser beam, but in the meantime the neutron polarizability has to be estimated from information obtained from Compton-effect experiments, particularly on the deuteron. The difficulties in the analysis of these experiments; as in the analysis of experiments on electron scattering on deuterons, are due primarily to the lack of a strict relativistic theory of the deuteron. The fact that the neutron is in motion in the deuteron must be taken into account. At the present stage of relativistic deuteron theory the problem of the effect of motion can be solved only approximately. Several assumptions have to be made and these ultimately reduce the reliability of the results.

The analysis of deuteron Compton-effect data is usually performed within the framework of the impulse approximation [41]. In this approximation the scattering amplitude on the deuteron is considered to be equal to the sum of the scattering amplitudes on the free proton and neutron, and the momentum distribution of these particles is taken to be the same as in the deuteron. However, as was shown in [29, 42, 43], the experimental data for photon scattering on deuterons [44, 45] do not fit into the framework of the impulse approximation. Photodisintegration of the deuteron has a pronounced effect on the scattering amplitude of the considered process in the energy range 50-100 MeV. This inelastic process, together with pion photoproduction at high energies, makes the impulse approximation invalid for the examination of the Compton effect on the deuteron in a wide energy range. The ultimate result is that no reliable conclusions regarding the polarizability of the neutron can be derived from experiments on photon scattering on deuterons.

In a recent investigation [42] the authors examined experimental deuteron Compton-effect data and came to the conclusion that the electric polarizability of the deuteron cannot exceed the corresponding value for the proton by more than 40%. This conclusion, however, was based on the impulse approximation and, as the authors themselves state, the consideration of some exchange effects can lead to a much higher value of α_n.

In [29] attention was given to the fact that a study of $\gamma + d \to d + \gamma$ and $\gamma + He^4 \to He^4 + \gamma$ processes can simplify the interpretation of the experimental data. As regards the first process, it is evident that at energies of the order of several tens of megaelectron-volts it is difficult to separate from the $\gamma + d \to p + n + \gamma$ process, while the polarizability is proportional to the square of the photon frequency, and at very low photon energies the effect is small. The possibility of evaluating α_n from an investigation of the $\gamma + He^4 \to He^4 + \gamma$ reaction was considered in [37]. In this case the contribution of inelastic processes to the scattering amplitude is much less than in the case of the deuteron, since the helium photodisintegration threshold is fairly high ($E \sim 20$ MeV). As was shown in [37], photon scattering on helium at an angle of 90° can be measured and the value of $\bar{\alpha}$ obtained from:

$$\bar{\alpha} = \alpha + \frac{1}{3}\left(\frac{e^2}{M}\right)\langle r_e^2 \rangle, \tag{24}$$

where $\langle r_e^2 \rangle$ is the rms radius of helium determined from experiments on $e - He^4$ scattering, and the helium polarizability $\alpha = \alpha_N + 2\alpha_p + 2\alpha_n$. The value of α_N can be determined from helium photodisintegration experiments by using the expression known from dispersion relations:

$$\alpha_N = \frac{1}{2\pi^2}\int \frac{\sigma_{E1}\, d\omega}{\omega^2}, \tag{25}$$

where σ_{E_1} is the total dipole absorption cross section (with meson photoproduction neglected). The integral (25) can be evaluated by replacing σ_{E_1} by σ_{tot}. This gives $\alpha_N = (70 \pm 4) \cdot 10^{-42}$ cm^3. The accuracy of this evaluation determines the upper limit of α_n. This method can be used to determine the value of α_n, if it is more than, or of the order of, $5 \cdot 10^{-42}$ cm^3. There have not yet been any experiments on photon scattering on helium.

An indirect estimate of the neutron polarizability can be obtained from meson photoproduction data [29, 38]: $\gamma + p \to \pi^+ + n$, $\gamma + n \to \pi^- + p$. From dispersion relations we have [38]:

$$\alpha = \frac{1}{2\pi^2}\int_{\omega_t}^{\infty} \frac{d\omega}{\omega^2}\left\{|E_1|^2 + 2|E_3|^2 + \frac{1}{3}|E_2|^2 - \frac{1}{6}|M_2|^2\right\}; \tag{26}$$

$$\beta = \frac{1}{2\pi^2}\int_{\omega_t}^{\infty} \frac{d\omega}{\omega^2}\left\{|M_1|^2 + 2|M_3|^2 + \frac{1}{3}|M_2|^2 - \frac{1}{6}|E_2|^2\right\}, \tag{27}$$

where E_i and M_i are the partial amplitudes of pion production on the nucleon (electric and magnetic types). It is known from experiments on pion photoproduction on protons and deuterons that the ratio of the cross section $\sigma_{\pi^-}/\sigma_{\pi^+}$ near the threshold is ~ 1.3 [45]. Since electric dipole pion production dominates at these energies, it can be assumed that $|E_1^{\pi^-}|^2 \sim 1.3|E_1^{\pi^+}|^2$. Hence, taking into account the value of α_p, we can obtain for the neutron $\alpha_n \sim 1.2 \cdot 10^{-42}$ cm^3. A similar value was found in [46]. The true value of α_n is evidently close to the obtained values. There have been no experimental estimates of β_n.

Neutron Scattering by Heavy Nuclei. An upper estimate of the electric polarizability of the neutron can be obtained by investigating neutron scattering by heavy nuclei. The concept of electric polarizability of the neutron was introduced in connection with the question of neutron scattering in [30, 31]. In the Hamiltonian for neutron-nucleus interaction there appears an additional term of the form $\frac{1}{2}\alpha_n E^2$, where

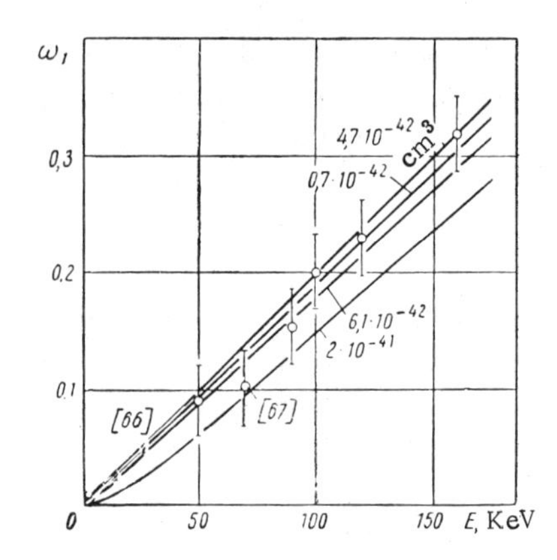

Fig. 10. ω_1 as function of neutron energy. Curves calculated from formulas (31) and (32) with fixed $\alpha = 1.91 \cdot 10^{-3}$ and indicated values of α_n.

E is the electric field of the nucleus. In [31] the amplitude of polarization scattering of the neutron in the Coulomb field of the nucleus was calculated in the Born approximation:

$$f(\theta) = \frac{M\alpha_n}{2R}\left(\frac{Ze}{\hbar}\right)^2 KR\left(\frac{\sin KR}{K^2R^2} + \frac{\cos KR}{KR} + \sin KR\right) \tag{28}$$

where $K = (2\sin\theta/2)/\lambda$; R is the radius of the nucleus; M is the reduced neutron mass.

On the assumption that the energy of nuclear interaction is independent of the spin the following expression was obtained for the differential cross section for elastic scattering of unpolarized neutrons:

$$\frac{d\sigma}{d\Omega} = |f_0(\theta)|^2 + \frac{1}{4}\varkappa_n^2\left(\frac{\hbar}{Mc}\right)^2\left(\frac{Ze^2}{\hbar c}\right)^2 \operatorname{ctg}^2\frac{\theta}{2} + 2\operatorname{Re} f_0(\theta) f(\theta) + f^2(\theta), \tag{29}$$

where $f_0(\theta)$ is the nuclear scattering amplitude. The second term of expression (29) represents the Schwinger scattering resulting from the interaction of the neutron magnetic moment $\mu_n = \varkappa_n e\hbar/2Mc$ with the Coulomb field of the nucleus. Since polarization scattering is due to long-range forces the effects caused by them at neutron energies of the order of several megaelectron-volts must be sought at small angles (< 10°). Besides the effect due to neutron polarizability, there will also be Schwinger scattering in the small-angle region, but the effect of this can be taken into account sufficiently accurately by calculation.

The main difficulty in interpretation of the experimental data lies in the assessment of nuclear interaction. Since there is no appropriate rigorous theory, the consideration of the effect of nuclear interaction necessitates model ideas of a different kind. In a first approximation nuclear scattering at angles < 10° at energies above 1 MeV can be regarded as isotropic. In fact, a neutron which has undergone interaction in a region of radius R (radius of nucleus) has an uncertainty in momentum of $\Delta p \sim \hbar/R$, which leads to uncertainty in the scattering angle θ: $\tan\theta = \Delta p/p \sim \hbar/pR = 1/kR$. If $R \sim 10^{-12}$ cm, the kinetic energy of the neutron $E_n \sim 2$ MeV, then $\tan\theta \sim 1/3$ and $\theta \sim 18°$, i.e., down to angles of 20° (or less) scattering is equally probable. In the first investigations of low-angle scattering of neutrons with energy of the order of several megaelectron volts [47-49], the results of the experiment after correction for Schwinger scattering were compared either with the relationship for nuclear scattering of the form $\sigma(\theta) = A + B\cos\theta$ [47, 48] or with the diffraction formula [49]. In both cases the behavior of the differential cross section in the low-angle region showed deviations from the above relationships for uranium, thorium, and plutonium. There were no deviations in the case of lighter nuclei. If these deviations are attributed to the effect of the electric polarizability of the neutron,the value of α_n obtained ($\alpha_n \sim 10^{-40}$ cm³) is too high, which contradicts not only the experimental value of the proton electric polarizability, from which α_n cannot differ too greatly, but also a whole series of theoretical calculations made for nucleons [50-54].

In later experimental investigations of low-angle scattering [55-60] covering the neutron energy range from 0.5 to 14 MeV, the results of the measurements were compared with calculations based on the optical model of the nucleus. It is difficult at present to speak of any regularity in the differences between the calculated and experimental value. Nevertheless, the purely classical optical model is not satisfactory for small angles –the inclusion of the long-range potential improves the agreement between calculations and experiment. Anikin came to this conclusion after a thorough analysis of the experimental data of [47-49, 55-60] by using for the computer calculations a program which included a long-range potential of the form $1/r^4$ and which allowed a direct selection of the parameters of the optical model during calculation so that the best agreement with the experimental data for angular scattering distributions and total cross sections could be secured.

In [61-64] attempts were made to include effects capable of causing various kinds of anomalies in the low-angle region, but they were not successful. The presence of uncertainty in calculations of the behavior of the nuclear scattering cross section greatly reduces the accuracy of the upper limit for the neutron electric polarizability derived from the discussed experiments. All that one can say is that $|\alpha_n| < (2\text{-}3) \cdot 10^{-40}$ cm³.

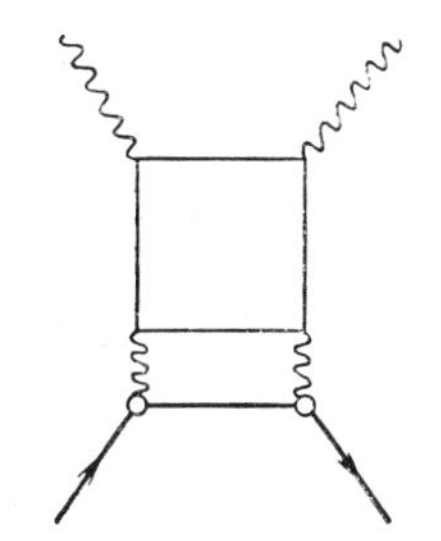

Fig. 11. Scattering due to polarizability of electron-positron cloud of charged particle in Coulomb field.

Another series of experiments from which an upper estimate for α_n can be obtained are experiments on neutron scattering by heavy nuclei in the low-energy region (less than 100 keV) [65, 66]. If the differential scattering cross section is put in the form

$$\sigma(\theta) = \frac{\sigma_0}{4\pi}\left[1 + \sum_{l=0}^{\infty} \omega_l P_l(\cos\theta)\right], \tag{30}$$

where σ_0 is the total potential scattering cross section and the well-known approximate relationship $\delta_l \sim (kR)^{2H}$ is used for the scattering phases on the short-range potential of the nucleus, it is easy to show that in the case of purely nuclear interaction the coefficient ω_1 is a linear function of the neutron energy E. If the interference of nuclear scattering and polarization scattering is taken into account ω_1 will contain a term proportional to k, so that

$$\omega_1 = aE + bE^{1/2}, \tag{31}$$

where

$$b = -2.5 \times 10^{-4} \frac{M^{3/2} e^2}{\hbar^3} \cdot \frac{\alpha_n Z^2}{\sigma_0^{1/2}}. \tag{32}$$

If the energy E in formula (31) is expressed in kiloelectron-volts. Such an analysis was performed in [65] for scattering of neutrons with energies of 50-300 keV on uranium nuclei and led to the estimate $|\alpha_n| < 20 \cdot 10^{-42}$ cm^3. A more suitable region for investigations is the energy region below 50 keV, since in this energy region the nonlinear variation of ω_1 with E is much more appreciable. The authors of [66] carried out experiments on neutron scattering in the energy region 0.6 – 26 keV (time-of-flight method, pulsed reactor). The material chosen for the scatterer was lead, which has no strong neutron resonances in the investigated energy region. This ensures that there is no uncertainty in the analysis due to neglect of the role of resonances. The obtained data were treated by the least-squares method in conjunction with the data of Langsdorf et al. in [67]. This treatment led to the conclusion that with a probability of 70% the value of α_n lies in the following range (Fig. 10): $-4.7 \cdot 10^{-42}$ cm$^3 \lesssim \alpha_n \lesssim 6.1 \cdot 10^{-42}$ cm^3. This result is the best direct experimental estimate of α_n so far.

Polarizability of Nucleons due to Nonlinear Electrodynamic Effects. So far we have been dealing with the polarizability of the nucleon due to the cloud of virtual pions surrounding it. The nucleon, however, is always surrounded by a cloud of virtual electron-positron pairs of radius $\hbar/2m_e c \sim 10^{-11}$ cm. This was considered in [68]. As was reported in [68], it is possible that an appreciable role in the scattering of nucleons on nuclei is played by sixth-order diagrams in e with a photon scattering block due to the Coulomb field of the nucleus. The corresponding diagram is shown in Fig. 11. The total coefficient for the quadratic frequency terms in this case can be written in the form [68]

$$\left.\begin{aligned} \alpha_t &\sim \int_{\omega_t}^{\infty} \frac{\sigma_t(\omega)\, d\omega}{\omega^2}, \\ \alpha_t &= \alpha_m + \alpha_e, \end{aligned}\right\} \tag{33}$$

where α_m and α_e are the contributions from mesonic and electrodynamic processes.

Because of the special nature of the angular dependence of γp scattering only the mesonic part, i.e., α_m, plays a significant role in large-angle scattering at energies of the order of 50-100 MeV. In the case of small-angle scattering or at very low energies the effect of electron-positron pair production becomes significant. Substituting in (33) the cross section for pair production on the proton we find that $\alpha_{ep} = 0.7 \cdot 10^{-39}$ cm^3. Such a large value could certainly be detected in the case of proton scattering by heavy nuclei. For a similar estimate for the neutron we need to know the cross section for e^+e^- pair production on the neutron. In [68] the cross section for e^+e^- pair production on the anomalous magnetic moment of the neutron was calculated. The results of calculation from formula (33) give $\alpha_{en} \sim 10^{-44}$ cm^3. No neutron or photon experiments have shown such a small value so far.

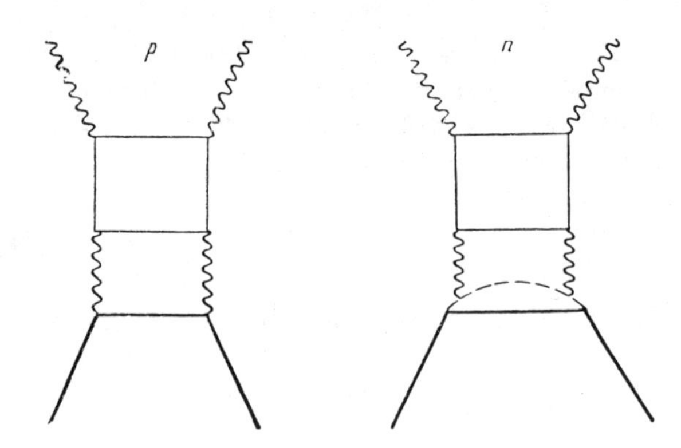

Fig. 12. Scattering due to polarizability of electron-positron cloud of proton and nucleon in Coulomb field.

It was reported in [69] that the main contribution to α_{en} could be made by pair production due to the electrically charged pion cloud, and not to the anomalous magnetic moment of the neutron. Diagrams representing the electromagnetic interactions of nucleons by their polarizability due to the cloud of virtual electron-positron pairs are shown in Fig. 12 for the proton and neutron. The polarizability of the electron-positron cloud in the proton and neutron [70] has the following order:

$$\alpha_{ep} \sim \overline{G}_{Ep}\left(\frac{e^2}{\hbar c}\right)\left(\frac{\hbar}{2m_e c}\right)^3; \tag{34}$$

$$\alpha_{en} \sim \overline{G}_{En}\left(\frac{g^2}{\hbar c}\right)\left(\frac{e^2}{\hbar c}\right)\left(\frac{\hbar}{2m_e c}\right)^3, \tag{35}$$

where $\bar{G}_{Ep}$ and $\bar{G}_{En}$ are the mean values of the charge form factor of the proton and neutron; g is the strong coupling constant. $\bar{G}_{Ep}$ for the proton is known, and the substitution of the values of the constants in [34] gives $\alpha_{Ep} \sim 1.5 \cdot 10^{-39}$ cm^3, which is close to the value obtained from expression (33). The neutron charge form factor, according to the data of [25, 71, 72], in the region of $q^2 = (0-2)f^{-2}$ *does not exceed 0.03-0.05. Hence, the value of α_{en} will be small and can hardly have a significant effect on neutron scattering by heavy nuclei at the considered energies, although more accurate calculations are required for a definitive conclusion, since different values of q^2 make a contribution to the cross section in the case of interaction of the neutron with the Coulomb field of the nucleus. In addition, as was reported in [68], there may be a strong energy dependence at the threshold for pair production by collision of a neutron with the nucleus. It is possible that effects of this kind can cause the anomalous scattering of neutrons by heavy nuclei at small angles at energies of several megaelectron-volts and have no effect in the very low-energy region. At any rate, this question requires further investigation.

Within the frameworks of SU_3 and SU_6, relationships between the electric and magnetic polarizabilities of baryons and mesons [73] can be obtained. The electric polarizabilities of baryons are connected by the relationships:

$$\left.\begin{array}{l} \alpha_p = \alpha_{\Sigma^+}; \quad \alpha_n = \alpha_{\Xi^0}; \quad \alpha_{\Sigma^-} = \alpha_{\Xi^-}; \\ \alpha_{\Sigma^0} - \alpha_{\Lambda^0} = 1\,(\alpha_{\Lambda^0} - \alpha_n) = \frac{2\sqrt{3}}{3}\,\alpha_{\Sigma^0\Lambda^0}, \end{array}\right\} \tag{36}$$

where $\alpha_{\Sigma^0\Lambda^0}$ is the matrix element of two-photon decay

$$\Sigma^0 \rightarrow \Lambda^0 + 2\gamma.$$

The baryon magnetic polarizability β satisfies similar relationships.

In [74] relationships between the polarizabilities of baryons and mesons were obtained on the basis of the nonrelativistic quark model of elementary particles possessing SU_6 symmetry. In particular, within the framework of this model $\alpha_p = \alpha_n$, $\beta_p = 1.5 \cdot 10^{-43}$ cm^3, $\beta_n = 1.2 \cdot 10^{-43}$ cm^3.

*The momentum $1f \approx 10^{13}$ cm^{-1} = 197 MeV/c.

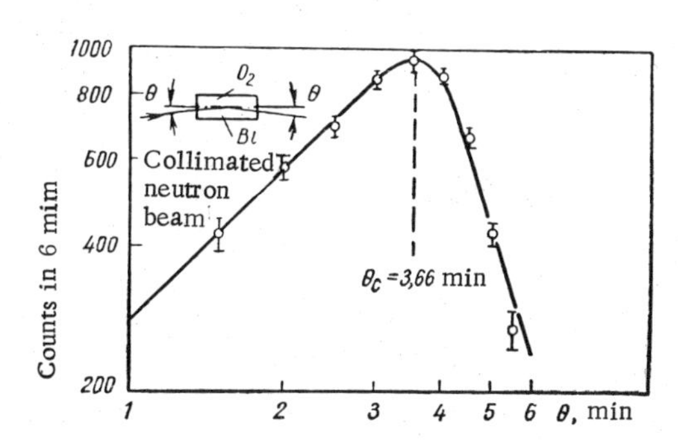

Fig. 13. Setup and results of experiment of Hughes et al. [85, 86].

Thus, electrons and photons are mutually supplementary means of investigating the electromagnetic structure of nucleons. The form factors G_E and G_M, which characterize the distribution of charge and magnetic moment of the nucleon, are determined from electron scattering. Polarizability – the ability of the meson cloud of the nucleon to undergo deformation – is manifested in interaction with photons. There is evidently a close link between the form factors and the polarizability, but this question has not been examined yet.

4. Neutron-Electron (ne) Interaction

Introduction. Since the neutron has a magnetic moment, there is a spin-dependent interaction of the dipole moments of the neutron and electron, and also a velocity-dependent interaction of the magnetic moment of the neutron with the electric field of the moving electron. These interactions, which lead to magnetic scattering of slow neutrons on atoms, are investigated by the techniques of neutron optics: small-angle scattering, diffraction, reflection, etc. [75, 76]. Spin- and velocity-independent interaction of the neutron and electron can occur if the neutron has a spatial electromagnetic structure. In this case there may be regions with non-zero charge density inside the neutron. Any charged particle getting "inside" the neutron will be subjected in this case to the action of electromagnetic forces. Any other charged particle could be taken as the object of investigation in place of the electron. In the case of the proton, however, the nuclear interaction between it and the neutron is not known well enough for us to identify the small electromagnetic contribution. For other particles, μ mesons for instance, such experiments are impossible owing to inadequate experimental technique.

The sought cross section for scattering of the neutron by the electron is eight orders less than the nuclear interaction cross section. Nevertheless, this effect can be detected from the interference between the elastic scattering of slow neutrons by the nucleus and electrons of the atom. The differential cross section for coherent scattering of slow neutrons with wavelength λ of the order of the atomic radius is described (absorption being ignored) by the relationship

$$\sigma(\theta) = \left| a + Zf\left(\frac{\sin\theta}{\lambda}\right) a_{ne} \right|^2, \tag{37}$$

where a is the coherent nuclear scattering amplitude; a_{ne} is the scattering amplitude of the neutron on the electron; $f(\sin\theta/\lambda)$ is the atomic form factor for electrons, which is known from x-ray scattering [77]. Estimates show that the relative magnitude of the additional contribution to the scattering

$$\frac{\Delta\sigma}{\sigma} \approx 2Z\,\Delta f\,\frac{a_{ne}}{a} \tag{38}$$

is about 1%. An effect of this magnitude can be measured. In neutron physics the ne interaction is usually described by a constant equivalent potential V_0, connected with the scattering amplitude by the expression

$$a_{ne} = \frac{2}{3}\cdot\frac{MR^3}{\hbar^2}V_0. \tag{39}$$

For the radius R of the potential well we take the classical electron radius $e^2/mc^2 = 2.8\cdot10^{-13}$ cm. Of course, the potential V_0 is of a purely arbitrary nature, since in this problem the classical electron radius does not play a fundamental role.

Experimental Methods of Determining ne Interaction. Neutron-electron interaction has been investigated by three methods which have now become classical. One of them [78] consists in observing the very small asymmetry of the scattering of thermal neutrons due to the variation of $f(\sin\theta/\lambda)$ with scattering angle. In this and similar experiments the scattering of thermal neutrons with a Maxwellian velocity distribution at angles of 45° and 135° is compared. Effects due to magnetic scattering and molecular diffraction are avoided by using noble gases with filled electron shells, particularly xenon (Z=54), as scatterers.

Table 2

Method	V_0, eV	Literature
Scattering on Xe	(300±5000)	[78]
Total cross section for Pb	Less than 5000	[82[
Total cross section for Bi	—(5300±1000)	[83]
Reflection from O_2-Bi mirror	—(3860±370)	[85, 86]
Scattering on Xe, Kr, Ar	—(4100±1000)	[79]
Measurement of amplitudes of Xe, Kr	—(3900±800)	]80]*
Total cross section for Bi	—(4340±140)	[84]
Scattering on Xe, Kr, Ar, Ne	—(3720±90)	[81]
Reflection from Bi mirror	—(4100±100)	[88]

*The authors used the data of [79].

The values $f(45°)=0.78$ and $f(135°)=0.52$ for thermal neutrons are responsible for the asymmetry of ne scattering, which is superimposed on the strong isotropic nuclear scattering. A correction must be introduced into the measured value of the asymmetry for the different geometric parameters of the detectors and to allow for the assymmetry due to the thermal motion of the gas atoms, which is many times greater than the sought effect. The measurements described in [75] and subsequent more accurate experiments [79, 80] led to a value $V_0=-(3900\pm800)$ eV; the minus sign corresponds to attraction between the neutron and electron. The main fault of experiments of such type is the large value of the correction for the effect due to thermal motion of the gas atoms. The main contribution to the correction comes from neutrons which have a very large wavelength in precisely that region where deviations from a Maxwellian distribution can be expected. This can lead to incorrect calculation of the correction. In the most accurate experiments this correction was determined experimentally from measurements with argon or neon, for which ne scattering is insignificant.

Accurate experiments by the method described in [75] were those carried out [81] in the Argonne National Laboratory in 1965-1966. Xenon, krypton, and argon were used in the experiments. Neon was used to check the calculated correction for asymmetry due to thermal motion of the gas. The magnitude of this correction exceeded the required effect for xenon by a factor of 4, for krypton by a factor of 10, and for argon by a factor of 18. The ne scattering amplitude was determined from the relationship

$$R=1+8\pi\frac{aa_{ne}}{\sigma_s}Z\,\Delta f\,(1+\delta), \qquad (40)$$

where $R=\sigma(45°)/\sigma(135°)$ is the measured asymmetry, including a correction for "false" asymmetry; a is the nuclear scattering length; σ_s is the scattering cross section; $\Delta f=\langle f(45°)-f(135°)\rangle$ is the form factor difference averaged over the neutron spectrum,

$$\delta\approx\frac{a_{ne}}{2a\,\Delta f}\langle f^2\,45°)-f^2(135°)\rangle-\frac{8\pi aa_{ne}}{\sigma_s}\langle f\,(135°)\rangle.$$

The value of σ did not exceed 0.01. The values of a for Xe and Kr were obtained by measurement of the critical angles of total reflection of neutrons from the surfaces of liquefied gases. The value of a for Xe was also obtained from experiments on neutron diffraction on XeF_4.

The final result of the experiment was as follows:

$$a_{ne}=(-1.34\pm0.03)\cdot10^{-16}\ \text{cm}\quad\text{or}\quad V_0=(-3720\pm90)\ \text{eV}. \qquad (41)$$

The second method of investigating ne interaction was used by Havens, Rainwater, and Rabi [82, 81]. It consists in observing the variation of the total scattering cross section with neutron wavelength in the $\lambda\sim$ 1 Å region. The nuclear scattering amplitude is constant, whereas the form factor $f(\sin\theta/\lambda)$ is responsible for the variation of the total cross section with λ. In [82, 83] the authors used liquid lead, and then liquid bismuth, as scatterers. They measured the total cross section for liquid bismuth in the range $\lambda=0.3-1.3$ Å. In this wavelength range the change in the form factor due to ne interaction causes a change of 0.1 b in the total cross section. Corrections for other effects must be introduced into the measurements. For instance, the correction for interatomic interference effects is 0.2 b in this wavelength range. A correction has to be introduced for the relative velocity due to thermal motion of the target atoms, and also for the

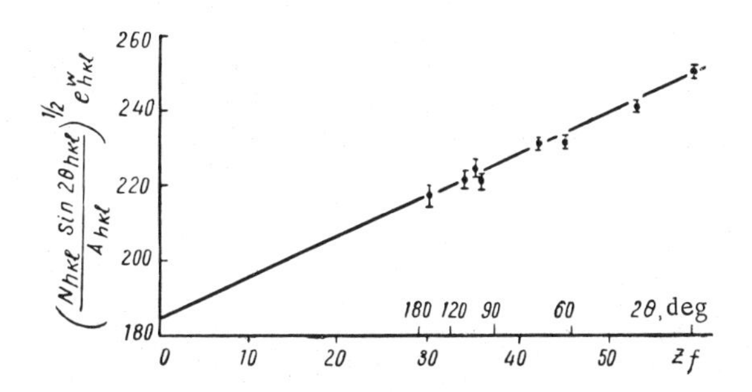

Fig. 14. $(N_{hkl} \sin 2\theta_{hkl} / A_{hkl})^{1/2} e^{W_{hkl}}$ as function of Zf.

effect of neutron capture in bismuth and impurities. Hence, despite the high statistical accuracy of this method, its over-all accuracy is low. The most accurate value $V_0 = -(4340 \pm 140)$ eV was obtained by this method in [84]; the cited error is the statistical error.

A more accurate experimental method of investigating ne interaction is the method of balancing the nuclear scattering amplitudes by reflection from a mirror. This method was used in [85, 86]. It is known that the refractive index of a substance for neutrons with wavelength λ is given by the relationship:

$$n^2 = 1 + \frac{\lambda^2}{\pi} \sum a_i N_i, \tag{42}$$

where N_i is the number of particles of type i per cm³ of substance; a_i is the coherent (forward) scattering length for neutrons by particles of type i. For the ne interaction case of interest to us

$$n^2 = 1 + \frac{\lambda^2}{\pi} N (a + Z a_{ne})^*. \tag{43}$$

For accurate determination of the relative refractive index of two substances the critical angle θ_c for total reflection at their interface is measured, and

$$\theta_C^2 = n_A^2 - n_B^2 \ (\text{for } \theta_C \ll 1), \tag{44}$$

where n_A and n_B are the refractive indices of the two substances. For liquid oxygen and bismuth we obtain from relationships (43) and (44)

$$\frac{\pi}{\lambda^2} \theta_C^2 = N_{\text{Bi}} a_{\text{Bi}} \left(\frac{N_{\text{O}} a_{\text{O}}}{N_{\text{Bi}} a_{\text{Bi}}} - 1 \right) - (N_{\text{Bi}} Z_{\text{Bi}} - N_{\text{O}} Z_{\text{O}}) a_{ne}, \tag{45}$$

where a_O and a_{Bi} are the coherent scattering amplitudes for oxygen and bismuth. Nuclear scattering on oxygen is only 2% more than on bismuth, whereas electronic scattering on bismuth, owing to its high Z, is much greater than on oxygen. Consequently, the measured critical angle θ_c is determined approximately equally by the unbalanced nuclear scattering and ne interaction. This angle is a few minutes.

The setup of the experiment carried out by Hughes et al. [85, 86] is shown in Fig. 13. The amplitudes a_O and a_{Bi} (or, more accurately, a_{Bi} and the ratio a_O/a_{Bi}) were determined from measurements of the free-atom cross sections at energies of the order of 10 eV. At these energies there is no ne interaction, since its form factor is practically zero. For conversion to coherent amplitudes a_{Bi} and a_O it was necessary to determine the spin-dependent incoherent scattering in bismuth. This was done by means of separate experiments with long-wavelength neutrons, for which there is only incoherent scattering. This led to the following value of V_0:

$$V_0 = -(3860 \pm 370) \text{ eV}. \tag{46}$$

The reflection method has an advantage over the other methods, since the measured effect is largely due to ne interaction. However, as in [82-84] the final result of the experiments [85, 86] rests on the assumption that in the energy region from 10 eV to thermal energies the amplitude is independent of the

*Since we are dealing with small-angle (forward scattering the form factor $f(\sin \theta/\lambda) = 1$.

neutron energy. Yet, as Halpern [87] reported, at energies of the order of 10 eV inelastic scattering can occur on bismuth and on oxygen. If this is so the values of a_O and a_{Bi} will differ from the values used by Hughes and the final results of the experiment would be altered accordingly. It is also essential to ensure that there is no effect due to resonances in this energy region. Nevertheless, the closeness of the results obtained by the three different experimental methods indicates that they are probably close to the true values.

Recently Koester et al. in Munich [88] made very accurate measurements of the coherent scattering amplitude for bismuth on a gravitational neutron refractometer [89] and found $a_{Bi} = (8.5234 \pm 0.0014) \cdot 10^{-13}$ cm. This amplitude was compared with scattering cross section data at neutron energy 5.2 eV, which led to the value

$$V_0 = -(4100 \pm 100)\,. \tag{47}$$

Table 2 gives the results of measurements of ne interaction during the period from 1947 through 1968.

Theoretical Analysis of ne Interaction. The experimentally obtained value of V_0 was explained by Foldy [90-93]. Foldy showed that in addition to the effect due to the existence of a cloud of charged virtual particles around the neutron there must be a magnetic effect, which is to be expected simply on the basis that the free neutron satisfies the Dirac equation and has an anomalous magnetic moment. A free Dirac particle does not move along a straight line, but "dances" with the velocity of light around a point which moves uniformly with velocity v. The "dance" of the particle covers a region of radius $\hbar/mc$. In the case of an electron, for instance, with charge e this movement is equivalent to a small loop with a current and in the presence of a magnetic field the electron behaves as if it had a normal magnetic moment of $e\hbar/2mc$. We can also expect effects due to the fact that the motion of the charge is not the same as that expected for a point particle, but resembles the motion of a charge distributed over a finite volume. This effect leads to a further shift of the S-electron levels in the hydrogen atom and is provided for in the Dirac theory by the term proposed by Darwin [94]. Thus, the "vibration" of a point electron leads to an apparent finite extension of its charge distribution and to a normal magnetic moment.

If the particle has an internal electromagnetic structure the apparent spatial extension of the charge is made up of the internal extension and the additional "smearing" due to the vibration. If we are to obtain information about the internal structure of the neutron from experimental ne-interaction data we have to separate the contribution made by vibration.

The scattering amplitude of a Dirac particle on a weak, slowly varying, purely electrostatic potential $\varphi(\mathbf{r})$ was obtained by Foldy [93] from the generalized Dirac equation [91, 95, 96]:

$$\gamma_\mu \frac{\partial \Psi}{\partial x_\mu} + \frac{Mc}{\hbar}\psi - \frac{i}{\hbar c}\sum_{n=0}^{\infty}\left[\varepsilon_n \gamma_\mu \square^n A_\mu + \frac{1}{2}\mu_n \gamma_\mu \gamma_\nu \square^n \left(\frac{\partial A_\mu}{\partial x_\nu} - \frac{\partial A_\nu}{\partial x_\mu}\right)\right]\psi = 0, \tag{48}$$

where ε_0 is the total charge of the Dirac particle; μ_0 is the anomalous magnetic dipole moment of the Dirac particle in the form introduced by Pauli [97].

The remaining terms in these series are the higher radial moments of the internal charge distribution and the current due to the particle. In particular, the term with coefficient ε_1 represents the radial extension of the internal charge distribution, while the coefficient ε_1 is connected with the rms electric radius by:

$$\varepsilon_1 \sim \frac{1}{6}\int r^2 \rho(\mathbf{r})\,d\mathbf{r} = \frac{e}{6}\langle r_e^2\rangle. \tag{49}$$

The term with μ_1 is due to the radial extension of the magnetic moment of the particle.

Thus, in the first Born approximation the scattering amplitude will have the form [93]

$$a(\mathbf{k}) = -\frac{M}{2\pi\hbar^2}\int e^{-i\mathbf{kr}}\sum_{n=0}^{\infty}\left[\varepsilon_n + \frac{\hbar}{2Mc}\mu_{n-1} + \frac{1}{2}\left(\frac{\hbar}{2Mc}\right)^2 \varepsilon_{n-1} + \ldots\right]\Delta^n \varphi(\mathbf{r})\,d\mathbf{r}, \tag{50}$$

where $\hbar\mathbf{k}$ is the momentum transfer in scattering.

In the case of small momentum transfers there is only the term with n=0, which gives the electrostatic scattering on a point charge ε_0:

$$a_0(\mathbf{k}) = -\frac{M\varepsilon_0}{2\pi\hbar^2}\int e^{i\,\mathbf{kr}}\,\varphi(\mathbf{r})\,d\mathbf{r} \tag{51}$$

and n=1;

$$a_1(\mathbf{k}) = -\frac{M}{2\pi\hbar^2}\left[\varepsilon_1 + \frac{\hbar}{2Mc}\mu_0 + \frac{1}{2}\left(\frac{\hbar}{2Mc}\right)^2\varepsilon_0\right]\int e^{i\,\mathbf{kr}}\,\nabla^2\varphi(\mathbf{r})\,d\mathbf{r}. \tag{52}$$

In the case of the neutron $\varepsilon_0=0$,and $a_1(\mathrm{k})$ is entirely responsible for the observed ne interaction. For the required ne scattering amplitude we obtain ($\mathrm{k}\to 0$)

$$a_{ne} = -\frac{2Me}{\hbar^2}\left(\varepsilon_1 + \frac{\hbar}{2Mc}\mu_0\right). \tag{53}$$

In this expression the term containing ε_1 is due to the radial extension of the charge distribution in the neutron. The term with μ_0 is the magnetic contribution and is due to the vibration of a particle having an anomalous magnetic moment μ_0.

As was related in [98], Weisskopf gave a simple semiqualitative interpretation of the effect. The trajectory of the moving neutron is a spiral along which it moves with the velocity of light c so that the transport velocity is equal to the velocity ν. When the neutron is at a distance $R \leq \hbar/Mc$ from the electron, magnetic spin-orbit interaction between the electron current and the magnetic moment of the neutron will occur. A calculation of this interaction shows that it agrees, aside from the factor 3/4, with Foldy's result.

By using (39) we can obtain from (53)

$$V_0 = 3e\left(\frac{mc^2}{e^2}\right)^3\left(\varepsilon_1 + \frac{\hbar}{2Mc}\mu_0\right) = V_{0\varepsilon_1} + V_{0\mu_0}. \tag{54}$$

Substitution of the numerical values gives $V_{0\mu_0}=-4080$ eV.

A comparison of this value with the experimental data of Table 2 shows that the majority of measurements, apart from those of [81, 84], which strongly contradict one another, agree within the limits of error with the value of $V_{0\mu_0}$. Thus, the determination of the contribution of $V_{0\varepsilon_1}$ or, in other words, the measurement of $\langle r_e^2\rangle_n$, will necessitate an increase in the accuracy of the measurements. It would be even better, however, to use some new, more sensitive technique. All that we can say at present about the value of $\langle r_e^2\rangle_n^{1/2}$ is that it is small and evidently does not exceed $0.1\cdot 10^{-13}$ cm.* It is of interest to note that if the neutron has the same $\langle r_e^2\rangle_n$ as the proton, then, as the calculations of [99] show, for $V_{0\varepsilon_1}$ we would obtain a value of $V_{0\varepsilon_1}\sim 16{,}000$ eV.

The amplitude a_{ne} (or the equivalent potential V_0) can be expressed in terms of the neutron electric form factor G_{En}. Differentiating expression (21) with respect to q^2 and using (19) we obtain

$$\left(\frac{\partial G_{En}}{\partial q^2}\right)_{q^2=0} = \frac{\langle r_e^2\rangle_n}{6} + \varkappa\frac{\hbar^2}{4M^2c^2}. \tag{55}$$

Taking (49) into account and comparing (55) and (54) we obtain

$$\left(\frac{\partial G_{En}}{\partial q^2}\right)_{q^2=0} = \frac{V_0}{3e^2}\left(\frac{e^2}{mc^2}\right)^3. \tag{56}$$

Thus, by investigating ne interaction we can obtain the value of $\left(\frac{\partial G_{En}}{\partial q^2}\right)_{q^2=0}$. The main contribution to $\left(\frac{\partial G_{En}}{\partial q^2}\right)_{q^2=0}$ is due to the magnetic term, equal to $\varkappa\hbar/4M^2c^2 = 0.0210f^2$. The contribution of the term containing $\langle r_e^2\rangle_n$ is still obscure.

Investigation of ne Interaction from Neutron Diffraction by a Tungsten Crystal. As already mentioned, if any conclusions regarding the contribution of the term containing ε_1 to the ne scattering amplitude

*The corresponding value for the proton is $\sim 0.8\cdot 10^{-13}$ cm [99].

are to be made, the accuracy of the measurements will have to be increased from 10% to 1–2%. Measurements made to an accuracy of 3% by various methods [81, 84, 88], however, lead to results which differ from one another by more than four errors and give values of ε_1 which differ even in sign.

The main defect of these methods is the very small value of the observed effect in comparison with the strong neutron-nucleus interaction. Hence, there is always the risk of some unconsidered nuclear effect affecting the results. For instance, in the case of measurements with bismuth a change of only 1/1000 in the nuclear cross section of bismuth between 0 and 10 eV leads to a 10% change in the measured ne interaction amplitude. In measurements of a half per cent effect to an accuracy of better than 3% on noble gases one must be absolutely sure that there are no other effects (e.g., p resonances, admixtures of light gases, etc.) which cause false asymmetry. In view of this it is of interest to find a new method of investigating ne interaction in which the measured effect is greater. Since the relative contribution to the scattering cross section from ne interaction is $\Delta\sigma/\sigma \sim Z\Delta f\, a_{ne}/a_N$, the measurements should be made on a heavy nucleus with small a_N.

It was reported in [100] that owing to the interference of potential and resonance scattering the isotope ^{186}W must have an anomalously small nuclear scattering amplitude in the region of neutron thermal energies. It was suggested that a_{ne} should be determined from an investigation of neutron diffraction in a mixture enriched with this isotope. Since tungsten metal is a paramagnetic [76], magnetic scattering should not make any contribution to the diffraction peaks [75, 76]. In [101] the single-crystal neutron diffraction technique was used to determine the energy dependence (in the range 0.008–0.13 eV) of the nuclear scattering amplitude of a mixture containing 90.7% ^{186}W. In the analysis of the results of these measurements the authors had to take into account ne scattering, the contribution of which to the total amplitude was about 20%.

Measurements of the intensities of the Bragg reflections of monochromatic neutrons with wavelength 1.15 Å from a single crystal containing 90.7% ^{186}W* have been made. The intensity of each reflection is

$$N_{hkl}\left(\frac{\sin\theta}{\lambda}\right)=k\left\{\left[a_{я}+Zf\left(\frac{\sin\theta}{\lambda}\right)_{hkl}a_{ne}\right]^2+\left[1-f\left(\frac{\sin\theta}{\lambda}\right)_{hkl}\right]^2\gamma^2\operatorname{ctg}^2\theta\right\}A_{hkl}\frac{e^{-2W_{hkl}}}{\sin 2\theta_{hkl}}, \quad (57)$$

where k is a constant for all the measured reflections; a_N is the nuclear scattering amplitude; A_{hkl} is the absorption factor determined by calculation; e^{-2W} is the Debye-Waller factor, which takes into account the thermal vibrations of the atoms in the lattice; $W=B(\sin\theta/\lambda)^2$, where B=const for all the measured reflections; θ_{hkl} is the glancing angle; and lastly, the term containing $\gamma^2\cot^2\theta$, where $\gamma=1/2x_n(-\hbar/Mc)(Ze^2/\hbar c)$, represents the Schwinger scattering.

The intensities of eight reflections [(110), (200), (220), (310), (400), (330), (420), and (510)] were determined experimentally. The values of the Debye-Waller factor were determined from a calculation based on [102, 103]. The analysis of the phonon spectrum of tungsten made in these investigations led to a value B=0.147 Å^2, which corresponds to the Debye temperature θ_D=355°K. Since the contribution of Schwinger scattering to (57) does not exceed 3% it can be neglected in a first approximation. Then the quantity

$$\left(\frac{N_{hkl}\sin 2\theta_{hkl}}{A_{hkl}}\right)^{1/2}e^{W_{hkl}}=k^{1/2}(a_{я}+Zf_{hkl}a_{ne})$$

will depend linearly on Zf. In Fig. 14 this relationship is illustrated by one of the series of measurements. Figure 14 shows that a_N and a_{ne} have the same sign. This agrees with the results of measurements of [10], which showed that a_N is negative for this isotopic mixture. The observed scattering asymmetry is due to ne interaction and is

$$R-1=\frac{(a_N+f_{(110)}Za_{ne})^2}{(a_N+f_{(510)}Za_{ne})^2}\approx .019.$$

i.e., almost 40 times greater than the corresponding effect in the case of Xe.

*This work was carried out by Yu. A. Aleksandrov, A. M. Balagurov, T. A. Machekhina, G. S. Samosvat, L. N. Sedlakova (JINR), and N V. Rannev and L. E. Fykin (Physicochemical Institute).

A least-squares treatment of equations of the type (57) for the eight measured reflections gave the value of the ratio a_{ne}/a_N. Specially conducted experiments with W powders (with natural and investigated isotope mixtures) gave the total scattering amplitude for the (110) reflection. These experiments were carried out on two different apparatuses with two sets of specimens prepared by different methods and gave similar results. The value of a_{ne} is determined by comparing the total amplitude for the (110) reflection with the ratio a_{ne}/a_N.

The obtained value of a_{ne} differed excessively from the results of previous measurements. Hence, attempts were made to find the reason for this large discrepancy. In particular, the authors carried out experiments in which the dimensions of the neutron beam reflected from the crystal were measured and their variation from reflection to reflection was determined. The dimensions of the beam were much less than the angular opening of the detector and the variation with Bragg angle was insignificant. Since the crystal of enriched tungsten was grown from a natural seed the natural tungsten might have diffused into the enriched tungsten and led to variation of a over the height of the crystal. This hypothesis was tested by measuring the intensities of the reflections from the upper and lower halves of the investigated specimen (the specimen was oriented along the direction of growth of the crystal). No differences between the slopes of the straight lines (see Fig. 14), which determine the ratio a_{ne}/a_N, were found. Nevertheless, it is intended to carry out a further analysis of the isotopic composition of the crystal used. The results of the measurements might also have been affected by effects associated with magnetic scattering of neutrons, effects which were absent in previous experiments. These effects are now being assessed.

It is possible that B was incorrectly determined: this would have a significant effect on the final result. In particular, the value $a_{ne} = -1.45 \cdot 10^{-16}$ cm (V_0=4080 eV) could be obtained from the measurements if B is taken as 0.26 Å^2 ($\theta_D \approx 270$°K). In view of this, direct experimental determination of B for W is extremely desirable.

The effect of magnetic scattering was investigated in experiments aimed at detecting ordered magnetic moments in W atoms. Such experiments were undertaken earlier and consisted in a search for the (100) reflection from powdered samples of natural W [104]. These measurements indicated that $\mu_W < 0.3$ μ_B, where μ_B is the Bohr magneton. Similar measurements made on a single crystal of a mixture enriched with ^{186}W (90.7%) at room temperature gave $\mu_W \lesssim 0.004$ μ_B.

LITERATURE CITED

1. R. Marshak and E. Sudarshan, Introduction to Elementary Particle Physics, Interscience Publ., New York (1961).
2. G. Feinberg and M. Goldhaber, Proc. Nat. Acad. Sci., U.S.A., 45, 1301 (1959).
3. P. Dee, Proc. Roy. Soc., (London), A136, 727 (1932).
4. I. S. Shapiro and I. V. Éstulin, Zh. Éksp. Teor. Fiz., 30, 579 (1956).
5. J. King, Phys. Rev. Lett., 5, 562 (1960).
6. V. Hughes, Phys. Rev., 105, 170 (1957).
7. J. C. Zorn et el., Phys. Rev., 129, 2566 (1963).
8. C. G. Shull et al., Phys. Rev., 153, 1415 (1967).
9. R. A. Lyttleton and H. Bondi, Proc. Roy. Soc. (London), A252, 313 (1959); A257, 442 (1960).
10. A. Piccard and E. Kessler, Arch. Sci. Phys. Nat., 7, 340 (1925); P. Blackett, Nature, 159, 658 (1947); V. A. J. Bailey, Proc. Roy. Soc. (London), 97, 77 (1960).
11. M. A. Markov, Zh. Éksp. Teor. Fiz., 51, 878 (1966); High-Energy Physics and Elementary Particle Theory [in Russian], Naukova Dumka, Kiev (1967), p. 671; in: Proceedings of International Seminar on Elementary Particle Theory, Varna, May 6-9, 1968; Preprint JINR R2-4050, Dubna (1968); Preprint JINR R2-4534, Dubna (1969).
12. L. D. Landau and E. M. Lifshits, Field Theory [in Russian], Nauka, Moscow (1967), p. 434.
13. Yu. B. Rumer, Investigations on 5-Optics [in Russian], Gostekhteorizdat, Moscow (1956).
14. L. D. Landau, Zh. Éksp. Teor. Fiz., 32, 405 (1957).
15. J. Smith et al., Phys. Rev., 108, 120 (1957).
16. P. D. Miller et al., Phys. Rev. Lett., 19, 381 (1967).
17. P. Miller, Usp. Fiz. Nauk, 95, 470 (1968).
18. P. D. Miller et al., 14th Internat. Conf. on High-Energy Physics, Vienna (1968), p. 300.
19. C. G. Shull and P. Nathans, Phys. Rev. Lett., 19, 384 (1967).
20. Yu. A. Aleksandrov et al., Preprint JINR R3-4121, Dubna (1968).

21. F. L. Shapiro, Usp. Fiz. Nauk, 95, 145 (1968).
22. S. D. Drell and F. Zachariasen, Electromagnetic Structure of Nucleons, O.U.P., London (1961).
23. V. K. Fedyanin, Electromagnetic Structure of Nuclei and Nucleons [in Russian], Vysshaya Shkola, Moscow (1967).
24. E. J. Ernst et al., Phys. Rev., 119, 1105; R. G. Sachs, Phys. Rev., 126, 2256 (1962); L. N. Hand et al., Phys. Rev. Lett., 8, 110 (1962).
25. Proc. Internat. Symp. on Electron and Proton Interactions at High Energies, Stanford, California, September 5-9, 1967, p. 70-71.
26. G. Berger et al., 14th Internat. Conf. on High-Energy Physics, Vienna, 1968, p. 516.
27. G. Höhler et al., 14th Internat. Conf. on High-Energy Physics, Vienna, 1968, p. 233.
28. A. Klein, Phys. Rev., 99, 998 (1955).
29. A. M. Baldin, Nucl. Phys., 18, 310 (1960).
30. Yu. A. Aleksandrov and I. I. Bondarenko, Zh. Éksp. Teor. Fiz., 31, 726 (1956).
31. V. S. Barashenkov et al., Zh. Éksp. Teor. Fiz., 32, 154 (1957).
32. V. I. Gol'danskii et al., Zh. Éksp. Teor. Fiz., 38, 1965 (1960).
33. I. L. Powell, Phys. Rev., 75, 32 (1949).
34. M. Gell-Mann and M. Goldberger, Phys. Rev., 96, 1433 (1954).
35. V. A. Petrun'kin, Zh. Éksp. Teor. Fiz., 40, 1148 (1960).
36. V. S. Barashenkov et al., Preprint JINR R-1348, Dubna (1963).
37. V. A. Petrunkin, Nucl. Phys., 55, 197 (1964).
38. L. I. Lapidus, Preprint JINR R-967, Dubna (1962).
39. Yu. A. Aleksandrov and G. S. Samosvat, Preprint JINR R-2495, Dubna (1965).
40. V. S. Barashenkov et al., Nucl. Phys., 50, 684 (1964).
41. G. F. Chew, Phys. Rev., 80, 196 (1950); G. F. Chew and M. Goldberger, Phys. Rev., 87, 778 (1952); R. Capps, Phys. Rev., 106, 1031 (1957).
42. A. Tenore and A. Verganelakis, Nuovo Cimento, 35, 261 (1965).
43. L. I. Lapidus and Chou Kuang-chao, Zh. Éksp. Teor. Fiz., 39, 1286 (1960).
44. R. S. Jones et al., Phys. Rev., 128, 1357 (1962); J. Fox et al., Bull. Amer. Phys. Soc., 9, 69 (1964).
45. G. Bernardini, 9th Internat. Annual Conf. on High-Energy Physics, Kiev, 1959.
46. L. G. Moroz and V. N. Tret'yakov, Dokl. Akad. Nauk BSSR, 8, 575 (1964).
47. Yu. A. Aleksandrov, Zh. Éksp. Teor. Fiz., 33, 294 (1957).
48. Yu. A. Aleksandrov et al., Zh. Éksp. Teor. Fiz., 40, 1878 (1961).
49. Yu. V. Dukarevich and A. N. Dyumin, Zh. Éksp. Teor. Fiz., 44, 130 (1963).
50. V. S. Barashenkov and B. M. Barbashov, Nucl. Phys., 9, 426 (1958).
51. V. S. Barashenkov and G. Yu. Kaizer, Preprint JINR R-771, Dubna (1961).
52. G. Breit and M. Rustgi, Phys. Rev., 114, 830 (1959).
53. T. Vela and M. Samawara, Progr. Theoret. Phys., 24, 519 (1960).
54. A. Kazanawa, Nucl. Phys., 24, 524 (1961).
55. M. Walt and D. Fossan, Phys. Rev., 137, B629 (1965).
56. G. V. Anikin et al., Nuclear Structure Study with Neutrons, Antwerp, 19-23 July, 1965, p. 574.
57. J. Monahan et al., Nuclear Structure Study with Neutrons, Antwerp, 19-23 July, 1965, p. 588; A. J. Elwyn et al., Phys. Rev., 142, 758 (1966).
58. G. V. Gorlov et al., Yadernaya Fizika, 8, 1086 (1968).
59. A. Adam et al., Nuclear Structure Symposium, Dubna, July 4-11, 1968.
60. F. T. Kuchnir, et al., Phys. Rev., 176, 1405 (1968).
61. Yu. A. Aleksandrov, Dissertation, FÉI (1959).
62. V. M. Koprov and L. N. Usachev, "Nuclear reactions at low and medium energies," in: Proceedings of All-Union Conference, 1960 [in Russian], Izd-vo AN SSSR, Moscow.
63. V. M. Agranovich and D. D. Odintsov, Proceedings of All-Union Conference, 1960 [in Russian] Izd-vo AN SSSR, Moscow.
64. A. N. Dyumin, Dissertation, Physicotechnical Institute (1964).
65. R. M. Thaler, Phys. Rev., 114, 827 (1959).
66. Yu. A. Aleksandrov et al., ZhÉTF Pis. Red., 4, 196 (1966).
67. M. Goldberg et al., BNL-400, Second edition, Vol. 11 (1962).
68. S. B. Gerasimov et al., Zh. Éksp. Teor. Fiz., 43, 1872 (1962).

69. V. S. Barashenkov and H. J. Kaiser, Nucleon Structure, Proc. of Internat. Conf. at Stanford Univ., June 24-27, 1963, p. 263.
70. V. S. Barashenkov, Internat. Winter School of Theoretical Physics in JINR, Vol. 3 [in Russian], JINR, Dubna (1964), p. 86.
71. G. de Vries et al., Preprint Stanford Univ. (1963).
72. W. K. H. Panofsky, 14th Internat. Conf. on High-Energy Physics, Vienna, August 28-September 5, 1968, p. 23.
73. V. S. Barashenkov et al., Preprint JINR R-2894, Dubna (1966).
74. R. P. Zaikov, Preprint JINR R2-3073, Dubna (1966).
75. D. Hughes, Neutron Optics, Interscience Publ., New York (1955).
76. G. Bacon, Neutron Diffraction, Clarendon Press, Oxford (1962).
77. A. H. Compton and S. K. Allison, X Rays in Theory and Experiment, New York (1957).
78. E. Fermi and L. Marshall, Phys. Rev., 72, 1139 (1947).
79. M. Hamermesh et al., Phys. Rev., 85, 483 (1952).
80. M. F. Crouch et al., Phys. Rev., 102, 1321 (1956).
81. V. Krohn and G. Ringo, Phys. Rev., 148, 1303 (1966).
82. W. Havens et al., Phys. Rev., 72, 634 (1947).
83. W. Havens et al., Phys. Rev., 82, 345 (1951).
84. E. Melkonian et al., Phys. Rev., 114, 1571 (1959).
85. J. Harvey et al., Phys. Rev., 87, 220 (1952).
86. D. Hughes et al., Phys. Rev., 90, 497 (1953).
87. O. Halpern, Phys. Rev., 133, B581 (1963).
88. L. Koester, Technische Hochschule, Munich (private communication) (1969).
89. L. Koester, Z. Phys., 182, 328 (1965); L. Koester, Z. Phys., 198, 187 (1967).
90. L. Foldy, Phys. Rev., 83, 688 (1951).
91. L. Foldy, Phys. Rev., 87, 688 (1952).
92. L. Foldy, Phys. Rev., 87, 693 (1952).
93. L. Foldy, Rev. Mod. Phys., 30, 471 (1958).
94. C. G. Darwin, Proc. Roy. Soc. (London), A118, 654 (1928).
95. G. Salzman, Phys. Rev., 99, 973 (1955).
96. A. C. Zemach, Phys. Rev., 104, 1771 (1957).
97. W. Pauli, Relativistic Theory of Elementary Particles [Russian translation], Izd-vo Inostr. Lit., Moscow (1947).
98. E. Segré (Ed.), Experimental Nuclear Physics, John Wiley & Sons, New York (1953).
99. D. R. Yenny et al., Rev. Mod. Phys., 29, 144 (1957).
100. Yu. A. Aleksandrov, Preprint JINR R3-3442, Dubna (1967).
101. Yu. A. Aleksandrov et al., Preprint JINR R3-4121, Dubna (1968).
102. P. S. Mahesh and B. Dayal, Phys. Rev., 143, 443 (1966).
103. F. H. Chem and B. N. Brockhouse, Solid State Commun., 2, 73 (1964).
104. C. G. Shull and M. K. Wilkinson, Rev. Mod. Phys., 25, 100 (1953).